Bureau of Flora and Fauna, Canberra

ZOOLOGICAL CATALOGUE OF AUSTRALIA

Volume 2

HYMENOPTERA:

FORMICOIDEA, VESPOIDEA AND SPHECOIDEA

Australian Government Publishing Service
Canberra 1985

© Commonwealth of Australia 1985
ISBN for series: 0 644 02840 8
ISBN for Volume 2: 0 644 03922 1

Zoological Catalogue of Australia

The compilation of the *Zoological Catalogue of Australia* is conducted under the auspices of the Bureau of Flora and Fauna, B. J. Richardson, Assistant Director (Fauna).

D. W. Walton, Executive Editor

Vol. 2

HYMENOPTERA : Formicoidea

by

Robert W. Taylor
Australian National Insect Collection
C.S.I.R.O. Division of Entomology
Canberra, A.C.T.

and

D. R. Brown
Bureau of Flora and Fauna
Department of Arts, Heritage and Environment
Canberra, A.C.T.

HYMENOPTERA : Vespoidea and Sphecoidea

by

Josephine C. Cardale
Australian National Insect Collection
C.S.I.R.O. Division of Entomology
Canberra, A.C.T.

Printed by Brown Prior Anderson Pty Limited, Victoria

CONTENTS

Editorial Preface	v
Formicoidea: Introduction	1
Formicidae	5
Nothomyrmeciinae	5
Myrmeciinae	6
Pseudomyrmecinae	17
Ponerinae	18
Dorylinae	52
Leptanillinae	53
Myrmicinae	53
Dolichoderinae	92
Formicinae	107
Vespoidea and Sphecoidea: Introduction	150
Masaridae	154
Eumenidae	160
Vespidae	208
Sphecidae	218
Appendix I Museum Acronyms	304
Appendix II Abbreviations	305
Taxonomic Index: Formicoidea	306
Taxonomic Index: Vespoidea and Sphecoidea	349
Publication date of the previous volume	381

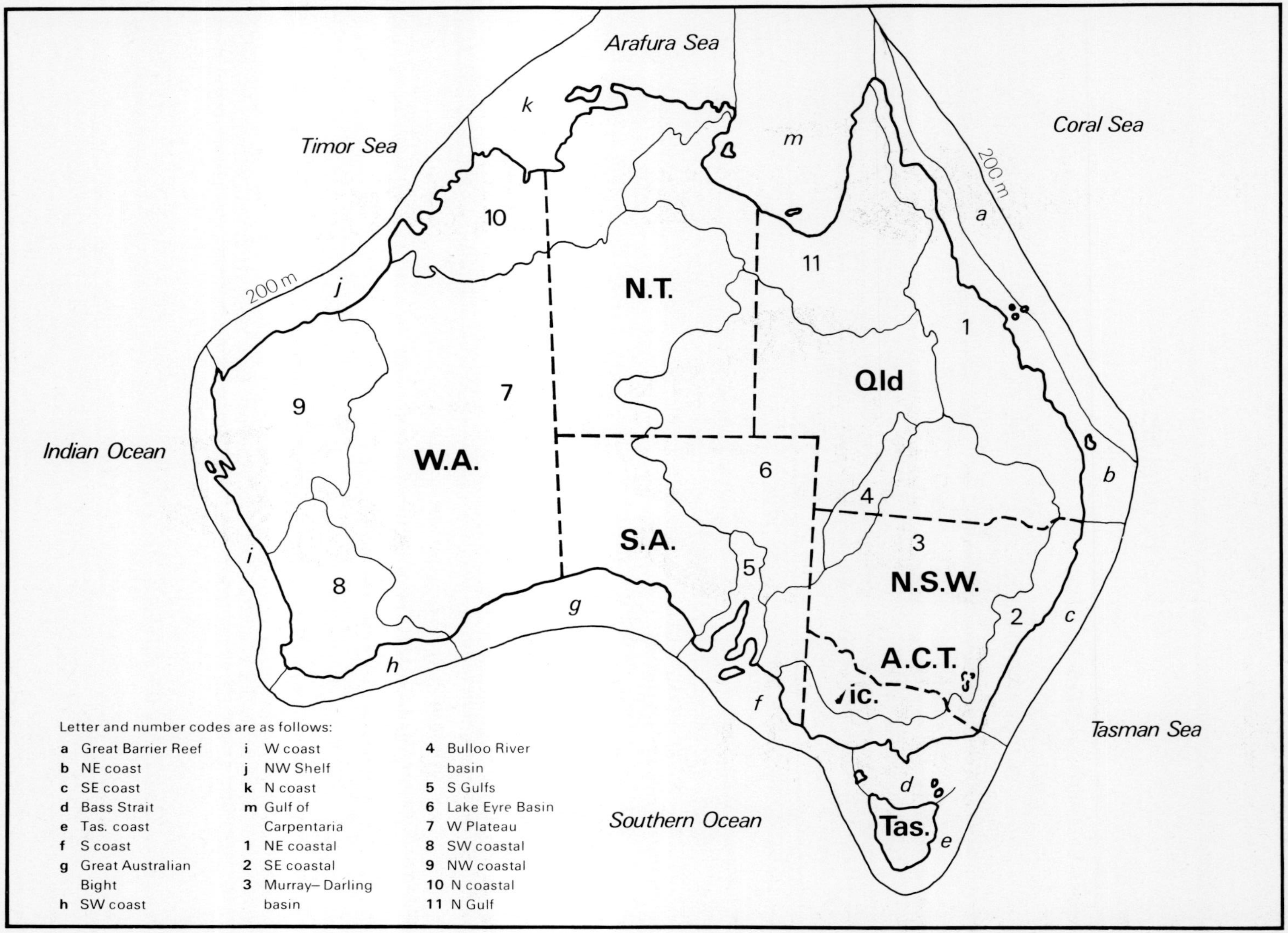
Arafura Sea
Coral Sea
Timor Sea
Indian Ocean
Southern Ocean
Tasman Sea
200 m
200 m
N.T.
Qld
W.A.
S.A.
N.S.W.
A.C.T.
Vic.
Tas.
Letter and number codes are as follows:
a Great Barrier Reef
b NE coast
c SE coast
d Bass Strait
e Tas. coast
f S coast
g Great Australian Bight
h SW coast
i W coast
j NW Shelf
k N coast
m Gulf of Carpentaria
1 NE coastal
2 SE coastal
3 Murray–Darling basin
4 Bulloo River basin
5 S Gulfs
6 Lake Eyre Basin
7 W Plateau
8 SW coastal
9 NW coastal
10 N coastal
11 N Gulf

Map 1. States, catchment areas and coastal zones of Australia

EDITORIAL PREFACE

INTRODUCTION

An objective of the Australian Biological Resources Study is to stimulate research and publications on the taxonomy and distribution of Australian fauna and flora. Consistent with this aim, the *Zoological Catalogue of Australia* was conceived as a concise, computer-based data bank consisting of current taxonomic and biological knowledge of the Australian fauna, accessible to all interested in such information. As the project developed, the advantages of publication of this information were recognised.

Data for inclusion in the *Catalogue* are assembled in four separate files: a genus taxonomic arrangement file, a species taxonomic arrangement file, a genus available name file and a species available name file. The contents of appropriate files are then integrated by computer. This methodology yields a standard format which will be maintained throughout the volumes of the *Catalogue* and provides consistency in the data. The format and style of presentation are, therefore, the responsibility of the Bureau of Flora and Fauna. The authors are responsible only for the information content.

Each volume of the *Catalogue*, treating specific taxa, will cite by name and original reference all species known from Australia. The species are arranged taxonomically by family and genus. Information for each species includes synonymy, literature citation, location/status of the type material and type locality for each available name in synonymy, a brief summary of geographical distribution and ecological attributes, and important references on various aspects of the biology. It is designed to serve primarily as a bibliographic directory to the most comprehensive and recent information available on each species.

This data base is intended to provide a substantial assessment of current knowledge and to stimulate and provide a starting point for future investigations. It is estimated that the Australian fauna exceeds 150 000 species of which about half have yet to be recognised and described. As knowledge of the Australian fauna advances, the data base will be updated and expanded.

TAXONOMIC INFORMATION

Nomenclature in the *Catalogue* adheres to the provisions established in the International Code of Zoological Nomenclature. The author and date of all names appearing in the *Catalogue* are presented so that the user may understand the nature and relationships of the names and all names appear in their legitimate form, not as they appeared in their original presentation. The valid genus and species group names and their allocation to families are determined by the contributors. No new genus or species group names are introduced in the *Catalogue* although new combinations may be established. Synonymies do not include new combinations. Treatment of family group names is not included.

ECOLOGY AND DISTRIBUTION

Information on ecology and distribution is given with each valid species. The ecological descriptors are general terms derived from a list prepared by the Bureau of Flora and Fauna. These descriptors act as computer search terms for use with the data base.

Distribution data are based on a standardised list of computer search terms established by the Bureau of Flora and Fauna. Both political and geographical region descriptors are included (see map). Political areas include the adjacent waters. Terrestrial geographical terms are based on the drainage systems of continental Australia, while marine terms are self explanatory except as follows: the boundary between the coastal and the oceanic zones is the 200 m contour; the Arafura Sea extends from Cape York to 124°E longitude; and the boundary between the Tasman and Coral Seas is considered to be the latitude of Fraser Island, also regarded as the southern terminus of the Great Barrier Reef. Ecological or distributional terms in parentheses imply that the information is unconfirmed but, in the opinion of the contributor, likely to be correct. Terms for terrestrial habitat or vegetation type follow Specht, R.L. (1970). Vegetation. pp. 44–67 *in* Leeper, G.W. (ed.) *The Australian Environment*. 4th edn. Melbourne : CSIRO-Melbourne Univ. Press.

BIBLIOGRAPHIC INFORMATION

Where possible, selected references are provided as an introduction to the biology of a species. Literature citations throughout the *Catalogue* are given in full. Older works, with extended subtitles, in some cases have been shortened but only if their identity is preserved. Serial titles are abbreviated in a manner designed to facilitate library research. The number and variety of sources for serial abbreviations employed by workers of different nationalities and among the various taxa precluded use of a standard guide. References or titles originally issued in a script other than Roman and lacking a Romanised translation are transliterated with the original language shown in brackets. Common abbreviations are listed in Appendix I. Acronyms of museums or collections, given as part of the Type data, are defined in Appendix II.

ACKNOWLEDGEMENTS

Within the Bureau of Flora and Fauna, Richard Longmore, Janet Godsell, Barry Richardson and Keith Houston assisted in the editing of the volume. David Berman and Chris Curtis managed the data base and Wendy Riley, Cindy Wolter and Cindy Warhurst the entry and revisions of the data.

To all those involved, grateful acknowledgement is extended.

D. W. Walton

FORMICOIDEA

Robert W. Taylor and D.R. Brown

INTRODUCTION

Ants are among the most ubiquitous, abundant and familiar of insects. They are a group of great ecological importance in most habitats found in Australia, ranging from rainforests to deserts, from the cold mountains of the southeast to the tropical plains of the far north. The fauna is estimated to include at least four thousand species, possibly many more. This is about three times the number of scientific names available in the literature, and more than twice the number of species currently recognized in collections (Taylor 1979,1983).

Ants were well represented among the first Australian insects returned to Europe for scientific study. A number of species in the collections of Joseph Banks and Daniel Solander were collected in 1770, during Captain James Cook's first voyage of discovery to eastern Australia. These were described by J.C. Fabricius in 1775, in the first publication ever to have contained scientific descriptions of Australian endemic animals. The Fabrician insect species, in fact, were described several years before any Australian endemic vertebrates were named.

There have been several checklists of the names available for Australian Formicidae. All were incomplete for their time and all are now out of date. They include the works of Dalla Torre (1893), Gustaf Mayr (1876) and W.W. Froggatt (1905), along with the rather more satisfactory coverage of the fauna in the world checklists of Carlo Emery, published in Wytsman's monumental *Genera Insectorum* (Emery 1910,1911,1912,1921,1922,1925). In most genera, the tally of species accumulated in a piecemeal fashion, and most species have never been the subjects of critical, let alone modern, synthetic monographic studies. Moreover, many of the species-group names of the past were first proposed with subspecific status, so that an infrageneric classification is implicit in the nomenclature. Overall, this arrangement will bear little resemblance to the structured products of future revisionary studies in which it is probable that most "subspecies" will be elevated to full species rank, and the remainder will become junior synonyms, often under names with which they have had no previous close association. The specific and subspecific arrangement in most genera, especially large ones like *Iridomyrmex* and *Camponotus*, evidences more disorder than order, and disorder will prevail until comprehensive revisionary monographs, based upon more representative collections and improved biological knowledge, can be completed.

It must be emphasized that this catalogue is preliminary in many aspects. We believe that all species names are correctly assigned to the genera currently recognized by ant taxonomists, and that future surprises in generic re-assignment of the names presented here are unlikely. However, in genera which have not been recently monographed, the status of individual names as specific or subspecific epithets usually follows the last published assignment. The final arrangement must be considered a piecemeal development, as discussed above. This has been unavoidable, but it has allowed us to place each species name in a logically identifiable place relative to other names, even if the taxonomic implications of the arrangement might be untenable in the light of future comprehensive taxonomic studies.

ORGANISATION OF THE CATALOGUE

Classification
The classification used here is primarily that of Brown (1973), with the

Nothomyrmeciinae raised to subfamily status separate from the Myrmeciinae, following Taylor (1978).

Citation of Taxon Names

All generic and specific names are listed in their currently legitimate form, without diacritic marks, capitilization, hyphenation, etc., even if these were present in the original or other subsequent references.

Taxonomic Arrangement of Subfamilies and Genera

Subfamily headings have been included and the genera are arranged in separate alphabetical cohorts for each subfamily. The order of generic listing is thus partly "taxonomic" and partly alphabetical.

In the present arrangement the subfamily Nothomyrmeciinae begins the listing with *Nothomyrmecia*; this is followed by Myrmeciinae, with *Myrmecia*; Pseudomyrmecinae, with *Tetraponera*; Ponerinae, with *Amblyopone*; Dorylinae, with *Aenictus*; Leptanillinae, with *Leptanilla*; Myrmicinae, with *Adlerzia*; Dolichoderinae, with *Bothriomyrmex*; and Formicinae, with *Acropyga*.

Synonymies

Within the limits prescribed above, the generic and specific synonymies are as complete as we have been able to achieve. We have proposed very few new synonyms, even though we are aware of likely future changes, occasionally at generic level and frequently at species level. All synonyms are listed in order of date of publication.

Taxonomic Arrangment at Species Level

Species names without synonyms, and those accepted as senior synonyms, are presented in alphabetical order within genera. Subspecies are listed alphabetically after the nominate subspecies name. The synonyms listed at generic level include only those names of which the type species is represented in Australia.

Subspecies

Names of the species group assigned subspecific status at the time of their most recent published citation are listed here as subspecies.

Formicid nomenclature has been burdened by the past use of the subspecies category. Much effort has been made by those engaged in modern revisionary studies to eliminate old subspecies names from the nomenclature, either by elevating them to species status or by submerging them as junior synonyms. Despite this, we have proposed few changes of status among names of the species group, although we would not expect the subspecies category as used here to be accepted in any modern taxonomic synthesis of an Australian ant genus. Editorial procedure has required the citation of nominate subspecies in the listings and this has sometimes introduced previously implicit but unpublished trinominal combinations into the literature.

Infrasubspecific Taxa

The treatment of infrasubspecific names follows Art. 45 of the ICZN. Other organizational matters involving nomenclature have followed the procedures laid down in the editorial code of the *Zoological Catalogue of Australia*, much as reviewed in the first volume of this series (Cogger *et al.* 1983).

Keywords

Because of the paucity of published information, we consider this work to be a beginning and not a definitive statement. As with all sections of the *Catalogue*, the computer data base files will be updated to refine, not only the nomenclature and classification of Australian ants, but also the knowledge of their distribution and biological attributes. We believe, and hope, that the nomenclatural and bibliographic components of this work will prove useful to others interested in the Australian Formicidae. We caution users that we have considerable reservation about the reliability of the keywords at this point in time. Our selection of distributional and biological keywords has been based largely on published data, with little reference to the data on the labels of specimens housed in public collections, specimens whose records have never been published. The next phase of this project will involve such a synthesis and we hope that updating of the existing computer data base will begin immediately.

Because of these constraints, many species are assigned a distribution limited to the prescribed geographic region which contains their type locality. This means that such areas as the Australian Capital Territory and Tasmania would appear to have ant faunas much less rich than is the case in nature. On the other hand, some regions, such as "NE coastal Queensland", apparently contain many species found nowhere else, so that there are likely to be few additions to the distributional keywords of species listed from them.

The biological keywords given for each species are based upon a prescription designed originally for the genus concerned, which has been repeated for each species, even though authoritative documentation is not available for all. For example, all *Pheidole* species are said to be harvesters of seeds, even though we have no proof of this for many of them. This section will become more useful as further biological information becomes available, as species are placed taxonomically and as data on their labels are added to the *Catalogue* data base.

Biological References
We are aware that many references have been omitted from the individual species entries under this heading. A few "key" references have been given to access the literature on some of the more extensively studied taxa (such as *Nothomyrmecia*, some *Myrmecia* species and *Iridomyrmex* species of the *purpureus* group). Several recent general works could not easily be accommodated in this way. They include, the karyological survey of Imai, Crozier & Taylor (1977), and Greenslade's *A Guide to Ants of South Australia* (1979); the latter usefully surveys the genera present in that State and provides keys to a large subset of the ant genera known from Australia. A number of ecological titles are also excluded, most notably Berg's (1975) milestone study on the relations between myrmecochorus plants and ants, along with the many papers which his work has inspired.

Tramp Species
There are a number of essentially pantropical "tramp" or "vagrant" ant species, some of which have been introduced by human agency into northern Australia and some southern cities. Some of these species have not been included in this catalogue. We expect to add them to the data base shortly. There is some confusion as to just which "tramp" species are present on the Australian continent, and the extent of their distribution is often unclear. One of us (RWT) has been progressively surveying these matters, but the work was incomplete at the time of publication. There are some species, including various *Tetramorium* spp., *Quadristruma emmae* (Emery), *Technomyrmex albipes* (Smith), *Iridomyrmex glaber* (Mayr), and *Anoplolepis longipes* (Jerdon), which are known to be vagrant in places peripheral to their main distributional areas, and are generally considered "tramp" species for this reason. In our opinion such species, if listed below, are likely native species, which have dispersed onto the Australian continent from Papuasian source areas in a late stage of the northwards drift of the continent.

ACKNOWLEDGEMENTS

We are grateful for assistance in the compilation, checking and recording of data gathered for this catalogue by Elizabeth Lockie, Renate Sadler, Patricia Hoyle, Marie-Louise Johnson and Timothy Wace. Valuable advice on myrmecological matters was given by Rev. B.B. Lowery, S.J., and on the computer data-logging procedures by Janet Pyke. Drs Dan Walton and Barry Richardson of the Bureau of Flora and Fauna are both thanked and complimented on their excellent and detailed attention to editorial matters. The compilation was supported by a grant from the Australian Biological Resources Study.

R.W.T. & D.R.B.

References

Brown, W.L. jr. (1973). A comparison of the Hylean and Congo-West African rain forest ant faunas. pp. 161–185 *in* Meggers, B.J., Ayensu, E.S. & Duckworth, W.D. (eds.) *Tropical forest ecosystems in Africa and South America: a comparative review.* Washington : Smithsonian Institution Press

Berg, R.Y. (1975). Myrmecochorous plants in Australia and their dispersal by ants. *Aust. J. Bot.* **23**: 475–508

Cogger, H.G., Cameron, E.E. & Cogger, H.M. (1983). Amphibia and Reptilia. *in* Walton, D.W. (ed.) *Zoological Catalogue of Australia.* Canberra : Australian Government Publishing Service Vol. 1 313 pp.

Dalla Torre, K.W. von. (1893). Formicidae. *Cat. Hymenoptera* 7 Leipzig : W. Engelmann

Emery, C. (1910). Formicidae subfam. Dorylinae *in* Wytsman, P. (ed.) *Genera Insectorum.* Fasc. 102 Brussells : Verteneuil & Desmet

Emery, C. (1911). Formicidae subfam. Ponerinae *ibid.* Fasc. 118

Emery, C. (1912). Formicidae subfam. Dolichoderinae *ibid.* Fasc. 137

Emery, C. (1921). Formicidae subfam. Myrmicinae *ibid.* Fasc. 174A

Emery, C. (1922). Formicidae subfam. Myrmicinae *ibid.* Fasc. 174B

Emery, C. (1922). Formicidae subfam. Myrmicinae *ibid.* Fasc. 174C

Emery, C. (1925). Formicidae subfam. Formicinae *ibid.* Fasc. 183

Froggatt, W.W. (1905). Domestic insects: ants. *Agric. Gaz. N.S.W.* **16**: 861–866. [Reprinted by Agric. Dept. N.S.W. as *Miscellaneous Publication No. 889*, with a catalogue of Australian species, 1906]

Greenslade, P.J.M. (1979). *A Guide to Ants of South Australia.* Adelaide : South Australian Museum 44 pp.

Imai, H.T., Crozier, R.H. & Taylor, R.W. (1977). Karyotype evolution in Australian ants. *Chromosoma (Berl.)* **59**: 341–393

Mayr, G. (1876). Die australischen Formiciden. *J. Mus. Godeffroy* **12**: 56–115

Taylor, R.W. (1978). *Nothomyrmecia macrops*: a living-fossil ant rediscovered. *Science* **201**: 979–985

Taylor, R.W. (1979). Some statistics relevant to Australian insect taxonomy. *CSIRO Aust. Div. Entomol. Rep. No. 8* pp. 1–9

Taylor, R.W. (1983). Descriptive taxonomy: past, present, and future. pp. 93–134 *in* Highley, E. & Taylor, R.W. (eds.) *Australian Systematic Entomology: a Bicentenary Perspective.* Melbourne : CSIRO

FORMICIDAE

INTRODUCTION

The family Formicidae accommodates all known true ants. Almost all species are fully eusocial. The exceptions are a few derived and sometimes highly specialised workerless parasites which are inquilines in the nests of other, usually closely related, ant species. Most formicid species have winged, wasp-like males, deciduously winged or wingless females, and a wingless neuter-female worker caste. The vast majority of individual ants are workers; the ants familiar to casual observers are usually members of this caste. Virgin winged females and males are abroad only during a limited, usually annual, season when they take part in mating flights. After these flights the males disperse and die, and the females, as the foundress queens of new colonies, shed their wings, secrete themselves in the soil or elsewhere, and begin to lay worker-producing eggs. With few exceptions, mature ant colonies include a single or very few coeval mated queens along with a large force of daughter workers. In addition, alate virgin males and females may be present during the weeks or months prior to their release for the mating flight. Eggs, larvae and pupae are usually also present in the nests, though the brood composition can vary seasonally, and broods may be absent during winter.

Ants have a distinctive habitus, though they may be confused (among non-mimics) with wingless females of the families Mutillidae and Thynnidae and certain other wingless Hymenoptera. All ants have a nodiform, binodal or scale-like "waist" consisting of the modified true abdominal segments II or II+III. The antennae of females are usually elbowed, with the basal segment or "scape" much longer than any of the succeeding "funicular" segments. With few exceptions, ants have a large "metapleural gland" with a small external orifice which opens on each side of the metathorax, at the lower posterior corners of the mesosoma, above the hind coxae.

The family Formicidae is treated as coextensive with the superfamily Formicoidea in the classification followed here. Some European authors tend to elevate the sub-families recognized here to family status. The recently proposed classification of the Hymenoptera by D.J. Brothers (1975) reduces the previously and commonly accepted seven superfamilies of aculeate Hymenoptera to three. The family Formicidae is placed in superfamily Vespoidea, along with eleven other families. Of the two informal groups included in the Vespoidea, the "Formiciformes" contains only the family Formicidae. It is thus equivalent as a taxon to the traditionally recognized superfamily Formicoidea, as used here.

References

Brothers, D.J. (1975). Phylogeny and classification of the aculeate Hymenoptera, with special reference to Mutillidae. *Univ. Kansas Sci. Bull.* **50**: 483–648

NOTHOMYRMECIINAE

Nothomyrmecia Clark, 1934

Nothomyrmecia Clark, J. (1934). Notes on Australian ants, with descriptions of new species and a new genus. *Mem. Natl. Mus. Vict.* **8**: 5–20 [17 pl 1]. Type species *Nothomyrmecia macrops* Clark, 1934 by original designation.

Nothomyrmecia macrops Clark, 1934

Nothomyrmecia macrops Clark, J. (1934). Notes on Australian ants, with descriptions of new species and a new genus. *Mem. Natl. Mus. Vict.* **8**: 5-20 [19 pl 1]. Type data: syntypes, NMV *W, from Russell Range, W.A.

Distribution: W plateau, S.A., W.A. Ecology: terrestrial, nocturnal, predator, woodland; nest in soil. Biological references: Taylor, R.W. (1978). *Nothomyrmecia macrops*: a living-fossil ant rediscovered. *Science* **201**: 979-985 (phylogeny, bionomics).

MYRMECIINAE

Myrmecia Fabricius, 1804

Myrmecia Fabricius, J.C. (1804). *Systema Piezatorum*. Brunsvigae [423]. Type species *Formica gulosa* Fabricius, 1775 by subsequent designation, see Shuckard, W.E. (1840). *Hist. and Nat. Arrang. Ins.* [173]. Compiled from secondary source: Wheeler, W.M. (1913). Corrections to "List of type species of the genera and subgenera of Formicidae". *Ann. N.Y. Acad. Sci.* **23**: 77-83 [29 May 1913].
Promyrmecia Emery, C. (1911). Hymenoptera Fam. Formicidae subfam. Ponerinae. *in* Wytsman, P. (ed.) *Genera Insectorum*. Fasc. 118 Brussels 125 pp. 3 pls [19] [proposed with subgeneric rank in *Myrmecia* Fabricius, 1804]. Type species *Myrmecia aberrans* Forel, 1900 by original designation.
Pristomyrmecia Emery, C. (1911). Hymenoptera. Fam. Formicidae, subfam. Ponerinae. *in* Wytsman, P. (ed.) *Genera Insectorum*. Fasc. 118 Brussels 125 pp. 3 pls [21] [proposed with subgeneric rank in *Myrmecia* Fabricius, 1804]. Type species *Myrmecia mandibularis* F. Smith, 1858 by original designation.
Halmamyrmecia Wheeler, W.M. (1922). Observations on *Gigantiops destructor* Fabricius and other leaping ants. *Biol. Bull. Mar. Biol. Lab., Woods Hole* **42**: 185-201 [195] [proposed with subgeneric rank in *Myrmecia* Fabricius, 1804]. Type species *Myrmecia pilosula* F. Smith, 1858 by original designation.

Synonymy that of Clark, J. (1951). *The Formicidae of Australia*. Subfamily Myrmeciinae. Melbourne : CSIRO. Vol. 1 230 pp. [119]; Brown, W.L. jr. (1953). Characters and synonymies among the genera of ants. Part I. *Breviora* **11**: 1-13 [20 Mar. 1953] [1].

This group is also found in New Caledonia (one endemic species) and New Zealand (one introduced species).

Myrmecia aberrans Forel, 1900

Myrmecia aberrans Forel, A. (1900). Ponerinae et Dorylinae d'Australie récoltées par MM. Turner, Froggatt, Nugent, Chase, Rothney, J.J. Walker, etc. *Ann. Soc. Entomol. Belg.* **44**: 54-77 [54]. Type data: syntypes, GMNH W, from Gawlertown, S.A.

Distribution: S Gulfs, S.A. Ecology: terrestrial, noctidiurnal, predator; nest in soil.

Myrmecia analis Mayr, 1862

Myrmecia analis Mayr, G.L. (1862). Myrmecologische Studien. *Verh. Zool.-Bot. Ges. Wien* **12**: Abhand. 649-776 [725,728 pl 19]. Type data: holotype, NHMW W, from Australia (as New Holland).
Myrmecia atriscapa Crawley, W.C. (1925). New ants from Australia. II. *Ann. Mag. Nat. Hist. (9)* **16**: 577-598 [580]. Type data: syntypes, OUM *W, from Albany, W.A.

Synonymy that of Clark, J. (1951). *The Formicidae of Australia*. Subfamily Myrmeciinae. Melbourne : CSIRO Vol. 1 230 pp. [54].

Distribution: SW coastal, SE coastal, NE coastal, Vic., N.S.W., Qld., W.A. Ecology: terrestrial, noctidiurnal, predator, open scrub, woodland, open forest; nest in soil.

Myrmecia arnoldi Clark, 1951

Myrmecia arnoldi Clark, J. (1951). *The Formicidae of Australia*. Subfamily Myrmeciinae. Melbourne : CSIRO Vol. 1 230 pp. [36]. Type data: holotype, ANIC W, from Emu Rock, W.A.

Distribution: SW coastal, W plateau, W.A. Ecology: terrestrial, noctidiurnal, predator, woodland, open forest; nest in soil.

Myrmecia atrata Clark, 1951

Myrmecia atrata Clark, J. (1951). *The Formicidae of Australia*. Subfamily Myrmeciinae. Melbourne : CSIRO Vol. 1 230 pp. [77]. Type data: holotype, ANIC W, from Ravensthorpe, W.A.

Distribution: SW coastal, W.A. Ecology: terrestrial, noctidiurnal, predator, woodland, open forest; nest in soil.

Myrmecia auriventris Mayr, 1870

Myrmecia auriventris Mayr, G.L. (1870). Neue Formiciden. *Verh. Zool.-Bot. Ges. Wien* **20**: Abhand. 939-996 [31 Dec. 1870] [968]. Type data: syntypes, NHMW W, from Port Mackay and Cape York, Qld.
Myrmecia auriventris athertonensis Forel, A. (1915). Results of Dr. E. Mjöbergs Swedish Scientific Expeditions to Australia 1910-1913. 2. Ameisen. *Ark. Zool.* **9**: 1-119 pls 1-3 [4 Dec. 1915] [8]. Type data: syntypes, GMNH W,M, ANIC W, other syntypes may exist, from Atherton, Qld.

Synonymy that of Brown, W.L. jr. (1953). Revisionary notes on the ant genus *Myrmecia* of Australia. *Bull. Mus. Comp. Zool.* **111**: 1-35 [10].

Distribution: NE coastal, Qld. Ecology: terrestrial, noctidiurnal, predator, open forest; nest in soil.

Myrmecia brevinoda Forel, 1910

Myrmecia forficata brevinoda Forel, A. (1910). Formicides australiens reçus de MM. Froggatt et Rowland Turner. *Rev. Suisse Zool.* **18**: 1-94 [2]. Type data: syntypes, GMNH W,F, ANIC W, from N.S.W. and Gisborne, Vic.

Myrmecia pyriformis gigas Forel, A. (1913). Formicides du Congo Belge récoltées par MM. Bequaert, Luja, etc. *Rev. Zool. Afr.* **2**: 306–351 [30 May 1913] [310]. Type data: syntypes, GMNH,RMB *W, from Qld.

Myrmecia forficata eudoxia Forel, A. (1915). Results of Dr. E. Mjöbergs Swedish Scientific Expeditions to Australia 1910–1913. 2. Ameisen. *Ark. Zool.* **9**: 1–119 pls 1–3 [4 Dec. 1915] [8]. Type data: syntypes, GMNH W, other syntypes may exist, from Atherton, Qld.

Synonymy that of Clark, J. (1951). *The Formicidae of Australia*. Subfamily Myrmeciinae. Melbourne : CSIRO Vol. 1 230 pp. [104]; Brown, W.L. jr. (1953). Revisionary notes on the ant genus *Myrmecia* of Australia. *Bull. Mus. Comp. Zool.* **111**: 1–35 [22].

Distribution: NE coastal, SE coastal, Murray-Darling basin, Qld., N.S.W. Ecology: terrestrial, noctidiurnal, predator, woodland, open forest; nest in soil.

Myrmecia callima (Clark, 1943)

Promyrmecia callima Clark, J. (1943). A revision of the genus *Promyrmecia* Emery (Formicidae). *Mem. Natl. Mus. Vict.* **13**: 83–149 pls 12–17 [125]. Type data: syntypes, NMV *W, from Kiata, Vic.

Distribution: Murray-Darling basin, Vic. Ecology: terrestrial, noctidiurnal, predator, desert, open forest; nest in soil.

Myrmecia cardigaster Brown, 1953

Myrmecia cordata Clark, J. (1951). *The Formicidae of Australia*. Subfamily Myrmeciinae. Melbourne : CSIRO Vol. 1 230 pp. [116] [*non Myrmecia cordata* Fabricius, 1805 = *Daceton armigerum* Latreille, 1802]. Type data: holotype, ANIC W, from Malanda, Qld.

Myrmecia cardigaster Brown, W.L. jr. (1953). Revisionary notes on the ant genus *Myrmecia* of Australia. *Bull. Mus. Comp. Zool.* **111**: 1–35 [28] [*nom. nov.* for *Myrmecia cordata* Clark, 1951].

Distribution: NE coastal, Qld. Ecology: terrestrial, noctidiurnal, predator, (open forest), (closed forest); nest in soil.

Myrmecia celaena (Clark, 1943)

Promyrmecia celaena Clark, J. (1943). A revision of the genus *Promyrmecia* Emery (Formicidae). *Mem. Natl. Mus. Vict.* **13**: 83–149 pls 12–17 [120]. Type data: syntypes, NMV *W, from Pilliga and Narrabri, N.S.W. and Millmerran, Qld.

Distribution: Murray-Darling basin, N.S.W., Qld. Ecology: terrestrial, noctidiurnal, predator, woodland, open forest; nest in soil.

Myrmecia cephalotes (Clark, 1943)

Promyrmecia cephalotes Clark, J. (1943). A revision of the genus *Promyrmecia* Emery (Formicidae). *Mem. Natl. Mus. Vict.* **13**: 83–149 pls 12–17 [123]. Type data: syntypes, NMV *W,F,M, from Cooper's Creek and Killalpaninna, S.A.

Distribution: Lake Eyre basin, S.A. Ecology: terrestrial, noctidiurnal, predator, desert, woodland; nest in soil.

Myrmecia chasei Forel, 1894

Myrmecia chasei chasei Forel, 1894

Myrmecia chasei Forel, A. (1894). Quelques fourmis de Madagascar (récoltées par M. le Dr. Völtzkow); de Nouvelle Zélande (récoltées par M. W.W. Smith); de Nouvelle Calédonie (récoltées par M. Sommer); de Queensland (Australie) récoltées par M. Wiederkehr; et de Perth (Australie occidentale) récoltées par M. Chase. *Ann. Soc. Entomol. Belg.* **38**: 226–237 [235]. Type data: holotype, GMNH W, from Perth, W.A.

Myrmecia pilosula mediorubra Forel, A. (1910). Formicides australiens reçus de MM. Froggatt et Rowland Turner. *Rev. Suisse Zool.* **18**: 1–94 [7]. Type data: holotype, GMNH W, from King George Sound, W.A.

Synonymy that of Clark, J. (1951). *The Formicidae of Australia*. Subfamily Myrmeciinae. Melbourne : CSIRO Vol. 1 230 pp. [212].

Distribution: SW coastal, W.A. Ecology: terrestrial, noctidiurnal, predator, woodland, open forest; nest in soil.

Myrmecia chasei ludlowi Crawley, 1922

Myrmecia chasei ludlowi Crawley, W.C. (1922). New ants from Australia. *Ann. Mag. Nat. Hist. (9)* **9**: 427–448 [431]. Type data: syntypes, OUM *W, from Ludlow, W.A.

Distribution: SW coastal, W.A. Ecology: terrestrial, noctidiurnal, predator, woodland, open forest; nest in soil.

Myrmecia chrysogaster (Clark, 1943)

Promyrmecia chrysogaster Clark, J. (1943). A revision of the genus *Promyrmecia* Emery (Formicidae). *Mem. Natl. Mus. Vict.* **13**: 83–149 pls 12–17 [114]. Type data: syntypes (probable), NMV *W, from Brisbane, Qld.

Distribution: NE coastal, Qld. Ecology: terrestrial, noctidiurnal, predator, woodland, open forest; nest in soil.

Myrmecia clarki Crawley, 1922

Myrmecia clarki Crawley, W.C. (1922). New ants from Australia. *Ann. Mag. Nat. Hist. (9)* **9**: 427–448 [432]. Type data: syntypes, OUM *W, from Mundaring Weir, W.A.

Distribution: SW coastal, W.A. Ecology: terrestrial, noctidiurnal, predator, woodland, open forest; nest in soil.

Myrmecia comata Clark, 1951

Myrmecia comata Clark, J. (1951). *The Formicidae of Australia*. Subfamily Myrmeciinae. Melbourne : CSIRO Vol. 1 230 pp. [43]. Type data: holotype, ANIC W, from Bunya Mts., Qld.

Distribution: NE coastal, Qld. Ecology: terrestrial, noctidiurnal, predator, open forest; nest in soil.

Myrmecia cydista (Clark, 1943)

Promyrmecia cydista Clark, J. (1943). A revision of the genus *Promyrmecia* Emery (Formicidae). *Mem. Natl. Mus. Vict.* **13**: 83–149 pls 12–17 [115]. Type data: syntypes, NMV *W, from Lismore, Dorrigo, Sydney, and Wahroonga, N.S.W.

Distribution: SE coastal, N.S.W. Ecology: terrestrial, noctidiurnal, predator, woodland, open forest; nest in soil.

Myrmecia decipians Clark, 1951

Myrmecia decipians Clark, J. (1951). *The Formicidae of Australia.* Subfamily Myrmeciinae. Melbourne : CSIRO Vol. 1 230 pp. [86]. Type data: holotype, ANIC W, from Quirindi, N.S.W.

Distribution: Murray-Darling basin, N.S.W. Ecology: terrestrial, noctidiurnal, predator, woodland; nest in soil.

Myrmecia desertorum Wheeler, 1915

Myrmecia vindex desertorum Wheeler, W.M. (1915). Hymenoptera. *Trans. R. Soc. S. Aust.* **39**: 805–823 pls 64–66 [Dec. 1915] [805]. Type data: syntypes, MCZ *W, from Todmorden, S.A.

Myrmecia lutea Crawley, W.C. (1922). New ants from Australia. *Ann. Mag. Nat. Hist.* (9) **9**: 427–448 [429]. Type data: syntypes, OUM *W, from Ludlow, W.A.

Myrmecia princeps Clark, J. (1951). *The Formicidae of Australia.* Subfamily Myrmeciinae. Melbourne : CSIRO Vol. 1 230 pp. [46]. Type data: holotype, ANIC W, from Tarcoola, S.A.

Myrmecia fuscipes Clark, J. (1951). *The Formicidae of Australia.* Subfamily Myrmeciinae. Melbourne : CSIRO Vol. 1 230 pp. [62]. Type data: holotype, ANIC W, from Port Lincoln, S.A.

Synonymy that of Brown, W.L. jr. (1953). Revisionary notes on the ant genus *Myrmecia* of Australia. *Bull. Mus. Comp. Zool.* **111**: 1–35 [25].

Distribution: W plateau, Lake Eyre basin, SW coastal, S.A., W.A. Ecology: terrestrial, noctidiurnal, predator, desert, woodland; nest in soil. Biological references: Gray, B. (1971). Notes on the field behaviour of two ant species *Myrmecia desertorum* Wheeler and *Myrmecia dispar* (Clark) (Hymenoptera : Formicidae). *Insectes Soc.* **18**: 81–94 (foraging behaviour).

Myrmecia dichospila Clark, 1938

Myrmecia (Promyrmecia) dichospila Clark, J. (1938). Reports of the McCoy Society for Field Investigation and Research. No. 2. Sir Joseph Bank Islands. Part I. Formicidae (Hymenoptera). *Proc. R. Soc. Vict.* **50**: 356–382 [359]. Type data: syntypes, NMV *W,F,M, from Reevesby Is., S.A.

Distribution: S Gulfs, S.A. Ecology: terrestrial, noctidiurnal, predator, desert, woodland; nest in soil.

Myrmecia dimidiata Clark, 1951

Myrmecia dimidiata Clark, J. (1951). *The Formicidae of Australia.* Subfamily Myrmeciinae. Melbourne : CSIRO Vol. 1 230 pp. [71]. Type data: holotype, ANIC W, from Stanthorpe, Qld.

Distribution: Murray-Darling basin, Qld. Ecology: terrestrial, noctidiurnal, predator, woodland; nest in soil.

Myrmecia dispar (Clark, 1951)

Promyrmecia dispar Clark, J. (1951). *The Formicidae of Australia.* Subfamily Myrmeciinae. Melbourne : CSIRO Vol. 1 230 pp. [226]. Type data: syntypes, ANIC W, from Cowra and Junee, N.S.W.

Distribution: Murray-Darling basin, Vic. Ecology: terrestrial, noctidiurnal, predator, desert, woodland; nest in soil. Biological references: Gray, B. (1971). Notes on the field behaviour of two ant species *Myrmecia desertorum* Wheeler and *Myrmecia dispar* (Clark) (Hymenoptera : Formicidae). *Insectes Soc.* **18**: 81–94 (foraging behaviour).

Myrmecia dixoni (Clark, 1943)

Promyrmecia dixoni Clark, J. (1943). A revision of the genus *Promyrmecia* Emery (Formicidae). *Mem. Natl. Mus. Vict.* **13**: 83–149 pls 12–17 [135]. Type data: syntypes, NMV *W,F, from Eltham, Vic., Albury, N.S.W. and Canberra, A.C.T.

Distribution: SE coastal, Murray-Darling basin, N.S.W., Vic., A.C.T. Ecology: terrestrial, noctidiurnal, predator, woodland, open forest; nest in soil.

Myrmecia elegans (Clark, 1943)

Promyrmecia elegans Clark, J. (1943). A revision of the genus *Promyrmecia* Emery (Formicidae). *Mem. Natl. Mus. Vict.* **13**: 83–149 pls 12–17 [122]. Type data: syntypes, NMV *W,F, from Hovea, Mt. Dale and Mundaring, W.A.

Distribution: SW coastal, W.A. Ecology: terrestrial, noctidiurnal, predator, woodland; nest in soil.

Myrmecia esuriens Fabricius, 1804

Myrmecia esuriens Fabricius, J.C. (1804). *Systema Piezatorum.* Brunsvigae [424]. Type data: uncertain, whereabouts unknown, from Australia, see Roger, J. (1861). Die *Ponera*-Artigen Ameisen. *Berl. Entomol. Z.* **5**: 1–54 [35].

Myrmecia tasmaniensis Smith, F. (1858). *Catalogue of hymenopterous insects in the collection of the British Museum.* Part 6. Formicidae. London : British Museum 216 pp. 14 pls [27 Mar. 1858] [147]. Publication date established from Donisthorpe, H. (1932). On the identity of Smith's types of Formicidae (Hymenoptera) collected by Alfred Russell Wallace in the Malay Archipelago, with descriptions of two new species. *Ann. Mag. Nat. Hist. (10)* **10**: 441–476. Type data: syntypes (probable), BMNH *W, from Tas.

Myrmecia walkeri Forel, A. (1893). Nouvelles fourmis d'Australie et des Canaries. *Ann. Soc. Entomol. Belg.* **37**: 454–466 [456]. Type data: syntypes, GMNH W, from Hobart, Tas.

Synonymy that of Emery, C. (1911). Hymenoptera Fam. Formicidae subfam. Ponerinae. *in* Wytsman, P. (ed.) *Genera Insectorum.* Fasc. 118 Brussels 125 pp. 3 pls [20].

Distribution: Tas. Ecology: terrestrial, noctidiurnal, predator, woodland, open forest; nest in ground layer.

Myrmecia eupoecila (Clark, 1943)

Promyrmecia eupoecila Clark, J. (1943). A revision of the genus *Promyrmecia* Emery (Formicidae). *Mem. Natl. Mus. Vict.* **13**: 83–149 pls 12–17 [98]. Type data: syntypes (probable), NMV *F, from Adelaide, S.A.

Distribution: S Gulfs, S.A. Ecology: terrestrial, noctidiurnal, predator, woodland, open forest; nest in soil.

Myrmecia excavata (Clark, 1951)

Promyrmecia excavata Clark, J. (1951). *The Formicidae of Australia.* Subfamily Myrmeciinae. Melbourne : CSIRO Vol. 1 230 pp. [137]. Type data: holotype, ANIC W, from Bundarra, N.S.W.

Distribution: Murray-Darling basin, N.S.W. Ecology: terrestrial, noctidiurnal, predator, woodland; nest in soil.

Myrmecia exigua (Clark, 1943)

Promyrmecia exigua Clark, J. (1943). A revision of the genus *Promyrmecia* Emery (Formicidae). *Mem. Natl. Mus. Vict.* **13**: 83–149 pls 12–17 [107]. Type data: syntypes, NMV *W, from Lake Hattah, Vic.

Distribution: Murray-Darling basin, Vic. Ecology: terrestrial, noctidiurnal, predator, woodland; nest in soil.

Myrmecia fasciata Clark, 1951

Myrmecia fasciata Clark, J. (1951). *The Formicidae of Australia.* Subfamily Myrmeciinae. Melbourne : CSIRO Vol. 1 230 pp. [63]. Type data: holotype, ANIC W, from Pilliga, N.S.W.

Distribution: Murray-Darling basin, N.S.W. Ecology: terrestrial, noctidiurnal, predator, woodland; nest in soil.

Myrmecia ferruginea Mayr, 1876

Myrmecia nigriceps ferruginea Mayr, G.L. (1876). Die australischen Formiciden. *J. Mus. Godeffroy* **5**: 56–115 [95]. Type data: syntypes, NHMW W, from Peak Downs, Qld.

Distribution: NE coastal, Qld. Ecology: terrestrial, noctidiurnal, predator, woodland, open forest; nest in soil. Biological references: Brown, W.L. jr. (1953). Revisionary notes on the ant genus *Myrmecia* of Australia. *Bull. Mus. Comp. Zool.* **111**: 1–35 (raised to species).

Myrmecia flammicollis Brown, 1953

Myrmecia flammicollis Brown, W.L. jr. (1953). Revisionary notes on the ant genus *Myrmecia* of Australia. *Bull. Mus. Comp. Zool.* **111**: 1–35 [23]. Type data: holotype, MCZ *W, from The Rocky Scrub around the headwaters of the Rocky River, in the McIlwraith Range, NE of Coen, Cape York Peninsula, Qld.

Distribution: NE coastal, Qld. Ecology: terrestrial, noctidiurnal, predator, open forest, closed forest; nest in soil.

Myrmecia flavicoma Roger, 1861

Myrmecia flavicoma flavicoma Roger, 1861

Myrmecia flavicoma Roger, J. (1861). Myrmicologische Nachlese. *Berl. Entomol. Z.* **5**: 163–174 [171]. Type data: syntypes, MNHP *W, from Australia.

Distribution: NE coastal, Qld. Ecology: terrestrial, noctidiurnal, predator, woodland, open forest; nest in soil.

Myrmecia flavicoma minuscula Forel, 1915

Myrmecia flavicoma minuscula Forel, A. (1915). Results of Dr. E. Mjöbergs Swedish Scientific Expeditions to Australia 1910–1913. 2. Ameisen. *Ark. Zool.* **9**: 1–119 pls 1–3 [4 Dec. 1915] [8]. Type data: syntypes, GMNH W, ANIC W, other syntypes may exist, from Malanda and Cedar Creek, Qld.

Distribution: NE coastal, Qld. Ecology: terrestrial, noctidiurnal, predator, (open forest); nest in soil.

Myrmecia forceps Roger, 1861

Myrmecia forceps Roger, J. (1861). Die *Ponera*-Artigen Ameisen. *Berl. Entomol. Z.* **5**: 1–54 [34]. Type data: syntypes (probable), BMN *W, from Australia (as New Holland).

Myrmecia forceps obscuriceps Viehmeyer, H. (1924). Formiciden der australischen Faunenregion. *Entomol. Mitt.* **13**: 219–229 [222]. Type data: syntypes (probable), ZMB *W, from Liverpool, N.S.W.

Myrmecia singularis Clark, J. (1951). *The Formicidae of Australia.* Subfamily Myrmeciinae. Melbourne : CSIRO Vol. 1 230 pp. [26]. Type data: holotype, ANIC W, from Kangaroo Is., S.A.

Synonymy that of Clark, J. (1951). *The Formicidae of Australia.* Subfamily Myrmeciinae. Melbourne : CSIRO Vol. 1 230 pp. [24]; Brown, W.L. jr. (1953). Revisionary notes on the ant genus *Myrmecia* of Australia. *Bull. Mus. Comp. Zool.* **111**: 1–35 [7].

Distribution: SE coastal, S Gulfs, N.S.W., S.A. Ecology: terrestrial, noctidiurnal, predator, woodland, open forest; nest in soil. Biological references: Freeland, J. (1958). Biological and social patterns in the Australian bulldog ants of the genus *Myrmecia*. *Aust. J. Zool.* **6**: 1–18 (social behaviour).

Myrmecia forficata (Fabricius, 1787)

Formica forficata Fabricius, J.C. (1787). *Mantissa Insectorum sistens eorum species nuper detectas adiectis characteribus genericis, differentiis specificis, emendationibus, observationibus.* Hafniae Vol. 1 [310]. Type data: holotype (probable), BMNH W, from Tas.
Myrmecia lucida Forel, A. (1893). Nouvelles fourmis d'Australie et des Canaries. *Ann. Soc. Entomol. Belg.* **37**: 454–466 [457]. Type data: syntypes (probable), GMNH W, from Hobart, Tas.
Myrmecia forficata rubra Forel, A. (1910). Formicides australiens reçus de MM. Froggatt et Rowland Turner. *Rev. Suisse Zool.* **18**: 1–94 [3]. Type data: syntypes, GMNH W, from Jarra distr., Vic.

Synonymy that of Clark, J. (1951). *The Formicidae of Australia.* Subfamily Myrmeciinae. Melbourne : CSIRO Vol. 1 230 pp. [93]; Brown, W.L. jr. (1953). Revisionary notes on the ant genus *Myrmecia* of Australia. *Bull. Mus. Comp. Zool.* **111**: 1–35 [28].

Distribution: SE coastal, Vic., Tas. Ecology: terrestrial, noctidiurnal, predator, alpine, woodland, open forest; nest in ground layer.

Myrmecia froggatti Forel, 1910

Myrmecia froggatti Forel, A. (1910). Formicides australiens reçus de MM. Froggatt et Rowland Turner. *Rev. Suisse Zool.* **18**: 1–94 [9] [introduced as *froggati*, incorrect spelling of collector, Froggatt]. Type data: holotype, GMNH W, from Manilla, N.S.W.
Myrmecia (Promyrmecia) aberrans taylori Wheeler, W.M. (1933). *Colony-founding among ants with an account of some primitive Australian species.* Cambridge : Harvard Univ. Press 179 pp. [53]. Type data: holotype, MCZ *W, from Roma distr., Qld.
Myrmecia (Promyrmecia) aberrans sericata Wheeler, W.M. (1933). *Colony-founding among ants with an account of some primitive Australian species.* Cambridge : Harvard Univ. Press 179 pp. [53]. Type data: holotype, MCZ *W, from Wagga Wagga, N.S.W.

Synonymy that of Brown, W.L. jr. (1953). Revisionary notes on the ant genus *Myrmecia* of Australia. *Bull. Mus. Comp. Zool.* **111**: 1–35 [17].

Distribution: Murray-Darling basin, N.S.W., Qld. Ecology: terrestrial, noctidiurnal, predator, woodland, open forest; nest in soil.

Myrmecia fucosa Clark, 1934

Myrmecia (Promyrmecia) fucosa Clark, J. (1934). Notes on Australian ants, with descriptions of new species and a new genus. *Mem. Natl. Mus. Vict.* **8**: 5–20 [15 pl 1].

Type data: syntypes, NMV *W,F, from Lake Hattah, Ouyen, Sea Lake, Wyperfield, Vic. and Murray Bridge, S.A.

Distribution: Murray-Darling basin, Vic., S.A. Ecology: terrestrial, noctidiurnal, predator, desert, woodland, open forest; nest in soil.

Myrmecia fulgida Clark, 1951

Myrmecia fulgida Clark, J. (1951). *The Formicidae of Australia.* Subfamily Myrmeciinae. Melbourne : CSIRO Vol. 1 230 pp. [73]. Type data: holotype, ANIC W, from Parker's Range, W.A.

Distribution: W plateau, W.A. Ecology: terrestrial, noctidiurnal, predator, desert, woodland; nest in soil.

Myrmecia fulviculis Forel, 1913

Myrmecia (Pristomyrmecia) fulvipes fulviculis Forel, A. (1913). Fourmis de Tasmanie et d'Australie récoltées par MM. Lea, Froggatt etc. *Bull. Soc. Vaud. Sci. Nat.* **49**: 173–196 pl 2 [174]. Type data: syntypes, GMNH W, from Sydney, N.S.W., see Clark, J. (1943). A revision of the genus *Promyrmecia* Emery (Formicidae). *Mem. Natl. Mus. Vict.* **13**: 83–149 pls 12–17.

Distribution: Tas. Ecology: terrestrial, noctidiurnal, predator, woodland, open forest; nest in soil. Biological references: Clark, J. (1943). A revision of the genus *Promyrmecia* Emery (Formicidae). *Mem. Natl. Mus. Vict.* **13**: 83–149 (raised to species).

Myrmecia fulvipes Roger, 1861

Myrmecia fulvipes Roger, J. (1861). Die *Ponera*-Artigen Ameisen. *Berl. Entomol. Z.* **5**: 1–54 [36]. Type data: holotype, MNHP *W, from Australia.
Myrmecia (Pristomyrmecia) piliventris femorata Santschi, F. (1928). Nouvelles fourmis d'Australie. *Bull. Soc. Vaud. Sci. Nat.* **56**: 465–483 [30 Aug. 1928] [466]. Type data: syntypes, BNHM W, from Franktown (=Frankston), Vic.
Myrmecia (Promyrmecia) fulvipes barbata Wheeler, W.M. (1933). *Colony-founding among ants with an account of some primitive Australian species.* Cambridge : Harvard Univ. Press 179 pp. [71]. Type data: syntypes, MCZ *W,F, from Dorrigo, N.S.W. and Belgrade (=Belgrave) Vic.

Synonymy that of Brown, W.L. jr. (1953). Revisionary notes on the ant genus *Myrmecia* of Australia. *Bull. Mus. Comp. Zool.* **111**: 1–35 [21].

Distribution: SE coastal, N.S.W., Vic. Ecology: terrestrial, noctidiurnal, predator, open heath, woodland; nest in soil.

Myrmecia gilberti Forel, 1910

Myrmecia fulvipes gilberti Forel, A. (1910). Formicides australiens reçus de MM. Froggatt et Rowland Turner. *Rev. Suisse Zool.* **18**: 1–94 [6]. Type data: syntypes, GMNH W, ANIC W, from Mackay, Qld.

Myrmecia (Pristomyrmecia) regina Santschi, F. (1928). Nouvelles fourmis d'Australie. *Bull. Soc. Vaud. Sci. Nat.* **56**: 465-483 [30 Aug. 1928] [465]. Type data: syntypes, BNHM W, from Townsville, Qld.

Synonymy that of Clark, J. (1951). *The Formicidae of Australia.* Subfamily Myrmeciinae. Melbourne : CSIRO Vol. 1 230 pp. [169].

Distribution: NE coastal, Qld. Ecology: terrestrial, noctidiurnal, predator, woodland, open forest; nest in soil.

Myrmecia gratiosa Clark, 1951

Myrmecia gratiosa Clark, J. (1951). *The Formicidae of Australia.* Subfamily Myrmeciinae. Melbourne : CSIRO Vol. 1 230 pp. [66]. Type data: holotype, ANIC W, from Bendering, W.A.

Distribution: SW coastal, W.A. Ecology: terrestrial, noctidiurnal, predator, woodland; nest in soil.

Myrmecia greavesi (Clark, 1943)

Promyrmecia greavesi Clark, J. (1943). A revision of the genus *Promyrmecia* Emery (Formicidae). *Mem. Natl. Mus. Vict.* **13**: 83-149 pls 12-17 [99]. Type data: syntypes (probable), NMV *F, from Mareeba, Qld.

Distribution: NE coastal, Qld. Ecology: terrestrial, noctidiurnal, predator, woodland; nest in soil.

Myrmecia gulosa (Fabricius, 1775)

Formica gulosa Fabricius, J.C. (1775). *Systema Entomologiae,* sistens insectorum classes, ordines, genera, species, adiectis synonymis, locis, descriptionibus, observationibus. Flensburgi et Lipsiae [395]. Type data: uncertain, BMNH W, from Australia (as New Holland).

Myrmecia gulosa obscurior Forel, A. (1922). Glanures myrmécologiques en 1922. *Rev. Suisse Zool.* **30**: 87-102 [87]. Type data: syntypes, GMNH W, from Australia.

Synonymy that of Clark, J. (1951). *The Formicidae of Australia.* Subfamily Myrmeciinae. Melbourne : CSIRO Vol. 1 230 pp. [49].

Distribution: NE coastal, SE coastal, Qld., N.S.W. Ecology: terrestrial, noctidiurnal, predator, open heath, woodland, open forest; nest in soil. Biological references: Freeland, J. (1958). Biological and social patterns in the Australian bulldog ants of the genus *Myrmecia. Aust. J. Zool.* **6**: 1-18 (social behaviour).

Myrmecia harderi Forel, 1910

Myrmecia harderi Forel, A. (1910). Formicides australiens reçus de MM. Froggatt et Rowland Turner. *Rev. Suisse Zool.* **18**: 1-94 [8]. Type data: syntypes, GMNH W, ANIC W, from Gundah, N.S.W.

Promyrmecia scabra Clark, J. (1943). A revision of the genus *Promyrmecia* Emery (Formicidae). *Mem. Natl. Mus. Vict.* **13**: 83-149 pls 12-17 [119]. Type data: syntypes, NMV *W,F, from Leigh Creek, S.A.

Promyrmecia maloni Clark, J. (1943). A revision of the genus *Promyrmecia* Emery (Formicidae). *Mem. Natl. Mus. Vict.* **13**: 83-149 pls 12-17 [121]. Type data: syntypes, NMV *W, from Inglewood, Vic.

Synonymy that of Brown, W.L. jr. (1953). Revisionary notes on the ant genus *Myrmecia* of Australia. *Bull. Mus. Comp. Zool.* **111**: 1-35 [16].

Distribution: Murray-Darling basin, S Gulfs, N.S.W., Vic., S.A. Ecology: terrestrial, noctidiurnal, predator, woodland; nest in soil.

Myrmecia hilli (Clark, 1943)

Promyrmecia hilli Clark, J. (1943). A revision of the genus *Promyrmecia* Emery (Formicidae). *Mem. Natl. Mus. Vict.* **13**: 83-149 pls 12-17 [125]. Type data: syntypes (probable), NMV *W, from Finke River, N.T.

Distribution: Lake Eyre basin, N.T. Ecology: terrestrial, noctidiurnal, predator, desert, woodland; nest in soil.

Myrmecia hirsuta Clark, 1951

Myrmecia hirsuta Clark, J. (1951). *The Formicidae of Australia.* Subfamily Myrmeciinae. Melbourne : CSIRO Vol. 1 230 pp. [109]. Type data: holotype, ANIC W, from Stawell, Vic.

Distribution: Murray-Darling basin, Vic. Ecology: terrestrial, noctidiurnal, predator, woodland, open forest; nest in soil, probably a social parasite of other *Myrmecia* species.

Myrmecia infima Forel, 1900

Myrmecia picta infima Forel, A. (1900). Ponerinae et Dorylinae d'Australie récoltées par MM. Turner, Froggatt, Nugent, Chase, Rothney, J.J. Walker, etc. *Ann. Soc. Entomol. Belg.* **44**: 54-77 [54]. Type data: holotype, GMNH W, from Perth, W.A.

Distribution: SW coastal, W.A. Ecology: terrestrial, noctidiurnal, predator, woodland, open forest; nest in soil. Biological references: Wheeler, W.M. (1933). *Colony-founding among ants with an account of some primitive Australian species.* Cambridge : Harvard Univ. Press 179 pp. (raised to species).

Myrmecia inquilina Douglas and Brown, 1959

Myrmecia inquilina Douglas, A. & Brown, W.L. jr. (1959). *Myrmecia inquilina* new species: the first parasite among the lower ants. *Insectes Soc.* **6**: 13-19 [13]. Type data: holotype, WAM 64-38 *F, from Badjanning Rocks, 4 mi NW of Wagin, W.A.

Distribution: SW coastal, W.A. Ecology: terrestrial, noctidiurnal, predator, woodland; nest in soil, workerless social parasite of other *Myrmecia* species.

Myrmecia longinodis Clark, 1951

Myrmecia longinodis Clark, J. (1951). *The Formicidae of Australia*. Subfamily Myrmeciinae. Melbourne : CSIRO Vol. 1 230 pp. [87]. Type data: holotype, ANIC W, from Kiama, N.S.W.

Distribution: SE coastal, N.S.W. Ecology: terrestrial, noctidiurnal, predator, woodland, open forest; nest in soil.

Myrmecia luteiforceps (Clark, 1943)

Promyrmecia luteiforceps Clark, J. (1943). A revision of the genus *Promyrmecia* Emery (Formicidae). *Mem. Natl. Mus. Vict.* **13**: 83–149 pls 12–17 [143] [introduced as a quadranomen by Forel, 1915]. Type data: syntypes, GMNH W, ANIC W, from Herberton, Qld.

Distribution: NE coastal, Qld. Ecology: terrestrial, noctidiurnal, predator, open forest, (closed forest); nest in soil.

Myrmecia mandibularis F. Smith, 1858

Myrmecia mandibularis Smith, F. (1858). *Catalogue of hymenopterous insects in the collection of the British Museum*. Part 6. Formicidae. London : British Museum 216 pp. 14 pls [27 Mar. 1858] [145]. Publication date established from Donisthorpe, H. (1932). On the identity of Smith's types of Formicidae (Hymenoptera) collected by Alfred Russell Wallace in the Malay Archipelago, with descriptions of two new species. *Ann. Mag. Nat. Hist.* (10) **10**: 441–476. Type data: syntypes (probable), BMNH *W, from Adelaide, S.A.

Myrmecia mandibularis aureorufa Forel, A. (1910). Formicides australiens reçus de MM. Froggatt et Rowland Turner. *Rev. Suisse Zool.* **18**: 1–94 [6]. Type data: holotype, GMNH W, from Australia.

Myrmecia (Promyrmecia) mandibularis postpetiolaris Wheeler, W.M. (1933). *Colony-founding among ants with an account of some primitive Australian species*. Cambridge : Harvard Univ. Press 179 pp. [65]. Type data: syntypes, MCZ *W,M, from Mt. Lofty, S.A., Ballarat, Vic. and Warren River, W.A.

Myrmecia (Promyrmecia) fulvipes caelatinoda Wheeler, W.M. (1933). *Colony-founding among ants with an account of some primitive Australian species*. Cambridge : Harvard Univ. Press 179 pp. [72]. Type data: holotype, lost, from Belair, S.A.

Promyrmecia laevinodis Clark, J. (1943). A revision of the genus *Promyrmecia* Emery (Formicidae). *Mem. Natl. Mus. Vict.* **13**: 83–149 pls 12–17 [139]. Type data: syntypes, NMV *W,F, from Armadale, Albany, and Bunbury, W.A., Lucindale, Melrose and Kangaroo Is., S.A. and Mallee, Vic.

Synonymy that of Clark, J. (1951). *The Formicidae of Australia*. Subfamily Myrmeciinae. Melbourne : CSIRO Vol. 1 230 pp. [151]; Brown, W.L. jr. (1953). Revisionary notes on the ant genus *Myrmecia* of Australia. *Bull. Mus. Comp. Zool.* **111**: 1–35 [4].

Distribution: SE coastal, S Gulfs, SW coastal, Vic., S.A., W.A. Ecology: terrestrial, noctidiurnal, predator; nest in soil.

Myrmecia maura Wheeler, 1933

Myrmecia maura maura Wheeler, 1933

Myrmecia (Promyrmecia) aberrans maura Wheeler, W.M. (1933). *Colony-founding among ants with an account of some primitive Australian species*. Cambridge : Harvard Univ. Press 179 pp. [51]. Type data: syntypes, MCZ *W, from Bathurst, N.S.W.

Distribution: Murray-Darling basin, N.S.W. Ecology: terrestrial, noctidiurnal, predator, woodland, open forest; nest in soil. Biological references: Brown, W.L. jr. (1953). Revisionary notes on the ant genus *Myrmecia* of Australia. *Bull. Mus. Comp. Zool.* **111**: 1–35 (raised to species).

Myrmecia maura formosa Wheeler, 1933

Myrmecia (Promyrmecia) aberrans formosa Wheeler, W.M. (1933). *Colony-founding among ants with an account of some primitive Australian species*. Cambridge : Harvard Univ. Press 179 pp. [52]. Type data: syntypes, MCZ *W, from Uralla, N.S.W.

Myrmecia (Promyrmecia) aberrans haematosticta Wheeler, W.M. (1933). *Colony-founding among ants with an account of some primitive Australian species*. Cambridge : Harvard Univ. Press 179 pp. [51]. Type data: syntypes, MCZ *W, from Uralla, N.S.W.

Synonymy that of Brown, W.L. jr. (1953). Revisionary notes on the ant genus *Myrmecia* of Australia. *Bull. Mus. Comp. Zool.* **111**: 1–35 [19].

Distribution: Murray-Darling basin, N.S.W. Ecology: terrestrial, noctidiurnal, predator, woodland, open forest; nest in soil.

Myrmecia michaelseni Forel, 1907

Myrmecia michaelseni michaelseni Forel, 1907

Myrmecia michaelseni Forel, A. (1907). Formicidae. pp. 263–310 *in* Michaelsen, W. & Hartmeyer, R. (eds.) *Die Fauna Südwest-Australiens*. Jena : G. Fischer Vol. 1 [267]. Type data: syntypes, GMNH W, ANIC W, from NE of Albany, W.A.

Myrmecia michaelseni perthensis Crawley, W.C. (1922). New ants from Australia. *Ann. Mag. Nat. Hist.* (9) **9**: 427–448 [431]. Type data: syntypes (probable), OUM *W, from Perth, W.A.

Synonymy that of Clark, J. (1951). *The Formicidae of Australia*. Subfamily Myrmeciinae. Melbourne : CSIRO Vol. 1 230 pp. [204].

Distribution: SW coastal, W.A. Ecology: terrestrial, noctidiurnal, predator, woodland, open forest; nest in soil.

Myrmecia michaelseni queenslandica Forel, 1915

Myrmecia michaelseni queenslandica Forel, A. (1915). Results of Dr. E. Mjöbergs Swedish Scientific Expeditions to Australia 1910–1913. 2. Ameisen. *Ark. Zool.* **9**: 1–119 pls 1–3 [4 Dec. 1915] [4]. Type data: holotype, SMNH *W, from Lamington Plateau, Qld.

Myrmecia michaelseni overbecki Viehmeyer, H. (1924). Formiciden der australischen Faunenregion. *Entomol. Mitt.* **13**: 219-229 [222]. Type data: syntypes, ZMB *W,F, from Trial Bay, N.S.W.
Synonymy that of Clark, J. (1951). *The Formicidae of Australia*. Subfamily Myrmeciinae. Melbourne : CSIRO Vol. 1 230 pp. [206].

Distribution: NE coastal, SE coastal, Qld., Vic. Ecology: terrestrial, noctidiurnal, predator, woodland, open forest; nest in soil.

Myrmecia midas Clark, 1951

Myrmecia midas Clark, J. (1951). *The Formicidae of Australia*. Subfamily Myrmeciinae. Melbourne : CSIRO Vol. 1 230 pp. [55]. Type data: holotype, ANIC W, from Dorrigo, N.S.W.

Distribution: SE coastal, N.S.W. Ecology: terrestrial, noctidiurnal, predator, open forest, closed forest; nest in soil.

Myrmecia mjobergi Forel, 1915

Myrmecia mjobergi Forel, A. (1915). Results of Dr. E. Mjöbergs Swedish Scientific Expeditions to Australia 1910-1913. 2. Ameisen. *Ark. Zool.* **9**: 1-119 pls 1-3 [4 Dec. 1915] [5]. Type data: syntypes, GMNH W,F, ANIC W, other syntypes may exist, from Atherton and Malanda, Qld.

Distribution: NE coastal, Qld. Ecology: terrestrial, nocturnal, predator, closed forest; nest arboreal (in epiphytes), occasionally in ground layer.

Myrmecia nigra Forel, 1907

Myrmecia picta nigra Forel, A. (1907). Formicidae. pp. 263-310 *in* Michaelsen, W. & Hartmeyer, R. (eds.) *Die Fauna Südwest-Australiens*. Jena : G. Fischer Vol. 1 [267]. Type data: holotype, probably destroyed in ZMH in WW II, from East Fremantle, W.A.

Distribution: SW coastal, W.A. Ecology: terrestrial, noctidiurnal, predator, woodland, open forest; nest in soil. Biological references: Clark, J. (1943). A revision of the genus *Promyrmecia* Emery (Formicidae). *Mem. Natl. Mus. Vict.* **13**: 83-149 pls 12-17 (raised to species).

Myrmecia nigriceps Mayr, 1862

Myrmecia nigriceps Mayr, G.L. (1862). Myrmecologische Studien. *Verh. Zool.-Bot. Ges. Wien* **12**: Abhand. 649-776 [725,728 pl 19]. Type data: syntypes, NHMW W, from Australia (as New Holland).

Distribution: SW coastal, W plateau, S Gulfs, SE coastal, Murray-Darling basin, N.S.W., A.C.T., Vic., S.A., W.A. Ecology: terrestrial, noctidiurnal, predator, desert, woodland, open forest; nest in soil.

Myrmecia nigriscapa Roger, 1861

Myrmecia nigriscapa Roger, J. (1861). Die Ponera-Artigen Ameisen. *Berl. Entomol. Z.* **5**: 1-54 [33]. Type data: syntypes, BMN *W, from Australia (as New Holland).

Distribution: SW coastal, W plateau, S Gulfs, SE coastal, NE coastal, Qld., N.S.W., Vic., S.A., W.A. Ecology: terrestrial, noctidiurnal, predator, woodland, open forest; nest in soil.

Myrmecia nigrocincta F. Smith, 1858

Myrmecia nigrocincta Smith, F. (1858). *Catalogue of hymenopterous insects in the collection of the British Museum. Part 6. Formicidae*. London : British Museum 216 pp. 14 pls [27 Mar. 1858] [147]. Publication date established from Donisthorpe, H. (1932). On the identity of Smith's types of Formicidae (Hymenoptera) collected by Alfred Russell Wallace in the Malay Archipelago, with descriptions of two new species. *Ann. Mag. Nat. Hist. (10)* **10**: 441-476. Type data: syntypes (probable), BMNH *W, from Australia.

Distribution: SE coastal, Murray-Darling basin, NE coastal, Qld., N.S.W., Vic. Ecology: terrestrial, noctidiurnal, predator, woodland, open forest, closed forest; nest in soil.

Myrmecia nobilis (Clark, 1943)

Promyrmecia nobilis Clark, J. (1943). A revision of the genus *Promyrmecia* Emery (Formicidae). *Mem. Natl. Mus. Vict.* **13**: 83-149 pls 12-17 [97]. Type data: syntypes, NMV *W,F,M, from Altona, Bacchus Marsh, Coburg, Broadmeadows, Geelong and Patho, Vic.

Distribution: Murray-Darling basin, SE coastal, Vic. Ecology: terrestrial, noctidiurnal, predator, woodland, open forest; nest in soil.

Myrmecia occidentalis (Clark, 1943)

Promyrmecia occidentalis Clark, J. (1943). A revision of the genus *Promyrmecia* Emery (Formicidae). *Mem. Natl. Mus. Vict.* **13**: 83-149 pls 12-17 [119]. Type data: syntypes, NMV *W,F, from Tammin, Eradu, Merredin and Beverley, W.A.

Distribution: SW coastal, W plateau, W.A. Ecology: terrestrial, noctidiurnal, predator, woodland; nest in soil.

Myrmecia opaca (Clark, 1943)

Promyrmecia opaca Clark, J. (1943). A revision of the genus *Promyrmecia* Emery (Formicidae). *Mem. Natl. Mus. Vict.* **13**: 83-149 pls 12-17 [123]. Type data: syntypes, NMV *W,F, from Tammin, Eradu and Dowerin, W.A.

Distribution: SW coastal, W plateau, W.A. Ecology: terrestrial, noctidiurnal, predator, woodland; nest in soil.

Myrmecia pavida Clark, 1951

Myrmecia pavida Clark, J. (1951). *The Formicidae of Australia.* Subfamily Myrmeciinae. Melbourne : CSIRO Vol. 1 230 pp. [76]. Type data: holotype, ANIC W, from Mt. Barker, W.A.

Distribution: SW coastal, W.A. Ecology: terrestrial, noctidiurnal, predator, woodland; nest in soil.

Myrmecia petiolata Emery, 1895

Myrmecia petiolata Emery, C. (1895). Descriptions de quelques fourmis nouvelles d'Australie. *Ann. Soc. Entomol. Belg.* **39**: 345-358 [345]. Type data: holotype, MCG W, from Mt. Bellenden Ker, Qld.

Distribution: NE coastal, Qld. Ecology: terrestrial, noctidiurnal, predator, closed forest; nest in soil.

Myrmecia picta F. Smith, 1858

Myrmecia picta Smith, F. (1858). *Catalogue of hymenopterous insects in the collection of the British Museum.* Part 6. Formicidae. London : British Museum 216 pp. 14 pls [27 Mar. 1858] [146]. Publication date established from Donisthorpe, H. (1932). On the identity of Smith's types of Formicidae (Hymenoptera) collected by Alfred Russell Wallace in the Malay Archipelago, with descriptions of two new species. *Ann. Mag. Nat. Hist. (10)* **10**: 441-476. Type data: syntypes, BMNH *W,F, from Adelaide, S.A.

Distribution: S Gulfs, S.A. Ecology: terrestrial, noctidiurnal, predator, woodland, open forest; nest in soil.

Myrmecia picticeps Clark, 1951

Myrmecia picticeps Clark, J. (1951). *The Formicidae of Australia.* Subfamily Myrmeciinae. Melbourne : CSIRO Vol. 1 230 pp. [47]. Type data: holotype, ANIC W, from Albany, W.A.

Distribution: SW coastal, W.A. Ecology: terrestrial, noctidiurnal, predator, woodland; nest in soil.

Myrmecia piliventris F. Smith, 1858

Myrmecia piliventris Smith, F. (1858). *Catalogue of hymenopterous insects in the collection of the British Museum.* Part 6. Formicidae. London : British Museum 216 pp. 14 pls [27 Mar. 1858] [146]. Publication date established from Donisthorpe, H. (1932). On the identity of Smith's types of Formicidae (Hymenoptera) collected by Alfred Russell Wallace in the Malay Archipelago, with descriptions of two new species. *Ann. Mag. Nat. Hist. (10)* **10**: 441-476. Type data: syntypes (probable), BMNH *W, from Australia.

Myrmecia piliventris rectidens Forel, A. (1910). Formicides australiens reçus de MM. Froggatt et Rowland Turner. *Rev. Suisse Zool.* **18**: 1-94 [5]. Type data: syntypes, GMNH W, from Kingstown, Australia".

Synonymy that of Brown, W.L. jr. (1953). Revisionary notes on the ant genus *Myrmecia* of Australia. *Bull. Mus. Comp. Zool.* **111**: 1-35 [20].

Distribution: SE coastal, N.S.W., Vic., Tas. Ecology: terrestrial, noctidiurnal, predator, open heath, woodland, open forest; nest in soil.

Myrmecia pilosula F. Smith, 1858

Myrmecia pilosula Smith, F. (1858). *Catalogue of hymenopterous insects in the collection of the British Museum.* Part 6. Formicidae. London : British Museum 216 pp. 14 pls [27 Mar. 1858] [146]. Publication date established from Donisthorpe, H. (1932). On the identity of Smith's types of Formicidae (Hymenoptera) collected by Alfred Russell Wallace in the Malay Archipelago, with descriptions of two new species. *Ann. Mag. Nat. Hist. (10)* **10**: 441-476. Type data: syntypes, BMNH *M,F,W, from Australia and Tas.".

Ponera ruginoda Smith, F. (1858). *Catalogue of hymenopterous insects in the collection of the British Museum.* Part 6. Formicidae. London : British Museum 216 pp. 14 pls [27 Mar. 1858] [93]. Publication date established from Donisthorpe, H. (1932). On the identity of Smith's types of Formicidae (Hymenoptera) collected by Alfred Russell Wallace in the Malay Archipelago, with descriptions of two new species. *Ann. Mag. Nat. Hist. (10)* **10**: 441-476. Type data: syntypes (probable), BMNH *M, from Australia.

Synonymy that of Brown, W.L. jr. (1953). Revisionary notes on the ant genus *Myrmecia* of Australia. *Bull. Mus. Comp. Zool.* **111**: 1-35 [6].

Distribution: SW coastal, S Gulfs, SE coastal, NE coastal, Murray-Darling basin, Qld., N.S.W., A.C.T., Vic., Tas., S.A., W.A. Ecology: terrestrial, noctidiurnal, predator, alpine, woodland, open forest; nest in soil. Biological references: Craig, R. & Crozier, R.H. (1979). Relatedness in the polygynous ant *Myrmecia pilosula*. *Evolution* **33**: 335-341 (social genetics).

Myrmecia potteri (Clark, 1951)

Promyrmecia potteri Clark, J. (1951). *The Formicidae of Australia.* Subfamily Myrmeciinae. Melbourne : CSIRO Vol. 1 230 pp. [168]. Type data: holotype, ANIC W, from Patho, Vic.

Distribution: Murray-Darling basin, Vic. Ecology: terrestrial, noctidiurnal, predator, desert, woodland; nest in soil.

Myrmecia pulchra Clark, 1929

Myrmecia pulchra Clark, J. (1929). Results of a collecting trip to the Cann River, East Gippsland. *Vict. Nat.* **46**: 115-123 [4 Oct. 1929] [119]. Type data: syntypes, NMV *W,F, from Cann River, Vic.

Myrmecia crassinoda Clark, J. (1934). Ants from the Otway Ranges. *Mem. Natl. Mus. Vict.* **8**: 48-73 [50 pl 4]. Type data: syntypes, NMV *W,F, from Gellibrand, Vic.

Myrmecia fallax Clark, J. (1951). *The Formicidae of Australia.* Subfamily Myrmeciinae. Melbourne : CSIRO Vol. 1 230 pp. [79]. Type data: holotype, ANIC W, from Kerrie, Vic.

Myrmecia murina Clark, J. (1951). *The Formicidae of Australia.* Subfamily Myrmeciinae. Melbourne : CSIRO Vol. 1 230 pp. [80]. Type data: holotype, ANIC W, from Belgrave, Vic.

Synonymy that of Brown, W.L. jr. (1953). Revisionary notes on the ant genus *Myrmecia* of Australia. *Bull. Mus. Comp. Zool.* **111**: 1–35 [27].

Distribution: SE coastal, Vic. Ecology: terrestrial, noctidiurnal, predator, alpine, woodland, open forest; nest in soil.

Myrmecia pyriformis F. Smith, 1858

Myrmecia pyriformis Smith, F. (1858). *Catalogue of hymenopterous insects in the collection of the British Museum. Part 6. Formicidae.* London : British Museum 216 pp. 14 pls [27 Mar. 1858] [144]. Publication date established from Donisthorpe, H. (1932). On the identity of Smith's types of Formicidae (Hymenoptera) collected by Alfred Russell Wallace in the Malay Archipelago, with descriptions of two new species. *Ann. Mag. Nat. Hist.* (10) **10**: 441–476. Type data: syntypes, BMNH *W,F,M, from Melbourne, Vic. and Hunter River, N.S.W.

Myrmecia sanguinea Smith, F. (1858). *Catalogue of hymenopterous insects in the collection of the British Museum. Part 6. Formicidae.* London : British Museum 216 pp. 14 pls [27 Mar. 1858] [148]. Publication date established from Donisthorpe, H. (1932). On the identity of Smith's types of Formicidae (Hymenoptera) collected by Alfred Russell Wallace in the Malay Archipelago, with descriptions of two new species. *Ann. Mag. Nat. Hist.* (10) **10**: 441–476. Type data: syntypes (probable), BMNH *W, from Tas.

Synonymy that of Brown, W.L. jr. (1953). Revisionary notes on the ant genus *Myrmecia* of Australia. *Bull. Mus. Comp. Zool.* **111**: 1–35 [9].

Distribution: SE coastal, N.S.W., Vic., Tas. Ecology: terrestrial, noctidiurnal, predator, woodland, open forest; nest in soil. Biological references: Wheeler, W.M. (1916). The marriage flight of a bull-dog ant (*Myrmecia sanguinea* F. Smith). *J. Anim. Behav.* **6**: 70–73 (reproductive behaviour).

Myrmecia regularis Crawley, 1925

Myrmecia regularis Crawley, W.C. (1925). New ants from Australia. II. *Ann. Mag. Nat. Hist.* (9) **16**: 577–598 [579]. Type data: syntypes, OUM *W, from Albany, W.A.

Distribution: SW coastal, W.A. Ecology: terrestrial, noctidiurnal, predator, woodland, open forest; nest in soil.

Myrmecia rowlandi Forel, 1910

Myrmecia tarsata rowlandi Forel, A. (1910). Formicides australiens reçus de MM. Froggatt et Rowland Turner. *Rev. Suisse Zool.* **18**: 1–94 [4]. Type data: syntypes, GMNH W, from Curanda (=Kuranda) and Cairns, Qld.

Myrmecia tarsata malandensis Forel, A. (1915). Results of Dr. E. Mjöbergs Swedish Scientific Expeditions to Australia 1910–1913. 2. Ameisen. *Ark. Zool.* **9**: 1–119 pls 1–3 [4 Dec. 1915] [9]. Type data: syntypes, GMNH W,M, ANIC W, other syntypes may exist, from Malanda, Cedar Creek and Atherton, Qld.

Synonymy that of Brown, W.L. jr. (1953). Revisionary notes on the ant genus *Myrmecia* of Australia. *Bull. Mus. Comp. Zool.* **111**: 1–35 [10].

Distribution: NE coastal, Qld. Ecology: terrestrial, noctidiurnal, predator, closed forest; nest in soil.

Myrmecia rubicunda (Clark, 1943)

Promyrmecia rubicunda Clark, J. (1943). A revision of the genus *Promyrmecia* Emery (Formicidae). *Mem. Natl. Mus. Vict.* **13**: 83–149 pls 12–17 [107]. Type data: syntypes, NMV *W, from Ooldea, S.A.

Distribution: W plateau, S.A. Ecology: terrestrial, noctidiurnal, predator, desert, woodland; nest in soil.

Myrmecia rubripes Clark, 1951

Myrmecia rubripes Clark, J. (1951). *The Formicidae of Australia.* Subfamily Myrmeciinae. Melbourne : CSIRO Vol. 1 230 pp. [34]. Type data: syntypes, specimens in ANIC may be syntypes, other syntypes may exist in NMV, from Ongerup, W.A.

Distribution: SW coastal, W.A. Ecology: terrestrial, noctidiurnal, predator, woodland; nest in soil.

Myrmecia rufinodis F. Smith, 1858

Myrmecia rufinodis Smith, F. (1858). *Catalogue of hymenopterous insects in the collection of the British Museum. Part 6. Formicidae.* London : British Museum 216 pp. 14 pls [27 Mar. 1858] [145]. Publication date established from Donisthorpe, H. (1932). On the identity of Smith's types of Formicidae (Hymenoptera) collected by Alfred Russell Wallace in the Malay Archipelago, with descriptions of two new species. *Ann. Mag. Nat. Hist.* (10) **10**: 441–476. Type data: syntypes (probable), BMNH *W, from Adelaide, S.A.

Myrmecia gracilis Emery, C. (1898). Descrizioni di formiche nuove Malesi e Australiane. Note sinonimiche. *Rec. Sess. Accad. Sci. Ist. Bologna* (ns) **2**: 231–245 [232]. Type data: holotype, MCG W, from Kingskate (=Kingscote), S.A.

Synonymy that of Brown, W.L. jr. (1953). Revisionary notes on the ant genus *Myrmecia* of Australia. *Bull. Mus. Comp. Zool.* **111**: 1–35 [8].

Distribution: S Gulfs, S.A. Ecology: terrestrial, noctidiurnal, predator, woodland, open forest; nest in soil.

Myrmecia rugosa Wheeler, 1933

Myrmecia (Promyrmecia) michaelseni rugosa Wheeler, W.M. (1933). *Colony-founding among ants with an account of some primitive Australian species.* Cambridge : Harvard Univ. Press 179 pp. [60]. Type data: syntypes, MCZ *W, ANIC W, from Ludlow, W.A.

Promyrmecia ruginodis Clark, J. (1943). A revision of the genus *Promyrmecia* Emery (Formicidae). *Mem. Natl. Mus. Vict.* **13**: 83–149 pls 12–17 [113] [*non Ponera ruginoda* F. Smith, 1858 = *Myrmecia ruginoda* (F. Smith, 1858)]. Type data: syntypes, NMV *W,F,M, from Perth, Armadale and Ludlow, W.A.

Synonymy that of Brown, W.L. jr. (1953). Revisionary notes on the ant genus *Myrmecia* of Australia. *Bull. Mus. Comp. Zool.* **111**: 1–35 [5].

Distribution: SW coastal, W.A. Ecology: terrestrial, noctidiurnal, predator, woodland, open forest; nest in soil.

Myrmecia simillima F. Smith, 1858

Myrmecia simillima Smith, F. (1858). *Catalogue of hymenopterous insects in the collection of the British Museum.* Part 6. Formicidae. London : British Museum 216 pp. 14 pls [27 Mar. 1858] [144]. Publication date established from Donisthorpe, H. (1932). On the identity of Smith's types of Formicidae (Hymenoptera) collected by Alfred Russell Wallace in the Malay Archipelago, with descriptions of two new species. *Ann. Mag. Nat. Hist. (10)* **10**: 441–476. Type data: syntypes (probable), BMNH *W, from Australia.

Myrmecia crudelis Smith, F. (1858). *Catalogue of hymenopterous insects in the collection of the British Museum.* Part 6. Formicidae. London : British Museum 216 pp. 14 pls [27 Mar. 1858] [147]. Publication date established from Donisthorpe, H. (1932). On the identity of Smith's types of Formicidae (Hymenoptera) collected by Alfred Russell Wallace in the Malay Archipelago, with descriptions of two new species. *Ann. Mag. Nat. Hist. (10)* **10**: 441–476. Type data: syntypes, BMNH *W,F, from Adelaide, S.A.

Myrmecia nigriventris Mayr, G.L. (1862). Myrmecologische Studien. *Verh. Zool.-Bot. Ges. Wien* **12**: Abhand. 649–776 [724,727 pl 19]. Type data: holotype, NHMW W, from Australia (as New Holland).

Myrmecia spadicea Mayr, G.L. (1862). Myrmecologische Studien. *Verh. Zool.-Bot. Ges. Wien* **12**: Abhand. 649–776 [724,728 pl 19]. Type data: status uncertain, NHMW F, from Sidney (=Sydney), N.S.W. and Adelaide, S.A.

Myrmecia affinis Mayr, G.L. (1862). Myrmecologische Studien. *Verh. Zool.-Bot. Ges. Wien* **12**: Abhand. 649–776 [725,728 pl 19]. Type data: syntypes, NHMW W, from Australia (as New Holland).

Myrmecia tricolor Mayr, G.L. (1862). Myrmecologische Studien. *Verh. Zool.-Bot. Ges. Wien* **12**: Abhand. 649–776 [724,728 pl 19]. Type data: syntypes (probable), NHMW W, from Sidney (=Sydney), N.S.W.

Myrmecia paucidens Forel, A. (1910). Formicides australiens reçus de MM. Froggatt et Rowland Turner. *Rev. Suisse Zool.* **18**: 1–94 [5]. Type data: syntypes, GMNH W, from Tas.

Myrmecia tricolor rogeri Emery, C. (1914). Formiche d'Australia e di Samoa raccolte dal Prof. Silvestri nel 1913. *Boll. Lab. Zool. Gen. Agr. R. Scuola Agric. Portici* **8**: 179–186 [30 Jan. 1914] [181]. Type data: uncertain, MCG *W, from N.S.W.

Synonymy that of Clark, J. (1951). *The Formicidae of Australia.* Subfamily Myrmeciinae. Melbourne : CSIRO Vol. 1 230 pp. [89]; Brown, W.L. jr. (1953). Revisionary notes on the ant genus *Myrmecia* of Australia. *Bull. Mus. Comp. Zool.* **111**: 1–35 [12].

Distribution: S Gulfs, SE coastal, S.A., Vic., N.S.W. Ecology: terrestrial, noctidiurnal, predator, woodland, open forest; nest in soil.

Myrmecia subfasciata Viehmeyer, 1924

Myrmecia subfasciata Viehmeyer, H. (1924). Formiciden der australischen Faunenregion. *Entomol. Mitt.* **13**: 219–229 [221]. Type data: holotype, ZMB *W, from Liverpool, N.S.W.

Distribution: SE coastal, N.S.W. Ecology: terrestrial, noctidiurnal, predator, (woodland); nest in soil.

Myrmecia suttoni Clark, 1951

Myrmecia suttoni Clark, J. (1951). *The Formicidae of Australia.* Subfamily Myrmeciinae. Melbourne : CSIRO Vol. 1 230 pp. [72]. Type data: holotype, ANIC W, from Fletcher, Qld.

Distribution: Murray-Darling basin, Qld. Ecology: terrestrial, noctidiurnal, predator, (woodland); nest in soil.

Myrmecia swalei Crawley, 1922

Myrmecia harderi swalei Crawley, W.C. (1922). New ants from Australia. *Ann. Mag. Nat. Hist. (9)* **9**: 427–448 [429]. Type data: holotype, OUM *W, from Albany, W.A.

Distribution: SW coastal, W.A. Ecology: terrestrial, noctidiurnal, predator, woodland, open forest; nest in soil. Biological references: Clark, J. (1943). A revision of the genus *Promyrmecia* Emery (Formicidae). *Mem. Natl. Mus. Vict.* **13**: 83–149 (raised to species).

Myrmecia tarsata F. Smith, 1858

Myrmecia tarsata Smith, F. (1858). *Catalogue of hymenopterous insects in the collection of the British Museum.* Part 6. Formicidae. London : British Museum 216 pp. 14 pls [27 Mar. 1858] [145]. Publication date established from Donisthorpe, H. (1932). On the identity of Smith's types of Formicidae (Hymenoptera) collected by Alfred Russell Wallace in the Malay Archipelago, with descriptions of two new species. *Ann. Mag. Nat. Hist. (10)* **10**: 441–476. Type data: syntypes (probable), BMNH *W, from Australia (Hunter River, &c) [*sic*].

Distribution: NE coastal, SE coastal, Murray-Darling basin, Qld., N.S.W., A.C.T., Vic. Ecology: terrestrial, noctidiurnal, predator, woodland, open forest, closed forest; nest in soil. Biological references: McAreavey, J.J. (1948). Some observations on *Myrmecia tarsata* Smith. *Proc. Linn. Soc. N.S.W.* **73**: 137–141 (colony-founding).

Myrmecia tepperi Emery, 1898

Myrmecia tepperi Emery, C. (1898). Descrizioni di formiche nuove Malesi e Australiane. Note sinonimiche. *Rec. Sess. Accad. Sci. Ist. Bologna (ns)* **2**: 231-245 [231]. Type data: syntypes, whereabouts unknown, from S.A.

Distribution: SW coastal, W plateau, S Gulfs, Murray-Darling basin, W.A., S.A., N.S.W., A.C.T., Vic. Ecology: terrestrial, noctidiurnal, predator, open heath, woodland, open forest; nest in soil.

Myrmecia testaceipes (Clark, 1943)

Promyrmecia testaceipes Clark, J. (1943). A revision of the genus *Promyrmecia* Emery (Formicidae). *Mem. Natl. Mus. Vict.* **13**: 83-149 pls 12-17 [134]. Type data: syntypes, NMV *W, from Albany, W.A.

Distribution: SW coastal, W.A. Ecology: terrestrial, noctidiurnal, predator, (woodland); nest in soil.

Myrmecia urens Lowne, 1865

Myrmecia urens Lowne, B.T. (1865). Contributions to the natural history of Australian ants. *Entomologist* **2**: 331-336 [336]. Type data: syntypes (probable), BMNH (probable) *W, from Sidney (=Sydney), N.S.W.

Myrmecia pumilio Mayr, G.L. (1866). Diagnosen neuer and wenig gekannter Formiciden. *Verh. Zool.-Bot. Ges. Wien* **16**: Abhand. 885-908 [896 pl 20]. Type data: syntypes (probable), NHMW (probable) *W, from Sidney (=Sydney), N.S.W.

Synonymy that of Clark, J. (1951). *The Formicidae of Australia*. Subfamily Myrmeciinae. Melbourne : CSIRO Vol. 1 230 pp. [190].

Distribution: SE coastal, N.S.W. Ecology: terrestrial, noctidiurnal, predator, open heath, woodland, closed forest; nest in soil.

Myrmecia varians Mayr, 1876

Myrmecia varians Mayr, G.L. (1876). Die australischen Formiciden. *J. Mus. Godeffroy* **5**: 56-115 [94]. Type data: syntypes, NHMW W, from Peak Downs and Rockhampton, Qld.

Myrmecia rufonigra Crawley, W.C. (1921). New and little-known species of ants from various localities. *Ann. Mag. Nat. Hist.* (9) **7**: 87-97 [87]. Type data: syntypes, OUM *W, from Townsville, Qld.

Promyrmecia wilsoni Clark, J. (1943). A revision of the genus *Promyrmecia* Emery (Formicidae). *Mem. Natl. Mus. Vict.* **13**: 83-149 pls 12-17 [127]. Type data: syntypes, NMV *W, from Mutchilba, Qld.

Promyrmecia shepherdi Clark, J. (1943). A revision of the genus *Promyrmecia* Emery (Formicidae). *Mem. Natl. Mus. Vict.* **13**: 83-149 pls 12-17 [128]. Type data: syntypes, NMV *W,F,M, from Broken Hill and Dubbo, N.S.W., "Finke River" and Murray Bridge, S.A. and Nhill, Vic.

Promyrmecia goudiei Clark, J. (1943). A revision of the genus *Promyrmecia* Emery (Formicidae). *Mem. Natl. Mus. Vict.* **13**: 83-149 pls 12-17 [129]. Type data: syntypes, NMV *W,F, from Sea Lake, Redcliffs, Hattah and Lake Hattah, Vic.

Promyrmecia marmorata Clark, J. (1951). *The Formicidae of Australia*. Subfamily Myrmeciinae. Melbourne : CSIRO Vol. 1 230 pp. [188]. Type data: holotype, ANIC W, from Patho, Vic.

Synonymy that of Clark, J. (1951). *The Formicidae of Australia*. Subfamily Myrmeciinae. Melbourne : CSIRO Vol. 1 230 pp. [181]; Brown, W.L. jr. (1953). Revisionary notes on the ant genus *Myrmecia* of Australia. *Bull. Mus. Comp. Zool.* **111**: 1-35 [14].

Distribution: NE coastal, Murray-Darling basin, Lake Eyre basin, Qld., N.S.W., Vic., S.A., N.T. Ecology: terrestrial, noctidiurnal, predator, desert, woodland, open forest; nest in soil.

Myrmecia vindex F. Smith, 1858

Myrmecia vindex vindex F. Smith, 1858

Myrmecia vindex Smith, F. (1858). *Catalogue of hymenopterous insects in the collection of the British Museum. Part 6. Formicidae.* London : British Museum 216 pp. 14 pls [27 Mar. 1858] [144]. Publication date established from Donisthorpe, H. (1932). On the identity of Smith's types of Formicidae (Hymenoptera) collected by Alfred Russell Wallace in the Malay Archipelago, with descriptions of two new species. *Ann. Mag. Nat. Hist.* (10) **10**: 441-476. Type data: syntypes (probable), BMNH *W, from W.A.

Distribution: SW coastal, W.A. Ecology: terrestrial, noctidiurnal, predator, woodland, open forest; nest in soil.

Myrmecia vindex basirufa Forel, 1907

Myrmecia vindex basirufa Forel, A. (1907). Formicidae. pp. 263-310 *in* Michaelsen, W. & Hartmeyer, R. (eds.) *Die Fauna Südwest-Australiens.* Jena : G. Fischer Vol.1 [264]. Type data: syntypes, GMNH W, ANIC W, from Subiaco, W.A.

Distribution: SW coastal, W.A. Ecology: terrestrial, noctidiurnal, predator, woodland, open forest; nest in soil.

PSEUDOMYRMECINAE

Tetraponera F. Smith, 1852

Tetraponera Smith, F. (1852). Descriptions of some hymenopterous insects captured in India, with notes on their economy, by Ezra T. Downes, Esq., who presented them to the Honourable the East India Company. *Ann. Mag. Nat. Hist.* (2) **9**: 44-50 [44] [redefined in Wheeler, W.M. (1922). Ants of the American Museum Congo Expedition. A contribution to the myrmecology of Africa Part II. The ants collected by the American Museum Congo Expedition. *Bull. Am. Mus. Nat. Hist.* **45**: 39-269 pls 2-23 (10 Feb. 1922)]. Type species *Eciton nigrum* Jerdon, 1851 (as *Tetraponera atrata* F. Smith, 1852) by subsequent designation, see Wheeler, W.M. (1911). A list of the type species of the genera and subgenera of Formicidae. *Ann. N.Y. Acad. Sci.* **21**: 157-175 [17 Oct. 1911].

This group is also found in the south Palearctic, Ethiopian, Malagasy and Oriental regions; New Guinea and east Melanesia in the Australian Region, see Brown, W.L. jr. (1973). A comparison of the Hylean and Congo-West African rain forest ant faunas. pp. 161-185 *in* Meggers, B.J., Ayensu, E.S. & Duckworth, W.D. (eds.) *Tropical forest ecosystems in Africa and South America: a comparative review*. Washington : Smithsonian Institution Press.

Tetraponera laeviceps (F. Smith, 1859)

Pseudomyrma laeviceps Smith, F. (1859). Catalogue of hymenopterous insects collected by Mr A.R. Wallace at the islands of Aru and Key. *J. Linn. Soc. Zool.* **3**: 132-178 [1 Feb. 1859] [145]. Publication date established from Donisthorpe, H. (1932). On the identity of Smith's types of Formicidae (Hymenoptera) collected by Alfred Russell Wallace in the Malay Archipelago, with descriptions of two new species. *Ann. Mag. Nat. Hist. (10)* **10**: 441-476. Type data: syntypes (probable), BMNH *W, from Aru Ils., Indonesia.

Distribution: N coastal, N Gulf, NE coastal, N.T., Qld.; also in New Guinea. Ecology: terrestrial, diurnal, predator, open forest, closed forest; nest arboreal.

Tetraponera punctulata F. Smith, 1877

Tetraponera punctulata punctulata F. Smith, 1877

Tetraponera punctulata Smith, F. (1877). Descriptions of new species of the genera *Pseudomyrma* and *Tetraponera*, belonging to the family Myrmicidae. *Trans. R. Entomol. Soc. Lond.* **25**: 57-72 [72]. Type data: holotype (probable), BMNH? *F, from Champion Bay, W.A.

Distribution: NW coastal, W.A. Ecology: terrestrial, diurnal, predator, open forest, closed forest; nest arboreal.

Tetraponera punctulata kimberleyensis (Forel, 1915)

Sima punctulata kimberleyensis Forel, A. (1915). Results of Dr. E. Mjöbergs Swedish Scientific Expeditions to Australia 1910-1913. 2. Ameisen. *Ark. Zool.* **9**: 1-119 pls 1-3 [4 Dec. 1915] [37]. Type data: syntypes, GMNH W, ANIC W, other syntypes may exist, from Kimberley distr., W.A. and Colosseum, Qld.

Distribution: NE coastal, N coastal, Qld., W.A. Ecology: terrestrial, diurnal, predator, open forest, closed forest; nest arboreal.

PONERINAE

Amblyopone Erichson, 1842

Amblyopone Erichson, W.F. (1842). Beitrag zur Fauna von Vandiemansland mit besonderer rucksicht auf die geographische Verbreitung der Insecten. *Arch. Naturg.* **8**: 83-287 [260]. Type species *Amblyopone australis* Erichson, 1842 by monotypy.

Neoamblyopone Wheeler, W.M. (1927). Ants of the genus *Amblyopone* Erichson. *Proc. Am. Acad. Arts Sci.* **62**: 1-29 [1] [proposed with subgeneric rank in *Amblyopone* Erichson, 1842]. Type species *Amblyopone clarki* Wheeler, 1927 by monotypy.

Protamblyopone Wheeler, W.M. (1927). Ants of the genus *Amblyopone* Erichson. *Proc. Am. Acad. Arts Sci.* **62**: 1-29 [1] [proposed with subgeneric rank in *Amblyopone* Erichson, 1842]. Type species *Amblyopone aberrans* Wheeler, 1927 by monotypy.

Lithomyrmex Clark, J. (1928). Australian Formicidae. *J. R. Soc. West. Aust.* **14**: 29-41 pl 1 [24 April 1928] [30]. Type species *Lithomyrmex glauerti* Clark, 1928 by original designation.

Synonymy that of Brown, W.L. jr. (1960). Contributions toward a reclassification of the Formicidae. III. Tribe Amblyoponini (Hymenoptera). *Bull. Mus. Comp. Zool.* **122**: 143-230 [155].

This group is also found in the Neotropical, Nearctic, south Palearctic, north Ethiopian, Oriental regions; New Guinea, east Melanesia, New Caledonia, New Zealand and Hawaii in the Australian Region, see Brown, W.L. jr. (1973). A comparison of the Hylean and Congo-West African rain forest ant faunas. pp. 161-185 *in* Meggers, B.J., Ayensu, E.S. & Duckworth, W.D. (eds.) *Tropical forest ecosystems in Africa and South America: a comparative review*. Washington : Smithsonian Institution Press.

Amblyopone aberrans Wheeler, 1927

Amblyopone aberrans Wheeler, W.M. (1927). Ants of the genus *Amblyopone* Erichson. *Proc. Am. Acad. Arts Sci.* **62**: 1-29 [26]. Type data: syntypes, MCZ *W,F,M, from Mundaring, W.A.

Distribution: SW coastal, W.A. Ecology: terrestrial, noctidiurnal, predator, woodland, open forest; nest in soil.

Amblyopone australis Erichson, 1842

Amblyopone australis Erichson, W.F. (1842). Beitrag zur Fauna von Vandiemansland mit besonderer rucksicht auf die geographische Verbreitung der Insecten. *Arch. Naturg.* **8**: 83-287 [261]. Type data: holotype (probable), ZMB *W, from Tas.

Amblyopone obscura Smith, F. (1858). *Catalogue of hymenopterous insects in the collection of the British Museum*. Part 6. Formicidae. London : British Museum 216 pp. 14 pls [27 Mar. 1858] [109]. Publication date established from Donisthorpe, H. (1932). On the identity of Smith's types of Formicidae (Hymenoptera) collected by Alfred Russell Wallace in the Malay Archipelago, with descriptions of two new species. *Ann. Mag. Nat. Hist. (10)* **10**: 441-476. Type data: syntypes, BMNH *W,F, from Australia.

Amblyopone australis fortis Forel, A. (1910). Formicides australiens reçus de MM. Froggatt et Rowland Turner. *Rev. Suisse Zool.* **18**: 1–94 [1]. Type data: syntypes, GMNH W, from Kuranda and Cairns, Qld.

Amblyopone australis minor Forel, A. (1915). Results of Dr. E. Mjöbergs Swedish Scientific Expeditions to Australia 1910–1913. 2. Ameisen. *Ark. Zool.* **9**: 1–119 pls 1–3 [4 Dec. 1915] [1]. Type data: syntypes, GMNH W,F,M, ANIC W, other syntypes may exist, from Mt. Tambourine (=Tamborine Mt.), Qld.

Amblyopone australis foveolata Wheeler, W.M. (1927). Ants of the genus *Amblyopone* Erichson. *Proc. Am. Acad. Arts Sci.* **62**: 1–29 [9]. Type data: syntypes, MCZ *W,F,M, from Denmark, W.A.

Synonymy that of Brown, W.L. jr. (1960). Contributions toward a reclassification of the Formicidae. III. Tribe Amblyoponini (Hymenoptera). *Bull. Mus. Comp. Zool.* **122**: 143–230 [167].

Distribution: NE coastal, SW coastal, Murray-Darling basin, SE coastal, S Gulfs, W plateau, N.S.W., Vic., S.A., Qld., Tas., W.A. Ecology: terrestrial, noctidiurnal, predator, alpine, shrubland, woodland, open forest, closed forest; nest in ground layer. Biological references: Taylor, R.W. (1979). Melanesian ants of the genus *Amblyopone* (Hymenoptera : Formicidae). *Aust. J. Zool.* **26**: 823–839 (bionomics).

Amblyopone clarki Wheeler, 1927

Amblyopone clarki Wheeler, W.M. (1927). Ants of the genus *Amblyopone* Erichson. *Proc. Am. Acad. Arts Sci.* **62**: 1–29 [24]. Type data: syntypes, MCZ *W,F, from Ludlow, W.A.

Distribution: SW coastal, W. plateau, W.A. Ecology: terrestrial, noctidiurnal, predator, desert, shrubland, woodland, open forest; nest in soil.

Amblyopone exigua Clark, 1928

Amblyopone exigua Clark, J. (1928). Australian Formicidae. *J. R. Soc. West. Aust.* **14**: 29–41 pl 1 [24 Apr. 1928] [35]. Type data: syntypes (probable), NMV *F, from Belgrave, Vic.

Distribution: SE coastal, Vic. Ecology: terrestrial, noctidiurnal, predator, shrubland, woodland, open forest; nest in ground layer.

Amblyopone ferruginea F. Smith, 1858

Amblyopone ferruginea Smith, F. (1858). *Catalogue of hymenopterous insects in the collection of the British Museum.* Part 6. Formicidae. London : British Museum 216 pp. 14 pls [27 Mar. 1858] [110]. Publication date established from Donisthorpe, H. (1932). On the identity of Smith's types of Formicidae (Hymenoptera) collected by Alfred Russell Wallace in the Malay Archipelago, with descriptions of two new species. *Ann. Mag. Nat. Hist.* (10) **10**: 441–476. Type data: syntypes (probable), BMNH *W, from Melbourne, Vic.

Amblyopone mandibularis Clark, J. (1928). Australian Formicidae. *J. R. Soc. West. Aust.* **14**: 29–41 pl 1 [24 Apr. 1928] [33]. Type data: syntypes, NMV *W, from Belgrave, Vic.

Synonymy that of Brown, W.L. jr. (1952). The status of some Australian *Amblyopone* species (Hymenoptera : Formicidae). *Entomol. News* **63**: 265–267 [265].

Distribution: SE coastal, Murray-Darling basin, N.S.W., Vic. Ecology: terrestrial, noctidiurnal, predator, woodland, open forest; nest in soil.

Amblyopone gingivalis Brown, 1960

Amblyopone gingivalis Brown, W.L. jr. (1960). Contributions toward a reclassification of the Formicidae. III. Tribe Amblyoponini (Hymenoptera). *Bull. Mus. Comp. Zool.* **122**: 143–230 [205]. Type data: holotype, ANIC W, from Calga, N.S.W.

Distribution: SE coastal, N.S.W. Ecology: terrestrial, noctidiurnal, predator, woodland; nest in soil.

Amblyopone glauerti (Clark, 1928)

Lithomyrmex glauerti Clark, J. (1928). Australian Formicidae. *J. R. Soc. West. Aust.* **14**: 29–41 pl 1 [24 Apr. 1928] [31]. Type data: syntypes, WAM 26–605a to 26–605d *W,F,M, from Irwin River, W.A.

Distribution: NW coastal, W plateau, W.A. Ecology: terrestrial, noctidiurnal, predator, desert, shrubland, woodland; nest in soil.

Amblyopone gracilis Clark, 1934

Amblyopone (Fulakora) gracilis Clark, J. (1934). Ants from the Otway Ranges. *Mem. Natl. Mus. Vict.* **8**: 48–73 [52 pl 4]. Type data: syntypes, NMV *W,F, from Beech Forest, Vic.

Distribution: SE coastal, Murray-Darling basin, N.S.W., Vic. Ecology: terrestrial, noctidiurnal, predator, woodland, open forest, closed forest; nest in soil.

Amblyopone hackeri Wheeler, 1927

Amblyopone hackeri Wheeler, W.M. (1927). Ants of the genus *Amblyopone* Erichson. *Proc. Am. Acad. Arts Sci.* **62**: 1–29 [22]. Type data: syntypes, MCZ *W, from the "National Park of Qld."

Distribution: NE coastal, Qld. Ecology: terrestrial, noctidiurnal, predator, open forest, closed forest; nest in ground layer.

Amblyopone leae Wheeler, 1927

Amblyopone leae Wheeler, W.M. (1927). Ants of the genus *Amblyopone* Erichson. *Proc. Am. Acad. Arts Sci.* **62**: 1–29 [16]. Type data: syntypes, MCZ *W, from Lord Howe Is.

Distribution: Lord Howe Is. Ecology: terrestrial, noctidiurnal, predator, closed forest; nest in soil.

Amblyopone longidens Forel, 1910

Amblyopone ferruginea longidens Forel, A. (1910). Formicides australiens reçus de MM. Froggatt et Rowland Turner. *Rev. Suisse Zool.* **18**: 1–94 [1]. Type data: syntypes, GMNH W, ANIC W, from Bombala, N.S.W.

Distribution: SE coastal, Murray-Darling basin, A.C.T., Vic., N.S.W. Ecology: terrestrial, noctidiurnal, predator, alpine, woodland, open forest; nest in ground layer. Biological references: Brown, W.L. jr. (1952). The status of some Australian *Amblyopone* species (Hymenoptera : Formicidae). *Entomol. News* **63**: 265–267 (raised to species).

Amblyopone lucida Clark, 1934

Amblyopone (Fulakora) lucida Clark, J. (1934). New Australian ants. *Mem. Natl. Mus. Vict.* **8**: 21–47 [27 pls 2–3]. Type data: syntypes, NMV *W, from Corrie Creek, A.C.T.

Distribution: Murray-Darling basin, N.S.W., A.C.T. Ecology: terrestrial, noctidiurnal, predator, alpine, woodland, open forest; nest in soil.

Amblyopone mercovichi Brown, 1960

Amblyopone mercovichi Brown, W.L. jr. (1960). Contributions toward a reclassification of the Formicidae. III. Tribe Amblyoponini (Hymenoptera). *Bull. Mus. Comp. Zool.* **122**: 143–230 [201]. Type data: holotype, ANIC W, from Kinglake West, Vic.

Distribution: SE coastal, Vic. Ecology: terrestrial, noctidiurnal, predator, woodland; nest in soil.

Amblyopone michaelseni Forel, 1907

Amblyopone michaelseni Forel, A. (1907). Formicidae. pp. 263–310 *in* Michaelsen, W. & Hartmeyer, R. (eds.) *Die Fauna Südwest-Australiens.* Jena : G. Fischer Vol. 1 [263]. Type data: holotype, probably destroyed in ZMH in WW II, from Jarrahdale, W.A.

Distribution: SW coastal, SE coastal, Vic., W.A. Ecology: terrestrial, noctidiurnal, predator, woodland, open forest; nest in ground layer.

Amblyopone punctulata Clark, 1934

Amblyopone (Fulakora) punctulata Clark, J. (1934). New Australian ants. *Mem. Natl. Mus. Vict.* **8**: 21–47 [28 pls 2–3]. Type data: syntypes, NMV *W,F, from Trevallyn, Tas.

Distribution: Tas. Ecology: terrestrial, noctidiurnal, predator, woodland, open forest; nest in ground layer.

Amblyopone smithi Brown, 1960

Amblyopone smithi Brown, W.L. jr. (1960). Contributions toward a reclassification of the Formicidae. III. Tribe Amblyoponini (Hymenoptera). *Bull. Mus. Comp. Zool.* **122**: 143–230 [211]. Type data: holotype, MCZ *W, from Aldgate near Mt. Lofty, Lofty Ranges, S.A.

Distribution: S Gulfs, S.A. Ecology: terrestrial, noctidiurnal, predator, open forest; nest in ground layer.

Amblyopone wilsoni Clark, 1928

Amblyopone wilsoni Clark, J. (1928). Australian Formicidae. *J. R. Soc. West. Aust.* **14**: 29–41 pl 1 [24 Apr. 1928] [34]. Type data: syntypes (probable), NMV *W, from Barrington Tops, N.S.W.

Distribution: SE coastal, N.S.W. Ecology: terrestrial, noctidiurnal, predator, woodland, open forest; nest in ground layer.

Anochetus Mayr, 1861

Anochetus Mayr, G.L. (1861). *Die europëischen Formiciden. (Ameisen.) Nach der analytischen Methode bearbeitet.* Vienna : Carl Gerolds Sohn 80 pp. 1 pl [53]. Type species *Odontomachus ghilianii* Spinola, 1853 by monotypy.

This group is also found in the Neotropical, south Nearctic, south Palearctic, Ethiopian, Malagasy and Oriental regions; throughout the Australian Region except New Zealand.

Anochetus armstrongi McAreavey, 1949

Anochetus armstrongi McAreavey, J.J. (1949). Australian Formicidae. New genera and species. *Proc. Linn. Soc. N.S.W.* **74**: 1–25 [15 June 1949] [1]. Type data: holotype, ANIC W,F, from Nyngan, N.S.W.

Distribution: Murray-Darling basin, N.S.W. Ecology: terrestrial, noctidiurnal, predator, shrubland, woodland; nest in soil.

Anochetus graeffei Mayr, 1870

Anochetus graeffei Mayr, G.L. (1870). Neue Formiciden. *Verh. Zool.-Bot. Ges. Wien* **20**: Abhand. 939–996 [31 Dec. 1870] [961]. Type data: syntypes (probable), NHMW (probable) *W, from Upolu Is., Samoa.

Distribution: N coastal, N Gulf, NE coastal, SE coastal, N.T., Qld., N.S.W. Ecology: terrestrial, noctidiurnal, predator, shrubland, woodland, open forest, closed forest; nest in ground layer.

Anochetus paripungens Brown, 1978

Anochetus paripungens Brown, W.L. jr. (1978). Contributions toward a reclassification of the Formicidae. Part VI. Ponerinae, tribe Ponerini, subtribe Odontomachiti. Section B. Genus *Anochetus* and Bibliography. *Studia Entomol. (ns)* **20**: 549–638 pls 1–12 [30 Aug. 1978] [596]. Type data: holotype, MCZ *W, from Howard Springs, Darwin area, N.T.

Distribution: N coastal, N.T. Ecology: terrestrial, noctidiurnal, predator, woodland; nest in soil.

Anochetus rectangularis Mayr, 1876

Anochetus rectangularis Mayr, G.L. (1876). Die australischen Formiciden. *J. Mus. Godeffroy* **5**: 56-115 [86]. Type data: holotype, NHMW W, from Rockhampton, Qld.

Anochetus rectangularis diabolus Forel, A. (1915). Results of Dr. E. Mjöbergs Swedish Scientific Expeditions to Australia 1910-1913. 2. Ameisen. *Ark. Zool.* **9**: 1-119 pls 1-3 [4 Dec. 1915] [35]. Type data: holotype, SMNH *W, from Christmas Creek, Qld.

Synonymy that of Brown, W.L. jr. (1978). Contributions toward a reclassification of the Formicidae. Part VI. Ponerinae, tribe Ponerini, subtribe Odontomachiti. Section B. Genus *Anochetus* and Bibliography. *Studia Entomol. (ns)* **20**: 549-638 pls 1-12 [558].

Distribution: NE coastal, Qld. Ecology: terrestrial, noctidiurnal, predator, woodland, open forest, closed forest; nest in ground layer.

Anochetus turneri Forel, 1900

Anochetus turneri Forel, A. (1900). Ponerinae et Dorylinae d'Australie récoltées par MM. Turner, Froggatt, Nugent, Chase, Rothney, J.J. Walker, etc. *Ann. Soc. Entomol. Belg.* **44**: 54-77 [55]. Type data: syntypes, GMNH W, ANIC W, from Mackay, Qld.

Anochetus turneri latunei Forel, A. (1915). Results of Dr. E. Mjöbergs Swedish Scientific Expeditions to Australia 1910-1913. 2. Ameisen. *Ark. Zool.* **9**: 1-119 pls 1-3 [4 Dec. 1915] [35]. Type data: holotype, SMNH *W, from Yarrabah, Qld.

Synonymy that of Brown, W.L. jr. (1978). Contributions toward a reclassification of the Formicidae. Part VI. Ponerinae, tribe Ponerini, subtribe Odontomachiti. Section B. Genus *Anochetus* and Bibliography. *Studia Entomol. (ns)* **20**: 549-638 pls 1-12 [559].

Distribution: NE coastal, Qld. Ecology: terrestrial, noctidiurnal, predator, open forest, closed forest; nest in ground layer.

Bothroponera Mayr, 1862

Bothroponera Mayr, G.L. (1862). Myrmecologische Studien. *Verh. Zool.-Bot. Ges. Wien* **12**: Abhand. 649-776 [717 pl 19] [redefined in Wheeler, W.M. (1922). Ants of the American Museum Congo Expedition. A contribution to the myrmecology of Africa. Part II. The ants collected by the American Museum Congo Expedition. *Bull. Am. Mus. Nat. Hist.* **45**: 39-269 pls 2-23 (10 Feb. 1922)]. Type species *Ponera pumicosa* Roger, 1860 by monotypy.

This group is also found in the Ethiopian, Malagasy and Oriental regions; New Guinea and east Melanesia in the Australian Region.

Bothroponera astuta (F. Smith, 1858)

Pachycondyla astuta Smith, F. (1858). Catalogue of hymenopterous insects in the collection of the British Museum. Part 6. Formicidae. London : British Museum 216 pp. 14 pls [27 Mar. 1858] [107]. Publication date established from Donisthorpe, H. (1932). On the identity of Smith's types of Formicidae (Hymenoptera) collected by Alfred Russell Wallace in the Malay Archipelago, with descriptions of two new species. *Ann. Mag. Nat. Hist. (10)* **10**: 441-476. Type data: syntypes (probable), BMNH *W, from Australia.

Distribution: NE coastal, N coastal, Qld., N.T. Ecology: terrestrial, noctidiurnal, predator, open forest, closed forest; nest in ground layer.

Bothroponera barbata (Stitz, 1911)

Pachycondyla (Bothroponera) barbata Stitz, H. (1911). Australische Ameisen (Neu-Guinea und Salomons-Inseln, Festland, Neu-Seeland). *Sber. Ges. Naturf. Freunde Berl.* **1911**: 351-381 [355]. Type data: syntypes, ZMB *W, from Adelaide, S.A.

Distribution: S Gulfs, S.A. Ecology: terrestrial, noctidiurnal, predator, woodland, open forest; nest in soil.

Bothroponera denticulata W.F. Kirby, 1896

Bothroponera denticulata Kirby, W.F. (1896). Hymenoptera. pp. 203-209 *in* Spencer, B. (ed.) *Report on the work of the Horn Scientific Expedition to Central Australia*. Melbourne : Melville, Mullen & Slade Pt. 1 supplement [206]. Type data: syntypes, BMNH (probable) *W, from Blood Creek, S.A.

Distribution: Lake Eyre basin, S.A. Ecology: terrestrial, noctidiurnal, predator, woodland; nest in soil.

Bothroponera dubitata (Forel, 1900)

Ponera (Bothroponera) dubitata Forel, A. (1900). Ponerinae et Dorylinae d'Australie récoltées par MM. Turner, Froggatt, Nugent, Chase, Rothney, J.J. Walker, etc. *Ann. Soc. Entomol. Belg.* **44**: 54-77 [63]. Type data: syntypes (probable), GMNH *W, from northern Australia.

Distribution: N Gulf, Qld. Ecology: terrestrial, noctidiurnal, predator, woodland, open forest; nest in soil.

Bothroponera excavata Emery, 1893

Bothroponera excavata excavata Emery, 1893

Bothroponera excavata Emery, C. (1893). Formicides de l'Archipel Malais. *Rev. Suisse Zool.* **1**: 187-229 [200 pl 8]. Type data: holotype, MCG *W, from Australia.

Distribution: N Gulf, Qld. Ecology: terrestrial, noctidiurnal, predator, woodland, open forest; nest in soil.

Bothroponera excavata acuticostata (Forel, 1900)

Ponera (Bothroponera) excavata acuticostata Forel, A. (1900). Ponerinae et Dorylinae d'Australie récoltées par MM. Turner, Froggatt, Nugent, Chase, Rothney, J.J. Walker, etc. *Ann. Soc. Entomol. Belg.* **44**: 54-77 [64]. Type data: holotype (probable), GMNH W, from Qld.

Distribution: N Gulf, NE coastal, Qld. Ecology: terrestrial, noctidiurnal, predator, woodland, open forest; nest in soil.

Bothroponera mayri Emery, 1887

Bothroponera mayri Emery, C. (1887). Catalogo delle formiche esistenti nelle collezioni del Museo Civico di Genova. Parte terza. Formiche della regione Indo-Malese e dell'Australia. *Ann. Mus. Civ. Stor. Nat. Giacomo Doria* **25**: 427–473 pls 1–2 [442]. Type data: syntypes (probable), NHMW *W, from Peak Downs, Rockhampton and Brisbane, Qld.

Distribution: NE coastal, Qld. Ecology: terrestrial, noctidiurnal, predator, woodland, open forest; nest in soil.

Bothroponera piliventris (F. Smith, 1858)

Bothroponera piliventris piliventris (F. Smith, 1858)

Pachycondyla piliventris Smith, F. (1858). *Catalogue of hymenopterous insects in the collection of the British Museum. Part 6. Formicidae.* London : British Museum 216 pp. 14 pls [27 Mar. 1858] [107]. Publication date established from Donisthorpe, H. (1932). On the identity of Smith's types of Formicidae (Hymenoptera) collected by Alfred Russell Wallace in the Malay Archipelago, with descriptions of two new species. *Ann. Mag. Nat. Hist. (10)* **10**: 441–476. Type data: syntypes (probable), BMNH *W, from Adelaide, S.A.

Distribution: S Gulfs, S.A. Ecology: terrestrial, noctidiurnal, predator, woodland, open forest; nest in soil.

Bothroponera piliventris intermedia (Forel, 1900)

Ponera (Bothroponera) piliventris intermedia Forel, A. (1900). Ponerinae et Dorylinae d'Australie récoltées par MM. Turner, Froggatt, Nugent, Chase, Rothney, J.J. Walker, etc. *Ann. Soc. Entomol. Belg.* **44**: 54–77 [63]. Type data: syntypes, GMNH W,M, ANIC W, from Mackay, Qld.

Distribution: NE coastal, Qld. Ecology: terrestrial, noctidiurnal, predator, woodland, open forest; nest in soil.

Bothroponera piliventris regularis (Forel, 1907)

Pachycondyla (Bothroponera) piliventris regularis Forel, A. (1907). Formicidae. pp. 263–310 *in* Michaelsen, W. & Hartmeyer, R. (eds.) *Die Fauna Südwest-Australiens.* Jena : G. Fischer Vol. 1 [271]. Type data: syntypes, GMNH W, ANIC W, from Tamala, W.A.

Distribution: NW coastal, W.A. Ecology: terrestrial, noctidiurnal, predator, woodland, open forest; nest in soil.

Bothroponera porcata (Emery, 1897)

Ponera (Bothroponera) porcata Emery, C. (1897). Viaggio do Lamberto Loria nella Papuasia orientale 18. Formiche raccolte nelle Nuova Guinea. *Ann. Mus. Civ. Stor. Nat. Giacomo Doria* **38**: 546–594 [22 Nov. 1897] [552 pl 1]. Type data: syntypes, MCG W, ANIC W, from N.S.W.

Distribution: NE coastal, SE coastal, Murray-Darling basin, N.S.W., Qld. Ecology: terrestrial, noctidiurnal, predator, woodland, open forest; nest in soil.

Bothroponera sublaevis Emery, 1887

Bothroponera sublaevis sublaevis Emery, 1887

Bothroponera sublaevis Emery, C. (1887). Catalogo delle formiche esistenti nelle collezioni del Museo Civico di Genova. Parte terza. Formiche della regione Indo-Malese e dell'Australia. *Ann. Mus. Civ. Stor. Nat. Giacomo Doria (2)* **5**: 427–473 pls 1–2 [442]. Type data: syntypes, MCG W, from Somerset, Qld.

Distribution: NE coastal, Qld. Ecology: terrestrial, noctidiurnal, predator, woodland, open forest; nest in soil.

Bothroponera sublaevis kurandensis (Forel, 1910)

Pachycondyla (Bothroponera) sublaevis kurandensis Forel, A. (1910). Formicides australiens reçus de MM. Froggatt et Rowland Turner. *Rev. Suisse Zool.* **18**: 1–94 [16]. Type data: syntypes, GMNH W, from Kuranda near Cairns, Qld.

Distribution: NE coastal, Qld. Ecology: terrestrial, noctidiurnal, predator, woodland, open forest; nest in soil.

Bothroponera sublaevis murina (Forel, 1910)

Pachycondyla (Bothroponera) sublaevis murina Forel, A. (1910). Formicides australiens reçus de MM. Froggatt et Rowland Turner. *Rev. Suisse Zool.* **18**: 1–94 [17]. Type data: syntypes, GMNH W, ANIC W, from Cape York, Qld.

Distribution: NE coastal, Qld. Ecology: terrestrial, noctidiurnal, predator, woodland, open forest; nest in soil.

Bothroponera sublaevis reticulata (Forel, 1900)

Ponera (Bothroponera) sublaevis reticulata Forel, A. (1900). Ponerinae et Dorylinae d'Australie récoltées par MM. Turner, Froggatt, Nugent, Chase, Rothney, J.J. Walker, etc. *Ann. Soc. Entomol. Belg.* **44**: 54–77 [62]. Type data: syntypes, GMNH W,M, ANIC W, from Mackay, Qld.

Distribution: NE coastal, Qld. Ecology: terrestrial, noctidiurnal, predator, woodland, open forest; nest in soil.

Bothroponera sublaevis rubicunda Emery, 1893

Bothroponera sublaevis rubicunda Emery, C. (1893). Formicides de l'Archipel Malais. *Rev. Suisse Zool.* **1**: 187–229 [201 pl 8]. Type data: holotype, MCG *W, from Qld.

Distribution: NE coastal, Qld. Ecology: terrestrial, noctidiurnal, predator, woodland, open forest; nest in soil.

Brachyponera Emery, 1901

Brachyponera Emery, C. (1901). Notes sur les sous-familles des Dorylines et Ponérines (famille des Formicides). *Ann. Soc. Entomol. Belg.* **45**: 32-54 [43] [proposed with subgeneric rank in *Euponera* Forel, 1891; raised to genus and redefined in Brown, W.L. jr. (1958). A review of the ants of New Zealand (Hymenoptera). *Acta Hymenopt.* **1**: 1-50]. Type species *Ponera sennaarensis* Mayr, 1862 by original designation.

This group is also found in the Ethiopian and Oriental regions; New Guinea and east Melanesia in Australian Region, see Brown, W.L. jr. (1973). A comparison of the Hylean and Congo-West African rain forest ant faunas. pp. 161-185 *in* Meggers, B.J., Ayensu, E.S. & Duckworth, W.D. (eds.) *Tropical forest ecosystems in Africa and South America: a comparative review.* Washington : Smithsonian Institution Press.

Brachyponera croceicornis (Emery, 1900)

Euponera (Brachyponera) luteipes croceicornis Emery, C. (1900). Formicidarum species novae vel minus cognitae in collectione Musaei Nationalis Hungarici, quas in Nova-Guinea, Colonia Germanica, collegit L. Biró. *Termész. Füz.* **23**: 310-338 pl 8 [315]. Type data: syntypes, probably MCG or MNH *W,F, from New Guinea.

Euponera (Brachyponera) luteipes inops Forel, A. (1910). Formicides australiens reçus de MM. Froggatt et Rowland Turner. *Rev. Suisse Zool.* **18**: 1-94 [17]. Type data: syntypes, GMNH W, from Kuranda near Cairns, Qld.

Synonymy that of Wilson, E.O. (1958). Studies on the ant fauna of Melanesia. III. *Rhytidoponera* in Western Melanesia and the Moluccas. IV. The tribe Ponerini. *Bull. Mus. Comp. Zool.* **119**: 301-371 [347].

Distribution: NE coastal, Qld.; also on New Guinea. Ecology: terrestrial, noctidiurnal, predator, closed forest; nest in ground layer.

Brachyponera lutea (Mayr, 1862)

Brachyponera lutea lutea (Mayr, 1862)

Ponera lutea Mayr, G.L. (1862). Myrmecologische Studien. *Verh. Zool.-Bot. Ges. Wien* **12**: Abhand. 649-776 [721 pl 19]. Type data: syntypes, NHMW W, from Sidney (=Sydney), N.S.W.

Ectatomma socialis MacLeay, W.J. (1873). Miscellanea entomologica. *Trans. Entomol. Soc. N.S.W.* **2**: 319-370 [369]. Type data: syntypes, ANIC W, from Mundarlo, N.S.W.

Synonymy that of Taylor, R.W. & Brown, D.R., this work.

Distribution: SE coastal, Murray-Darling basin, S Gulfs, Bulloo River basin, Lake Eyre basin, W plateau, SW coastal, NW coastal, N coastal, N Gulf, NE coastal, Vic., S.A., W.A., N.T., Qld., N.S.W. Ecology: terrestrial, noctidiurnal, predator, woodland, open forest; nest in soil.

Brachyponera lutea clara (Crawley, 1915)

Euponera (Brachyponera) lutea clara Crawley, W.C. (1915). Ants from north and central Australia, collected by G.F. Hill. Part I. *Ann. Mag. Nat. Hist. (8)* **15**: 130-136 [133]. Type data: syntypes (probable), BMNH *W, from Stapleton, N.T.

Distribution: N coastal, N.T. Ecology: terrestrial, noctidiurnal, predator, woodland, open forest; nest in soil.

Cerapachys F. Smith, 1857

Cerapachys Smith, F. (1857). Catalogue of the hymenopterous insects collected at Sarawak, Borneo, Mount Ophir, Malacca; and at Singapore by A. R. Wallace. *J. Linn. Soc. Zool.* **2**: 42-130 [2 Nov. 1857] [74 pls 1-2]. Publication date established from Donisthorpe, H. (1932). On the identity of Smith's types of Formicidae (Hymenoptera) collected by Alfred Russell Wallace in the Malay Archipelago, with descriptions of two new species. *Ann. Mag. Nat. Hist. (10)* **10**: 441-476. Type species *Cerapachys antennatus* Smith, 1857 by monotypy.

Neophyracaces Clark, J. (1941). Australian Formicidae. Notes and new species. *Mem. Natl. Mus. Vict.* **12**: 71-94 [76 pl 13]. Type species *Phyracaces princeps* Clark, 1934 (as *Phyracaces clarus* Clark, 1930) by original designation.

Synonymy that of Brown, W.L. jr. (1975). Contributions toward a reclassification of the Formicidae. V. Ponerinae, tribes Platythyreini, Cerapachyini, Cylindromyrmecini, Acanthostichini, and Aenictogitini. *Search Agric.* **5**: 1-116 [18].

This group is also found in the Neotropical, south Nearctic, south Palearctic, Ethiopian, Malagasy and Oriental regions; New Guinea, east Melanesia, New Caledonia and parts of Polynesia in Australian Region, see Brown, W.L. jr. (1973). A comparison of the Hylean and Congo-West African rain forest ant faunas. pp. 161-185 *in* Meggers, B.J., Ayensu, E.S. & Duckworth, W.D. (eds.) *Tropical forest ecosystems in Africa and South America: a comparative review.* Washington : Smithsonian Institution Press.

Cerapachys aberrans (Clark, 1934)

Phyracaces aberrans Clark, J. (1934). New Australian ants. *Mem. Natl. Mus. Vict.* **8**: 21-47 [25 pls 2-3]. Type data: syntypes (probable), SAMA *W, from Kuranda, Qld.

Distribution: NE coastal, Qld. Ecology: terrestrial, noctidiurnal, predator, desert, tussock grassland, shrubland, woodland, open forest, closed forest; nest in ground layer.

Cerapachys adamus Forel, 1910

Cerapachys (Phyracaces) adamus Forel, A. (1910). Formicides australiens reçus de MM. Froggatt et Rowland Turner. *Rev. Suisse Zool.* **18**: 1-94 [19]. Type data: syntypes, GMNH W, ANIC W, from Kuranda near Cairns, Qld.

Distribution: NE coastal, Qld. Ecology: terrestrial, noctidiurnal, predator, closed forest; nest in soil.

Cerapachys angustatus (Clark, 1924)

Phyracaces angustatus Clark, J. (1924). Australian Formicidae. *J. R. Soc. West. Aust.* **10**: 75-89 pls 6-7 [30 Apr. 1924] [76]. Type data: holotype, NMV *F, from National Park, W.A."

Distribution: SW coastal, W.A. Ecology: terrestrial, noctidiurnal, predator, shrubland, woodland; nest in soil.

Cerapachys bicolor (Clark, 1924)

Phyracaces bicolor Clark, J. (1924). Australian Formicidae. *J. R. Soc. West. Aust.* **10**: 75-89 pls 6-7 [30 Apr. 1924] [77]. Type data: syntypes, NMV *W,F, from Armadale, W.A.

Distribution: SW coastal, W.A. Ecology: terrestrial, noctidiurnal, predator, shrubland, woodland; nest in soil.

Cerapachys binodis Forel, 1910

Cerapachys (Phyracaces) binodis Forel, A. (1910). Formicides australiens reçus de MM. Froggatt et Rowland Turner. *Rev. Suisse Zool.* **18**: 1-94 [20]. Type data: syntypes, GMNH W, ANIC W, from Kuranda near Cairns, Qld.

Distribution: NE coastal, Qld. Ecology: terrestrial, noctidiurnal, predator, open forest, closed forest; nest in ground layer.

Cerapachys brevicollis (Clark, 1923)

Phyracaces brevicollis Clark, J. (1923). Australian Formicidae. *J. R. Soc. West. Aust.* **9**: 72-89 [78]. Type data: holotype, NMV *W, from Kelmscott, W.A.

Distribution: SW coastal, W.A. Ecology: terrestrial, noctidiurnal, predator, shrubland, woodland; nest in soil.

Cerapachys brevis (Clark, 1924)

Phyracaces brevis Clark, J. (1924). Australian Formicidae. *J. R. Soc. West. Aust.* **10**: 75-89 pls 6-7 [30 Apr. 1924] [78]. Type data: syntypes, NMV *W, from Hovea, W.A.

Distribution: SW coastal, W.A. Ecology: terrestrial, noctidiurnal, predator, shrubland, woodland; nest in soil.

Cerapachys clarki (Crawley, 1922)

Phyracaces clarki Crawley, W.C. (1922). New ants from Australia. *Ann. Mag. Nat. Hist. (9)* **9**: 427-448 [433]. Type data: syntypes, OUM *W, from Darlington, W.A.

Phyracaces castaneus Clark, J. (1924). Australian Formicidae. *J. R. Soc. West. Aust.* **10**: 75-89 pls 6-7 [30 Apr. 1924] [79]. Type data: syntypes, NMV *W,F,M, from Hovea, W.A.

Synonymy that of Brown, W.L. jr. (1975). Contributions toward a reclassification of the Formicidae. V. Ponerinae, tribes Platythyreini, Cerapachyini, Cylindromyrmecini, Acanthostichini, and Aenictogitini. *Search Agric.* **5**: 1-116 [22].

Distribution: SW coastal, W.A. Ecology: terrestrial, noctidiurnal, predator, shrubland, woodland; nest in soil.

Cerapachys constrictus (Clark, 1923)

Phyracaces constricta Clark, J. (1923). Australian Formicidae. *J. R. Soc. West. Aust.* **9**: 72-89 [79]. Type data: holotype, NMV *F, from Armadale, W.A.

Distribution: SW coastal, W.A. Ecology: terrestrial, noctidiurnal, predator, desert, hummock grassland, shrubland, woodland; nest in soil.

Cerapachys crassus (Clark, 1941)

Phyracaces crassus Clark, J. (1941). Australian Formicidae. Notes and new species. *Mem. Natl. Mus. Vict.* **12**: 71-94 [74 pl 13]. Type data: syntypes, NMV *W, from Hattah, Vic.

Distribution: Murray-Darling basin, Vic., N.S.W. Ecology: terrestrial, noctidiurnal, predator, shrubland, woodland; nest in soil.

Cerapachys edentatus (Forel, 1900)

Syscia australis Forel, A. (1900). Ponerinae et Dorylinae d'Australie récoltées par MM. Turner, Froggatt, Nugent, Chase, Rothney, J.J. Walker, etc. *Ann. Soc. Entomol. Belg.* **44**: 54-77 [68] [introduced as *autralis* but used by the original author in 1902 as *australis*; non *Lioponera longitarsus australis* Forel, 1895 = *Lioponera longitarsus* Mayr, 1878 = *Cerapachys longitarsus* (Mayr, 1878)]. Type data: syntypes, GMNH W, ANIC W, from Mackay, Qld.

Syscia australis edentata Forel, A. (1900). Ponerinae et Dorylinae d'Australie récoltées par MM. Turner, Froggatt, Nugent, Chase, Rothney, J.J. Walker, etc. *Ann. Soc. Entomol. Belg.* **44**: 54-77 [69] [introduced as *autralis* but used by the original author in 1902 as *australis*]. Type data: syntypes, GMNH W, ANIC W, from Mackay, Qld.

Synonymy that of Brown, W.L. jr. (1975). Contributions toward a reclassification of the Formicidae. V. Ponerinae, tribes Platythyreini, Cerapachyini, Cylindromyrmecini, Acanthostichini, and Aenictogitini. *Search Agric.* **5**: 1-116 [22].

Distribution: NE coastal, Murray-Darling basin, SE coastal, N.S.W., A.C.T., Qld. Ecology: terrestrial, noctidiurnal, predator, shrubland, woodland, open forest; nest in soil.

Cerapachys elegans (Wheeler, 1918)

Phyracaces elegans Wheeler, W.M. (1918). The Australian ants of the ponerine tribe Cerapachyini. *Proc. Am. Acad. Arts Sci.* **53**: 213-265 [254]. Type data: syntypes, MCZ *W,F, from ·Southerland (=Sutherland), N.S.W.

Distribution: SE coastal, N.S.W. Ecology: terrestrial, noctidiurnal, predator, woodland, open forest, closed forest; nest in ground layer.

Cerapachys emeryi (Viehmeyer, 1913)

Phyracaces emeryi Viehmeyer, H. (1913). Neue und unvollständig bekannte Ameisen der Alten Welt. *Arch. Naturg.* **79A**(12): 24-60 [26]. Type data: holotype, ZMB *W, from Killalpaninna, S.A.

Distribution: Lake Eyre basin, S.A., N.T., Qld. Ecology: terrestrial, noctidiurnal, predator, desert, shrubland; nest in soil.

Cerapachys fervidus (Wheeler, 1918)

Phyracaces fervidus Wheeler, W.M. (1918). The Australian ants of the ponerine tribe Cerapachyini. *Proc. Am. Acad. Arts Sci.* **53**: 213-265 [245]. Type data: syntypes, MCZ *W, from Cairns, Qld.

Phyracaces leae Wheeler, W.M. (1918). The Australian ants of the ponerine tribe Cerapachyini. *Proc. Am. Acad. Arts Sci.* **53**: 213-265 [243]. Type data: holotype, SAMA *W, from Townsville, Qld.

Phyracaces scrutator Wheeler, W.M. (1918). The Australian ants of the ponerine tribe Cerapachyini. *Proc. Am. Acad. Arts Sci.* **53**: 213-265 [247]. Type data: syntypes, MCZ *W, from Toowong near Brisbane, Qld.

Phyracaces newmani Clark, J. (1923). Australian Formicidae. *J. R. Soc. West. Aust.* **9**: 72-89 [82]. Type data: syntypes, NMV *W, from Mundaring, W.A.

Phyracaces fici Viehmeyer, H. (1924). Formiciden der australischen Faunenregion. *Entomol. Mitt.* **13**: 219-229 [222]. Type data: syntypes, ZMB *W,F, from Trial Bay, N.S.W.

Phyracaces flavescens Clark, J. (1930). New Formicidae, with notes on some little-known species. *Proc. R. Soc. Vict.* **43**: 2-25 [30 Aug. 1930] [5]. Type data: syntypes, NMV *W.F, from Eradu, W.A.

Phyracaces dromus Clark, J. (1941). Australian Formicidae. Notes and new species. *Mem. Natl. Mus. Vict.* **12**: 71-94 [75 pl 13]. Type data: syntypes, NMV *W,F, from Patho, Vic.

Synonymy that of Brown, W.L. jr. (1975). Contributions toward a reclassification of the Formicidae. V. Ponerinae,

tribes Platythyreini, Cerapachyini, Cylindromyrmecini, Acanthostichini, and Aenictogitini. *Search Agric.* **5**: 1-116 [22].

Distribution: NE coastal, SE coastal, Murray-Darling basin, SW coastal, NW coastal, N.S.W., Vic., Qld., W.A. Ecology: terrestrial, noctidiurnal, predator, shrubland, woodland, open forest; nest in ground layer.

Cerapachys ficosus (Wheeler 1918)

Phyracaces ficosus Wheeler, W.M. (1918). The Australian ants of the ponerine tribe Cerapachyini. *Proc. Am. Acad. Arts Sci.* **53**: 213-265 [252]. Type data: syntypes, MCZ *W, from Bulli Pass, N.S.W.

Distribution: SE coastal, N.S.W. Ecology: terrestrial, noctidiurnal, predator, open forest, closed forest; nest in ground layer.

Cerapachys flammeus (Clark, 1930)

Phyracaces flammeus Clark, J. (1930). New Formicidae, with notes on some little-known species. *Proc. R. Soc. Vict.* **43**: 2-25 [30 Aug. 1930] [4]. Type data: syntypes, NMV *W,F, from Lesmurdie Falls, W.A.

Distribution: SW coastal, W.A. Ecology: terrestrial, noctidiurnal, predator, woodland, open forest; nest in soil.

Cerapachys gilesi (Clark, 1923)

Phyracaces gilesi Clark, J. (1923). Australian Formicidae. *J. R. Soc. West. Aust.* **9**: 72-89 [81]. Type data: syntypes, NMV *W,F, from Mundaring, W.A.

Distribution: SW coastal, W.A. Ecology: terrestrial, noctidiurnal, predator, shrubland, woodland, open forest; nest in soil.

Cerapachys grandis (Clark, 1934)

Phyracaces grandis Clark, J. (1934). New Australian ants. *Mem. Natl. Mus. Vict.* **8**: 21-47 [22 pls 2-3]. Type data: syntypes, NMV *W, from South Australia.

Distribution: W plateau, Lake Eyre basin, S Gulfs, S.A. Ecology: terrestrial, noctidiurnal, predator, hummock grassland, tussock grassland, shrubland, woodland; nest in soil.

Cerapachys greavesi (Clark, 1934)

Phyracaces greavesi Clark, J. (1934). New Australian ants. *Mem. Natl. Mus. Vict.* **8**: 21-47 [25 pls 2-3]. Type data: syntypes (probable), NMV *W, from Bungulla, W.A.

Distribution: SW coastal, W.A. Ecology: terrestrial, noctidiurnal, predator, shrubland, woodland, open forest; nest in soil.

Cerapachys gwynethae (Clark, 1941)

Neophyracaces gwynethae Clark, J. (1941). Australian Formicidae. Notes and new species. *Mem. Natl. Mus. Vict.* **12**: 71–94 [77 pl 13]. Type data: syntypes, NMV *W,M, from Red Cliffs, Vic.

Distribution: Murray-Darling basin, Vic. Ecology: terrestrial, noctidiurnal, predator, desert, shrubland, woodland; nest in soil.

Cerapachys heros (Wheeler, 1918)

Phyracaces heros Wheeler, W.M. (1918). The Australian ants of the ponerine tribe Cerapachyini. *Proc. Am. Acad. Arts Sci.* **53**: 213–265 [240]. Type data: holotype, MCZ *W, from Qld.

Distribution: NE coastal, Qld. Ecology: terrestrial, noctidiurnal, predator, open forest, closed forest; nest in ground layer.

Cerapachys incontentus Brown, 1975

Phyracaces inconspicuus Clark, J. (1924). Australian Formicidae. *J. R. Soc. West. Aust.* **10**: 75–89 pls 6–7 [30 Apr. 1924] [82] (*non Cerapachys inconspicuus* Emery, 1902). Type data: syntypes, NMV *W,F, from National Park, W.A."

Cerapachys incontentus Brown, W.L. jr. (1975). Contributions toward a reclassification of the Formicidae. V. Ponerinae, tribes Platythyreini, Cerapachyini, Cylindromyrmecini, Acanthostichini, and Aenictogitini. *Search Agric.* **5**: 1–116 [23] [*nom. nov.* for *Phyracaces inconspicuus* Clark, 1924].

Distribution: SW coastal, W.A. Ecology: terrestrial, noctidiurnal, predator, shrubland, woodland, open forest; nest in ground layer.

Cerapachys jovis Forel, 1915

Cerapachys (Phyracaces) jovis Forel, A. (1915). Results of Dr. E. Mjöbergs Swedish Scientific Expeditions to Australia 1910–1913. 2. Ameisen. *Ark. Zool.* **9**: 1–119 pls 1–3 [4 Dec. 1915] [20]. Type data: syntypes, GMNH W, ANIC W, other syntypes may exist, from Alice River, Qld.

Distribution: NE coastal, Qld. Ecology: terrestrial, noctidiurnal, predator, tussock grassland, shrubland, woodland; nest in soil.

Cerapachys larvatus (Wheeler, 1918)

Phyracaces larvatus Wheeler, W.M. (1918). The Australian ants of the ponerine tribe Cerapachyini. *Proc. Am. Acad. Arts Sci.* **53**: 213–265 [257]. Type data: syntypes, MCZ *W, from Katoomba, N.S.W.

Distribution: SE coastal, N.S.W. Ecology: terrestrial, noctidiurnal, predator, open forest, closed forest; nest in ground layer.

Cerapachys latus Brown, 1975

Phyracaces reticulatus Clark, J. (1926). Australian Formicidae. *J. R. Soc. West. Aust.* **12**: 43–51 pl 6 [25 Jan. 1926] [45] [*non Cerapachys reticulatus* Emery, 1893]. Type data: syntypes, NMV *W, from National Park, W.A.

Cerapachys latus Brown, W.L. jr. (1975). Contributions toward a reclassification of the Formicidae. V. Ponerinae, tribes Platythreini, Cerapachyini, Cylindromyrmecini, Acanthostichini, and Aenictogitini. *Search Agric.* **5**: 1–116 [23] [*nom. nov.* for *Phyracaces reticulatus* Clark, 1926].

Distribution: SW coastal, W.A. Ecology: terrestrial, noctidiurnal, predator, shrubland, woodland, open forest; nest in soil.

Cerapachys longitarsus (Mayr, 1878)

Lioponera longitarsus Mayr, G.L. (1878). Beiträge zur Amesien-Fauna Asiens. *Verh. Zool.-Bot. Ges. Wien* **28**: 645–686 [667]. Type data: syntypes, NHMW *W,F, from Calcutta, India.

Lioponera longitarsus australis Forel, A. (1895). Nouvelles fourmis d'Australie, récoltée à The Ridge, Mackay, Queensland par M. Gilbert Turner. *Ann. Soc. Entomol. Belg.* **39**: 417–428 [422]. Type data: syntypes (probable), GMNH (probable) *W, from Mackay, Qld.

Phyracaces pygmaeus Clark, J. (1934). New Australian ants. *Mem. Natl. Mus. Vict.* **8**: 21–47 [26 pls 2–3]. Type data: syntypes, NMV *W, from Kuranda, Qld.

Synonymy that of Brown, W.L. jr. (1975). Contributions toward a reclassification of the Formicidae. V. Ponerinae, tribes Platythyreini, Cerapachyini, Cylindromyrmecini, Acanthostichini, and Aenictogitini. *Search Agric.* **5**: 1–116 [23].

Distribution: N coastal, N Gulf, NE coastal, N.T., Qld. Ecology: terrestrial, noctidiurnal, predator, woodland, open forest, closed forest; nest in ground layer.

Cerapachys macrops (Clark, 1941)

Neophyracaces macrops Clark, J. (1941). Australian Formicidae. Notes and new species. *Mem. Natl. Mus. Vict.* **12**: 71–94 [79 pl 13]. Type data: syntypes, NMV *W, from Patho, Vic.

Distribution: Murray-Darling basin, Vic. Ecology: terrestrial, noctidiurnal, predator, desert, hummock grassland, tussock grassland, shrubland; nest in soil.

Cerapachys mjobergi Forel, 1915

Cerapachys (Phyracaces) mjobergi Forel, A. (1915). Results of Dr. E. Mjöbergs Swedish Scientific Expeditions to Australia 1910–1913. 2. Ameisen. *Ark. Zool.* **9**: 1–119 pls 1–3 [4 Dec. 1915] [18]. Type data: holotype, SMNH *W, from Derby, W.A.

Distribution: N coastal, W.A. Ecology: terrestrial, noctidiurnal, predator, desert, shrubland, woodland; nest in soil.

Cerapachys mullewanus (Wheeler, 1918)

Phyracaces mullewanus Wheeler, W.M. (1918). The Australian ants of the ponerine tribe Cerapachyini. *Proc.*

Am. Acad. Arts Sci. **53**: 213–265 [251] [name based on male specimens only]. Type data: holotype, MCZ *M, from Mullewa, W.A.

Distribution: NW coastal, W.A. Ecology: terrestrial, noctidiurnal, predator, desert, hummock grassland, tussock grassland, shrubland, woodland.

Cerapachys nigriventris (Clark, 1924)

Phyracaces nigriventris Clark, J. (1924). Australian Formicidae. *J. R. Soc. West. Aust.* **10**: 75–89 pls 6–7 [30 Apr. 1924] [84]. Type data: syntypes, NMV *W,F, from National Park, W.A.

Distribution: SW coastal, W.A. Ecology: terrestrial, noctidiurnal, predator, woodland, open forest; nest in ground layer.

Cerapachys picipes (Clark, 1924)

Phyracaces picipes Clark, J. (1924). Australian Formicidae. *J. R. Soc. West. Aust.* **10**: 75–89 pls 6–7 [30 Apr. 1924] [86]. Type data: syntypes, NMV *W, from Tammin, W.A.

Distribution: SW coastal, W.A. Ecology: terrestrial, noctidiurnal, predator, shrubland, woodland, open forest; nest in soil.

Cerapachys pictus (Clark, 1934)

Phyracaces pictus Clark, J. (1934). New Australian ants. *Mem. Natl. Mus. Vict.* **8**: 21–47 [23 pls 2–3]. Type data: syntypes (probable), NMV *W, from Western distr., Vic.

Distribution: Murray-Darling basin, N.S.W., Vic. Ecology: terrestrial, noctidiurnal, predator, hummock grassland, tussock grassland, shrubland, woodland; nest in soil.

Cerapachys piliventris (Clark, 1941)

Neophyracaces piliventris Clark, J. (1941). Australian Formicidae. Notes and new species. *Mem. Natl. Mus. Vict.* **12**: 71–94 [80 pl 13]. Type data: syntypes, NMV *W, from Brisbane, Qld.

Distribution: NE coastal, Murray-Darling basin, Qld. Ecology: terrestrial, noctidiurnal, predator, shrubland, woodland; nest in soil.

Cerapachys potteri (Clark, 1941)

Neophyracaces potteri Clark, J. (1941). Australian Formicidae. Notes and new species. *Mem. Natl. Mus. Vict.* **12**: 71–94 [76 pl 13]. Type data: syntypes, NMV *W,M, from Patho, Vic.

Distribution: Murray-Darling basin, Vic., N.S.W. Ecology: terrestrial, noctidiurnal, predator, shrubland, woodland; nest in soil.

Cerapachys princeps (Clark, 1934)

Phyracaces clarus Clark, J. (1930). New Formicidae, with notes on some little-known species. *Proc. R. Soc. Vict.* **43**: 2–25 [30 Aug. 1930] [3] [*non Cerapachys emeryi clarus* Forel, 1893 = *Sphinctomyrmex clarus* (Forel, 1893)]. Type data: syntypes, NMV *W,F,M, from Cannington, Mundaring, Kalamunda and "National Park", W.A.

Phyracaces princeps Clark, J. (1934). New Australian ants. *Mem. Natl. Mus. Vict.* **8**: 21–47 [24 pls 2–3]. Type data: syntypes, SAMA *W, from Minnie Downs, S.A.

Synonymy that of Brown, W.L. jr. (1975). Contributions toward a reclassification of the Formicidae. V. Ponerinae, tribes Platythyreini, Cerapachyini, Cylindromyrmecini, Acanthostichini, and Aenictogitini. *Search Agric.* **5**: 1–116 [23].

Distribution: SW coastal, W plateau, S Gulfs, W.A., S.A. Ecology: terrestrial, noctidiurnal, predator, desert, shrubland, woodland; nest in soil.

Cerapachys punctatissimus (Clark, 1923)

Phyracaces punctatissima Clark, J. (1923). Australian Formicidae. *J. R. Soc. West. Aust.* **9**: 72–89 [84]. Type data: syntypes, NMV *W,F, from Mundaring, W.A.

Distribution: SW coastal, W.A. Ecology: terrestrial, noctidiurnal, predator, shrubland, woodland; nest in soil.

Cerapachys ruficornis (Clark, 1923)

Phyracaces ruficornis Clark, J. (1923). Australian Formicidae. *J. R. Soc. West. Aust.* **9**: 72–89 [86]. Type data: syntypes, NMV *W, from Mundaring, W.A.

Distribution: SW coastal, W.A. Ecology: terrestrial, noctidiurnal, predator, shrubland, woodland; nest in soil.

Cerapachys rugulinodis (Wheeler, 1918)

Phyracaces rugulinodis Wheeler, W.M. (1918). The Australian ants of the ponerine tribe Cerapachyini. *Proc. Am. Acad. Arts Sci.* **53**: 213–265 [249] [name based on male specimens only]. Type data: lectotype, MCZ *M, from Murat Bay, S.A., designation by Brown, W.L. jr. (1975). Contributions toward a reclassification of the Formicidae. V. Ponerinae, tribes Platythyreini, Cerapachyini, Cylindromyrmecini, Acanthostichini, and Aenictogitini. *Search Agric.* **5**: 1–116.

Distribution: W plateau, S.A. Ecology: terrestrial, noctidiurnal, predator, woodland, open forest.

Cerapachys senescens (Wheeler, 1918)

Phyracaces senescens Wheeler, W.M. (1918). The Australian ants of the ponerine tribe Cerapachyini. *Proc. Am. Acad. Arts Sci.* **53**: 213–265 [259]. Type data: syntypes, MCZ *W, from Salisbury Court near Uralla, N.S.W.

Distribution: Murray-Darling basin, N.S.W. Ecology: terrestrial, noctidiurnal, predator, hummock grassland, tussock grassland, shrubland, woodland; nest in soil.

Cerapachys simmonsae (Clark, 1923)

Phyracaces simmonsae Clark, J. (1923). Australian Formicidae. *J. R. Soc. West. Aust.* **9**: 72-89 [87]. Type data: syntypes, NMV *W,F, from Mundaring and Denmark, W.A.

Distribution: SW coastal, W.A. Ecology: terrestrial, noctidiurnal, predator, shrubland, woodland; nest in soil.

Cerapachys singularis Forel, 1900

Cerapachys singularis Forel, A. (1900). Ponerinae et Dorylinae d'Australie récoltées par MM. Turner, Froggatt, Nugent, Chase, Rothney, J.J. Walker, etc. *Ann. Soc. Entomol. Belg.* **44**: 54-77 [69]. Type data: syntypes, GMNH W, from S.A.

Cerapachys (Phyracaces) singularis rotula Forel, A. (1910). Formicides australiens reçus de MM. Froggatt et Rowland Turner. *Rev. Suisse Zool.* **18**: 1-94 [21]. Type data: syntypes, GMNH W, ANIC W, from Reedy Creek, Inverell, N.S.W.

Synonymy that of Brown, W.L. jr. (1975). Contributions toward a reclassification of the Formicidae. V. Ponerinae, tribes Platythyreini, Cerapachyini, Cylindromyrmecini, Acanthostichini, and Aenictogitini. *Search Agric.* **5**: 1-116 [23].

Distribution: Murray-Darling basin, SE coastal, S Gulfs, N.S.W., Vic., S.A. Ecology: terrestrial, noctidiurnal, predator, shrubland, woodland, open forest; nest in ground layer.

Cerapachys sjostedti Forel, 1915

Cerapachys (Phyracaces) sjostedti Forel, A. (1915). Results of Dr. E. Mjöbergs Swedish Scientific Expeditions to Australia 1910-1913. 2. Ameisen. *Ark. Zool.* **9**: 1-119 pls 1-3 [4 Dec. 1915] [19]. Type data: syntypes, GMNH W, ANIC W, other syntypes may exist, from NW Australia.

Distribution: N coastal, W.A. Ecology: terrestrial, noctidiurnal, predator, shrubland, woodland; nest in soil.

Cerapachys turneri Forel, 1902

Cerapachys (Phyracaces) turneri Forel, A. (1902). Fourmis nouvelles d'Australie. *Rev. Suisse Zool.* **10**: 405-548 [405]. Type data: syntypes, GMNH W,F, ANIC W, from Mackay, Qld.

Distribution: NE coastal, Qld. Ecology: terrestrial, noctidiurnal, predator, closed forest; nest in ground layer.

Cerapachys varians (Clark, 1924)

Phyracaces varians Clark, J. (1924). Australian Formicidae. *J. R. Soc. West. Aust.* **10**: 75-89 pls 6-7 [30 Apr. 1924] [87]. Type data: syntypes, NMV *W, from Lion Mill, W.A.

Distribution: SW coastal, W.A. Ecology: terrestrial, noctidiurnal, predator, shrubland, woodland, open forest; nest in soil.

Cryptopone Emery, 1892

Cryptopone Emery, C. (1892). Diagnoses de cinq nouveaux genres de Formicides. *Bull. Soc. Entomol. Fr.* **61**: 275-277 [275] [redefined in Brown, W.L., jr. (1963). Characters and synonymies among the genera of ants. Part III. Some members of the tribe Ponerini (Ponerinae, Formicidae). *Breviora* 190: 1-10 (30 Sept. 1963)]. Type species *Amblyopone testacea* Motschoulsky, 1863 by monotypy.

This group is also found in the north Neotropical, south Nearctic, south Palearctic, south Ethiopian and east Oriental regions; New Guinea, east Melanesia, New Caledonia and southwest Polynesia in the Australian Region, see Brown, W.L. jr. (1973). A comparison of the Hylean and Congo-West African rain forest ant faunas. pp. 161-185 *in* Meggers, B.J., Ayensu, E.S. & Duckworth, W.D. (eds.) *Tropical forest ecosystems in Africa and South America: a comparative review.* Washington : Smithsonian Institution Press.

Cryptopone rotundiceps (Emery, 1914)

Euponera (Trachymesopsus) rotundiceps Emery, C. (1914). Les fourmis de la Nouvelle-Calédonie et des îles Loyalty. *in* Sarasin, F. & Roux, J. (1914-1921). *Forschungen in Neu-Caledonien und auf den Loyalty-Inseln.* Zoologie 1: 393-437 pl 13 [397]. Type data: holotype, BNHM *F, from Mt. Canala, New Caledonia.

Ponera mjobergi Forel, A. (1915). Results of Dr. E. Mjöbergs Swedish Scientific Expeditions to Australia 1910-1913. 2. Ameisen. *Ark. Zool.* **9**: 1-119 pls 1-3 [4 Dec. 1915] [22]. Type data: syntypes, GMNH W,F, ANIC W, other syntypes may exist, from Blackal (=Blackall) Range and Mt. Tambourine (=Tamborine Mt.), Qld.

Synonymy that of Brown, W.L. jr. (1963). Characters and synonymies among the genera of ants. Part III. Some members of the tribe Ponerini (Ponerinae, Formicidae). *Breviora* 190: 1-10 [6].

Distribution: NE coastal, SE coastal, N.S.W., Qld. Ecology: terrestrial, noctidiurnal, predator, open forest, closed forest; nest in ground layer.

Diacamma Mayr, 1862

Diacamma Mayr, G.L. (1862). Myrmecologische Studien. *Verh. Zool.-Bot. Ges. Wien* **12**: Abhand. 649-776 [718 pl 19]. Type species *Ponera rugosa* Le Guillou, 1841 by subsequent designation, see Bingham, C.T. (1903). *The Fauna of British India, including Ceylon and Burma.* Hymenoptera. Vol. 2 Ants and cuckoo-wasps. London : Taylor & Francis [75].

This group is also found in the east Palearctic and Oriental regions; New Guinea and east Melanesia in Australian Region.

Diacamma australe (Fabricius, 1775)

Formica australis Fabricius, J.C. (1775). *Systema Entomologiae, sistens insectorum classes, ordines, genera, species, adiectis synonymis, locis, descriptionibus, observationibus.* Flensburgi et Lipsiae [393]. Type data: holotype (probable), BMNH W, from Australia (as New Holland).

Diacamma australe colosseensis Forel, A. (1915). Results of Dr. E. Mjöbergs Swedish Scientific Expeditions to Australia 1910–1913. 2. Ameisen. *Ark. Zool.* 9: 1–119 pls 1–3 [4 Dec. 1915] [26]. Type data: syntypes, GMNH W, other syntypes may exist, from Colosseum, Chillagoe and Atherton, Qld.

Diacamma australe levis Crawley, W.C. (1915). Ants from north and central Australia, collected by G.F. Hill. Part I. *Ann. Mag. Nat. Hist.* (8) **15**: 130–136 [134]. Type data: syntypes, BMNH *W, GMNH W, from Near Adelaide Plains, N.T.

Synonymy that of Taylor, R.W. and Brown, D.R., this work.

Distribution: NE coastal, Qld. Ecology: terrestrial, diurnal, predator, tussock grassland, woodland, open forest; nest in soil.

Discothyrea Roger, 1863

Discothyrea Roger, J. (1863). Die neu aufgeführten Gattungen und Arten meines Formiciden-Verzeichnisses. *Berl. Entomol. Z.* **7**: 129–214 [June 1863] [176]. Type species *Discothyrea testacea* Roger, 1863 by monotypy.

Prodiscothyrea Wheeler, W.M. (1916). *Prodiscothyrea*, a new genus of ponerine ants from Queensland. *Trans. R. Soc. S. Aust.* **40**: 33–37 [23 Dec. 1916] [33 pl 4]. Type species *Prodiscothyrea velutina* Wheeler, 1916 by monotypy.

Synonymy that of Brown, W.L. jr. (1958). Contributions toward a reclassification of the Formicidae. II. Tribe Ectatommini (Hymenoptera). *Bull. Mus. Comp. Zool.* **118**: 173–362 [248].

This group is also found in the Neotropical, north Nearctic, Ethiopian and east Oriental regions; New Guinea, east Melanesia, New Caledonia and New Zealand in the Australian Region, see Brown, W.L. jr. (1973). A comparison of the Hylean and Congo-West African rain forest ant faunas. pp. 161–185 *in* Meggers, B.J., Ayensu, E.S. & Duckworth, W.D. (eds.) *Tropical forest ecosystems in Africa and South America: a comparative review.* Washington : Smithsonian Institution Press.

Discothyrea bidens Clark, 1928

Discothyrea bidens Clark, J. (1928). Australian Formicidae. *J. R. Soc. West. Aust.* **14**: 29–41 pl 1 [24 Apr. 1928] [38]. Type data: syntypes (probable), NMV *W, from Warburton, Vic.

Distribution: SE coastal, Murray-Darling basin, N.S.W., A.C.T., Vic. Ecology: terrestrial, noctidiurnal, predator, closed forest; nest in ground layer.

Discothyrea crassicornis Clark, 1926

Discothyrea crassicornis Clark, J. (1926). Australian Formicidae. *J. R. Soc. West. Aust.* **12**: 43–51 pl 6 [25 Jan. 1926] [46]. Type data: syntypes, NMV *W, from Manjimup, W.A.

Distribution: SW coastal, W.A. Ecology: terrestrial, noctidiurnal, predator, open forest, closed forest; nest in ground layer.

Discothyrea leae Clark, 1934

Discothyrea leae Clark, J. (1934). New Australian ants. *Mem. Natl. Mus. Vict.* **8**: 21–47 [29 pls 2–3]. Type data: syntypes (probable), SAMA *W, from Mt. Lofty, S.A.

Distribution: S Gulfs, S.A. Ecology: terrestrial, noctidiurnal, predator, open forest, closed forest; nest in ground layer.

Discothyrea turtoni Clark, 1934

Discothyrea turtoni Clark, J. (1934). Ants from the Otway Ranges. *Mem. Natl. Mus. Vict.* **8**: 48–73 [53 pl 4]. Type data: syntypes, NMV *W,F, from Beech Forest, Vic.

Distribution: SE coastal, Murray-Darling basin, A.C.T., N.S.W., Vic. Ecology: terrestrial, noctidiurnal, predator, open forest, closed forest; nest in ground layer.

Discothyrea velutina (Wheeler, 1916)

Prodiscothyrea velutina Wheeler, W.M. (1916). *Prodiscothyrea*, a new genus of ponerine ants from Queensland. *Trans. R. Soc. S. Aust.* **40**: 33–37 [23 Dec. 1916] [34 pl 4]. Type data: syntypes, MCZ *W,F, from Kuranda, Qld.

Distribution: NE coastal, Qld. Ecology: terrestrial, noctidiurnal, predator, closed forest; nest in ground layer.

Ectomomyrmex Mayr, 1867

Ectomomyrmex Mayr, G.L. (1867). Adnotationes in Monographiam formicidarum Indo-Neerlandicarum. *Tijdschr. Entomol.* **10**: 33–117 [83 pl 2] [redefined in Brown, W.L. jr. (1963). Characters and synonymies among the genera of ants. Part III. Some members of the tribe Ponerini (Ponerinae, Formicidae). *Breviora* 190: 1–10 (30 Sept. 1963)]. Type species *Ectomomyrmex javanus* Mayr, 1867 by subsequent designation, see Bingham, C.T. (1903). *The Fauna of British India, including Ceylon and Burma. Hymenoptera.* Vol. 2 Ants and cuckoo-wasps. London : Taylor & Francis [85].

This group is also found in the west Palearctic and Oriental regions; New Guinea, east Melanesia and south Polynesia in the Australian Region.

Ectomomyrmex ruficornis Clark, 1934

Ectomomyrmex ruficornis Clark, J. (1934). New Australian ants. *Mem. Natl. Mus. Vict.* **8**: 21–47 [31 pls 2–3]. Type data: holotype, NMV *W, from Cairns, Qld.

Distribution: NE coastal, Qld. Ecology: terrestrial, noctidiurnal, predator, open forest, closed forest; nest in ground layer.

Gnamptogenys Roger, 1863

Gnamptogenys Roger, J. (1863). Die neu aufgeführten Gattungen und Arten meines Formiciden-Verzeichnisses. *Berl. Entomol. Z.* **7**: 129–214 [June 1863] [174] [redefined in Brown, W.L. jr. (1958). Contributions toward a reclassification of the Formicidae. II. Tribe Ectatommini (Hymenoptera). *Bull. Mus. Comp. Zool.* **118**: 173–362]. Type species *Ponera tornata* Roger, 1861 by subsequent designation, see Emery, C. (1911). Hymenoptera. Fam. Formicidae. subfam. Ponerinae. *in* Wytsman, P. (ed.) *Genera Insectorum*. Fasc. 118 Brussels 125 pp. 3 pls [44].

This group is also found in the Neotropical, south Nearctic and Oriental regions; New Guinea, east Melanesia (to Fiji) in the Australian Region, see Brown, W.L. jr. (1973). A comparison of the Hylean and Congo-West African rain forest ant faunas. pp. 161–185 *in* Meggers, B.J., Ayensu, E.S. & Duckworth, W.D. (eds.) *Tropical forest ecosystems in Africa and South America: a comparative review*. Washington : Smithsonian Institution Press.

Gnamptogenys biroi (Emery, 1902)

Stictoponera biroi Emery, C. (1902). Formicidarum species novae vel minus cognitae in collectione Musaei Nationalis Hungarici, quas in Nova-Guinea, Colonia Germanica, collegit L. Biró. *Termész. Füz.* **25**: 152–160 [154]. Type data: holotype, probably MCG or MNH *W, from Sattleburg, New Guinea.

Distribution: NE coastal, Qld.; also on New Guinea. Ecology: terrestrial, noctidiurnal, predator, closed forest; nest in ground layer.

Heteroponera Mayr, 1887

Heteroponera Mayr, G.L. (1887). Südamerikanische Formiciden. *Verh. Zool.-Bot. Ges. Wien* **37**: Abhand. 511–632 [532]. Type species *Heteroponera carinifrons* Mayr, 1887 by monotypy.

Paranomopone Wheeler, W.M. (1915). *Paranomopone*, a new genus of ponerine ants from Queensland. *Psyche Camb.* **22**: 117–120 pl 8 [117]. Type species *Paranomopone relicta* Wheeler, 1915 by monotypy.

Synonymy that of Brown, W.L. jr. (1958). Contributions toward a reclassification of the Formicidae. II. Tribe Ectatommini (Hymenoptera). *Bull. Mus. Comp. Zool.* **118**: 173–362 [194].

This group is also found in the Neotropical Region; New Zealand in the Australian Region, see Brown, W.L. jr. (1973). A comparison of the Hylean and Congo-West African rain forest ant faunas. pp. 161–185 *in* Meggers, B.J., Ayensu, E.S. & Duckworth, W.D. (eds.) *Tropical forest ecosystems in Africa and South America: a comparative review*. Washington : Smithsonian Institution Press.

Heteroponera imbellis (Emery, 1895)

Acanthoponera imbellis Emery, C. (1895). Descriptions de quelques fourmis nouvelles d'Australie. *Ann. Soc. Entomol. Belg.* **39**: 345–358 [346]. Type data: holotype, MCG *W, from Kamerunga, Qld.

Ectatomma (Acanthoponera) imbellis hilare Forel, A. (1895). Nouvelles fourmis d'Australie, récoltée à The Ridge, Mackay, Queensland par M. Gilbert Turner. *Ann. Soc. Entomol. Belg.* **39**: 417–428 [421]. Type data: syntypes (probable), GMNH (probable) *W, from Mackay, Qld.

Acanthoponera (Anacanthoponera) imbellis scabra Wheeler, W.M. (1923). Ants of the genera *Myopias* and *Acanthoponera*. *Psyche Camb.* **30**: 175–192 [181]. Type data: syntypes, MCZ *W, from Sydney, N.S.W.

Acanthoponera occidentalis Clark, J. (1926). Australian Formicidae. *J. R. Soc. West. Aust.* **12**: 43–51 pl 6 [25 Jan. 1926] [47]. Type data: syntypes, NMV *W, from National Park, W.A.

Acanthoponera nigra Clark, J. (1930). New Formicidae, with notes on some little-known species. *Proc. R. Soc. Vict.* **43**: 2–25 [30 Aug. 1930] [6]. Type data: syntypes, NMV *W, from Mt. William, Grampians, Vic.

Synonymy that of Brown, W.L. jr. (1958). Contributions toward a reclassification of the Formicidae. II. Tribe Ectatommini (Hymenoptera). *Bull. Mus. Comp. Zool.* **118**: 173–362 [195].

Distribution: NE coastal, Murray-Darling basin, S Gulfs, W plateau, SW coastal, Qld., Vic., W.A., A.C.T., S.A. Ecology: terrestrial, noctidiurnal, predator, woodland, open forest, closed forest; nest in ground layer.

Heteroponera leae (Wheeler, 1923)

Acanthoponera (Anacanthoponera) leae Wheeler, W.M. (1923). Ants of the genera *Myopias* and *Acanthoponera*. *Psyche Camb.* **30**: 175–192 [181]. Type data: syntypes, MCZ *W, from The National Park, near Sydney, N.S.W.

Distribution: SE coastal, N.S.W. Ecology: terrestrial, noctidiurnal, predator, closed forest; nest in ground layer.

Heteroponera relicta (Wheeler, 1915)

Paranomopone relicta Wheeler, W.M. (1915). *Paranomopone*, a new genus of ponerine ants from Queensland. *Psyche Camb.* **22**: 117–120 [118 pl 8]. Type data: syntypes, MCZ *W,F, from Kuranda, Qld.

Distribution: NE coastal, Qld. Ecology: terrestrial, noctidiurnal, predator, closed forest; nest in ground layer.

Hypoponera Santschi, 1938

Hypoponera Santschi, F. (1938). Notes sur quelques *Ponera* Latr. *Bull. Soc. Entomol. Fr.* **43**: 78–80 [15 Apr. 1938]. [79] [proposed with subgeneric rank in *Ponera* Latreille, 1804; raised to genus and redefined in Taylor, R.W. (1967). A monographic revision of the ant genus *Ponera* Latreille; (Hymenoptera : Formicidae). *Pac. Insects Monogr.* 13: 1–112]. Type species *Ponera abeillei* E. André, 1881 by original designation.

This group is also found in the Neotropical, south Nearctic, south Palearctic, Ethiopian, Malagasy and Oriental regions; widespread in the Australian Region, see Brown, W.L. jr. (1973). A comparison of the Hylean and Congo-West African rain forest ant faunas. pp. 161–185 *in* Meggers, B.J., Ayensu, E.S. & Duckworth, W.D. (eds.) *Tropical forest ecosystems in Africa and South America: a comparative review.* Washington : Smithsonian Institution Press.

Hypoponera congrua (Wheeler, 1934)

Ponera congrua Wheeler, W.M. (1934). Contributions to the fauna of Rottnest Island, Western Australia No. IX. The ants. *J. R. Soc. West. Aust.* **20**: 137–163 [5 Oct. 1934] [142]. Type data: syntypes, MCZ *W,F, from White Hill, Rottnest Is., W.A.

Distribution: SW coastal, W.A. Ecology: terrestrial, noctidiurnal, predator, open forest, closed forest; nest in ground layer.

Hypoponera convexiuscula (Forel, 1900)

Ponera trigona convexiuscula Forel, A. (1900). Ponerinae et Dorylinae d'Australie récoltées par MM. Turner, Froggatt, Nugent, Chase, Rothney, J.J. Walker, etc. *Ann. Soc. Entomol. Belg.* **44**: 54–77 [60]. Type data: syntypes, GMNH W, ANIC W, from Mackay, Qld.

Distribution: NE coastal, Qld. Ecology: terrestrial, noctidiurnal, predator, closed forest; nest in ground layer.

Hypoponera decora (Clark, 1934)

Ponera decora Clark, J. (1934). Ants from the Otway Ranges. *Mem. Natl. Mus. Vict.* **8**: 48–73 [56 pl 4]. Type data: syntypes, NMV *W,F, from Gellibrand, Vic.

Distribution: SE coastal, Murray-Darling basin, Vic., N.S.W., Tas. Ecology: terrestrial, noctidiurnal, predator, open forest, closed forest; nest in ground layer.

Hypoponera elliptica (Forel, 1900)

Ponera truncata elliptica Forel, A. (1900). Ponerinae et Dorylinae d'Australie récoltées par MM. Turner, Froggatt, Nugent, Chase, Rothney, J.J. Walker, etc. *Ann. Soc. Entomol. Belg.* **44**: 54–77 [62]. Type data: syntypes, GMNH W,F, ANIC W, from unknown locality.

Distribution: NE coastal, Qld. Ecology: terrestrial, noctidiurnal, predator, closed forest; nest in ground layer.

Hypoponera herbertonensis (Forel, 1915)

Ponera pruinosa herbertonensis Forel, A. (1915). Results of Dr. E. Mjöbergs Swedish Scientific Expeditions to Australia 1910–1913. 2. Ameisen. *Ark. Zool.* **9**: 1–119 pls 1–3 [4 Dec. 1915] [24]. Type data: syntypes, GMNH W,F,M, other syntypes may exist, from Herberton and Malanda, Qld.

Distribution: NE coastal, Qld. Ecology: terrestrial, noctidiurnal, predator, closed forest; nest in ground layer.

Hypoponera mackayensis (Forel, 1900)

Ponera coarctata mackayensis Forel, A. (1900). Ponerinae et Dorylinae d'Australie récoltées par MM. Turner, Froggatt, Nugent, Chase, Rothney, J.J. Walker, etc. *Ann. Soc. Entomol. Belg.* **44**: 54–77 [61]. Type data: syntypes, GMNH W, ANIC W, from Mackay, Qld.

Distribution: NE coastal, Qld. Ecology: terrestrial, noctidiurnal, predator, closed forest; nest in ground layer.

Hypoponera mina (Wheeler, 1927)

Ponera mina Wheeler, W.M. (1927). The ants of Lord Howe Island and Norfolk Island. *Proc. Am. Acad. Arts Sci.* **62**: 121–153 [131]. Type data: syntypes, MCZ *W,F,M, from Norfolk Is.

Distribution: Norfolk Is. Ecology: terrestrial, noctidiurnal, predator, closed forest; nest in ground layer.

Hypoponera queenslandensis (Forel, 1900)

Ponera queenslandensis Forel, A. (1900). Ponerinae et Dorylinae d'Australie récoltées par MM. Turner, Froggatt, Nugent, Chase, Rothney, J.J. Walker, etc. *Ann. Soc. Entomol. Belg.* **44**: 54–77 [61]. Type data: syntypes, GMNH W,F, from Mackay, Qld.

Distribution: NE coastal, Qld. Ecology: terrestrial, noctidiurnal, predator, closed forest; nest in ground layer.

Hypoponera rectidens (Clark, 1934)

Ponera rectidens Clark, J. (1934). Ants from the Otway Ranges. *Mem. Natl. Mus. Vict.* **8**: 48–73 [57 pl 4]. Type data: syntypes (probable), NMV *W, from Gellibrand, Vic.

Distribution: SE coastal, Murray-Darling basin, N.S.W., Tas., Vic. Ecology: terrestrial, noctidiurnal, predator, open forest, closed forest; nest in ground layer.

Hypoponera scitula (Clark, 1934)

Ponera scitula Clark, J. (1934). Ants from the Otway Ranges. *Mem. Natl. Mus. Vict.* **8**: 48–73 [55 pl 4]. Type data: syntypes, NMV *W,F, from Turton's Track, Otway Range, Vic.

Distribution: SE coastal, Murray-Darling basin, N.S.W., Vic. Ecology: terrestrial, noctidiurnal, predator, woodland, open forest, closed forest; nest in ground layer.

Hypoponera sulciceps (Clark, 1928)

Ponera sulciceps Clark, J. (1928). Entomological Reports. Formicidae. *in* Report of the Victorian Field Naturalists' expedition through the Western District of Victoria. *Vict. Nat.* **45** suppl.: 39–44 [40]. Type data: syntypes, NMV *W, from Mt. Arapiles, Vic.

Distribution: SE coastal, Murray-Darling basin, N.S.W., Vic. Ecology: terrestrial, noctidiurnal, predator, open forest, closed forest; nest in ground layer.

Leptogenys Roger, 1861

Leptogenys Roger, J. (1861). Die *Ponera*-Artigen Ameisen. *Berl. Entomol. Z.* **5**: 1–54 [41]. Type species *Leptogenys falcigera* Roger, 1861 by subsequent designation, see Bingham, C.T. (1903). *The Fauna of British India, including Ceylon and Burma.* Hymenoptera. Vol. 2 Ants and cuckoo-wasps. London : Taylor & Francis [52].
Odontopelta Emery, C. (1911). Hymenoptera Fam. Formicidae subfam. Ponerinae. *in* Wytsman, P. (ed.) *Genera Insectorum.* Fasc. 118 Brussels 125 pp. 3 pls [101] [proposed with subgeneric rank in *Leptogenys* Roger, 1861]. Type species *Leptogenys turneri* Forel, 1900 by monotypy.
Dorylozelus Forel, A. (1915). Results of Dr. E. Mjöbergs Swedish Scientific Expeditions to Australia. 1910-1913. 2. Ameisen. *Ark. Zool.* **9**: 1–119 [4 Dec. 1915] [24 pls 1–3]. Type species *Leptogenys tricosa* Taylor, 1969 (as *Dorylozelus mjobergi* Forel, 1915) by monotypy.

Synonymy that of Bolton, B. (1975). A revision of the ant genus *Leptogenys* Roger (Hymenoptera : Formicidae) in the Ethiopian region with a review of the Malagasy species. *Bull. Br. Mus. Nat. Hist. (Entomol.)* **75**: 237–305 [5 Feb. 1975] [239].

This group is also found in the Neotropical, south Nearctic, Ethiopian, Malagasy and Oriental regions; New Guinea, east Melanesia and New Caledonia in the Australian Region, see Brown, W.L. jr. (1973). A comparison of the Hylean and Congo-West African rain forest ant faunas. pp. 161–185 *in* Meggers, B.J., Ayensu, E.S. & Duckworth, W.D. (eds.) *Tropical forest ecosystems in Africa and South America: a comparative review.* Washington : Smithsonian Institution Press.

Leptogenys angustinoda Clark, 1934

Leptogenys (Lobopelta) angustinoda Clark, J. (1934). New Australian ants. *Mem. Natl. Mus. Vict.* **8**: 21–47 [34 pls 2–3]. Type data: syntypes, NMV *W,F, from Armidale, N.S.W.

Distribution: Murray-Darling basin, N.S.W., A.C.T. Ecology: terrestrial, noctidiurnal, nomadic, predator, woodland, open forest; nest in ground layer.

Leptogenys anitae Forel, 1915

Leptogenys (Lobopelta) anitae Forel, A. (1915). Results of Dr. E. Mjöbergs Swedish Scientific Expeditions to Australia 1910-1913. 2. Ameisen. *Ark. Zool.* **9**: 1–119 pls 1–3 [4 Dec. 1915] [29]. Type data: holotype, SMNH *W, from Mt. Tambourine (=Tamborine Mt.), Qld.

Distribution: NE coastal, Qld. Ecology: terrestrial, noctidiurnal, nomadic, predator, closed forest; nest in ground layer.

Leptogenys bidentata Forel, 1900

Leptogenys (Lobopelta) bidentata Forel, A. (1900). Ponerinae et Dorylinae d'Australie récoltées par MM. Turner, Froggatt, Nugent, Chase, Rothney, J.J. Walker, etc. *Ann. Soc. Entomol. Belg.* **44**: 54–77 [66]. Type data: syntypes (probable), GMNH W, from Mackay, Qld.

Distribution: NE coastal, Qld. Ecology: terrestrial, noctidiurnal, nomadic, predator, closed forest; nest in ground layer.

Leptogenys chelifer (Santschi, 1928)

Pseudoponera chelifer Santschi, F. (1928). Nouvelles fourmis d'Australie. *Bull. Soc. Vaud. Sci. Nat.* **56**: 465–483 [30 Aug. 1928] [466]. Type data: syntypes, BNHM W, from Beyfield (=Byfield), Qld.

Distribution: NE coastal, Qld. Ecology: terrestrial, noctidiurnal, nomadic, predator, open forest, closed forest; nest in ground layer.

Leptogenys clarki Wheeler, 1933

Leptogenys clarki Wheeler, W.M. (1933). *Colony-founding among ants with an account of some primitive Australian species.* Cambridge : Harvard Univ. Press 179 pp. [82]. Type data: syntypes, MCZ *W, from Geraldton, W.A.

Distribution: NW coastal, W.A. Ecology: terrestrial, noctidiurnal, nomadic, predator, woodland, open forest; nest in ground layer.

Leptogenys conigera (Mayr, 1876)

Leptogenys conigera conigera (Mayr, 1876)

Lobopelta conigera Mayr, G.L. (1876). Die australischen Formiciden. *J. Mus. Godeffroy* **5**: 56–115 [89]. Type data: syntypes, NHMW W, from Peak Downs and Gayndah, Qld.

Distribution: NE coastal, Qld. Ecology: terrestrial, noctidiurnal, nomadic, predator, woodland, open forest; nest in ground layer.

Leptogenys conigera adlerzi Forel, 1900

Leptogenys (Lobopelta) conigera adlerzi Forel, A. (1900). Ponerinae et Dorylinae d'Australie récoltées par MM. Turner, Froggatt, Nugent, Chase, Rothney, J.J. Walker, etc. *Ann. Soc. Entomol. Belg.* **44**: 54–77 [65]. Type data: syntypes, GMNH W, ANIC W, from Townsville and Charters Towers, Qld.

Distribution: NE coastal, Qld. Ecology: terrestrial, noctidiurnal, nomadic, predator, woodland, open forest; nest in ground layer.

Leptogenys conigera centralis Wheeler, 1915

Leptogenys (Lobopelta) conigera centralis Wheeler, W.M. (1915). Hymenoptera. *Trans. R. Soc. S. Aust.* **39**: 805–823 pls 64–66 [Dec. 1915] [805]. Type data: syntypes, MCZ *W,M, from Moorilyanna, S.A.

Distribution: Lake Eyre basin, S.A. Ecology: terrestrial, noctidiurnal, nomadic, predator, woodland; nest in ground layer.

Leptogenys conigera exigua Crawley, 1921

Leptogenys (Lobopelta) conigera exigua Crawley, W.C. (1921). New and little-known species of ants from various localities. *Ann. Mag. Nat. Hist. (9)* **7**: 87–97 [89]. Type data: syntypes (probable), BMNH *W, from Darwin, N.T.

Distribution: N coastal, N.T. Ecology: terrestrial, noctidiurnal, nomadic, predator, woodland, open forest; nest in ground layer.

Leptogenys conigera mutans Forel, 1900

Leptogenys (Lobopelta) conigera mutans Forel, A. (1900). Ponerinae et Dorylinae d'Australie récoltées par MM. Turner, Froggatt, Nugent, Chase, Rothney, J.J. Walker, etc. *Ann. Soc. Entomol. Belg.* **44**: 54–77 [65]. Type data: syntypes, GMNH W, ANIC W, from Mackay, Qld.

Distribution: NE coastal, Qld. Ecology: terrestrial, noctidiurnal, nomadic, predator, woodland, open forest; nest in ground layer.

Leptogenys darlingtoni Wheeler, 1933

Leptogenys (Lobopelta) darlingtoni Wheeler, W.M. (1933). *Colony-founding among ants with an account of some primitive Australian species.* Cambridge : Harvard Univ. Press 179 pp. [90]. Type data: syntypes, MCZ *W,F, from near Mullewa, W.A.

Distribution: NW coastal, W.A. Ecology: terrestrial, noctidiurnal, nomadic, predator, woodland, open forest; nest in ground layer.

Leptogenys diminuta (F. Smith, 1854)

Ponera diminuta Smith, F. (1854). Catalogue of the hymenopterous insects collected at Sarawak, Borneo; Mount Ophir, Malacca; and at Singapore, by A.R. Wallace. *J. Linn. Soc. Zool.* **2**: 42–130 [69]. Type data: status unknown, ?BMNH, from Borneo (Sarawak).

Leptogenys diminuta yarrabahna Forel, 1915

Leptogenys (Lobopelta) diminuta yarrabahna Forel, A. (1915). Results of Dr. E. Mjöbergs Swedish Scientific Expeditions to Australia 1910-1913. 2. Ameisen. *Ark. Zool.* **9**: 1–119 pls 1–3 [4 Dec. 1915] [29]. Type data: syntypes, GMNH W, ANIC W, other syntypes may exist, from Yarrabah and Mt. Bellenden Ker, Qld.

Distribution: NE coastal, Qld. Ecology: terrestrial, noctidiurnal, nomadic, predator, woodland, open forest, closed forest; nest in ground layer.

Leptogenys ebenina Forel, 1915

Leptogenys (Lobopelta) ebenina Forel, A. (1915). Results of Dr. E. Mjöbergs Swedish Scientific Expeditions to Australia 1910-1913. 2. Ameisen. *Ark. Zool.* **9**: 1–119 pls 1–3 [4 Dec. 1915] [30]. Type data: syntypes, GMNH W, ANIC W, other syntypes may exist, from Malanda, Qld.

Distribution: NE coastal, Qld. Ecology: terrestrial, noctidiurnal, nomadic, predator, closed forest; nest in ground layer.

Leptogenys excisa (Mayr, 1876)

Leptogenys excisa excisa (Mayr, 1876)

Lobopelta excisa Mayr, G.L. (1876). Die australischen Formiciden. *J. Mus. Godeffroy* **5**: 56–115 [89]. Type data: syntypes, NHMW W, from Rockhampton, Qld.

Distribution: NE coastal, Qld. Ecology: terrestrial, noctidiurnal, nomadic, predator, open forest, closed forest; nest in ground layer.

Leptogenys excisa major Forel, 1910

Leptogenys (Lobopelta) excisa major Forel, A. (1910). Formicides australiens reçus de MM. Froggatt et Rowland Turner. *Rev. Suisse Zool.* **18**: 1–94 [18]. Type data: syntypes, GMNH W, from Tweed River, N.S.W.

Distribution: SE coastal, NE coastal, Qld., N.S.W. Ecology: terrestrial, noctidiurnal, nomadic, predator, open forest, closed forest; nest in ground layer.

Leptogenys fallax (Mayr, 1876)

Leptogenys fallax fallax (Mayr, 1876)

Lobopelta fallax Mayr, G.L. (1876). Die australischen Formiciden. *J. Mus. Godeffroy* **5**: 56–115 [88]. Type data: syntypes, NHMW W,M, from Cape York, Rockhampton, Gayndah and Peak Downs, Qld.

Distribution: NE coastal, Qld. Ecology: terrestrial, noctidiurnal, nomadic, predator, woodland, open forest; nest in ground layer.

Leptogenys fallax fortior Forel, 1900

Leptogenys (Lobopelta) fallax fortior Forel, A. (1900). Ponerinae et Dorylinae d'Australie récoltées par MM. Turner, Froggatt, Nugent, Chase, Rothney, J.J. Walker, etc. *Ann. Soc. Entomol. Belg.* **44**: 54–77 [64]. Type data: syntypes, GMNH W,M, ANIC W,M, from Cairns, Qld.

Distribution: NE coastal, Qld. Ecology: terrestrial, noctidiurnal, nomadic, predator, woodland, open forest; nest in ground layer.

Leptogenys hackeri Clark, 1934

Leptogenys (Lobopelta) hackeri Clark, J. (1934). New Australian ants. *Mem. Natl. Mus. Vict.* **8**: 21–47 [35 pls 2–3]. Type data: syntypes, NMV *W, from Cascade, N.S.W. and "National Park", Qld.

Distribution: NE coastal, SE coastal, Qld., N.S.W. Ecology: terrestrial, noctidiurnal, nomadic, predator, closed forest; nest in ground layer.

Leptogenys intricata Viehmeyer, 1924

Leptogenys (Lobopelta) intricata Viehmeyer, H. (1924). Formiciden der australischen Faunenregion. *Entomol. Mitt.* **13**: 219–229 [228]. Type data: syntypes, ZMB *W,M, from Trial Bay, N.S.W.

Distribution: SE coastal, N.S.W. Ecology: terrestrial, noctidiurnal, nomadic, predator, closed forest; nest in ground layer.

Leptogenys magna Forel, 1900

Leptogenys (Lobopelta) magna Forel, A. (1900). Ponerinae et Dorylinae d'Australie récoltées par MM. Turner, Froggatt, Nugent, Chase, Rothney, J.J. Walker, etc. *Ann. Soc. Entomol. Belg.* **44**: 54–77 [65]. Type data: syntypes, GMNH W,M, ANIC W, from Mackay, Qld.

Distribution: NE coastal, Qld. Ecology: terrestrial, noctidiurnal, nomadic, predator, closed forest; nest in ground layer.

Leptogenys mjobergi Forel, 1915

Leptogenys (Lobopelta) mjobergi Forel, A. (1915). Results of Dr. E. Mjöbergs Swedish Scientific Expeditions to Australia 1910–1913. 2. Ameisen. *Ark. Zool.* **9**: 1–119 pls 1–3 [4 Dec. 1915] [32]. Type data: syntypes, GMNH W, ANIC W, other syntypes may exist, from Blackal (=Blackall) Range, Qld.

Distribution: NE coastal, Qld. Ecology: terrestrial, noctidiurnal, nomadic, predator, closed forest; nest in ground layer.

Leptogenys neutralis Forel, 1907

Leptogenys (Lobopelta) neutralis Forel, A. (1907). Formicidae. pp. 263–310 *in* Michaelsen, W. & Hartmeyer, R. (eds.) *Die Fauna Südwest-Australiens.* Jena : G. Fischer Vol. 1 [271]. Type data: holotype, probably destroyed in ZMH in WW II, from Pickering Brook, W.A.

Distribution: SW coastal, W.A. Ecology: terrestrial, noctidiurnal, nomadic, predator, closed forest; nest in ground layer.

Leptogenys sjostedti Forel, 1915

Leptogenys sjostedti Forel, A. (1915). Results of Dr. E. Mjöbergs Swedish Scientific Expeditions to Australia 1910–1913. 2. Ameisen. *Ark. Zool.* **9**: 1–119 pls 1–3 [4 Dec. 1915] [27]. Type data: syntypes, GMNH W, ANIC W, other syntypes may exist, from Lamington Plateau and Malanda, Qld.

Distribution: NE coastal, Qld. Ecology: terrestrial, noctidiurnal, nomadic, predator, open forest, closed forest; nest in ground layer.

Leptogenys tricosa Taylor, 1969

Dorylozelus mjobergi Forel, A. (1915). Results of Dr. E. Mjöbergs Swedish Scientific Expeditions to Australia 1910–1913. 2. Ameisen. *Ark. Zool.* **9**: 1–119 pls 1–3 [4 Dec. 1915] [25] [*non Leptogenys mjobergi* Forel, 1915]. Type data: holotype, SMNH W, from Blackal (=Blackall) Range, Qld.

Leptogenys tricosa Taylor, R.W. (1969). The identity of *Dorylozelus mjobergi* Forel (Hymenoptera : Formicidae). *J. Aust. Entomol. Soc.* **8**: 131–133 [132] [*nom. nov.* for *Dorylozelus mjobergi* Forel, 1915].

Distribution: NE coastal, Qld. Ecology: terrestrial, noctidiurnal, nomadic, predator, woodland, open forest, closed forest; nest in ground layer.

Leptogenys turneri Forel, 1900

Leptogenys turneri turneri Forel, 1900

Leptogenys turneri Forel, A. (1900). Ponerinae et Dorylinae d'Australie récoltées par MM. Turner, Froggatt, Nugent, Chase, Rothney, J.J. Walker, etc. *Ann. Soc. Entomol. Belg.* **44**: 54–77 [67]. Type data: syntypes, GMNH W, ANIC W, from Mackay, Qld.

Distribution: NE coastal, Qld. Ecology: terrestrial, noctidiurnal, nomadic, predator, closed forest; nest in ground layer.

Leptogenys turneri longensis Forel, 1915

Leptogenys (Odontopelta) turneri longensis Forel, A. (1915). Results of Dr. E. Mjöbergs Swedish Scientific Expeditions to Australia 1910–1913. 2. Ameisen. *Ark. Zool.* **9**: 1–119 pls 1–3 [4 Dec. 1915] [33]. Type data: syntypes, GMNH W, ANIC W, other syntypes may exist, from Malanda, Qld.

Distribution: NE coastal, Qld. Ecology: terrestrial, noctidiurnal, nomadic, predator, closed forest; nest in ground layer.

Mesoponera Emery, 1901

Mesoponera Emery, C. (1901). Notes sur les sous-familles des Dorylines et Ponérines (famille des Formicides). *Ann. Soc. Entomol. Belg.* **45**: 32–54 [43] [proposed with subgeneric rank in *Euponera* Forel, 1891; raised to genus and redefined in Brown, W.L. jr. (1958).

A review of the ants of New Zealand (Hymenoptera). *Acta Hymenopt.* **1**: 1–50]. Type species *Ponera caffraria* F. Smith, 1858 by original designation.

This group is also found in the Ethiopian, Malagasy and Oriental regions; New Guinea, east Melanesia and New Zealand in the Australian Region.

Mesoponera australis (Forel, 1900)

Ponera melanaria australis Forel, A. (1900). Ponerinae et Dorylinae d'Australie récoltées par MM. Turner, Froggatt, Nugent, Chase, Rothney, J.J. Walker, etc. *Ann. Soc. Entomol. Belg.* **44**: 54–77 [62]. Type data: syntypes, GMNH W, ANIC W, from Mackay, Qld.

Distribution: NE coastal, SE coastal, N.S.W., Qld. Ecology: terrestrial, noctidiurnal, predator, woodland, open forest, closed forest; nest in soil. Biological references: Wilson, E.O. (1958). Studies of the ant fauna of Melanesia. III. *Rhytidoponera* in Western Melanesia and the Moluccas. IV. The tribe Ponerini. *Bull. Mus. Comp. Zool.* **119**: 301–371 (taxonomy, raised to species).

Myopias Roger, 1861

Myopias Roger, J. (1861). Die *Ponera*-Artigen Ameisen. *Berl. Entomol. Z.* **5**: 1–54 [39]. Type species *Myopias amblyops* Roger, 1861 by monotypy.

This group is also found in the Oriental Region; New Guinea and east Melanesia in the Australian Region.

Myopias tasmaniensis Wheeler, 1923

Myopias tasmaniensis Wheeler, W.M. (1923). Ants of the genera *Myopias* and *Acanthoponera*. *Psyche Camb.* **30**: 175–192 [177]. Type data: syntypes, MCZ *W, from Hobart, Tas.

Trapeziopelta diadela Clark, J. (1934). Ants from the Otway Ranges. *Mem. Natl. Mus. Vict.* **8**: 48–73 [54 pl 4]. Type data: syntypes, NMV *W,F, from Turton's Track, Beech Forest, Vic.

Synonymy that of Brown, W.L. jr. (1953). An Australian *Trapeziopelta* (Hymenoptera : Formicidae). *Psyche Camb.* **60**: 51.

Distribution: SE coastal, Vic., Tas. Ecology: terrestrial, noctidiurnal, predator, closed forest; nest in ground layer.

Myopopone Roger, 1861

Myopopone Roger, J. (1861). Die *Ponera*-Artigen Ameisen. *Berl. Entomol. Z.* **5**: 1–54 [49] [redefined in Brown, W.L. jr. (1960). Contributions toward a reclassification of the Formicidae III Tribe Ambyloponini (Hymenoptera). *Bull. Mus. Comp. Zool.* **122**: 143–230]. Type species *Amblyopone castaneus* F. Smith, 1860 (as *Myopopone maculata* Roger, 1861) by subsequent designation, see Bingham, C.T. (1903). *The fauna of British India, including Ceylon and Burma.* Hymenoptera. Vol. 2 Ants and cuckoo-wasps. London : Taylor & Francis [33].

This group is also found in the east Oriental Region; New Guinea and east Melanesia in the Australian Region, see Brown, W.L. jr. (1973). A comparison of the Hylean and Congo-West African rain forest ant faunas. pp. 161–185 *in* Meggers, B.J., Ayensu, E.S. & Duckworth, W.D. (eds.) *Tropical forest ecosystems in Africa and South America: a comparative review.* Washington : Smithsonian Institution Press.

Myopopone castanea (F. Smith, 1860)

Amblyopone castaneus Smith, F. (1860). Catalogue of hymenopterous insects collected by Mr A.R. Wallace in the islands of Bachian, Kaisaa, Amboyne, Gilolo, and at Dory in New Guinea. *J. Linn. Soc. Zool.* **5**: 93–143 pl 1 [18 July 1860] [105]. Publication date established from Donisthorpe, H. (1932). On the identity of Smith's types of Formicidae (Hymenoptera) collected by Alfred Russell Wallace in the Malay Archipelago, with descriptions of two new species. *Ann. Mag. Nat. Hist. (10)* **10**: 441–476. Type data: syntypes (probable), BMNH *W, from Bachian, Indonesia.

Distribution: NE coastal, Qld. Ecology: terrestrial, noctidiurnal, predator, closed forest; nest in ground layer.

Mystrium Roger, 1862

Mystrium Roger, J. (1862). Einige neue exotische Ameisen - Gattungen und Arten. *Berl. Entomol. Z.* **6**: 233–254 [245 pl 1] [redefined in Brown, W.L. jr. (1960). Contributions toward a reclassification of the Formicidae. III Tribe Amblyoponini (Hymenoptera). *Bull. Mus. Comp. Zool.* **122**: 143–230]. Type species *Mystrium mysticum* Roger, 1862 by monotypy.

This group is also found in the north Ethiopian, Malagasy and east Oriental regions; New Guinea in the Australian Region, see Brown, W.L. jr. (1973). A comparison of the Hylean and Congo-West African rain forest ant faunas. pp. 161–185 *in* Meggers, B.J., Ayensu, E.S. & Duckworth, W.D. (eds.) *Tropical forest ecosystems in Africa and South America: a comparative review.* Washington : Smithsonian Institution Press.

Mystrium camillae Emery, 1889

Mystrium camillae Emery, C. (1889). Viaggio di Leonardo Fea in Birmania e regioni vicine. XX. Formiche di Birmania e del Tenasserim racolte de Leonardo Fea (1885–87). *Ann. Mus. Civ. Stor. Nat. Giacomo Doria* **27**: 485–520 pls 10–11 [491]. Type data: syntypes, MCG *W,F, from Bhamo, Burma.

Distribution: N coastal, N.T. Ecology: terrestrial, noctidiurnal, predator, closed forest; nest in ground layer.

Odontomachus Latreille, 1804

Odontomachus Latreille, P.A. (1804). *Nouveau Dictionnaire d'Histoire Naturelle.* Paris Vol. 24 [179]. Type species *Formica haematoda* Linnaeus, 1758 by monotypy. Compiled from secondary source: Donisthorpe, H. (1943). A list of the type-species of the genera and subgenera of the Formicidae. *Ann. Mag. Nat. Hist. (11)* **10**: 649–688.

This group is also found in the Neotropical, south Nearctic, south Palearctic, Ethiopian, Malagasy and Oriental regions; New Guinea, east Melanesia, New Caledonia and parts of Polynesia in the Australian Region, see Brown, W.L. jr. (1973). A comparison of the Hylean and Congo-West African rain forest ant faunas. pp. 161–185 *in* Meggers, B.J., Ayensu, E.S. & Duckworth, W.D. (eds.) *Tropical forest ecosystems in Africa and South America: a comparative review.* Washington : Smithsonian Institution Press.

Odontomachus cephalotes F. Smith, 1863

Odontomachus cephalotes Smith, F. (1863). Catalogue of hymenopterous insects collected by Mr A.R. Wallace in the islands of Mysol, Ceram, Waigiou, Bouru and Timor. *J. Linn. Soc. Zool.* **7**: 6–48 [4 Mar. 1863] [19]. Publication date established from Donisthorpe, H. (1932). On the identity of Smith's types of Formicidae (Hymenoptera) collected by Alfred Russell Wallace in the Malay Archipelago, with descriptions of two new species. *Ann. Mag. Nat. Hist. (10)* **10**: 441–476. Type data: syntypes (probable), BMNH *W, from Ceram, Indonesia.

Distribution: N coastal, N Gulf, NE coastal, N.T., Qld.; also in New Guinea, the Moluccas and other parts of Indonesia. Ecology: terrestrial, diurnal, predator, closed forest; nest in ground layer.

Odontomachus ruficeps F. Smith, 1858

Odontomachus ruficeps Smith, F. (1858). *Catalogue of hymenopterous insects in the collection of the British Museum.* Part 6. Formicidae. London : British Museum 216 pp. 14 pls [27 Mar. 1858] [81]. Publication date established from Donisthorpe, H. (1932). On the identity of Smith's types of Formicidae (Hymenoptera) collected by Alfred Russell Wallace in the Malay Archipelago, with descriptions of two new species. *Ann. Mag. Nat. Hist. (10)* **10**: 441–476. Type data: syntypes (probable), BMNH *W, from Australia.

Odontomachus coriarius Mayr, G.L. (1876). Die australischen Formiciden. *J. Mus. Godeffroy* **5**: 56–115 [85]. Type data: syntypes, NHMW W,M, from Rockhampton, Qld.

Odontomachus coriarius semicircularis Mayr, G.L. (1876). Die australischen Formiciden. *J. Mus. Godeffroy* **5**: 56–115 [85]. Type data: syntypes, NHMW W, from Peak Downs and Gayndah, Qld.

Odontomachus coriarius magnus Mayr, G.L. (1876). Die australischen Formiciden. *J. Mus. Godeffroy* **5**: 56–115 [85]. Type data: syntypes, NHMW W, from Rockhampton, Qld.

Odontomachus sharpei Forel, A. (1893). Nouvelles fourmis d'Australie et des Canaries. *Ann. Soc. Entomol. Belg.* **37**: 454–466 [458]. Type data: syntypes (probable), GMNH F, from Adelaide River, N.T.

Odontomachus ruficeps acutidens Forel, A. (1900). Ponerinae et Dorylinae d'Australie récoltées par MM. Turner, Froggatt, Nugent, Chase, Rothney, J.J. Walker, etc. *Ann. Soc. Entomol. Belg.* **44**: 54–77 [56]. Type data: holotype (probable), GMNH W, from Adelaide River, N.T.

Odontomachus ruficeps rubriceps Forel, A. (1915). Results of Dr. E. Mjöbergs Swedish Scientific Expeditions to Australia 1910–1913. 2. Ameisen. *Ark. Zool.* **9**: 1–119 pls 1–3 [4 Dec. 1915] [33]. Type data: syntypes, GMNH W, ANIC W, other syntypes may exist, from Kimberley distr., Noonkanbah and Broome, W.A.

Odontomachus ruficeps rufescens Forel, A. (1915). Results of Dr. E. Mjöbergs Swedish Scientific Expeditions to Australia 1910–1913. 2. Ameisen. *Ark. Zool.* **9**: 1–119 pls 1–3 [4 Dec. 1915] [34]. Type data: syntypes, GMNH W, other syntypes may exist, from Kimberley distr., W.A.

Odontomachus septentrionalis Crawley, W.C. (1915). Ants from north and central Australia, collected by G.F. Hill. Part I. *Ann. Mag. Nat. Hist. (8)* **15**: 130–136 [130]. Type data: holotype, BMNH *W, from Stapleton, N.T.

Odontomachus coriarius obscura Crawley, W.C. (1922). New ants from Australia. *Ann. Mag. Nat. Hist. (9)* **9**: 427–448 [437]. Type data: syntypes (probable), OUM *W, from W.A.

Synonymy that of Brown, W.L. jr. (1976). Contributions toward a reclassification of the Formicidae. Part VI. Ponerinae, tribe Ponerini, subtribe Odontomachiti. Section A. Introduction, subtribal characters. Genus *Odontomachus. Studia Entomol.* **19**: 67–171 [105].

Distribution: NE coastal, N coastal, Qld., W.A., N.T. Ecology: terrestrial, diurnal, predator, woodland, open forest; nest in ground layer.

Odontomachus turneri Forel, 1900

Odontomachus ruficeps turneri Forel, A. (1900). Ponerinae et Dorylinae d'Australie récoltées par MM. Turner, Froggatt, Nugent, Chase, Rothney, J.J. Walker, etc. *Ann. Soc. Entomol. Belg.* **44**: 54–77 [56]. Type data: syntypes, GMNH W, ANIC W, from Townsville, Qld.

Distribution: NE coastal, N coastal, N Gulf, N.T., Qld. Ecology: terrestrial, diurnal, predator, woodland, open forest; nest in ground layer. Biological references: Brown, W.L. jr. (1978). A supplement to the world revision of *Odontomachus* (Hymenoptera : Formicidae). *Psyche Camb.* **83**: 281–285 (reinstated from synonymy).

Onychomyrmex Emery, 1895

Onychomyrmex Emery, C. (1895). Descriptions de quelques fourmis nouvelles d'Australie. *Ann. Soc. Entomol. Belg.* **39**: 345–358 [349] [redefined in Brown, W.L. jr. (1960). Contributions toward a reclassification of the Formicidae. III. Tribe Amblyoponini (Hymenoptera)

Bull. Mus. Comp. Zool. **122**: 143–230]. Type species *Onychomyrmex hedleyi* Emery, 1895 by monotypy.

Onychomyrmex doddi Wheeler, 1916

Onychomyrmex doddi Wheeler, W.M. (1916). The Australian ants of the genus *Onychomyrmex*. *Bull. Mus. Comp. Zool.* **60**: 45–54 pls 1–2 [53]. Type data: syntypes, MCZ *W,F, from Kuranda, Qld.

Distribution: NE coastal, Qld. Ecology: terrestrial, noctidiurnal, nomadic, predator, closed forest; nest in ground layer, army ant.

Onychomyrmex hedleyi Emery, 1895

Onychomyrmex hedleyi Emery, C. (1895). Descriptions de quelques fourmis nouvelles d'Australie. *Ann. Soc. Entomol. Belg.* **39**: 345–358 [350]. Type data: syntypes, MCG W, ANIC W, from Mt. Bellenden Ker, Qld.

Distribution: NE coastal, Qld. Ecology: terrestrial, noctidiurnal, nomadic, predator, closed forest; nest in ground layer, army ant.

Onychomyrmex mjobergi Forel, 1915

Onychomyrmex mjobergi Forel, A. (1915). Results of Dr. E. Mjöbergs Swedish Scientific Expeditions to Australia 1910–1913. 2. Ameisen. *Ark. Zool.* **9**: 1–119 pls 1–3 [4 Dec. 1915] [2]. Type data: syntypes, GMNH W, other syntypes may exist, from Herberton, Atherton and Cedar Creek, Qld.

Distribution: NE coastal, Qld. Ecology: terrestrial, noctidiurnal, nomadic, predator, open forest, closed forest; nest in ground layer, army ant.

Platythyrea Roger, 1863

Platythyrea Roger, J. (1863). Die neu aufgeführten Gattungen und Arten meines Formiciden-Verzeichnisses. *Berl. Entomol. Z.* **7**: 129–214 [June 1863] [172]. Type species *Pachycondyla punctata* F. Smith, 1858 by subsequent designation, see Bingham, C.T. (1903). *The Fauna of British India, including Ceylon and Burma*. Hymenoptera. Vol. 2 Ants and cuckoo-wasps. London : Taylor & Francis [73].
Eubothroponera Clark, J. (1930). New Formicidae, with notes on some little-known species. *Proc. R. Soc. Vict.* **43**: 2–25 [30 Aug. 1930] [8]. Type species *Eubothroponera dentinodis* Clark, 1930 by original designation.
Synonymy that of Brown, W.L. jr. (1975). Contributions toward a reclassification of the Formicidae. V. Ponerinae, tribes Platythyreini, Cerapachyini, Cylindromyrmecini, Acanthostichini, and Aenictogitini. *Search Agric.* **5**: 1–116 [6].

This group is also found in the Neotropical, north Nearctic, Ethiopian and Oriental regions; New Guinea and east Melanesia in the Australian Region, see Brown, W.L. jr. (1973). A comparison of the Hylean and Congo-West African rain forest ant faunas. pp. 161–185 *in* Meggers, B.J., Ayensu, E.S. & Duckworth, W.D. (eds.) *Tropical forest ecosystems in Africa and South America: a comparative review.* Washington : Smithsonian Institution Press.

Platythyrea brunnipes (Clark, 1938)

Eubothroponera brunnipes Clark, J. (1938). Reports of the McCoy Society for Field Investigation and Research. No. 2. Sir Joseph Bank Islands. Part I. Formicidae (Hymenoptera). *Proc. R. Soc. Vict.* **50**: 356–382 [361]. Type data: syntypes (probable), NMV *W, from Reevesby Is., S.A.

Distribution: S Gulfs, SW coastal, W plateau, W.A., S.A. Ecology: terrestrial, noctidiurnal, predator, woodland, open forest; nest in soil.

Platythyrea dentinodis (Clark, 1930)

Eubothroponera dentinodis Clark, J. (1930). New Formicidae, with notes on some little-known species. *Proc. R. Soc. Vict.* **43**: 2–25 [30 Aug. 1930] [9]. Type data: syntypes, NMV *W, from Bungulla, W.A.

Distribution: SW coastal, W.A. Ecology: terrestrial, noctidiurnal, predator, open forest, closed forest; nest in ground layer.

Platythyrea micans (Clark, 1930)

Eubothroponera micans Clark, J. (1930). New Formicidae, with notes on some little-known species. *Proc. R. Soc. Vict.* **43**: 2–25 [30 Aug. 1930] [10]. Type data: syntypes, NMV *W, from Mundaring, W.A.

Distribution: SW coastal, W.A. Ecology: terrestrial, noctidiurnal, predator, woodland, open forest; nest in ground layer.

Platythyrea parallela (F. Smith, 1859)

Ponera parallela Smith, F. (1859). Catalogue of hymenopterous insects collected by Mr A.R. Wallace at the islands of Aru and Key. *J. Linn. Soc. Zool.* **3**: 132–178 [1 Feb. 1859] [143]. Publication date established from Donisthorpe, H. (1932). On the identity of Smith's types of Formicidae (Hymenoptera) collected by Alfred Russell Wallace in the Malay Archipelago, with descriptions of two new species. *Ann. Mag. Nat. Hist. (10)* **10**: 441–476. Type data: syntypes, BMNH *W, from Aru Ils., Indonesia.
Platythyrea pusilla australis Forel, A. (1915). Results of Dr. E. Mjöbergs Swedish Scientific Expeditions to Australia 1910–1913. 2. Ameisen. *Ark. Zool.* **9**: 1–119 pls 1–3 [4 Dec. 1915] [10]. Type data: syntypes, GMNH W, ANIC W, other syntypes may exist, from Blackal (=Blackall) Range and Mt. Tambourine (=Tamborine Mt.), Qld.
Platythyrea parva Crawley, W.C. (1915). Ants from north and central Australia, collected by G.F. Hill. Part I. *Ann. Mag. Nat. Hist. (8)* **15**: 130–136 [133]. Type data: syntypes, BMNH *W, from Darwin, N.T.
Platythyrea cephalotes Viehmeyer, H. (1924). Formiciden der australischen Faunenregion. *Entomol. Mitt.* **13**: 219–229 [224]. Type data: holotype, ZMB *W, from Trial Bay, N.S.W.

Synonymy that of Brown, W.L. jr. (1975). Contributions toward a reclassification of the Formicidae. V. Ponerinae, tribes Platythyreini, Cerapachyini, Cylindromyrmecini, Acanthostichini, and Aenictogitini. *Search Agric.* **5**: 1-116 [8].

Distribution: N coastal, N Gulf, NE coastal, SE coastal, Murray-Darling basin, N.T., Qld., N.S.W., A.C.T. Ecology: terrestrial, noctidiurnal, predator, woodland, open forest, closed forest; nest in ground layer.

Platythyrea turneri Forel, 1895

Platythyrea turneri Forel, A. (1895). Nouvelles fourmis d'Australie, récoltée à The Ridge, Mackay, Queensland par M. Gilbert Turner. *Ann. Soc. Entomol. Belg.* **39**: 417-428 [420]. Type data: syntypes, GMNH W, from Mackay, Qld.

Pachycondyla (Bothroponera) tasmaniensis Forel, A. (1913). Fourmis de Tasmanie et d'Australie récoltées par MM. Lea, Froggatt etc. *Bull. Soc. Vaud. Sci. Nat.* **49**: 173-196 pl 2 [176]. Type data: syntypes, GMNH W, from Hobart, Tas.

Eubothroponera bicolor Clark, J. (1930). New Formicidae, with notes on some little-known species. *Proc. R. Soc. Vict.* **43**: 2-25 [30 Aug. 1930] [11]. Type data: syntypes, NMV *W, from Ludlow, W.A.

Eubothroponera reticulata Clark, J. (1934). New Australian ants. *Mem. Natl. Mus. Vict.* **8**: 21-47 [33 pls 2-3]. Type data: syntypes (probable), NMV *W, from Sutherland, N.S.W.

Eubothroponera septentrionalis Clark, J. (1934). New Australian ants. *Mem. Natl. Mus. Vict.* **8**: 21-47 [34 pls 2-3]. Type data: syntypes (probable), QM *W, from Townsville, Qld.

Synonymy that of Brown, W.L. jr. (1975). Contributions toward a reclassification of the Formicidae. V. Ponerinae, tribes Platythyreini, Cerapachyini, Cylindromyrmecini, Acanthostichini, and Aenictogitini. *Search Agric.* **5**: 1-116 [9].

Distribution: NE coastal, SE coastal, SW coastal, Qld., N.S.W., Tas., W.A. Ecology: terrestrial, noctidiurnal, predator, woodland, open forest; nest in ground layer.

Ponera Latreille, 1804

Ponera Latreille, P.A. (1804). *Nouveau Dictionnaire d'Histoire Naturelle.* Paris Vol. 24 [179]. Type species *Formica coarctata* Latreille, 1802 (as *Formica contracta* Latreille, 1802) by subsequent designation, see Westwood, J.O. (1840). *An Introduction to the Modern Classification of Insects*; founded on the natural habits and corresponding organisation of the different families. Vol. 2. Synopsis of the genera of British Insects. London : Longman [Synopsis 83]. Compiled from secondary source: Taylor, R.W. (1967). A monographic revision of the ant genus *Ponera* Latreille. (Hymenoptera : Formicidae). *Pac. Insects Monogr.* **13**: 1-112 [30 May 1967].

This group is also found in the north Neotropical, Nearctic, Palearctic and east Oriental regions; New Guinea, east Melanesia, New Caledonia, New Zealand (N. Is.) and parts of Polynesia in the Australian Region, see Brown, W.L. jr. (1973). A comparison of the Hylean and Congo-West African rain forest ant faunas. pp. 161-185 *in* Meggers, B.J., Ayensu, E.S. & Duckworth, W.D. (eds.) *Tropical forest ecosystems in Africa and South America: a comparative review.* Washington : Smithsonian Institution Press.

Ponera clavicornis Emery, 1900

Ponera clavicornis Emery, C. (1900). Formicidarum species novae vel minus cognitae in collectione Musaei Nationalis Hungarici, quas in Nova-Guinea, Colonia Germanica, collegit L. Biró. *Termész. Füz.* **23**: 310-338 [1 Aug. 1900] [317 pl 8]. Type data: syntypes (probable), probably MCG* or MNH, from Friedrich-Wilhelmshafen (=Madang), New Guinea.

Distribution: NE coastal, Qld.; also in New Guinea. Ecology: terrestrial, noctidiurnal, predator, closed forest; nest in ground layer.

Ponera leae Forel, 1913

Ponera leae Forel, A. (1913). Fourmis de Tasmanie et d'Australie récoltées par MM. Lea, Froggatt etc. *Bull. Soc. Vaud. Sci. Nat.* **49**: 173-196 pl 2 [174]. Type data: holotype, GMNH W, from Tas.

Ponera leae oculata Wheeler, W.M. (1927). The ants of Lord Howe Island and Norfolk Island. *Proc. Am. Acad. Arts Sci.* **62**: 121-153 [130] [*non Ponera oculata* F. Smith, 1858]. Type data: syntypes, MCZ *W,F, from Norfolk Is.

Ponera leae norfolkensis Wheeler, W.M. (1935). Check list of the ants of Oceania. *Occ. Pap. Bernice P. Bishop Mus.* **11**(11): 1-56 [13] [*nom. nov.* for *Ponera leae oculata* Wheeler, 1927].

Ponera exedra Wilson, E.O. (1957). The *tenuis* and *selenophora* groups of the ant genus *Ponera* (Hymenoptera : Formicidae). *Bull. Mus. Comp. Zool.* **116**: 353-386 [364]. Type data: holotype, MCZ *W, from Arthurs Seat at McCrae, Vic.

Synonymy that of Taylor, R.W. (1967). A monographic revision of the ant genus *Ponera* Latreille (Hymenoptera : Formicidae). *Pac. Insects Monogr.* **13**: 1-112 [88].

Distribution: NE coastal, SE coastal, Qld., N.S.W., Vic., S.A., Tas., Norfolk Is. Ecology: terrestrial, noctidiurnal, predator, open forest, closed forest; nest in ground layer.

Ponera selenophora Emery, 1900

Ponera selenophora Emery, C. (1900). Formicidarum species novae vel minus cognitae in collectione Musaei Nationalis Hungarici, quas in Nova-Guinea, Colonia Germanica, collegit L. Biró. *Termész. Füz.* **23**: 310-338 [1 Aug. 1900] [317 pl 8]. Type data: syntypes, probably MCG* or MNH*, from Lemien, New Guinea.

Distribution: NE coastal, Qld.; also in New Guinea. Ecology: terrestrial, noctidiurnal, predator, open forest, closed forest; nest in ground layer.

Prionogenys Emery, 1895

Prionogenys Emery, C. (1895). Descriptions de quelques fourmis nouvelles d'Australie. *Ann. Soc. Entomol. Belg.* **39**: 345-358 [348]. Type species *Prionogenys podenzanai* Emery, 1895 by monotypy.

This group is also found in New Caledonia.

Prionogenys podenzanai Emery, 1895

Prionogenys podenzanai podonzanai Emery, 1895

Prionogenys podenzanai Emery, C. (1895). Descriptions de quelques fourmis nouvelles d'Australie. *Ann. Soc. Entomol. Belg.* **39**: 345-358 [349]. Type data: syntypes, MCG *W, from Mt. Bellenden Ker, Qld.

Distribution: NE coastal, Qld. Ecology: terrestrial, nocturnal, nomadic, predator, closed forest; nest in in ground layer.

Prionogenys podenzanai malandensis Forel, 1915

Prionogenys podenzanai malandensis Forel, A. (1915). Results of Dr. E. Mjöbergs Swedish Scientific Expeditions to Australia 1910-1913. 2. Ameisen. *Ark. Zool.* **9**: 1-119 pls 1-3 [4 Dec. 1915] [27]. Type data: syntypes, GMNH W, other syntypes may exist, from Malanda, Qld.

Distribution: NE coastal, Qld. Ecology: terrestrial, nocturnal, nomadic, predator, closed forest; nest in in ground layer.

Prionopelta Mayr, 1866

Prionopelta Mayr, G.L. (1866). Myrmecologische beiträge. *Sber. Akad. Wiss. Wien* **53**: Abt. 1 484-517 [503] [redefined in Brown, W.L. jr. (1960). Contributions toward a reclassification of the Formicidae. III. Tribe Amblyoponini (Hymenoptera). *Bull. Mus. Comp. Zool.* **122**: 143-230]. Type species *Prionopelta punctulata* Mayr, 1866 by monotypy.

This group is also found in the Neotropical, south Nearctic, Ethiopian, Malagasy and east Oriental regions; New Guinea, east Melanesia, New Caledonia and parts of Polynesia in the Australian Region, see Brown, W.L. jr. (1973). A comparison of the Hylean and Congo-West African rain forest ant faunas. pp. 161-185 *in* Meggers, B.J., Ayensu, E.S. & Duckworth, W.D. (eds.) *Tropical forest ecosystems in Africa and South America: a comparative review.* Washington : Smithsonian Institution Press.

Prionopelta opaca Emery, 1897

Prionopelta opaca Emery, C. (1897). Formicidarum species novae vel minus cognitae in collectione Musaei Nationalis Hungarici, quas in Nova-Guinea, Colonia Germanica, collegit L. Biró. *Termész. Füz.* **20**: 571-599 pl 14-15 [596]. Type data: syntypes, probably MCG or MNH *W,M,F, from New Guinea.

Distribution: NE coastal, SE coastal, N.S.W., Vic., Qld.; also in New Guinea and Micronesia. Ecology: terrestrial, noctidiurnal, predator, open forest, closed forest; nest in ground layer.

Probolomyrmex Mayr, 1901

Probolomyrmex Mayr, G.L. (1901). Südafrikanische Formiciden, gesammelt von Dr. Hans Brauns. *Ann. Natl. Mus. Wien* **16**: 1-30 pls 1-2 [2] [redefined in Brown, W.L. jr. (1975). Contributions toward a reclassification of the Formicidae. V. Ponerinae, Tribes Platythyreini, Cerapachyini, Cylindromyrmecini, Acanthostichini, and Aenictogitini. *Search Agric.* **5**: 1-116]. Type species *Probolomyrmex filiformis* Mayr, 1901 by monotypy.

This group is also found in the north Neotropical, Ethiopian and east Oriental regions, New Guinea and east Melanesia in the Australian Region, see Brown, W.L. jr. (1973). A comparison of the Hylean and Congo-West African rain forest ant faunas. pp. 161-185 *in* Meggers, B.J., Ayensu, E.S. & Duckworth, W.D. (eds.) *Tropical forest ecosystems in Africa and South America: a comparative review.* Washington : Smithsonian Institution Press.

Probolomyrmex greavesi Taylor, 1965

Probolomyrmex greavesi Taylor, R.W. (1965). A monographic revision of the rare tropicopolitan ant genus *Probolomyrmex* Mayr (Hymenoptera : Formicidae). *Trans. R. Entomol. Soc. Lond.* **117**: 345-365 [31 Dec. 1965] [358]. Type data: holotype, ANIC W, from Mt. Stromlo, A.C.T.

Distribution: Murray-Darling basin, A.C.T. Ecology: terrestrial, noctidiurnal, predator, woodland; nest in ground layer.

Proceratium Roger, 1863

Proceratium Roger, J. (1863). Die neu aufgeführten Gattungen und Arten meines Formiciden-Verzeichnisses. *Berl. Entomol. Z.* **7**: 129-214 [June 1863] [171]. Type species *Proceratium silaceum* Roger, 1863 by monotypy.

This group is also found in the Neotropical, Nearctic, Palearctic, north Ethiopian, Malagasy, east Oriental regions; New Guinea and east Melanesia in the Australian Region, see Brown, W.L. jr. (1973). A comparison of the Hylean and Congo-West African rain forest ant faunas. pp. 161-185 *in* Meggers, B.J., Ayensu, E.S. & Duckworth, W.D. (eds.) *Tropical forest ecosystems in Africa and South America: a comparative review.* Washington : Smithsonian Institution Press.

Proceratium papuanum Emery, 1897

Proceratium papuanum Emery, C. (1897). Formicidarum species novae vel minus cognitae in collectione Musaei Nationalis Hungarici, quas in Nova-Guinea, Colonia Germanica, collegit L. Biró. *Termész. Füz.* **20**: 571–599 pls 14–15 [592]. Type data: holotype, MCG *F, from New Guinea.

Distribution: NE coastal, SE coastal, N.S.W., Qld., Lord Howe Is. Ecology: terrestrial, noctidiurnal, predator, woodland, open forest; nest in ground layer.

Proceratium stictum Brown, 1958

Proceratium stictum Brown, W.L. jr. (1958). Contributions toward a reclassification of the Formicidae. II. Tribe Ectatommini (Hymenoptera). *Bull. Mus. Comp. Zool.* **118**: 173–362 [336]. Type data: holotype, MCZ *W, from Kuranda, Qld.

Distribution: NE coastal, Qld. Ecology: terrestrial, noctidiurnal, predator, closed forest; nest in ground layer.

Rhytidoponera Mayr, 1862

Rhytidoponera Mayr, G.L. (1862). Myrmecologische Studien. *Verh. Zool.-Bot. Ges. Wien* **12**: Abhand. 649–776 [731 pl 19] [proposed with subgeneric rank in *Ectatomma* F. Smith, 1858]. Type species *Ponera araneoides* Le Guillou, 1841 by subsequent designation, see Emery, C. (1911). Hymenoptera Fam. Formicidae subfam. Ponerinae. *in* Wytsman, P. (ed.) *Genera Insectorum*. Fasc. 118 Brussels 125 pp. 3 pls [37].

Chalcoponera Emery, C. (1897). Viaggio di Lamberto Loria nella Papuasia orientale 18. Formiche raccolte nella Nuova Guinea. *Ann. Mus. Civ. Stor. Nat. Giacomo Doria* **38**: 546–594 [22 Nov. 1897] [548 pl 1] [proposed with subgeneric rank in *Rhytidoponera* Mayr, 1862]. Type species *Ponera metallica* F. Smith, 1858 by subsequent designation, see Emery, C. (1911). Hymenoptera Fam. Formicidae subfam. Ponerinae. *in* Wytsman, P. (ed.) *Genera Insectorum*. Fasc. 118 Brussels 125 pp. 3 pls [38].

Synonymy that of Brown, W.L. jr. (1953). Characters and synonymies among the genera of ants. Part I. *Breviora* **11**: 1–13 [20 Mar. 1953] [2].

This group is also found in the east Oriental Region; New Guinea, east Melanesia, New Caledonia and Timor in the Australian Region, see Brown, W.L. jr. (1973). A comparison of the Hylean and Congo-West African rain forest ant faunas. pp. 161–185 *in* Meggers, B.J., Ayensu, E.S. & Duckworth, W.D. (eds.) *Tropical forest ecosystems in Africa and South America: a comparative review*. Washington : Smithsonian Institution Press.

Rhytidoponera aciculata (F. Smith, 1858)

Ectatomma aciculata Smith, F. (1858). *Catalogue of hymenopterous insects in the collection of the British Museum.* Part 6. Formicidae. London : British Museum 216 pp. 14 pls [27 Mar. 1858] [104]. Publication date established from Donisthorpe, H. (1932). On the identity of Smith's types of Formicidae (Hymenoptera) collected by Alfred Russell Wallace in the Malay Archipelago, with descriptions of two new species. *Ann. Mag. Nat. Hist. (10)* **10**: 441–476. Type data: syntypes (probable), BMNH *W, from Hunter River, N.S.W.

Ectatomma (Rhytidoponera) cristatum caro Forel, A. (1910). Formicides australiens reçus de MM. Froggatt et Rowland Turner. *Rev. Suisse Zool.* **18**: 1–94 [11]. Type data: syntypes, GMNH W, ANIC W, from N.S.W.

Synonymy that of Clark, J. (1936). A revision of Australian species of *Rhytidoponera* Mayr (Formicidae). *Mem. Natl. Mus. Vict.* **9**: 14–89 pls 3–6 [55].

Distribution: SE coastal, N.S.W., Vic. Ecology: terrestrial, noctidiurnal, omnivore, woodland, open forest, closed forest; nest in soil.

Rhytidoponera anceps Emery, 1898

Rhytidoponera anceps Emery, C. (1898). Descrizioni di formiche nuove Malesi e Australiane. Note sinonimiche. *Rec. Sess. Accad. Sci. Ist. Bologna (ns)* **2**: 231–245 [233]. Type data: holotype, MCG W, from Qld.

Distribution: NE coastal, SW coastal, Qld., W.A. Ecology: terrestrial, noctidiurnal, predator, woodland, open forest; nest in soil.

Rhytidoponera araneoides (Le Guillou, 1841)

Ponera araneoides Le Guillou, E.J.F. (1841). Catalogue raisonné des insectes hyménoptères recueillis dans le voyage de circumnavigation des corvettes l'*Astrolabe* et la *Zélée*. *Ann. Soc. Entomol. Fr.* **10**: 311–324 [317]. Type data: syntypes (probable), MNHP (probable) *W, from Salomon (=Solomon) Ils.

Rhytidoponera araneoides arcuata Stitz, H. (1911). Australische Ameisen (Neu-Guinea und Salomons-Inseln, Festland, Neu-Seeland). *Sber. Ges. Naturf. Freunde Berl.* **1911**: 351–381 [352]. Type data: syntypes, ZMB *W, from Cape York, Qld.

Synonymy that of Brown, W.L. jr. (1958). Contributions toward a reclassification of the Formicidae. II. Tribe Ectatommini (Hymenoptera). *Bull. Mus. Comp. Zool.* **118**: 173–362 [202].

Distribution: NE coastal, Qld.; also in New Guinea and Solomon Ils. Ecology: terrestrial, noctidiurnal, predator, open forest, closed forest; nest in ground layer.

Rhytidoponera aspera (Roger, 1860)

Ponera metallica aspera Roger, J. (1860). Die *Ponera*-Artigen Ameisen. *Berl. Entomol. Z.* **4**: 278–312 [308]. Type data: holotype, BMN (probable) *W, from Australia (as New Holland).

Rhytidoponera (Chalcoponera) arnoldi Forel, A. (1915). Results of Dr. E. Mjöbergs Swedish Scientific Expeditions to Australia 1910-1913. 2. Ameisen. *Ark. Zool.* **9**: 1–119 pls 1–3 [4 Dec. 1915] [14]. Type data: syntypes, GMNH W, other syntypes may exist, from Healesville, Vic.

Synonymy that of Brown, W.L. jr. (1954). Systematic and other notes on some of the smaller species of the ant genus *Rhytidoponera* Mayr. *Breviora* 33: 1-11 [9].

Distribution: SE coastal, Murray-Darling basin, N.S.W., Vic. Ecology: terrestrial, noctidiurnal, predator, open forest, closed forest; nest in ground layer.

Rhytidoponera aurata (Roger, 1861)

Ponera (Ectatomma) aurata Roger, J. (1861). Myrmicologische Nachlese. *Berl. Entomol. Z.* **5**: 163-174 [169]. Type data: holotype, whereabouts unknown, from Australia.

Rhytidoponera flava Crawley, W.C. (1915). Ants from north and south-west Australia (G.F. Hill, Rowland Turner) and Christmas Island, Straits Settlements. Part II. *Ann. Mag. Nat. Hist. (8)* **15**: 232-239 [232]. Type data: syntypes, BMNH *M, from Darwin, N.T.

Synonymy that of Clark, J. (1936). A revision of Australian species of *Rhytidoponera* Mayr (Formicidae). *Mem. Natl. Mus. Vict.* **9**: 14-89 pls 3-6 [27].

Distribution: N coastal, N Gulf, NE coastal, N.T., Qld. Ecology: terrestrial, noctidiurnal, predator, woodland, open forest; nest in soil.

Rhytidoponera barnardi Clark, 1936

Rhytidoponera barnardi Clark, J. (1936). A revision of Australian species of *Rhytidoponera* Mayr (Formicidae). *Mem. Natl. Mus. Vict.* **9**: 14-89 pls 3-6 [54]. Type data: syntypes, NMV *W, from Cape York, Qld.

Distribution: N Gulf, Qld. Ecology: terrestrial, noctidiurnal, predator, woodland, open forest; nest in soil.

Rhytidoponera barretti Clark, 1941

Rhytidoponera barretti Clark, J. (1941). Australian Formicidae. Notes and new species. *Mem. Natl. Mus. Vict.* **12**: 71-94 [81 pl 13]. Type data: syntypes, NMV *W, from Harts Range, N.T.

Distribution: Lake Eyre basin, N.T. Ecology: terrestrial, noctidiurnal, predator, desert, woodland; nest in soil.

Rhytidoponera borealis Crawley, 1918

Rhytidoponera (Chalcoponera) numeensis borealis Crawley, W.C. (1918). Some new Australian ants. *Entomol. Rec. J. Var.* **30**: 86-92 [88]. Type data: syntypes (probable), possibly OUM, from Stapleton, N.T.

Chalcoponera brunnea Clark, J. (1941). Australian Formicidae. Notes and new species. *Mem. Natl. Mus. Vict.* **12**: 71-94 [86 pl 13]. Type data: syntypes, NMV *W, from Koolpinyah, N.T.

Synonymy that of Brown, W.L. jr. (1958). Contributions toward a reclassification of the Formicidae. II. Tribe Ectatommini (Hymenoptera). *Bull. Mus. Comp. Zool.* **118**: 173-362 [202].

Distribution: N coastal, N.T. Ecology: terrestrial, noctidiurnal, predator, woodland; nest in soil.

Rhytidoponera carinata Clark, 1936

Rhytidoponera carinata Clark, J. (1936). A revision of Australian species of *Rhytidoponera* Mayr (Formicidae). *Mem. Natl. Mus. Vict.* **9**: 14-89 pls 3-6 [54]. Type data: syntypes (probable), NMV *W, from Borroloola, N.T.

Distribution: N Gulf, N.T. Ecology: terrestrial, noctidiurnal, predator, desert, woodland; nest in soil.

Rhytidoponera cerastes Crawley, 1925

Rhytidoponera cerastes Crawley, W.C. (1925). New ants from Australia. II. *Ann. Mag. Nat. Hist. (9)* **16**: 577-598 [584]. Type data: syntypes, OUM *W, from Derby, W.A.

Distribution: N coastal, W.A. Ecology: terrestrial, noctidiurnal, predator, desert, woodland; nest in soil.

Rhytidoponera chalybaea Emery, 1901

Rhytidoponera impressa chalybaea Emery, C. (1901). Notes sur les sous-familles des Dorylines et Ponérines (famille des Formicides). *Ann. Soc. Entomol. Belg.* **45**: 32-54 [51]. Type data: holotype (probable), MCG W, from N.S.W.

Ectatomma (Rhytidoponera) cyrus Forel, A. (1910). Formicides australiens reçus de MM. Froggatt et Rowland Turner. *Rev. Suisse Zool.* **18**: 1-94 [13]. Type data: syntypes, GMNH W,F, ANIC W, from Ballina, N.S.W.

Synonymy that of Brown, W.L. jr. (1954). Systematic and other notes on some of the smaller species of the ant genus *Rhytidoponera* Mayr. *Breviora* 33: 1-11 [4].

Distribution: NE coastal, SE coastal, Qld., N.S.W. Ecology: terrestrial, noctidiurnal, predator, open forest, closed forest; nest in ground layer. Biological references: Ward, P.S. (1980). Genetic variation and population differentiation in the *Rhytidoponera impressa* group, a species complex of ponerine ants (Hymenoptera : Formicidae). *Evolution* **34**: 1060-1076 (genetic variation).

Rhytidoponera chnoopyx Brown, 1958

Rhytidoponera chnoopyx Brown, W.L. jr. (1958). Contributions toward a reclassification of the Formicidae. II. Tribe Ectatommini (Hymenoptera). *Bull. Mus. Comp. Zool.* **118**: 173-362 [269]. Type data: holotype, MCZ *W, from Millaa Millaa, Atherton Tableland, Qld.

Distribution: NE coastal, Qld. Ecology: terrestrial, noctidiurnal, predator, closed forest; nest in ground layer.

Rhytidoponera clarki Donisthorpe, 1943

Ectatomma (Rhytidoponera) metallicum obscurum Forel, A. (1900). Ponerinae et Dorylinae d'Australie récoltées par MM. Turner, Froggatt, Nugent, Chase, Rothney, J.J. Walker, etc. *Ann. Soc. Entomol. Belg.* **44**: 54-77 [60] [*non Ectatomma (Holcoponera) obscurum*

Emery, 1869 = *Holcoponera obscura* (Emery, 1869)]. Type data: syntypes, GMNH W,F,M, ANIC W, from Mackay, Qld.

Chalcoponera hilli Clark, J. (1941). Australian Formicidae. Notes and new species. *Mem. Natl. Mus. Vict.* **12**: 71-94 [85 pl 13] [*non Rhytidoponera hilli* Crawley, 1915]. Type data: syntypes, NMV *W, from Palm Is., Qld.

Rhytidoponera (Chalcoponera) clarki Donisthorpe, H. (1943). Myrmecological gleanings. *Proc. R. Entomol. Soc. Lond. (B)* **12**: 115-116 [115] [*nom. nov.* for *Chalcoponera hilli* Clark, 1941].

Synonymy that of Brown, W.L. jr. (1958). Contributions toward a reclassification of the Formicidae. II. Tribe Ectatommini (Hymenoptera). *Bull. Mus. Comp. Zool.* **118**: 173-362 [203].

Distribution: NE coastal, Great Barrier Reef, Qld. Ecology: terrestrial, noctidiurnal, predator, woodland; nest in ground layer.

Rhytidoponera confusa Ward, 1980

Rhytidoponera confusa Ward, P.S. (1980). A systematic revision of the *Rhytidoponera impressa* group (Hymenoptera : Formicidae) in Australia and New Guinea. *Aust. J. Zool.* **28**: 475-498 [26 Aug. 1980] [482]. Type data: holotype, ANIC W, from Royal Natl. Park, N.S.W.

Distribution: SE coastal, NE coastal, Qld., Vic., N.S.W. Ecology: terrestrial, noctidiurnal, predator, open forest, closed forest; nest in ground layer. Biological references: Ward, P.S. (1980). Genetic variation and population differentiation in the *Rhytidoponera impressa* group, a species complex of ponerine ants (Hymenoptera : Formicidae). *Evolution* **34**: 1060-1076 (genetic variation).

Rhytidoponera convexa (Mayr, 1876)

Ectatomma convexum Mayr, G.L. (1876). Die australischen Formiciden. *J. Mus. Godeffroy* **5**: 56-115 [92]. Type data: syntypes, NHMW W,M, from Rockhampton, Gayndah and Peak Downs, Qld.

Rhytidoponera nigra Clark, J. (1936). A revision of Australian species of *Rhytidoponera* Mayr (Formicidae). *Mem. Natl. Mus. Vict.* **9**: 14-89 pls 3-6 [81]. Type data: syntypes, SAMA *W, from Mt. Serle and Owieandana, S.A.

Synonymy that of Brown, W.L. jr. (1958). Contributions toward a reclassification of the Formicidae. II. Tribe Ectatommini (Hymenoptera). *Bull. Mus. Comp. Zool.* **118**: 173-362 [272].

Distribution: NE coastal, S Gulfs, Qld., S.A. Ecology: terrestrial, noctidiurnal, predator, woodland, open forest; nest in soil.

Rhytidoponera cornuta (Emery, 1895)

Ectatomma (Rhytidoponera) cornutum Emery, C. (1895). Descriptions de quelques fourmis nouvelles d'Australie. *Ann. Soc. Entomol. Belg.* **39**: 345-358 [347]. Type data: holotype (probable), MCG W, from Cooktown, Qld.

Distribution: NE coastal, N Gulf, N.T., Qld. Ecology: terrestrial, noctidiurnal, predator, woodland; nest in soil.

Rhytidoponera crassinodis (Forel, 1907)

Ectatomma (Rhytidoponera) crassinode Forel, A. (1907). Formicidae. pp. 263-310 *in* Michaelsen, W. & Hartmeyer, R. (eds.) *Die Fauna Südwest-Australiens*. Jena : G. Fischer Vol. 1 [270]. Type data: holotype, probably destroyed in ZMH in WW II, from Day Dawn, W.A.

Distribution: NW coastal, W.A. Ecology: terrestrial, noctidiurnal, predator, woodland; nest in ground layer.

Rhytidoponera cristata (Mayr, 1876)

Ectatomma cristatum Mayr, G.L. (1876). Die australischen Formiciden. *J. Mus. Godeffroy* **5**: 56-115 [91]. Type data: syntypes, NHMW W, from Gayndah, Qld.

Distribution: NE coastal, Qld. Ecology: terrestrial, noctidiurnal, predator, woodland, open forest; nest in ground layer.

Rhytidoponera croesus Emery, 1901

Rhytidoponera croesus Emery, C. (1901). Notes sur les sous-familles des Dorylines et Ponérines (famille des Formicides). *Ann. Soc. Entomol. Belg.* **45**: 32-54 [50]. Type data: syntypes, MCG W, ANIC W, from N.S.W.

Rhytidoponera (Chalcoponera) fastuosa Santschi, F. (1916). Deux nouvelles fourmis d'Australie. *Bull. Soc. Entomol. Fr.* **1916**: 174-175 [174]. Type data: syntypes, BNHM W,F,M, from Australia.

Chalcoponera victoriae andrei Wheeler, W.M. & Chapman, J.W. (1925). The ants of the Philippine Islands. *Philipp. J. Sci.* **28**: 47-73 pls 1-2 [21 Sept. 1925] [59]. Type data: syntypes, MCZ *W, from Dorrigo, N.S.W., see Brown, W.L. jr. (1954). Systematic and other notes on some of the smaller species of the ant genus *Rhytidoponera* Mayr. *Breviora* 33: 1-11.

Synonymy that of Brown, W.L. jr. (1954). Systematic and other notes on some of the smaller species of the ant genus *Rhytidoponera* Mayr. *Breviora* 33: 1-11 [10].

Distribution: NE coastal, SE coastal, Qld., N.S.W., Vic. Ecology: terrestrial, noctidiurnal, predator, closed forest; nest in ground layer.

Rhytidoponera douglasi Brown, 1952

Rhytidoponera punctata levior Crawley, W.C. (1925). New ants from Australia. II. *Ann. Mag. Nat. Hist.* (9) **16**: 577-598 [581] [*non Rhytidoponera mayri glabrius laevior* Stitz, 1911]. Type data: syntypes (probable), OUM *W, from Rottnest Is., W.A.

Rhytidoponera douglasi Brown, W.L. jr. (1952). Notes on two well-known Australian ant species. *West. Aust. Nat.* **3**: 137-138 [15 Sept. 1952] [137] [*nom. nov.* for *Rhytidoponera punctata levior* Crawley, 1925].

Distribution: SW coastal, W.A. Ecology: terrestrial, noctidiurnal, predator, open forest; nest in soil.

Rhytidoponera dubia Crawley, 1915

Rhytidoponera (Chalcoponera) dubia Crawley, W.C. (1915). Ants from north and central Australia, collected by G.F. Hill. Part I. *Ann. Mag. Nat. Hist. (8)* **15**: 130-136 [132]. Type data: holotype, BMNH *W, from Stapleton, N.T.

Distribution: N coastal, N.T. Ecology: terrestrial, noctidiurnal, predator, woodland; nest in ground layer.

Rhytidoponera enigmatica Ward, 1980

Rhytidoponera enigmatica Ward, P.S. (1980). A systematic revision of the *Rhytidoponera impressa* group (Hymenoptera : Formicidae) in Australia and New Guinea. *Aust. J. Zool.* **28**: 475-498 [26 Aug. 1980] [484]. Type data: holotype, ANIC W, from Stringy Bark Creek, Lane Cove West, N.S.W.

Distribution: SE coastal, N.S.W. Ecology: terrestrial, noctidiurnal, predator, woodland; nest in ground layer. Biological references: Ward, P.S. (1980). Genetic variation and population differentiation in the *Rhytidoponera impressa* group, a species complex of ponerine ants (Hymenoptera : Formicidae). *Evolution* **34**: 1060-1076 (genetic variation).

Rhytidoponera eremita Clark, 1936

Rhytidoponera eremita Clark, J. (1936). A revision of Australian species of *Rhytidoponera* Mayr (Formicidae). *Mem. Natl. Mus. Vict.* **9**: 14-89 pls 3-6 [78]. Type data: syntypes, NMV *W, from Tennant Creek, Powell's Creek and Newcastle Waters, N.T.

Distribution: N Gulf, N coastal, W plateau, N.T. Ecology: terrestrial, noctidiurnal, predator, woodland; nest in ground layer.

Rhytidoponera ferruginea Clark, 1936

Rhytidoponera ferruginea Clark, J. (1936). A revision of Australian species of *Rhytidoponera* Mayr (Formicidae). *Mem. Natl. Mus. Vict.* **9**: 14-89 pls 3-6 [48]. Type data: syntypes, NMV *W, from Longreach, Qld.

Distribution: Lake Eyre basin, Qld. Ecology: terrestrial, noctidiurnal, predator, woodland; nest in ground layer.

Rhytidoponera flavicornis Clark, 1936

Rhytidoponera flavicornis Clark, J. (1936). A revision of Australian species of *Rhytidoponera* Mayr (Formicidae). *Mem. Natl. Mus. Vict.* **9**: 14-89 pls 3-6 [64]. Type data: syntypes, WAM *W, from Mundi Windi, W.A.

Distribution: NW coastal, W.A. Ecology: terrestrial, noctidiurnal, predator, woodland; nest in ground layer.

Rhytidoponera flavipes (Clark, 1941)

Chalcoponera flavipes Clark, J. (1941). Australian Formicidae. Notes and new species. *Mem. Natl. Mus. Vict.* **12**: 71-94 [84 pl 13]. Type data: syntypes, NMV *W, from Ooldea, S.A.

Distribution: W plateau, S.A. Ecology: terrestrial, noctidiurnal, predator, woodland; nest in ground layer.

Rhytidoponera flindersi Clark, 1936

Rhytidoponera flindersi Clark, J. (1936). A revision of Australian species of *Rhytidoponera* Mayr (Formicidae). *Mem. Natl. Mus. Vict.* **9**: 14-89 pls 3-6 [60]. Type data: syntypes (probable), NMV *W, from Flinders Is., S.A.

Distribution: W plateau, W.A., S.A. Ecology: terrestrial, noctidiurnal, predator, desert, woodland; nest in ground layer.

Rhytidoponera foreli Crawley, 1918

Rhytidoponera foreli Crawley, W.C. (1918). Some new Australian ants. *Entomol. Rec. J. Var.* **30**: 86-92 [87]. Type data: syntypes (probable), possibly OUM, from Koolpinyah, N.T.

Distribution: N coastal, N.T. Ecology: terrestrial, noctidiurnal, predator, woodland; nest in soil.

Rhytidoponera foveolata Crawley, 1925

Rhytidoponera foveolata Crawley, W.C. (1925). New ants from Australia. II. *Ann. Mag. Nat. Hist. (9)* **16**: 577-598 [581]. Type data: syntypes (probable), OUM *W, from Perth, W.A.

Distribution: SW coastal, W.A. Ecology: terrestrial, noctidiurnal, predator, woodland, open forest; nest in ground layer.

Rhytidoponera fuliginosa Clark, 1936

Rhytidoponera fuliginosa Clark, J. (1936). A revision of Australian species of *Rhytidoponera* Mayr (Formicidae). *Mem. Natl. Mus. Vict.* **9**: 14-89 pls 3-6 [79]. Type data: syntypes, NMV *W, from Birdum and Johnston's Lagoon, N.T.

Distribution: N Gulf, N.T. Ecology: terrestrial, noctidiurnal, predator, woodland; nest in ground layer.

Rhytidoponera greavesi Clark, 1941

Rhytidoponera greavesi Clark, J. (1941). Australian Formicidae. Notes and new species. *Mem. Natl. Mus. Vict.* **12**: 71-94 [81 pl 13]. Type data: syntypes, NMV *W, from Julia Creek, Qld.

Distribution: N Gulf, Qld., N.T. Ecology: terrestrial, noctidiurnal, predator, woodland; nest in ground layer.

Rhytidoponera gregoryi Clark, 1936

Rhytidoponera gregoryi Clark, J. (1936). A revision of Australian species of *Rhytidoponera* Mayr (Formicidae). *Mem. Natl. Mus. Vict.* **9**: 14–89 pls 3–6 [47]. Type data: syntypes, NMV *W, from Lake Killalpaninna, S.A.

Distribution: Lake Eyre basin, S.A. Ecology: terrestrial, noctidiurnal, predator, desert, woodland; nest in ground layer.

Rhytidoponera haeckeli (Forel, 1910)

Ectatomma (Rhytidoponera) haeckeli Forel, A. (1910). Formicides australiens reçus de MM. Froggatt et Rowland Turner. *Rev. Suisse Zool.* **18**: 1–94 [15]. Type data: syntypes, GMNH W, ANIC W, from Cape York, Qld.

Distribution: N Gulf, Qld. Ecology: terrestrial, noctidiurnal, predator, woodland; nest in ground layer.

Rhytidoponera hilli Crawley, 1915

Rhytidoponera hilli Crawley, W.C. (1915). Ants from north and central Australia, collected by G.F. Hill. Part I. *Ann. Mag. Nat. Hist. (8)* **15**: 130–136 [131]. Type data: syntypes, BMNH *W, from Stapleton, N.T.

Distribution: N coastal, N.T. Ecology: terrestrial, noctidiurnal, predator, woodland; nest in ground layer.

Rhytidoponera impressa (Mayr, 1876)

Ectatomma impressum Mayr, G.L. (1876). Die australischen Formiciden. *J. Mus. Godeffroy* **5**: 56–115 [92]. Type data: syntypes, NHMW W,F, from Gayndah, Qld.

Distribution: NE coastal, Qld. Ecology: terrestrial, noctidiurnal, predator, closed forest; nest in ground layer. Biological references: Ward, P.S. (1980). Genetic variation and population differentiation in the *Rhytidoponera impressa* group, a species complex of ponerine ants (Hymenoptera : Formicidae). *Evolution* **34**: 1060–1076 (genetic variation).

Rhytidoponera incisa Crawley, 1915

Rhytidoponera incisa Crawley, W.C. (1915). Ants from north and central Australia, collected by G.F. Hill. Part I. *Ann. Mag. Nat. Hist. (8)* **15**: 130–136 [132]. Type data: syntypes, BMNH *W, from Alice Springs, N.T.

Distribution: Lake Eyre basin, N.T. Ecology: terrestrial, noctidiurnal, predator, desert, woodland; nest in ground layer.

Rhytidoponera inornata Crawley, 1922

Rhytidoponera (Chalcoponera) metallica inornata Crawley, W.C. (1922). New ants from Australia. *Ann. Mag. Nat. Hist. (9)* **9**: 427–448 [436]. Type data: syntypes, OUM *W, from Perth, W.A.

Chalcoponera metallica carbonaria Wheeler, W.M. (1934). Contributions to the fauna of Rottnest Island, Western Australia No. IX. The ants. *J. R. Soc. West. Aust.* **20**: 137–163 [5 Oct. 1934] [139]. Type data: syntypes, MCZ *W, from White Hill, Tourists' Camp Reserve and west end of Rottnest Is., W.A.

Synonymy that of Brown, W.L. jr. (1958). Contributions toward a reclassification of the Formicidae. II. Tribe Ectatommini (Hymenoptera). *Bull. Mus. Comp. Zool.* **118**: 173–362 [203].

Distribution: SW coastal, W.A. Ecology: terrestrial, noctidiurnal, predator, woodland, open forest; nest in ground layer.

Rhytidoponera kurandensis Brown, 1958

Rhytidoponera kurandensis Brown, W.L. jr. (1958). Contributions toward a reclassification of the Formicidae. II. Tribe Ectatommini (Hymenoptera). *Bull. Mus. Comp. Zool.* **118**: 173–362 [267]. Type data: holotype, MCZ *W, from Kuranda near Cairns, Qld.

Distribution: NE coastal, Qld. Ecology: terrestrial, noctidiurnal, predator, closed forest; nest in ground layer.

Rhytidoponera lamellinodis Santschi, 1919

Rhytidoponera (Chalcoponera) lamellinodis Santschi, F. (1919). Cinq notes myrmécologiques. *Bull. Soc. Vaud. Sci. Nat.* **52**: 325–350 [327]. Type data: syntypes, BNHM W, from Townsville, Qld.

Distribution: NE coastal, Qld. Ecology: terrestrial, noctidiurnal, predator, woodland, open forest; nest in ground layer.

Rhytidoponera laticeps Forel, 1915

Rhytidoponera laticeps Forel, A. (1915). Results of Dr. E. Mjöbergs Swedish Scientific Expeditions to Australia 1910–1913. 2. Ameisen. *Ark. Zool.* **9**: 1–119 pls 1–3 [4 Dec. 1915] [12]. Type data: syntypes, GMNH W,M, other syntypes may exist, from Mt. Bellenden Ker, Qld.

Distribution: NE coastal, Qld. Ecology: terrestrial, noctidiurnal, predator, closed forest; nest in ground layer.

Rhytidoponera maledicta Forel, 1915

Rhytidoponera (Chalcoponera) victoriae maledicta Forel, A. (1915). Results of Dr. E. Mjöbergs Swedish Scientific Expeditions to Australia 1910–1913. 2. Ameisen. *Ark. Zool.* **9**: 1–119 pls 1–3 [4 Dec. 1915] [15]. Type data: syntypes, GMNH W,M,F, other syntypes may exist, from Malanda and Cedar Creek, Qld.

Distribution: NE coastal, Qld. Ecology: terrestrial, noctidiurnal, predator, open forest, closed forest; nest in ground layer. Biological references: Brown, W.L. jr. (1958). Contributions toward a reclassification of the Formicidae. II. Tribe Ectatommini (Hymenoptera). *Bull. Mus. Comp. Zool.* **118**: 173–362 (raised to species).

Rhytidoponera maniae (Forel, 1900)

Ectatomma (Rhytidoponera) maniae Forel, A. (1900). Ponerinae et Dorylinae d'Australie récoltées par MM. Turner, Froggatt, Nugent, Chase, Rothney, J.J. Walker, etc. *Ann. Soc. Entomol. Belg.* **44**: 54-77 [57]. Type data: syntypes, GMNH W, from Adelaide, S.A.

Ectatomma (Rhytidoponera) convexum spatiatum Forel, A. (1900). Ponerinae et Dorylinae d'Australie récoltées par MM. Turner, Froggatt, Nugent, Chase, Rothney, J.J. Walker, etc. *Ann. Soc. Entomol. Belg.* **44**: 54-77 [58]. Type data: syntypes, GMNH W, ANIC W, from S.A.

Synonymy that of Brown, W.L. jr. (1958). Contributions toward a reclassification of the Formicidae. II. Tribe Ectatommini (Hymenoptera). *Bull. Mus. Comp. Zool.* **118**: 173-362 [203].

Distribution: S Gulfs, Murray-Darling basin, N.S.W., Vic., S.A. Ecology: terrestrial, noctidiurnal, predator, desert, woodland, open forest; nest in ground layer.

Rhytidoponera mayri (Emery, 1883)

Ectatomma mayri Emery, C. (1883). Alcune formiche della Nuova Caledonia. *Boll. Soc. Entomol. Ital.* **15**: 145-151 [150]. Type data: syntypes, MCG *W, from eastern Australia.

Ectatomma (Rhytidoponera) mayri glabrius Forel, A. (1907). Formicidae. pp. 263-310 *in* Michaelsen, W. & Hartmeyer, R. (eds.) *Die Fauna Südwest-Australiens.* Jena : G. Fischer Vol. 1 [268]. Type data: syntypes, GMNH W, from Day Dawn and Yalgoo, W.A.

Rhytidoponera quadriceps Clark, J. (1936). A revision of Australian species of *Rhytidoponera* Mayr (Formicidae). *Mem. Natl. Mus. Vict.* **9**: 14-89 pls 3-6 [30]. Type data: syntypes, NMV *W, from Tennant Creek, N.T.

Rhytidoponera stridulator Clark, J. (1936). A revision of Australian species of *Rhytidoponera* Mayr (Formicidae). *Mem. Natl. Mus. Vict.* **9**: 14-89 pls 3-6 [37]. Type data: syntypes missing, originally lodged in ANIC, from 20 mi N of Bourke, N.S.W.

Rhytidoponera occidentalis Clark, J. (1936). A revision of Australian species of *Rhytidoponera* Mayr (Formicidae). *Mem. Natl. Mus. Vict.* **9**: 14-89 pls 3-6 [39]. Type data: syntypes, WAM *W, from Wadgingarra, N of Yalgoo, W.A.

Rhytidoponera petiolata Clark, J. (1936). A revision of Australian species of *Rhytidoponera* Mayr (Formicidae). *Mem. Natl. Mus. Vict.* **9**: 14-89 pls 3-6 [41]. Type data: syntypes, NMV *W, from Lake Killalpaninna, S.A.

Rhytidoponera dixoni Clark, J. (1936). A revision of Australian species of *Rhytidoponera* Mayr (Formicidae). *Mem. Natl. Mus. Vict.* **9**: 14-89 pls 3-6 [46]. Type data: syntypes, NMV *W,M, from Lake Hattah, Wyperfeld Natl. Park and Pomonal, Vic.

Synonymy that of Brown, W.L. jr. (1958). Contributions toward a reclassification of the Formicidae. II. Tribe Ectatommini (Hymenoptera). *Bull. Mus. Comp. Zool.* **118**: 173-362 [203].

Distribution: Murray-Darling basin, Lake Eyre basin, W plateau, NW coastal, N.S.W., Vic., S.A., W.A., N.T. Ecology: terrestrial, noctidiurnal, predator, desert, woodland, open forest; nest in ground layer.

Rhytidoponera metallica (F. Smith, 1858)

Ponera metallica Smith, F. (1858). *Catalogue of hymenopterous insects in the collection of the British Museum.* Part 6. Formicidae. London : British Museum 216 pp. 14 pls [27 Mar. 1858] [94]. Publication date established from Donisthorpe, H. (1932). On the identity of Smith's types of Formicidae (Hymenoptera) collected by Alfred Russell Wallace in the Malay Archipelago, with descriptions of two new species. *Ann. Mag. Nat. Hist. (10)* **10**: 441-476. Type data: lectotype, BMNH *W,F, from Adelaide, S.A., designation by Brown, W.L. jr. (1958). Contributions toward a reclassification of the Formicidae. II. Tribe Ectatommini (Hymenoptera). *Bull. Mus. Comp. Zool.* **118**: 173-362 [275].

Rhytidoponera (Chalcoponera) metallica purpurascens Wheeler, W.M. (1915). Hymenoptera. *Trans. R. Soc. S. Aust.* **39**: 805-823 pls 64-66 [Dec. 1915] [805]. Type data: holotype, MCZ *W, from Moorilyanna, S.A.

Rhytidoponera (Chalcoponera) metallica varians Crawley, W.C. (1922). New ants from Australia. *Ann. Mag. Nat. Hist. (9)* **9**: 427-448 [436]. Type data: syntypes, OUM *W, from Darlington, W.A.

Rhytidoponera (Chalcoponera) caeciliae Viehmeyer, H. (1924). Formiciden der australischen Faunenregion. *Entomol. Mitt.* **13**: 219-229 [227]. Type data: syntypes, ZMB *W,F, from Kilolpanino (=Killalpaninna), S.A.

Chalcoponera pulchra Clark, J. (1941). Australian Formicidae. Notes and new species. *Mem. Natl. Mus. Vict.* **12**: 71-94 [86 pl 13]. Type data: syntypes, NMV *W, from Forrest, W.A.

Synonymy that of Brown, W.L. jr. (1958). Contributions toward a reclassification of the Formicidae. II. Tribe Ectatommini (Hymenoptera). *Bull. Mus. Comp. Zool.* **118**: 173-362 [204].

Distribution: S Gulfs, Lake Eyre basin, W plateau, NW coastal, SW coastal, Bulloo River basin, Murray-Darling basin, NE coastal, SE coastal, Tas., N.S.W., Vic., A.C.T., Qld., N.T., S.A., W.A. Ecology: terrestrial, noctidiurnal, predator, desert, woodland, open forest; nest in ground layer. Biological references: Crozier, R.H. (1969). Chromosome number polymorphism in an Australian ponerine ant. *Can. J. Genet. Cytol.* **11**: 333-339 (genetics).

Rhytidoponera micans Clark, 1936

Rhytidoponera micans Clark, J. (1936). A revision of Australian species of *Rhytidoponera* Mayr (Formicidae). *Mem. Natl. Mus. Vict.* **9**: 14-89 pls 3-6 [62]. Type data: syntypes, NMV *W,M, from Eradu and Mullewa, W.A.

Distribution: NW coastal, W.A. Ecology: terrestrial, noctidiurnal, predator, woodland; nest in ground layer.

Rhytidoponera mirabilis Clark, 1936

Rhytidoponera mirabilis Clark, J. (1936). A revision of Australian species of *Rhytidoponera* Mayr (Formicidae). *Mem. Natl. Mus. Vict.* **9**: 14–89 pls 3–6 [29]. Type data: syntypes, NMV *W, from Alice Springs, N.T.

Distribution: Lake Eyre basin, N.T. Ecology: terrestrial, noctidiurnal, predator, desert, woodland; nest in ground layer.

Rhytidoponera nitida Clark, 1936

Rhytidoponera nitida Clark, J. (1936). A revision of Australian species of *Rhytidoponera* Mayr (Formicidae). *Mem. Natl. Mus. Vict.* **9**: 14–89 pls 3–6 [45]. Type data: syntypes, NMV *W, from Bourke, N.S.W.

Distribution: Murray-Darling basin, N.S.W. Ecology: terrestrial, noctidiurnal, predator, desert, woodland; nest in ground layer.

Rhytidoponera nodifera (Emery, 1895)

Ectatomma (Rhytidoponera) convexum nodiferum Emery, C. (1895). Descriptions de quelques fourmis nouvelles d'Australie. *Ann. Soc. Entomol. Belg.* **39**: 345–358 [348]. Type data: syntypes, MCG W, from Laidily (=Laidley) and Kamerunga, Qld.

Ectatomma (Rhytidoponera) rothneyi Forel, A. (1900). Ponerinae et Dorylinae d'Australie récoltées par MM. Turner, Froggatt, Nugent, Chase, Rothney, J.J. Walker, etc. *Ann. Soc. Entomol. Belg.* **44**: 54–77 [56]. Type data: syntypes, GMNH W, from Brisbane, Qld.

Rhytidoponera rothneyi mediana Viehmeyer, H. (1924). Formiciden der australischen Faunenregion. *Entomol. Mitt.* **13**: 219–229 [224]. Type data: syntypes, ZMB *W,M, from Trial Bay, N.S.W.

Rhytidoponera pronotalis Crawley, W.C. (1925). New ants from Australia. II. *Ann. Mag. Nat. Hist. (9)* **16**: 577–598 [588]. Type data: syntypes, OUM *W, from Lismore, N.S.W.

Synonymy that of Clark, J. (1936). A revision of Australian species of *Rhytidoponera* Mayr (Formicidae). *Mem. Natl. Mus. Vict.* **9**: 14–89 pls 3–6 [68].

Distribution: NE coastal, SE coastal, N.S.W., Qld. Ecology: terrestrial, noctidiurnal, predator, woodland, open forest; nest in ground layer.

Rhytidoponera nudata (Mayr, 1876)

Ectatomma nudatum Mayr, G.L. (1876). Die australischen Formiciden. *J. Mus. Godeffroy* **5**: 56–115 [91]. Type data: holotype, NHMW W, from Gayndah, Qld.

Distribution: NE coastal, Qld. Ecology: terrestrial, noctidiurnal, predator, woodland; nest in ground layer.

Rhytidoponera peninsularis Brown, 1958

Rhytidoponera peninsularis Brown, W.L. jr. (1958). Contributions toward a reclassification of the Formicidae. II. Tribe Ectatommini (Hymenoptera). *Bull. Mus. Comp. Zool.* **118**: 173–362 [280]. Type data: holotype, MCZ *W, from Rocky Scrub in the McIlwraith Range, NE of Coen, Cape York Peninsula, Qld.

Distribution: NE coastal, Qld. Ecology: terrestrial, noctidiurnal, predator, open forest, closed forest; nest in ground layer.

Rhytidoponera pilosula Clark, 1936

Rhytidoponera pilosula Clark, J. (1936). A revision of Australian species of *Rhytidoponera* Mayr (Formicidae). *Mem. Natl. Mus. Vict.* **9**: 14–89 pls 3–6 [80]. Type data: syntypes, NMV *W, from Bourke, N.S.W.

Distribution: Murray-Darling basin, N.S.W. Ecology: terrestrial, noctidiurnal, predator, woodland; nest in ground layer.

Rhytidoponera punctata (F. Smith, 1858)

Ectatomma punctata Smith, F. (1858). *Catalogue of hymenopterous insects in the collection of the British Museum.* Part 6. Formicidae. London : British Museum 216 pp. 14 pls [27 Mar. 1858] [104]. Publication date established from Donisthorpe, H. (1932). On the identity of Smith's types of Formicidae (Hymenoptera) collected by Alfred Russell Wallace in the Malay Archipelago, with descriptions of two new species. *Ann. Mag. Nat. Hist. (10)* **10**: 441–476. Type data: syntypes (probable), BMNH *W, from Port Lincoln, S.A.

Distribution: S Gulfs, S.A. Ecology: terrestrial, noctidiurnal, predator, woodland; nest in ground layer.

Rhytidoponera punctigera Crawley, 1925

Rhytidoponera punctigera Crawley, W.C. (1925). New ants from Australia. II. *Ann. Mag. Nat. Hist. (9)* **16**: 577–598 [582]. Type data: syntypes (probable), OUM *W, from Manjimup, W.A.

Distribution: SW coastal, W.A. Ecology: terrestrial, noctidiurnal, predator, woodland; nest in ground layer.

Rhytidoponera punctiventris (Forel, 1900)

Ectatomma (Rhytidoponera) cristatum punctiventris Forel, A. (1900). Ponerinae et Dorylinae d'Australie récoltées par MM. Turner, Froggatt, Nugent, Chase, Rothney, J.J. Walker, etc. *Ann. Soc. Entomol. Belg.* **44**: 54–77 [56]. Type data: syntypes (probable), GMNH W, from Sydney, N.S.W.

Distribution: SE coastal, N.S.W. Ecology: terrestrial, noctidiurnal, predator, woodland, open forest; nest in ground layer. Biological references: Brown, W.L. jr. (1958). Contributions toward a reclassification of the Formicidae. II. Tribe Ectatommini (Hymenoptera). *Bull. Mus. Comp. Zool.* **118**: 173–362 (raised to species).

Rhytidoponera purpurea (Emery, 1887)

Ectatomma impressum purpureum Emery, C. (1887). Catalogo delle formiche esistenti nelle collezioni del

Museo Civico di Genova. Parte terza. Formiche della regione Indo-Malese e dell'Australia. *Ann. Mus. Civ. Stor. Nat. Giacomo Doria (2)* **5**: 427–473 pls 1–2 [444]. Type data: syntypes, MCG W,F, from Hatam, New Guinea.

Ectatomma (Rhytidoponera) impressum splendidum Forel, A. (1910). Formicides australiens reçus de MM. Froggatt et Rowland Turner. *Rev. Suisse Zool.* **18**: 1–94 [12]. Type data: syntypes, GMNH W, from Kuranda and Cairns, Qld.

Synonymy that of Brown, W.L. jr. (1954). Systematic and other notes on some of the smaller species of the ant genus *Rhytidoponera* Mayr. *Breviora* **33**: 1–11 [7].

Distribution: NE coastal, Qld.; also in New Guinea. Ecology: terrestrial, noctidiurnal, predator, closed forest; nest in ground layer. Biological references: Ward, P.S. (1980). Genetic variation and population differentiation in the *Rhytidoponera impressa* group, a species complex of ponerine ants (Hymenoptera : Formicidae). *Evolution* **34**: 1060–1076 (genetic variation).

Rhytidoponera reflexa Clark, 1936

Rhytidoponera reflexa Clark, J. (1936). A revision of Australian species of *Rhytidoponera* Mayr (Formicidae). *Mem. Natl. Mus. Vict.* **9**: 14–89 pls 3–6 [76]. Type data: syntypes, NMV *W, from Koolpinyah and Bathurst Is., N.T.

Distribution: N coastal, N.T. Ecology: terrestrial, noctidiurnal, predator, woodland; nest in ground layer.

Rhytidoponera reticulata (Forel, 1893)

Ectatomma (Rhytidoponera) reticulatum Forel, A. (1893). Nouvelles fourmis d'Australie et des Canaries. *Ann. Soc. Entomol. Belg.* **37**: 454–466 [459]. Type data: syntypes, GMNH W,F, from Port Darwin, N.T.

Distribution: N coastal, N.T. Ecology: terrestrial, noctidiurnal, predator, woodland; nest in ground layer.

Rhytidoponera rufescens (Forel, 1900)

Ectatomma (Rhytidoponera) convexum rufescens Forel, A. (1900). Ponerinae et Dorylinae d'Australie récoltées par MM. Turner, Froggatt, Nugent, Chase, Rothney, J.J. Walker, etc. *Ann. Soc. Entomol. Belg.* **44**: 54–77 [58]. Type data: syntypes, GMNH W, from Charter (=Charters) Towers and Townsville, Qld.

Distribution: NE coastal, Qld. Ecology: terrestrial, noctidiurnal, predator, woodland; nest in ground layer. Biological references: Brown, W.L. jr. (1958). Contributions toward a reclassification of the Formicidae. II. Tribe Ectatommini (Hymenoptera). *Bull. Mus. Comp. Zool.* **118**: 173–362 (raised to species).

Rhytidoponera rufithorax Clark, 1941

Rhytidoponera rufithorax Clark, J. (1941). Australian Formicidae. Notes and new species. *Mem. Natl. Mus. Vict.* **12**: 71–94 [82 pl 13]. Type data: syntypes, NMV *W, from Alexandria Station, N.T.

Distribution: N Gulf, N.T. Ecology: terrestrial, noctidiurnal, predator, woodland; nest in ground layer.

Rhytidoponera rufiventris Forel, 1915

Rhytidoponera convexa rufiventris Forel, A. (1915). Results of Dr. E. Mjöbergs Swedish Scientific Expeditions to Australia 1910–1913. 2. Ameisen. *Ark. Zool.* **9**: 1–119 pls 1–3 [4 Dec. 1915] [11]. Type data: syntypes, GMNH W,M, ANIC W, other syntypes may exist, from Herberton, Atherton, Evelyne, Malanda and Cedar Creek, Qld.

Rhytidoponera castanea Crawley, W.C. (1925). New ants from Australia. II. *Ann. Mag. Nat. Hist. (9)* **16**: 577–598 [589]. Type data: syntypes, OUM *W, from Derby, N.S.W.

Synonymy that of Clark, J. (1936). A revision of Australian species of *Rhytidoponera* Mayr (Formicidae). *Mem. Natl. Mus. Vict.* **9**: 14–89 pls 3–6 [83].

Distribution: N coastal, N gulf, NE coastal, W.A., N.T., Qld. Ecology: terrestrial, noctidiurnal, predator, woodland, open forest; nest in ground layer.

Rhytidoponera rufonigra Clark, 1936

Rhytidoponera rufonigra Clark, J. (1936). A revision of Australian species of *Rhytidoponera* Mayr (Formicidae). *Mem. Natl. Mus. Vict.* **9**: 14–89 pls 3–6 [58]. Type data: syntypes, NMV *W,M, from Perth, Mundaring and Armadale, W.A.

Distribution: SW coastal, W.A. Ecology: terrestrial, noctidiurnal, predator, woodland, open forest; nest in ground layer.

Rhytidoponera scaberrima (Emery, 1895)

Ectatomma (Rhytidoponera) scaberrimum Emery, C. (1895). Descriptions de quelques fourmis nouvelles d'Australie. *Ann. Soc. Entomol. Belg.* **39**: 345–358 [347]. Type data: holotype, MCG W, from Mt. Bellenden Ker, Qld.

Rhytidoponera laciniosa malandensis Forel, A. (1915). Results of Dr. E. Mjöbergs Swedish Scientific Expeditions to Australia 1910–1913. 2. Ameisen. *Ark. Zool.* **9**: 1–119 pls 1–3 [4 Dec. 1915] [10]. Type data: syntypes, GMNH W,M, ANIC W, other syntypes may exist, from Malanda, Qld.

Synonymy that of Brown, W.L. jr. (1958). Contributions toward a reclassification of the Formicidae. II. Tribe Ectatommini (Hymenoptera). *Bull. Mus. Comp. Zool.* **118**: 173–362 [204].

Distribution: NE coastal, Qld. Ecology: terrestrial, noctidiurnal, predator, closed forest; nest in ground layer.

Rhytidoponera scabra (Mayr, 1876)

Ectatomma scabrum Mayr, G.L. (1876). Die australischen Formiciden. *J. Mus. Godeffroy* **5**: 56–115 [90]. Type data: syntypes, NHMW W,M, from Port Mackay, Rockhampton and Peak Downs, Qld.

Distribution: NE coastal, Qld. Ecology: terrestrial, noctidiurnal, predator, woodland, open forest; nest in ground layer.

Rhytidoponera scabrior Crawley, 1925

Rhytidoponera (Chalcoponera) aspera scabrior Crawley, W.C. (1925). New ants from Australia. II. *Ann. Mag. Nat. Hist. (9)* **16**: 577–598 [590]. Type data: syntypes, OUM *W, from Lismore, N.S.W.

Distribution: NE coastal, SE coastal, Qld., N.S.W. Ecology: terrestrial, noctidiurnal, predator, open forest, closed forest; nest in ground layer. Biological references: Ward, P.S. (1980). A systematic revision of the *Rhytidoponera impressa* group (Hymenoptera : Formicidae) in Australia and New Guinea. *Aust. J. Zool.* **28**: 475–498 (raised to species).

Rhytidoponera socrus (Forel, 1894)

Ectatomma (Rhytidoponera) socrus Forel, A. (1894). Quelques fourmis de Madagascar (récoltées par M. le Dr. Völtzkow); de Nouvelle Zélande (récoltées par M. W.W. Smith); de Nouvelle Calédonie (récoltées par M. Sommer); de Queensland (Australie) récoltées par M. Wiederkehr; et de Perth (Australie occidentale) récoltées par M. Chase. *Ann. Soc. Entomol. Belg.* **38**: 226–237 [236]. Type data: syntypes, GMNH W, ANIC W, from Charters Towers, Qld.

Distribution: NE coastal, Qld. Ecology: terrestrial, noctidiurnal, predator, desert, woodland; nest in soil.

Rhytidoponera spoliata (Emery, 1895)

Ectatomma (Rhytidoponera) spoliatum Emery, C. (1895). Descriptions de quelques fourmis nouvelles d'Australie. *Ann. Soc. Entomol. Belg.* **39**: 345–358 [348]. Type data: syntypes, MCG W, ANIC W, from Mt. Bellenden Ker, Qld.

Distribution: NE coastal, Qld. Ecology: terrestrial, noctidiurnal, predator, closed forest; nest in ground layer.

Rhytidoponera tasmaniensis Emery, 1898

Rhytidoponera metallica tasmaniensis Emery, C. (1898). Descrizioni di formiche nuove Malesi e Australiane. Note sinonimiche. *Rec. Sess. Accad. Sci. Ist. Bologna (ns)* **2**: 231–245 [232]. Type data: syntypes, MCG W, ANIC W, from Tas.

Ectatomma (Rhytidoponera) metallicum cristulatum Forel, A. (1900). Ponerinae et Dorylinae d'Australie récoltées par MM. Turner, Froggatt, Nugent, Chase, Rothney, J.J. Walker, etc. *Ann. Soc. Entomol. Belg.* **44**: 54–77 [59]. Type data: syntypes, GMNH W, ANIC W, from Australia.

Synonymy that of Brown, W.L. jr. (1958). Contributions toward a reclassification of the Formicidae. II. Tribe Ectatommini (Hymenoptera). *Bull. Mus. Comp. Zool.* **118**: 173–362 [205].

Distribution: S Gulfs, Murray-Darling basin, SE coastal, N.S.W., Vic., Tas. Ecology: terrestrial, noctidiurnal, predator, woodland, open forest; nest in ground layer.

Rhytidoponera taurus (Forel, 1910)

Ectatomma (Rhytidoponera) cornutum taurus Forel, A. (1910). Formicides australiens reçus de MM. Froggatt et Rowland Turner. *Rev. Suisse Zool.* **18**: 1–94 [12]. Type data: syntypes, GMNH W, ANIC W, from Tennant Creek, N.T.

Rhytidoponera cornuta fusciventris Stitz, H. (1911). Australische Ameisen (Neu-Guinea und Salomons-Inseln, Festland, Neu-Seeland). *Sber. Ges. Naturf. Freunde Berl.* **1911**: 351–381 [352]. Type data: syntypes, ZMB *W, from Adelaide, S.A.

Rhytidoponera cerastes brevior Crawley, W.C. (1925). New ants from Australia. II. *Ann. Mag. Nat. Hist. (9)* **16**: 577–598 [586]. Type data: syntypes, OUM *W,M, from Derby, W.A.

Synonymy that of Clark, J. (1936). A revision of Australian species of *Rhytidoponera* Mayr (Formicidae). *Mem. Natl. Mus. Vict.* **9**: 14–89 pls 3–6 [25].

Distribution: S Gulfs, W plateau, N coastal, N Gulf, Lake Eyre basin, S.A., W.A., N.T. Ecology: terrestrial, noctidiurnal, predator, desert, woodland, open forest; nest in soil.

Rhytidoponera tenuis (Forel, 1900)

Ectatomma (Rhytidoponera) tenue Forel, A. (1900). Ponerinae et Dorylinae d'Australie récoltées par MM. Turner, Froggatt, Nugent, Chase, Rothney, J.J. Walker, etc. *Ann. Soc. Entomol. Belg.* **44**: 54–77 [58]. Type data: syntypes, GMNH W, ANIC W, from Mackay, Qld.

Distribution: NE coastal, Qld. Ecology: terrestrial, noctidiurnal, predator, woodland, open forest; nest in ground layer.

Rhytidoponera trachypyx Brown, 1958

Rhytidoponera trachypyx Brown, W.L. jr. (1958). Contributions toward a reclassification of the Formicidae. II. Tribe Ectatommini (Hymenoptera). *Bull. Mus. Comp. Zool.* **118**: 173–362 [281]. Type data: holotype, MCZ *W, from river bank at Katherine, N.T.

Distribution: N coastal, N.T. Ecology: terrestrial, noctidiurnal, predator, woodland; nest in ground layer.

Rhytidoponera turneri (Forel, 1910)

Ectatomma (Rhytidoponera) turneri Forel, A. (1910). Formicides australiens reçus de MM. Froggatt et Rowland Turner. *Rev. Suisse Zool.* **18**: 1–94 [14]. Type data: syntypes, GMNH W, from Cape York, Qld.

Distribution: N Gulf, Qld. Ecology: terrestrial, noctidiurnal, predator, woodland, open forest; nest in ground layer.

Rhytidoponera tyloxys Brown and Douglas, 1958

Rhytidoponera tyloxys Brown, W.L. & Douglas, A.M. (1958). *in* Brown, W.L. jr. (1958). Contributions toward a reclassification of the Formicidae. II. Tribe Ectatommini (Hymenoptera). *Bull. Mus. Comp. Zool.* **118**: 173–362 [282]. Type data: holotype, WAM 64–37 *W, from Woodstock Station, 900 mi N of Perth, W.A.

Distribution: NW coastal, W.A. Ecology: terrestrial, noctidiurnal, predator, desert, woodland; nest in ground layer.

Rhytidoponera victoriae (E. André, 1896)

Ectatomma (Rhytidoponera) victoriae André, E. (1896). Fourmis nouvelles d'Asie et d'Australie. *Rev. Entomol.* **15**: 251–265 [261]. Type data: syntypes, MNHP W, ANIC W, from Victorian Alps.

Ectatomma (Rhytidoponera) metallicum modestum Emery, C. (1895). Descriptions de quelques fourmis nouvelles d'Australie. *Ann. Soc. Entomol. Belg.* **39**: 345–358 [348]. Type data: syntypes, MCG W, from Kamerunga, Qld.

Ectatomma (Rhytidoponera) metallicum scrobiculatum Forel, A. (1900). Ponerinae et Dorylinae d'Australie récoltées par MM. Turner, Froggatt, Nugent, Chase, Rothney, J.J. Walker, etc. *Ann. Soc. Entomol. Belg.* **44**: 54–77 [59]. Type data: syntypes, GMNH W,M,F, ANIC W, from Richmond, N.S.W.

Rhytidoponera (Chalcoponera) victoriae cedarensis Forel, A. (1915). Results of Dr. E. Mjöbergs Swedish Scientific Expeditions to Australia 1910–1913. 2. Ameisen. *Ark. Zool.* **9**: 1–119 pls 1–3 [4 Dec. 1915] [15]. Type data: syntypes, GMNH (probable) *W,M, from Cedar Creek, Qld.

Synonymy that of Brown, W.L. jr. (1958). Contributions toward a reclassification of the Formicidae. II. Tribe Ectatommini (Hymenoptera). *Bull. Mus. Comp. Zool.* **118**: 173–362 [205].

Distribution: NE coastal, SE coastal, Murray-Darling basin, Qld., N.S.W., Vic. Ecology: terrestrial, noctidiurnal, predator, woodland, open forest, closed forest; nest in ground layer.

Rhytidoponera violacea (Forel, 1907)

Ectatomma (Rhytidoponera) convexum violaceum Forel, A. (1907). Formicidae. pp. 263–310 *in* Michaelsen, W. & Hartmeyer, R. (eds.) *Die Fauna Südwest-Australiens.* Jena : G. Fischer Vol. 1 [269]. Type data: syntypes, GMNH W, ANIC W, from Northampton, Eradu, Wooroloo, Lion Mill, Mundaring Weir, South Perth, Subiaco, Jarrahdale and York, W.A.

Ectatomma (Rhytidoponera) convexum gemma Forel, A. (1907). Formicidae. pp. 263–310 *in* Michaelsen, W. & Hartmeyer, R. (eds.) *Die Fauna Südwest-Australiens.* Jena : G. Fischer Vol. 1 [269]. Type data: syntypes, GMNH W, ANIC W, from Yarloop, Gooseberry Hill and York, W.A.

Rhytidoponera convexa opacior Clark, J. (1936). A revision of Australian species of *Rhytidoponera* Mayr (Formicidae). *Mem. Natl. Mus. Vict.* **9**: 14–89 pls 3–6 [86] [introduced as a quadranomen by Crawley, 1925]. Type data: syntypes, OUM W, from Jigalong, W.A.

Synonymy that of Clark, J. (1936). A revision of Australian species of *Rhytidoponera* Mayr (Formicidae). *Mem. Natl. Mus. Vict.* **9**: 14–89 pls 3–6 [87]; Brown, W.L. jr. (1958). Contributions toward a reclassification of the Formicidae. II. Tribe Ectatommini (Hymenoptera). *Bull. Mus. Comp. Zool.* **118**: 173–362 [205].

Distribution: SW coastal, NW coastal, W plateau, W.A. Ecology: terrestrial, noctidiurnal, predator, woodland, open forest; nest in soil.

Rhytidoponera viridis (Clark, 1941)

Chalcoponera viridis Clark, J. (1941). Australian Formicidae. Notes and new species. *Mem. Natl. Mus. Vict.* **12**: 71–94 [83 pl 13]. Type data: syntypes, NMV *W, from Kalamurina, Lake Eyre, S.A.

Distribution: Lake Eyre basin, S.A. Ecology: terrestrial, noctidiurnal, predator, desert, woodland; nest in soil.

Rhytidoponera yorkensis Forel, 1915

Rhytidoponera cristata yorkensis Forel, A. (1915). Results of Dr. E. Mjöbergs Swedish Scientific Expeditions to Australia 1910–1913. 2. Ameisen. *Ark. Zool.* **9**: 1–119 pls 1–3 [4 Dec. 1915] [12]. Type data: syntypes, GMNH W, other syntypes may exist, from Cape York, Qld.

Distribution: NE coastal, Qld. Ecology: terrestrial, noctidiurnal, predator, woodland, open forest; nest in ground layer. Biological references: Clark, J. (1936). A revision of Australian species of *Rhytidoponera* Mayr (Formicidae). *Mem. Natl. Mus. Vict.* **9**: 14–89 pls 3–6 (raised to species).

Sphinctomyrmex Mayr, 1866

Sphinctomyrmex Mayr, G.L. (1866). Diagnosen neuer und wenig gekannter Formiciden. *Verh. Zool.-Bot. Ges. Wien* **16**: Abhand. 885–908 [895 pl 20]. Type species *Sphinctomyrmex stali* Mayr, 1866 by monotypy.

Nothosphinctus Wheeler, W.M. (1918). The Australian ants of the ponerine tribe Cerapachyini. *Proc. Am. Acad. Arts Sci.* **53**: 215–265 [219] [proposed with subgeneric rank in *Eusphinctus* Emery, 1893]. Type species *Sphinctomyrmex froggatti* Forel, 1900 by subsequent designation, see Donisthorpe, H. (1943). A list of the type-species of the genera and subgenera of the Formicidae. *Ann. Mag. Nat. Hist. (11)* **10**: 649–688.

Zasphinctus Wheeler, W.M. (1918). The Australian ants of the ponerine tribe Cerapachyini. *Proc. Am. Acad. Arts Sci.* **53**: 215–265 [219] [proposed with subgeneric rank in

Eusphinctus Emery, 1893]. Type species *Sphinctomyrmex turneri* Forel, 1900 by monotypy.

Synonymy that of Brown, W.L. jr. (1975). Contributions toward a reclassification of the Formicidae. V. Ponerinae, tribes Platythyreini, Cerapachyini, Cylindromyrmecini, Acanthostichini, and Aenictogitini. *Search Agric.* **5**: 1–116 [31].

This group is also found in the south Neotropical, north Ethiopian and Oriental regions; New Guinea, east Melanesia and New Caledonia in the Australian Region, see Brown, W.L. jr. (1973). A comparison of the Hylean and Congo-West African rain forest ant faunas. pp. 161–185 *in* Meggers, B.J., Ayensu, E.S. & Duckworth, W.D. (eds.) *Tropical forest ecosystems in Africa and South America: a comparative review.* Washington : Smithsonian Institution Press.

Sphinctomyrmex asper Brown, 1975

Sphinctomyrmex asper Brown, W.L. jr. (1975). Contributions toward a reclassification of the Formicidae. V. Ponerinae, tribes Platythyreini, Cerapachyini, Cylindromyrmecini, Acanthostichini, and Aenictogitini. *Search Agric.* **5**: 1–116 [78]. Type data: holotype, MCZ W, from Halifax, Qld.

Distribution: NE coastal, Qld. Ecology: terrestrial, noctidiurnal, predator, open forest, closed forest; nest in soil.

Sphinctomyrmex cedaris Forel, 1915

Sphinctomyrmex (Eusphinctus) fallax cedaris Forel, A. (1915). Results of Dr. E. Mjöbergs Swedish Scientific Expeditions to Australia 1910–1913. 2. Ameisen. *Ark. Zool.* **9**: 1–119 pls 1–3 [4 Dec. 1915] [16]. Type data: syntypes, GMNH W, ANIC W, other syntypes may exist, from Cedar Creek, Qld.

Distribution: NE coastal, Qld. Ecology: terrestrial, noctidiurnal, predator, open forest, closed forest; nest in soil. Biological references: Brown, W.L. jr. (1975). Contributions toward a reclassification of the Formicidae. V. Ponerinae, tribes Platythyreini, Cerapachyini, Cylindromyrmecini, Acanthostichini, and Aenictogitini. *Search Agric.* **5**: 1–116 (raised to species).

Sphinctomyrmex clarus (Forel, 1893)

Cerapachys emeryi clarus Forel, A. (1893). Nouvelles fourmis d'Australie et des Canaries. *Ann. Soc. Entomol. Belg.* **37**: 454–466 [462]. Type data: syntypes, GMNH W, from Adelaide River, N.T.

Distribution: N coastal, N.T., W.A. Ecology: terrestrial, noctidiurnal, predator, woodland, open forest; nest in soil. Biological references: Brown, W.L. jr. (1975). Contributions toward a reclassification of the Formicidae. V. Ponerinae, tribes Platythyreini, Cerapachyini, Cylindromyrmecini, Acanthostichini, and Aenictogitini. *Search Agric.* **5**: 1–116 (raised to species).

Sphinctomyrmex duchaussoyi (E. André, 1905)

Eusphinctus duchaussoyi André, E. (1905). Description d'un genre nouveau et de deux espèces nouvelles de fourmis d'Australie. *Rev. Entomol.* **24**: 205–208 [205]. Type data: syntypes, MNHP *W,F, from Sydney, N.S.W.

Eusphinctus (Eusphinctus) hackeri Wheeler, W.M. (1918). The Australian ants of the ponerine tribe Cerapachyini. *Proc. Am. Acad. Arts Sci.* **53**: 213–265 [229]. Type data: syntypes, MCZ *W,F, from Bribie Is. near Brisbane, Qld.

Synonymy that of Brown, W.L. jr. (1975). Contributions toward a reclassification of the Formicidae. V. Ponerinae, tribes Platythyreini, Cerapachyini, Cylindromyrmecini, Acanthostichini, and Aenictogitini. *Search Agric.* **5**: 1–116 [33].

Distribution: NE coastal, SE coastal, Qld., N.S.W., Vic. Ecology: terrestrial, noctidiurnal, predator, woodland, open forest; nest in ground layer.

Sphinctomyrmex froggatti Forel, 1900

Sphinctomyrmex froggatti Forel, A. (1900). Ponerinae et Dorylinae d'Australie récoltées par MM. Turner, Froggatt, Nugent, Chase, Rothney, J.J. Walker, etc. *Ann. Soc. Entomol. Belg.* **44**: 54–77 [71]. Type data: syntypes, GMNH W, ANIC W, from N.S.W.

Distribution: Murray-Darling basin, SE coastal, N.S.W. Ecology: terrestrial, noctidiurnal, predator, woodland, open forest; nest in soil.

Sphinctomyrmex imbecilis Forel, 1907

Sphinctomyrmex froggatti imbecilis Forel, A. (1907). Formicidae. pp. 263–310 *in* Michaelsen, W. & Hartmeyer, R. (eds.) *Die Fauna Südwest-Australiens.* Jena : G. Fischer Vol. 1 [272]. Type data: syntypes, GMNH W, ANIC W, from Lion Mill, W.A.

Eusphinctus (Nothosphinctus) manni Wheeler, W.M. (1918). The Australian ants of the ponerine tribe Cerapachyini. *Proc. Am. Acad. Arts Sci.* **53**: 213–265 [236]. Type data: syntypes, MCZ *W,F, from Leura in the Blue Mts., N.S.W.

Eusphinctus (Nothosphinctus) fulvidus Clark, J. (1923). Australian Formicidae. *J. R. Soc. West. Aust.* **9**: 72–89 [75]. Type data: syntypes, NMV *W,F, from Mundaring, W.A.

Eusphinctus (Nothosphinctus) silaceus Clark, J. (1923). Australian Formicidae. *J. R. Soc. West. Aust.* **9**: 72–89 [77]. Type data: syntypes, NMV *W, from Armadale, W.A.

Eusphinctus (Nothosphinctus) brunnicornis Clark, J. (1930). New Formicidae, with notes on some little-known species. *Proc. R. Soc. Vict.* **43**: 2–25 [30 Aug. 1930] [2]. Type data: syntypes, NMV *W, from Collie, W.A.

Synonymy that of Brown, W.L. jr. (1975). Contributions toward a reclassification of the Formicidae. V. Ponerinae, tribes Platythyreini, Cerapachyini, Cylindromyrmecini, Acanthostichini, and Aenictogitini. *Search Agric.* **5**: 1–116 [33].

Distribution: SE coastal, SW coastal, S Gulfs, S.A., Vic., N.S.W., W.A. Ecology: terrestrial, noctidiurnal, predator, woodland, open forest, closed forest; nest in soil.

Sphinctomyrmex mjobergi Forel, 1915

Sphinctomyrmex clarus mjobergi Forel, A. (1915). Results of Dr. E. Mjöbergs Swedish Scientific Expeditions to Australia 1910–1913. 2. Ameisen. *Ark. Zool.* **9**: 1–119 pls 1–3 [4 Dec. 1915] [16]. Type data: syntypes, GMNH W, other syntypes may exist, from Mt. Tambourine (=Tamborine Mt.), Qld.

Distribution: NE coastal, Qld. Ecology: terrestrial, noctidiurnal, predator, open forest, closed forest; nest in soil. Biological references: Brown, W.L. jr. (1975). Contributions toward a reclassification of the Formicidae. V. Ponerinae, tribes Platythyreini, Cerapachyini, Cylindromyrmecini, Acanthostichini, and Aenictogitini. *Search Agric.* **5**: 1–116 (raised to species).

Sphinctomyrmex myops Forel, 1895

Sphinctomyrmex emeryi myops Forel, A. (1895). Nouvelles fourmis d'Australie, récoltée à The Ridge, Mackay, Queensland par M. Gilbert Turner. *Ann. Soc. Entomol. Belg.* **39**: 417–428 [421]. Type data: syntypes, GMNH W, from Mackay, Qld.

Distribution: NE coastal, Qld. Ecology: terrestrial, noctidiurnal, predator, woodland, open forest; nest in soil. Biological references: Brown, W.L. jr. (1975). Contributions toward a reclassification of the Formicidae. V. Ponerinae, tribes Platythyreini, Cerapachyini, Cylindromyrmecini, Acanthostichini, and Aenictogitini. *Search Agric.* **5**: 1–116 (raised to species).

Sphinctomyrmex nigricans (Clark, 1926)

Eusphinctus (Nothosphinctus) nigricans Clark, J. (1926). Australian Formicidae. *J. R. Soc. West. Aust.* **12**: 43–51 pl 6 [25 Jan. 1926] [44]. Type data: syntypes, NMV *W, from Lismore, N.S.W.

Distribution: SE coastal, N.S.W. Ecology: terrestrial, noctidiurnal, predator, woodland, open forest; nest in soil.

Sphinctomyrmex occidentalis (Clark, 1923)

Eusphinctus (Eusphinctus) occidentalis Clark, J. (1923). Australian Formicidae. *J. R. Soc. West. Aust.* **9**: 72–89 [74]. Type data: syntypes, NMV *W,F, from Mundaring, W.A.

Distribution: SW coastal, W.A. Ecology: terrestrial, noctidiurnal, predator, woodland, open forest; nest in soil.

Sphinctomyrmex perstictus Brown, 1975

Cerapachys emeryi Forel, A. (1893). Nouvelles fourmis d'Australie et des Canaries. *Ann. Soc. Entomol. Belg.* **37**: 454–466 [461] [*non Cerapachys (Simopone) emeryi* Forel, 1892 =*Simopone emeryi* (Forel, 1892)]. Type data: syntypes, GMNH W, from Baudin Is., W.A.

Sphinctomyrmex perstictus Brown, W.L. jr. (1975). Contributions toward a reclassification of the Formicidae. V. Ponerinae, tribes Platythyreini, Cerapachyini, Cylindromyrmecini, Acanthostichini, and Aenictogitini. *Search Agric.* **5**: 1–116 [33] [*nom. nov.* for *Cerapachys emeryi* Forel, 1893].

Distribution: NW coastal, W.A. Ecology: terrestrial, noctidiurnal, predator, woodland, open forest; nest in soil.

Sphinctomyrmex septentrionalis (Crawley, 1925)

Eusphinctus (Nothosphinctus) septentrionalis Crawley, W.C. (1925). New ants from Australia. II. *Ann. Mag. Nat. Hist. (9)* **16**: 577–598 [577]. Type data: syntypes, OUM *W, BMNH *W, from Darwin, N.T.

Distribution: N coastal, N.T. Ecology: terrestrial, noctidiurnal, predator, woodland, open forest; nest in soil.

Sphinctomyrmex steinheili Forel, 1900

Sphinctomyrmex (Eusphinctus) steinheili Forel, A. (1900). Ponerinae et Dorylinae d'Australie récoltées par MM. Turner, Froggatt, Nugent, Chase, Rothney, J.J. Walker, etc. *Ann. Soc. Entomol. Belg.* **44**: 54–77 [72]. Type data: syntypes, GMNH W, ANIC W, from Mackay, Qld.

Sphinctomyrmex (Eusphinctus) fallax Forel, A. (1900). Ponerinae et Dorylinae d'Australie récoltées par MM. Turner, Froggatt, Nugent, Chase, Rothney, J.J. Walker, etc. *Ann. Soc. Entomol. Belg.* **44**: 54–77 [73]. Type data: syntypes, GMNH W, ANIC W, from Mackay, Qld.

Sphinctomyrmex (Eusphinctus) fallax hedwigae Forel, A. (1910). Formicides australiens reçus de MM. Froggatt et Rowland Turner. *Rev. Suisse Zool.* **18**: 1–94 [22]. Type data: syntypes, GMNH W, ANIC W, from N.S.W.

Eusphinctus hirsutus Clark, J. (1929). Results of a collecting trip to the Cann River, East Gippsland. *Vict. Nat.* **46**: 115–123 [4 Oct. 1929] [118]. Type data: syntypes, NMV *W,F, from Cann River, Vic.

Eusphinctus fulvipes Clark, J. (1934). Ants from the Otway Ranges. *Mem. Natl. Mus. Vict.* **8**: 48–73 [49 pl 4]. Type data: syntypes, NMV *W,F, from Gellibrand, Vic.

Synonymy that of Brown, W.L. jr. (1975). Contributions toward a reclassification of the Formicidae. V. Ponerinae, tribes Platythyreini, Cerapachyini, Cylindromyrmecini, Acanthostichini, and Aenictogitini. *Search Agric.* **5**: 1–116 [33].

Distribution: NE coastal, SE coastal, Murray-Darling basin, Qld., N.S.W., Vic. Ecology: terrestrial, noctidiurnal, predator, woodland, open forest, closed forest; nest in soil.

Sphinctomyrmex trux Brown, 1975

Sphinctomyrmex trux Brown, W.L. jr. (1975). Contributions toward a reclassification of the Formicidae. V. Ponerinae, tribes Platythyreini, Cerapachyini, Cylindromyrmecini, Acanthostichini, and Aenictogitini. *Search Agric.* **5**: 1–116 [77]. Type data: holotype, MCZ *W, from near Ravenshoe, on the Atherton Tableland, Qld.

Distribution: NE coastal, Qld. Ecology: terrestrial, noctidiurnal, predator, open forest, closed forest; nest in ground layer.

Sphinctomyrmex turneri Forel, 1900

Sphinctomyrmex turneri Forel, A. (1900). Ponerinae et Dorylinae d'Australie récoltées par MM. Turner, Froggatt, Nugent, Chase, Rothney, J.J. Walker, etc. *Ann. Soc. Entomol. Belg.* **44**: 54–77 [70]. Type data: syntypes, GMNH W, ANIC W, from Mackay, Qld.

Distribution: NE coastal, Qld. Ecology: terrestrial, noctidiurnal, predator, closed forest; nest in ground layer.

Trachymesopus Emery, 1911

Trachymesopus Emery, C. (1911). Hymenoptera Fam. Formicidae subfam. Ponerinae. *in* Wytsman, P. (ed.) *Genera Insectorum.* Fasc. 118 Brussels 125 pp. 3 pls [84] [proposed with subgeneric rank in *Euponera* Forel, 1891; raised to genus in Wilson, E.O. (1958). Studies on the ant fauna of Melanesia III. *Rhytidoponera* in Western Melanesia and the Moluccas. IV. The tribe Ponerini. *Bull. Mus. Comp. Zool.* **119**: 301–371]. Type species *Formica stigma* Fabricius, 1804 by original designation.

This group is also found in the Neotropical, south Nearctic, Ethiopian and Oriental regions; New Guinea, eastern Melanesia and parts of Polynesia in the Australian Region.

Trachymesopus clarki (Wheeler, 1934)

Euponera (Trachymesopus) clarki Wheeler, W.M. (1934). Contributions to the fauna of Rottnest Island, Western Australia No. IX. The ants. *J. R. Soc. West. Aust.* **20**: 137–163 [5 Oct. 1934] [140]. Type data: syntypes, MCZ *W,F, from Serpentine Lake, Rottnest Is. and Margaret River, W.A.

Distribution: SW coastal, W.A. Ecology: terrestrial, noctidiurnal, predator, open forest, closed forest; nest in ground layer.

Trachymesopus darwinii (Forel, 1893)

Belonopelta darwinii Forel, A. (1893). Nouvelles fourmis d'Australie et des Canaries. *Ann. Soc. Entomol. Belg.* **37**: 454–466 [460]. Type data: holotype (probable), GMNH F, from Port Darwin, N.T.

Distribution: N coastal, N Gulf, NE coastal, Qld., N.T. Ecology: terrestrial, noctidiurnal, predator, open forest, closed forest; nest in ground layer.

Trachymesopus pachynoda (Clark, 1930)

Euponera (Trachymesopus) pachynoda Clark, J. (1930). New Formicidae, with notes on some little-known species. *Proc. R. Soc. Vict.* **43**: 2–25 [30 Aug. 1930] [7]. Type data: syntypes (probable), NMV *W, from Ferntree Gully, Vic.

Distribution: SE coastal, N.S.W., Vic. Ecology: terrestrial, noctidiurnal, predator, open forest, closed forest; nest in ground layer.

Trachymesopus rufonigra (Clark, 1934)

Euponera (Brachyponera) rufonigra Clark, J. (1934). New Australian ants. *Mem. Natl. Mus. Vict.* **8**: 21–47 [30 pls 2–3]. Type data: syntypes, NMV *W,F, from Perth, Armadale, Mundaring, Busselton and Albany, W.A.

Distribution: SW coastal, W.A. Ecology: terrestrial, noctidiurnal, predator, open forest, closed forest; nest in ground layer.

DORYLINAE

Aenictus Shuckard, 1840

Aenictus Shuckard, W.E. (1840). Monograph of the Dorylidae, a family of the Hymenoptera Heterogyna. *Ann. Mag. Nat. Hist. (1)* **5**: 258–271 [266]. Type species *Aenictus ambiguus* Shuckard, 1840 by original designation.

This group is also found in the south Palearctic, Ethiopian and Oriental regions; New Guinea and east Melanesia in the Australian Region, see Brown, W.L. jr. (1973). A comparison of the Hylean and Congo-West African rain forest ant faunas. pp. 161–185 *in* Meggers, B.J., Ayensu, E.S. & Duckworth, W.D. (eds.) *Tropical forest ecosystems in Africa and South America: a comparative review.* Washington : Smithsonian Institution Press.

Aenictus aratus Forel, 1900

Aenictus aratus Forel, A. (1900). Ponerinae et Dorylinae d'Australie récoltées par MM. Turner, Froggatt, Nugent, Chase, Rothney, J.J. Walker, etc. *Ann. Soc. Entomol. Belg.* **44**: 54–77 [74]. Type data: syntypes, GMNH W, ANIC W, from Mackay, Qld.

Distribution: NE coastal, Qld. Ecology: terrestrial, noctidiurnal, nomadic, predator, open forest, closed forest; nest in ground layer, army ant.

Aenictus ceylonicus (Mayr, 1866)

Typhlatta ceylonica Mayr, G.L. (1866). Myrmecologische beiträge. *Sber. Akad. Wiss. Wien* **53** Abt. 1: 484–517 [505]. Type data: syntypes, NHMW *W, from Sri Lanka (as Ceylon).

Aenictus turneri Forel, A. (1900). Ponerinae et Dorylinae d'Australie récoltées par MM. Turner,

Froggatt, Nugent, Chase, Rothney, J.J. Walker, etc. *Ann. Soc. Entomol. Belg.* **44**: 54–77 [75]. Type data: syntypes, GMNH W, ANIC W, from Mackay, Qld.

Aenictus deuqueti Crawley, W.C. (1923). Myrmecological notes - new Australian Formicidae. *Entomol. Rec. J. Var.* **35**: 177–179 [177]. Type data: syntypes, OUM *W, from Lismore, N.S.W.

Aenictus exiguus Clark, J. (1934). New Australian ants. *Mem. Natl. Mus. Vict.* **8**: 21–47 [21 pls 2–3]. Type data: syntypes (probable), SAMA *W, from Cairns district, Qld.

Synonymy that of Brown, W.L. jr. (1952). New synonymy in the army ant genus *Aenictus* Shuckard. *Psyche Camb.* **58**: 123; Wilson, E.O. (1964). The true army ants of the Indo-Australian area (Hymenoptera : Formicidae : Dorylinae). *Pac. Insects Monogr.* **6**: 427–483 [452].

Distribution: NE coastal, Murray-Darling basin, Qld., N.S.W. Ecology: terrestrial, noctidiurnal, nomadic, predator, open heath, woodland, open forest, closed forest; nest in ground layer, army ant.

Aenictus hilli Clark, 1928

Aenictus hilli Clark, J. (1928). Australian Formicidae. *J. R. Soc. West. Aust.* **14**: 29–41 pl 1 [24 Apr. 1928] [38] [this name is based on males, which are rarely observed in this genus, and it may be synonymous with *Aenictus ceylonicus* (Mayr, 1866)]. Type data: syntypes (probable), NMV *M, from Malanda, Qld.

Distribution: NE coastal, Qld. Ecology: terrestrial, noctidiurnal, nomadic, predator, woodland; nest in ground layer, army ant.

Aenictus philiporum Wilson, 1964

Aenictus philiporum Wilson, E.O. (1964). The true army ants of the Indo-Australian area (Hymenoptera : Formicidae : Dorylinae). *Pac. Insects Monogr.* **6**: 427–483 [10 Nov. 1964] [473]. Type data: holotype, MCZ *W, from Iron Range, Cape York, Qld.

Distribution: NE coastal, Qld. Ecology: terrestrial, noctidiurnal, nomadic, predator, open forest, closed forest; nest in ground layer, army ant.

LEPTANILLINAE

Leptanilla Emery, 1870

Leptanilla Emery, C. (1870). Studi mirmecologici. *Boll. Soc. Entomol. Ital.* **2**: 193–201 [196 pl 2] [redefined in Urbani, C. Baroni (1977). Materiali per una revisione della sottofamiglia Leptanillinae Emery (Hymenoptera : Formicidae). V. *Entomologica Bas.* **2**: 427–488]. Type species *Leptanilla revelierii* Emery, 1870 by monotypy.

This group is also found in the south Palearctic and east Oriental regions, see Brown, W.L. jr. (1973). A comparison of the Hylean and Congo-West African rain forest ant faunas. pp. 161–185 *in* Meggers, B.J., Ayensu, E.S. & Duckworth, W.D. (eds.) *Tropical forest ecosystems in Africa and South America: a comparative review.* Washington : Smithsonian Institution Press.

Leptanilla swani Wheeler, 1932

Leptanilla swani Wheeler, W.M. (1932). An Australian *Leptanilla*. *Psyche Camb.* **39**: 53–58 [54]. Type data: syntypes, WAM 32-1252 to 32-1254 *W, MCZ *W,F, from Goyamin Pool, Chittering, W.A.

Distribution: SW coastal, NE coastal, N Gulf, N coastal, Qld., W.A. Ecology: terrestrial, nomadic, predator, woodland, open forest; nest in ground layer.

MYRMICINAE

Adlerzia Forel, 1902

Adlerzia Forel, A. (1902). Fourmis nouvelles d'Australie. *Rev. Suisse Zool.* **10**: 405–548 [445] [proposed with subgeneric rank in *Monomorium* Mayr, 1855]. Type species *Monomorium (Adlerzia) froggatti* Forel, 1902 by original designation.

Stenothorax McAreavey, J.J. (1949). Australian Formicidae. New genera and species. *Proc. Linn. Soc. N.S.W.* **74**: 1–25 [15 June 1949] [3]. Type species *Stenothorax katerinae* McAreavey, 1949 by original designation.

Synonymy that of Brown, W.L. jr. (1952). *Adlerzia froggatti* Forel and some new synonymy (Hymenoptera : Formicidae). *Psyche Camb.* **58**: 110 [7 Apr. 1952].

Adlerzia froggatti (Forel, 1902)

Monomorium (Adlerzia) froggatti Forel, A. (1902). Fourmis nouvelles d'Australie. *Rev. Suisse Zool.* **10**: 405–548 [445]. Type data: holotype, GMNH W, from Bendigo, Vic.

Machomyrma silvestrii Emery, C. (1914). Formiche d'Australia e di Samoa raccolte dal Prof. Silvestri nel 1913. *Boll. Lab. Zool. Gen. Agr. R. Scuola Agric. Portici* **8**: 179–186 [30 Jan. 1914] [182]. Type data: holotype, MCG *W, from Mt. Lofty, Adelaide, S.A.

Stenothorax katerinae McAreavey, J.J. (1949). Australian Formicidae. New genera and species. *Proc. Linn. Soc. N.S.W.* **74**: 1–25 [15 June 1949] [3]. Type data: holotype, whereabouts unknown, from Greensborough, Vic.

Synonymy that of Brown, W.L. jr. (1952). *Adlerzia froggatti* Forel and some new synonymy (Hymenoptera : Formicidae). *Psyche Camb.* **58**: 110.

Distribution: Murray-Darling basin, S Gulfs, SE coastal, Vic., S.A. Ecology: terrestrial, noctidiurnal, predator, tall open shrubland, woodland, open forest; nest in ground layer.

Anisopheidole Forel, 1914

Anisopheidole Forel, A. (1914). Einige amerikanische Ameisen. *Dtsch. Entomol. Zeit.* **1914**: 615–620 [10 Dec. 1914] [616] [proposed with subgeneric rank in *Pheidole* Westwood, 1841]. Type species *Pheidole froggatti* Forel, 1902 by monotypy.

Anisopheidole antipodum (F. Smith, 1858)

Atta antipodum Smith, F. (1858). *Catalogue of hymenopterous insects in the collection of the British Museum.* Part 6. Formicidae. London : British Museum 216 pp. 14 pls [27 Mar. 1858] [166]. Publication date established from Donisthorpe, H. (1932). On the identity of Smith's types of Formicidae (Hymenoptera) collected by Alfred Russell Wallace in the Malay Archipelago, with descriptions of two new species. *Ann. Mag. Nat. Hist. (10)* **10**: 441–476. Type data: syntypes (probable), BMNH *F, from Swan River, W.A.

Pheidole froggatti Forel, A. (1902). Fourmis nouvelles d'Australie. *Rev. Suisse Zool.* **10**: 405–548 [414]. Type data: syntypes, GMNH W,M, ANIC W, from Kalgoorlie, W.A.

Pheidole myops Forel, A. (1902). Fourmis nouvelles d'Australie. *Rev. Suisse Zool.* **10**: 405–548 [421]. Type data: syntypes, whereabouts unknown., from Native Dog Bore, Darling River, N.S.W.

Monomorium lippulum Wheeler, W.M. (1927). Ants collected by Professor F. Silvestri in Indochina. *Boll. Lab. Zool. Gen. Agr. R. Scuola Agric. Portici* **20**: 83–106 [6 May 1927] [89]. Type data: syntypes, MCZ *W, from Port Lincoln, S.A. and McDonnel (=McDonnell Range), N.T.

Synonymy that of Ettershank, G. (1966). A generic revision of the world Myrmicinae related to *Solenopsis* and *Pheidologeton* (Hymenoptera : Formicidae). *Aust. J. Zool.* **14**: 73–171 [132].

Distribution: Murray-Darling basin, Lake Eyre basin, W plateau, S Gulfs, N.S.W., S.A., W.A., N.T. Ecology: terrestrial, noctidiurnal, predator, shrubland, open forest; nest in soil.

Aphaenogaster Mayr, 1853

Aphaenogaster Mayr, G.L. (1853) Beiträge der Kenntniss der Ameisen. *Verh. Zool.-Bot. Ges. Wien* **3**: Abhand. 105–114 [107]. Type species *Aphaenogaster sardoa* Mayr, 1853 by subsequent designation, see Bingham, C.T. (1903). *The Fauna of British India, including Ceylon and Burma.* Hymenoptera. Vol. 2 Ants and cuckoo-wasps. London : Taylor & Francis 507 pp.

Nystalomyrma Wheeler, W. M. (1916). The Australian ants of the genus *Aphaenogaster* Mayr. *Trans. R. Soc. S. Aust.* **40**: 213–223 [23 Dec. 1916] [215 pls 21–22] [proposed with subgeneric rank in *Aphaenogaster* Mayr, 1853]. Type species *Myrmica longiceps* F. Smith, 1858 by original designation.

Synonymy that of Brown, W.L. jr. (1973). A comparison of the Hylean and Congo-West African rain forest ant faunas. pp. 161–185 *in* Meggers, B.J., Ayensu, E.S. & Duckworth, W.D. (eds.) *Tropical forest ecosystems in Africa and South America: a comparative review.* Washington : Smithsonian Institution Press [177].

This group is also found in the south Neotropical, Nearctic, Palearctic, Malagasy and Oriental regions; New Guinea in the Australian Region, see Brown, W.L. jr. (1973). A comparison of the Hylean and Congo-West African rain forest ant faunas. pp. 161–185 *in* Meggers, B.J., Ayensu, E.S. & Duckworth, W.D. (eds.) *Tropical forest ecosystems in Africa and South America: a comparative review.* Washington : Smithsonian Institution Press.

Aphaenogaster barbigula Wheeler, 1916

Aphaenogaster (Nystalomyrma) barbigula Wheeler, W.M. (1916). The Australian ants of the genus *Aphaenogaster* Mayr. *Trans. R. Soc. S. Aust.* **40**: 213–223 pls 21–22 [23 Dec. 1916] [221]. Type data: syntypes, MCZ *W,F, from Adelaide, Meningie, Gawler, Karoonda to Peebinga, S.A. and Dongarra, Gooseberry Hill, Wallaby Is., Beverley, W.A. and Sea Lake, Vic. and Yanco, N.S.W.

Distribution: Murray-Darling basin, S Gulfs, SW coastal, W plateau, Lake Eyre basin, N.S.W., Vic., S.A., W.A. Ecology: terrestrial, noctidiurnal, predator, desert, shrubland, woodland; nest in soil.

Aphaenogaster longiceps (F. Smith, 1858)

Myrmica longiceps Smith, F. (1858). *Catalogue of hymenopterous insects in the collection of the British Museum.* Part 6. Formicidae. London : British Museum 216 pp. 14 pls [27 Mar. 1858] [128]. Publication date established from Donisthorpe, H. (1932). On the identity of Smith's types of Formicidae (Hymenoptera) collected by Alfred Russell Wallace in the Malay Archipelago, with descriptions of two new species. *Ann. Mag. Nat. Hist. (10)* **10**: 441–476. Type data: syntypes (probable), BMNH *W, from N.S.W.

Stenamma (Ischnomyrmex) longiceps ruginota Forel, A. (1902). Fourmis nouvelles d'Australie. *Rev. Suisse Zool.* **10**: 405–548 [439]. Type data: syntypes, whereabouts unknown., from N.S.W. and Yarra distr., Vic.

Synonymy that of Wheeler, W.M. (1916). The Australian ants of the genus *Aphaenogaster* Mayr. *Trans. R. Soc. S. Aust.* **40**: 213–223 pls 21–22 [216]; Ettershank, G. (1966). A generic revision of the world Myrmecinae related to *Solenopsis* and *Pheidologeton* (Hymenoptera : Formicidae). *Aust. J. Zool.* **14**: 73–171 [132].

Distribution: SE coastal, SW coastal, N.S.W., Vic., W.A. Ecology: terrestrial, noctidiurnal, predator, woodland, open forest; nest in soil.

Aphaenogaster poultoni Crawley, 1922

Aphaenogaster poultoni Crawley, W.C. (1922). New ants from Australia. *Ann. Mag. Nat. Hist. (9)* **10**: 16–36 [17]. Type data: syntypes (probable), OUM *W, from Beenup, W.A.

Distribution: SW coastal, W.A. Ecology: terrestrial, noctidiurnal, predator, woodland, open forest; nest in soil.

Aphaenogaster pythia Forel, 1915

Aphaenogaster pythia Forel, A. (1915). Results of Dr. E. Mjöbergs Swedish Scientific Expeditions to Australia 1910-1913. 2. Ameisen. *Ark. Zool.* **9**: 1-119 pls 1-3 [4 Dec. 1915] [76]. Type data: syntypes, GMNH *W,M,F. from Herberton, Qld.

Distribution: NE coastal, Qld. Ecology: terrestrial, noctidiurnal, predator, open forest; nest in soil.

Calyptomyrmex Emery, 1887

Calyptomyrmex Emery, C. (1887). Catalogo delle Formiche esistenti nelle collezioni del Museo Civico di Genova. Parte terza. Formiche della regione Indo-Malese e dell'Australia. *Ann. Mus. Civ. Stor. Nat. Giacomo Doria* **25**: 427-473 [471 pls 1-2] [redefined in Bolton, B. (1981). A revision of the ant genera *Meranoplus* F. Smith, *Dicroaspis* Emery and *Calyptomyrmex* Emery (Hymenoptera : Formicidae) in the Ethiopian zoogeographical region. *Bull. Br. Mus. Nat. Hist. (Entomol.)* **42**: 43-81 (2 Feb. 1981)]. Type species *Calyptomyrmex beccarii* Emery, 1887 by monotypy.

This group is also found in the Ethiopian and east Oriental regions; New Guinea in the Australian Region, see Brown, W.L. jr. (1973). A comparison of the Hylean and Congo-West African rain forest ant faunas. pp. 161-185 *in* Meggers, B.J., Ayensu, E.S. & Duckworth, W.D. (eds.) *Tropical forest ecosystems in Africa and South America: a comparative review*. Washington : Smithsonian Institution Press.

Calyptomyrmex schraderi Forel, 1901

Calyptomyrmex schraderi Forel, A. (1901). Formiciden des Naturhistorischen Museums zu Hamburg. Neue *Calyptomyrmex-*, *Dacryon-*, *Podomyrma-*, und *Echinopla*-Arten. *Mitt. Naturh. Mus. Hamb.* **18**: 45-82 [50]. Type data: syntypes, probably destroyed in ZMH in W.W. II, from Australia.

Distribution: NE coastal, Qld. Ecology: terrestrial, noctidiurnal, predator, closed forest; nest in ground layer.

Cardiocondyla Emery, 1869

Cardiocondyla Emery, C. (1869). Enumerazione dei formicidi che rinvengonsi nei contorni di Napoli con descrizione di specie nuove o meno conosciute. *Ann. Accad. Asp. Nat. Napoli (era 2)* **2**: 1-26 [20]. Type species *Cardiocondyla elegans* Emery, 1869 by monotypy. Compiled from secondary source: Wheeler, W.M. (1911). A list of the type species of the genera and subgenera of Formicidae. *Ann. N.Y. Acad. Sci.* **21**: 157-175 [17 Oct. 1911].

This group is also found in the south Palearctic, Ethiopian, Malagasy and Oriental regions; New Guinea, east Melanesia, New Caledonia and parts of Polynesia in the Australian Region, see Brown, W.L. jr. (1973). A comparison of the Hylean and Congo-West African rain forest ant faunas. pp. 161-185 *in* Meggers, B.J., Ayensu, E.S. & Duckworth, W.D. (eds.) *Tropical forest ecosystems in Africa and South America: a comparative review*. Washington : Smithsonian Institution Press.

Cardiocondyla nuda (Mayr, 1866)

Leptothorax nudus Mayr, G.L. (1866). Myrmecologische Beiträge. *Sber. Akad. Wiss. Wien* **53**(1): 484-517 [508]. Type data: status unknown, ?NHMW, from Ovalau, Viti, Fiji.

Cardiocondyla nuda atalanta Forel, 1915

Cardiocondyla nuda atalanta Forel, A. (1915). Results of Dr. E. Mjöbergs Swedish Scientific Expeditions to Australia 1910-1913. 2. Ameisen. *Ark. Zool.* **9**: 1-119 pls 1-3 [4 Dec. 1915] [75]. Type data: syntypes, GMNH W, other syntypes may exist, from Kimberley distr., W.A.

Distribution: N coastal, W.A. Ecology: terrestrial, noctidiurnal, predator, woodland, open forest; nest in ground layer.

Cardiocondyla nuda nereis Wheeler, 1927

Cardiocondyla nuda nereis Wheeler, W.M. (1927). The ants of Lord Howe Island and Norfolk Island. *Proc. Am. Acad. Arts Sci.* **62**: 121-153 [140]. Type data: syntypes, MCZ *W,F, from Norfolk Is.

Distribution: Norfolk Is. Ecology: terrestrial, noctidiurnal, predator, woodland, open forest; nest in ground layer.

Chelaner Emery, 1914

Chelaner Emery, C. (1914). Les fourmis de la Nouvelle-Calédonie et des Îles Loyalty. *in* Sarasin, F. & Roux, J. (1914-1921). *Forschungen in Neu-Caledonien und auf den Loyalty-Inseln*. Zoologie 1: 393-437 pl 13 [410] [proposed with subgeneric rank in *Monomorium* Mayr, 1855]. Type species *Monomorium (Chelaner) forcipatum* Emery, 1914 by subsequent designation, see Emery, C. (1921). Hymenoptera. Fam. Formicidae. subfam. Myrmecinae. *in* Wytsman, P. (ed.) *Genera Insectorum*. Fasc. 174C pp. 207-397 7 pls.

Protholcomyrmex Wheeler, W.M. (1922). Ants of the American Museum Congo Expedition. A contribution to the myrmecology of Africa. II. The ants collected by the American Museum Congo Expedition. *Bull. Am. Mus. Nat. Hist.* **45**: 39-269 pls 2-23 [10 Feb. 1922] [162] [proposed with subgeneric rank in *Monomorium* Mayr, 1855]. Type species *Monomorium rothsteini* Forel, 1902 by original designation.

Schizopelta McAreavey, J.J. (1949). Australian Formicidae. New genera and species. *Proc. Linn. Soc. N.S.W.* **74**: 1-25 [15 June 1949] [14]. Type species

Schizopelta falcata McAreavey, 1949 by original designation.

Synonymy that of Ettershank, G. (1966). A generic revision of the world Myrmicinae related to *Solenopsis* and *Pheidologeton* (Hymenoptera : Formicidae). *Aust. J. Zool.* **14**: 73–171 [93].

This group is also found in New Guinea, New Caledonia, New Zealand, Kermadec Ils. and Rapa in Polynesia.

Chelaner armstrongi (McAreavey, 1949)

Monomorium (Holcomyrmex) armstrongi McAreavey, J.J. (1949). Australian Formicidae. New genera and species. *Proc. Linn. Soc. N.S.W.* **74**: 1–25 [15 June 1949] [10]. Type data: holotype, ANIC W, from Nyngan, N.S.W.

Distribution: NE coastal, N coastal, Qld., W.A. Ecology: terrestrial, noctidiurnal, predator, desert, woodland, open forest; nest in soil.

Chelaner bicornis (Forel, 1907)

Monomorium bicorne Forel, A. (1907). Formicidae. pp. 263–310 *in* Michaelsen, W. & Hartmeyer, R. (eds.) *Die Fauna Südwest-Australiens.* Jena : G. Fischer Vol. 1 [276]. Type data: holotype, probably destroyed in ZMH in WW II, from Grooseberry (=Gooseberry) Hill, W.A.

Distribution: SW coastal, W.A. Ecology: terrestrial, noctidiurnal, predator, woodland, open forest; nest in soil.

Chelaner centralis (Forel, 1910)

Monomorium centrale Forel, A. (1910). Formicides australiens reçus de MM. Froggatt et Rowland Turner. *Rev. Suisse Zool.* **18**: 1–94 [28]. Type data: holotype, GMNH W, from Tennant Creek, N.T.

Distribution: W plateau, N.T. Ecology: terrestrial, noctidiurnal, predator, desert, woodland; nest in soil.

Chelaner falcatus (McAreavey, 1949)

Schizopelta falcata McAreavey, J.J. (1949). Australian Formicidae. New genera and species. *Proc. Linn. Soc. N.S.W.* **74**: 1–25 [15 June 1949] [15]. Type data: holotype, ANIC W, from Nyngan, N.S.W.

Distribution: Murray-Darling basin, N.S.W. Ecology: terrestrial, noctidiurnal, predator, desert, woodland, open forest; nest in soil.

Chelaner flavigaster (Clark, 1938)

Xiphomyrmex flavigaster Clark, J. (1938). Reports of the McCoy Society for Field Investigation and Research. No. 2. Sir Joseph Banks Islands. Part I. Formicidae (Hymenoptera). *Proc. R. Soc. Vict.* **50**: 356–382 [366]. Type data: syntypes, NMV *W, from Reevesby Is., S.A.

Distribution: S Gulfs, S.A. Ecology: terrestrial, noctidiurnal, predator, woodland, open forest; nest in soil. Biological references: Bolton, B. (1976). The ant tribe Tetramoriini (Hymenoptera : Formicidae). Constituent genera, review of small genera and revision of *Triglyphothrix* Forel. *Bull. Br. Mus. Nat. Hist. (Entomol.)* **34**: 283–379 (transferred to *Chelaner*).

Chelaner flavipes (Clark, 1938)

Monomorium (Notomyrmex) flavipes Clark, J. (1938). Reports of the McCoy Society for Field Investigation and Research. No. 2. Sir Joseph Banks Islands. Part I. Formicidae (Hymenoptera). *Proc. R. Soc. Vict.* **50**: 356–382 [369]. Type data: syntypes, NMV *W,F, from N end of Reevesby Is., S.A.

Distribution: S Gulfs, S.A. Ecology: terrestrial, noctidiurnal, predator, woodland; nest in soil.

Chelaner foreli (Viehmeyer, 1913)

Monomorium (Holcomyrmex) foreli Viehmeyer, H. (1913). Neue und unvollständig bekannte Ameisen der Alten Welt. *Arch. Naturg.* **79A**(12): 24–60 [32]. Type data: syntypes, ZMB *W, ANIC W, from Killalpaninna, S.A.

Distribution: Lake Eyre basin, S.A. Ecology: terrestrial, noctidiurnal, predator, desert, woodland; nest in soil.

Chelaner gilberti (Forel, 1902)

Chelaner gilberti gilberti (Forel, 1902)

Monomorium gilberti Forel, A. (1902). Fourmis nouvelles d'Australie. *Rev. Suisse Zool.* **10**: 405–548 [440]. Type data: syntypes, GMNH W, from Mackay, Qld.

Distribution: NE coastal, Qld. Ecology: terrestrial, noctidiurnal, predator, open forest, closed forest; nest in ground layer.

Chelaner gilberti mediorubrus (Forel, 1915)

Monomorium gilberti mediorubra Forel, A. (1915). Results of Dr. E. Mjöbergs Swedish Scientific Expeditions to Australia 1910–1913. 2. Ameisen. *Ark. Zool.* **9**: 1–119 pls 1–3 [4 Dec. 1915] [72]. Type data: syntypes, GMNH W, ANIC W, other syntypes may exist, from Malanda, Qld.

Distribution: NE coastal, Qld. Ecology: terrestrial, noctidiurnal, predator, open forest, closed forest; nest in ground layer.

Chelaner howensis (Wheeler, 1927)

Monomorium (Notomyrmex) howense Wheeler, W.M. (1927). The ants of Lord Howe Island and Norfolk Island. *Proc. Am. Acad. Arts Sci.* **62**: 121–153 [138]. Type data: syntypes, MCZ *W,F, from Lord Howe Is.

Distribution: Lord Howe Is. Ecology: terrestrial, noctidiurnal, predator, woodland, open forest; nest in ground layer.

Chelaner insolescens (Wheeler, 1934)

Monomorium (Notomyrmex) insolescens Wheeler, W.M. (1934). Contributions to the fauna of Rottnest Island, Western Australia No. IX. The ants. *J. R. Soc. West. Aust.* **20**: 137–163 [5 Oct. 1934] [145]. Type data: syntypes, MCZ *W,M, from Derby, W.A.

Distribution: N coastal, W.A. Ecology: terrestrial, noctidiurnal, predator, desert, woodland; nest in soil.

Chelaner insularis (Clark, 1938)

Monomorium (Notomyrmex) insularis Clark, J. (1938). Reports of the McCoy Society for Field Investigation and Research. No. 2. Sir Joseph Bank Islands. Part I. Formicidae (Hymenoptera). *Proc. R. Soc. Vict.* **50**: 356–382 [368]. Type data: syntypes, NMV *W,F, from Reevesby Is., S.A.

Distribution: S Gulfs, S.A. Ecology: terrestrial, noctidiurnal, predator, woodland, open forest; nest in soil.

Chelaner kiliani (Forel, 1902)

Chelaner kiliani kiliani (Forel, 1902)

Monomorium kiliani Forel, A. (1902). Fourmis nouvelles d'Australie. *Rev. Suisse Zool.* **10**: 405–548 [441]. Type data: syntypes, GMNH W, ANIC W, from Bong Bong, N.S.W.

Distribution: SE coastal, N.S.W. Ecology: terrestrial, noctidiurnal, predator, woodland, open forest; nest in ground layer.

Chelaner kiliani obscurellus (Viehmeyer, 1925)

Monomorium kiliani obscurella Viehmeyer, H. (1925). Formiciden der australischen Faunenregion. *Entomol. Mitt.* **14**: 25–39 [27]. Type data: syntypes (probable), ZMB *W, from Liverpool, N.S.W.

Distribution: NE coastal, Qld. Ecology: terrestrial, noctidiurnal, predator, woodland, open forest; nest in ground layer.

Chelaner kiliani tambourinensis (Forel, 1915)

Monomorium kiliani tambourinensis Forel, A. (1915). Results of Dr. E. Mjöbergs Swedish Scientific Expeditions to Australia 1910–1913. 2. Ameisen. *Ark. Zool.* **9**: 1–119 pls 1–3 [4 Dec. 1915] [71]. Type data: syntypes, GMNH W, ANIC W, other syntypes may exist, from Mt. Tambourine (=Tamborine Mt.), Qld.

Distribution: NE coastal, Qld. Ecology: terrestrial, noctidiurnal, predator, open forest, closed forest; nest in ground layer.

Chelaner leae (Forel, 1913)

Monomorium leae Forel, A. (1913). Fourmis de Tasmanie et d'Australie récoltées par MM. Lea, Froggatt etc. *Bull. Soc. Vaud. Sci. Nat.* **49**: 173–196 pl 2 [185]. Type data: syntypes, GMNH W,F, ANIC W, from Tas.

Monomorium (Notomyrmex) hemiphaeum Clark, J. (1934). Ants from the Otway Ranges. *Mem. Natl. Mus. Vict.* **8**: 48–73 [61 pl 4]. Type data: syntypes, NMV *W,F, from Beech Forest and Gellibrand, Vic.

Synonymy that of Ettershank, G. (1966). A generic revision of the world Myrmicinae related to *Solenopsis* and *Pheidologeton* (Hymenoptera : Formicidae). *Aust. J. Zool.* **14**: 73–171 [97].

Distribution: SE coastal, Vic., Tas. Ecology: terrestrial, noctidiurnal, predator, open forest, closed forest; nest in ground layer.

Chelaner longiceps (Wheeler, 1934)

Monomorium (Notomyrmex) longiceps Wheeler, W.M. (1934). Contributions to the fauna of Rottnest Island, Western Australia No. IX. The ants. *J. R. Soc. West. Aust.* **20**: 137–163 [5 Oct. 1934] [146]. Type data: syntypes, MCZ *W, from Lady Edeline Beach, Rottnest Is. and Ludlow, W.A.

Distribution: SW coastal, W.A. Ecology: terrestrial, noctidiurnal, predator, woodland, open forest; nest in ground layer.

Chelaner macareaveyi Ettershank, 1966

Monomorium (Holcomyrmex) niger McAreavey, J.J. (1949). Australian Formicidae. New genera and species. *Proc. Linn. Soc. N.S.W.* **74**: 1–25 [15 June 1949] [12] [*non Holcomyrmex criniceps nigrum* Forel, 1902]. Type data: holotype, ANIC W, from Nyngan, N.S.W.

Chelaner macareaveyi Ettershank, G. (1966). A generic revision of the world Myrmicinae related to *Solenopsis* and *Pheidologeton* (Hymenoptera : Formicidae). *Aust. J. Zool.* **14**: 73–171 [97] [*nom. nov.* for *Monomorium (Holcomyrmex) niger* McAreavey, 1949].

Distribution: Murray-Darling basin, N.S.W. Ecology: terrestrial, noctidiurnal, predator, desert, woodland; nest in soil.

Chelaner occidaneus (Crawley, 1922)

Monomorium occidaneus Crawley, W.C. (1922). New ants from Australia. *Ann. Mag. Nat. Hist. (9)* **9**: 427–448 [447]. Type data: syntypes, OUM *W,F, from Swan River, W.A.

Distribution: SW coastal, W.A. Ecology: terrestrial, noctidiurnal, predator, woodland, open forest; nest in ground layer.

Chelaner rothsteini (Forel, 1902)

Chelaner rothsteini rothsteini (Forel, 1902)

Monomorium rothsteini Forel, A. (1902). Fourmis nouvelles d'Australie. *Rev. Suisse Zool.* **10**: 405-548 [444]. Type data: syntypes, GMNH W, ANIC W, from Charters Towers, Qld.

Distribution: NE coastal, Qld. Ecology: terrestrial, noctidiurnal, predator, woodland; nest in soil.

Chelaner rothsteini humilior (Forel, 1910)

Monomorium rothsteini humilior Forel, A. (1910). Formicides australiens reçus de MM. Froggatt et Rowland Turner. *Rev. Suisse Zool.* **18**: 1–94 [27]. Type data: holotype, GMNH W, from Tennant Creek, N.T.

Distribution: W plateau, N.T. Ecology: terrestrial, noctidiurnal, predator, desert, woodland; nest in soil.

Chelaner rothsteini leda (Forel, 1915)

Monomorium rothsteini leda Forel, A. (1915). Results of Dr. E. Mjöbergs Swedish Scientific Expeditions to Australia 1910–1913. 2. Ameisen. *Ark. Zool.* **9**: 1–119 pls 1–3 [4 Dec. 1915] [71]. Type data: syntypes, GMNH W,M, ANIC W, other syntypes may exist, from Kimberley distr. and Noonkanbah, W.A. and Laura and Alice River, Qld.

Distribution: NE coastal, Qld. Ecology: terrestrial, noctidiurnal, predator, woodland; nest in soil.

Chelaner rothsteini tostum (Wheeler, 1915)

Monomorium rothsteini tostum Wheeler, W.M. (1915). Hymenoptera. *Trans. R. Soc. S. Aust.* **39**: 805-823 pls 64-66 [Dec. 1915] [806]. Type data: syntypes, MCZ *W, from Everard Range, S.A.

Distribution: W plateau, S.A. Ecology: terrestrial, noctidiurnal, predator, desert, woodland; nest in soil.

Chelaner rothsteini doddi (Santschi, 1919)

Monomorium (Paraholcomyrmex) rothsteini doddi Santschi, F. (1919). Cinq notes myrmécologiques. *Bull. Soc. Vaud. Sci. Nat.* **52**: 325-350 [328]. Type data: syntypes, BNHM W, from Townsville, Qld.

Distribution: NE coastal, Qld. Ecology: terrestrial, noctidiurnal, predator, woodland; nest in soil.

Chelaner rothsteini squamigena (Viehmeyer, 1925)

Monomorium rothsteini squamigena Viehmeyer, H. (1925). Formiciden der australischen Faunenregion. *Entomol. Mitt.* **14**: 25-39 [28]. Type data: holotype, ZMB *W, from Trial Bay, N.S.W.

Distribution: SE coastal, N.S.W. Ecology: terrestrial, noctidiurnal, predator, woodland, open forest; nest in soil.

Chelaner rubriceps (Mayr, 1876)

Chelaner rubriceps rubriceps (Mayr, 1876)

Monomorium rubriceps Mayr, G.L. (1876). Die australischen Formiciden. *J. Mus. Godeffroy* **5**: 56–115 [101]. Type data: syntypes, NHMW W,M, from Cape York and Rockhampton, Qld. and Sidney (=Sydney), N.S.W.

Distribution: NE coastal, SE coastal, Qld., N.S.W. Ecology: terrestrial, noctidiurnal, predator, woodland, open forest, closed forest; nest in ground layer.

Chelaner rubriceps cinctus (Wheeler, 1917)

Monomorium rubriceps cinctum Wheeler, W.M. (1917). The phylogenetic development of subapterous and apterous castes in the Formicidae. *Proc. Nat. Acad. Sci. U.S.A.* **3**: 109–117 [113]. Type data: syntypes, MCZ *W,F, from Vic.

Distribution: Murray-Darling basin, N.S.W. Ecology: terrestrial, noctidiurnal, predator, desert, woodland; nest in soil.

Chelaner rubriceps extreminigrus (Forel, 1915)

Monomorium rubriceps extreminigrum Forel, A. (1915). Results of Dr. E. Mjöbergs Swedish Scientific Expeditions to Australia 1910–1913. 2. Ameisen. *Ark. Zool.* **9**: 1–119 pls 1–3 [4 Dec. 1915] [73]. Type data: holotype, SMNH *W, from Cedar Creek, Qld.

Distribution: NE coastal, Qld. Ecology: terrestrial, noctidiurnal, predator, woodland, open forest, closed forest; nest in ground layer.

Chelaner rubriceps rubrus (Forel, 1915)

Monomorium rubriceps rubra Forel, A. (1915). Results of Dr. E. Mjöbergs Swedish Scientific Expeditions to Australia 1910–1913. 2. Ameisen. *Ark. Zool.* **9**: 1–119 pls 1–3 [4 Dec. 1915] [72]. Type data: syntypes, GMNH W,M,F, ANIC W, other syntypes may exist, from N.S.W.

Distribution: SE coastal, N.S.W. Ecology: terrestrial, noctidiurnal, predator, woodland, open forest, closed forest; nest in ground layer.

Chelaner sanguinolentus (Wheeler, 1927)

Monomorium (Notomyrmex) sanguinolentum Wheeler, W.M. (1927). The ants of Lord Howe Island and Norfolk Island. *Proc. Am. Acad. Arts Sci.* **62**: 121-153 [135]. Type data: syntypes, MCZ *W,M, from Norfolk Is.

Distribution: Norfolk Is. Ecology: terrestrial, noctidiurnal, predator, woodland, open forest; nest in ground layer.

Chelaner sculpturatus (Clark, 1934)

Monomorium (Notomyrmex) sculpturatum Clark, J. (1934). Ants from the Otway Ranges. *Mem. Natl. Mus. Vict.* **8**: 48-73 [59 pl 4]. Type data: syntypes, NMV *W,F, from Beech Forest, Vic.

Distribution: SE coastal, Vic. Ecology: terrestrial, noctidiurnal, predator, open forest, closed forest; nest in ground layer.

Chelaner sordidus (Forel, 1902)

Chelaner sordidus sordidus (Forel, 1902)

Monomorium sordidum Forel, A. (1902). Fourmis nouvelles d'Australie. *Rev. Suisse Zool.* **10**: 405–548 [443]. Type data: syntypes, GMNH W, ANIC W, from Queanbeyan, N.S.W.

Distribution: Murray-Darling basin, N.S.W. Ecology: terrestrial, noctidiurnal, predator, woodland, open forest; nest in ground layer.

Chelaner sordidus nigriventris (Forel, 1910)

Monomorium sordidum nigriventris Forel, A. (1910). Formicides australiens reçus de MM. Froggatt et Rowland Turner. *Rev. Suisse Zool.* **18**: 1–94 [29]. Type data: syntypes, GMNH W,F, ANIC W, from Howlong, N.S.W.

Distribution: Murray-Darling basin, N.S.W. Ecology: terrestrial, noctidiurnal, predator, woodland, open forest; nest in ground layer.

Chelaner subapterus (Wheeler, 1917)

Chelaner subapterus subapterus (Wheeler, 1917)

Monomorium subapterum Wheeler, W.M. (1917). The phylogenetic development of subapterous and apterous castes in the Formicidae. *Proc. Nat. Acad. Sci. U.S.A.* **3**: 109–117 [112]. Type data: syntypes, MCZ *W,F,M, from Harding River and Derby, W.A.

Distribution: NW coastal, W.A. Ecology: terrestrial, noctidiurnal, predator, desert, woodland; nest in soil.

Chelaner subapterus bogischi (Wheeler, 1917)

Monomorium subapterum bogischi Wheeler, W.M. (1917). The phylogenetic development of subapterous and apterous castes in the Formicidae. *Proc. Nat. Acad. Sci. U.S.A.* **3**: 109–117 [112]. Type data: syntypes, MCZ *W,F, from Point (=Port) Wakefield, S.A.

Distribution: S Gulfs, S.A. Ecology: terrestrial, noctidiurnal, predator, desert, woodland; nest in soil.

Chelaner turneri (Forel, 1910)

Vollenhovia turneri Forel, A. (1910). Formicides australiens reçus de MM. Froggatt et Rowland Turner. *Rev. Suisse Zool.* **18**: 1–94 [26]. Type data: syntypes, GMNH W, ANIC W, from Kuranda near Cairns, Qld.

Distribution: NE coastal, Qld. Ecology: terrestrial, noctidiurnal, predator, closed forest; nest in ground layer.

Chelaner whitei (Wheeler, 1915)

Monomorium (Holcomyrmex) whitei Wheeler, W.M. (1915). Hymenoptera. *Trans. R. Soc. S. Aust.* **39**: 805–823 pls 64–66 [Dec. 1915] [807]. Type data: syntypes, MCZ *W, from Flat Rock Hole in the Musgrave Ranges, S.A.

Distribution: W plateau, S.A. Ecology: terrestrial, noctidiurnal, predator, desert, woodland; nest in soil.

Colobostruma Wheeler, 1927

Colobostruma Wheeler, W.M. (1927). The physiognomy of insects. *Q. Rev. Biol.* **2**: 1–36 [32] [proposed with subgeneric rank in *Epopostruma* Forel, 1895]. Type species *Epopostruma (Colobostruma) leae* Wheeler, 1927 by monotypy.

Clarkistruma Brown, W.L. jr. (1948). A preliminary generic revision of the higher Dacetini (Hymenoptera : Formicidae). *Trans. Am. Entomol. Soc.* **74**: 101–129 [27 July 1948] [124]. Type species *Epopostruma alinodis* Forel, 1913 by original designation.

Synonymy that of Brown, W.L. jr. (1973). A comparison of the Hylean and Congo-West African rain forest ant faunas. pp. 161–185 *in* Meggers, B.J., Ayensu, E.S. & Duckworth, W.D. (eds.) *Tropical forest ecosystems in Africa and South America: a comparative review.* Washington : Smithsonian Institution Press. [177].

This group is also found in New Guinea and east Melanesia.

Colobostruma alinodis (Forel, 1913)

Epopostruma alinodis Forel, A. (1913). Fourmis de Tasmanie et d'Australie récoltées par MM. Lea, Froggatt etc. *Bull. Soc. Vaud. Sci. Nat.* **49**: 173–196 pl 2 [179]. Type data: syntypes, GMNH W, ANIC W, from Railton, Tas.

Distribution: SE coastal, Murray-Darling basin, Vic., N.S.W., Tas. Ecology: terrestrial, noctidiurnal, predator, open forest; nest in ground layer.

Colobostruma australis Brown, 1959

Colobostruma australis Brown, W.L. jr. (1959). Some new species of dacetine ants. *Breviora* **108**: 1–11 [7 May 1959] [4]. Type data: holotype, MCZ *W, from Kallista in the Dandenong Range, Vic.

Distribution: SE coastal, Vic. Ecology: terrestrial, noctidiurnal, predator, woodland, open forest; nest in ground layer.

Colobostruma cerornata Brown, 1959

Colobostruma cerornata Brown, W.L. jr. (1959). Some new species of dacetine ants. *Breviora* **108**: 1–11 [7 May 1959] [1]. Type data: holotype, MCZ *W, from Dempster Head (=Telegraph Hill), Esperance, W.A.

Distribution: SW coastal, W.A. Ecology: terrestrial, noctidiurnal, predator, tall shrubland; nest in ground layer.

Colobostruma elliotti (Clark, 1928)

Epitritus elliotti Clark, J. (1928). Entomological Reports. Formicidae. *in*, Report of the Victorian Field Naturalists' expedition through the Western District of Victoria. *Vict. Nat.* **45** suppl.: 39–44 [42]. Type data: syntypes, NMV *W,F, from Mt. Arapiles, Vic.

Distribution: Murray-Darling basin, N.S.W., Vic. Ecology: terrestrial, noctidiurnal, predator, woodland, open forest; nest in ground layer.

Colobostruma froggatti (Forel, 1913)

Epopostruma froggatti Forel, A. (1913). Fourmis de Tasmanie et d'Australie récoltées par MM. Lea, Froggatt etc. *Bull. Soc. Vaud. Sci. Nat.* **49**: 173–196 pl 2 [177]. Type data: syntypes, GMNH W, from New Norfolk, Tas.

Distribution: SE coastal, Murray-Darling basin, Vic., N.S.W., Tas. Ecology: terrestrial, noctidiurnal, predator, woodland, open forest; nest in ground layer.

Colobostruma leae (Wheeler, 1927)

Epopostruma (Colobostruma) leae Wheeler, W.M. (1927). The physiognomy of insects. *Q. Rev. Biol.* **2**: 1–36 [32 fig 4]. Type data: holotype, MCZ *F, from Cairns district, Qld., see Brown, W.L. jr. (1948). A preliminary generic revision of the higher Dacetini (Hymenoptera : Formicidae). *Trans. Am. Entomol. Soc.* **74**: 101–129 [27 July 1948] [118].

Distribution: NE coastal, Qld. Ecology: terrestrial, noctidiurnal, predator, open forest; nest in ground layer.

Colobostruma nancyae Brown, 1965

Colobostruma nancyae Brown, W.L. jr. (1965). *Colobostruma nancyae* species nov. Pilot Register of Zoology, Cornell University, Ithaca, New York, Card no. 22 [5 Apr. 1965]. Type data: holotype, MCZ *W, from 8 km NE of (old) Thomas River Station, about 100 km E of Esperance, W.A.

Distribution: SW coastal, W.A. Ecology: terrestrial, noctidiurnal, predator, tall shrubland; nest in ground layer.

Colobostruma papulata Brown, 1965

Colobostruma papulata Brown, W.L. jr. (1965). *Colobostruma papulata* species nov. Insecta : Hymenoptera : Formicidae. Pilot Register of Zoology, Cornell University, Ithaca, New York, Card no. 21 [5 Apr. 1965]. Type data: holotype, MCZ *W, from Dempster Head (=Telegraph Hill) at Esperance, W.A.

Distribution: SW coastal, W.A. Ecology: terrestrial, noctidiurnal, predator, tall shrubland; nest in ground layer.

Crematogaster Lund, 1831

Crematogaster Lund, M. (1831). Lettre sur les habitudes de quelques fourmis de Brésil, adressée à M. Audouin. *Ann. Sci. Nat.* **23**: 113–138 [132]. Type species *Formica scutellaris* Olivier, 1791 by subsequent designation, see Bingham, C.T. (1903). *The Fauna of British India, including Ceylon and Burma.* Hymenoptera. Vol. 2 Ants and cuckoo wasps. London : Taylor & Francis [124].

Cremastogaster Mayr, G.L. (1861). *Die europëischen Formiciden. (Ameisen.) Nach der analytischen Methode bearbeitet.* Vienna : Carl Gerolds Sohn 80 pp. 1 pl [74] [invalid emend. of *Crematogaster* Lund, 1831].

This group is also found in the Neotropical, Nearctic, south Palearctic, Ethiopian, Malagasy and Oriental regions; widespread in the Australian Region except New Zealand and Polynesia, see Brown, W.L. jr. (1973). A comparison of the Hylean and Congo-West African rain forest ant faunas. pp. 161–185 *in* Meggers, B.J., Ayensu, E.S. & Duckworth, W.D. (eds.) *Tropical forest ecosystems in Africa and South America: a comparative review.* Washington : Smithsonian Institution Press.

Crematogaster australis Mayr, 1876

Crematogaster australis australis Mayr, 1876

Crematogaster australis Mayr, G.L. (1876). Die australischen Formiciden. *J. Mus. Godeffroy* **5**: 56–115 [108]. Type data: syntypes, NHMW W,F,M, from Peak Downs, Qld.

Distribution: NE coastal, Qld. Ecology: terrestrial, diurnal, predator, woodland, open forest; nest in ground layer.

Crematogaster australis chillagoensis Forel, 1915

Crematogaster australis chillagoensis Forel, A. (1915). Results of Dr. E. Mjöbergs Swedish Scientific Expeditions to Australia 1910–1913. 2. Ameisen. *Ark. Zool.* **9**: 1–119 pls 1–3 [4 Dec. 1915] [57]. Type data: syntypes, GMNH W, ANIC W, other syntypes may exist, from Chillagoe, Qld.

Distribution: NE coastal, Qld. Ecology: terrestrial, diurnal, predator, woodland, open forest; nest in ground layer.

Crematogaster australis sycites Forel, 1916

Crematogaster australis sycites Forel, A. (1916). Fourmis du Congo et d'autres provenances récoltées par MM. Hermann, Kohl, Luja, Mayné, etc. *Rev. Suisse Zool.* **24**: 397–460 [406]. Type data: syntypes (probable), GMNH (probable) *W, from Townsville, Qld.

Distribution: NE coastal, Qld. Ecology: terrestrial, diurnal, predator, woodland, open forest; nest in ground layer.

Crematogaster cornigera Forel, 1902

Cremastogaster cornigera Forel, A. (1902). Fourmis nouvelles d'Australie. *Rev. Suisse Zool.* **10**: 405–548 [407]. Type data: syntypes, GMNH W,F, ANIC W, from Mackay, Qld.

Distribution: NE coastal, Qld. Ecology: terrestrial, diurnal, predator, woodland, open forest; nest in ground layer.

Crematogaster dispar Forel, 1902

Cremastogaster sordidula dispar Forel, A. (1902). Fourmis nouvelles d'Australie. *Rev. Suisse Zool.* **10**: 405–548 [412]. Type data: syntypes, GMNH W,F,M, ANIC W, from Bendigo, Vic.

Distribution: Murray-Darling basin, Vic. Ecology: terrestrial, diurnal, predator, woodland, open forest; nest in ground layer. Biological references: Emery, C. (1922). Hymenoptera Fam. Formicidae subfam. Myrmicinae. *in* Wytsman, P. (ed.) *Genera Insectorum.* Fasc. 174B 112 pp. (raised to species).

Crematogaster eurydice Forel, 1915

Cremastogaster (Atopogyne) eurydice Forel, A. (1915). Results of Dr. E. Mjöbergs Swedish Scientific Expeditions to Australia 1910–1913. 2. Ameisen. *Ark. Zool.* **9**: 1–119 pls 1–3 [4 Dec. 1915] [56]. Type data: syntypes, GMNH F, other syntypes may exist, from Noonkanbah, W.A.

Distribution: N coastal, W.A. Ecology: terrestrial, diurnal, predator, woodland, open forest; nest in ground layer.

Crematogaster frivola Forel, 1902

Crematogaster frivola frivola Forel, 1902

Cremastogaster frivolus Forel, A. (1902). Fourmis nouvelles d'Australie. *Rev. Suisse Zool.* **10**: 405–548 [412]. Type data: syntypes, GMNH W, ANIC W, from Kalgoorlie, W.A.

Distribution: W plateau, W.A. Ecology: terrestrial, diurnal, predator, woodland, open forest; nest in ground layer.

Crematogaster frivola sculpticeps Forel, 1907

Cremastogaster frivola sculpticeps Forel, A. (1907). Formicidae. pp. 263–310 *in* Michaelsen, W. & Hartmeyer, R. (eds.) *Die Fauna Südwest-Australiens.* Jena : G. Fischer Vol.1 [279]. Type data: syntypes, GMNH W, from Kalgoorlie, W.A.

Distribution: W plateau, W.A. Ecology: terrestrial, diurnal, predator, woodland; nest in ground layer.

Crematogaster fusca Mayr, 1876

Cremastogaster fusca Mayr, G.L. (1876). Die australischen Formiciden. *J. Mus. Godeffroy* **5**: 56–115 [107]. Type data: syntypes, NHMW W, from Rockhampton, Qld.

Distribution: NE coastal, Qld. Ecology: terrestrial, diurnal, predator, woodland, open forest; nest in ground layer.

Crematogaster kutteri Viehmeyer, 1924

Cremastogaster kutteri Viehmeyer, H. (1924). Formiciden der australischen Faunenregion. *Entomol. Mitt.* **13**: 310–319 [314]. Type data: syntypes, ZMB *W, from Liverpool and Trial Bay, N.S.W.

Distribution: SE coastal, N.S.W. Ecology: terrestrial, diurnal, predator, woodland, open forest; nest in ground layer.

Crematogaster laeviceps F. Smith, 1858

Crematogaster laeviceps laeviceps F. Smith, 1858

Crematogaster laeviceps Smith, F. (1858). *Catalogue of hymenopterous insects in the collection of the British Museum. Part 6. Formicidae.* London : British Museum 216 pp. 14 pls [27 Mar. 1858] [138]. Publication date established from Donisthorpe, H. (1932). On the identity of Smith's types of Formicidae (Hymenoptera) collected by Alfred Russell Wallace in the Malay Archipelago, with descriptions of two new species. *Ann. Mag. Nat. Hist. (10)* **10**: 441–476. Type data: syntypes, BMNH *W,F, from Melbourne, Vic.

Distribution: SE coastal, Vic. Ecology: terrestrial, diurnal, predator, woodland, open forest; nest in ground layer.

Crematogaster laeviceps broomensis Forel, 1915

Crematogaster laeviceps broomensis Forel, A. (1915). Results of Dr. E. Mjöbergs Swedish Scientific Expeditions to Australia 1910–1913. 2. Ameisen. *Ark. Zool.* **9**: 1–119 pls 1–3 [4 Dec. 1915] [56]. Type data: syntypes, GMNH (probable) *W, from Broome, W.A.

Distribution: N coastal, W.A. Ecology: terrestrial, diurnal, predator, woodland, open forest; nest in ground layer.

Crematogaster laeviceps chasei Forel, 1902

Cremastogaster laeviceps chasei Forel, A. (1902). Fourmis nouvelles d'Australie. *Rev. Suisse Zool.* **10**: 405–548 [413]. Type data: syntypes, GMNH W, ANIC W, from Perth, W.A.

Distribution: SW coastal, W.A. Ecology: terrestrial, diurnal, predator, woodland, open forest; nest in ground layer.

Crematogaster laeviceps clarior Forel, 1902

Cremastogaster laeviceps clarior Forel, A. (1902). Fourmis nouvelles d'Australie. *Rev. Suisse Zool.* **10**: 405–548 [414]. Type data: syntypes, GMNH W, ANIC W, from Mackay, Qld.

Distribution: NE coastal, Qld. Ecology: terrestrial, diurnal, predator, woodland, open forest; nest in ground layer.

Crematogaster longiceps Forel, 1910

Crematogaster longiceps longiceps Forel, 1910

Cremastogaster longiceps Forel, A. (1910). Formicides australiens reçus de MM. Froggatt et Rowland Turner. *Rev. Suisse Zool.* **18**: 1-94 [32]. Type data: syntypes, GMNH W,F, ANIC W, from Tennant Creek, N.T.

Distribution: W plateau, N.T. Ecology: terrestrial, diurnal, predator, desert, woodland; nest in ground layer.

Crematogaster longiceps curticeps Wheeler, 1915

Crematogaster longiceps curticeps Wheeler, W.M. (1915). Hymenoptera. *Trans. R. Soc. S. Aust.* **39**: 805-823 pls 64-66 [Dec. 1915] [809]. Type data: syntypes, MCZ *W, from Ellery Creek in the MacDonnell Ranges, N.T.

Distribution: Lake Eyre basin, N.T. Ecology: terrestrial, diurnal, predator, desert, woodland; nest in ground layer.

Crematogaster mjobergi Forel, 1915

Cremastogaster mjobergi Forel, A. (1915). Results of Dr. E. Mjöbergs Swedish Scientific Expeditions to Australia 1910-1913. 2. Ameisen. *Ark. Zool.* **9**: 1-119 pls 1-3 [4 Dec. 1915] [54]. Type data: syntypes, GMNH W, ANIC W, other syntypes may exist, from Kimberley distr., W.A.

Distribution: N coastal, W.A. Ecology: terrestrial, diurnal, predator, woodland, open forest; nest in ground layer.

Crematogaster pallida Lowne, 1865

Crematogaster pallidus Lowne, B.T. (1865). Contributions to the natural history of Australian ants. *Entomologist* **2**: 331-336 [335]. Type data: syntypes, BMNH (probable) *W,F, from Sidney (=Sydney), N.S.W.

Distribution: SE coastal, N.S.W. Ecology: terrestrial, diurnal, predator, woodland, open forest; nest in ground layer.

Crematogaster pallipes Mayr, 1862

Cremastogaster pallipes Mayr, G.L. (1862). Myrmecologische Studien. *Verh. Zool.-Bot. Ges. Wien* **12**: Abhand. 649-776 [768 pl 19]. Type data: syntypes, NHMW W, from Sidney (=Sydney), N.S.W.

Cremastogaster piceus Lowne, B.T. (1865). Contributions to the natural history of Australian ants. *Entomologist* **2**: 331-336 [335]. Type data: syntypes (probable), BMNH (probable) *W, from Sidney (=Sydney), N.S.W.

Cremastogaster pallidipes Dalla Torre, C.G. De (1893). *Catalogus hymenopterorum hucusque descriptorum systematicus et synonymicus*. Vol. 7 Formicidae (Heterogyna). Lipsiae : G. Engelmann 289 pp. [84] [invalid emend. of *Cremastogaster pallipes* Mayr, 1862].

Synonymy that of Emery, C. (1922). Hymenoptera Fam. Formicidae subfam. Myrmicinae. *in* Wytsman, P. (ed.) *Genera Insectorum*. Fasc. 174B Brussels pp. 95-206 [133].

Distribution: SE coastal, N.S.W. Ecology: terrestrial, diurnal, predator, woodland, open forest; nest in ground layer.

Crematogaster perthensis Crawley, 1922

Crematogaster perthensis Crawley, W.C. (1922). New ants from Australia. *Ann. Mag. Nat. Hist. (9)* **10**: 16-36 [21]. Type data: syntypes, OUM *W,M, BMNH *W, from Perth, W.A.

Distribution: SW coastal, W.A. Ecology: terrestrial, diurnal, predator, woodland, open forest; nest in ground layer.

Crematogaster pythia Forel, 1915

Cremastogaster pythia Forel, A. (1915). Results of Dr. E. Mjöbergs Swedish Scientific Expeditions to Australia 1910-1913. 2. Ameisen. *Ark. Zool.* **9**: 1-119 pls 1-3 [4 Dec. 1915] [53]. Type data: syntypes, GMNH W, ANIC W, other syntypes may exist, from Yarrabah, Qld.

Distribution: NE coastal, Qld. Ecology: terrestrial, diurnal, predator, woodland, open forest; nest in ground layer.

Crematogaster queenslandica Forel, 1902

Crematogaster queenslandica queenslandica Forel, 1902

Cremastogaster sordidula queenslandica Forel, A. (1902). Fourmis nouvelles d'Australie. *Rev. Suisse Zool.* **10**: 405-548 [410]. Type data: syntypes, GMNH W,F, ANIC W, from Mackay, Qld.

Distribution: NE coastal, Qld. Ecology: terrestrial, diurnal, predator, woodland, open forest; nest in ground layer. Biological references: Emery, C. (1921). Hymenoptera. Fam. Formicidae subfam. Myrmicinae. *in* Wytsman, P. (ed.) *Genera Insectorum*. Fasc. 174A Brussels pp. 1-94 (raised to species).

Crematogaster queenslandica froggatti Forel, 1902

Cremastogaster sordidula froggatti Forel, A. (1902). Fourmis nouvelles d'Australie. *Rev. Suisse Zool.* **10**: 405-548 [410]. Type data: syntypes, GMNH W,F, ANIC W, from Sydney, N.S.W.

Distribution: SE coastal, N.S.W. Ecology: terrestrial, diurnal, predator, woodland, open forest; nest in ground layer.

Crematogaster queenslandica gilberti Forel, 1910

Cremastogaster sordidula gilberti Forel, A. (1910). Formicides australiens reçus de MM. Froggatt et Rowland Turner. *Rev. Suisse Zool.* **18**: 1-94 [32]. Type data: syntypes, GMNH W, ANIC W, from Mackay, Qld.

Distribution: NE coastal, Qld. Ecology: terrestrial, diurnal, predator, woodland, open forest; nest in ground layer.

Crematogaster queenslandica rogans Forel, 1902

Cremastogaster sordidula rogans Forel, A. (1902). Fourmis nouvelles d'Australie. *Rev. Suisse Zool.* **10**: 405–548 [411]. Type data: syntypes, GMNH W,F, ANIC W, from Sydney, N.S.W.

Distribution: SE coastal, N.S.W. Ecology: terrestrial, diurnal, predator, woodland, open forest; nest in ground layer.

Crematogaster queenslandica scabrula Emery, 1914

Crematogaster froggatti scabrula Emery, C. (1914). Formiche d'Australia e di Samoa raccolte dal Prof. Silvestri nel 1913. *Boll. Lab. Zool. Gen. Agr. R. Scuola Agric. Portici* **8**: 179–186 [30 Jan. 1914] [184]. Type data: syntypes (probable), MCG *W, from Mt. Lofty, Adelaide, S.A.

Distribution: S Gulfs, S.A. Ecology: terrestrial, diurnal, predator, woodland, open forest; nest in ground layer.

Crematogaster rufotestacea Mayr, 1876

Crematogaster rufotestacea rufotestacea Mayr, 1876

Cremastogaster rufotestacea Mayr, G.L. (1876). Die australischen Formiciden. *J. Mus. Godeffroy* **5**: 56–115 [109]. Type data: holotype, NHMW W, from Sidney (=Sydney), N.S.W.

Distribution: SE coastal, N.S.W. Ecology: terrestrial, diurnal, predator, woodland, open forest; nest in ground layer.

Crematogaster rufotestacea dentinasis Santschi, 1929

Crematogaster (Orthocrema) rufotestacea dentinasis Santschi, F. (1929). Mélange myrmécologique. *Wien Entomol. Ztg.* **46**: 84–93 [15 Sept. 1929] [89]. Type data: syntypes, BNHM W,F,M, from Mittagong, N.S.W.

Distribution: SE coastal, N.S.W. Ecology: terrestrial, diurnal, predator, woodland, open forest; nest in ground layer.

Crematogaster scita Forel, 1902

Crematogaster scita scita Forel, 1902

Cremastogaster scita Forel, A. (1902). Fourmis nouvelles d'Australie. *Rev. Suisse Zool.* **10**: 405–548 [409]. Type data: syntypes, GMNH W, from Mackay, Qld.

Distribution: NE coastal, Qld. Ecology: terrestrial, diurnal, predator, woodland, open forest; nest in ground layer.

Crematogaster scita mixta Forel, 1902

Cremastogaster scita mixta Forel, A. (1902). Fourmis nouvelles d'Australie. *Rev. Suisse Zool.* **10**: 405–548 [409]. Type data: syntypes, GMNH W, ANIC W, from Mackay, Qld.

Distribution: NE coastal, Qld. Ecology: terrestrial, diurnal, predator, woodland, open forest; nest in ground layer.

Crematogaster whitei Wheeler, 1915

Crematogaster whitei Wheeler, W.M. (1915). Hymenoptera. *Trans. R. Soc. S. Aust.* **39**: 805–823 pls 64–66 [Dec. 1915] [808]. Type data: holotype, MCZ *W, from Everard Range, S.A.

Distribution: W plateau, S.A. Ecology: terrestrial, diurnal, predator, desert, woodland; nest in ground layer.

Crematogaster xerophila Wheeler, 1915

Crematogaster xerophila xerophila Wheeler, 1915

Crematogaster xerophila Wheeler, W.M. (1915). Hymenoptera. *Trans. R. Soc. S. Aust.* **39**: 805–823 pls 64–66 [Dec. 1915] [810]. Type data: syntypes, MCZ *W, from Moorilyanna, S.A.

Distribution: Lake Eyre basin, S.A. Ecology: terrestrial, diurnal, predator, desert, woodland; nest in ground layer.

Crematogaster xerophila exigua Wheeler, 1915

Crematogaster xerophila exigua Wheeler, W.M. (1915). Hymenoptera. *Trans. R. Soc. S. Aust.* **39**: 805–823 pls 64–66 [Dec. 1915] [811]. Type data: syntypes, MCZ *W, from Moorilyanna, S.A.

Distribution: Lake Eyre basin, S.A. Ecology: terrestrial, diurnal, predator, desert, woodland; nest in ground layer.

Epopostruma Forel, 1895

Epopostruma Forel, A. (1895). Nouvelles fourmis d'Australie, récoltées à The Ridge, Mackay, Queensland par M. Gilbert Turner. *Ann. Soc. Entomol. Belg.* **39**: 417–428 [422] [proposed with subgeneric rank in *Strumigenys* F. Smith, 1860]. Type species *Strumigenys (Epopostruma) quadrispinosa* Forel, 1895 by subsequent designation, see Wheeler, W.M. (1911). A list of the type species of the genera and subgenera of Formicidae. *Ann. N.Y. Acad. Sci.* **21**: 157–175 [7 Oct. 1911].

Hexadaceton Brown, W.L. jr. (1948). A preliminary generic revision of the higher Dacetini (Hymenoptera : Formicidae). *Trans. Am. Entomol. Soc.* **74**: 101–129 [27 July 1948] [120]. Type species *Hexadaceton frosti* Brown, 1948 by original designation.

Synonymy that of Brown, W.L. jr. (1973). A comparison of the Hylean and Congo-West African rain forest ant faunas. pp. 161–185 *in* Meggers, B.J., Ayensu, E.S. & Duckworth, W.D. (eds.) *Tropical forest ecosystems in Africa and South America: a comparative review.* Washington : Smithsonian Institution Press [177].

Epopostruma frosti (Brown, 1948)

Hexadaceton frosti Brown, W.L. jr. (1948). A preliminary generic revision of the higher Dacetini (Hymenoptera : Formicidae). *Trans. Am. Entomol. Soc.* **74**: 101–129 [27 July 1948] [120]. Type data: holotype, MCZ No. 27838 *W, from N Mecklenburg, S.A."

Distribution: S Gulfs, W plateau, S.A. Ecology: terrestrial, noctidiurnal, predator, woodland, open forest; nest in ground layer.

Epopostruma monstrosa Viehmeyer, 1925

Epopostruma monstrosa Viehmeyer, H. (1925). Formiciden der australischen Faunenregion. *Entomol. Mitt.* **14**: 25–39 [30]. Type data: holotype, ZMB *F, from Trial Bay, N.S.W.

Distribution: SE coastal, N.S.W. Ecology: terrestrial, noctidiurnal, predator, open forest; nest in ground layer.

Epopostruma quadrispinosa (Forel, 1895)

Epopostruma quadrispinosa quadrispinosa (Forel, 1895)

Strumigenys (Epopostruma) quadrispinosa Forel, A. (1895). Nouvelles fourmis d'Australie, récoltée à The Ridge, Mackay, Queensland par M. Gilbert Turner. *Ann. Soc. Entomol. Belg.* **39**: 417–428 [422]. Type data: holotype, GMNH W, from Mackay, Qld.

Distribution: NE coastal, Qld. [51]. Type data: holotype, GMNH F, from N.S.W.

Distribution: N.S.W.; State only specified. Ecology: terrestrial, noctidiurnal, predator, open forest; nest in ground layer.

Eurhopalothrix Brown and Kempf, 1961

Eurhopalothrix Brown, W.L. jr. & Kempf, W.W. (1961). The type species of the ant genus *Eurhopalothrix*. *Psyche Camb.* **67**: 44 [16 Feb. 1961]. Type species *Rhopalothrix bolaui* Mayr, 1870 by original designation.

This group is also found in the Neotropical, south Nearctic and east Oriental regions; New Guinea, east Melanesia, New Caledonia and Samoa in the Australian Region, see Brown, W.L. jr. (1973). A comparison of the Hylean and Congo-West African rain forest ant faunas. pp. 161–185 *in* Meggers, B.J., Ayensu, E.S. & Duckworth, W.D. (eds.) *Tropical forest ecosystems in Africa and South America: a comparative review*. Washington : Smithsonian Institution Press. Species now known not to occur in Australia: *Rhopalothrix emeryi*, see Brown, W.L. jr. & Kempf, W.W. (1960). A world revision of the ant tribe Basicerotini (Hym. Formicidae). *Studia Entomol.* **3**: 161–250 [as *Eurhopalothrix emeryi* (Forel, 1912)].

Eurhopalothrix australis Brown and Kempf, 1960

Eurhopalothrix australis Brown, W.L. jr. & Kempf, W.W. (1960). A world revision of the ant tribe Basicerotini (Hym. Formicidae). *Studia Entomol.* **3**: 161–250 [218]. Type data: holotype, MCZ *W, from near Crawford's Lookout by the Beatrice River, on the Millaa-Millaa-Innisfail Highway descending from the Atherton Tableland, Qld.

Distribution: NE coastal, SE coastal, N.S.W., Qld. Ecology: terrestrial, noctidiurnal, predator, closed forest; nest in ground layer.

Eurhopalothrix procera (Emery, 1897)

Rhopalothrix procera Emery, C. (1897). Formicidarum species novae vel minus cognitae in collectione Musaei Nationalis Hungarici, quas in Nova-Guinea, Colonia Germanica, collegit L. Biró. *Termész. Füz.* **20**: 571–599 pls 14–15 [572]. Type data: syntypes, MCG *W,F, from Berlinhafen (=Aitape), Seleo Is. and Friedrich-Wilhelmshafen (=Madang), New Guinea.

Distribution: NE coastal, Qld. Ecology: terrestrial, noctidiurnal, predator, closed forest; nest in ground layer.

Glamyromyrmex Wheeler, 1915

Glamyromyrmex Wheeler, W.M. (1915). Two new genera of myrmicine ants from Brazil. *Bull. Mus. Comp. Zool.* **59**: 483–491 [487]. Type species *Glamyromyrmex beebei* Wheeler, 1915 by monotypy.

This group is also found in the Neotropical and north Ethiopian regions; New Guinea in the Australian Region, see Brown, W.L. jr. (1973). A comparison of the Hylean and Congo-West African rain forest ant faunas. pp. 161–185 *in* Meggers, B.J., Ayensu, E.S. & Duckworth, W.D. (eds.) *Tropical forest ecosystems in Africa and South America: a comparative review*. Washington : Smithsonian Institution Press.

Glamyromyrmex flagellatus (Taylor, 1962)

Codiomyrmex flagellatus Taylor, R.W. (1962). New Australian dacetine ants of the genera *Mesostruma* Brown and *Codiomyrmex* Wheeler (Hymenoptera : Formicidae). *Breviora* 152: 1–10 [15 Jan. 1962] [7]. Type data: holotype, QM *W, from Clump Point near Mourilyan, Qld.

Distribution: NE coastal, Qld. Ecology: terrestrial, noctidiurnal, predator, closed forest; nest in ground layer.

Glamyromyrmex semicomptus (Brown, 1959)

Codiomyrmex semicomptus Brown, W.L. jr. (1959). Some new species of dacetine ants. *Breviora* 108: 1–11 [7 May 1959] [9]. Type data: holotype, MCZ *W, from Shipton's Flat, about 20–25 mi S of Cooktown, Qld.

Distribution: NE coastal, Qld. Ecology: terrestrial, noctidiurnal, predator, closed forest; nest in ground layer.

Leptothorax Mayr, 1855

Leptothorax Mayr, G.L. (1855). Formicina Austriaca. *Verh. Zool.-Bot. Ges. Wien* **5**: Abhand. 273–478 [431]. Type species *Myrmica clypeata* Mayr, 1853 by subsequent designation, see Emery, C. (1912). Les espèces-type des genres et sous-genres de la famille des Formicides. *Ann. Soc. Entomol. Belg.* **56**: 271–273 [271].

This group is also found in the Neotropical, Nearctic, Palearctic, Ethiopian, Malagasy and west Oriental regions, see Brown, W.L. jr. (1973). A comparison of the Hylean and Congo-West African rain forest ant faunas. pp. 161–185 *in* Meggers, B.J., Ayensu, E.S. & Duckworth, W.D. (eds.) *Tropical forest ecosystems in Africa and South America: a comparative review.* Washington : Smithsonian Institution Press.

Leptothorax australis Wheeler, 1934

Leptothorax (Goniothorax) australis Wheeler, W.M. (1934). An Australian ant of the genus *Leptothorax* Mayr. *Psyche Camb.* **41**: 60–62 [60]. Type data: syntypes, MCZ *W, from Cairns distr., Qld.

Distribution: NE coastal, Qld. Ecology: terrestrial, noctidiurnal, predator, closed forest; nest in ground layer.

Lordomyrma Emery, 1897

Lordomyrma Emery, C. (1897). Formicidarum species novae vel minus cognitae in collectione Musaei Nationalis Hungarici, quas in Nova-Guinea, Colonia Germanica, collegit L. Biró. *Termész. Füz.* **20**: 571–599 [591 pls 14–15]. Type species *Lordomyrma furcifera* Emery, 1897 by subsequent designation, see Wheeler, W.M. (1911). A list of the type species of the genera and subgenera of Formicidae. *Ann. N.Y. Acad. Sci.* **21**: 157–175 [17 Oct. 1911].

This group is also found in the north Ethiopian and Oriental regions; New Guinea, east Melanesia and New Caledonia in the Australian Region (apparent "species flock" on New Caledonia and Fiji), see Brown, W.L. jr. (1973). A comparison of the Hylean and Congo-West African rain forest ant faunas. pp. 161–185 *in* Meggers, B.J., Ayensu, E.S. & Duckworth, W.D. (eds.) *Tropical forest ecosystems in Africa and South America: a comparative review.* Washington : Smithsonian Institution Press.

Lordomyrma leae Wheeler, 1919

Lordomyrma leae Wheeler, W.M. (1919). The ant genus *Lordomyrma* Emery. *Psyche Camb.* **26**: 97–106 [102]. Type data: syntypes, MCZ *W,M, from Lord Howe Is.

Distribution: Lord Howe Is. Ecology: terrestrial, noctidiurnal, predator, closed forest; nest in ground layer.

Lordomyrma punctiventris Wheeler, 1919

Lordomyrma punctiventris Wheeler, W.M. (1919). The ant genus *Lordomyrma* Emery. *Psyche Camb.* **26**: 97–106 [105]. Type data: syntypes, MCZ *W, from Kuranda, Qld.

Distribution: NE coastal, Qld. Ecology: terrestrial, noctidiurnal, predator, closed forest; nest in ground layer.

Machomyrma Forel, 1895

Machomyrma Forel, A. (1895). Nouvelles fourmis d'Australie, récoltées à The Ridge, Mackay, Queensland par M. Gilbert Turner. *Ann. Soc. Entomol. Belg.* **39**: 417–428 [425] [proposed with subgeneric rank in *Liomyrmex* Mayr, 1865]. Type species *Liomyrmex (Machomyrma) dispar* Forel, 1895 by monotypy.

Machomyrma dispar (Forel, 1895)

Liomyrmex (Machomyrma) dispar Forel, A. (1895). Nouvelles fourmis d'Australie, récoltée à The Ridge, Mackay, Queensland par M. Gilbert Turner. *Ann. Soc. Entomol. Belg.* **39**: 417–428 [425]. Type data: syntypes, GMNH W,F, ANIC W, from Mackay, Qld.

Distribution: NE coastal, Qld. Ecology: terrestrial, noctidiurnal, predator, woodland, open forest; nest in ground layer.

Mayriella Forel, 1902

Mayriella Forel, A. (1902). Fourmis nouvelles d'Australie. *Rev. Suisse Zool.* **10**: 405–548 [425]. Type species *Mayriella abstinens* Forel, 1902 by monotypy.

This group is also found in the east Oriental region; New Guinea in the Australian Region, see Brown, W.L. jr. (1973). A comparison of the Hylean and Congo-West African rain forest ant faunas. pp. 161–185 *in* Meggers, B.J., Ayensu, E.S. & Duckworth, W.D. (eds.) *Tropical forest ecosystems in Africa and South America: a comparative review.* Washington : Smithsonian Institution Press.

Mayriella abstinens Forel, 1902

Mayriella abstinens abstinens Forel, 1902

Mayriella abstinens Forel, A. (1902). Fourmis nouvelles d'Australie. *Rev. Suisse Zool.* **10**: 405–548 [452]. Type data: syntypes, GMNH W, from Mackay, Qld.

Mayriella overbecki Viehmeyer, H. (1925). Formiciden der australischen Faunenregion. *Entomol. Mitt.* **14**: 25–39 [26]. Type data: syntypes, ZMB *W,F, from Trial Bay, N.S.W.

Synonymy that of Wheeler, W.M. (1935). The Australian ant genus *Mayriella* Forel. *Psyche Camb.* **42**: 151–160 [157].

Distribution: NE coastal, SE coastal, Qld., N.S.W., Vic. Ecology: terrestrial, noctidiurnal, predator, woodland, open forest; nest in ground layer.

Mayriella abstinens hackeri Wheeler, 1935

Mayriella abstinens hackeri Wheeler, W.M. (1935). The Australian ant genus *Mayriella* Forel. *Psyche Camb.* **42**: 151–160 [157]. Type data: syntypes, MCZ *W,F, from Brisbane, Qld.

Distribution: NE coastal, Qld. Ecology: terrestrial, noctidiurnal, predator, woodland, open forest; nest in ground layer.

Mayriella abstinens venustula Wheeler, 1935

Mayriella abstinens venustula Wheeler, W.M. (1935). The Australian ant genus *Mayriella* Forel. *Psyche Camb.* **42**: 151–160 [158]. Type data: holotype, MCZ *W, from Mt. Tambourine (=Tamborine Mt.), Qld.

Distribution: NE coastal, Qld. Ecology: terrestrial, noctidiurnal, predator, woodland, open forest; nest in ground layer.

Mayriella spinosior Wheeler, 1935

Mayriella spinosior Wheeler, W.M. (1935). The Australian ant genus *Mayriella* Forel. *Psyche Camb.* **42**: 151–160 [159]. Type data: holotype, MCZ *W, from Cairns distr., Qld.

Distribution: NE coastal, Qld. Ecology: terrestrial, noctidiurnal, predator, woodland, open forest; nest in ground layer.

Meranoplus F. Smith, 1854

Meranoplus Smith, F. (1854). Monograph of the genus *Cryptocerus*, belonging to the group Cryptoceridae-Family Myrmicidae-Division Hymenoptera Heterogyna. *Trans. R. Entomol. Soc. Lond.* **7**: 213–228 pls 19–21 [224] [redefined in Bolton, B. (1981). A revision of the ant genera *Meranoplus* F. Smith, *Dicroaspis* Emery and *Calyptomyrmex* Emery (Hymenoptera : Formicidae) in the Ethiopian zoogeographic region. *Bull. Br. Mus. Nat. Hist. (Entomol.)* **42**: 43–81 (26 Feb. 1981)]. Type species *Cryptocerus bicolor* Guérin, 1845 by subsequent designation, see Bingham, C.T. (1903). *The Fauna of British India, including Ceylon and Burma.* Hymenoptera. Vol. 2 Ants and cuckoo-wasps. London : Taylor & Francis [116].

This group is also found in the Ethiopian, Malagasy and Oriental regions; New Guinea, east Melanesia and New Caledonia in the Australian Region, see Brown, W.L. jr. (1973). A comparison of the Hylean and Congo-West African rain forest ant faunas. pp. 161–185 *in* Meggers, B.J., Ayensu, E.S. & Duckworth, W.D. (eds.) *Tropical forest ecosystems in Africa and South America: a comparative review*. Washington : Smithsonian Institution Press.

Meranoplus aureolus Crawley, 1921

Meranoplus aureolus aureolus Crawley, 1921

Meranoplus aureolus Crawley, W.C. (1921). New and little-known species of ants from various localities. *Ann. Mag. Nat. Hist. (9)* **7**: 87–97 [91]. Type data: syntypes (probable), possibly OUM, from Koolpinyah, N.T.

Distribution: N coastal, N.T. Ecology: terrestrial, noctidiurnal, predator, granivore, woodland, open forest; nest in soil.

Meranoplus aureolus doddi Santschi, 1928

Meranoplus aureolus doddi Santschi, F. (1928). Nouvelles fourmis d'Australie. *Bull. Soc. Vaud. Sci. Nat.* **56**: 465–483 [30 Aug. 1928] [469]. Type data: syntypes, BNHM W, from Townsville, Qld.

Distribution: NE coastal, Qld. Ecology: terrestrial, noctidiurnal, predator, granivore, woodland, open forest; nest in soil.

Meranoplus aureolus linae Santschi, 1928

Meranoplus aureolus linae Santschi, F. (1928). Nouvelles fourmis d'Australie. *Bull. Soc. Vaud. Sci. Nat.* **56**: 465–483 [30 Aug. 1928] [469]. Type data: syntypes, BNHM W, from Townsville, Qld.

Distribution: NE coastal, Qld. Ecology: terrestrial, noctidiurnal, predator, granivore, woodland, open forest; nest in soil.

Meranoplus barretti Santschi, 1928

Meranoplus barretti Santschi, F. (1928). Nouvelles fourmis d'Australie. *Bull. Soc. Vaud. Sci. Nat.* **56**: 465–483 [30 Aug. 1928] [468]. Type data: syntypes, BNHM W, from Elsternwick, Vic.

Distribution: SE coastal, Vic. Ecology: terrestrial, noctidiurnal, predator, woodland, open forest; nest in soil.

Meranoplus dichrous Forel, 1907

Meranoplus dichrous Forel, A. (1907). Formicidae. pp. 263–310 *in* Michaelsen, W. & Hartmeyer, R. (eds.) *Die Fauna Südwest-Australiens*. Jena : G. Fischer Vol. 1 [274]. Type data: holotype, probably destroyed in ZMH in WW II, from Yalgoo, W.A.

Distribution: NW coastal, W.A. Ecology: terrestrial, noctidiurnal, predator, granivore, woodland, open forest; nest in soil.

Meranoplus dimidiatus F. Smith, 1867

Meranoplus dimidiatus Smith, F. (1867). Descriptions of new species of Cryptoceridae. *Trans. R. Entomol. Soc. Lond.* **15**: 523–528 [527 pl 26]. Type data: holotype (probable), BMNH *W, from Champion Bay, W.A.

Distribution: NW coastal, W.A. Ecology: terrestrial, noctidiurnal, predator, granivore, woodland, open forest; nest in soil.

Meranoplus diversus F. Smith, 1867

Meranoplus diversus diversus F. Smith, 1867

Meranoplus diversus Smith, F. (1867). Descriptions of new species of Cryptoceridae. *Trans. R. Entomol. Soc. Lond.* **15**: 523–528 [527 pl 26]. Type data: holotype (probable), BMNH *W, from Champion Bay, W.A.

Distribution: NW coastal, W.A. Ecology: terrestrial, noctidiurnal, predator, granivore, woodland, open forest; nest in soil.

Meranoplus diversus duyfkeni Forel, 1915

Meranoplus diversus duyfkeni Forel, A. (1915). Results of Dr. E. Mjöbergs Swedish Scientific Expeditions to Australia 1910–1913. 2. Ameisen. *Ark. Zool.* **9**: 1–119 pls 1–3 [4 Dec. 1915] [45]. Type data: syntypes, GMNH W, ANIC W, other syntypes may exist, from Kimberley distr., W.A.

Distribution: N coastal, W.A. Ecology: terrestrial, noctidiurnal, predator, granivore, woodland, open forest; nest in soil.

Meranoplus diversus oxleyi Forel, 1915

Meranoplus diversus oxleyi Forel, A. (1915). Results of Dr. E. Mjöbergs Swedish Scientific Expeditions to Australia 1910–1913. 2. Ameisen. *Ark. Zool.* **9**: 1–119 pls 1–3 [4 Dec. 1915] [45]. Type data: syntypes, GMNH W, ANIC W, other syntypes may exist, from Kimberley distr., W.A.

Distribution: N coastal, W.A. Ecology: terrestrial, noctidiurnal, predator, granivore, woodland, open forest; nest in soil.

Meranoplus diversus unicolor Forel, 1902

Meranoplus diversus unicolor Forel, A. (1902). Fourmis nouvelles d'Australie. *Rev. Suisse Zool.* **10**: 405–548 [455]. Type data: syntypes, GMNH W, ANIC W, from King's Sound (?=King Sound), W.A.

Distribution: N coastal, W.A. Ecology: terrestrial, noctidiurnal, predator, granivore, woodland, open forest; nest in soil.

Meranoplus excavatus Clark, 1938

Meranoplus excavatus Clark, J. (1938). Reports of the McCoy Society for Field Investigation and Research. No. 2. Sir Joseph Bank Islands. Part I. Formicidae (Hymenoptera). *Proc. R. Soc. Vict.* **50**: 356–382 [367]. Type data: syntypes, NMV *W, from Reevesby Is., S.A.

Distribution: S Gulfs, S.A. Ecology: terrestrial, noctidiurnal, predator, granivore, woodland, open forest; nest in soil.

Meranoplus fenestratus F. Smith, 1867

Meranoplus fenestratus Smith, F. (1867). Descriptions of new species of Cryptoceridae. *Trans. R. Entomol. Soc. Lond.* **15**: 523–528 [526 pl 26]. Type data: holotype (probable), BMNH *W, from Champion Bay, W.A.

Distribution: NW coastal, W.A. Ecology: terrestrial, noctidiurnal, predator, granivore, woodland, open forest; nest in soil.

Meranoplus ferrugineus Crawley, 1922

Meranoplus ferrugineus Crawley, W.C. (1922). New ants from Australia. *Ann. Mag. Nat. Hist. (9)* **9**: 427–448 [444]. Type data: syntypes, OUM *W, from Serpentine River, W.A.

Distribution: SW coastal, W.A. Ecology: terrestrial, noctidiurnal, predator, granivore, woodland, open forest; nest in soil.

Meranoplus froggatti Forel, 1913

Meranoplus froggatti Forel, A. (1913). Fourmis de Tasmanie et d'Australie récoltées par MM. Lea, Froggatt etc. *Bull. Soc. Vaud. Sci. Nat.* **49**: 173–196 pl 2 [183]. Type data: syntypes, GMNH W, from Vic.

Distribution: SE coastal, Vic. Ecology: terrestrial, noctidiurnal, predator, granivore, woodland, open forest; nest in soil.

Meranoplus hilli Crawley, 1922

Meranoplus hilli Crawley, W.C. (1922). New ants from Australia. *Ann. Mag. Nat. Hist. (9)* **9**: 427–448 [445]. Type data: syntypes (probable), OUM *W, from Seaford, Vic.

Distribution: SE coastal, Vic. Ecology: terrestrial, noctidiurnal, predator, granivore, woodland, open forest; nest in soil.

Meranoplus hirsutus Mayr, 1876

Meranoplus hirsutus hirsutus Mayr, 1876

Meranoplus hirsutus Mayr, G.L. (1876). Die australischen Formiciden. *J. Mus. Godeffroy* **5**: 56–115 [112]. Type data: syntypes, NHMW W, from Gayndah, Qld.

Distribution: NE coastal, Qld. Ecology: terrestrial, noctidiurnal, predator, granivore, woodland, open forest, closed forest; nest in ground layer.

Meranoplus hirsutus minor Forel, 1902

Meranoplus hirsutus minor Forel, A. (1902). Fourmis nouvelles d'Australie. *Rev. Suisse Zool.* **10**: 405–548 [457]. Type data: syntypes, GMNH W,F, ANIC W, from Sydney and Thornleigh, N.S.W.

Distribution: SE coastal, N.S.W. Ecology: terrestrial, noctidiurnal, predator, granivore, woodland, open forest; nest in soil.

Meranoplus hirsutus rugosa Crawley, 1922

Meranoplus hirsutus rugosa Crawley, W.C. (1922). New ants from Australia. *Ann. Mag. Nat. Hist. (9)* **9**: 427–448 [443]. Type data: syntypes (probable), OUM *W, from Parkerville, W.A.

Distribution: SW coastal, W.A. Ecology: terrestrial, noctidiurnal, predator, granivore, woodland, open forest; nest in soil.

Meranoplus hospes Forel, 1910

Meranoplus hospes Forel, A. (1910). Formicides australiens reçus de MM. Froggatt et Rowland Turner. *Rev. Suisse Zool.* **18**: 1–94 [48]. Type data: syntypes, GMNH W,M, from Howlong, N.S.W.

Distribution: Murray-Darling basin, N.S.W. Ecology: terrestrial, noctidiurnal, predator, granivore, woodland, open forest; nest in soil.

Meranoplus mars Forel, 1902

Meranoplus mars mars Forel, 1902

Meranoplus mars Forel, A. (1902). Fourmis nouvelles d'Australie. *Rev. Suisse Zool.* **10**: 405–548 [454]. Type data: syntypes, GMNH W, ANIC W, from Charters Towers, Qld.

Distribution: NE coastal, Qld. Ecology: terrestrial, noctidiurnal, predator, granivore, woodland; nest in soil.

Meranoplus mars ajax Forel, 1915

Meranoplus mars ajax Forel, A. (1915). Results of Dr. E. Mjöbergs Swedish Scientific Expeditions to Australia 1910–1913. 2. Ameisen. *Ark. Zool.* **9**: 1–119 pls 1–3 [4 Dec. 1915] [44]. Type data: holotype, SMNH *W, from Kimberley distr., W.A.

Distribution: N coastal, W.A. Ecology: terrestrial, noctidiurnal, predator, granivore, desert, woodland, open forest; nest in soil.

Meranoplus minimus Crawley, 1922

Meranoplus minor Crawley, W.C. (1918). Some new Australian ants. *Entomol. Rec. J. Var.* **30**: 86–92 [89] [*non Meranoplus hirsutus minor* Forel, 1902]. Type data: syntypes, possibly OUM, from Koolpinyah, N.T.

Meranoplus minimus Crawley, W.C. (1922). New ants from Australia. *Ann. Mag. Nat. Hist.* (9) **9**: 427–448 [445] [*nom. nov.* for *Meranoplus minor* Crawley, 1918].

Meranoplus crawleyi Viehmeyer, H. (1925). Formiciden der australischen Faunenregion. *Entomol. Mitt.* **14**: 25–39 [27] [*nom. nov.* for *Meranoplus minor* Crawley, 1918].

Distribution: N coastal, N.T. Ecology: terrestrial, noctidiurnal, predator, granivore, woodland, open forest; nest in soil.

Meranoplus mjobergi Forel, 1915

Meranoplus mjobergi Forel, A. (1915). Results of Dr. E. Mjöbergs Swedish Scientific Expeditions to Australia 1910–1913. 2. Ameisen. *Ark. Zool.* **9**: 1–119 pls 1–3 [4 Dec. 1915] [46]. Type data: syntypes, GMNH W, ANIC W, other syntypes may exist, from Noonkanbah, W.A.

Distribution: N coastal, W.A. Ecology: terrestrial, noctidiurnal, predator, granivore, woodland, open forest; nest in soil.

Meranoplus oceanicus F. Smith, 1862

Meranoplus oceanicus Smith, F. (1862). A list of the genera and species belonging to the family Cryptoceridae, with descriptions of new species; also a list of the species of the genus *Echinopla*. *Trans. R. Entomol. Soc. Lond.* **11**: 407–416 pls 12–13 [414]. Type data: holotype (probable), BMNH *W, from Moreton Bay, Qld.

Distribution: NE coastal, Qld. Ecology: terrestrial, noctidiurnal, predator, granivore, woodland, open forest; nest in soil.

Meranoplus pubescens (F. Smith, 1854)

Cryptocerus pubescens Smith, F. (1854). Monograph of the genus *Cryptocerus* belonging to the group Cryptoceridae - Family Myrmicidae - Division Hymenoptera Heterogyna. *Trans. R. Entomol. Soc. Lond.* **7**: 213–228 pls 19–21 [223]. Type data: syntypes (probable), BMNH *F, from Adelaide, N.S.W. (*sic*).

Distribution: S Gulfs, S.A. Ecology: terrestrial, noctidiurnal, predator, granivore, woodland, open forest; nest in soil.

Meranoplus puryi Forel, 1902

Meranoplus puryi puryi Forel, 1902

Meranoplus puryi Forel, A. (1902). Fourmis nouvelles d'Australie. *Rev. Suisse Zool.* **10**: 405–548 [456]. Type data: syntypes, GMNH W, ANIC W, from Yarra distr., Vic.

Distribution: SE coastal, Vic. Ecology: terrestrial, noctidiurnal, predator, granivore, woodland, open forest; nest in soil.

Meranoplus puryi curvispina Forel, 1910

Meranoplus puryi curvispina Forel, A. (1910). Formicides australiens reçus de MM. Froggatt et Rowland Turner. *Rev. Suisse Zool.* **18**: 1–94 [47]. Type data: syntypes, GMNH W, ANIC W, from N.S.W.

Distribution: SE coastal, N.S.W. Ecology: terrestrial, noctidiurnal, predator, granivore, woodland, open forest; nest in soil.

Meranoplus similis Viehmeyer, 1922

Meranoplus similis Viehmeyer, H. (1922). Neue Ameisen. *Arch. Naturg.* **88A**(7): 203–220 [208]. Type data: syntypes, ZMB *W, ANIC W, from Killapaninno (=Killalpaninna), S.A.

Distribution: Lake Eyre basin, S.A. Ecology: terrestrial, noctidiurnal, predator, granivore, desert, woodland; nest in soil.

Meranoplus testudineus McAreavey, 1956

Meranoplus testudineus McAreavey, J.J. (1956). A new species of the genus *Meranoplus*. *Mem. Qd. Mus.* **13**: 148–150 [26 Apr. 1956] [148]. Type data: holotype, QM T5319 *W, from Port George the Fourth, W.A.

Distribution: N coastal, W.A. Ecology: terrestrial, noctidiurnal, predator, granivore, desert, woodland, open forest; nest in soil.

Mesostruma Brown, 1948

Mesostruma Brown, W.L. jr. (1948). A preliminary generic revision of the higher Dacetini (Hymenoptera : Formicidae). *Trans. Am. Entomol. Soc.* **74**: 101–129 [27 July 1948] [118]. Type species *Strumigenys (Epopostruma) turneri* Forel, 1895 by original designation.

Mesostruma browni Taylor, 1962

Mesostruma browni Taylor, R.W. (1962). New Australian dacetine ants of the genera *Mesostruma* Brown and *Codiomyrmex* Wheeler (Hymenoptera : Formicidae). *Breviora* 152: 1–10 [15 Jan. 1962] [1]. Type data: holotype, ANIC W, from 2 mi E of Berry, N.S.W.

Distribution: SE coastal, N.S.W. Ecology: terrestrial, noctidiurnal, predator, woodland, open forest; nest in ground layer.

Mesostruma eccentrica Taylor, 1973

Mesostruma eccentrica Taylor, R.W. (1973). Ants of the Australian genus *Mesostruma* Brown (Hymenoptera : Formicidae). *J. Aust. Entomol. Soc.* **12**: 24–38 [31]. Type data: holotype, ANIC Type no. 7513 W, from 14 km W of Balranald, N.S.W.

Distribution: S Gulfs, Murray-Darling basin, S.A., Vic., N.S.W. Ecology: terrestrial, noctidiurnal, predator, woodland, open forest; nest in ground layer.

Mesostruma exolympica Taylor, 1973

Mesostruma exolympica Taylor, R.W. (1973). Ants of the Australian genus *Mesostruma* Brown (Hymenoptera : Formicidae). *J. Aust. Entomol. Soc.* **12**: 24–38 [35]. Type data: holotype, ANIC Type no. 7515 W, from Mt. Ainslie, A.C.T.

Distribution: Murray-Darling basin, S Gulfs, S.A., A.C.T. Ecology: terrestrial, noctidiurnal, predator, woodland, open forest; nest in ground layer.

Mesostruma laevigata Brown, 1952

Mesostruma laevigata Brown, W.L. jr. (1952). The dacetine ant genus *Mesostruma* Brown. *Trans. R. Soc. S. Aust.* **75**: 9–13 [12]. Type data: holotype, ANIC W, from Sea Lake, Vic.

Distribution: Murray-Darling basin, S Gulfs, N.S.W., S.A., Vic. Ecology: terrestrial, noctidiurnal, predator, woodland, open forest; nest in ground layer.

Mesostruma loweryi Taylor, 1973

Mesostruma loweryi Taylor, R.W. (1973). Ants of the Australian genus *Mesostruma* Brown (Hymenoptera : Formicidae). *J. Aust. Entomol. Soc.* **12**: 24–38 [35]. Type data: holotype, ANIC Type no. 7514 W, from Willaston near Gawler, S.A.

Distribution: S Gulfs, S.A. Ecology: terrestrial, noctidiurnal, predator, woodland; nest in ground layer.

Mesostruma turneri (Forel, 1895)

Strumigenys (Epopostruma) turneri Forel, A. (1895). Nouvelles fourmis d'Australie, récoltée à The Ridge, Mackay, Queensland par M. Gilbert Turner. *Ann. Soc. Entomol. Belg.* **39**: 417–428 [424]. Type data: syntypes, GMNH W, ANIC W, from Mackay, Qld.

Distribution: NE coastal, Qld. Ecology: terrestrial, noctidiurnal, predator, woodland, open forest; nest in ground layer.

Metapone Forel, 1911

Metapone Forel, A. (1911). Sur le genre *Metapone* n.g. nouveau groupe des Formicides et sur quelques autres formes nouvelles. *Rev. Suisse Zool.* **19**: 445–459 [447 pl 14]. Type species *Metapone greeni* Forel, 1911 by monotypy.

This group is also found in the Malagasy and Oriental regions; New Guinea in the Australian Region, see Brown, W.L. jr. (1973). A comparison of the Hylean and Congo-West African rain forest ant faunas. pp. 161–185 *in* Meggers, B.J., Ayensu, E.S. & Duckworth, W.D. (eds.) *Tropical forest ecosystems in Africa and South America: a comparative review.* Washington : Smithsonian Institution Press.

Metapone leae Wheeler, 1919

Metapone leae Wheeler, W.M. (1919). The ants of the genus *Metapone* Forel. *Ann. Entomol. Soc. Am.* **12**: 173–191 [21 Oct. 1919] [183]. Type data: syntypes, MCZ *F, from Mt. Tambourine (=Tamborine Mt.), Qld.

Distribution: NE coastal, Qld. Ecology: terrestrial, noctidiurnal, omnivore, open forest, closed forest; nest in soil.

Metapone mjobergi Forel, 1915

Metapone mjobergi Forel, A. (1915). Results of Dr. E. Mjöbergs Swedish Scientific Expeditions to Australia 1910-1913. 2. Ameisen. *Ark. Zool.* **9**: 1–119 pls 1–3 [4 Dec. 1915] [36]. Type data: syntypes, GMNH W, other syntypes may exist, from Malanda, Qld.

Distribution: NE coastal, Qld. Ecology: terrestrial, noctidiurnal, omnivore, open forest, closed forest; nest in soil.

Metapone tillyardi Wheeler, 1919

Metapone tillyardi Wheeler, W.M. (1919). The ants of the genus *Metapone* Forel. *Ann. Entomol. Soc. Am.* **12**: 173–191 [21 Oct. 1919] [187]. Type data: syntypes, MCZ *W, from Dorrigo, N.S.W.

Distribution: SE coastal, N.S.W. Ecology: terrestrial, noctidiurnal, omnivore, open forest, closed forest; nest in soil.

Metapone tricolor McAreavey, 1949

Metapone tricolor McAreavey, J.J. (1949). Australian Formicidae. New genera and species. *Proc. Linn. Soc. N.S.W.* **74**: 1–25 [15 June 1949] [4]. Type data: holotype, ANIC F, from Nyngan, N.S.W.

Distribution: Murray-Darling basin, N.S.W. Ecology: terrestrial, noctidiurnal, omnivore, woodland, open forest; nest in soil.

Monomorium Mayr, 1855

Monomorium Mayr, G.L. (1855). Formicina Austriaca. *Verh. Zool.-Bot. Ges. Wien* **5**: Abhand. 273–478 [452]. Type species *Monomorium minutum* Mayr, 1855 by monotypy.

Mitara Emery, C. (1913). Études sur les Myrmicinae. *Ann. Soc. Entomol. Belg.* **57**: 250–262 [261] [proposed with subgeneric rank in *Monomorium* Mayr, 1855]. Type species *Monomorium laeve* Mayr, 1876 by original designation.

Synonymy that of Emery, C. (1922). Hymenoptera Fam. Formicidae subfam. Myrmicinae *in* Wytsman, P. (ed.) *Genera Insectorum*. Fasc. 174B Brussels pp. 95–206 [183]; Ettershank, G. (1966). A generic revision of the world Myrmicinae related to *Solenopsis* and *Pheidologeton* (Hymenoptera : Formicidae). *Aust. J. Zool.* **14**: 73–171 [82].

This group is also found in the north Neotropical, Nearctic, south Palearctic, Ethiopian, Malagasy and Oriental regions; widespread in the Australian Region, see Brown, W.L. jr. (1973). A comparison of the Hylean and Congo-West African rain forest ant faunas. pp. 161–185 *in* Meggers, B.J., Ayensu, E.S. & Duckworth, W.D. (eds.) *Tropical forest ecosystems in Africa and South America: a comparative review.* Washington : Smithsonian Institution Press.

Monomorium australicum Forel, 1907

Monomorium subcoecum australicum Forel, A. (1907). Formicides du Musée National Hongrois. *Ann. Hist.-Nat. Mus. Natl. Hung.* **5**: 1–42 [30 June 1907] [20]. Type data: syntypes (probable), probably in GMNH or MNH, from Mt. Victoria, Blue Mts., N.S.W.

Distribution: SE coastal, N.S.W. Ecology: terrestrial, noctidiurnal, predator, open forest; nest in ground layer. Biological references: Ettershank, G. (1966). A generic revision of the world Myrmicinae related to *Solenopsis* and *Pheidologeton* (Hymenoptera : Formicidae) *Aust. J. Zool.* **14**: 73–171 (raised to species).

Monomorium broomense Forel, 1915

Monomorium (Mitara) laeve broomense Forel, A. (1915). Results of Dr. E. Mjöbergs Swedish Scientific Expeditions to Australia 1910–1913. 2. Ameisen. *Ark. Zool.* **9**: 1–119 pls 1–3 [4 Dec. 1915] [74] [introduced as *leve*]. Type data: syntypes, GMNH W, ANIC W, other syntypes may exist, from Broome, W.A.

Distribution: N coastal, W.A. Ecology: terrestrial, noctidiurnal, predator, woodland, open forest; nest in ground layer. Biological references: Taylor, R.W. and Brown, D.R., this work, raised to species level.

Monomorium donisthorpei Crawley, 1915

Monomorium (Mitara) donisthorpei Crawley, W.C. (1915). Ants from north and central Australia, collected by G.F. Hill. Part I. *Ann. Mag. Nat. Hist. (8)* **15**: 130–136 [134]. Type data: syntypes (probable), BMNH *W, from Darwin, N.T.

Distribution: N coastal, N.T. Ecology: terrestrial, noctidiurnal, predator, woodland, open forest, closed forest; nest in ground layer.

Monomorium fieldi Forel, 1910

Monomorium (Martia) fieldi Forel, A. (1910). Formicides australiens reçus de MM. Froggatt et Rowland Turner. *Rev. Suisse Zool.* **18**: 1–94 [30]. Type data: syntypes, GMNH W,M, ANIC W, from Tennant Creek, N.T.

Distribution: W plateau, Lake Eyre basin, S.A., Qld., N.T. Ecology: terrestrial, noctidiurnal, predator, desert, woodland; nest in ground layer.

Monomorium fraterculus Santschi, 1919

Monomorium fraterculus fraterculus Santschi, 1919

Monomorium (Mitara) laeve fraterculus Santschi, F. (1919). Cinq notes myrmécologiques. *Bull. Soc. Vaud. Sci. Nat.* **52**: 325–350 [328]. Type data: syntypes, BNHM W, from Townsville, Qld.

Distribution: NE coastal, Qld. Ecology: terrestrial, noctidiurnal, predator, woodland, open forest; nest in ground layer. Biological references: Santschi, F. (1928). Nouvelles fourmis d'Australie. *Bull. Soc. Vaud. Sci. Nat.* **56**: 465–483 (raised to species).

Monomorium fraterculus barretti Santschi, 1928

Monomorium (Lampromyrmex) fraterculus barretti Santschi, F. (1928). Nouvelles fourmis d'Australie. *Bull. Soc. Vaud. Sci. Nat.* **56**: 465–483 [30 Aug. 1928] [467]. Type data: syntypes, BNHM W, from Elsternwick, Vic.

Distribution: SE coastal, Vic. Ecology: terrestrial, noctidiurnal, predator, woodland, open forest; nest in ground layer.

Monomorium ilia Forel, 1907

Monomorium ilia ilia Forel, 1907

Monomorium (Martia) ilia Forel, A. (1907). Formicidae. pp. 263–310 *in* Michaelsen, W. & Hartmeyer, R. (eds.) *Die Fauna Südwest-Australiens.* Jena : G. Fischer Vol. 1 [277]. Type data: syntypes, GMNH W, ANIC W, from Day Dawn and Guildford, W.A.

Distribution: SW coastal, NW coastal, W.A. Ecology: terrestrial, noctidiurnal, predator, woodland, open forest; nest in ground layer.

Monomorium ilia lamingtonense Forel, 1915

Monomorium (Mitara) ilia lamingtonensis Forel, A. (1915). Results of Dr. E. Mjöbergs Swedish Scientific Expeditions to Australia 1910–1913. 2. Ameisen. *Ark. Zool.* **9**: 1–119 pls 1–3 [4 Dec. 1915] [73]. Type data: syntypes, GMNH W,F, other syntypes may exist, from Glen Lamington, Qld.

Distribution: NE coastal, Qld. Ecology: terrestrial, noctidiurnal, predator, open forest, closed forest; nest in ground layer.

Monomorium laeve Mayr, 1876

Monomorium laeve laeve Mayr, 1876

Monomorium laeve Mayr, G.L. (1876). Die australischen Formiciden. *J. Mus. Godeffroy* **5**: 56–115 [101]. Type data: syntypes, NHMW W, from Rockhampton, Qld.

Distribution: NE coastal, Qld. Ecology: terrestrial, noctidiurnal, predator, woodland, open forest; nest in ground layer.

Monomorium laeve nigrius Forel, 1915

Monomorium (Mitara) laeve nigrius Forel, A. (1915). Results of Dr. E. Mjöbergs Swedish Scientific Expeditions to Australia 1910–1913. 2. Ameisen. *Ark. Zool.* **9**: 1–119 pls 1–3 [4 Dec. 1915] [74]. Type data: syntypes, GMNH W,F, ANIC W, other syntypes may exist, from Mt. Tambourine (=Tamborine Mt.), Cedar Creek and Alice River, Qld.

Distribution: NE coastal, Qld. Ecology: terrestrial, noctidiurnal, predator, woodland, open forest; nest in ground layer.

Monomorium micron Crawley, 1925

Monomorium micron Crawley, W.C. (1925). New ants from Australia. II. *Ann. Mag. Nat. Hist. (9)* **16**: 577–598 [593]. Type data: syntypes, OUM *W,F, from W.A.

Distribution: NW coastal, W.A. Ecology: terrestrial, noctidiurnal, predator, woodland, open forest; nest in ground layer.

Monomorium sydneyense Forel, 1902

Monomorium sydneyense sydneyense Forel, 1902

Monomorium sydneyense Forel, A. (1902). Fourmis nouvelles d'Australie. *Rev. Suisse Zool.* **10**: 405–548 [442]. Type data: syntypes, GMNH W, ANIC W, from Sydney, N.S.W.

Distribution: SE coastal, N.S.W. Ecology: terrestrial, noctidiurnal, predator, open forest; nest in soil.

Monomorium sydneyense nigellum Emery, 1914

Monomorium (Mitara) sydneyense nigella Emery, C. (1914). Formiche d'Australia e di Samoa raccolte dal Prof. Silvestri nel 1913. *Boll. Lab. Zool. Gen. Agr. R. Scuola Agric. Portici* **8**: 179–186 [30 Jan. 1914] [184]. Type data: syntypes (probable), MCG *W, from Loftus, N.S.W.

Distribution: SE coastal, N.S.W. Ecology: terrestrial, noctidiurnal, predator, open forest; nest in soil.

Myrmecina Curtis, 1829

Myrmecina Curtis, J. (1829). *British Entomology; or illustrations and descriptions of the genera of insects found in Great Britain and Ireland, etc.* London Vol. 6 [226]. Type species *Formica graminicola* Latreille, 1802 (as *Myrmecina latreillei* Curtis, 1829) by monotypy. Compiled from secondary source: Donisthorpe, H. (1943). A list of the type-species of the genera and subgenera of the Formicidae. *Ann. Mag. Nat. Hist. (11)* **10**: 649–688.

This group is also found in the north Neotropical, south Nearctic, Palearctic and Oriental regions; New Guinea and east Melanesia in the Australian Region, see Brown, W.L. jr. (1973). A comparison of the Hylean and Congo-West African rain forest ant faunas. pp. 161–185 *in* Meggers, B.J., Ayensu, E.S. & Duckworth, W.D. (eds.) *Tropical forest ecosystems in Africa and South America: a comparative review.* Washington : Smithsonian Institution Press.

Myrmecina rugosa Forel, 1902

Myrmecina rugosa Forel, A. (1902). Fourmis nouvelles d'Australie. *Rev. Suisse Zool.* **10**: 405–548 [438]. Type data: syntypes, GMNH W,M, ANIC W, from Mackay, Qld.

Distribution: NE coastal, Qld. Ecology: terrestrial, noctidiurnal, predator, open forest, closed forest; nest in ground layer.

Oligomyrmex Mayr, 1867

Oligomyrmex Mayr, G.L. (1867). Adnotationes in Monographiam formicidarum Indo-Neerlandicarum. *Tijdschr. Entomol.* **10**: 33–117 [110 pl 2]. Type species *Oligomyrmex concinnus* Mayr, 1867 by monotypy.

Octella Forel, A. (1915). Results of Dr. E. Mjöbergs Swedish Scientific Expeditions to Australia. 1910–1913. 2. Ameisen. *Ark. Zool.* **9**: 1–119 [4 Dec. 1915] [69 pls 1–3] [proposed with subgeneric rank in *Oligomyrmex* Mayr, 1867]. Type species *Oligomyrmex (Octella) pachycerus* Forel, 1915 by monotypy.

Synonymy that of Ettershank, G. (1966). A generic revision of the world Myrmicinae related to *Solenopsis* and *Pheidologeton* (Hymenoptera : Formicidae). *Aust. J. Zool.* **14**: 73–171 [119].

This group is also found in the Neotropical, south Nearctic, south Palearctic, Ethiopian, Malagasy Oriental regions; New Guinea, east Melanesia, New Caledonia and southwest Polynesia in the Australian Region, see Brown, W.L. jr. (1973). A comparison of the Hylean and Congo-West African rain forest ant faunas. pp. 161–185 *in* Meggers, B.J., Ayensu, E.S. & Duckworth, W.D. (eds.) *Tropical forest ecosystems in Africa and South America: a comparative review.* Washington : Smithsonian Institution Press.

Oligomyrmex corniger Forel, 1902

Oligomyrmex corniger corniger Forel, 1902

Oligomyrmex corniger Forel, A. (1902). Fourmis nouvelles d'Australie. *Rev. Suisse Zool.* **10**: 405–548 [449]. Type data: syntypes, GMNH W,F,M, ANIC W, from Mackay, Qld.

Distribution: NE coastal, Qld. Ecology: terrestrial, noctidiurnal, predator, granivore, open forest, closed forest; nest in ground layer.

Oligomyrmex corniger parvicornis Forel, 1915

Oligomyrmex corniger parvicornis Forel, A. (1915). Results of Dr. E. Mjöbergs Swedish Scientific Expeditions to Australia 1910–1913. 2. Ameisen. *Ark. Zool.* **9**: 1–119 pls 1–3 [4 Dec. 1915] [70]. Type data: syntypes, GMNH W,M,F, ANIC W, other syntypes may exist, from Malanda, Herberton and Cedar Creek, Qld.

Distribution: NE coastal, Qld. Ecology: terrestrial, noctidiurnal, predator, granivore, open forest, closed forest; nest in ground layer.

Oligomyrmex mjobergi Forel, 1915

Oligomyrmex mjobergi Forel, A. (1915). Results of Dr. E. Mjöbergs Swedish Scientific Expeditions to Australia 1910–1913. 2. Ameisen. *Ark. Zool.* **9**: 1–119 pls 1–3 [4 Dec. 1915] [69]. Type data: syntypes, GMNH W, ANIC W, other syntypes may exist, from Malanda, Qld.

Distribution: NE coastal, Qld. Ecology: terrestrial, noctidiurnal, predator, granivore, open forest, closed forest; nest in ground layer.

Oligomyrmex norfolkensis Donisthorpe, 1941

Oligomyrmex manni norfolkensis Donisthorpe, H. (1941). The ants of Norfolk Island. *Entomol. Mon. Mag.* **77**: 90–93 [2 Apr. 1941] [92]. Type data: syntypes, BMNH *W,F, from Norfolk Is.

Distribution: Norfolk Is. Ecology: terrestrial, noctidiurnal, predator, granivore, open forest; nest in ground layer. Biological references: Taylor, R.W. and Brown, D.R., this work, raised to species.

Oligomyrmex pachycerus Forel, 1915

Oligomyrmex (Octella) pachycerus Forel, A. (1915). Results of Dr. E. Mjöbergs Swedish Scientific Expeditions to Australia 1910–1913. 2. Ameisen. *Ark. Zool.* **9**: 1–119 pls 1–3 [4 Dec. 1915] [69]. Type data: holotype, SMNH *W, from Cedar Creek, Qld.

Distribution: NE coastal, Qld. Ecology: terrestrial, noctidiurnal, predator, granivore, woodland, open forest; nest in ground layer.

Orectognathus F. Smith, 1854

Orectognathus Smith, F. (1854). Monograph of the genus *Cryptocerus*, belonging to the group Cryptoceridae-Family Myrmicidae-Division Hymenoptera Heterogyna. *Trans. R. Entomol. Soc. Lond.* **7**: 213–228 pls 19–21 [227]. Type species *Orectognathus antennatus* F. Smith, 1854 by monotypy.

This group is also found in New Guinea, New Zealand (North Island).

Orectognathus alligator Taylor, 1980

Orectognathus alligator Taylor, R.W. (1980). New Australian ants of the genus *Orectognathus*, with summary description of the twenty-nine known species (Hymenoptera : Formicidae). *Aust. J. Zool.* **27**: 773–788 [15 Feb. 1980] [778]. Type data: holotype, ANIC Type no. 7528 W, from Spencer Gap, 20 km SW of Walkerston, Qld.

Distribution: NE coastal, Qld. Ecology: terrestrial, noctidiurnal, predator, woodland; nest in ground layer.

Orectognathus antennatus F. Smith, 1854

Orectognathus antennatus Smith, F. (1854). Monograph of the genus *Cryptocerus* belonging to the group Cryptoceridae - Family Myrmicidae - Division Hymenoptera Heterogyna. *Trans. R. Entomol. Soc. Lond.* **7**: 213–228 pls 19–21 [228]. Type data: syntypes (probable), BMNH *W, from New Zealand.

Orectognathus antennatus septentrionalis Forel, A. (1910). Formicides australiens reçus de MM. Froggatt et Rowland Turner. *Rev. Suisse Zool.* **18**: 1–94 [51]. Type data: holotype (probable), whereabouts unknown, from Wollongbar, Richmond River, N.S.W.

Synonymy that of Brown, W.L. jr. (1953). A revision of the dacetine ant genus *Orectognathus*. *Mem. Qd. Mus.* **13**: 84–104 [99].

Distribution: SE coastal, NE coastal, N.S.W., Qld. Ecology: terrestrial, noctidiurnal, predator, open forest, closed forest; nest in ground layer.

Orectognathus clarki Brown, 1953

Orectognathus clarki Brown, W.L. jr. (1953). A revision of the dacetine ant genus *Orectognathus*. *Mem. Qd. Mus.* **13**: 84–104 [14 Dec. 1953] [94]. Type data: holotype, ANIC W, from Fern Tree Gully, Vic.

Distribution: SE coastal, NE coastal, Murray-Darling basin, S Gulfs, Qld., N.S.W., Tas., S.A., Vic. Ecology: terrestrial, noctidiurnal, predator, woodland, open forest, closed forest; nest in ground layer.

Orectognathus coccinatus Taylor, 1980

Orectognathus coccinatus Taylor, R.W. (1980). New Australian ants of the genus *Orectognathus*, with summary description of the twenty-nine known species (Hymenoptera : Formicidae). *Aust. J. Zool.* **27**: 773–788 [15 Feb. 1980] [779]. Type data: holotype, ANIC Type no. 7529 W, from Byfield, near Yeppoon, Qld.

Distribution: NE coastal, Qld. Ecology: terrestrial, noctidiurnal, predator, closed forest; nest in ground layer.

Orectognathus darlingtoni Taylor, 1977

Orectognathus darlingtoni Taylor, R.W. (1977). New ants of the Australasian genus *Orectognathus*, with a key to the known species (Hymenoptera : Formicidae). *Aust. J. Zool.* **25**: 581–612 [5 Aug. 1977] [606]. Type data: holotype, ANIC Type no. 7517 W, from Lake Eacham Natl. Park, near Yungaburra, Qld.

Distribution: NE coastal, Qld. Ecology: terrestrial, noctidiurnal, predator, closed forest; nest in ground layer.

Orectognathus elegantulus Taylor, 1977

Orectognathus elegantulus Taylor, R.W. (1977). New ants of the Australasian genus *Orectognathus*, with a key to the known species (Hymenoptera : Formicidae). *Aust. J. Zool.* **25**: 581–612 [5 Aug. 1977] [589]. Type data: holotype, ANIC Type no. 7504 W, from Lamington Natl. Park, Qld.

Distribution: NE coastal, SE coastal, N.S.W., Qld. Ecology: terrestrial, noctidiurnal, predator, closed forest; nest in ground layer.

Orectognathus howensis Wheeler, 1927

Orectognathus antennatus howensis Wheeler, W.M. (1927). The ants of Lord Howe Island and Norfolk Island. *Proc. Am. Acad. Arts Sci.* **62**: 121–153 [145]. Type data: holotype, ANIC Type no. 7518 W, from Howe Is. (=Lord Howe Is.).

Distribution: Lord Howe Is. Ecology: terrestrial, noctidiurnal, predator, open forest, closed forest; nest in ground layer. Biological references: Brown, W.L. jr. (1953). A revision of the dacetine ant genus *Orectognathus*. *Mem. Qd. Mus.* **13**: 84–104 (raised to species).

Orectognathus kanangra Taylor, 1980

Orectognathus kanangra Taylor, R.W. (1980). New Australian ants of the genus *Orectognathus*, with summary description of the twenty-nine known species (Hymenoptera : Formicidae). *Aust. J. Zool.* **27**: 773–788 [15 Feb. 1980] [776]. Type data: holotype, ANIC Type no. 7527 W, from Gingra Range, near Kanangra Tops, N.S.W.

Distribution: NE coastal, Qld. Ecology: terrestrial, noctidiurnal, predator, woodland, open forest; nest in ground layer.

Orectognathus mjobergi Forel, 1915

Orectognathus mjobergi Forel, A. (1915). Results of Dr. E. Mjöbergs Swedish Scientific Expeditions to Australia 1910–1913. 2. Ameisen. *Ark. Zool.* **9**: 1–119 pls 1–3 [4 Dec. 1915] [38]. Type data: syntypes, GMNH W, ANIC W, other syntypes may exist, from Cedar Creek, Qld.

Orectognathus mjobergi unicolor Forel, A. (1915). Results of Dr. E. Mjöbergs Swedish Scientific Expeditions to Australia 1910–1913. 2. Ameisen. *Ark. Zool.* **9**: 1–119 pls 1–3 [4 Dec. 1915] [39]. Type data: holotype, whereabouts uncertain, from Malanda, Qld.

Synonymy that of Brown, W.L. jr. (1953). A revision of the dacetine ant genus *Orectognathus*. *Mem. Qd. Mus.* **13**: 84–104 [98].

Distribution: NE coastal, SE coastal, N.S.W., Qld. Ecology: terrestrial, noctidiurnal, predator, closed forest; nest in ground layer.

Orectognathus nanus Taylor, 1977

Orectognathus nanus Taylor, R.W. (1977). New ants of the Australasian genus *Orectognathus*, with a key to the known species (Hymenoptera : Formicidae). *Aust. J. Zool.* **25**: 581–612 [5 Aug. 1977] [605]. Type data: holotype, ANIC Type no. 7509 W, from Seymour Range, about 5 km N of Innisfail, Qld.

Distribution: NE coastal, Qld. Ecology: terrestrial, noctidiurnal, predator, closed forest; nest in ground layer.

Orectognathus nigriventris Mercovich, 1958

Orectognathus nigriventris Mercovich, T.C. (1958). A new species of the genus *Orectognathus*. *Mem. Qd. Mus.* **13**: 195–198 [28 July 1958] [195]. Type data: holotype, QM *W, from Dora Creek, Martinville, near Morisset, N.S.W.

Distribution: SE coastal, N.S.W. Ecology: terrestrial, noctidiurnal, predator, open forest; nest in ground layer.

Orectognathus parvispinus Taylor, 1977

Orectognathus parvispinus Taylor, R.W. (1977). New ants of the Australasian genus *Orectognathus*, with a key to the known species (Hymenoptera : Formicidae). *Aust. J. Zool.* **25**: 581–612 [5 Aug. 1977] [603]. Type data: holotype, ANIC Type no. 7508 W, from Eungella Natl. Park, about 3 km S of Eungella, Qld.

Distribution: NE coastal, Qld. Ecology: terrestrial, noctidiurnal, predator, closed forest; nest in ground layer.

Orectognathus phyllobates Brown, 1958

Orectognathus phyllobates Brown, W.L. jr. (1958). A supplement to the revisions of the dacetine ant genera *Orectognathus* and *Arnoldidris*, with keys to the species. *Psyche Camb.* **64**: 17–29 [10 Jan. 1958] [25]. Type data: holotype, MCZ *W, from Joalah Natl. Park, near the top of Tamborine Mt., Qld.

Distribution: NE coastal, SE coastal, N.S.W., Qld. Ecology: terrestrial, noctidiurnal, predator, closed forest; nest in ground layer.

Orectognathus robustus Taylor, 1977

Orectognathus robustus Taylor, R.W. (1977). New ants of the Australasian genus *Orectognathus*, with a key to the known species (Hymenoptera : Formicidae). *Aust. J. Zool.* **25**: 581–612 [5 Aug. 1977] [599]. Type data: holotype, ANIC Type no. 7507 W, from Lake Eacham Natl. Park near Yungaburra, Qld.

Distribution: NE coastal, Qld. Ecology: terrestrial, noctidiurnal, predator, closed forest; nest in ground layer.

Orectognathus rostratus Lowery, 1967

Orectognathus rostratus Lowery, B.B. (1967). A new ant of the dacetine genus *Orectognathus* (Hymenoptera : Formicidae). *J. Aust. Entomol. Soc.* **6**: 137–140 [31 Dec. 1967] [137]. Type data: holotype, ANIC Type no. 7501 W, from Karrumbyn Creek (=Breakfast Creek), Mt. Warning State Park, 10 mi W of Murwillumbah, N.S.W.

Distribution: SE coastal, NE coastal, Qld., N.S.W. Ecology: terrestrial, noctidiurnal, predator, closed forest; nest in ground layer.

Orectognathus satan Brown, 1953

Orectognathus satan Brown, W.L. jr. (1953). A revision of the dacetine ant genus *Orectognathus*. *Mem. Qd. Mus.* **13**: 84–104 [14 Dec. 1953] [102]. Type data: holotype, MCZ *W, from Malanda Falls, Malanda, Qld.

Distribution: NE coastal, Qld. Ecology: terrestrial, noctidiurnal, predator, closed forest; nest in ground layer.

Orectognathus sexspinosus Forel, 1915

Orectognathus sexspinosus Forel, A. (1915). Results of Dr. E. Mjöbergs Swedish Scientific Expeditions to Australia 1910–1913. 2. Ameisen. *Ark. Zool.* **9**: 1–119 pls 1–3 [4 Dec. 1915] [39]. Type data: syntypes, GMNH W,M, ANIC W, other syntypes may exist, from Cedar Creek, Qld.

Distribution: NE coastal, Qld. Ecology: terrestrial, noctidiurnal, predator, closed forest; nest in ground layer.

Orectognathus versicolor Donisthorpe, 1940

Orectognathus versicolor Donisthorpe, H. (1940). Descriptions of new species of ants (Hym., Formicidae) from various localities. *Ann. Mag. Nat. Hist. (11)* **5** : 39–48 [46]. Type data: holotype, BMNH *W, from Tambourine (=Tamborine) Mt., Qld.

Distribution: NE coastal, SE coastal, N.S.W., Qld. Ecology: terrestrial, noctidiurnal, predator, open forest, closed forest; nest in ground layer.

Peronomyrmex Viehmeyer, 1922

Peronomyrmex Viehmeyer, H. (1922). Neue Ameisen. *Arch. Naturg.* **88A**(7): 203–220 [212]. Type species *Peronomyrmex overbecki* Viehmeyer, 1922 by monotypy.

Peronomyrmex overbecki Viehmeyer, 1922

Peronomyrmex overbecki Viehmeyer, H. (1922). Neue Ameisen. *Arch. Naturg.* **88A**(7): 203–220 [213]. Type data: holotype, ZMB W, from Trial Bay, N.S.W.

Distribution: SE coastal, N.S.W. Ecology: terrestrial, diurnal, predator, (closed forest); (nest arboreal). Biological references: Taylor, R.W. (1970). Characterization of the Australian endemic ant genus *Peronomyrmex* Viehmeyer (Hymenoptera : Formicidae). *J. Aust. Entomol. Soc.* **9**: 209–211 (systematics).

Pheidole Westwood, 1841

Pheidole Westwood, J.O. (1841). Observations on the genus *Typhlopone*, with descriptions of several exotic species of ants. *Ann. Mag. Nat. Hist. (1)* **6**: 81–89 [87 pl 2]. Type species *Atta providens* Sykes, 1835 by monotypy.

This group is found world-wide, no native species in New Zealand, see Brown, W.L. jr. (1973). A comparison of the Hylean and Congo-West African rain forest ant faunas. pp. 161–185 *in* Meggers, B.J., Ayensu, E.S. & Duckworth, W.D. (eds.) *Tropical forest ecosystems in Africa and South America: a comparative review.* Washington : Smithsonian Institution Press.

Pheidole ampla Forel, 1893

Pheidole ampla ampla Forel, 1893

Pheidole variabilis ampla Forel, A. (1893). Nouvelles fourmis d'Australie et des Canaries. *Ann. Soc. Entomol. Belg.* **37**: 454–466 [462]. Type data: syntypes, GMNH W, from East Wallaby Is., W.A.

Distribution: NW coastal, W.A. Ecology: terrestrial, noctidiurnal, predator, granivore; nest in ground layer. Biological references: Forel, A. (1902). Fourmis nouvelles d'Australie. *Rev. Suisse Zool.* **10**: 405–548 (raised to species).

Pheidole ampla mackayensis Forel, 1902

Pheidole ampla mackayensis Forel, A. (1902). Fourmis nouvelles d'Australie. *Rev. Suisse Zool.* **10**: 405–548 [436]. Type data: syntypes, GMNH W, ANIC W, from Mackay, Qld.

Distribution: NE coastal, Qld. Ecology: terrestrial, noctidiurnal, predator, granivore; nest in ground layer.

Pheidole ampla parviceps Forel, 1915

Pheidole ampla parviceps Forel, A. (1915). Results of Dr. E. Mjöbergs Swedish Scientific Expeditions to Australia 1910–1913. 2. Ameisen. *Ark. Zool.* **9**: 1–119 pls 1–3 [4 Dec. 1915] [57]. Type data: holotype, SMNH *W, from Herberton, Qld.

Distribution: NE coastal, Qld. Ecology: terrestrial, noctidiurnal, predator, granivore; nest in ground layer.

Pheidole ampla perthensis Crawley, 1922

Pheidole ampla perthensis Crawley, W.C. (1922). New ants from Australia. *Ann. Mag. Nat. Hist. (9)* **10**: 16–36 [24]. Type data: syntypes, OUM *W,F, from Perth, W.A.

Distribution: SW coastal, W.A. Ecology: terrestrial, noctidiurnal, predator, granivore; nest in ground layer.

Pheidole anthracina Forel, 1902

Pheidole anthracina anthracina Forel, 1902

Pheidole anthracina Forel, A. (1902). Fourmis nouvelles d'Australie. *Rev. Suisse Zool.* **10**: 405–548 [419]. Type data: syntypes, GMNH W,F, ANIC W, from The Ridge, Mackay, Qld.

Distribution: NE coastal, Qld. Ecology: terrestrial, noctidiurnal, predator, granivore; nest in ground layer.

Pheidole anthracina grandii Emery, 1914

Pheidole anthracina grandii Emery, C. (1914). Formiche d'Australia e di Samoa raccolte dal Prof. Silvestri nel 1913. *Boll. Lab. Zool. Gen. Agr. R. Scuola Agric. Portici* **8**: 179–186 [30 Jan. 1914] [183]. Type data: syntypes, MCG *W, from Gosford, N.S.W.

Distribution: SE coastal, N.S.W. Ecology: terrestrial, noctidiurnal, predator, granivore; nest in ground layer.

Pheidole anthracina orba Forel, 1902

Pheidole anthracina orba Forel, A. (1902). Fourmis nouvelles d'Australie. *Rev. Suisse Zool.* **10**: 405–548 [421]. Type data: syntypes, GMNH W,F, ANIC W, from Wollongbar, Richmond River, N.S.W.

Distribution: SE coastal, N.S.W. Ecology: terrestrial, noctidiurnal, predator, granivore; nest in ground layer.

Pheidole athertonensis Forel, 1915

Pheidole athertonensis athertonensis Forel, 1915

Pheidole athertonensis Forel, A. (1915). Results of Dr. E. Mjöbergs Swedish Scientific Expeditions to Australia 1910–1913. 2. Ameisen. *Ark. Zool.* **9**: 1–119 pls 1–3 [4 Dec. 1915] [62]. Type data: syntypes, GMNH W,F, ANIC W, other syntypes may exist, from Atherton, Qld.

Distribution: NE coastal, Qld. Ecology: terrestrial, noctidiurnal, predator, granivore; nest in ground layer.

Pheidole athertonensis cedarensis Forel, 1915

Pheidole athertonensis cedarensis Forel, A. (1915). Results of Dr. E. Mjöbergs Swedish Scientific Expeditions to Australia 1910–1913. 2. Ameisen. *Ark. Zool.* **9**: 1–119 pls 1–3 [4 Dec. 1915] [64]. Type data: syntypes, GMNH W,M, ANIC W, other syntypes may exist, from Cedar Creek, Qld.

Distribution: NE coastal, Qld. Ecology: terrestrial, noctidiurnal, predator, granivore; nest in ground layer.

Pheidole athertonensis tambourinensis Forel, 1915

Pheidole athertonensis tambourinensis Forel, A. (1915). Results of Dr. E. Mjöbergs Swedish Scientific Expeditions to Australia 1910–1913. 2. Ameisen. *Ark. Zool.* **9**: 1–119 pls 1–3 [4 Dec. 1915] [65]. Type data: syntypes, GMNH W,M,F, other syntypes may exist, from Mt. Tambourine (=Tamborine Mt.), Qld.

Distribution: NE coastal, Qld. Ecology: terrestrial, noctidiurnal, predator, granivore; nest in ground layer.

Pheidole bos Forel, 1893

Pheidole bos bos Forel, 1893

Pheidole bos Forel, A. (1893). Nouvelles fourmis d'Australie et des Canaries. *Ann. Soc. Entomol. Belg.* **37**: 454–466 [463]. Type data: syntypes, GMNH W, from Fremantle, W.A.

Distribution: SW coastal, W.A. Ecology: terrestrial, noctidiurnal, predator, granivore; nest in ground layer.

Pheidole bos baucis Forel, 1910

Pheidole bos baucis Forel, A. (1910). Formicides australiens reçus de MM. Froggatt et Rowland Turner. *Rev. Suisse Zool.* **18**: 1–94 [37]. Type data: syntypes, GMNH W,F, ANIC W, from N.S.W.

Distribution: N.S.W. Ecology: terrestrial, noctidiurnal, predator, granivore; nest in ground layer.

Pheidole bos eubos Forel, 1915

Pheidole bos eubos Forel, A. (1915). Results of Dr. E. Mjöbergs Swedish Scientific Expeditions to Australia

1910–1913. 2. Ameisen. *Ark. Zool.* **9**: 1–119 pls 1–3 [4 Dec. 1915] [62]. Type data: syntypes, GMNH W, ANIC W, other syntypes may exist, from Cedar Creek, Atherton, Laura and Cape York, Qld.

Distribution: NE coastal, Qld. Ecology: terrestrial, noctidiurnal, predator, granivore; nest in ground layer.

Pheidole brevicornis Mayr, 1876

Pheidole brevicornis Mayr, G.L. (1876). Die australischen Formiciden. *J. Mus. Godeffroy* **5**: 56–115 [106]. Type data: syntypes, whereabouts unknown, from Rockhampton, Qld.

Distribution: NE coastal, Qld. Ecology: terrestrial, noctidiurnal, predator, granivore; nest in ground layer.

Pheidole cairnsiana Forel, 1902

Pheidole javana cairnsiana Forel, A. (1902). Fourmis nouvelles d'Australie. *Rev. Suisse Zool.* **10**: 405–548 [438]. Type data: syntypes, GMNH (probable) *W, from Cairns, Qld.

Distribution: NE coastal, Qld. Ecology: terrestrial, noctidiurnal, predator, granivore; nest in ground layer. Biological references: Taylor, R.W. and Brown, D.R., this work, raised to species.

Pheidole concentrica Forel, 1902

Pheidole concentrica concentrica Forel, 1902

Pheidole concentrica Forel, A. (1902). Fourmis nouvelles d'Australie. *Rev. Suisse Zool.* **10**: 405–548 [416]. Type data: syntypes, GMNH W, ANIC W, from N.S.W.

Distribution: N.S.W. Ecology: terrestrial, noctidiurnal, predator, granivore; nest in ground layer.

Pheidole concentrica recurva Forel, 1910

Pheidole concentrica recurva Forel, A. (1910). Formicides australiens reçus de MM. Froggatt et Rowland Turner. *Rev. Suisse Zool.* **18**: 1–94 [39]. Type data: syntypes, GMNH W,F,M, from Launceston, Tas.

Distribution: Tas. Ecology: terrestrial, noctidiurnal, predator, granivore; nest in ground layer.

Pheidole conficta Forel, 1902

Pheidole conficta Forel, A. (1902). Fourmis nouvelles d'Australie. *Rev. Suisse Zool.* **10**: 405–548 [417]. Type data: syntypes, GMNH W, ANIC W, from N.S.W.

Distribution: N.S.W. Ecology: terrestrial, noctidiurnal, predator, granivore; nest in ground layer.

Pheidole deserticola Forel, 1910

Pheidole deserticola deserticola Forel, 1910

Pheidole deserticola Forel, A. (1910). Formicides australiens reçus de MM. Froggatt et Rowland Turner. *Rev. Suisse Zool.* **18**: 1–94 [34]. Type data: syntypes, GMNH W,F, ANIC W, from Tennant Creek, N.T.

Distribution: W plateau, N.T. Ecology: terrestrial, noctidiurnal, predator, granivore; nest in ground layer.

Pheidole deserticola foveifrons Viehmeyer, 1924

Pheidole deserticola foveifrons Viehmeyer, H. (1924). Formiciden der australischen Faunenregion. *Entomol. Mitt.* **13**: 310–319 [312]. Type data: syntypes, ZMB *W, from Killalpanino (=Killalpaninna), S.A.

Distribution: Lake Eyre basin, S.A. Ecology: terrestrial, noctidiurnal, predator, granivore; nest in ground layer.

Pheidole gellibrandi Clark, 1934

Pheidole gellibrandi Clark, J. (1934). Ants from the Otway Ranges. *Mem. Natl. Mus. Vict.* **8**: 48–73 [58 pl 4]. Type data: syntypes, NMV *W, from Gellibrand, Vic.

Distribution: SE coastal, Vic. Ecology: terrestrial, noctidiurnal, predator, granivore; nest in ground layer.

Pheidole hartmeyeri Forel, 1907

Pheidole hartmeyeri Forel, A. (1907). Formicidae. pp. 263–310 *in* Michaelsen, W. & Hartmeyer, R. (eds.) *Die Fauna Südwest-Australiens.* Jena : G. Fischer Vol. 1 [280]. Type data: syntypes, GMNH W, ANIC W, from Buckland Hill near Fremantle and Broome Hill, W.A.

Distribution: SW coastal, W.A. Ecology: terrestrial, noctidiurnal, predator, granivore; nest in ground layer.

Pheidole impressiceps Mayr, 1876

Pheidole impressiceps Mayr, G.L. (1876). Die australischen Formiciden. *J. Mus. Godeffroy* **5**: 56–115 [105]. Type data: syntypes, NHMW W, from Rockhampton, Qld.

Distribution: NE coastal, Qld. Ecology: terrestrial, noctidiurnal, predator, granivore; nest in ground layer.

Pheidole incurvata Viehmeyer, 1924

Pheidole incurvata Viehmeyer, H. (1924). Formiciden der australischen Faunenregion. *Entomol. Mitt.* **13**: 310–319 [313]. Type data: syntypes, ZMB *W, from Liverpool, N.S.W.

Distribution: SE coastal, N.S.W. Ecology: terrestrial, noctidiurnal, predator, granivore; nest in ground layer.

Pheidole liteae Forel, 1910

Pheidole liteae Forel, A. (1910). Formicides australiens reçus de MM. Froggatt et Rowland Turner. *Rev. Suisse Zool.* **18**: 1–94 [41]. Type data: syntypes, GMNH W, ANIC W, from Tas.

Distribution: Tas. Ecology: terrestrial, noctidiurnal, predator, granivore; nest in ground layer.

Pheidole longiceps Mayr, 1876

Pheidole longiceps longiceps Mayr, 1876

Pheidole longiceps Mayr, G.L. (1876). Die australischen Formiciden. *J. Mus. Godeffroy* **5**: 56–115 [106]. Type data: syntypes, NHMW W, from Rockhampton, Qld.

Distribution: NE coastal, Qld. Ecology: terrestrial, noctidiurnal, predator, granivore; nest in ground layer.

Pheidole longiceps doddi Forel, 1910

Pheidole longiceps doddi Forel, A. (1910). Formicides australiens reçus de MM. Froggatt et Rowland Turner. *Rev. Suisse Zool.* **18**: 1–94 [38]. Type data: syntypes, GMNH W,F, ANIC W, from Bunderbury, Qld.

Distribution: NE coastal, Qld. Ecology: terrestrial, noctidiurnal, predator, granivore; nest in ground layer.

Pheidole longiceps frontalis Forel, 1902

Pheidole longiceps frontalis Forel, A. (1902). Fourmis nouvelles d'Australie. *Rev. Suisse Zool.* **10**: 405–548 [436]. Type data: syntypes, GMNH W,F, ANIC W, from Mackay, Qld.

Distribution: NE coastal, Qld. Ecology: terrestrial, noctidiurnal, predator, granivore; nest in ground layer.

Pheidole mjobergi Forel, 1915

Pheidole (Pheidolacanthinus) mjobergi Forel, A. (1915). Results of Dr. E. Mjöbergs Swedish Scientific Expeditions to Australia 1910–1913. 2. Ameisen. *Ark. Zool.* **9**: 1–119 pls 1–3 [4 Dec. 1915] [66]. Type data: syntypes, GMNH W, ANIC W, other syntypes may exist, from Kimberley distr., W.A.

Distribution: N coastal, W.A. Ecology: terrestrial, noctidiurnal, predator, granivore; nest in ground layer.

Pheidole opaciventris Mayr, 1876

Pheidole opaciventris Mayr, G.L. (1876). Die australischen Formiciden. *J. Mus. Godeffroy* **5**: 56–115 [105]. Type data: syntypes, NHMW W, from Rockhampton, Qld.

Distribution: NE coastal, Qld. Ecology: terrestrial, noctidiurnal, predator, granivore; nest in ground layer.

Pheidole platypus Crawley, 1915

Pheidole platypus Crawley, W.C. (1915). Ants from north and south-west Australia (G.F. Hill, Rowland Turner) and Christmas Island, Straits Settlements. Part II. *Ann. Mag. Nat. Hist.* (8) **15**: 232–239 [234]. Type data: syntypes, BMNH *W, from Stapleton, N.T.

Distribution: N coastal, N.T. Ecology: terrestrial, noctidiurnal, predator, granivore; nest in ground layer.

Pheidole proxima Mayr, 1876

Pheidole proxima proxima Mayr, 1876

Pheidole proxima Mayr, G.L. (1876). Die australischen Formiciden. *J. Mus. Godeffroy* **5**: 56–115 [104]. Type data: syntypes, NHMW W, from Peak Downs, Qld.

Distribution: NE coastal, Qld. Ecology: terrestrial, noctidiurnal, predator, granivore; nest in ground layer.

Pheidole proxima bombalensis Forel, 1910

Pheidole proxima bombalensis Forel, A. (1910). Formicides australiens reçus de MM. Froggatt et Rowland Turner. *Rev. Suisse Zool.* **18**: 1–94 [43]. Type data: syntypes, GMNH W, ANIC W, from Bombala, N.S.W.

Distribution: SE coastal, N.S.W. Ecology: terrestrial, noctidiurnal, predator, granivore; nest in ground layer.

Pheidole proxima transversa Forel, 1902

Pheidole proxima transversa Forel, A. (1902). Fourmis nouvelles d'Australie. *Rev. Suisse Zool.* **10**: 405–548 [428]. Type data: syntypes, GMNH W,F,M, ANIC W, from Mackay, Qld.

Distribution: NE coastal, Qld. Ecology: terrestrial, noctidiurnal, predator, granivore; nest in ground layer.

Pheidole pyriformis Clark, 1938

Pheidole pyriformis Clark, J. (1938). Reports of the McCoy Society for Field Investigation and Research. No. 2. Sir Joseph Bank Islands. Part I. Formicidae (Hymenoptera). *Proc. R. Soc. Vict.* **50**: 356–382 [371]. Type data: syntypes, NMV *W, from Reevesby Is., Winceby Is. and English Is., S.A.

Distribution: S Gulfs, S.A. Ecology: terrestrial, noctidiurnal, predator, granivore; nest in ground layer.

Pheidole spinoda (F. Smith, 1858)

Atta spinoda Smith, F. (1858). *Catalogue of hymenopterous insects in the collection of the British Museum. Part 6. Formicidae*. London : British Museum 216 pp. 14 pls [27 Mar. 1858] [166]. Publication date established from Donisthorpe, H. (1932). On the identity of Smith's types of Formicidae (Hymenoptera) collected by Alfred Russell Wallace in the Malay Archipelago, with descriptions of two new species. *Ann. Mag. Nat. Hist. (10)* **10**: 441–476. Type data: syntypes (probable), BMNH *F, from Adelaide, S.A.

Distribution: S Gulfs, S.A. Ecology: terrestrial, noctidiurnal, predator, granivore; nest in ground layer.

Pheidole tasmaniensis Mayr, 1866

Pheidole tasmaniensis tasmaniensis Mayr, 1866

Pheidole tasmaniensis Mayr, G.L. (1866). Myrmecologische Beiträge. *Sber. Akad. Wiss. Wien* **53**(1): 484–517 [511]. Type data: syntypes, NHMW *W, from Tas.

Distribution: Tas. Ecology: terrestrial, noctidiurnal, predator, granivore; nest in ground layer.

Pheidole tasmaniensis continentis Forel, 1902

Pheidole tasmaniensis continentis Forel, A. (1902). Fourmis nouvelles d'Australie. *Rev. Suisse Zool.* **10**: 405–548 [437]. Type data: syntypes, GMNH W,F, ANIC W, from Ballarat, Vic.

Distribution: SE coastal, Vic. Ecology: terrestrial, noctidiurnal, predator, granivore; nest in ground layer.

Pheidole trapezoidea Viehmeyer, 1913

Pheidole trapezoidea Viehmeyer, H. (1913). Neue und unvollständig bekannte Ameisen der Alten Welt. *Arch. Naturg.* **79A**(12): 24–60 [36]. Type data: syntypes (probable), ZMB *W, from Killalpaninna, S.A.

Distribution: Lake Eyre basin, S.A. Ecology: terrestrial, noctidiurnal, predator, granivore; nest in ground layer.

Pheidole turneri Forel, 1902

Pheidole turneri Forel, A. (1902). Fourmis nouvelles d'Australie. *Rev. Suisse Zool.* **10**: 405–548 [430]. Type data: syntypes, GMNH W, ANIC W, from Mackay, Qld.

Distribution: NE coastal, Qld. Ecology: terrestrial, noctidiurnal, predator, granivore; nest in ground layer.

Pheidole variabilis Mayr, 1876

Pheidole variabilis variabilis Mayr, 1876

Pheidole variabilis Mayr, G.L. (1876). Die australischen Formiciden. *J. Mus. Godeffroy* **5**: 56–115 [103]. Type data: syntypes, NHMW W,F,M, from Rockhampton, Qld.

Distribution: NE coastal, Qld. Ecology: terrestrial, noctidiurnal, predator, granivore; nest in ground layer.

Pheidole variabilis latigena Forel, 1907

Pheidole variabilis latigena Forel, A. (1907). Formicidae. pp. 263–310 *in* Michaelsen, W. & Hartmeyer, R. (eds.) *Die Fauna Südwest-Australiens.* Jena : G. Fischer Vol. 1 [279]. Type data: syntypes, GMNH W, from Day Dawn, W.A.

Distribution: NW coastal, W.A. Ecology: terrestrial, noctidiurnal, predator, granivore; nest in ground layer.

Pheidole variabilis mediofusca Forel, 1902

Pheidole variabilis mediofusca Forel, A. (1902). Fourmis nouvelles d'Australie. *Rev. Suisse Zool.* **10**: 405–548 [425]. Type data: syntypes, GMNH W, ANIC W, from Wollongbar, Richmond River, N.S.W.

Distribution: SE coastal, N.S.W. Ecology: terrestrial, noctidiurnal, predator, granivore; nest in ground layer.

Pheidole variabilis ocior Forel, 1915

Pheidole variabilis ocior Forel, A. (1915). Results of Dr. E. Mjöbergs Swedish Scientific Expeditions to Australia 1910–1913. 2. Ameisen. *Ark. Zool.* **9**: 1–119 pls 1–3 [4 Dec. 1915] [58]. Type data: syntypes, GMNH W, other syntypes may exist, from Malanda and Tolga, Qld.

Distribution: NE coastal, Qld. Ecology: terrestrial, noctidiurnal, predator, granivore; nest in ground layer.

Pheidole variabilis ocyma Forel, 1915

Pheidole variabilis ocyma Forel, A. (1915). Results of Dr. E. Mjöbergs Swedish Scientific Expeditions to Australia 1910–1913. 2. Ameisen. *Ark. Zool.* **9**: 1–119 pls 1–3 [4 Dec. 1915] [59]. Type data: syntypes, GMNH W,F, ANIC W, other syntypes may exist, from Christmas Creek, Qld.

Distribution: NE coastal, Qld. Ecology: terrestrial, noctidiurnal, predator, granivore; nest in ground layer.

Pheidole variabilis parvispina Forel, 1902

Pheidole variabilis parvispina Forel, A. (1902). Fourmis nouvelles d'Australie. *Rev. Suisse Zool.* **10**: 405–548 [424]. Type data: syntypes, GMNH W,M, ANIC W, from Mackay, Qld.

Distribution: NE coastal, Qld. Ecology: terrestrial, noctidiurnal, predator, granivore; nest in ground layer.

Pheidole variabilis praedo Forel, 1902

Pheidole variabilis praedo Forel, A. (1902). Fourmis nouvelles d'Australie. *Rev. Suisse Zool.* **10**: 405–548 [426]. Type data: syntypes, GMNH W, ANIC W, from Wollongbar, Richmond River, N.S.W.

Distribution: SE coastal, N.S.W. Ecology: terrestrial, noctidiurnal, predator, granivore; nest in ground layer.

Pheidole variabilis redunca Crawley, 1915

Pheidole variabilis redunca Crawley, W.C. (1915). Ants from north and south-west Australia (G.F. Hill, Rowland

Turner) and Christmas Island, Straits Settlements. Part II. *Ann. Mag. Nat. Hist. (8)* **15**: 232-239 [235]. Type data: syntypes, possibly OUM, from Darwin, N.T.

Distribution: N coastal, N.T. Ecology: terrestrial, noctidiurnal, predator, granivore; nest in ground layer.

Pheidole variabilis rugocciput Forel, 1902

Pheidole variabilis rugocciput Forel, A. (1902). Fourmis nouvelles d'Australie. *Rev. Suisse Zool.* **10**: 405-548 [423]. Type data: syntypes, GMNH W, ANIC W, from Mackay, Qld.

Distribution: NE coastal, Qld. Ecology: terrestrial, noctidiurnal, predator, granivore; nest in ground layer.

Pheidole variabilis rugosula Forel, 1902

Pheidole variabilis rugosula Forel, A. (1902). Fourmis nouvelles d'Australie. *Rev. Suisse Zool.* **10**: 405-548 [423]. Type data: syntypes, GMNH W, ANIC W, from Bong Bong, N.S.W.

Distribution: SE coastal, N.S.W. Ecology: terrestrial, noctidiurnal, predator, granivore; nest in ground layer.

Pheidole vigilans (F. Smith, 1858)

Atta vigilans Smith, F. (1858). *Catalogue of hymenopterous insects in the collection of the British Museum*. Part 6. Formicidae. London : British Museum 216 pp. 14 pls [27 Mar. 1858] [166]. Publication date established from Donisthorpe, H. (1932). On the identity of Smith's types of Formicidae (Hymenoptera) collected by Alfred Russell Wallace in the Malay Archipelago, with descriptions of two new species. *Ann. Mag. Nat. Hist. (10)* **10**: 441-476. Type data: syntypes (probable), BMNH *W, from Melbourne, Vic.

Pheidole dolichocephala André, E. (1896). Fourmis nouvelles d'Asie et d'Australie. *Rev. Entomol.* **15**: 251-265 [262]. Type data: syntypes, MNHP W, from W.A.

Pheidole ampla yarrensis Forel, A. (1902). Fourmis nouvelles d'Australie. *Rev. Suisse Zool.* **10**: 405-548 [434]. Type data: syntypes, GMNH W,F, from Yarra distr., Vic.

Pheidole ampla parallela Forel, A. (1902). Fourmis nouvelles d'Australie. *Rev. Suisse Zool.* **10**: 405-548 [435]. Type data: syntypes, GMNH W,M, ANIC W, from N.S.W.

Pheidole ampla norfolkensis Wheeler, W.M. (1927). The ants of Lord Howe Island and Norfolk Island. *Proc. Am. Acad. Arts Sci.* **62**: 121-153 [134]. Type data: syntypes, MCZ *W, from Norfolk Is.

Synonymy that of Brown, W.L. jr. (1971). The identity and synonymy of *Pheidole vigilans* a common ant of Southeastern Australia (Hymenoptera : Formicidae). *Aust. J. Zool.* **10**: 13-14 [13].

Distribution: Murray-Darling basin, SE coastal, S Gulfs, N.S.W., Vic., S.A., Tas., Norfolk Is. Ecology: terrestrial, noctidiurnal, predator, granivore; nest in ground layer.

Pheidole wiesei Forel, 1910

Pheidole wiesei Forel, A. (1910). Formicides australiens reçus de MM. Froggatt et Rowland Turner. *Rev. Suisse Zool.* **18**: 1-94 [40]. Type data: syntypes, GMNH W,F,M, ANIC W, from N.S.W.

Distribution: N.S.W. Ecology: terrestrial, noctidiurnal, predator, granivore; nest in ground layer.

Pheidologeton Mayr, 1862

Pheidologeton Mayr, G.L. (1862). Myrmecologische Studien. *Verh. Zool.-Bot. Ges. Wien* **12**: Abhand. 649-776 pl 19 [750] [redefined in Ettershank, G. (1966). A generic revision of the world Myrmicinae related to *Solenopsis* and *Pheidologeton* (Hymenoptera : Formicidae). *Aust. J. Zool.* **14**: 73-171]. Type species *Oecodoma diversa* Jerdon, 1851 by subsequent designation, see Bingham, C.T. (1903). *The Fauna of British India, including Ceylon and Burma.* Hymenoptera. Vol. 2 Ants and cuckoo-wasps. London : Taylor & Francis [160].

This group is also found in the Ethiopian and Oriental regions; New Guinea and east Melanesia in the Australian Region, see Brown, W.L. jr. (1973). A comparison of the Hylean and Congo-West African rain forest ant faunas. pp. 161-185 *in* Meggers, B.J., Ayensu, E.S. & Duckworth, W.D. (eds.) *Tropical forest ecosystems in Africa and South America: a comparative review*. Washington : Smithsonian Institution Press.

Pheidologeton australis Forel, 1915

Pheidologeton australis australis Forel, 1915

Pheidologeton affinis australis Forel, A. (1915). Results of Dr. E. Mjöbergs Swedish Scientific Expeditions to Australia 1910-1913. 2. Ameisen. *Ark. Zool.* **9**: 1-119 pls 1-3 [4 Dec. 1915] [68]. Type data: syntypes, GMNH W, other syntypes may exist, from Cedar Creek, Herberton and Atherton, Qld.

Distribution: NE coastal, Qld. Ecology: terrestrial, noctidiurnal, nomadic, predator, open forest, closed forest; nest in ground layer. Biological references: Forel, A. (1918). Études myrmécologiques en 1917. *Bull. Soc. Vaud. Sci. Nat.* **51**: 717-727 (raised to species).

Pheidologeton australis mjobergi Forel, 1918

Pheidologeton australis mjobergi Forel, A. (1918). Études myrmécologiques en 1917. *Bull. Soc. Vaud. Sci. Nat.* **51**: 717-727 [5 Apr. 1918] [723]. Type data: syntypes, GMNH F, from Atherton, Qld.

Distribution: NE coastal, Qld. Ecology: terrestrial, noctidiurnal, nomadic, predator, open forest, closed forest; nest in ground layer.

Podomyrma F. Smith, 1859

Podomyrma Smith, F. (1859). Catalogue of hymenopterous insects collected by Mr. A.R. Wallace at the islands of Aru and Key. *J. Linn. Soc. Zool.* **3**: 132–178 [1 Feb. 1859] [145]. Publication date established from Donisthorpe, H. (1932). On the identity of Smith's types of Formicidae (Hymenoptera) collected by Alfred Russell Wallace in the Malay Archipelago, wtih descriptions of two species. *Ann. Mag. Nat. Hist. (10)* **10**: 441–476. Type species *Podomyrma femorata* F. Smith, 1859 by subsequent designation, see Wheeler, W.M. (1911). A list of the type species of the genera and subgenera of Formicidae. *Ann. N.Y. Acad. Sci.* **21**: 157–175 [17 Oct. 1911].

Dacryon Forel, A. (1895). Nouvelles fourmis d'Australie, récoltées à The Ridge, Mackay, Queensland par M. Gilbert Turner. *Ann. Soc. Entomol. Belg.* **39**: 417–428 [421]. Type species *Dacryon omniparens* Forel, 1895 by monotypy.

Pseudopodomyrma Crawley, W.C. (1925). Formicidae. A new genus. *Entomol. Rec. J. Var.* **37**: 40–41 [40]. Type species *Pseudopodomyrma clarki* Crawley, 1925 by monotypy.

Synonymy that of Brown, W.L. jr. (1973). A comparison of the Hylean and Congo-West African rain forest ant faunas. pp. 161–185 *in* Meggers, B.J., Ayensu, E.S. & Duckworth, W.D. (eds.) *Tropical forest ecosystems in Africa and South America: a comparative review.* Washington : Smithsonian Institution Press [177].

This group is also found in New Guinea and east Melanesia in the Australian Region.

Podomyrma abdominalis Emery, 1887

Podomyrma abdominalis Emery, C. (1887). Catalogue delle Formiche esistenti nelle collezioni del Museo Civico di Genova, Parte terza. Formiche della regione Indo-Malese e dell'Australia. *Ann. Mus. Civ. Stor. Nat. Giacomo Doria (2)* **5**: 427–473 [459]. Type data: status unknown, ?MGB, from Ternate, Indonesia.

Podomyrma abdominalis pulchra Forel, 1901

Podomyrma abdominalis pulchra Forel, A. (1901). Formiciden des Naturhistorischen Museums zu Hamburg. Neue *Calyptomyrmex-, Dacryon-, Podomyrma-,* und *Echinopla*-Arten. *Mitt. Naturh. Mus. Hamb.* **18**: 45–82 [54]. Type data: syntypes, GMNH W, ANIC W, from Cairns, Qld.

Distribution: NE coastal, Qld. Ecology: terrestrial, arboreal, noctidiurnal, predator, open forest, closed forest; nest arboreal.

Podomyrma adelaidae (F. Smith, 1858)

Podomyrma adelaidae adelaidae (F. Smith, 1858)

Myrmica adelaidae Smith, F. (1858). *Catalogue of hymenopterous insects in the collection of the British Museum.* Part 6. Formicidae. London : British Museum 216 pp. 14 pls [27 Mar. 1858] [128]. Publication date established from Donisthorpe, H. (1932). On the identity of Smith's types of Formicidae (Hymenoptera) collected by Alfred Russell Wallace in the Malay Archipelago, with descriptions of two new species. *Ann. Mag. Nat. Hist. (10)* **10**: 441–476. Type data: holotype, BMNH *W, from Adelaide, S.A.

Podomyrma micans sericeiventris Emery, C. (1898). Descrizioni di formiche nuove Malesi e Australiane. Note sinonimiche. *Rec. Sess. Accad. Sci. Ist. Bologna (ns)* **2**: 231–245 [235]. Type data: syntypes, MCG *W,F, from unknown locality.

Podomyrma bimaculata Forel, A. (1901). Formiciden des Naturhistorischen Museums zu Hamburg. Neue *Calyptomyrmex-, Dacryon-, Podomyrma-,* und *Echinopla*-Arten. *Mitt. Naturh. Mus. Hamb.* **18**: 45–82 [57]. Type data: syntypes, GMNH W,F, ANIC W, from Kalgoorlie, W.A.

Synonymy that of Emery, C. (1922). Hymenoptera Fam. Formicidae subfam. Myrmicinae. *in* Wytsman, P. (ed.) *Genera Insectorum.* Fasc. 174C pp. 207–397 [237].

Distribution: W plateau, S Gulfs, Murray-Darling basin, W.A., S.A., Vic., N.S.W. Ecology: terrestrial, arboreal, noctidiurnal, predator, woodland, open forest; nest arboreal.

Podomyrma adelaidae brevidentata Forel, 1915

Podomyrma bimaculata brevidentata Forel, A. (1915). Results of Dr. E. Mjöbergs Swedish Scientific Expeditions to Australia 1910–1913. 2. Ameisen. *Ark. Zool.* **9**: 1–119 pls 1–3 [4 Dec. 1915] [49]. Type data: syntypes, GMNH W, other syntypes may exist, from Kimberley distr., W.A.

Distribution: N coastal, W.A. Ecology: terrestrial, arboreal, noctidiurnal, predator, woodland, open forest; nest arboreal.

Podomyrma adelaidae obscurior Forel, 1915

Podomyrma bimaculata obscurior Forel, A. (1915). Results of Dr. E. Mjöbergs Swedish Scientific Expeditions to Australia 1910–1913. 2. Ameisen. *Ark. Zool.* **9**: 1–119 pls 1–3 [4 Dec. 1915] [50]. Type data: holotype, probably GMNH or SMNH, from Alice River, Qld.

Distribution: NE coastal, Qld. Ecology: terrestrial, arboreal, noctidiurnal, predator, woodland, open forest; nest arboreal.

Podomyrma basalis F. Smith, 1859

Podomyrma basalis Smith, F. (1859). Catalogue of hymenopterous insects collected by Mr A.R. Wallace at the islands of Aru and Key. *J. Linn. Soc. Zool.* **3**: 132–178 [1 Feb. 1859] [147]. Publication date established from Donisthorpe, H. (1932). On the identity of Smith's types of Formicidae (Hymenoptera) collected by Alfred Russell Wallace in the Malay Archipelago, with descriptions of two new species. *Ann. Mag. Nat. Hist. (10)* **10**: 441–476. Type data: syntypes (probable), BMNH *W, from Aru Ils., Indonesia.

Distribution: N coastal, N Gulf, NE coastal, N.T., Qld.; also in New Guinea. Ecology: terrestrial, arboreal, noctidiurnal, predator, open forest, closed forest; nest arboreal.

Podomyrma bispinosa Forel, 1901

Podomyrma bispinosa Forel, A. (1901). Formiciden des Naturhistorischen Museums zu Hamburg. Neue *Calyptomyrmex-*, *Dacryon-*, *Podomyrma-*, und *Echinopla*-Arten. *Mitt. Naturh. Mus. Hamb.* **18**: 45–82 [56]. Type data: syntypes, GMNH W, from Mackay, Qld.

Distribution: NE coastal, Qld. Ecology: terrestrial, arboreal, noctidiurnal, predator, woodland, open forest; nest arboreal.

Podomyrma chasei Forel, 1901

Podomyrma chasei Forel, A. (1901). Formiciden des Naturhistorischen Museums zu Hamburg. Neue *Calyptomyrmex-*, *Dacryon-*, *Podomyrma-*, und *Echinopla*-Arten. *Mitt. Naturh. Mus. Hamb.* **18**: 45–82 [58]. Type data: syntypes, GMNH W,M, ANIC W, from Perth, W.A.

Distribution: SW coastal, W.A. Ecology: terrestrial, arboreal, noctidiurnal, predator, woodland, open forest; nest arboreal.

Podomyrma christae (Forel, 1907)

Dacryon christae Forel, A. (1907). Formicides du Musée National Hongrois. *Ann. Hist.- Nat. Mus. Natl. Hung.* **5**: 1–42 [30 June 1907] [16]. Type data: syntypes (probable), probably in GMNH or MNH, from Sydney, Botany Bay, N.S.W.

Distribution: SE coastal, N.S.W. Ecology: terrestrial, arboreal, noctidiurnal, predator, woodland, open forest; nest arboreal.

Podomyrma clarki (Crawley, 1925)

Pseudopodomyrma clarki Crawley, W.C. (1925). Formicidae. A new genus. *Entomol. Rec. J. Var.* **37**: 40–41 [40]. Type data: syntypes (probable), OUM *W, from Swan River, W.A.

Distribution: SW coastal, W.A. Ecology: terrestrial, arboreal, noctidiurnal, predator, woodland, open forest; nest arboreal.

Podomyrma convergens Forel, 1895

Podomyrma convergens Forel, A. (1895). Nouvelles fourmis d'Australie, récoltée à The Ridge, Mackay, Queensland par M. Gilbert Turner. *Ann. Soc. Entomol. Belg.* **39**: 417–428 [427]. Type data: holotype, GMNH W, from Mackay, Qld.

Distribution: NE coastal, Qld. Ecology: terrestrial, arboreal, noctidiurnal, predator, woodland, open forest; nest arboreal.

Podomyrma delbruckii Forel, 1901

Podomyrma delbruckii Forel, A. (1901). Formiciden des Naturhistorischen Museums zu Hamburg. Neue *Calyptomyrmex-*, *Dacryon-*, *Podomyrma-*, und *Echinopla*-Arten. *Mitt. Naturh. Mus. Hamb.* **18**: 45–82 [58]. Type data: syntypes, GMNH W, ANIC W, from Mackay, Qld.

Distribution: NE coastal, Qld. Ecology: terrestrial, arboreal, noctidiurnal, predator, woodland, open forest; nest arboreal.

Podomyrma densestrigosa Viehmeyer, 1924

Podomyrma densestrigosa densestrigosa Viehmeyer, 1924

Podomyrma densestrigosa Viehmeyer, H. (1924). Formiciden der australischen Faunenregion. *Entomol. Mitt.* **13**: 310–319 [316]. Type data: syntypes, ZMB *W, from Liverpool, N.S.W.

Distribution: SE coastal, N.S.W. Ecology: terrestrial, arboreal, noctidiurnal, predator, woodland, open forest; nest arboreal.

Podomyrma densestrigosa teres Viehmeyer, 1924

Podomyrma densestrigosa teres Viehmeyer, H. (1924). Formiciden der australischen Faunenregion. *Entomol. Mitt.* **13**: 310–319 [317]. Type data: syntypes, ZMB *W, from Liverpool and Trial Bay, N.S.W.

Distribution: SE coastal, N.S.W. Ecology: terrestrial, arboreal, noctidiurnal, predator, woodland, open forest; nest arboreal.

Podomyrma elongata Forel, 1895

Podomyrma elongata Forel, A. (1895). Nouvelles fourmis d'Australie, récoltée à The Ridge, Mackay, Queensland par M. Gilbert Turner. *Ann. Soc. Entomol. Belg.* **39**: 417–428 [428]. Type data: syntypes, GMNH W, ANIC W, from Mackay, Qld.

Podomyrma parva Crawley, W.C. (1925). New ants from Australia. II. *Ann. Mag. Nat. Hist. (9)* **16**: 577–598 [592]. Type data: syntypes (probable), OUM *W, from W.A.

Synonymy that of Brown, W.L. jr. (1953). Notes on Australian *Podomyrma* (Hymenoptera : Formicidae). *N. Qd. Nat.* **21**: 3.

Distribution: NE coastal, SW coastal, W.A., Qld. Ecology: terrestrial, arboreal, noctidiurnal, predator, woodland, open forest; nest arboreal.

Podomyrma femorata F. Smith, 1859

Podomyrma femorata Smith, F. (1859). Catalogue of hymenopterous insects collected by Mr A.R. Wallace at the islands of Aru and Key. *J. Linn. Soc. Zool.* **3**: 132–178 [1 Feb. 1859] [145]. Publication date established from Donisthorpe, H. (1932). On the identity of Smith's types of Formicidae (Hymenoptera) collected by Alfred Russell Wallace in the Malay Archipelago, with descriptions of two new species. *Ann. Mag. Nat. Hist. (10)* **10**: 441–476. Type data: syntypes, BMNH *W,F, from Aru Ils., Indonesia.

Distribution: N coastal, N Gulf, NE coastal, W.A., N.T., Qld.; also in New Guinea. Ecology: terrestrial, arboreal, noctidiurnal, predator, woodland, open forest; nest arboreal.

Podomyrma ferruginea (Clark, 1934)

Dacryon ferruginea Clark, J. (1934). New Australian ants. *Mem. Natl. Mus. Vict.* **8**: 21–47 [37 pls 2–3]. Type data: syntypes, NMV *W, from Bombala, N.S.W. and Canberra, A.C.T.

Distribution: SE coastal, Murray-Darling basin, N.S.W. Ecology: terrestrial, arboreal, noctidiurnal, predator, woodland, open forest; nest arboreal.

Podomyrma formosa (F. Smith, 1858)

Myrmica formosa Smith, F. (1858). *Catalogue of hymenopterous insects in the collection of the British Museum. Part 6. Formicidae.* London : British Museum 216 pp. 14 pls [27 Mar. 1858] [128]. Publication date established from Donisthorpe, H. (1932). On the identity of Smith's types of Formicidae (Hymenoptera) collected by Alfred Russell Wallace in the Malay Archipelago, with descriptions of two new species. *Ann. Mag. Nat. Hist. (10)* **10**: 441–476. Type data: syntypes (probable), BMNH *W, from Adelaide, S.A.

Distribution: S Gulfs, S.A. Ecology: terrestrial, arboreal, noctidiurnal, predator, woodland, open forest; nest arboreal.

Podomyrma fortirugis Viehmeyer, 1924

Podomyrma fortirugis Viehmeyer, H. (1924). Formiciden der australischen Faunenregion. *Entomol. Mitt.* **13**: 310–319 [315]. Type data: syntypes, ZMB *W,F,M, from Trial Bay, N.S.W.

Distribution: SE coastal, N.S.W. Ecology: terrestrial, arboreal, noctidiurnal, predator, woodland, open forest; nest arboreal.

Podomyrma gracilis Emery, 1887

Podomyrma gracilis Emery, C. (1887). Cataloge delle Formiche esistenti nelle collezioni del Museo Civico di Genova, Parte terza. Formiche della regione Indo-Malese e dell'Australia. *Ann. Mus. Civ. Stor. Nat. Giacomo Doria (2)* **5**: 427–473 [460]. Type data: status unknown, ?MCG, from Ramoi, New Guinea.

Podomyrma gracilis nugenti Forel, 1901

Podomyrma gracilis nugenti Forel, A. (1901). Formiciden des Naturhistorischen Museums zu Hamburg. Neue *Calyptomyrmex-, Dacryon-, Podomyrma-,* und *Echinopla*-Arten. *Mitt. Naturh. Mus. Hamb.* **18**: 45–82 [54]. Type data: syntypes, GMNH W, ANIC W, from Cairns, Qld.

Distribution: NE coastal, Qld. Ecology: terrestrial, arboreal, noctidiurnal, predator, woodland, open forest; nest arboreal.

Podomyrma gratiosa (F. Smith, 1858)

Myrmecina gratiosa Smith, F. (1858). *Catalogue of hymenopterous insects in the collection of the British Museum. Part 6. Formicidae.* London : British Museum 216 pp. 14 pls [27 Mar. 1858] [133]. Publication date established from Donisthorpe, H. (1932). On the identity of Smith's types of Formicidae (Hymenoptera) collected by Alfred Russell Wallace in the Malay Archipelago, with descriptions of two new species. *Ann. Mag. Nat. Hist. (10)* **10**: 441–476. Type data: syntypes, BMNH *W,F, from Adelaide, S.A.

Distribution: S Gulfs, S.A. Ecology: terrestrial, arboreal, noctidiurnal, predator, woodland, open forest; nest arboreal.

Podomyrma grossestriata Forel, 1915

Podomyrma elongata grossestriata Forel, A. (1915). Results of Dr. E. Mjöbergs Swedish Scientific Expeditions to Australia 1910–1913. 2. Ameisen. *Ark. Zool.* **9**: 1–119 pls 1–3 [4 Dec. 1915] [50]. Type data: holotype, probably GMNH or SMNH, from Malanda, Qld.

Distribution: NE coastal, Qld. Ecology: terrestrial, arboreal, noctidiurnal, predator, woodland, open forest; nest arboreal. Biological references: Brown, W.L. jr. (1953). Notes on Australian *Podomyrma* (Hymenoptera : Formicidae). *N. Qd. Nat.* **21**: 3 (raised to species).

Podomyrma inermis Mayr, 1876

Podomyrma inermis Mayr, G.L. (1876). Die australischen Formiciden. *J. Mus. Godeffroy* **5**: 56–115 [111]. Type data: syntypes (probable), whereabouts unknown, from Peak Downs, Qld.

Distribution: NE coastal, Qld. Ecology: terrestrial, arboreal, noctidiurnal, predator, woodland, open forest; nest arboreal.

Podomyrma kitschneri (Forel, 1915)

Dacryon kitschneri Forel, A. (1915). Results of Dr. E. Mjöbergs Swedish Scientific Expeditions to Australia 1910–1913. 2. Ameisen. *Ark. Zool.* **9**: 1–119 pls 1–3 [4 Dec. 1915] [52]. Type data: syntypes, GMNH W, other syntypes may exist, from Cedar Creek, Qld.

Distribution: NE coastal, Qld. Ecology: terrestrial, arboreal, noctidiurnal, predator, woodland, open forest; nest arboreal.

Podomyrma kraepelini Forel, 1901

Podomyrma kraepelini Forel, A. (1901). Formiciden des Naturhistorischen Museums zu Hamburg. Neue *Calyptomyrmex-, Dacryon-, Podomyrma-,* und *Echinopla*-Arten. *Mitt. Naturh. Mus. Hamb.* **18**: 45–82 [59]. Type data: holotype (probable), probably destroyed in ZMH in W.W. II, from Australia.

Distribution: (NE coastal), (Qld.). Ecology: terrestrial, arboreal, noctidiurnal, predator, woodland, open forest; nest arboreal.

Podomyrma laevissima F. Smith, 1863

Podomyrma laevissima Smith, F. (1863). Catalogue of hymenopterous insects collected by Mr A.R. Wallace in the islands of Mysol, Ceram, Waigiou, Bouru and Timor. *J. Linn. Soc. Zool.* **7**: 6–48 [4 Mar. 1863] [20]. Publication date established from Donisthorpe, H. (1932). On the identity of Smith's types of Formicidae (Hymenoptera) collected by Alfred Russell Wallace in the Malay Archipelago, with descriptions of two new species. *Ann. Mag. Nat. Hist. (10)* **10**: 441–476. Type data: syntypes (probable), BMNH *W, from Mysol, Indonesia.

Distribution: N coastal, N Gulf, NE coastal, N.T., Qld.; also in Papua New Guinea. Ecology: terrestrial, arboreal, noctidiurnal, predator, woodland, open forest; nest arboreal.

Podomyrma lampros Viehmeyer, 1924

Podomyrma lampros Viehmeyer, H. (1924). Formiciden der australischen Faunenregion. *Entomol. Mitt.* **13**: 310–319 [317]. Type data: syntypes, ZMB *W, from Trial Bay, N.S.W.

Distribution: SE coastal, N.S.W. Ecology: terrestrial, arboreal, noctidiurnal, predator, woodland, open forest; nest arboreal.

Podomyrma libra (Forel, 1907)

Dacryon liber Forel, A. (1907). Formicidae. pp. 263–310 in Michaelsen, W. & Hartmeyer, R. (eds.) *Die Fauna Südwest-Australiens.* Jena : G. Fischer Vol. 1 [275]. Type data: holotype, probably destroyed in ZMH in WW II, from Eradu, W.A.

Distribution: NW coastal, W.A. Ecology: terrestrial, arboreal, noctidiurnal, predator, woodland, open forest; nest arboreal.

Podomyrma macrophthalma Viehmeyer, 1925

Podomyrma macrophthalma Viehmeyer, H. (1925). Formiciden der australischen Faunenregion. *Entomol. Mitt.* **14**: 25–39 [25]. Type data: holotype, ZMB *W, from Trial Bay, N.S.W.

Distribution: SE coastal, N.S.W. Ecology: terrestrial, arboreal, noctidiurnal, predator, woodland, open forest; nest arboreal.

Podomyrma marginata (McAreavey, 1949)

Dacryon marginatus McAreavey, J.J. (1949). Australian Formicidae. New genera and species. *Proc. Linn. Soc. N.S.W.* **74**: 1–25 [15 June 1949] [8]. Type data: holotype, ANIC W, from Nyngan, N.S.W.

Distribution: Murray-Darling basin, N.S.W. Ecology: terrestrial, arboreal, noctidiurnal, predator, woodland, open forest; nest arboreal.

Podomyrma micans Mayr, 1876

Podomyrma micans micans Mayr, 1876

Podomyrma micans Mayr, G.L. (1876). Die australischen Formiciden. *J. Mus. Godeffroy* **5**: 56–115 [111]. Type data: syntypes, NHMW W, from Rockhampton, Qld.

Distribution: NE coastal, Qld. Ecology: terrestrial, arboreal, noctidiurnal, predator, woodland, open forest; nest arboreal.

Podomyrma micans maculiventris Emery, 1887

Podomyrma micans maculiventris Emery, C. (1887). Catalogo delle formiche esistenti nelle collezioni del Museo Civico di Genova. Parte terza. Formiche della regione Indo-Malese e dell'Australia. *Ann. Mus. Civ. Stor. Nat. Giacomo Doria (2)* **5**: 427–473 pls 1–2 [459]. Type data: syntypes, MCG *W, from Somerset, Qld.

Distribution: NE coastal, Qld. Ecology: terrestrial, arboreal, noctidiurnal, predator, woodland, open forest; nest arboreal.

Podomyrma mjobergi (Forel, 1915)

Dacryon mjobergi Forel, A. (1915). Results of Dr. E. Mjöbergs Swedish Scientific Expeditions to Australia 1910–1913. 2. Ameisen. *Ark. Zool.* **9**: 1–119 pls 1–3 [4 Dec. 1915] [51]. Type data: syntypes, GMNH W, other syntypes may exist, from Cedar Creek and Mt. Bellenden Ker, Qld.

Distribution: NE coastal, Qld. Ecology: terrestrial, arboreal, noctidiurnal, predator, woodland, open forest; nest arboreal.

Podomyrma muckeli Forel, 1910

Podomyrma muckeli Forel, A. (1910). Formicides australiens reçus de MM. Froggatt et Rowland Turner. *Rev. Suisse Zool.* **18**: 1–94 [25]. Type data: holotype (probable), GMNH (probable) W, from Kuranda near Cairns, Qld.

Distribution: NE coastal, Qld. Ecology: terrestrial, arboreal, noctidiurnal, predator, woodland, open forest; nest arboreal.

Podomyrma nitida (Clark, 1938)

Dacryon nitida Clark, J. (1938). Reports of the McCoy Society for Field Investigation and Research. No. 2. Sir Joseph Bank Islands. Part I. Formicidae (Hymenoptera). *Proc. R. Soc. Vict.* **50**: 356–382 [364]. Type data: syntypes, NMV *W,F,M, from Reevesby Is., S.A.

Distribution: S Gulfs, S.A. Ecology: terrestrial, arboreal, noctidiurnal, predator, woodland, open forest; nest arboreal.

Podomyrma novemdentata Forel, 1901

Podomyrma novemdentata Forel, A. (1901). Formiciden des Naturhistorischen Museums zu Hamburg. Neue *Calyptomyrmex-*, *Dacryon-*, *Podomyrma-*, und

Echinopla-Arten. *Mitt. Naturh. Mus. Hamb.* **18**: 45–82 [55]. Type data: syntypes, GMNH W,F, from Mackay, Qld.

Distribution: NE coastal, Qld. Ecology: terrestrial, arboreal, noctidiurnal, predator, woodland, open forest; nest arboreal.

Podomyrma nuda Crawley, 1922

Podomyrma nuda Crawley, W.C. (1922). New ants from Australia. *Ann. Mag. Nat. Hist. (9)* **9**: 427–448 [441]. Type data: holotype, OUM *W, from Murray River, W.A.

Distribution: SW coastal, W.A. Ecology: terrestrial, arboreal, noctidiurnal, predator, woodland, open forest; nest arboreal.

Podomyrma obscura Stitz, 1911

Podomyrma obscura Stitz, H. (1911). Australische Ameisen (Neu-Guinea und Salomons-Inseln, Festland, Neu-Seeland). *Sber. Ges. Naturf. Freunde Berl.* **1911**: 351–381 [362]. Type data: holotype, ZMB *W, from Newcastle, N.S.W.

Distribution: SE coastal, N.S.W. Ecology: terrestrial, arboreal, noctidiurnal, predator, woodland, open forest; nest arboreal.

Podomyrma octodentata Forel, 1901

Podomyrma octodentata Forel, A. (1901). Formiciden des Naturhistorischen Museums zu Hamburg. Neue *Calyptomyrmex*-, *Dacryon*-, *Podomyrma*-, und *Echinopla*-Arten. *Mitt. Naturh. Mus. Hamb.* **18**: 45–82 [54]. Type data: holotype (probable), GMNH W, from Mackay, Qld.

Distribution: NE coastal, Qld. Ecology: terrestrial, aboreal, noctidiurnal, predator, woodland, open forest; nest arboreal.

Podomyrma odae Forel, 1910

Podomyrma odae Forel, A. (1910). Formicides australiens reçus de MM. Froggatt et Rowland Turner. *Rev. Suisse Zool.* **18**: 1–94 [23]. Type data: syntypes, GMNH W, ANIC W, from Kuranda near Cairns, Qld.

Distribution: NE coastal, Qld. Ecology: terrestrial, arboreal, noctidiurnal, predator, woodland, open forest; nest arboreal.

Podomyrma omniparens (Forel, 1895)

Dacryon omniparens Forel, A. (1895). Nouvelles fourmis d'Australie, récoltée à The Ridge, Mackay, Queensland par M. Gilbert Turner. *Ann. Soc. Entomol. Belg.* **39**: 417–428 [421]. Type data: holotype, GMNH W, from Mackay, Qld.

Distribution: NE coastal, Qld. Ecology: terrestrial, arboreal, noctidiurnal, predator, woodland, open forest; nest arboreal.

Podomyrma overbecki Viehmeyer, 1924

Podomyrma overbecki overbecki Viehmeyer, 1924

Podomyrma overbecki Viehmeyer, H. (1924). Formiciden der australischen Faunenregion. *Entomol. Mitt.* **13**: 310–319 [318]. Type data: syntypes, ZMB *W, from Trial Bay, N.S.W.

Distribution: SE coastal, N.S.W. Ecology: terrestrial, arboreal, noctidiurnal, predator, woodland, open forest; nest arboreal.

Podomyrma overbecki varicolor Viehmeyer, 1925

Podomyrma overbecki varicolor Viehmeyer, H. (1925). Formiciden der australischen Faunenregion. *Entomol. Mitt.* **14**: 25–39 [25]. Type data: holotype, ZMB *W, from Liverpool, N.S.W.

Distribution: SE coastal, N.S.W. Ecology: terrestrial, arboreal, noctidiurnal, predator, woodland, open forest; nest arboreal.

Podomyrma rugosa (Clark, 1934)

Lordomyrma rugosa Clark, J. (1934). New Australian ants. *Mem. Natl. Mus. Vict.* **8**: 21–47 [38 pls 2–3]. Type data: syntypes, NMV *W,F, from Ferntree Gully, Vic.

Distribution: SE coastal, Vic. Ecology: terrestrial, arboreal, noctidiurnal, predator, woodland, open forest; nest arboreal.

Podomyrma striata F. Smith, 1859

Podomyrma striata Smith, F. (1859). Catalogue of hymenopterous insects collected by Mr A.R. Wallace at the islands of Aru and Key. *J. Linn. Soc. Zool.* **3**: 132–178 [1 Feb. 1859] [146]. Publication date established from Donisthorpe, H. (1932). On the identity of Smith's types of Formicidae (Hymenoptera) collected by Alfred Russell Wallace in the Malay Archipelago, with descriptions of two new species. *Ann. Mag. Nat. Hist. (10)* **10**: 441–476. Type data: syntypes (probable), BMNH *W, from Aru Ils., Indonesia.

Podomyrma castanea Stitz, H. (1911). Australische Ameisen (Neu-Guinea und Salomons-Inseln, Festland, Neu-Seeland). *Sber. Ges. Naturf. Freunde Berl.* **1911**: 351–381 [358]. Type data: syntypes, ZMB *W, from Cape York, Qld.

Synonymy that of Emery, C. (1922). Hymenoptera Fam. Formicidae subfam. Myrmicinae. *in* Wytsman, P. (ed.) *Genera Insectorum.* Fasc. 174C pp. 207–397 [238].

Distribution: NE coastal, Qld. Ecology: terrestrial, arboreal, noctidiurnal, predator, woodland, open forest; nest arboreal.

Podomyrma tricolor Clark, 1934

Podomyrma tricolor Clark, J. (1934). New Australian ants. *Mem. Natl. Mus. Vict.* **8**: 21–47 [36 pls 2–3]. Type data: syntypes, NMV *W, from Claudie River, Qld.

Distribution: NE coastal, Qld. Ecology: terrestrial, arboreal, noctidiurnal, predator, woodland, open forest; nest arboreal.

Podomyrma turneri (Forel, 1901)

Dacryon turneri Forel, A. (1901). Formiciden des Naturhistorischen Museums zu Hamburg. Neue *Calyptomyrmex*-, *Dacryon*-, *Podomyrma*-, und *Echinopla*-Arten. *Mitt. Naturh. Mus. Hamb.* **18**: 45–82 [60]. Type data: syntypes, GMNH W,F, ANIC W, from Mackay, Qld.

Distribution: NE coastal, Qld. Ecology: terrestrial, arboreal, noctidiurnal, predator, woodland, open forest; nest arboreal.

Pristomyrmex Mayr, 1866

Pristomyrmex Mayr, G.L. (1866). Diagnosen neuer und wenig gekannter Formiciden. *Verh. Zool.-Bot. Ges. Wien* **16**: Abhand. 885–908 [903 pl 20]. Type species *Pristomyrmex pungens* Mayr, 1866 by monotypy.

Odontomyrmex André, E. (1905). Description d'un genre nouveau et de deux espèces nouvelles de fourmis d'Australie. *Rev. Entomol.* **24**: 205–208 [207]. Type species *Odontomyrmex quadridentatus* E. André, 1905 by monotypy.

Synonymy that of Bolton, B. (1981). A revision of six minor genera of Myrmicinae (Hymenoptera : Formicidae) in the Ethiopian zoogeographical region. *Bull. Br. Mus. Nat. Hist. (Entomol.)* **43**: 245–307 [26 Nov. 1981] [282].

This group is also found in the Ethiopian, Malagasy and Oriental regions; New Guinea and east Melanesia in the Australian Region, see Brown, W.L. jr. (1973). A comparison of the Hylean and Congo-West African rain forest ant faunas. pp. 161–185 *in* Meggers, B.J., Ayensu, E.S. & Duckworth, W.D. (eds.) *Tropical forest ecosystems in Africa and South America: a comparative review.* Washington : Smithsonian Institution Press.

Pristomyrmex erythropygus Taylor, 1968

Pristomyrmex erythropygus Taylor, R.W. (1968). A supplement to the revision of Australian *Pristomyrmex* species (Hymenoptera : Formicidae). *J. Aust. Entomol. Soc.* **7**: 63–66 [30 June 1968] [65]. Type data: holotype, MCZ Type no. 31325 *W, from Acacia Plateau, near Old Koreelah, N.S.W.

Distribution: SE coastal, N.S.W. Ecology: terrestrial, nocturnal, predator, closed forest; nest in soil.

Pristomyrmex foveolatus Taylor, 1965

Pristomyrmex foveolatus Taylor, R.W. (1965). The Australian ants of the genus *Pristomyrmex*, with a case of apparent character displacement. *Psyche Camb.* **72**: 35–54 [26 June 1965] [38]. Type data: holotype, MCZ Type no. 31152 *W, from Clump Point W of Tully, Qld.

Distribution: NE coastal, Qld. Ecology: terrestrial, nocturnal, predator, closed forest; nest in ground layer.

Pristomyrmex quadridentatus (E. André, 1905)

Odontomyrmex quadridentatus André, E. (1905). Description d'un genre nouveau et de deux espèces nouvelles de fourmis d'Australie. *Rev. Entomol.* **24**: 205–208 [208]. Type data: lectotype, MNHP W, from Sydney, N.S.W., designation by Taylor, R.W. (1965). The Australian ants of the genus *Pristomyrmex*, with a case of apparent character displacement. *Psyche Camb.* **72**: 35–54.

Pristomyrmex (Odontomyrmex) quadridentatus queenslandensis Forel, A. (1915). Results of Dr. E. Mjöbergs Swedish Scientific Expeditions to Australia 1910–1913. 2. Ameisen. *Ark. Zool.* **9**: 1–119 pls 1–3 [4 Dec. 1915] [53]. Type data: syntypes, GMNH W, other syntypes may exist, from Mt. Tambourine (=Tamborine Mt.), Qld.

Synonymy that of Taylor, R.W. (1965). The Australian ants of the genus *Pristomyrmex*, with a case of apparent character displacement. *Psyche Camb.* **72**: 35–54 [42].

Distribution: NE coastal, SE coastal, Qld., N.S.W. Ecology: terrestrial, nocturnal, predator, closed forest; nest in ground layer.

Pristomyrmex thoracicus Taylor, 1965

Pristomyrmex thoracicus Taylor, R.W. (1965). The Australian ants of the genus *Pristomyrmex*, with a case of apparent character displacement. *Psyche Camb.* **72**: 35–54 [26 June 1965] [41]. Type data: holotype, MCZ Type no. 31153 *W, from Vision Falls, Lake Eacham Natl. Park, Qld.

Distribution: NE coastal, Qld. Ecology: terrestrial, nocturnal, predator, closed forest; nest in ground layer.

Pristomyrmex wheeleri Taylor, 1965

Pristomyrmex wheeleri Taylor, R.W. (1965). The Australian ants of the genus *Pristomyrmex*, with a case of apparent character displacement. *Psyche Camb.* **72**: 35–54 [26 June 1965] [48]. Type data: holotype, MCZ Type no. 31154 *W, from vicinity of Binna Burra, Qld.

Distribution: NE coastal, SE coastal, N.S.W., Qld. Ecology: terrestrial, nocturnal, predator, closed forest; nest in soil.

Pristomyrmex wilsoni Taylor, 1968

Pristomyrmex wilsoni Taylor, R.W. (1968). A supplement to the revision of Australian *Pristomyrmex* species (Hymenoptera : Formicidae). *J. Aust. Entomol. Soc.* **7**: 63–66 [30 June 1968] [63]. Type data: holotype, ANIC Type no. 7502 W, from Mt. Lewis near Julatten, Qld.

Distribution: NE coastal, Qld. Ecology: terrestrial, nocturnal, predator, closed forest; nest in ground layer.

Quadristruma Brown, 1949

Quadristruma Brown, W.L. jr. (1949). Revision of the ant tribe Dacetini: 3. *Epitritus* Emery and *Quadristruma* new genus (Hymenoptera : Formicidae). *Trans. Am. Entomol. Soc.* **75**: 43–51 [6 July 1949] [47]. Type species *Epitritus emmae* Emery, 1890 by original designation.

This group is also found in New Guinea, east Melanesia and parts of Polynesia.

Quadristruma emmae (Emery, 1890)

Epitritus emmae Emery, C. (1890). Studii sulle formiche della fauna neotropica. *Boll. Soc. Entomol. Ital.* **22**: 38–80 pls 5–9 [70]. Type data: holotype, probably MCG *W, from St. Thomas Is., Virgin Ils.

Distribution: NW coastal, N coastal, N Gulf, NE coastal, W.A., N.T., Qld.; also in Africa, SE Asia, New Guinea, Micronesia and Polynesia, doubtfully native to Australia. Ecology: terrestrial, noctidiurnal, predator, woodland, open forest, closed forest; nest in ground layer.

Rhopalomastix Forel, 1900

Rhopalomastix Forel, A. (1900). Un nouveau genre et une nouvelle espèce de Myrmicide. *Ann. Soc. Entomol. Belg.* **44**: 24–26 [24]. Type species *Rhopalomastix rothneyi* Forel, 1900 by monotypy.

This group is also found in the Oriental Region; New Guinea in the Australian Region, see Brown, W.L. jr. (1973). A comparison of the Hylean and Congo-West African rain forest ant faunas. pp. 161–185 *in* Meggers, B.J., Ayensu, E.S. & Duckworth, W.D. (eds.) *Tropical forest ecosystems in Africa and South America: a comparative review.* Washington : Smithsonian Institution Press.

Rhopalomastix rothneyi Forel, 1900

Rhopalomastix rothneyi Forel, A. (1900). Un nouveau genre et une nouvelle espèce de Myrmicide. *Ann. Soc. Entomol. Belg.* **44**: 24–26 [24]. Type data: holotype, probably GMNH *F, from Barrackpore, India.

Distribution: NE coastal, Qld.; also in SE Asia and New Guinea, probably native to N Australia. Ecology: terrestrial, noctidiurnal, predator, closed forest; nest arboreal.

Rhopalothrix Mayr, 1870

Rhopalothrix Mayr, G.L. (1870) Formicidae Novogranadenses. *Sber. Akad. Wiss. Wien* Abt. 1 **61**: 370–417 pl [415]. Type species *Rhopalothrix ciliata* Mayr, 1870 by subsequent designation, see Wheeler, W.M. (1911). A list of the type species of the genera and subgenera of Formicidae. *Ann. N.Y. Acad. Sci.* **21**: 157–175 [17 Oct. 1911].

This group is also found in the Neotropical Region; New Guinea in the Australian Region, see Brown, W.L. jr. (1973). A comparison of the Hylean and Congo-West African rain forest ant faunas. pp. 161–185 *in* Meggers, B.J., Ayensu, E.S. & Duckworth, W.D. (eds.) *Tropical forest ecosystems in Africa and South America: a comparative review.* Washington : Smithsonian Institution Press.

Rhopalothrix orbis Taylor, 1968

Rhopalothrix orbis Taylor, R.W. (1968). Notes on the Indo-Australian basicerotine ants (Hymenoptera : Formicidae). *Aust. J. Zool.* **16**: 333–348 [336]. Type data: holotype, ANIC Type no. 7503 W, from Tamborine Mt., north side near Curtis Falls, Qld.

Distribution: NE coastal, Qld. Ecology: terrestrial, noctidiurnal, predator, closed forest; nest in ground layer.

Rhoptromyrmex Mayr, 1901

Rhoptromyrmex Mayr, G.L. (1901). Südafrikanische Formiciden, gesammelt von Dr. Hans Brauns. *Ann. Natl. Mus. Wien* **16**: 1–30 [18 pls 1–2] [redefined in Bolton, B. (1976). The ant tribe Tetramoriini (Hymenoptera : Formicidae). Constituent genera, review of smaller genera and revision of *Triglyphothrix* Forel. *Bull. Br. Mus. Nat. Hist. (Entomol.)* **34**: 283–379 (28 Oct. 1976)]. Type species *Rhoptromyrmex globulinodis* Mayr, 1901 by subsequent designation, see Wheeler, W.M. (1911). A list of the type of species of the genera and subgenera of Formicidae. *Ann. N.Y. Acad. Sci.* **21**: 157–175 [17 Oct. 1911].

This group is also found in the Ethiopian and Oriental regions; New Guinea in the Australian Region, see Brown, W.L. jr. (1973). A comparison of the Hylean and Congo-West African rain forest ant faunas. pp. 161–185 *in* Meggers, B.J., Ayensu, E.S. & Duckworth, W.D. (eds.) *Tropical forest ecosystems in Africa and South America: a comparative review.* Washington : Smithsonian Institution Press.

Rhoptromyrmex melleus (Emery, 1897)

Tetramorium melleum Emery, C. (1897). Formicidarum species novae vel minus cognitae in collectione Musaei Nationalis Hungarici, quas in Nova-Guinea, Colonia Germanica, collegit L. Biró. *Természz. Füz.* **20**: 571–599 pls 14–15 [586]. Type data: holotype, HMN *W, from Beliao Is. near Friedrich-Wilhelmshafen (=Madang), New Guinea.

Distribution: NE coastal, Qld. Ecology: terrestrial, noctidiurnal, predator, closed forest; nest in ground layer.

Rhoptromyrmex wroughtonii Forel, 1902

Rhoptromyrmex wroughtonii Forel, A. (1902). Myrmicinae nouveaux de l'Inde et de Ceylan. *Rev. Suisse Zool.* **10**: 165–249 [231]. Type data: syntypes, GMNH *W,M, from Kanara, India.

Distribution: NE coastal, Qld. Ecology: terrestrial, noctidiurnal, predator, closed forest; nest in ground layer.

Solenopsis Westwood, 1841

Solenopsis Westwood, J.O. (1841). Observations on the genus *Typhlopone*, with descriptions of several exotic species of ants. *Ann. Mag. Nat. Hist. (1)* **6**: 81–89 [86 pl 2] [redefined in Ettershank, G. (1966). A generic revision of the world Myrmicinae related to *Solenopsis* and *Pheidologeton* (Hymenoptera : Formicidae). *Aust. J. Zool.* **14**: 73–171]. Type species *Atta geminata* Fabricius, 1804 (as *Solenopsis mandibularis* Westwood, 1841) by monotypy.

This group is also found in the Neotropical, Nearctic, Palearctic, Ethiopian and Oriental regions; widespread in the Australian Region except New Zealand, see Brown, W.L. jr. (1973). A comparison of the Hylean and Congo-West African rain forest ant faunas. pp. 161–185 *in* Meggers, B.J., Ayensu, E.S. & Duckworth, W.D. (eds.) *Tropical forest ecosystems in Africa and South America: a comparative review.* Washington : Smithsonian Institution Press.

Solenopsis belisaria Forel, 1907

Solenopsis belisarius Forel, A. (1907). Formicidae. pp. 263–310 *in* Michaelsen, W. & Hartmeyer, R. (eds.) *Die Fauna Südwest-Australiens.* Jena : G. Fischer Vol. 1 [278]. Type data: syntypes, GMNH W,M, from Northampton, W.A.

Distribution: NW coastal, W.A. Ecology: terrestrial, noctidiurnal, predator, desert, woodland; nest in soil.

Solenopsis clarki Crawley, 1922

Solenopsis clarki Crawley, W.C. (1922). New ants from Australia. *Ann. Mag. Nat. Hist. (9)* **10**: 16–36 [16]. Type data: syntypes, OUM *W, from Byford, W.A.

Distribution: SW coastal, W.A. Ecology: terrestrial, noctidiurnal, predator, woodland, open forest; nest in soil.

Solenopsis froggatti Forel, 1913

Solenopsis froggatti Forel, A. (1913). Fourmis de Tasmanie et d'Australie récoltées par MM. Lea, Froggatt etc. *Bull. Soc. Vaud. Sci. Nat.* **49**: 173–196 pl 2 [187]. Type data: syntypes, GMNH W, from Hobart, Tas.

Distribution: Tas., SE coastal, Vic. Ecology: terrestrial, noctidiurnal, predator, woodland, open forest; nest in soil.

Solenopsis fusciventris Clark, 1934

Solenopsis fusciventris Clark, J. (1934). Ants from the Otway Ranges. *Mem. Natl. Mus. Vict.* **8**: 48–73 [62 pl 4]. Type data: syntypes, NMV *W, from Gellibrand, Vic.

Distribution: SE coastal, Vic., N.S.W. Ecology: terrestrial, noctidiurnal, predator, open forest; nest in ground layer.

Solenopsis insculpta Clark, 1938

Solenopsis insculptus Clark, J. (1938). Reports of the McCoy Society for Field Investigation and Research. No. 2. Sir Joseph Bank Islands. Part I. Formicidae (Hymenoptera). *Proc. R. Soc. Vict.* **50**: 356–382 [370]. Type data: syntypes, NMV *W, from Reevesby Is., S.A.

Distribution: S Gulfs, S.A. Ecology: terrestrial, noctidiurnal, predator, woodland; nest in soil.

Strumigenys F. Smith, 1860

Strumigenys Smith, F. (1860). Descriptions of new genera and species of exotic hymenoptera. *J. Entomol.* **1**: 65–84 [72 pl 4] [redefined in Brown, W.L. jr. (1948). A preliminary generic revision of the higher Dacetini (Hymenoptera : Formicidae). *Trans. Am. Entomol. Soc.* **74**: 101–129 (27 July 1948)]. Type species *Strumigenys mandibularis* F. Smith, 1860 by monotypy.

This group is also found in the Neotropical, south Nearctic, Ethiopian, Malagasy and Oriental regions; widespread in the Australian Region, see Brown, W.L. jr. (1973). A comparison of the Hylean and Congo-West African rain forest ant faunas. pp. 161–185 *in* Meggers, B.J., Ayensu, E.S. & Duckworth, W.D. (eds.) *Tropical forest ecosystems in Africa and South America: a comparative review.* Washington : Smithsonian Institution Press.

Strumigenys emdeni Forel, 1915

Strumigenys emdeni Forel, A. (1915). Results of Dr. E. Mjöbergs Swedish Scientific Expeditions to Australia 1910–1913. 2. Ameisen. *Ark. Zool.* **9**: 1–119 pls 1–3 [4 Dec. 1915] [41]. Type data: syntypes, GMNH W, ANIC W, other syntypes may exist, from Atherton, Qld.

Distribution: NE coastal, Qld. Ecology: terrestrial, noctidiurnal, predator, open forest, closed forest; nest in ground layer.

Strumigenys ferocior Brown, 1973

Strumigenys ferocior Brown, W.L. jr. (1973). The Indo-Australian species of the ant genus *Strumigenys*: groups of *horvathi*, *mayri* and *wallacei*. *Pac. Insects Monogr.* **15**: 259–269 [20 July 1973] [266]. Type data: holotype, ANIC Type no. 7516 W, from Iron Range, Cape York Peninsula, Qld.

Distribution: NE coastal, Qld. Ecology: terrestrial, noctidiurnal, predator, closed forest; nest in ground layer.

Strumigenys friedae Forel, 1915

Strumigenys friedae Forel, A. (1915). Results of Dr. E. Mjöbergs Swedish Scientific Expeditions to Australia 1910–1913. 2. Ameisen. *Ark. Zool.* **9**: 1–119 pls 1–3 [4 Dec. 1915] [42]. Type data: syntypes, GMNH W, ANIC W, other syntypes may exist, from Malanda, Qld.

Distribution: NE coastal, Qld. Ecology: terrestrial, noctidiurnal, predator, closed forest; nest in ground layer.

Strumigenys godeffroyi Mayr, 1866

Strumigenys godeffroyi Mayr, G.L. (1866). Myrmecologische Beiträge. *Sber. Akad. Wiss. Wien* **53**(1): 484–517 [516]. Type data: syntypes, NHMW *W, from Upolu, Samoa.

Distribution: NE coastal, Qld.; also in SE Asia, Micronesia, Melanesia, and S Polynesia. Ecology: terrestrial, noctidiurnal, predator, closed forest; nest in ground layer.

Strumigenys guttulata Forel, 1902

Strumigenys guttulata Forel, A. (1902). Fourmis nouvelles d'Australie. *Rev. Suisse Zool.* **10**: 405–548 [458]. Type data: syntypes, GMNH W, ANIC W, from Mackay, Qld.

Distribution: NE coastal, Qld. Ecology: terrestrial, noctidiurnal, predator, closed forest; nest in ground layer.

Strumigenys mayri Emery, 1897

Strumigenys mayri Emery, C. (1897). Formicidarum species novae vel minus cognitae in collectione Musaei Nationalis Hungarici, quas in Nova-Guinea, Colonia Germanica, collegit L. Biró. *Termész. Füz.* **20**: 571–599 pls 14–15 [579]. Type data: syntypes, MCG *W,F, MNH *W,F, from Friedrich-Wilhelmshafen (=Madang), New Guinea, see Brown, W.L. jr. (1973). The Indo-Australian species of the ant genus *Strumigenys*: groups of *horvathi*, *mayri* and *wallacei*. *Pac. Insects Monogr.* **15**: 259–269 [20 July 1973] [264].

Distribution: NE coastal, Qld.; also in Micronesia and New Guinea. Ecology: terrestrial, noctidiurnal, predator, closed forest; nest in ground layer.

Strumigenys opaca Brown, 1954

Strumigenys opaca Brown, W.L. jr. (1954). The Indo-Australian species of the ant genus *Strumigenys* Fr. Smith: *S. wallaci* Emery and relatives. *Psyche Camb.* **60**: 85–89. [8 Jan. 1954] [86]. Type data: holotype, MCZ *W, from Lankelly Creek in the McIlwraith Range, a few mi E of Coen, Qld.

Distribution: NE coastal, Qld. Ecology: terrestrial, noctidiurnal, predator, closed forest; nest in ground layer.

Strumigenys perplexa (F. Smith, 1876)

Orectognathus perplexus Smith, F. (1876). Descriptions of three new species of Hymenoptera (Formicidae) from New Zealand. *Trans. R. Entomol. Soc. Lond.* **24**: 489–492 [491]. Type data: syntypes, BMNH *W,F, from Tairua, near Mercury Bay, N.Z.

Strumigenys leae Forel, A. (1913). Fourmis de Tasmanie et d'Australie récoltées par MM. Lea, Froggatt etc. *Bull. Soc. Vaud. Sci. Nat.* **49**: 173–196 pl 2 [182]. Type data: syntypes, GMNH W, from Tas.

Synonymy that of Brown, W.L. jr. (1958). A review of the ants of New Zealand (Hymenoptera). *Acta Hymen.* **1**: 1–50 [38].

Distribution: S Gulfs, Murray-Darling basin, SE coastal, N.S.W., Vic., S.A., Tas.; also in New Zealand (N. Is.). Ecology: terrestrial, noctidiurnal, predator, open forest, closed forest; nest in ground layer.

Strumigenys quinquedentata Crawley, 1923

Strumigenys quinquedentata Crawley, W.C. (1923). Myrmecological notes - new Australian Formicidae. *Entomol. Rec. J. Var.* **35**: 177–179 [177]. Type data: syntypes, OUM *W, from Manjimup, W.A.

Distribution: SW coastal, W.A. Ecology: terrestrial, noctidiurnal, predator, open forest, closed forest; nest in ground layer.

Strumigenys szalayi Emery, 1897

Strumigenys szalayi Emery, C. (1897). Formicidarum species novae vel minus cognitae in collectione Musaei Nationalis Hungarici, quas in Nova-Guinea, Colonia Germanica, collegit L. Biró. *Termész. Füz.* **20**: 571–599 pls 14–15 [578]. Type data: syntypes, probably MCG* or MNH*, from Seleo Is. near Berlinhafen (=Aitape), New Guinea, see Brown, W.L. jr. (1971). The Indo-Australian species of the ant genus *Strumigenys*: group of *szalayi* (Hymenoptera : Formicidae). pp. 73–86 *in, Entomological Essays to Commemorate the Retirement of Professor K. Yasumatsu*. Tokyo : Hokuryukan.

Strumigenys szalayi australis Forel, A. (1910). Formicides australiens reçus de MM. Froggatt et Rowland Turner. *Rev. Suisse Zool.* **18**: 1–94 [50]. Type data: syntypes, GMNH W,M, from Kuranda near Cairns, Qld.

Synonymy that of Brown, W.L. jr. (1971). The Indo-Australian species of the ant genus *Strumigenys*: group of *szalayi* (Hymenoptera : Formicidae). pp. 73–86 *in, Entomological Essays to Commemorate the Retirement of Professor K. Yasumatsu*. Tokyo : Hokuryukan [75].

Distribution: NE coastal, Qld.; also in Phillipines, Micronesia, E Melanesia, and S Polynesia. Ecology: terrestrial, noctidiurnal, predator, closed forest; nest in ground layer.

Strumigenys xenos Brown, 1955

Strumigenys xenos Brown, W.L. jr. (1955). The first social parasite in the ant tribe Dacetini. *Insectes Soc.* **2**: 181–186 [182]. Type data: holotype, MCZ *F, from lower slopes of the Warburton Range, just above Warburton, Vic.

Distribution: SE coastal, Vic., N.S.W.; also in New Zealand (N. Is.). Ecology: terrestrial, noctidiurnal, predator, open forest, closed forest; nest in ground layer, social parasite of other ants.

Tetramorium Mayr, 1855

Tetramorium Mayr, G.L. (1855). Formicina Austriaca. *Verh. Zool.-Bot. Ges. Wien* **5**: Abhand. 273–478 [423] [redefined in Bolton, B. (1976). The ant tribe Tetramoriini (Hymenoptera : Formicidae). Constituent genera, review of smaller genera and revision of *Triglyphothrix* Forel. *Bull. Br. Mus. Nat. Hist. (Entomol.)* **34**: 283-379 (28 Oct. 1976)]. Type species *Formica caespita* Linnaeus, 1758 by subsequent designation, see Girard, M. (1879). *Les Insectes*. Traité elementaire d'entomologie, etc. Paris 3 vols [1016]. Compiled from secondary source: Wheeler, W.M. (1913). Corrections and additions to a "list of the type species of the genera and subgenera of Formicidae". *Ann. N.Y. Acad. Sci.* **23**: 77-83 [29 May 1913].

This group is also found in the Palearctic, Ethiopian, Malagasy and Oriental regions; New Guinea, east Melanesia, New Caledonia and parts of Polynesia in the Australian Region, see Brown, W.L. jr. (1973). A comparison of the Hylean and Congo-West African rain forest ant faunas. pp. 161-185 *in* Meggers, B.J., Ayensu, E.S. & Duckworth, W.D. (eds.) *Tropical forest ecosystems in Africa and South America: a comparative review*. Washington : Smithsonian Institution Press.

Tetramorium andrynicum Bolton, 1977

Tetramorium andrynicum Bolton, B. (1977). The ant tribe Tetramoriini (Hymenoptera : Formicidae). The genus *Tetramorium* Mayr in the Oriental and Indo-Australian regions, and in Australia. *Bull. Br. Mus. Nat. Hist. (Entomol.)* **36**: 67-151 [29 Sept. 1977] [142]. Type data: holotype, MCZ *W, from west slope, Mt. Bartle Frere, Qld.

Distribution: NE coastal, Qld. Ecology: terrestrial, noctidiurnal, predator, closed forest; nest in ground layer.

Tetramorium australe Bolton, 1977

Tetramorium australe Bolton, B. (1977). The ant tribe Tetramoriini (Hymenoptera : Formicidae). The genus *Tetramorium* Mayr in the Oriental and Indo-Australian regions, and in Australia. *Bull. Br. Mus. Nat. Hist. (Entomol.)* **36**: 67-151 [29 Sept. 1977] [146]. Type data: holotype, MCZ *W, from Tozer Gap, Cape York, Qld.

Distribution: NE coastal, Qld. Ecology: terrestrial, noctidiurnal, predator, open forest; nest in ground layer.

Tetramorium bicarinatum (Nylander, 1846)

Myrmica bicarinata Nylander, W. (1846). Additamentum adnotationum in monographiam formicarum borealium Europae. *Acta Soc. Sci. Fenn.* **2**: 1041-1062 [1061]. Type data: syntypes, lost, from California, U.S.A. Compiled from secondary source: Bolton, B. (1977). The ant tribe Tetramoriini (Hymerioptera : Formicidae). The genus *Tetramorium* Mayr in the Oriental and Indo-Australian regions, and in Australia. *Bull. Br. Mus. Nat. Hist. (Entomol.)* **36**: 67-151 [29 Sept. 1977].

Distribution: SE coastal, NE coastal, SW coastal, N coastal, Qld., N.S.W., W.A., N.T.; introduced from overseas into many areas of eastern Qld., N.S.W., SW W.A. and N.T. Ecology: terrestrial, noctidiurnal, peridomestic, predator, desert, woodland, open forest, closed forest; nest in ground layer.

Tetramorium capitale (McAreavey, 1949)

Xiphomyrmex capitalis McAreavey, J.J. (1949). Australian Formicidae. New genera and species. *Proc. Linn. Soc. N.S.W.* **74**: 1-25 [15 June 1949] [6]. Type data: holotype, ANIC W, from Nyngan, N.S.W.

Distribution: Murray-Darling basin, N.S.W. Ecology: terrestrial, noctidiurnal, predator, desert, woodland; nest in ground layer.

Tetramorium confusum Bolton, 1977

Tetramorium confusum Bolton, B. (1977). The ant tribe Tetramoriini (Hymenoptera : Formicidae). The genus *Tetramorium* Mayr in the Oriental and Indo-Australian regions, and in Australia. *Bull. Br. Mus. Nat. Hist. (Entomol.)* **36**: 67-151 [29 Sept. 1977] [143]. Type data: holotype, CAS *W, from Thegib (=The Gib) near Bowral, N.S.W.

Distribution: SE coastal, N.S.W. Ecology: terrestrial, noctidiurnal, predator, woodland, closed forest; nest in ground layer.

Tetramorium deceptum Bolton, 1977

Tetramorium deceptum Bolton, B. (1977). The ant tribe Tetramoriini (Hymenoptera : Formicidae). The genus *Tetramorium* Mayr in the Oriental and Indo-Australian regions, and in Australia. *Bull. Br. Mus. Nat. Hist. (Entomol.)* **36**: 67-151 [29 Sept. 1977] [146]. Type data: holotype, MCZ *W, from Shipton's Flat, Qld.

Distribution: NE coastal, Qld. Ecology: terrestrial, noctidiurnal, predator, open forest, closed forest; nest in ground layer.

Tetramorium fuscipes (Viehmeyer, 1925)

Xiphomyrmex turneri fuscipes Viehmeyer, H. (1925). Formiciden der australischen Faunenregion. *Entomol. Mitt.* **14**: 25-39 [29]. Type data: syntypes, ZMB *W, from Liverpool, N.S.W.

Distribution: SE coastal, N.S.W. Ecology: terrestrial, noctidiurnal, predator, woodland, open forest; nest in ground layer. Biological references: Bolton, B. (1977). The ant tribe Tetramoriini (Hymenoptera : Formicidae). The genus *Tetramorium* Mayr in the Oriental and Indo-Australian regions, and in Australia. *Bull. Br. Mus. Nat. Hist. (Entomol.)* **36**: 67-151 (raised to species).

Tetramorium impressum (Viehmeyer, 1925)

Xiphomyrmex impressus Viehmeyer, H. (1925). Formiciden der australischen Faunenregion. *Entomol. Mitt.* **14**: 25-39 [30]. Type data: holotype, ZMB *W, from Liverpool, N.S.W.

Distribution: SE coastal, N.S.W. Ecology: terrestrial, noctidiurnal, predator, woodland, open forest; nest in ground layer.

Tetramorium laticephalum Bolton, 1977

Tetramorium laticephalum Bolton, B. (1977). The ant tribe Tetramoriini (Hymenoptera : Formicidae). The genus *Tetramorium* Mayr in the Oriental and Indo-Australian regions, and in Australia. *Bull. Br. Mus. Nat. Hist. (Entomol.)* **36**: 67-151 [29 Sept. 1977] [139]. Type data: holotype, MCZ *W, from Patho, Vic.

Distribution: Murray-Darling basin, Vic. Ecology: terrestrial, noctidiurnal, predator, desert, woodland; nest in ground layer.

Tetramorium megalops Bolton, 1977

Tetramorium megalops Bolton, B. (1977). The ant tribe Tetramoriini (Hymenoptera : Formicidae). The genus *Tetramorium* Mayr in the Oriental and Indo-Australian regions, and in Australia. *Bull. Br. Mus. Nat. Hist. (Entomol.)* **36**: 67-151 [29 Sept. 1977] [139]. Type data: holotype, MCZ *W, from about 60 km NW of Balladonia, W.A.

Distribution: W plateau, W.A. Ecology: terrestrial, noctidiurnal, predator, woodland, closed forest; nest in ground layer.

Tetramorium ornatum Emery, 1897

Tetramorium ornatum Emery, C. (1897). Formicidarum species novae vel minus cognitae in collectione Musaei Nationalis Hungarici, quas in Nova-Guinea, Colonia Germanica, collegit L. Biró. *Termész. Füz.* **20**: 571-599 pls 14-15 [585]. Type data: syntypes, GMNH *W, from Berlinhafen (=Aitape), New Guinea.

Distribution: NE coastal, Qld. Ecology: terrestrial, noctidiurnal, predator, closed forest; nest in ground layer or arboreal.

Tetramorium pacificum Mayr, 1870

Tetramorium pacificum Mayr, G.L. (1870). Neue Formiciden. *Verh. Zool.-Bot. Ges. Wien* **20**: Abhand. 939-996 [31 Dec. 1870] [972,976]. Type data: syntypes, NHMW *W,F, from Tongatabu, Tonga.

Distribution: N coastal, NE coastal, SE coastal, N.T., Qld., N.S.W., Lord Howe Is. Ecology: terrestrial, noctidiurnal, predator, closed forest; nest in ground layer.

Tetramorium simillimum (F. Smith, 1851)

Myrmica simillima Smith, F. (1851). *List of the specimens of British animals in the collection of the British Museum.* Hymenoptera Aculeata. London : British Museum Vol. 6 [118]. Type data: syntypes, lost, presumed destroyed, from Dorset, England. Compiled from secondary source: Bolton, B. (1977). The ant tribe Tetramoriini (Hymenoptera : Formicidae). The genus *Tetramorium* Mayr in the Oriental and Indo-Australian regions, and in Australia. *Bull. Br. Mus. Nat. Hist. (Entomol.)* **36**: 67-151 [29 Sept. 1977].

Tetramorium antipodum Wheeler, W.M. (1927). The ants of Lord Howe Island and Norfolk Island. *Proc. Am. Acad. Arts Sci.* **62**: 121-153 [143]. Type data: syntypes, whereabouts unknown, from Norfolk Is.

Synonymy that of Bolton, B. (1977). The ant tribe Tetramoriini (Hymenoptera : Formicidae). The genus *Tetramorium* Mayr in the Oriental and Indo-Australian regions, and in Australia. *Bull. Br. Mus. Nat. Hist. (Entomol.)* **36**: 67-151 [131].

Distribution: N coastal, NE coastal, SE coastal, N.T., Qld., N.S.W., Lord Howe Is. Ecology: terrestrial, noctidiurnal, predator, open forest, closed forest; nest in ground layer.

Tetramorium sjostedti Forel, 1915

Tetramorium (Xiphomyrmex) sjostedti Forel, A. (1915). Results of Dr. E. Mjöbergs Swedish Scientific Expeditions to Australia 1910-1913. 2. Ameisen. *Ark. Zool.* **9**: 1-119 pls 1-3 [4 Dec. 1915] [48]. Type data: lectotype, SMNH *W, from Kimberley distr., W.A., designation by Bolton, B. (1977). The ant tribe Tetramoriini (Hymenoptera : Formicidae). The genus *Tetramorium* Mayr in the Oriental and Indo-Australian regions, and in Australia. *Bull. Br. Mus. Nat. Hist. (Entomol.)* **36**: 67-151 [140].

Distribution: N coastal, W.A. Ecology: terrestrial, noctidiurnal, predator, desert, woodland; nest in ground layer.

Tetramorium spininode Bolton, 1977

Tetramorium spininode Bolton, B. (1977). The ant tribe Tetramoriini (Hymenoptera : Formicidae). The genus *Tetramorium* Mayr in the Oriental and Indo-Australian regions, and in Australia. *Bull. Br. Mus. Nat. Hist. (Entomol.)* **36**: 67-151 [29 Sept. 1977] [140]. Type data: holotype, CAS *W, from Winjana Gorge, W.A.

Distribution: N coastal, W.A. Ecology: terrestrial, noctidiurnal, predator, desert, woodland, closed forest; nest in ground layer.

Tetramorium splendidior (Viehmeyer, 1925)

Xiphomyrmex striolatus splendidior Viehmeyer, H. (1925). Formiciden der australischen Faunenregion. *Entomol. Mitt.* **14**: 25-39 [29]. Type data: holotype, ZMB *W, from Liverpool, N.S.W.

Distribution: SE coastal, N.S.W. Ecology: terrestrial, noctidiurnal, predator, woodland, open forest; nest in ground layer. Biological references: Bolton, B. (1977). The ant tribe Tetramoriini (Hymenoptera : Formicidae). The genus *Tetramorium* Mayr in the Oriental and

Indo-Australian regions, and in Australia. *Bull. Br. Mus. Nat. Hist. (Entomol.)* **36**: 67–151 (raised to species).

Tetramorium strictum Bolton, 1977

Tetramorium strictum Bolton, B. (1977). The ant tribe Tetramoriini (Hymenoptera : Formicidae). The genus *Tetramorium* Mayr in the Oriental and Indo-Australian regions, and in Australia. *Bull. Br. Mus. Nat. Hist. (Entomol.)* **36**: 67–151 [29 Sept. 1977] [144]. Type data: holotype, MCZ *W, from Mt. Alexander (=Alexandra), NW of Daintree, Qld.

Distribution: NE coastal, Qld. Ecology: terrestrial, noctidiurnal, predator, closed forest; nest in ground layer.

Tetramorium striolatum Viehmeyer, 1913

Tetramorium (Xiphomyrmex) viehmeyeri striolatus Viehmeyer, H. (1913). Neue und unvollständig bekannte Ameisen der Alten Welt. *Arch. Naturg.* **79A**(12): 24–60 [39]. Type data: syntypes, ZMB *W, from Killalpaninna, S.A.

Distribution: Lake Eyre basin, S.A. Ecology: terrestrial, noctidiurnal, predator, desert, woodland; nest in ground layer or arboreal. Biological references: Bolton, B. (1977). The ant tribe Tetramoriini (Hymenoptera : Formicidae). The genus *Tetramorium* Mayr in the Oriental and Indo-Australian regions, and in Australia. *Bull. Br. Mus. Nat. Hist. (Entomol.)* **36**: 67–151 (raised to species).

Tetramorium thalidum Bolton, 1977

Tetramorium thalidum Bolton, B. (1977). The ant tribe Tetramoriini (Hymenoptera : Formicidae). The genus *Tetramorium* Mayr in the Oriental and Indo-Australian regions, and in Australia. *Bull. Br. Mus. Nat. Hist. (Entomol.)* **36**: 67–151 [29 Sept. 1977] [141]. Type data: holotype, MCZ *W, from Kuranda-Mareeba Rd., Davies Creek, Qld.

Distribution: NE coastal, Qld. Ecology: terrestrial, noctidiurnal, predator; nest in ground layer.

Tetramorium turneri Forel, 1902

Tetramorium (Xiphomyrmex) turneri Forel, A. (1902). Fourmis nouvelles d'Australie. *Rev. Suisse Zool.* **10**: 405–548 [447]. Type data: syntypes, GMNH W,F, ANIC W, from Mackay, Qld.

Distribution: NE coastal, SE coastal, N.S.W., Qld. Ecology: terrestrial, noctidiurnal, predator, open forest, closed forest; nest in ground layer or arboreal.

Tetramorium validiusculum Emery, 1897

Tetramorium pacificum validiusculum Emery, C. (1897). Formicidarum species novae vel minus cognitae in collectione Musaei Nationalis Hungarici, quas in Nova-Guinea, Colonia Germanica, collegit L. Biró.

Termész. Füz. **20**: 571–599 pls 14–15 [585]. Type data: syntypes, GMNH *W, from near Berlinhafen (=Aitape), New Guinea.

Distribution: NE coastal, Qld. Ecology: terrestrial, noctidiurnal, predator, closed forest; nest in ground layer. Biological references: Bolton, B. (1977). The ant tribe Tetramoriini (Hymenoptera : Formicidae). The genus *Tetramorium* Mayr in the Oriental and Indo-Australian regions, and in Australia. *Bull. Br. Mus. Nat. Hist. (Entomol.)* **36**: 67–151 (raised to species).

Tetramorium viehmeyeri Forel, 1907

Tetramorium (Xiphomyrmex) viehmeyeri Forel, A. (1907). Formicidae. pp. 263–310 *in* Michaelsen, W. & Hartmeyer, R. (eds.) *Die Fauna Südwest-Australiens*. Jena : G. Fischer Vol. 1 [275]. Type data: holotype, probably destroyed in ZMH in WW II, from Day Dawn, W.A.

Xiphomyrmex viehmeyeri venustus Wheeler, W.M. (1934). Contributions to the fauna of Rottnest Island, Western Australia No. IX. The ants. *J. R. Soc. West. Aust.* **20**: 137–163 [5 Oct. 1934] [147]. Type data: holotype lost, paratypes MCZ, from near Government House, Rottnest Is., W.A.

Synonymy that of Bolton, B. (1977). The ant tribe Tetramoriini (Hymenoptera : Formicidae). The genus *Tetramorium* Mayr in the Oriental and Indo-Australian regions, and in Australia. *Bull. Br. Mus. Nat. Hist. (Entomol.)* **36**: 67–151 [142].

Distribution: NW coastal, SW coastal, W.A. Ecology: terrestrial, noctidiurnal, predator, woodland, open forest; nest in ground layer.

Triglyphothrix Forel, 1890

Triglyphothrix Forel, A. (1890). Aenictus-Typhlatta découverte de M. Wroughton. Nouveaux genres de Formicides. *Ann. Soc. Entomol. Belg.* **34**: Bull. Compt.-Rend. Sci. 102–114 [106]. Type species *Triglyphothrix walshi* Forel, 1890 by monotypy.

This group is also found in the Ethiopian and Oriental regions; New Guinea, east Melanesia and parts of Polynesia in the Australian Region, see Brown, W.L. jr. (1973). A comparison of the Hylean and Congo-West African rain forest ant faunas. pp. 161–185 *in* Meggers, B.J., Ayensu, E.S. & Duckworth, W.D. (eds.) *Tropical forest ecosystems in Africa and South America: a comparative review*. Washington : Smithsonian Institution Press.

Triglyphothrix lanuginosa (Mayr, 1870)

Tetramorium lanuginosum Mayr, G.L. (1870). Neue Formiciden. *Verh. Zool.-Bot. Ges. Wien* **20**: Abhand., 939–996 [31 Dec. 1870] [972,976]. Type data: holotype, NHMW *W, from Batavia (=Djakarta), Java.

Triglyphothrix (Xiphomyrmex) striatidens australis Forel, A. (1902). Fourmis nouvelles d'Australie. *Rev. Suisse Zool.* **10**: 405-548 [449]. Type data: syntypes, GMNH W,F, ANIC W, from Mackay, Qld.

Synonymy that of Bolton, B. (1976). The ant tribe Tetramoriini (Hymenoptera : Formicidae). Constituent genera, review of small genera and revision of *Triglyphothrix* Forel. *Bull. Br. Mus. Nat. Hist. (Entomol.)* **34**: 283-379 [28 Oct. 1976] [350].

Distribution: N coastal, NE coastal, Qld., N.T. Ecology: terrestrial, noctidiurnal, predator, open forest, closed forest; nest in ground layer.

Vollenhovia Mayr, 1868

Vollenhovia Mayr, G.L. (1868). Formicidae. *in, Reise der österreichischen Fregatte Novara um die Erde in der Jahren 1857, 1858, 1859.* Zool. 2, Abth. IA3: 1-123 4 pls [21] [redefined in Ettershank, G. (1966). A generic revision of the world Myrmicinae related to *Solenopsis* and *Pheidologeton* (Hymenoptera : Formicidae). *Aust. J. Zool.* **14**: 73-171]. Type species *Vollenhovia punctatostriata* Mayr, 1868 by monotypy.

This group is also found in the Oriental Region; New Guinea, east Melanesia, New Caledonia and southwest Polynesia in the Australian Region, see Brown, W.L. jr. (1973). A comparison of the Hylean and Congo-West African rain forest ant faunas. pp. 161-185 *in* Meggers, B.J., Ayensu, E.S. & Duckworth, W.D. (eds.) *Tropical forest ecosystems in Africa and South America: a comparative review.* Washington : Smithsonian Institution Press; undescribed species are present in the Iron Range, Qld.

Vollenhovia oblonga (F. Smith, 1860)

Myrmica oblonga Smith, F. (1860). Catalogue of hymenopterous insects collected by Mr A.R. Wallace in the islands of Bachian, Kaisaa, Amboyna, Gilolo, and at Dory in New Guinea. *J. Linn. Soc. Zool.* **5**: 93-143 pl 1 [18 July 1860] [107]. Publication date established from Donisthorpe, H. (1932). On the identity of Smith's types of Formicidae (Hymenoptera) collected by Alfred Russell Wallace in the Malay Archipelago, with descriptions of two new species. *Ann. Mag. Nat. Hist. (10)* **10**: 441-476. Type data: syntypes (probable), BMNH *W, from Bachian, Indonesia.

Distribution: NE coastal, Qld. Ecology: terrestrial, noctidiurnal, predator, closed forest; nest in ground layer.

DOLICHODERINAE

Bothriomyrmex Emery, 1869

Bothriomyrmex Emery, C. (1869). Descrizione di una nuova Formica Italiana. *Annuar. R. Mus. Zool. R. Univ. Napoli* **5**: 117-118 [117]. Type species *Tapinoma meridionale* Roger, 1863 (as *Bothriomyrmex costae* Emery, 1869) by monotypy.

This group is also found in the south Palearctic and Oriental regions; New Guinea and east Melanesia in the Australian Region, see Brown, W.L. jr. (1973). A comparison of the Hylean and Congo-West African rain forest ant faunas. pp. 161-185 *in* Meggers, B.J., Ayensu, E.S. & Duckworth, W.D. (eds.) *Tropical forest ecosystems in Africa and South America: a comparative review.* Washington : Smithsonian Institution Press.

Bothriomyrmex flavus Crawley, 1922

Bothriomyrmex flavus Crawley, W.C. (1922). New ants from Australia. *Ann. Mag. Nat. Hist. (9)* **10**: 16-36 [27]. Type data: syntypes, OUM *W,F,M, from Mundaring Weir, W.A.

Distribution: SW coastal, W.A. Ecology: terrestrial, nocturnal, omnivore, open forest, closed forest; nest in ground layer.

Bothriomyrmex pusillus (Mayr, 1876)

Bothriomyrmex pusillus pusillus (Mayr, 1876)

Tapinoma pusillum Mayr, G.L. (1876). Die australischen Formiciden. *J. Mus. Godeffroy* **5**: 56-115 [83]. Type data: syntypes, NHMW *W,F,M, from Rockhampton, Qld. and Sidney (=Sydney), N.S.W.

Distribution: NE coastal, SE coastal, Qld., N.S.W. Ecology: terrestrial, nocturnal, omnivore, open forest, closed forest; nest in ground layer.

Bothriomyrmex pusillus aequalis Forel, 1902

Bothriomyrmex pusillus aequalis Forel, A. (1902). Fourmis nouvelles d'Australie. *Rev. Suisse Zool.* **10**: 405-548 [476]. Type data: syntypes, GMNH W,F,M, from Bendigo, Vic.

Distribution: SE coastal, Vic. Ecology: terrestrial, nocturnal, omnivore, open forest, closed forest; nest in ground layer.

Bothriomyrmex scissor Crawley, 1922

Bothriomyrmex scissor Crawley, W.C. (1922). New ants from Australia. *Ann. Mag. Nat. Hist. (9)* **10**: 16-36 [29]. Type data: syntypes, OUM *F, from Murray River, W.A.

Distribution: SW coastal, W.A. Ecology: terrestrial, nocturnal, omnivore, open forest, closed forest; nest in ground layer.

Bothriomyrmex wilsoni Clark, 1934

Bothriomyrmex wilsoni Clark, J. (1934). New Australian ants. *Mem. Natl. Mus. Vict.* **8**: 21-47 [39 pls 2-3]. Type data: syntypes, NMV *W, from Port Lincoln, S.A.

Distribution: S Gulfs, S.A. Ecology: terrestrial, nocturnal, omnivore, open forest; nest in ground layer.

Dolichoderus Lund, 1831

Dolichoderus Lund, M. (1831). Lettre sur les habitudes de quelques fourmis de Bresil, adressée à M. Audouin. *Ann. Sci. Nat.* **23**: 113-138 [130]. Type species *Formica attelaboides* Fabricius, 1775 by monotypy.

Acanthoclinea Wheeler, W.M. (1935). Myrmecological notes. *Psyche Camb.* **42**: 68-72 [69] [proposed with subgeneric rank in *Dolichoderus* Lund, 1831]. Type species *Dolichoderus doriae* Emery, 1887 by original designation.

Diceratoclinea Wheeler, W.M. (1935). Myrmecological notes. *Psyche Camb.* **42**: 68-72 [69] [proposed with subgeneric rank in *Dolichoderus* Lund, 1831]. Type species *Dolichoderus scabridus* Roger, 1862 by original designation.

Synonymy that of Brown, W.L. jr. (1973). A comparison of the Hylean and Congo-West African rain forest ant faunas. pp. 161-185 *in* Meggers, B.J., Ayensu, E.S. & Duckworth, W.D. (eds.) *Tropical forest ecosystems in Africa and South America: a comparative review.* Washington : Smithsonian Institution Press [177].

This group is also found in the Neotropical, Nearctic, Palearctic and Oriental regions; New Guinea and east Melanesia in the Australian Region, see Brown, W.L. jr. (1973). A comparison of the Hylean and Congo-West African rain forest ant faunas. pp. 161-185 *in* Meggers, B.J., Ayensu, E.S. & Duckworth, W.D. (eds.) *Tropical forest ecosystems in Africa and South America: a comparative review.* Washington : Smithsonian Institution Press.

Dolichoderus angusticornis Clark, 1930

Dolichoderus (Hypoclinea) angusticornis Clark, J. (1930). The Australian ants of the genus *Dolichoderus* (Formicidae). Subgenus *Hypoclinea* Mayr. *Aust. Zool.* **6**: 252-268 [20 Aug. 1930] [260]. Type data: syntypes, NMV *W, from Burracoppin, W.A.

Distribution: SW coastal, W.A. Ecology: terrestrial, noctidiurnal, omnivore, open forest, closed forest; nest in ground layer.

Dolichoderus armstrongi McAreavey, 1949

Dolichoderus (Hypoclinea) armstrongi McAreavey, J.J. (1949). Australian Formicidae. New genera and species. *Proc. Linn. Soc. N.S.W.* **74**: 1-25 [15 June 1949] [17]. Type data: holotype, ANIC W, from Nyngan, N.S.W.

Distribution: Murray-Darling basin, N.S.W. Ecology: terrestrial, noctidiurnal, omnivore, woodland; nest in ground layer.

Dolichoderus australis E. André, 1896

Dolichoderus australis André, E. (1896). Fourmis nouvelles d'Asie et d'Australie. *Rev. Entomol.* **15**: 251-265 [257]. Type data: syntypes, MNHP W, from Victorian Alps.

Distribution: Murray-Darling basin, A.C.T., N.S.W., Vic. Ecology: terrestrial, noctidiurnal, omnivore, alpine, woodland; nest in ground layer.

Dolichoderus clarki Wheeler, 1935

Dolichoderus (Hypoclinea) tristis Clark, J. (1930). The Australian ants of the genus *Dolichoderus* (Formicidae). Subgenus *Hypoclinea* Mayr. *Aust. Zool.* **6**: 252-268 [20 Aug. 1930] [254] [*non Dolichoderus (Monacis) tristis* Mann, 1916]. Type data: syntypes, NMV *W, from Bondi and Cooma, N.S.W.

Dolichoderus clarki Wheeler, W.M. (1935). Myrmecological notes. *Psyche Camb.* **42**: 68-72 [69] [*nom. nov.* for *Dolichoderus (Hypoclinea) tristis* Clark, 1930].

Distribution: SE coastal, Murray-Darling basin, N.S.W. Ecology: terrestrial, noctidiurnal, omnivore, woodland; nest in ground layer.

Dolichoderus clusor Forel, 1907

Dolichoderus clusor Forel, A. (1907). Formicidae. pp. 263-310 *in* Michaelsen, W. & Hartmeyer, R. (eds.) *Die Fauna Südwest-Australiens.* Jena : G. Fischer Vol. 1 [285]. Type data: holotype, probably destroyed in ZMH in WW II, from Fremantle, W.A.

Distribution: SW coastal, W.A. Ecology: terrestrial, noctidiurnal, omnivore, woodland, open forest; nest in ground layer.

Dolichoderus dentatus Forel, 1902

Dolichoderus doriae dentatus Forel, A. (1902). Fourmis nouvelles d'Australie. *Rev. Suisse Zool.* **10**: 405-548 [462]. Type data: syntypes, GMNH W, ANIC W, from Mackay, Qld.

Distribution: NE coastal, Qld. Ecology: terrestrial, noctidiurnal, omnivore, woodland, open forest; nest in ground layer. Biological references: Clark, J. (1930). The Australian ants of the genus *Dolichoderus* (Formicidae). Subgenus *Hypoclinea* Mayr. *Aust. Zool.* **6**: 252-268 [20 Aug. 1930] (raised to species).

Dolichoderus doriae Emery, 1887

Dolichoderus doriae Emery, C. (1887). Catalogo delle formiche esistenti nelle collezioni del Museo Civico di Genova. Parte terza. Formiche della regione Indo-Malese e dell'Australia. *Ann. Mus. Civ. Stor. Nat. Giacomo Doria* **25**: 209-258 pls 3-4 [253]. Type data: syntypes, MCG *W, from Blue Mts. and Mt. Victoria, N.S.W.

Distribution: SE coastal, N.S.W. Ecology: terrestrial, noctidiurnal, omnivore, open forest, closed forest; nest in ground layer.

Dolichoderus extensispinus Forel, 1915

Dolichoderus doriae extensispina Forel, A. (1915). Results of Dr. E. Mjöbergs Swedish Scientific Expeditions to Australia 1910-1913. 2. Ameisen. *Ark.*

Zool. **9**: 1-119 pls 1-3 [4 Dec. 1915] [76]. Type data: syntypes, GMNH W, ANIC W, other syntypes may exist, from Blackal (=Blackall) Range, Qld.

Distribution: NE coastal, Qld. Ecology: terrestrial, noctidiurnal, omnivore, woodland, open forest; nest in ground layer. Biological references: Clark, J. (1930). The Australian ants of the genus *Dolichoderus* (Formicidae). Subgenus *Hypoclinea* Mayr. *Aust. Zool.* **6**: 252-268 [20 Aug. 1930] (raised to species).

Dolichoderus formosus Clark, 1930

Dolichoderus (Hypoclinea) formosus Clark, J. (1930). The Australian ants of the genus *Dolichoderus* (Formicidae). Subgenus *Hypoclinea* Mayr. *Aust. Zool.* **6**: 252-268 [20 Aug. 1930] [265]. Type data: syntypes, NMV *W,F, from Armadale, Mundaring and Mt. Dale, W.A.

Distribution: SW coastal, W.A. Ecology: terrestrial, noctidiurnal, omnivore, woodland, open forest; nest in ground layer.

Dolichoderus glauerti Wheeler, 1934

Dolichoderus (Hypoclinea) glauerti Wheeler, W.M. (1934). Contributions to the fauna of Rottnest Island, Western Australia No. IX. The ants. *J. R. Soc. West. Aust.* **20**: 137-163 [5 Oct. 1934] [147]. Type data: syntypes, MCZ *W,M, from City of York Bay, Rottnest Is., W.A.

Distribution: SW coastal, W.A. Ecology: terrestrial, noctidiurnal, omnivore, woodland, open forest; nest in ground layer.

Dolichoderus goudiei Clark, 1930

Dolichoderus (Hypoclinea) goudiei Clark, J. (1930). The Australian ants of the genus *Dolichoderus* (Formicidae). Subgenus *Hypoclinea* Mayr. *Aust. Zool.* **6**: 252-268 [20 Aug. 1930] [264]. Type data: syntypes, NMV *W, from Maldon, Vic.

Distribution: Murray-Darling basin, Vic. Ecology: terrestrial, noctidiurnal, omnivore, woodland; nest in ground layer.

Dolichoderus nigricornis Clark, 1930

Dolichoderus (Hypoclinea) nigricornis Clark, J. (1930). The Australian ants of the genus *Dolichoderus* (Formicidae). Subgenus *Hypoclinea* Mayr. *Aust. Zool.* **6**: 252-268 [20 Aug. 1930] [265]. Type data: syntypes, NMV *W, from Tammin, W.A.

Distribution: SW coastal, W.A. Ecology: terrestrial, noctidiurnal, omnivore, woodland; nest in ground layer.

Dolichoderus occidentalis Clark, 1930

Dolichoderus (Hypoclinea) occidentalis Clark, J. (1930). The Australian ants of the genus *Dolichoderus* (Formicidae). Subgenus *Hypoclinea* Mayr. *Aust. Zool.* **6**: 252-268 [20 Aug. 1930] [268]. Type data: syntypes, NMV *W, from Albany, W.A.

Distribution: SW coastal, W.A. Ecology: terrestrial, noctidiurnal, omnivore, woodland; nest in ground layer.

Dolichoderus parvus Clark, 1930

Dolichoderus (Hypoclinea) parvus Clark, J. (1930). The Australian ants of the genus *Dolichoderus* (Formicidae). Subgenus *Hypoclinea* Mayr. *Aust. Zool.* **6**: 252-268 [20 Aug. 1930] [263]. Type data: syntypes, NMV *W, from Sea Lake, Vic.

Distribution: Murray-Darling basin, Vic. Ecology: terrestrial, noctidiurnal, omnivore, woodland; nest in ground layer.

Dolichoderus reflexus Clark, 1930

Dolichoderus (Hypoclinea) reflexus Clark, J. (1930). The Australian ants of the genus *Dolichoderus* (Formicidae). Subgenus *Hypoclinea* Mayr. *Aust. Zool.* **6**: 252-268 [20 Aug. 1930] [261]. Type data: syntypes, NMV *W, from Murray Bridge and Mt. Lofty, S.A.

Distribution: Murray-Darling basin, S.A. Ecology: terrestrial, noctidiurnal, omnivore, woodland; nest in ground layer.

Dolichoderus scabridus Roger, 1862

Dolichoderus scabridus scabridus Roger, 1862

Dolichoderus scabridus Roger, J. (1862). Einige neue exotische Ameisen-Gattungen und Arten. *Berl. Entomol. Z.* **6**: 233-254 pl 1 [244]. Type data: syntypes, BMN (probable) *W, from Australia.

Polyrhachis foveolatus Lowne, B.T. (1865). Contributions to the natural history of Australian ants. *Entomologist* **2**: 331-336 [334]. Type data: syntypes (probable), BMNH (probable) *W, from Sidney (=Sydney), N.S.W.

Synonymy that of Emery, C. (1912). Hymenoptera Fam. Formicidae subfam. Dolichoderinae *in* Wytsman, P. (ed.) *Genera Insectorum.* Fasc. 137 50 pp. 2 pls [13].

Distribution: SE coastal, N.S.W. Ecology: terrestrial, noctidiurnal, omnivore, woodland, open forest; nest in ground layer.

Dolichoderus scabridus ruficornis Santschi, 1916

Dolichoderus (Hypoclinea) scabridus ruficornis Santschi, F. (1916). Deux nouvelles fourmis d'Australie. *Bull. Soc. Entomol. Fr.* **1916**: 174-175 [175]. Type data: syntypes, BNHM W, from Australia.

Distribution: S Gulfs, SE coastal, S.A., Vic. Ecology: terrestrial, noctidiurnal, omnivore, woodland, open forest; nest in ground layer.

Dolichoderus scrobiculatus (Mayr, 1876)

Hypoclinea scrobiculata Mayr, G.L. (1876). Die australischen Formiciden. *J. Mus. Godeffroy* **5**: 56–115 [80]. Type data: syntypes, NHMW *W, from Peak Downs, Qld.

Distribution: NE coastal, Qld. Ecology: terrestrial, noctidiurnal, omnivore, woodland, open forest; nest in ground layer.

Dolichoderus turneri Forel, 1902

Dolichoderus turneri Forel, A. (1902). Fourmis nouvelles d'Australie. *Rev. Suisse Zool.* **10**: 405–548 [462]. Type data: syntypes, GMNH W, ANIC W, from Mackay, Qld.

Distribution: NE coastal, Qld. Ecology: terrestrial, noctidiurnal, omnivore, open forest, closed forest; nest in ground layer.

Dolichoderus ypsilon Forel, 1902

Dolichoderus ypsilon ypsilon Forel, 1902

Dolichoderus scabridus ypsilon Forel, A. (1902). Fourmis nouvelles d'Australie. *Rev. Suisse Zool.* **10**: 405–548 [461]. Type data: syntypes, GMNH W, ANIC W, from Perth, W.A.

Distribution: SW coastal, W.A. Ecology: terrestrial, noctidiurnal, omnivore, woodland, open forest; nest in ground layer. Biological references: Forel, A. (1907). Formicidae. pp. 263–310 *in* Michaelsen, W. & Hartmeyer, R. (eds.) *Die Fauna Südwest-Australiens.* Jena : G. Fischer Vol. 1 (raised to species).

Dolichoderus ypsilon nigra Crawley, 1922

Dolichoderus (Hypoclinea) ypsilon nigra Crawley, W.C. (1922). New ants from Australia. *Ann. Mag. Nat. Hist. (9)* **10**: 16–36 [25]. Type data: syntypes (probable), OUM *W, from Kelmscott, W.A.

Distribution: SW coastal, W.A. Ecology: terrestrial, noctidiurnal, omnivore, woodland, open forest; nest in ground layer.

Dolichoderus ypsilon rufotibialis Clark, 1930

Dolichoderus (Hypoclinea) ypsilon rufotibialis Clark, J. (1930). The Australian ants of the genus *Dolichoderus* (Formicidae). Subgenus *Hypoclinea* Mayr. *Aust. Zool.* **6**: 252–268 [20 Aug. 1930] [259]. Type data: syntypes, NMV *W, from Albany, W.A.

Distribution: SW coastal, W.A. Ecology: terrestrial, noctidiurnal, omnivore, woodland, open forest; nest in ground layer.

Froggattella Forel, 1902

Froggattella Forel, A. (1902). Fourmis nouvelles d'Australie. *Rev. Suisse Zool.* **10**: 405–548 [459]. Type species *Acantholepis kirbii* Lowne, 1865 by original designation.

Froggattella kirbii (Lowne, 1865)

Froggattella kirbii kirbii (Lowne, 1865)

Acantholepis kirbii Lowne, B.T. (1865). Contributions to the natural history of Australian ants. *Entomologist* **2**: 331–336 [333]. Type data: syntypes (probable), BMNH (probable) *W, from Sidney (=Sydney), N.S.W.

Dolichoderus kirbyi Dalla Torre, C.G. de (1893). *Catalogus hymenopterorum hucusque descriptorum systematicus et synonymicus.* Formicidae (Heterogyna). Lipsiae : G. Engelmann Vol. 7 [159] [invalid emend. of *Acantholepis kirbii* Lowne, 1865].

Distribution: SE coastal, SW coastal, W plateau, S Gulfs, Murray-Darling basin, NE coastal, W.A., S.A., Vic., Qld., N.S.W. Ecology: terrestrial, noctidiurnal, predator, woodland; nest in ground layer.

Froggattella kirbii bispinosa Forel, 1902

Froggattella kirbyi bispinosa Forel, A. (1902). Fourmis nouvelles d'Australie. *Rev. Suisse Zool.* **10**: 405–548 [460]. Type data: syntypes, GMNH W, A W, from Sydney and Oatley, N.S.W.

Distribution: SE coastal, N.S.W. Ecology: terrestrial, noctidiurnal, predator, woodland; nest in ground layer.

Froggattella kirbii ianthina Wheeler, 1936

Froggattella kirbyi ianthina Wheeler, W.M. (1936). The Australian ant genus *Froggattella. Am. Mus. Novit.* 842: 1–11 [13 Apr. 1936] [8]. Type data: syntypes, MCZ *W, from near Brisbane, Qld.

Distribution: NE coastal, Qld. Ecology: terrestrial, noctidiurnal, predator, woodland; nest in ground layer.

Froggattella kirbii laticeps Wheeler, 1936

Froggattella kirbyi laticeps Wheeler, W.M. (1936). The Australian ant genus *Froggattella. Am. Mus. Novit.* 842: 1–11 [13 Apr. 1936] [10]. Type data: syntypes, MCZ *W, from Lucindale, S.A.

Distribution: Murray-Darling basin, S.A. Ecology: terrestrial, noctidiurnal, predator, woodland; nest in ground layer.

Froggattella kirbii lutescens Wheeler, 1936

Froggattella kirbyi lutescens Wheeler, W.M. (1936). The Australian ant genus *Froggattella. Am. Mus. Novit.* 842: 1–11 [13 Apr. 1936] [9]. Type data: syntypes, MCZ *W, from near Sydney, N.S.W.

Distribution: SE coastal, N.S.W. Ecology: terrestrial, noctidiurnal, predator, woodland; nest in ground layer.

Froggattella kirbii nigripes Wheeler, 1936

Froggattella kirbyi nigripes Wheeler, W.M. (1936). The Australian ant genus *Froggattella*. *Am. Mus. Novit.* 842: 1–11 [13 Apr. 1936] [8]. Type data: syntypes, MCZ *W, from Coen, Cape York Peninsula, Qld.

Distribution: N Gulf, Qld. Ecology: terrestrial, noctidiurnal, predator, woodland; nest in ground layer.

Froggattella latispina Wheeler, 1936

Froggattella latispina Wheeler, W.M. (1936). The Australian ant genus *Froggattella*. *Am. Mus. Novit.* 842: 1–11 [13 Apr. 1936] [10]. Type data: syntypes, MCZ *W, from Port Lincoln, S.A.

Distribution: S Gulfs, S.A. Ecology: terrestrial, noctidiurnal, predator, woodland; nest in ground layer.

Iridomyrmex Mayr, 1862

Iridomyrmex Mayr, G.L. (1862). Myrmecologische Studien. *Verh. Zool.-Bot. Ges. Wien* **12**: Abhand. 649–776 [702 pl 19]. Type species *Formica purpurea* F. Smith, 1858 (as *Formica detecta* F. Smith, 1858) by subsequent designation, see Bingham, C.T. (1903). *The Fauna of British India, including Ceylon and Burma. Hymenoptera. Vol. 2 Ants and cuckoo-wasps.* London : Taylor & Francis [297].

Doleromyrma Forel, A. (1907). Formicides du Musée National Hongrois. *Ann. Hist.-Nat. Mus. Natl. Hung.* **5**: 1–42 [30 June 1907] [28] [proposed with subgeneric rank in *Tapinoma* Förster, 1850]. Type species *Tapinoma (Doleromyrma) darwinianum* Forel, 1907 by original designation.

Synonymy that of Emery, C. (1912). Hymenoptera Fam. Formicidae subfam. Dolichoderinae *in* Wytsman, P. (ed.) *Genera Insectorum.* Fasc. 137 Brussels 50 pp. 2 pls [21].

This group is also found in the Neotropical, Nearctic and east Oriental regions; New Guinea, east Melanesia and New Caledonia in the Australian Region, see Brown, W.L. jr. (1973). A comparison of the Hylean and Congo-West African rain forest ant faunas. pp. 161–185 *in* Meggers, B.J., Ayensu, E.S. & Duckworth, W.D. (eds.) *Tropical forest ecosystems in Africa and South America: a comparative review.* Washington : Smithsonian Institution Press.

Iridomyrmex agilis Forel, 1907

Iridomyrmex agilis Forel, A. (1907). Formicidae. pp. 263–310 *in* Michaelsen, W. & Hartmeyer, R. (eds.) *Die Fauna Südwest-Australiens.* Jena : G. Fischer Vol. 1 [295]. Type data: syntypes, GMNH W, ANIC W, from Yalgoo, W.A.

Distribution: NW coastal, W.A. Ecology: terrestrial, noctidiurnal, omnivore, woodland; nest in ground layer.

Iridomyrmex albitarsus Wheeler, 1927

Iridomyrmex albitarsus Wheeler, W.M. (1927). The ants of Lord Howe Island and Norfolk Island. *Proc. Am. Acad. Arts Sci.* **62**: 121–153 [147]. Type data: syntypes, MCZ *W,M,F, from Norfolk Is.

Distribution: Norfolk Is. Ecology: terrestrial, noctidiurnal, omnivore, open forest; nest in ground layer.

Iridomyrmex anceps (Roger, 1863)

Formica anceps Roger, J. (1863). Die neu aufgeführten Gattungen und Arten meines Formiciden-Verzeichnisses. *Berl. Entomol. Z.* **7**: 129–214 [164]. Type data: status unknown, ?ZMB, from Malacca (Malaysia?).

Distribution: NE coastal, SE coastal, Qld., N.S.W.; also India to Cook Ils. Ecology: terrestrial, noctidiurnal, omnivore, open forest; nest in ground layer.

Iridomyrmex arcadius Forel, 1915

Iridomyrmex arcadius Forel, A. (1915). Results of Dr. E. Mjöbergs Swedish Scientific Expeditions to Australia 1910–1913. 2. Ameisen. *Ark. Zool.* **9**: 1–119 pls 1–3 [4 Dec. 1915] [82]. Type data: syntypes, GMNH W, other syntypes may exist, from Malanda and Atherton, Qld.

Distribution: NE coastal, Qld. Ecology: terrestrial, noctidiurnal, omnivore, woodland; nest in ground layer.

Iridomyrmex bicknelli Emery, 1898

Iridomyrmex bicknelli bicknelli Emery, 1898

Iridomyrmex bicknelli Emery, C. (1898). Descrizioni di formiche nuove Malesi e Australiane. Note sinonimiche. *Rec. Sess. Accad. Sci. Ist. Bologna (ns)* **2**: 231–245 [236]. Type data: syntypes, MCG *W, from Tas.

Distribution: Tas. Ecology: terrestrial, noctidiurnal, omnivore, woodland; nest in ground layer.

Iridomyrmex bicknelli azureus Viehmeyer, 1913

Iridomyrmex bicknelli azureus Viehmeyer, H. (1913). Neue und unvollständig bekannte Ameisen der Alten Welt. *Arch. Naturg.* **79A**(12): 24–60 [41]. Type data: syntypes, ZMB *W, from Killalpaninna, S.A.

Distribution: Lake Eyre basin, S.A. Ecology: terrestrial, noctidiurnal, omnivore, woodland; nest in ground layer.

Iridomyrmex bicknelli brunneus Forel, 1902

Iridomyrmex bicknelli brunneus Forel, A. (1902). Fourmis nouvelles d'Australie. *Rev. Suisse Zool.* **10**: 405–548 [469]. Type data: syntypes, GMNH W, ANIC W, from Kalgoorlie, W.A.

Distribution: W plateau, W.A. Ecology: terrestrial, noctidiurnal, omnivore, woodland; nest in ground layer.

Iridomyrmex bicknelli lutea Forel, 1915

Iridomyrmex bicknelli lutea Forel, A. (1915). Results of Dr. E. Mjöbergs Swedish Scientific Expeditions to Australia 1910-1913. 2. Ameisen. *Ark. Zool.* **9**: 1-119 pls 1-3 [4 Dec. 1915] [77]. Type data: holotype (probable), whereabouts unknown, from Kimberley distr., W.A.

Distribution: N coastal, W.A. Ecology: terrestrial, noctidiurnal, omnivore, woodland; nest in ground layer.

Iridomyrmex bicknelli splendidus Forel, 1902

Iridomyrmex bicknelli splendidus Forel, A. (1902). Fourmis nouvelles d'Australie. *Rev. Suisse Zool.* **10**: 405-548 [468]. Type data: holotype (probable), GMNH W, from Perth, W.A.

Distribution: SW coastal, W.A. Ecology: terrestrial, noctidiurnal, omnivore, woodland; nest in ground layer.

Iridomyrmex biconvexus Santschi, 1928

Iridomyrmex biconvexus Santschi, F. (1928). Nouvelles fourmis d'Australie. *Bull. Soc. Vaud. Sci. Nat.* **56**: 465-483 [30 Aug. 1928] [471]. Type data: syntypes, BNHM *W, from Ringwood, Vic.

Iridomyrmex foetans Clark, J. (1929). Results of a collecting trip to the Cann River, East Gippsland. *Vict. Nat.* **46**: 115-123 [4 Oct. 1929] [122]. Type data: syntypes, NMV *W, from Cann River, Vic.

Synonymy that of Brown, W.L. jr. (1954). New synonymy of an Australian *Iridomyrmex* (Hymenoptera : Formicidae). *Psyche Camb.* **61**: 67.

Distribution: SE coastal, Vic. Ecology: terrestrial, noctidiurnal, omnivore, alpine, woodland; nest in ground layer.

Iridomyrmex chasei Forel, 1902

Iridomyrmex chasei chasei Forel, 1902

Iridomyrmex chasei Forel, A. (1902). Fourmis nouvelles d'Australie. *Rev. Suisse Zool.* **10**: 405-548 [467]. Type data: syntypes, GMNH W, ANIC W, from Perth, W.A.

Distribution: SW coastal, W.A. Ecology: terrestrial, noctidiurnal, omnivore, woodland; nest in ground layer.

Iridomyrmex chasei concolor Forel, 1902

Iridomyrmex chasei concolor Forel, A. (1902). Fourmis nouvelles d'Australie. *Rev. Suisse Zool.* **10**: 405-548 [468]. Type data: syntypes, GMNH W, ANIC W, from Kalgoorlie, W.A.

Distribution: W plateau, W.A. Ecology: terrestrial, noctidiurnal, omnivore, woodland; nest in ground layer.

Iridomyrmex chasei yalgooensis Forel, 1907

Iridomyrmex chasei yalgooensis Forel, A. (1907). Formicidae. pp. 263-310 *in* Michaelsen, W. & Hartmeyer, R. (eds.) *Die Fauna Südwest-Australiens*. Jena : G. Fischer Vol. 1 [288]. Type data: syntypes, GMNH W, ANIC W, from Geraldton, Day Dawn, Yalgoo and Coolgardie, W.A.

Distribution: W plateau, NW coastal, W.A. Ecology: terrestrial, noctidiurnal, omnivore, woodland; nest in ground layer.

Iridomyrmex conifer Forel, 1902

Iridomyrmex conifer Forel, A. (1902). Fourmis nouvelles d'Australie. *Rev. Suisse Zool.* **10**: 405-548 [463]. Type data: syntypes, GMNH W, ANIC W, from Perth, W.A.

Distribution: SW coastal, W.A. Ecology: terrestrial, noctidiurnal, omnivore, woodland; nest in ground layer.

Iridomyrmex cordatus (F. Smith, 1859)

Formica cordata Smith, F. (1859). Catalogue of hymenopterous insects collected by Mr A.R. Wallace at the islands of Aru and Key. *J. Linn. Soc. Zool.* **3**: 132-178 [137] [1 Feb. 1859]. Publication date established from Donisthorpe, H. (1932). On the identity of Smith's types of Formicidae (Hymenoptera) collected by Alfred Russell Wallace in the Malay Archipelago, with descriptions of two new species. *Ann. Mag. Nat. Hist. (10)* **10**: 441-476. Type data: status unknown, ?BMNH, from Aru Ils., Indonesia.

Iridomyrmex cordatus stewartii Forel, 1893

Iridomyrmex cordatus stewartii Forel, A. (1893). Nouvelles fourmis d'Australie et des Canaries. *Ann. Soc. Entomol. Belg.* **37**: 454-466 [456]. Type data: syntypes, GMNH W, ANIC W, from Torres Strait.

Distribution: Qld.; Torres Strait. Ecology: terrestrial, noctidiurnal, omnivore, open forest, closed forest; nest in ground layer.

Iridomyrmex cyaneus Wheeler, 1915

Iridomyrmex cyaneus Wheeler, W.M. (1915). Hymenoptera. *Trans. R. Soc. S. Aust.* **39**: 805-823 pls 64-66 [Dec. 1915] [812]. Type data: syntypes, MCZ *W, from Black Rock Hole in the Musgrave Ranges and Moorilyanna, S.A.

Distribution: W plateau, Lake Eyre basin, S.A. Ecology: terrestrial, noctidiurnal, omnivore, woodland; nest in ground layer.

Iridomyrmex darwinianus (Forel, 1907)

Iridomyrmex darwinianus darwinianus (Forel, 1907)

Tapinoma (Doleromyrma) darwinianum Forel, A. (1907). Formicides du Musée National Hongrois. *Ann. Hist.- Nat. Mus. Natl. Hung.* **5**: 1-42 [30 June 1907]

[28]. Type data: syntypes, GMNH W,M, ANIC W, other syntypes may exist in MNH, from Mt. Victoria, Blue Mts., N.S.W.

Distribution: SE coastal, N.S.W. Ecology: terrestrial, noctidiurnal, omnivore, woodland; nest in ground layer.

Iridomyrmex darwinianus fida (Forel, 1907)

Tapinoma (Doleromyrma) darwinianum fida Forel, A. (1907). Formicidae. pp. 263–310 *in* Michaelsen, W. & Hartmeyer, R. (eds.) *Die Fauna Südwest-Australiens.* Jena : G. Fischer Vol. 1 [286]. Type data: syntypes, GMNH W,F, ANIC W, from Guildford, Collie, Bunbury, Bridgetown, Donnybrook, Boyanup and Pickering Brook, W.A.

Distribution: SW coastal, W.A. Ecology: terrestrial, noctidiurnal, omnivore, woodland; nest in ground layer.

Iridomyrmex darwinianus leae Forel, 1913

Iridomyrmex darwinianus leae Forel, A. (1913). Fourmis de Tasmanie et d'Australie récoltées par MM. Lea, Froggatt etc. *Bull. Soc. Vaud. Sci. Nat.* **49**: 173–196 pl 2 [189]. Type data: syntypes, GMNH W, ANIC W, from Geelong, Vic.

Distribution: SE coastal, Vic. Ecology: terrestrial, noctidiurnal, omnivore, woodland; nest in ground layer.

Iridomyrmex discors Forel, 1902

Iridomyrmex discors discors Forel, 1902

Iridomyrmex discors Forel, A. (1902). Fourmis nouvelles d'Australie. *Rev. Suisse Zool.* **10**: 405–548 [464]. Type data: syntypes, GMNH W, ANIC W, from Charters Towers, Qld.

Distribution: NE coastal, Qld. Ecology: terrestrial, noctidiurnal, omnivore, woodland; nest in ground layer.

Iridomyrmex discors aeneogaster Wheeler, 1915

Iridomyrmex discors aeneogaster Wheeler, W.M. (1915). Hymenoptera. *Trans. R. Soc. S. Aust.* **39**: 805–823 pls 64–66 [Dec. 1915] [811]. Type data: holotype, MCZ *W, from Flat Rock Hole, Musgrave Ranges, .S.A.

Distribution: W plateau, S.A. Ecology: terrestrial, noctidiurnal, omnivore, woodland; nest in ground layer.

Iridomyrmex discors obscurior Forel, 1902

Iridomyrmex discors obscurior Forel, A. (1902). Fourmis nouvelles d'Australie. *Rev. Suisse Zool.* **10**: 405–548 [465]. Type data: syntypes, GMNH W, ANIC W, from Ballarat, Vic.

Distribution: SE coastal, Vic. Ecology: terrestrial, noctidiurnal, omnivore, woodland; nest in ground layer.

Iridomyrmex discors occipitalis Forel, 1907

Iridomyrmex discors occipitalis Forel, A. (1907). Formicidae. pp. 263–310 *in* Michaelsen, W. & Hartmeyer, R. (eds.) *Die Fauna Südwest-Australiens.* Jena : G. Fischer Vol. 1 [294]. Type data: syntypes, GMNH W, ANIC W, from Northampton, W.A.

Distribution: NW coastal, W.A. Ecology: terrestrial, noctidiurnal, omnivore, woodland; nest in ground layer.

Iridomyrmex dromus Clark, 1938

Iridomyrmex dromus Clark, J. (1938). Reports of the McCoy Society for Field Investigation and Research. No. 2. Sir Joseph Bank Islands. Part I. Formicidae (Hymenoptera). *Proc. R. Soc. Vict.* **50**: 356–382 [374]. Type data: syntypes (probable), NMV *W, from Reevesby Is., S.A.

Distribution: S Gulfs, S.A. Ecology: terrestrial, noctidiurnal, omnivore, woodland; nest in ground layer.

Iridomyrmex emeryi Crawley, 1918

Iridomyrmex emeryi Crawley, W.C. (1918). Some new Australian ants. *Entomol. Rec. J. Var.* **30**: 86–92 [90]. Type data: syntypes, possibly OUM, from Healesville, Vic.

Distribution: SE coastal, Vic. Ecology: terrestrial, noctidiurnal, omnivore, woodland; nest in ground layer.

Iridomyrmex exsanguis Forel, 1907

Iridomyrmex exsanguis Forel, A. (1907). Formicidae. pp. 263–310 *in* Michaelsen, W. & Hartmeyer, R. (eds.) *Die Fauna Südwest-Australiens.* Jena : G. Fischer Vol. 1 [296]. Type data: syntypes, GMNH W,F, from Denham, W.A.

Distribution: NW coastal, W.A. Ecology: terrestrial, noctidiurnal, omnivore, woodland; nest in ground layer.

Iridomyrmex flavipes (W.F. Kirby, 1896)

Hypoclinea flavipes Kirby, W.F. (1896). Hymenoptera. pp. 203–209 *in* Spencer, B. (ed.) *Report on the work of the Horn Scientific Expedition to Central Australia.* Melbourne : Melville, Mullen & Slade Pt. 1 supplement [206]. Type data: syntypes, BMNH (probable) *W, NMV *W, from Tempe Downs, N.T.

Iridomyrmex rostrinotus Forel, A. (1910). Formicides australiens reçus de MM. Froggatt et Rowland Turner. *Rev. Suisse Zool.* **18**: 1–94 [53]. Type data: syntypes, GMNH W,F,M, ANIC W, from Tennant Creek, N.T.

Synonymy that of Clark, J. (1930). The Australian ants of the genus *Dolichoderus* (Formicidae). Subgenus *Hypoclinea* Mayr. *Aust. Zool.* **6**: 252–268 [20 Aug. 1930] [268].

Distribution: Lake Eyre basin, W plateau, N.T. Ecology: terrestrial, noctidiurnal, omnivore, woodland; nest in ground layer.

Iridomyrmex flavus Mayr, 1868

Iridomyrmex flavus Mayr, G.L. (1868). Formicidae. *in, Reise der österreichischen fregatte Novara um die Erde in der Jahren 1857, 1858, 1859.* Zool. 2 Abth 1A3 1–123 pls 1–4 [60]. Type data: syntypes, NHMW (probable) *W, from Sidney (=Sydney), N.S.W.

Distribution: SE coastal, N.S.W. Ecology: terrestrial, noctidiurnal, omnivore, woodland; nest in ground layer.

Iridomyrmex fornicatus Emery, 1914

Iridomyrmex fornicatus Emery, C. (1914). Formiche d'Australia e di Samoa raccolte dal Prof. Silvestri nel 1913. *Boll. Lab. Zool. Gen. Agr. R. Scuola Agric. Portici* **8**: 179–186 [30 Jan. 1914] [185]. Type data: syntypes, whereabouts uncertain, probably MCG or MNHP, from Mt. Lofty, Adelaide, S.A.

Distribution: S Gulfs, S.A. Ecology: terrestrial, noctidiurnal, omnivore, woodland; nest in ground layer.

Iridomyrmex froggatti Forel, 1902

Iridomyrmex froggatti Forel, A. (1902). Fourmis nouvelles d'Australie. *Rev. Suisse Zool.* **10**: 405–548 [470]. Type data: holotype (probable), GMNH W, from Sydney, N.S.W.

Distribution: SE coastal, N.S.W. Ecology: terrestrial, noctidiurnal, omnivore, woodland; nest in ground layer.

Iridomyrmex gilberti Forel, 1902

Iridomyrmex gilberti Forel, A. (1902). Fourmis nouvelles d'Australie. *Rev. Suisse Zool.* **10**: 405–548 [470]. Type data: syntypes, GMNH W, ANIC W, from Cairns and Mackay, Qld.

Distribution: NE coastal, Qld. Ecology: terrestrial, noctidiurnal, omnivore, open forest, closed forest; nest in ground layer.

Iridomyrmex glaber (Mayr, 1862)

Iridomyrmex glaber glaber (Mayr, 1862)

Hypoclinea glabra Mayr, G.L. (1862). Myrmecologische Studien. *Verh. Zool.-Bot. Ges. Wien* **12**: Abhand. 649–776 [705 pl 19]. Type data: syntypes, NHMW *W,F, from Sidney (=Sydney), N.S.W.

Distribution: SE coastal, N.S.W. Ecology: terrestrial, noctidiurnal, omnivore, woodland, open forest; nest in ground layer.

Iridomyrmex glaber clarithorax Forel, 1902

Iridomyrmex glaber clarithorax Forel, A. (1902). Fourmis nouvelles d'Australie. *Rev. Suisse Zool.* **10**: 405–548 [473]. Type data: syntypes, GMNH W, ANIC W, from Brisbane, Qld. and Sydney, N.S.W.

Distribution: NE coastal, SE coastal, Qld., N.S.W. Ecology: terrestrial, noctidiurnal, omnivore, woodland, open forest; nest in ground layer.

Iridomyrmex gracilis (Lowne, 1865)

Iridomyrmex gracilis gracilis (Lowne, 1865)

Formica gracilis Lowne, B.T. (1865). Contributions to the natural history of Australian ants. *Entomologist* **2**: 275–280 [280]. Type data: syntypes (probable), BMNH (probable) *W, from Sidney (=Sydney), N.S.W.

Distribution: SE coastal, N.S.W. Ecology: terrestrial, noctidiurnal, omnivore, woodland; nest in ground layer.

Iridomyrmex gracilis fusciventris Forel, 1913

Iridomyrmex gracilis fusciventris Forel, A. (1913). Fourmis de Tasmanie et d'Australie récoltées par MM. Lea, Froggatt etc. *Bull. Soc. Vaud. Sci. Nat.* **49**: 173–196 pl 2 [188]. Type data: syntypes, GMNH W, from Mullewa, W.A. and Sea Lake, Vic.

Distribution: Murray-Darling basin, NW coastal, Vic., W.A. Ecology: terrestrial, noctidiurnal, omnivore, woodland; nest in ground layer.

Iridomyrmex gracilis mayri Forel, 1915

Iridomyrmex gracilis mayri Forel, A. (1915). Results of Dr. E. Mjöbergs Swedish Scientific Expeditions to Australia 1910–1913. 2. Ameisen. *Ark. Zool.* **9**: 1–119 pls 1–3 [4 Dec. 1915] [80]. Type data: syntypes, GMNH W, ANIC W, other syntypes may exist, from Blackal (=Blackall) Range, Glen Lamington and Cedar Creek, Qld.

Distribution: NE coastal, Qld. Ecology: terrestrial, noctidiurnal, omnivore, woodland; nest in ground layer.

Iridomyrmex gracilis minor Forel, 1915

Iridomyrmex gracilis minor Forel, A. (1915). Results of Dr. E. Mjöbergs Swedish Scientific Expeditions to Australia 1910–1913. 2. Ameisen. *Ark. Zool.* **9**: 1–119 pls 1–3 [4 Dec. 1915] [80]. Type data: syntypes, GMNH W,F, ANIC W, other syntypes may exist, from Atherton, Yarrabah, Cooktown and Cape York, Qld. and Perth, Noonkanbah, Kimberley distr. and Port Hedland, W.A.

Distribution: NE coastal, SW coastal, NW coastal, N coastal, Qld., W.A. Ecology: terrestrial, noctidiurnal, omnivore, woodland; nest in ground layer.

Iridomyrmex gracilis rubriceps Forel, 1902

Iridomyrmex gracilis rubriceps Forel, A. (1902). Fourmis nouvelles d'Australie. *Rev. Suisse Zool.* **10**: 405–548 [468]. Type data: syntypes, GMNH W, ANIC W, from Mackay, Qld.

Distribution: NE coastal, Qld. Ecology: terrestrial, noctidiurnal, omnivore, woodland; nest in ground layer.

Iridomyrmex gracilis spurcus Wheeler, 1915

Iridomyrmex gracilis spurcus Wheeler, W.M. (1915). Hymenoptera. *Trans. R. Soc. S. Aust.* **39**: 805–823 pls 64–66 [Dec. 1915] [813]. Type data: syntypes, MCZ *W, from Moorilyanna, S.A.

Distribution: Lake Eyre basin, S.A. Ecology: terrestrial, noctidiurnal, omnivore, woodland; nest in ground layer.

Iridomyrmex hartmeyeri Forel, 1907

Iridomyrmex hartmeyeri Forel, A. (1907). Formicidae. pp. 263–310 *in* Michaelsen, W. & Hartmeyer, R. (eds.) *Die Fauna Südwest-Australiens.* Jena : G. Fischer Vol. 1 [296]. Type data: syntypes, GMNH W, from Day Dawn, W.A.

Distribution: NW coastal, W.A. Ecology: terrestrial, noctidiurnal, omnivore, woodland; nest in ground layer.

Iridomyrmex innocens Forel, 1907

Iridomyrmex innocens innocens Forel, 1907

Iridomyrmex innocens Forel, A. (1907). Formicidae. pp. 263–310 *in* Michaelsen, W. & Hartmeyer, R. (eds.) *Die Fauna Südwest-Australiens.* Jena : G. Fischer Vol. 1 [292]. Type data: syntypes, GMNH W,M,F, from Yalgoo, Lion Mill, Midland and Yarloop, W.A.

Distribution: SW coastal, NW coastal, W.A. Ecology: terrestrial, noctidiurnal, omnivore, woodland; nest in ground layer.

Iridomyrmex innocens malandanus Forel, 1915

Iridomyrmex innocens malandanus Forel, A. (1915). Results of Dr. E. Mjöbergs Swedish Scientific Expeditions to Australia 1910–1913. 2. Ameisen. *Ark. Zool.* **9**: 1–119 pls 1–3 [4 Dec. 1915] [81]. Type data: syntypes, GMNH W, other syntypes may exist, from Mt. Bellenden Ker, Malanda and Chillagoe, Qld.

Distribution: NE coastal, Qld. Ecology: terrestrial, noctidiurnal, omnivore, woodland; nest in ground layer.

Iridomyrmex itinerans (Lowne, 1865)

Iridomyrmex itinerans itinerans (Lowne, 1865)

Formica itinerans Lowne, B.T. (1865). Contributions to the natural history of Australian ants. *Entomologist* **2**: 275–280 [278]. Type data: syntypes, BMNH (probable) *W, from Sidney (=Sydney), N.S.W.

Distribution: SE coastal, N.S.W. Ecology: terrestrial, noctidiurnal, omnivore, woodland; nest in ground layer.

Iridomyrmex itinerans ballaratensis Forel, 1902

Iridomyrmex itinerans ballaratensis Forel, A. (1902). Fourmis nouvelles d'Australie. *Rev. Suisse Zool.* **10**: 405–548 [472]. Type data: syntypes, GMNH W,M, from Ballarat, Vic.

Distribution: SE coastal, Vic. Ecology: terrestrial, noctidiurnal, omnivore, woodland; nest in ground layer.

Iridomyrmex itinerans depilis Forel, 1902

Iridomyrmex itinerans depilis Forel, A. (1902). Fourmis nouvelles d'Australie. *Rev. Suisse Zool.* **10**: 405–548 [471]. Type data: syntypes, GMNH W, ANIC W, from Mackay, Qld.

Distribution: NE coastal, Qld. Ecology: terrestrial, noctidiurnal, omnivore, woodland; nest in ground layer.

Iridomyrmex itinerans perthensis Forel, 1902

Iridomyrmex itinerans perthensis Forel, A. (1902). Fourmis nouvelles d'Australie. *Rev. Suisse Zool.* **10**: 405–548 [472]. Type data: syntypes, GMNH W, from Perth, W.A.

Distribution: SW coastal, W.A. Ecology: terrestrial, noctidiurnal, omnivore, woodland; nest in ground layer.

Iridomyrmex longiceps Forel, 1907

Iridomyrmex longiceps Forel, A. (1907). Formicides du Musée National Hongrois. *Ann. Hist.- Nat. Mus. Natl. Hung.* **5**: 1–42 [30 June 1907] [27]. Type data: syntypes (probable), probably in GMNH or MNH, from Mt. Victoria, Blue Mts., N.S.W.

Distribution: SE coastal, N.S.W. Ecology: terrestrial, noctidiurnal, omnivore, woodland; nest in ground layer.

Iridomyrmex macrocephalus (Erichson, 1842)

Formica macrocephala Erichson, W.F. (1842). Beitrag zur Fauna von Vandiemansland mit besonderer rucksicht auf die geographische Verbreitung der Insecten. *Arch. Naturg.* **8**: 83–287 [259]. Type data: holotype (probable), ZMB *F, from Tas.

Distribution: Tas. Ecology: terrestrial, noctidiurnal, omnivore, woodland; nest in ground layer.

Iridomyrmex mattiroloi Emery, 1898

Iridomyrmex mattiroloi mattiroloi Emery, 1898

Iridomyrmex mattiroloi Emery, C. (1898). Descrizioni di formiche nuove Malesi e Australiane. Note sinonimiche. *Rec. Sess. Accad. Sci. Ist. Bologna (ns)* **2**: 231–245 [236]. Type data: syntypes, MCG *W, from Tas.

Distribution: Tas. Ecology: terrestrial, noctidiurnal, omnivore, woodland; nest in ground layer.

Iridomyrmex mattiroloi continentis Forel, 1907

Iridomyrmex mattiroloi continentis Forel, A. (1907). Formicidae. pp. 263–310 *in* Michaelsen, W. & Hartmeyer, R. (eds.) *Die Fauna Südwest-Australiens.* Jena : G. Fischer Vol. 1 [290]. Type data: syntypes, GMNH W,F, from Denham and Kalgoorlie, W.A.

Distribution: SW coastal, W plateau, W.A. Ecology: terrestrial, noctidiurnal, omnivore, woodland; nest in ground layer.

Iridomyrmex mattiroloi parcens Forel, 1907

Iridomyrmex mattiroloi parcens Forel, A. (1907). Formicides du Musée National Hongrois. *Ann. Hist.-Nat. Mus. Natl. Hung.* **5**: 1–42 [30 June 1907] [27]. Type data: syntypes (probable), probably in GMNH or MNH, from Mt. Victoria, Blue Mts., N.S.W.

Distribution: SE coastal, N.S.W. Ecology: terrestrial, noctidiurnal, omnivore, woodland; nest in ground layer.

Iridomyrmex mattiroloi splendens Forel, 1907

Iridomyrmex mattiroloi splendens Forel, A. (1907). Formicidae. pp. 263–310 *in* Michaelsen, W. & Hartmeyer, R. (eds.) *Die Fauna Südwest-Australiens.* Jena : G. Fischer Vol. 1 [290]. Type data: syntypes, GMNH W, from Donnybrook and Albany, W.A.

Distribution: SW coastal, W.A. Ecology: terrestrial, noctidiurnal, omnivore, woodland; nest in ground layer.

Iridomyrmex mjobergi Forel, 1915

Iridomyrmex mjobergi Forel, A. (1915). Results of Dr. E. Mjöbergs Swedish Scientific Expeditions to Australia 1910–1913. 2. Ameisen. *Ark. Zool.* **9**: 1–119 pls 1–3 [4 Dec. 1915] [77]. Type data: syntypes, GMNH W, other syntypes may exist, from Kimberley distr., W.A. and Cedar Creek and Malanda, Qld.

Distribution: NE coastal, N coastal, Qld., W.A. Ecology: terrestrial, noctidiurnal, omnivore, woodland; nest in ground layer.

Iridomyrmex nitidiceps E. André, 1896

Iridomyrmex nitidiceps André, E. (1896). Fourmis nouvelles d'Asie et d'Australie. *Rev. Entomol.* **15**: 251–265 [258]. Type data: syntypes, MNHP W, ANIC W, from Victorian Alps.

Distribution: Murray-Darling basin, Vic. Ecology: terrestrial, noctidiurnal, omnivore, woodland; nest in ground layer.

Iridomyrmex nitidus Mayr, 1862

Iridomyrmex nitidus nitidus Mayr, 1862

Iridomyrmex nitida Mayr, G.L. (1862). Myrmecologische Studien. *Verh. Zool.-Bot. Ges. Wien* **12**: Abhand. 649–776 [702 pl 19]. Type data: syntypes (probable), NHMW (probable) *W, from Australia (as New Holland).

Acantholepis tuberculatus Lowne, B.T. (1865). Contributions to the natural history of Australian ants. *Entomologist* **2**: 331–336 [332]. Type data: syntypes (probable), BMNH (probable) *W, from Sidney (=Sydney), N.S.W.

Synonymy that of Dalla Torre, C.G. De (1893). *Catalogus hymenopterorum hucusque descriptorum systematicus et synonymicus.* Formicidae (Heterogyna). Lipsiae : G. Engelmann Vol. 7 289 pp. [169].

Distribution: SE coastal, N.S.W. Ecology: terrestrial, noctidiurnal, omnivore, woodland; nest in ground layer.

Iridomyrmex nitidus clitellarius Viehmeyer, 1925

Iridomyrmex nitidus clitellarius Viehmeyer, H. (1925). Formiciden der australischen Faunenregion. *Entomol. Mitt.* **14**: 25–39 [32]. Type data: syntypes (probable), ZMB *W, from Trial Bay, N.S.W.

Distribution: SE coastal, N.S.W. Ecology: terrestrial, noctidiurnal, omnivore, woodland; nest in ground layer.

Iridomyrmex nitidus queenslandensis Forel, 1901

Iridomyrmex nitidus queenslandensis Forel, A. (1901). Formiciden aus dem Bismarck-Archipel, auf Grundlage des von Prof. Dr. F. Dahl gesammelten Materials bearbeitet. *Mitt. Zool. Mus. Berl.* **2**: 1–37 [3 Apr. 1901] [21]. Type data: syntypes, GMNH W,F,M, ANIC W, from Mackay, Qld.

Distribution: NE coastal, Qld. Ecology: terrestrial, noctidiurnal, omnivore, woodland; nest in ground layer.

Iridomyrmex obscurus Crawley, 1921

Iridomyrmex obscurus Crawley, W.C. (1921). New and little-known species of ants from various localities. *Ann. Mag. Nat. Hist.* (9) **7**: 87–97 [92]. Type data: syntypes, BMNH *W, from Koolpinyah, N.T.

Distribution: N coastal, N.T. Ecology: terrestrial, noctidiurnal, omnivore, woodland; nest in ground layer.

Iridomyrmex prociduus (Erichson, 1842)

Formica procidua Erichson, W.F. (1842). Beitrag zur Fauna von Vandiemansland mit besonderer rucksicht auf die geographische Verbreitung der Insecten. *Arch. Naturg.* **8**: 83–287 [259]. Type data: holotype (probable), ZMB *F, from Tas.

Distribution: Tas. Ecology: terrestrial, noctidiurnal, omnivore, woodland; nest in ground layer.

Iridomyrmex punctatissimus Emery, 1887

Iridomyrmex punctatissimus Emery, C. (1887). Catalogo delle formiche esistenti nelle collezioni del Museo Civico di Genova. Parte terza. Formiche della regione Indo-Malese e dell'Australia. *Ann. Mus. Civ. Stor. Nat. Giacomo Doria* **25**: 209–258 pls 3–4 [251]. Type data: syntypes, MCG *W, from Mt. Victoria, N.S.W.

Distribution: SE coastal, N.S.W. Ecology: terrestrial, noctidiurnal, omnivore, woodland; nest in ground layer.

Iridomyrmex purpureus (F. Smith, 1858)

Iridomyrmex purpureus purpureus (F. Smith, 1858)

Formica purpurea Smith, F. (1858). *Catalogue of hymenopterous insects in the collection of the British Museum.* Part 6. Formicidae. London : British Museum 216 pp. 14 pls [27 Mar. 1858] [40]. Publication date established from Donisthorpe, H. (1932). On the identity of Smith's types of Formicidae (Hymenoptera) collected by Alfred Russell Wallace in the Malay Archipelago, with descriptions of two new species. *Ann. Mag. Nat. Hist. (10)* **10**: 441–476. Type data: syntypes (probable), BMNH *W, from Melbourne, Vic.

Formica detecta Smith, F. (1858). *Catalogue of hymenopterous insects in the collection of the British Museum.* Part 6. Formicidae. London : British Museum 216 pp. 14 pls [27 Mar. 1858] [36]. Publication date established from Donisthorpe, H. (1932). On the identity of Smith's types of Formicidae (Hymenoptera) collected by Alfred Russell Wallace in the Malay Archipelago, with descriptions of two new species. *Ann. Mag. Nat. Hist. (10)* **10**: 441–476. Type data: syntypes (probable), BMNH *F, from Hunter River, N.S.W.

Liometopum aeneum Mayr, G.L. (1862). Myrmecologische Studien. *Verh. Zool.-Bot. Ges. Wien* **12**: Abhand. 649–776 [704 pl 19]. Type data: syntypes (probable), NHMW *F, from Australia (as New Holland).

Formica smithii Lowne, B.T. (1865). Contributions to the natural history of Australian ants. *Entomologist* **2**: 275–280 [276]. Type data: syntypes (probable), BMNH (probable) *W, from Sidney (=Sydney), N.S.W.

Synonymy that of Dalla Torre, C.G. De (1893). *Catalogus hymenopterorum hucusque descriptorum systematicus et synonymicus.* Formicidae (Heterogyna). Lipsiae : G. Engelmann Vol. 7 289 pp. [168].

Distribution: SE coastal, N.S.W., Vic. Ecology: terrestrial, noctidiurnal, omnivore, woodland; nest in ground layer. Biological references: Greenslade, P.J.M. (1975). Dispersion and history of a population of the meat ant *Iridomyrmex purpureus* (Hymenoptera : Formicidae). *Aust. J. Zool.* **23**: 495–510.

Iridomyrmex purpureus castrae Viehmeyer, 1925

Iridomyrmex detectus castrae Viehmeyer, H. (1925). Formiciden der australischen Faunenregion. *Entomol. Mitt.* **14**: 25–39 [31]. Type data: syntypes, ZMB *W, from Liverpool, N.S.W.

Distribution: SE coastal, N.S.W. Ecology: terrestrial, noctidiurnal, omnivore, woodland; nest in ground layer.

Iridomyrmex purpureus sanguinea Forel, 1910

Iridomyrmex detectus sanguinea Forel, A. (1910). Formicides australiens reçus de MM. Froggatt et Rowland Turner. *Rev. Suisse Zool.* **18**: 1–94 [53]. Type data: syntypes, GMNH W, ANIC W, from Mackay and Townsville, Qld.

Distribution: NE coastal, Qld. Ecology: terrestrial, noctidiurnal, omnivore, woodland; nest in ground layer.

Iridomyrmex purpureus viridiaeneus Viehmeyer, 1913

Iridomyrmex detectus viridiaeneus Viehmeyer, H. (1913). Neue und unvollständig bekannte Ameisen der Alten Welt. *Arch. Naturg.* **79A**(12): 24–60 [41]. Type data: syntypes, ZMB *W, ANIC W, from Killalpaninna, S.A.

Distribution: Lake Eyre basin, S.A. Ecology: terrestrial, noctidiurnal, omnivore, woodland; nest in ground layer. Biological references: Greenslade, P. (1981). Temperature limits to trailing activity in the Australian arid-zone ant *Iridomyrmex purpureus* form *viridiaeneus*. *Aust. J. Zool.* **29**: 621–630 (foraging behaviour).

Iridomyrmex rufoniger (Lowne, 1865)

Iridomyrmex rufoniger rufoniger (Lowne, 1865)

Formica rufonigra Lowne, B.T. (1865). Contributions to the natural history of Australian ants. *Entomologist* **2**: 275–280 [279]. Type data: syntypes (probable), BMNH (probable) *W, from Sidney (=Sydney), N.S.W.

Acantholepis mamillatus Lowne, B.T. (1865). Contributions to the natural history of Australian ants. *Entomologist* **2**: 331–336 [333]. Type data: syntypes (probable), BMNH (probable) *W, from Sidney (=Sydney), N.S.W.

Synonymy that of Dalla Torre, C.G. De (1893). *Catalogus hymenopterorum hucusque descriptorum systematicus et synonymicus.* Formicidae (Heterogyna). Lipsiae : G. Engelmann Vol. 7 289 pp. [169].

Distribution: SE coastal, N.S.W. Ecology: terrestrial, noctidiurnal, omnivore, woodland; nest in ground layer.

Iridomyrmex rufoniger domestica Forel, 1910

Iridomyrmex rufoniger domestica Forel, A. (1910). Formicides australiens reçus de MM. Froggatt et Rowland Turner. *Rev. Suisse Zool.* **18**: 1–94 [51]. Type data: syntypes, GMNH W,F,M, ANIC W, from Howlong and Richmond near Sydney, N.S.W.

Distribution: SE coastal, N.S.W. Ecology: terrestrial, noctidiurnal, omnivore, woodland; nest in ground layer.

Iridomyrmex rufoniger incerta Forel, 1902

Iridomyrmex rufoniger incertus Forel, A. (1902). Fourmis nouvelles d'Australie. *Rev. Suisse Zool.* **10**: 405–548 [466]. Type data: syntypes, GMNH W, ANIC W, from Ralum, Bismarck Archipelago.

Distribution: NE coastal, Qld. Ecology: terrestrial, noctidiurnal, omnivore, woodland; nest in ground layer.

Iridomyrmex rufoniger pallidus Forel, 1901

Iridomyrmex rufoniger pallidus Forel, A. (1901). Formiciden aus dem Bismarck-Archipel, auf Grundlage des von Prof. Dr. F. Dahl gesammelten Materials bearbeitet. *Mitt. Zool. Mus. Berl.* 2: 1–37 [3 Apr. 1901] [22]. Type data: syntypes, GMNH W, ANIC W, from Mackay, Qld.

Distribution: NE coastal, Qld. Ecology: terrestrial, noctidiurnal, omnivore, woodland; nest in ground layer.

Iridomyrmex rufoniger septentrionalis Forel, 1902

Iridomyrmex rufoniger septentrionalis Forel, A. (1902). Fourmis nouvelles d'Australie. *Rev. Suisse Zool.* 10: 405–548 [465]. Type data: syntypes, GMNH W,F,M, ANIC W, from Mackay, Qld.

Distribution: NE coastal, Qld. Ecology: terrestrial, noctidiurnal, omnivore, woodland; nest in ground layer.

Iridomyrmex rufoniger suchieri Forel, 1907

Iridomyrmex rufoniger suchieri Forel, A. (1907). Formicidae. pp. 263–310 *in* Michaelsen, W. & Hartmeyer, R. (eds.) *Die Fauna Südwest-Australiens.* Jena : G. Fischer Vol. 1 [291]. Type data: syntypes, GMNH W,M,F, ANIC W, from Day Dawn, Yalgoo, Eradu, Dougarra (=Dongarra), Woorolloo and Subiaco, W.A.

Distribution: SW coastal, NW coastal, W.A. Ecology: terrestrial, noctidiurnal, omnivore, woodland; nest in ground layer.

Iridomyrmex rufoniger victorianus Forel, 1902

Iridomyrmex rufoniger victorianus Forel, A. (1902). Fourmis nouvelles d'Australie. *Rev. Suisse Zool.* 10: 405–548 [466]. Type data: syntypes, GMNH W,F, ANIC W, from Ballarat, Vic.

Distribution: SE coastal, Vic. Ecology: terrestrial, noctidiurnal, omnivore, woodland; nest in ground layer.

Iridomyrmex vicina Clark, 1934

Iridomyrmex vicina Clark, J. (1934). Ants from the Otway Ranges. *Mem. Natl. Mus. Vict.* 8: 48–73 [62 pl 4]. Type data: syntypes, NMV *W,F, from Beech Forest, Vic.

Distribution: SE coastal, Vic. Ecology: terrestrial, noctidiurnal, omnivore, woodland; nest in ground layer.

Iridomyrmex viridigaster Clark, 1941

Iridomyrmex viridigaster Clark, J. (1941). Australian Formicidae. Notes and new species. *Mem. Natl. Mus. Vict.* 12: 71–94 [87 pl 13]. Type data: syntypes (probable), NMV *W, from Patho, Vic.

Distribution: Murray-Darling basin, Vic. Ecology: terrestrial, noctidiurnal, omnivore, woodland; nest in ground layer.

Leptomyrmex Mayr, 1862

Leptomyrmex Mayr, G.L. (1862). Myrmecologische Studien. *Verh. Zool.-Bot. Ges. Wien* 12: Abhand. 649–776 [695 pl 19]. Type species *Formica erythrocephala* Fabricius, 1775 by monotypy.

This group is also found in New Guinea and New Caledonia.

Leptomyrmex darlingtoni Wheeler, 1934

Leptomyrmex darlingtoni darlingtoni Wheeler, 1934

Leptomyrmex darlingtoni Wheeler, W.M. (1934). A second revision of the ants of the genus *Leptomyrmex* Mayr. *Bull. Mus. Comp. Zool.* 77: 67–118 [104]. Type data: syntypes, MCZ *W,M, from Lankelly Creek in the McIlthwaite (=McIlwraith) Range, Cape York Peninsula, Qld.

Distribution: NE coastal, Qld. Ecology: terrestrial, noctidiurnal, omnivore, closed forest; nest in ground layer.

Leptomyrmex darlingtoni fascigaster Wheeler, 1934

Leptomyrmex darlingtoni fascigaster Wheeler, W.M. (1934). A second revision of the ants of the genus *Leptomyrmex* Mayr. *Bull. Mus. Comp. Zool.* 77: 67–118 [107]. Type data: syntypes, MCZ *W, from Coen, Cape York Peninsula, Qld.

Distribution: NE coastal, Qld. Ecology: terrestrial, noctidiurnal, omnivore, woodland, open forest; nest in ground layer.

Leptomyrmex darlingtoni jucundus Wheeler, 1934

Leptomyrmex darlingtoni jucundus Wheeler, W.M. (1934). A second revision of the ants of the genus *Leptomyrmex* Mayr. *Bull. Mus. Comp. Zool.* 77: 67–118 [107]. Type data: syntypes, MCZ *W, from Coen, Cape York Peninsula, Qld.

Distribution: NE coastal, Qld. Ecology: terrestrial, noctidiurnal, omnivore, open forest; nest in ground layer.

Leptomyrmex erythrocephalus (Fabricius, 1775)

Leptomyrmex erythrocephalus erythrocephalus (Fabricius, 1775)

Formica erythrocephala Fabricius, J.C. (1775). *Systema Entomologiae,* sistens insectorum classes, ordines, genera, species, adiectis synonymis, locis, descriptionibus, observationibus. Flensburgi et Lipsiae [391]. Type data: holotype (probable), BMNH W, from Australia (as New Holland).

Distribution: NE coastal, SE coastal, Qld., N.S.W. Ecology: terrestrial, noctidiurnal, omnivore, woodland, open forest; nest in ground layer.

Leptomyrmex erythrocephalus basirufus Wheeler, 1934

Leptomyrmex erythrocephalus basirufus Wheeler, W.M. (1934). A second revision of the ants of the genus *Leptomyrmex* Mayr. *Bull. Mus. Comp. Zool.* **77**: 67–118 [90]. Type data: syntypes, MCZ *W, from Buderim Mts. and Bundaberg, Qld.

Distribution: NE coastal, Qld. Ecology: terrestrial, noctidiurnal, omnivore, woodland, open forest; nest in ground layer.

Leptomyrmex erythrocephalus brunneiceps Wheeler, 1934

Leptomyrmex erythrocephalus brunneiceps Wheeler, W.M. (1934). A second revision of the ants of the genus *Leptomyrmex* Mayr. *Bull. Mus. Comp. Zool.* **77**: 67–118 [88]. Type data: syntypes, MCZ *W, from Mt. Wilson and Wentworth Falls, N.S.W.

Distribution: SE coastal, N.S.W. Ecology: terrestrial, noctidiurnal, omnivore, woodland, open forest; nest in ground layer.

Leptomyrmex erythrocephalus clarki Wheeler, 1934

Leptomyrmex erythrocephalus clarki Wheeler, W.M. (1934). A second revision of the ants of the genus *Leptomyrmex* Mayr. *Bull. Mus. Comp. Zool.* **77**: 67–118 [117]. Type data: syntypes, MCZ *W, from Fletcher, Qld.

Distribution: Murray-Darling basin, Qld. Ecology: terrestrial, noctidiurnal, omnivore, woodland, open forest; nest in ground layer.

Leptomyrmex erythrocephalus cnemidatus Wheeler, 1915

Leptomyrmex erythrocephalus cnemidatus Wheeler, W.M. (1915). The Australian honey-ants of the genus *Leptomyrmex* Mayr. *Proc. Am. Acad. Arts Sci.* **51**: 253–286 [268]. Type data: holotype, MCZ *W, from N.S.W.

Distribution: NE coastal, SE coastal, Qld., N.S.W. Ecology: terrestrial, noctidiurnal, omnivore, open forest, closed forest; nest in ground layer.

Leptomyrmex erythrocephalus decipiens Wheeler, 1915

Leptomyrmex erythrocephalus decipiens Wheeler, W.M. (1915). The Australian honey-ants of the genus *Leptomyrmex* Mayr. *Proc. Am. Acad. Arts Sci.* **51**: 253–286 [268]. Type data: syntypes, MCZ *W, from Gin-Gin, Qld.

Distribution: NE coastal, Qld. Ecology: terrestrial, noctidiurnal, omnivore, open forest, closed forest; nest in ground layer.

Leptomyrmex erythrocephalus mandibularis Wheeler, 1915

Leptomyrmex erythrocephalus mandibularis Wheeler, W.M. (1915). The Australian honey-ants of the genus *Leptomyrmex* Mayr. *Proc. Am. Acad. Arts Sci.* **51**: 253–286 [268]. Type data: holotype, MCZ *W, from vicinity of Sydney, N.S.W.

Distribution: SE coastal, N.S.W. Ecology: terrestrial, noctidiurnal, omnivore, woodland, open forest; nest in ground layer.

Leptomyrmex erythrocephalus rufithorax Forel, 1915

Leptomyrmex erythrocephalus rufithorax Forel, A. (1915). Results of Dr. E. Mjöbergs Swedish Scientific Expeditions to Australia 1910–1913. 2. Ameisen. *Ark. Zool.* **9**: 1–119 pls 1–3 [4 Dec. 1915] [83]. Type data: syntypes, GMNH W, ANIC W, other syntypes may exist, from Mt. Tambourine (=Tamborine Mt.) and Blackal (=Blackall) Range, Qld.

Distribution: NE coastal, Qld. Ecology: terrestrial, noctidiurnal, omnivore, open forest, closed forest; nest in ground layer.

Leptomyrmex erythrocephalus unctus Wheeler, 1934

Leptomyrmex erythrocephalus unctus Wheeler, W.M. (1934). A second revision of the ants of the genus *Leptomyrmex* Mayr. *Bull. Mus. Comp. Zool.* **77**: 67–118 [87]. Type data: syntypes, MCZ *W, from Condor Creek, near Canberra, A.C.T.

Distribution: Murray-Darling basin, A.C.T. Ecology: terrestrial, noctidiurnal, omnivore, woodland, open forest; nest in ground layer.

Leptomyrmex erythrocephalus venustus Wheeler, 1934

Leptomyrmex erythrocephalus venustus Wheeler, W.M. (1934). A second revision of the ants of the genus *Leptomyrmex* Mayr. *Bull. Mus. Comp. Zool.* **77**: 67–118 [87]. Type data: syntypes, MCZ *W,F, from Mt. Tomah, N.S.W.

Distribution: SE coastal, N.S.W. Ecology: terrestrial, noctidiurnal, omnivore, woodland, open forest; nest in ground layer.

Leptomyrmex froggatti Forel, 1910

Leptomyrmex froggatti Forel, A. (1910). Formicides australiens reçus de MM. Froggatt et Rowland Turner. *Rev. Suisse Zool.* **18**: 1–94 [57]. Type data: syntypes, GMNH W,M, ANIC W, from N.S.W.

Distribution: SE coastal, N.S.W. Ecology: terrestrial, noctidiurnal, omnivore, woodland, open forest; nest in ground layer.

Leptomyrmex mjobergi Forel, 1915

Leptomyrmex mjobergi Forel, A. (1915). Results of Dr. E. Mjöbergs Swedish Scientific Expeditions to Australia 1910–1913. 2. Ameisen. *Ark. Zool.* **9**: 1–119 pls 1–3 [4 Dec. 1915] [84]. Type data: syntypes, GMNH W, ANIC W, other syntypes may exist, from Colosseum, Tolga and Herberton, Qld.

Distribution: NE coastal, Qld. Ecology: terrestrial, noctidiurnal, omnivore, closed forest; nest in ground layer.

Leptomyrmex nigriventris (Guérin, 1831)

Leptomyrmex nigriventris nigriventris (Guérin, 1831)

Formica nigriventris Guérin-Meneville, F.E. (1831). Chapter 12, Insectes. in Duperrey, M.L.I. (1838). *Voyage autour du monde, exécuté par ordre du Roi, sur la corvette de La Majesté, La Coquille, pendant les années 1822, 1823, 1824 et 1825*. Vol. 2 part 2 division 1 : 57–302 Atlas (1830–1832) Ins pls 1–21 [205 pl 8 fig 4]. Publication date established from Bequaert, J. (1926). The date of publication of the Hymenoptera and Diptera described by Guérin in Duperrey's Voyage de La Coquille". *Entomol Mitt.* **15**: 186–195 [20 Mar. 1926]. Type data: uncerain, MNHP (probable) *W, from Port Jackson, N.S.W.

Distribution: SE coastal, N.S.W. Ecology: terrestrial, noctidiurnal, omnivore, woodland, open forest; nest in ground layer.

Leptomyrmex nigriventris hackeri Wheeler, 1934

Leptomyrmex nigriventris hackeri Wheeler, W.M. (1934). A second revision of the ants of the genus *Leptomyrmex* Mayr. *Bull. Mus. Comp. Zool.* **77**: 67–118 [99]. Type data: syntypes, MCZ *W, from Stradbroke Is., Qld.

Distribution: NE coastal, Qld. Ecology: terrestrial, noctidiurnal, omnivore, woodland, open forest; nest in ground layer.

Leptomyrmex nigriventris tibialis Emery, 1895

Leptomyrmex nigriventris tibialis Emery, C. (1895). Descriptions de quelques fourmis nouvelles d'Australie. *Ann. Soc. Entomol. Belg.* **39**: 345–358 [351]. Type data: syntypes, MCG *W, from N Qld.

Distribution: NE coastal, SE coastal, N.S.W., Qld. Ecology: terrestrial, noctidiurnal, omnivore, open forest, closed forest; nest in ground layer.

Leptomyrmex unicolor Emery, 1895

Leptomyrmex unicolor Emery, C. (1895). Descriptions de quelques fourmis nouvelles d'Australie. *Ann. Soc. Entomol. Belg.* **39**: 345–358 [352]. Type data: syntypes, MCG *W, from Cairus (=Cairns), Qld.

Distribution: NE coastal, Qld. Ecology: terrestrial, noctidiurnal, omnivore, closed forest; nest in ground layer.

Leptomyrmex varians Emery, 1895

Leptomyrmex varians varians Emery, 1895

Leptomyrmex varians Emery, C. (1895). Descriptions de quelques fourmis nouvelles d'Australie. *Ann. Soc. Entomol. Belg.* **39**: 345–358 [352]. Type data: syntypes, NHMW (probable) *W, from Rockhampton, Qld.

Distribution: NE coastal, Qld. Ecology: terrestrial, noctidiurnal, omnivore, woodland, open forest; nest in ground layer.

Leptomyrmex varians angusticeps Santschi, 1929

Leptomyrmex varians angusticeps Santschi, F. (1929). Mélange myrmécologique. *Wien Entomol. Ztg.* **46**: 84–93 [15 Sept. 1929] [93]. Type data: syntypes, BNHM M, from Beyfield (=Byfield), Qld.

Distribution: NE coastal, Qld. Ecology: terrestrial, noctidiurnal, omnivore, woodland, open forest; nest in ground layer.

Leptomyrmex varians quadricolor Wheeler, 1934

Leptomyrmex varians quadricolor Wheeler, W.M. (1934). A second revision of the ants of the genus *Leptomyrmex* Mayr. *Bull. Mus. Comp. Zool.* **77**: 67–118 [104]. Type data: syntypes, MCZ *W, from Lankelly Creek in the McIlthwaite (=McIlwraith) Range, Cape York Peninsula, Qld.

Distribution: NE coastal, Qld. Ecology: terrestrial, noctidiurnal, omnivore, woodland, open forest; nest in ground layer.

Leptomyrmex varians rothneyi Forel, 1902

Leptomyrmex varians rothneyi Forel, A. (1902). Fourmis nouvelles d'Australie. *Rev. Suisse Zool.* **10**: 405–548 [473]. Type data: syntypes, GMNH W, from Brisbane, Qld.

Distribution: NE coastal, Qld. Ecology: terrestrial, noctidiurnal, omnivore, woodland, open forest; nest in ground layer.

Leptomyrmex varians ruficeps Emery, 1895

Leptomyrmex varians ruficeps Emery, C. (1895). Descriptions de quelques fourmis nouvelles d'Australie. *Ann. Soc. Entomol. Belg.* **39**: 345–358 [352]. Type data: syntypes, MCG *W, from Mt. Bellenden Ker, Qld.

Distribution: NE coastal, Qld. Ecology: terrestrial, noctidiurnal, omnivore, woodland, open forest; nest in ground layer.

Leptomyrmex varians rufipes Emery, 1895

Leptomyrmex varians rufipes Emery, C. (1895). Descriptions de quelques fourmis nouvelles d'Australie. *Ann. Soc. Entomol. Belg.* **39**: 345–358 [352]. Type data: syntypes, MCG *W, from Laidely (=Laidley) and Brisbane, Qld.

Distribution: NE coastal, SE coastal, N.S.W., Qld. Ecology: terrestrial, noctidiurnal, omnivore, woodland, open forest; nest in ground layer.

Leptomyrmex wiburdi Wheeler, 1915

Leptomyrmex wiburdi wiburdi Wheeler, 1915

Leptomyrmex wiburdi Wheeler, W.M. (1915). The Australian honey-ants of the genus *Leptomyrmex* Mayr. *Proc. Am. Acad. Arts Sci.* **51**: 253–286 [272]. Type data: syntypes, MCZ *W, from Jenolan Caves, N.S.W.

Distribution: SE coastal, N.S.W. Ecology: terrestrial, noctidiurnal, omnivore, woodland, open forest; nest in ground layer.

Leptomyrmex wiburdi pictus Wheeler, 1915

Leptomyrmex wiburdi pictus Wheeler, W.M. (1915). The Australian honey-ants of the genus *Leptomyrmex* Mayr. *Proc. Am. Acad. Arts Sci.* **51**: 253–286 [274]. Type data: syntypes, MCZ *W, from Bulli Pass and Katoomba, N.S.W.

Distribution: SE coastal, N.S.W. Ecology: terrestrial, noctidiurnal, omnivore, woodland, open forest; nest in ground layer.

Tapinoma Förster, 1850

Tapinoma Förster, A. (1850). *Hymenopterologische Studien.* Formicariae. pp. 1–74 Aachen : Ernst ter Meer Vol. 1 [43]. Type species *Formica erratica* Latrielle, 1798 (as *Tapinoma collina* Förster, 1850) by monotypy.

This group is also found in the Neotropical, Nearctic, south Palearctic, Ethiopian, Malagasy and Oriental regions; New Guinea, east Melanesia, New Caledonia in the Australian Region, see Brown, W.L. jr. (1973). A comparison of the Hylean and Congo-West African rain forest ant faunas. pp. 161–185 *in* Meggers, B.J., Ayensu, E.S. & Duckworth, W.D. (eds.) *Tropical forest ecosystems in Africa and South America: a comparative review.* Washington : Smithsonian Institution Press.

Tapinoma minutum Mayr, 1862

Tapinoma minutum minutum Mayr, 1862

Tapinoma minutum Mayr, G.L. (1862). Myrmecologische Studien. *Verh. Zool.-Bot. Ges. Wien* **12**: Abhand. 649–776 [703 pl 19]. Type data: syntypes, NHMW *W, from Sidney (=Sydney), N.S.W.

Distribution: SE coastal, NE coastal, N coastal, Qld., N.T., N.S.W. Ecology: terrestrial, noctidiurnal, omnivore, open forest, closed forest; nest in ground layer or arboreal.

Tapinoma minutum broomense Forel, 1915

Tapinoma minutum broomensis Forel, A. (1915). Results of Dr. E. Mjöbergs Swedish Scientific Expeditions to Australia 1910–1913. 2. Ameisen. *Ark. Zool.* **9**: 1–119 pls 1–3 [4 Dec. 1915] [83]. Type data: syntypes, GMNH W, other syntypes may exist, from Broome, W.A.

Distribution: N coastal, W.A. Ecology: terrestrial, noctidiurnal, omnivore, open forest; nest in ground layer or arboreal.

Tapinoma minutum cephalicum Santschi, 1928

Tapinoma (Micromyrma) minutum cephalicum Santschi, F. (1928). Nouvelles fourmis d'Australie. *Bull. Soc. Vaud. Sci. Nat.* **56**: 465–483 [30 Aug. 1928] [472]. Type data: syntypes, BNHM *W,F,M, from Townsville, Qld.

Distribution: NE coastal, Qld. Ecology: terrestrial, noctidiurnal, omnivore, open forest, closed forest; nest in ground layer or arboreal.

Tapinoma minutum integrum Forel, 1902

Tapinoma minutum integrum Forel, A. (1902). Fourmis nouvelles d'Australie. *Rev. Suisse Zool.* **10**: 405–548 [476]. Type data: syntypes, GMNH W,M, ANIC W, from Mackay and Townsville, Qld.

Distribution: NE coastal, Qld. Ecology: terrestrial, noctidiurnal, omnivore, open forest, closed forest; nest in ground layer or arboreal.

Tapinoma rottnestense Wheeler, 1934

Tapinoma (Micromyrma) rottnestense Wheeler, W.M. (1934). Contributions to the fauna of Rottnest Island, Western Australia No. IX. The ants. *J. R. Soc. West. Aust.* **20**: 137–163 [5 Oct. 1934] [150]. Type data: syntypes, MCZ *W, from Lady Edeline Beach, Rottnest Is., W.A.

Distribution: SW coastal, W.A. Ecology: terrestrial, noctidiurnal, omnivore, woodland, open forest; nest in ground layer or arboreal.

Technomyrmex Mayr, 1872

Technomyrmex Mayr, G.L. (1872). Formicidae Borneenses collectae a J. Doria et O. Beccari in territorio Sarawak annis 1865–1867. *Ann. Mus. Civ. Stor. Nat. Giacomo Doria* **2**: 133–155 [147]. Type species *Technomyrmex strenuus* Mayr, 1872 by monotypy.

Aphantolepis Wheeler, W.M. (1930). Two new genera of ants from Australia and the Philippines. *Psyche Camb.* **37**: 41–47 [44]. Type species *Aphantolepis quadricolor* Wheeler, 1930 by monotypy.

Synonymy that of Brown, W.L. jr. (1953). Characters and synonymies among the genera of ants. Part II. *Breviora* **18**: 1–8 [23 Sept. 1953] [5].

This group is also found in the Ethiopian, Malagasy and Oriental regions; widespread in the Australian Region, see Brown, W.L. jr. (1973). A comparison of the Hylean and Congo-West African rain forest ant faunas. pp. 161–185 *in* Meggers, B.J., Ayensu, E.S. & Duckworth, W.D. (eds.) *Tropical forest ecosystems in Africa and South America: a comparative review.* Washington : Smithsonian Institution Press.

Technomyrmex albipes (F. Smith, 1861)

Formica (Tapinoma) albipes Smith, F. (1861). Catalogue of hymenopterous insects collected by Mr A.R. Wallace in the islands of Ceram, Celebes, Ternate and Gilolo. *J. Linn. Soc. Zool.* **6**: 36–66 [38]. Type data: status unknown, ?BMNH, from India.

Technomyrmex albipes cedarensis Forel, 1915

Technomyrmex albipes cedarensis Forel, A. (1915). Results of Dr. E. Mjöbergs Swedish Scientific Expeditions to Australia 1910–1913. 2. Ameisen. *Ark. Zool.* **9**: 1–119 pls 1–3 [4 Dec. 1915] [85]. Type data: syntypes, GMNH W,F, ANIC W, other syntypes may exist, from Cedar Creek, Qld.

Distribution: NE coastal, Qld. Ecology: terrestrial, noctidiurnal, omnivore, open forest, closed forest; nest in ground layer.

Technomyrmex bicolor Emery, 1893

Technomyrmex bicolor Emery, C. (1893). Voyage de M.E. Simon à l'Île de Ceylon (Janvier-Février 1892), 3ᵉ Mémoire(1), Formicides. *Ann. Soc. Entomol. Fr.* **62**: 239–258 [249]. Type data: status unknown, ?MCG, from Ceylon.

Technomyrmex bicolor antonii Forel, 1902

Technomyrmex bicolor antonii Forel, A. (1902). Fourmis nouvelles d'Australie. *Rev. Suisse Zool.* **10**: 405–548 [475]. Type data: syntypes, GMNH W,M, ANIC W, from Mackay, Qld.

Distribution: NE coastal, Qld. Ecology: terrestrial, noctidiurnal, omnivore, woodland, open forest; nest in ground layer.

Technomyrmex jocosus Forel, 1910

Technomyrmex jocosus Forel, A. (1910). Formicides australiens reçus de MM. Froggatt et Rowland Turner. *Rev. Suisse Zool.* **18**: 1–94 [56]. Type data: syntypes, GMNH W, ANIC W, from Yarra distr., Vic.

Distribution: SE coastal, Vic. Ecology: terrestrial, noctidiurnal, omnivore, woodland, open forest; nest in ground layer.

Technomyrmex quadricolor (Wheeler, 1930)

Aphantolepis quadricolor Wheeler, W.M. (1930). Two new genera of ants from Australia and the Philippines. *Psyche Camb.* **37**: 41–47 [44]. Type data: holotype, MCZ *W, from Cairns distr., Qld.

Distribution: NE coastal, Qld. Ecology: terrestrial, noctidiurnal, omnivore, closed forest; nest in ground layer.

Technomyrmex sophiae Forel, 1902

Technomyrmex sophiae Forel, A. (1902). Fourmis nouvelles d'Australie. *Rev. Suisse Zool.* **10**: 405–548 [474]. Type data: syntypes, GMNH W,F, ANIC W, from Mackay, Qld.

Distribution: NE coastal, Qld. Ecology: terrestrial, noctidiurnal, omnivore, open forest, closed forest; nest in ground layer.

Turneria Forel, 1895

Turneria Forel, A. (1895). Nouvelles fourmis d'Australie, récoltées à The Ridge, Mackay, Queensland par M. Gilbert Turner. *Ann. Soc. Entomol. Belg.* **39**: 417–428 [419]. Type species *Turneria bidentata* Forel, 1895 by monotypy.

This group is also found in New Guinea and east Melanesia.

Turneria bidentata Forel, 1895

Turnesia bidentata Forel, A. (1895). Nouvelles fourmis d'Australie, récoltée à The Ridge, Mackay, Queensland par M. Gilbert Turner. *Ann. Soc. Entomol. Belg.* **39**: 417–428 [419]. Type data: syntypes (probable), GMNH W, from Mackay, Qld.

Distribution: NE coastal, Qld. Ecology: terrestrial, diurnal, omnivore, closed forest; nest arboreal.

Turneria frenchi Forel, 1911

Turneria frenchi Forel, A. (1911). Ameisen aus Java beobachtet und gesammelt von Herrn Edward Jacobson. *Notes Leyden Mus.* **33**: 193–218 [29 Apr. 1911] [207]. Type data: syntypes (probable), RIB *W, from Australia.

Distribution: NE coastal, Qld. Ecology: terrestrial, diurnal, omnivore, closed forest; nest arboreal.

FORMICINAE

Acropyga Roger, 1862

Acropyga Roger, J. (1862). Einige neue exotische Ameisen - Gattungen und Arten. *Berl. Entomol. Z.* **6**: 233–254 [242 pl 1]. Type species *Acropya acutiventris* Roger, 1862 by monotypy.

This group is also found in the Neotropical, south Nearctic, north Ethiopian and Oriental regions; New Guinea and east Melanesia in the Australian Region, see Brown, W.L. jr. (1973). A comparison of the Hylean and Congo-West African rain forest ant faunas. pp. 161–185 *in* Meggers, B.J., Ayensu, E.S. & Duckworth, W.D. (eds.) *Tropical forest ecosystems in Africa and South America: a comparative review.* Washington : Smithsonian Institution Press.

Acropyga indistincta Crawley, 1923

Acropyga indistincta Crawley, W.C. (1923). Myrmecological notes - new Australian Formicidae. *Entomol. Rec. J. Var.* **35**: 177–179 [178]. Type data: syntypes, OUM *W, from Mundaring, W.A.

Distribution: SW coastal, W.A. Ecology: terrestrial, noctidiurnal, omnivore, woodland, open forest; nest in ground layer.

Acropyga moluccana Mayr, 1878

Acropyga moluccana Mayr, G.L. (1878). Beiträge zur Ameisen-Fauna Asiens. *Verh. Zool-Bot. Ges. Wien* **28**: 645–686 [658]. Type data: status unknown, ?NHMW, from Ceram Is., Indonesia.

Acropyga moluccana australis Forel, 1902

Acropyga moluccana australis Forel, A. (1902). Fourmis nouvelles d'Australie. *Rev. Suisse Zool.* **10**: 405–548 [477]. Type data: syntypes, GMNH W, ANIC W, from Mackay, Qld.

Distribution: NE coastal, N coast, N Gulf, W.A., N.T., Qld. Ecology: terrestrial, noctidiurnal, omnivore, open forest, closed forest; nest in ground layer.

Acropyga myops Forel, 1910

Acropyga myops Forel, A. (1910). Formicides australiens reçus de MM. Froggatt et Rowland Turner. *Rev. Suisse Zool.* **18**: 1–94 [59]. Type data: syntypes, GMNH W, ANIC W, from Bombala, N.S.W.

Distribution: SE coastal, N.S.W. Ecology: terrestrial, noctidiurnal, omnivore, open forest, closed forest; nest in ground layer.

Anoplolepis Santschi, 1914

Anoplolepis Santschi, F. (1914). Formicidae. *in, Voyage de Ch. Alluaud et R. Jeannel en Afrique orientale, 1911–1912.* Hymenoptera. **2**: 41–148 [25 Feb. 1914] [123 pls 2–3] [proposed with subgeneric rank in *Plagiolepis* Mayr, 1861]. Type species *Formica longipes* Jerdon, 1851 by original designation.

This group is also found in the Ethiopian and Oriental regions; New Guinea, east Melanesia and parts of Polynesia in the Australian Region, see Brown, W.L. jr. (1973). A comparison of the Hylean and Congo-West African rain forest ant faunas. pp. 161–185 *in* Meggers, B.J., Ayensu, E.S. & Duckworth, W.D. (eds.) *Tropical forest ecosystems in Africa and South America: a comparative review.* Washington : Smithsonian Institution Press.

Anoplolepis longipes (Jerdon, 1851)

Formica longipes Jerdon, T.C. (1851). A catalogue of the species of ants found in southern India. *Madras J. Lit. Sci.* **17**: 103–127 [122]. Type data: unknown, from India.

Distribution: N coastal, NE coastal, N Gulf, N.T., Qld.; widespread in SE Asia and Pacific, a "tramp" species of African origin. Ecology: terrestrial, arboreal, diurnal, omnivore, open forest, closed forest; nest in ground layer or aboreal.

Calomyrmex Emery, 1895

Calomyrmex Emery, C. (1895). Die Gattung *Dorylus* Fab. und die systematische Einteilung der Formiciden. *Zool. Jb. (Syst.)* **8**: 685–778 [8 Oct. 1895] [772 pls 14–17]. Type species *Formica laevissima* F. Smith, 1859 by subsequent designation, see Wheeler, W.M. (1911). A list of the type species of the genera and subgenera of Formicidae. *Ann. N.Y. Acad. Sci.* **21**: 157–175 [17 Oct. 1911].

This group is also found in New Guinea.

Calomyrmex albertisi (Emery, 1887)

Camponotus albertisi Emery, C. (1887). Catalogo delle formiche esistenti nelle collezioni del Museo Civico di Genova. Parte terza. Formiche della regione Indo-Malese e dell'Australia. *Ann. Mus. Civ. Stor. Nat. Giacomo Doria* **25**: 209–258 pls 3–4 [221]. Type data: holotype, MCG *W, from Fly River, New Guinea.

Distribution: NE coastal, Qld. Ecology: terrestrial, noctidiurnal, omnivore, woodland, open forest, closed forest; nest in soil.

Calomyrmex albopilosus (Mayr, 1876)

Calomyrmex albopilosus albopilosus (Mayr, 1876)

Camponotus albopilosus Mayr, G.L. (1876). Die australischen Formiciden. *J. Mus. Godeffroy* **5**: 56–115 [61]. Type data: syntypes, NHMW W,M,F, from Rockhampton, Peak Downs and Gayndah, Qld.

Distribution: NE coastal, Qld. Ecology: terrestrial, noctidiurnal, omnivore, open forest, closed forest; nest in soil.

Calomyrmex albopilosus wienandsi (Forel, 1910)

Camponotus (Calomyrmex) albopilosus wienandsi Forel, A. (1910). Formicides australiens reçus de MM. Froggatt et Rowland Turner. *Rev. Suisse Zool.* **18**: 1–94 [82]. Type data: syntypes, GMNH W,F, ANIC W, from Gunnedah, N.S.W.

Distribution: Murray-Darling basin, N.S.W. Ecology: terrestrial, noctidiurnal, omnivore, woodland, open forest; nest in soil.

Calomyrmex glauerti Clark, 1930

Calomyrmex glauerti Clark, J. (1930). Some new Australian Formicidae. *Proc. R. Soc. Vict.* **42**: 116–128 [10 Mar. 1930] [125]. Type data: holotype, WAM 22–391 *W, from Murchison River, W.A.

Distribution: NW coastal, W.A. Ecology: terrestrial, noctidiurnal, omnivore, woodland, open forest; nest in soil.

Calomyrmex impavidus (Forel, 1893)

Camponotus impavidus Forel, A. (1893). Nouvelles fourmis d'Australie et des Canaries. *Ann. Soc. Entomol. Belg.* **37**: 454–466 [455]. Type data: syntypes, GMNH W, from Port Darwin, N.T.

Distribution: N coastal, N.T. Ecology: terrestrial, noctidiurnal, omnivore, open forest; nest in soil.

Calomyrmex purpureus (Mayr, 1876)

Calomyrmex purpureus purpureus (Mayr, 1876)

Camponotus purpureus Mayr, G.L. (1876). Die australischen Formiciden. *J. Mus. Godeffroy* **5**: 56-115 [62]. Type data: syntypes, NHMW W, from Peak Downs, Qld.

Distribution: NE coastal, Qld. Ecology: terrestrial, noctidiurnal, omnivore, woodland, open forest; nest in soil.

Calomyrmex purpureus smaragdina Emery, 1898

Calomyrmex purpureus smaragdina Emery, C. (1898). Descrizioni di formiche nuove Malesi e Australiane. Note sinonimiche. *Rec. Sess. Accad. Sci. Ist. Bologna (ns)* **2**: 231-245 [238]. Type data: holotype, MCG *W, from Adelaide, S.A.

Distribution: S Gulfs, S.A. Ecology: terrestrial, noctidiurnal, omnivore, woodland, open forest; nest in soil.

Calomyrmex similis (Mayr, 1876)

Camponotus similis Mayr, G.L. (1876). Die australischen Formiciden. *J. Mus. Godeffroy* **5**: 56-115 [61]. Type data: syntypes, NHMW W, from Rockhampton and Gayndah, Qld.

Distribution: NE coastal, Qld. Ecology: terrestrial, noctidiurnal, omnivore, woodland, open forest; nest in soil.

Calomyrmex splendidus (Mayr, 1876)

Calomyrmex splendidus splendidus (Mayr, 1876)

Camponotus splendidus Mayr, G.L. (1876). Die australischen Formiciden. *J. Mus. Godeffroy* **5**: 56-115 [61]. Type data: syntypes, NHMW W, from Peak Downs, Qld.

Distribution: NE coastal, Qld. Ecology: terrestrial, noctidiurnal, omnivore, woodland, open forest; nest in soil.

Calomyrmex splendidus mutans (Forel, 1910)

Camponotus (Calomyrmex) splendidus mutans Forel, A. (1910). Formicides australiens reçus de MM. Froggatt et Rowland Turner. *Rev. Suisse Zool.* **18**: 1-94 [83]. Type data: syntypes, GMNH W,F, from Tennant Creek, N.T.

Distribution: W plateau, N.T. Ecology: terrestrial, noctidiurnal, omnivore, woodland; nest in soil.

Calomyrmex splendidus viridiventris Forel, 1915

Calomyrmex splendidus viridiventris Forel, A. (1915). Results of Dr. E. Mjöbergs Swedish Scientific Expeditions to Australia 1910-1913. 2. Ameisen. *Ark. Zool.* **9**: 1-119 pls 1-3 [4 Dec. 1915] [106]. Type data: syntypes, GMNH W, ANIC W, other syntypes may exist, from Kimberley distr., W.A. and Laura and Alice River, Qld.

Distribution: N coastal, NE coastal, W.A., Qld. Ecology: terrestrial, noctidiurnal, omnivore, woodland, open forest; nest in soil.

Camponotus Mayr, 1861

Camponotus Mayr, G.L. (1861). *Die europëischen Formiciden. (Ameisen.) Nach der analytischen Methode bearbeitet.* Vienna : Carl Gerolds Sohn 80 pp. 1 pl [35]. Type species *Formica ligniperda* Latreille, 1802 by subsequent designation, see Bingham, C.T. (1903). *The Fauna of British India, including Ceylon and Burma. Hymenoptera. Vol. 2 Ants and cuckoo-wasps.* London : Taylor & Francis [347].

Myrmophyma Forel, A. (1912). Formicides néotropiques. Part 6. 5me sous-famille Camponotinae Forel. *Mém. Soc. Entomol. Belg.* **20**: 59-92 [92] [proposed with subgeneric rank in *Camponotus* Mayr, 1861]. Type species *Camponotus capito* Mayr, 1876 by subsequent designation, see Wheeler, W.M. (1913). Corrections and additions to "List of type species of the genera and subgenera of Formicidae". *Ann. N.Y. Acad. Sci.* **23**: 77-83 [29 May 1913].

Myrmocamelus Forel, A. (1914). Le genre *Camponotus* Mayr and les genres voisins. *Rev. Suisse Zool.* **22**: 257-276 [261] [proposed with subgeneric rank in *Camponotus* Mayr, 1861; redefined in Ettershank, G. (1966). A generic revision of the world Myrmicinae related to *Solenopsis* and *Pheidologeton* (Hymenoptera : Formicidae). *Aust. J. Zool.* **14**: 73-171]. Type species *Formica ephippium* F. Smith, 1858 by original designation.

Thlipsepinotus Santschi, F. (1928). Nouvelles fourmis d'Australie. *Bull. Soc. Vaud. Sci. Nat.* **56**: 465-483 [483] [proposed with subgeneric rank in *Camponotus* Mayr, 1861]. Type species *Camponotus claripes* Mayr, 1876 by original designation.

Synonymy that of Brown, W.L. jr. (1973). A comparison of the Hylean and Congo-West African rain forest ant faunas. pp. 161-185 *in* Meggers, B.J., Ayensu, E.S. & Duckworth, W.D. (eds.) *Tropical forest ecosystems in Africa and South America: a comparative review.* Washington : Smithsonian Institution Press [177].

This group is found worldwide, see Brown, W.L. jr. (1973). A comparison of the Hylean and Congo-West African rain forest ant faunas. pp. 161-185 *in* Meggers, B.J., Ayensu, E.S. & Duckworth, W.D. (eds.) *Tropical forest ecosystems in Africa and South America: a comparative review.* Washington : Smithsonian Institution Press.

Camponotus adami Forel, 1910

Camponotus adami Forel, A. (1910). Formicides australiens reçus de MM. Froggatt et Rowland Turner. *Rev. Suisse Zool.* **18**: 1-94 [70]. Type data: syntypes, GMNH W, ANIC W, from Bombala, N.S.W.

Distribution: SE coastal, N.S.W. Ecology: terrestrial, noctidiurnal, omnivore, woodland, open forest; nest in ground layer.

Camponotus aeneopilosus Mayr, 1862

Camponotus aeneopilosus aeneopilosus Mayr, 1862

Camponotus aeneopilosus Mayr, G.L. (1862). Myrmecologische Studien. *Verh. Zool.-Bot. Ges. Wien* **12**: Abhand. 649–776 [665 pl 19]. Type data: syntypes, NHMW W, from Sidney (=Sydney), N.S.W.

Distribution: SE coastal, N.S.W. Ecology: terrestrial, noctidiurnal, omnivore, woodland, open forest; nest in ground layer.

Camponotus aeneopilosus flavidopubescens Forel, 1902

Camponotus aeneopilosus flavidopubescens Forel, A. (1902). Fourmis nouvelles d'Australie. *Rev. Suisse Zool.* **10**: 405–548 [504]. Type data: syntypes, GMNH W, ANIC W, from N.S.W.

Distribution: SE coastal, N.S.W. Ecology: terrestrial, noctidiurnal, omnivore, woodland, open forest; nest in ground layer.

Camponotus afflatus Viehmeyer, 1925

Camponotus (Myrmosaga) afflatus Viehmeyer, H. (1925). Formiciden der australischen Faunenregion. *Entomol. Mitt.* **14**: 139–149 [140]. Type data: syntypes (probable), ZMB *W, from Killalpaninno (=Killalpaninna), S.A.

Distribution: Lake Eyre basin, S.A. Ecology: terrestrial, noctidiurnal, omnivore, desert, woodland; nest in ground layer.

Camponotus arcuatus Mayr, 1876

Camponotus arcuatus arcuatus Mayr, 1876

Camponotus arcuatus Mayr, G.L. (1876). Die australischen Formiciden. *J. Mus. Godeffroy* **5**: 56–115 [63]. Type data: syntypes, NHMW W, from Rockhampton, Qld.

Distribution: NE coastal, Qld. Ecology: terrestrial, noctidiurnal, omnivore, hummock grassland, woodland, open forest; nest in ground layer.

Camponotus arcuatus aesopus Forel, 1907

Camponotus arcuatus aesopus Forel, A. (1907). Formicidae. pp. 263–310 *in* Michaelsen, W. & Hartmeyer, R. (eds.) *Die Fauna Südwest-Australiens.* Jena : G. Fischer Vol. 1 [302]. Type data: holotype, probably destroyed in ZMH in WW II, from Mt. Robinson near Kalgoorlie, W.A.

Distribution: W plateau, W.A. Ecology: terrestrial, noctidiurnal, omnivore, hummock grassland, woodland, open forest; nest in ground layer.

Camponotus armstrongi McAreavey, 1949

Camponotus (Myrmogonia) armstrongi McAreavey, J.J. (1949). Australian Formicidae. New genera and species. *Proc. Linn. Soc. N.S.W.* **74**: 1–25 [15 June 1949] [19]. Type data: holotype, ANIC W, from Nyngan, N.S.W.

Distribution: Murray-Darling basin, N.S.W. Ecology: terrestrial, noctidiurnal, omnivore, desert, woodland, open forest; nest in ground layer.

Camponotus aurocinctus (F. Smith, 1858)

Formica aurocincta Smith, F. (1858). *Catalogue of hymenopterous insects in the collection of the British Museum.* Part 6. Formicidae. London : British Museum 216 pp. 14 pls [27 Mar. 1858] [39]. Publication date established from Donisthorpe, H. (1932). On the identity of Smith's types of Formicidae (Hymenoptera) collected by Alfred Russell Wallace in the Malay Archipelago, with descriptions of two new species. *Ann. Mag. Nat. Hist. (10)* **10**: 441–476. Type data: syntypes (probable), BMNH *W, from Adelaide, S.A.

Distribution: S Gulfs, S.A. Ecology: terrestrial, noctidiurnal, omnivore, woodland, open forest; nest in ground layer.

Camponotus bigenus Santschi, 1919

Camponotus (Myrmocamelus) bigenus Santschi, F. (1919). Cinq notes myrmécologiques. *Bull. Soc. Vaud. Sci. Nat.* **52**: 325–350 [333]. Type data: syntypes, BNHM W,M, from Townsville, Qld.

Distribution: NE coastal, Qld. Ecology: terrestrial, noctidiurnal, omnivore, woodland, open forest; nest in ground layer.

Camponotus cameratus Viehmeyer, 1925

Camponotus (Myrmogonia) cameratus Viehmeyer, H. (1925). Formiciden der australischen Faunenregion. *Entomol. Mitt.* **14**: 139–149 [146]. Type data: syntypes, ZMB *W, from Trial Bay, N.S.W.

Distribution: SE coastal, N.S.W. Ecology: terrestrial, noctidiurnal, omnivore, woodland, open forest; nest in ground layer.

Camponotus capito Mayr, 1876

Camponotus capito capito Mayr, 1876

Camponotus capito Mayr, G.L. (1876). Die australischen Formiciden. *J. Mus. Godeffroy* **5**: 56–115 [64]. Type data: syntypes, NHMW W,F, from Peak Downs, Qld.

Distribution: NE coastal, Qld. Ecology: terrestrial, noctidiurnal, omnivore, woodland, open forest; nest in ground layer.

Camponotus capito ebeninithorax Forel, 1915

Camponotus (Myrmophyma) capito ebeninithorax Forel, A. (1915). Results of Dr. E. Mjöbergs Swedish Scientific Expeditions to Australia 1910–1913. 2. Ameisen. *Ark.*

Zool. **9**: 1-119 pls 1-3 [4 Dec. 1915] [100]. Type data: syntypes, GMNH W, other syntypes may exist, from Australia.

Distribution: NE coastal, Qld. Ecology: terrestrial, noctidiurnal, omnivore, woodland, open forest; nest in ground layer.

Camponotus ceriseipes Clark, 1938

Camponotus (Myrmophyma) ceriseipes Clark, J. (1938). Reports of the McCoy Society for Field Investigation and Research. No. 2. Sir Joseph Bank Islands. Part I. Formicidae (Hymenoptera). *Proc. R. Soc. Vict.* **50**: 356-382 [378]. Type data: syntypes, NMV *W, from N end of Reevesby Is., S.A.

Distribution: S Gulfs, S.A. Ecology: terrestrial, noctidiurnal, omnivore, woodland, open forest; nest in ground layer.

Camponotus chalceoides Clark, 1938

Camponotus (Myrmophyma) chalceoides Clark, J. (1938). Reports of the McCoy Society for Field Investigation and Research. No. 2. Sir Joseph Bank Islands. Part I. Formicidae (Hymenoptera). *Proc. R. Soc. Vict.* **50**: 356-382 [376]. Type data: syntypes, NMV *W, from Reevesby Is., S.A.

Distribution: S Gulfs, S.A. Ecology: terrestrial, noctidiurnal, omnivore, woodland, open forest; nest in ground layer.

Camponotus chalceus Crawley, 1915

Camponotus (Myrmosaga) chalceus Crawley, W.C. (1915). Ants from north and south-west Australia (G.F. Hill, Rowland Turner) and Christmas Island, Straits Settlements. Part II. *Ann. Mag. Nat. Hist. (8)* **15**: 232-239 [236]. Type data: syntypes, possibly OUM, from Yallingup, W.A.

Distribution: SW coastal, W.A. Ecology: terrestrial, noctidiurnal, omnivore, woodland, open forest; nest in ground layer.

Camponotus cinereus Mayr, 1876

Camponotus cinereus cinereus Mayr, 1876

Camponotus cinereus Mayr, G.L. (1876). Die australischen Formiciden. *J. Mus. Godeffroy* **5**: 56-115 [62]. Type data: syntypes, NHMW W, from Peak Downs, Qld.

Distribution: NE coastal, Qld. Ecology: terrestrial, noctidiurnal, omnivore, woodland, open forest; nest in ground layer.

Camponotus cinereus amperei Forel, 1913

Camponotus (Myrmocamelus) cinereus amperei Forel, A. (1913). Fourmis de Tasmanie et d'Australie récoltées par MM. Lea, Froggatt etc. *Bull. Soc. Vaud. Sci. Nat.* **49**: 173-196 pl 2 [192]. Type data: syntypes, GMNH W, from Sea Lake, Vic.

Distribution: Murray-Darling basin, Vic. Ecology: terrestrial, noctidiurnal, omnivore, woodland, open forest; nest in ground layer.

Camponotus cinereus notterae Forel, 1907

Camponotus cinereus notterae Forel, A. (1907). Formicidae. pp. 263-310 *in* Michaelsen, W. & Hartmeyer, R. (eds.) *Die Fauna Südwest-Australiens.* Jena : G. Fischer Vol. 1 [303]. Type data: holotype, probably destroyed in ZMH in WW II, from Grooseberry (=Gooseberry) Hill, W.A.

Distribution: SW coastal, W.A. Ecology: terrestrial, noctidiurnal, omnivore, woodland, open forest; nest in ground layer.

Camponotus claripes Mayr, 1876

Camponotus claripes claripes Mayr, 1876

Camponotus claripes Mayr, G.L. (1876). Die australischen Formiciden. *J. Mus. Godeffroy* **5**: 56-115 [64]. Type data: syntypes, whereabouts unknown, from Peak Downs and Gayndah, Qld.

Distribution: NE coastal, Qld. Ecology: terrestrial, noctidiurnal, omnivore, woodland, open forest; nest in ground layer.

Camponotus claripes elegans Forel, 1902

Camponotus claripes elegans Forel, A. (1902). Fourmis nouvelles d'Australie. *Rev. Suisse Zool.* **10**: 405-548 [496]. Type data: syntypes, GMNH W, ANIC W, from Wallsend, N.S.W.

Distribution: SE coastal, N.S.W. Ecology: terrestrial, noctidiurnal, omnivore, woodland, open forest; nest in ground layer.

Camponotus claripes inverallensis Forel, 1910

Camponotus claripes inverallensis Forel, A. (1910). Formicides australiens reçus de MM. Froggatt et Rowland Turner. *Rev. Suisse Zool.* **18**: 1-94 [72]. Type data: syntypes, GMNH W, from Reedy Creek, Inverell, N.S.W.

Distribution: Murray-Darling basin, N.S.W. Ecology: terrestrial, noctidiurnal, omnivore, woodland, open forest; nest in ground layer.

Camponotus claripes marcens Forel, 1907

Camponotus claripes marcens Forel, A. (1907). Formicidae. pp. 263-310 *in* Michaelsen, W. & Hartmeyer, R. (eds.) *Die Fauna Südwest-Australiens.* Jena : G. Fischer Vol. 1 [300]. Type data: syntypes, GMNH W, ANIC W, from Mundaring Weir and Guildford, W.A.

Distribution: SW coastal, W.A. Ecology: terrestrial, noctidiurnal, omnivore, woodland, open forest; nest in ground layer.

Camponotus claripes minimus Crawley, 1922

Camponotus (Myrmophyma) claripes minima Crawley, W.C. (1922). New ants from Australia. *Ann. Mag. Nat. Hist. (9)* **10**: 16–36 [31]. Type data: syntypes, OUM *W,F,M, from Mundaring, W.A.

Distribution: SW coastal, W.A. Ecology: terrestrial, noctidiurnal, omnivore, woodland, open forest; nest in ground layer.

Camponotus claripes nudimalis Forel, 1913

Camponotus claripes nudimalis Forel, A. (1913). Fourmis de Tasmanie et d'Australie récoltées par MM. Lea, Froggatt etc. *Bull. Soc. Vaud. Sci. Nat.* **49**: 173–196 pl 2 [191]. Type data: holotype, GMNH W, from Bridgetown, W.A.

Distribution: SW coastal, W.A. Ecology: terrestrial, noctidiurnal, omnivore, woodland, open forest; nest in ground layer.

Camponotus claripes orbiculatopunctatus Viehmeyer, 1925

Camponotus (Myrmophyma) claripes orbiculatopunctatus Viehmeyer, H. (1925). Formiciden der australischen Faunenregion. *Entomol. Mitt.* **14**: 139–149 [143]. Type data: syntypes, ZMB *W,F, from Liverpool, N.S.W.

Distribution: SE coastal, N.S.W. Ecology: terrestrial, noctidiurnal, omnivore, woodland, open forest; nest in ground layer.

Camponotus claripes piperatus Wheeler, 1933

Camponotus (Myrmophyma) claripes piperatus Wheeler, W.M. (1933). Mermis parasitism in some Australian and Mexican ants. *Psyche Camb.* **40**: 20–31 [26]. Type data: syntypes, MCZ *W,F,M, from Mt. Lofty, S.A.

Distribution: S Gulfs, S.A. Ecology: terrestrial, noctidiurnal, omnivore, woodland, open forest; nest in ground layer.

Camponotus consectator (F. Smith, 1858)

Formica consectator Smith, F. (1858). *Catalogue of hymenopterous insects in the collection of the British Museum. Part 6. Formicidae.* London : British Museum 216 pp. 14 pls [27 Mar. 1858] [38]. Publication date established from Donisthorpe, H. (1932). On the identity of Smith's types of Formicidae (Hymenoptera) collected by Alfred Russell Wallace in the Malay Archipelago, with descriptions of two new species. *Ann. Mag. Nat. Hist. (10)* **10**: 441–476. Type data: syntypes (probable), BMNH *F, from Australia.

Distribution: (SE coastal), (N.S.W.). Ecology: terrestrial, noctidiurnal, omnivore, woodland, open forest; nest in ground layer.

Camponotus consobrinus (Erichson, 1842)

Formica consobrina Erichson, W.F. (1842). Beitrag zur Fauna von Vandiemansland mit besonderer rucksicht auf die geographische Verbreitung der Insecten. *Arch. Naturg.* **8**: 83–287 [258]. Type data: holotype (probable), ZMB *F, from Tas.

Camponotus dimidiatus Roger, J. (1863). Verzeichniss der Formiciden-Gattungen und Arten. *Berl. Entomol. Z.* **7** appendix to vol.: 1–65 [4]. Type data: holotype, NHMW *W,F, from Australia (as New Holland).

Synonymy that of Clark, J. (1934). Ants from the Otway Ranges. *Mem. Natl. Mus. Vict.* **8**: 48–73 [70].

Distribution: SE coastal, Murray-Darling basin, NE coastal, S Gulfs, Qld., N.S.W., A.C.T., Vic., S.A., Tas. Ecology: terrestrial, noctidiurnal, omnivore, woodland, open forest; nest in soil.

Camponotus cowlei Froggatt, 1896

Camponotus cowlei Froggatt, W.W. (1896). Honey ants. pp. 385–392 *in* Spencer, B. (ed.) *Report on the work of the Horn Scientific Expedition to Central Australia.* Melbourne : Melville, Mullen & Slade Pt. 2 Zoology [387 pl 27]. Type data: syntypes, AM W,F,M, from Illamurta in the James Range and Spencer Gorge in the McDonnell Range, N.T.

Distribution: Lake Eyre basin, N.T. Ecology: terrestrial, noctidiurnal, omnivore, desert, woodland; nest in soil.

Camponotus crenatus Mayr, 1876

Camponotus crenatus Mayr, G.L. (1876). Die australischen Formiciden. *J. Mus. Godeffroy* **5**: 56–115 [64]. Type data: holotype (probable), NHMW W, from Rockhampton, Qld.

Distribution: NE coastal, Qld. Ecology: terrestrial, noctidiurnal, omnivore, woodland, open forest; nest in ground layer.

Camponotus cruentatus (Latreille, 1802)

Formica cruentata Latreille, P.A. (1802). *Histoire naturelle des fourmis, et recueil de mémoires et d'observations sur les abeilles, les araignées, les faucheurs, et autre insectes.* Paris : Crapelet 445 pp. 12 pls [116]. Type data: status unknown, ?MNHP, from Afrique.

Camponotus cruentatus aspera Menozzi, 1925

Camponotus (Myrmosericus) cruentatus aspera Menozzi, C. (1925). Qualche formica nuova od interessante del Deutsch. Entomol. Institut di Dahlem (Form.). *Entomol. Mitt.* **14**: 368–371 [371]. Type data: syntypes, probably BIE* or DEIB*, from Melbourne, Vic.

Distribution: SE coastal, Vic. Ecology: terrestrial, noctidiurnal, omnivore, woodland, open forest; nest in ground layer.

Camponotus denticulatus W.F. Kirby, 1896

Camponotus denticulatus Kirby, W.F. (1896). Hymenoptera. pp. 203–209 *in* Spencer, B. (ed.) *Report on the work of the Horn Scientific Expedition to Central Australia.* Melbourne : Melville, Mullen & Slade Pt. 1 supplement [204]. Type data: syntypes, BMNH (probable) *W, from McDonnell Range, N.T.

Distribution: Lake Eyre basin, N.T. Ecology: terrestrial, noctidiurnal, omnivore, desert, woodland; nest in soil.

Camponotus discors Forel, 1902

Camponotus discors discors Forel, 1902

Camponotus maculatus discors Forel, A. (1902). Fourmis nouvelles d'Australie. *Rev. Suisse Zool.* **10**: 405–548 [497]. Type data: syntypes, GMNH W, ANIC W, from Pera Bore, N.S.W.

Distribution: Murray-Darling basin, N.S.W. Ecology: terrestrial, noctidiurnal, omnivore, woodland, open forest; nest in ground layer. Biological references: Emery, C. (1920). Studi sui *Camponotus. Boll. Soc. Entomol. Ital.* **52**: 1–49 (raised to species).

Camponotus discors yarrabahensis Forel, 1915

Camponotus (Myrmoturba) maculatus yarrabahensis Forel, A. (1915). Results of Dr. E. Mjöbergs Swedish Scientific Expeditions to Australia 1910–1913. 2. Ameisen. *Ark. Zool.* **9**: 1–119 pls 1–3 [4 Dec. 1915] [98]. Type data: syntypes, GMNH W, other syntypes may exist, from Yarrabah and Malanda, Qld.

Distribution: NE coastal, Qld. Ecology: terrestrial, noctidiurnal, omnivore, open forest, closed forest; nest in ground layer.

Camponotus dorycus (F. Smith, 1860)

Formica dorycus Smith, F. (1860). Catalogue of hymenopterous insects collected by Mr A.R. Wallace in the Islands of Bachian, Kaisaa, Amboyne, Gilolo, and at Dory in New Guinea. *J. Linn. Soc. Zool.* **4** (suppl.): 93–143 [96]. Type data: status unknown, ?BMNH, from Dory.

Camponotus dorycus confusus Emery, 1887

Camponotus dorycus confusus Emery, C. (1887). Catalogo delle formiche esistenti nelle collezioni del Museo Civico di Genova. Parte terza. Formiche della regione Indo-Malese e dell'Australia. *Ann. Mus. Civ. Stor. Nat. Giacomo Doria* **25**: 209–258 pls 3–4 [215]. Type data: syntypes, MCG *W,F, from Katau, New Guinea, Percy Isles and Somerset, Qld.

Distribution: NE coastal, Qld. Ecology: terrestrial, noctidiurnal, omnivore, open forest, closed forest; nest in ground layer.

Camponotus dromas Santschi, 1919

Camponotus (Myrmocamelus) dromas Santschi, F. (1919). Cinq notes myrmécologiques. *Bull. Soc. Vaud. Sci. Nat.* **52**: 325–350 [332]. Type data: syntypes, BNHM W,M, from Townsville, Qld.

Distribution: NE coastal, Qld. Ecology: terrestrial, noctidiurnal, omnivore, woodland, open forest; nest in ground layer.

Camponotus ephippium (F. Smith, 1858)

Camponotus ephippium ephippium (F. Smith, 1858)

Formica ephippium Smith, F. (1858). *Catalogue of hymenopterous insects in the collection of the British Museum.* Part 6. Formicidae. London : British Museum 216 pp. 14 pls [27 Mar. 1858] [39]. Publication date established from Donisthorpe, H. (1932). On the identity of Smith's types of Formicidae (Hymenoptera) collected by Alfred Russell Wallace in the Malay Archipelago, with descriptions of two new species. *Ann. Mag. Nat. Hist. (10)* **10**: 441–476. Type data: syntypes (probable), BMNH *W, from Adelaide, S.A.

Distribution: S Gulfs, S.A. Ecology: terrestrial, noctidiurnal, omnivore, desert, woodland; nest in soil.

Camponotus ephippium narses Forel, 1915

Camponotus (Myrmocamelus) ephippium narses Forel, A. (1915). Results of Dr. E. Mjöbergs Swedish Scientific Expeditions to Australia 1910–1913. 2. Ameisen. *Ark. Zool.* **9**: 1–119 pls 1–3 [4 Dec. 1915] [103]. Type data: syntypes, GMNH W, other syntypes may exist, from Kimberley distr. and Broome, W.A.

Distribution: N coastal, W.A. Ecology: terrestrial, noctidiurnal, omnivore, desert, woodland, open forest; nest in soil.

Camponotus eremicus Wheeler, 1915

Camponotus (Myrmogonia) eremicus Wheeler, W.M. (1915). Hymenoptera. *Trans. R. Soc. S. Aust.* **39**: 805–823 pls 64–66 [Dec. 1915] [815]. Type data: syntypes, MCZ *W, from Everard Range, S.A.

Distribution: W plateau, S.A. Ecology: terrestrial, noctidiurnal, omnivore, desert, woodland; nest in soil.

Camponotus erythropus Viehmeyer, 1925

Camponotus (Myrmosaga) erythropus Viehmeyer, H. (1925). Formiciden der australischen Faunenregion. *Entomol. Mitt.* **14**: 139–149 [141]. Type data: syntypes, ZMB *W,F, from Liverpool, N.S.W.

Distribution: SE coastal, N.S.W. Ecology: terrestrial, noctidiurnal, omnivore, open forest; nest in ground layer.

Camponotus esau Forel, 1915

Camponotus (Myrmocamelus) esau Forel, A. (1915). Results of Dr. E. Mjöbergs Swedish Scientific Expeditions to Australia 1910–1913. 2. Ameisen. *Ark. Zool.* **9**: 1–119 pls 1–3 [4 Dec. 1915] [103]. Type data: syntypes, GMNH W, other syntypes may exist, from Cedar Creek, Qld.

Distribution: NE coastal, Qld. Ecology: terrestrial, noctidiurnal, omnivore, woodland, open forest; nest in ground layer.

Camponotus evae Forel, 1910

Camponotus evae evae Forel, 1910

Camponotus evae Forel, A. (1910). Formicides australiens reçus de MM. Froggatt et Rowland Turner. *Rev. Suisse Zool.* **18**: 1–94 [74]. Type data: syntypes, GMNH W, ANIC W, from Cape York, Qld.

Distribution: N Gulf, Qld. Ecology: terrestrial, noctidiurnal, omnivore, woodland; nest in ground layer.

Camponotus evae zeuxis Forel, 1915

Camponotus (Myrmogonia) evae zeuxis Forel, A. (1915). Results of Dr. E. Mjöbergs Swedish Scientific Expeditions to Australia 1910–1913. 2. Ameisen. *Ark. Zool.* **9**: 1–119 pls 1–3 [4 Dec. 1915] [101]. Type data: syntypes, GMNH W, other syntypes may exist, from Broome, W.A.

Distribution: N coastal, W.A. Ecology: terrestrial, noctidiurnal, omnivore, woodland, open forest; nest in ground layer.

Camponotus extensus Mayr, 1876

Camponotus extensus Mayr, G.L. (1876). Die australischen Formiciden. *J. Mus. Godeffroy* **5**: 56–115 [65]. Type data: syntypes, NHMW W, from Rockhampton, Qld.

Distribution: NE coastal, Qld. Ecology: terrestrial, noctidiurnal, omnivore, woodland, open forest; nest in ground layer.

Camponotus fictor Forel, 1902

Camponotus fictor fictor Forel, 1902

Camponotus (Colobopsis) fictor Forel, A. (1902). Fourmis nouvelles d'Australie. *Rev. Suisse Zool.* **10**: 405–548 [509]. Type data: syntypes, GMNH W, ANIC W, from New Castle (=Newcastle) and Native Dog Bore, N.S.W.

Distribution: SE coastal, Murray-Darling basin, N.S.W. Ecology: terrestrial, noctidiurnal, omnivore, woodland, open forest; nest arboreal.

Camponotus fictor augustulus Viehmeyer, 1925

Camponotus (Colobopsis) fictor augustulus Viehmeyer, H. (1925). Formiciden der australischen Faunenregion. *Entomol. Mitt.* **14**: 139–149 [145] [introduced as *victor*]. Type data: syntypes, ZMB *W,F, from Trial Bay, N.S.W.

Distribution: SE coastal, N.S.W. Ecology: terrestrial, noctidiurnal, omnivore, woodland, open forest; nest arboreal.

Camponotus fieldeae Forel, 1902

Camponotus fieldeae Forel, A. (1902). Fourmis nouvelles d'Australie. *Rev. Suisse Zool.* **10**: 405–548 [495]. Type data: syntypes, GMNH W, ANIC W, from Charters Towers, Qld.

Distribution: NE coastal, Qld. Ecology: terrestrial, noctidiurnal, omnivore, woodland, open forest; nest in ground layer.

Camponotus fieldellus Forel, 1910

Camponotus fieldellus Forel, A. (1910). Formicides australiens reçus de MM. Froggatt et Rowland Turner. *Rev. Suisse Zool.* **18**: 1–94 [79]. Type data: syntypes, GMNH W,F, ANIC W, from Tennant Creek, N.T.

Distribution: W plateau, N.T. Ecology: terrestrial, noctidiurnal, omnivore, desert, woodland; nest in ground layer.

Camponotus froggatti Forel, 1902

Camponotus froggatti Forel, A. (1902). Fourmis nouvelles d'Australie. *Rev. Suisse Zool.* **10**: 405–548 [504]. Type data: syntypes, GMNH W, ANIC W, from Wollongbar, Richmond River, N.S.W.

Distribution: SE coastal, N.S.W. Ecology: terrestrial, noctidiurnal, omnivore, woodland, open forest; nest in ground layer.

Camponotus gasseri (Forel, 1894)

Camponotus gasseri gasseri (Forel, 1894)

Colobopsis gasseri Forel, A. (1894). Quelques fourmis de Madagascar (récoltées par M. le Dr. Völtzkow); de Nouvelle Zélande (récoltées par M. W.W. Smith); de Nouvelle Calédonie (récoltées par M. Sommer); de Queensland (Australie) récoltées par M. Wiederkehr; et de Perth (Australie occidentale) récoltées par M. Chase. *Ann. Soc. Entomol. Belg.* **38**: 226–237 [233]. Type data: syntypes, GMNH W, from Perth, W.A.

Distribution: SW coastal, W.A. Ecology: terrestrial, noctidiurnal, omnivore, woodland, open forest; nest arboreal.

Camponotus gasseri caloratus Wheeler, 1934

Camponotus (Colobopsis) gasseri caloratus Wheeler, W.M. (1934). Contributions to the fauna of Rottnest Island, Western Australia No. IX. The ants. *J. R. Soc. West. Aust.* **20**: 137–163 [5 Oct. 1934] [162]. Type data: syntypes, MCZ *W,F,M, from near Government House, Rottnest Is., W.A.

Distribution: SW coastal, W.A. Ecology: terrestrial, noctidiurnal, omnivore, woodland, open forest; nest arboreal.

Camponotus gasseri lysias Forel, 1913

Camponotus (Colobopsis) gasseri lysias Forel, A. (1913). Fourmis de Tasmanie et d'Australie récoltées par MM. Lea, Froggatt etc. *Bull. Soc. Vaud. Sci. Nat.* **49**: 173–196 pl 2 [193]. Type data: syntypes, GMNH W, from Ulverstone, Tas.

Distribution: Tas. Ecology: terrestrial, noctidiurnal, omnivore, woodland, open forest; nest arboreal.

Camponotus gasseri obtusitruncatus Forel, 1902

Camponotus (Colobopsis) gasseri obtusitruncatus Forel, A. (1902). Fourmis nouvelles d'Australie. *Rev. Suisse Zool.* **10**: 405-548 [508]. Type data: syntypes, GMNH W,F,M, ANIC W, from Mackay, Qld.

Distribution: NE coastal, Qld. Ecology: terrestrial, noctidiurnal, omnivore, woodland, open forest; nest arboreal.

Camponotus gibbinotus Forel, 1902

Camponotus gibbinotus Forel, A. (1902). Fourmis nouvelles d'Australie. *Rev. Suisse Zool.* **10**: 405-548 [498]. Type data: syntypes, GMNH W, from Kalgoorlie, W.A.

Distribution: W plateau, W.A. Ecology: terrestrial, noctidiurnal, omnivore, desert, woodland; nest in soil.

Camponotus gouldianus Forel, 1922

Camponotus gouldianus Forel, A. (1922). Glanures myrmécologiques en 1922. *Rev. Suisse Zool.* **30**: 87-102 [100]. Type data: syntypes, GMNH W, from Sea Lake, Vic.

Distribution: Murray-Darling basin, Vic. Ecology: terrestrial, noctidiurnal, omnivore, woodland, open forest; nest in ground layer.

Camponotus hartogi Forel, 1902

Camponotus hartogi Forel, A. (1902). Fourmis nouvelles d'Australie. *Rev. Suisse Zool.* **10**: 405-548 [500]. Type data: holotype (probable), GMNH W, from Yarra distr., Vic.

Camponotus (Myrmosaga) ferruginipes Crawley, W.C. (1922). *in* Poulton, E.B. & Crawley, W.C. (1922). Notes on some Australian ants. *Entomol. Mon. Mag. (3)* **8**: 118-126 [125]. Type data: holotype, possibly OUM, from near Healesville, Vic.

Synonymy that of Brown, W.L. jr. (1956). Some synonymies in the ant genus *Camponotus*. *Psyche Camb.* **63**: 38-40 [40].

Distribution: SE coastal, Vic. Ecology: terrestrial, noctidiurnal, omnivore, woodland, open forest; nest in ground layer.

Camponotus horni W.F. Kirby, 1896

Camponotus horni Kirby, W.F. (1896). Hymenoptera. pp 203-209 *in* Spencer, B. (ed.) *Report on the work of the Horn Scientific Expedition to Central Australia.* Melbourne : Melville, Mullen & Slade Pt. 1 supplement [205]. Type data: syntypes, BMNH (probable) *W,F, from Palm Creek, N.T.

Distribution: Lake Eyre basin, N.T. Ecology: terrestrial, noctidiurnal, omnivore, woodland, open forest; nest in ground layer.

Camponotus howensis Wheeler, 1927

Camponotus (Colobopsis) howensis Wheeler, W.M. (1927). The ants of Lord Howe Island and Norfolk Island. *Proc. Am. Acad. Arts Sci.* **62**: 121-153 [152]. Type data: syntypes, MCZ *W, from Lord Howe Is.

Distribution: Lord Howe Is. Ecology: terrestrial, noctidiurnal, omnivore, woodland, open forest; nest arboreal.

Camponotus inflatus Lubbock, 1880

Camponotus inflatus Lubbock, J. (1880). Observations on Ants, Bees and Wasps; With a Description of a New Species of Honey-Ant. Part vii. Ants. *J. Linn. Soc. Zool.* **15**: 167-187 [3 Sept. 1880] [186 pl 8]. Type data: syntypes (probable), BMNH (probable) *W, from Adelaide, S.A.

Camponotus (Myrmamblys) aurofasciatus Wheeler, W.M. (1915). Hymenoptera. *Trans. R. Soc. S. Aust.* **39**: 805-823 pls 64-66 [Dec. 1915] [817]. Type data: syntypes, MCZ *W, from Musgrave Ranges and Moorilyanna, S.A.

Synonymy that of Emery, C. (1925). Hymenoptera Fam. Formicidae subfam. Formicinae. *in* Wytsman, P. (ed.) *Genera Insectorum.* Fasc. 183 302 pp. 4 pls [111].

Distribution: S Gulfs, W plateau, Lake Eyre basin, S.A. Ecology: terrestrial, noctidiurnal, omnivore, woodland; nest in soil.

Camponotus innexus Forel, 1902

Camponotus innexus Forel, A. (1902). Fourmis nouvelles d'Australie. *Rev. Suisse Zool.* **10**: 405-548 [499]. Type data: syntypes, GMNH W,F,M, ANIC W, from Bong Bong, N.S.W.

Distribution: SE coastal, N.S.W. Ecology: terrestrial, noctidiurnal, omnivore, woodland, open forest; nest in ground layer.

Camponotus insipidus Forel, 1893

Camponotus insipidus Forel, A. (1893). Nouvelles fourmis d'Australie et des Canaries. *Ann. Soc. Entomol. Belg.* **37**: 454-466 [454]. Type data: holotype (probable), GMNH W, from East Wallaby Is., W.A.

Distribution: W coast, W.A. Ecology: terrestrial, noctidiurnal, omnivore, woodland, open forest; nest in soil.

Camponotus intrepidus (W. Kirby, 1818)

Camponotus intrepidus intrepidus (W. Kirby, 1818)

Formica intrepida Kirby, W. (1818). A description of several new species of insects collected in New Holland by Robert Brown, Esq., F.R.S., Lib. Linn. Soc. *Trans. Linn. Soc. Lond.* **12**: 454-482 pls 21-23 [477]. Type data: uncertain, BMNH *W, from Port Jackson, N.S.W.

Formica agilis Smith, F. (1858). *Catalogue of hymenopterous insects in the collection of the British Museum.* Part 6. Formicidae. London : British Museum

216 pp. 14 pls [27 Mar. 1858] [37]. Publication date established from Donisthorpe, H. (1932). On the identity of Smith's types of Formicidae (Hymenoptera) collected by Alfred Russell Wallace in the Malay Archipelago, with descriptions of two new species. *Ann. Mag. Nat. Hist. (10)* **10**: 441–476. Type data: syntypes (probable), BMNH *W, from Australia (as New Holland).

Camponotus magnus Mayr, G.L. (1862). Myrmecologische Studien. *Verh. Zool.-Bot. Ges. Wien* **12**: Abhand. 649–776 [673 pl 19]. Type data: syntypes, NHMW *W, from Sidney (=Sydney) and Australia (as New Holland).

Synonymy that of Emery, C. (1925). Hymenoptera Fam. Formicidae subfam. Formicinae. *in* Wytsman, P. (ed.) *Genera Insectorum*. Fasc. 183 302 pp. 4 pls [114].

Distribution: SE coastal, N.S.W. Ecology: terrestrial, noctidiurnal, omnivore, woodland, open forest; nest in ground layer.

Camponotus intrepidus bellicosus Forel, 1902

Camponotus intrepidus bellicosus Forel, A. (1902). Fourmis nouvelles d'Australie. *Rev. Suisse Zool.* **10**: 405–548 [493]. Type data: syntypes, GMNH W, ANIC W, from Sydney, N.S.W.

Distribution: SE coastal, N.S.W. Ecology: terrestrial, noctidiurnal, omnivore, woodland, open forest; nest in ground layer.

Camponotus janeti Forel, 1895

Camponotus janeti Forel, A. (1895). Nouvelles fourmis d'Australie, récoltée à The Ridge, Mackay, Queensland par M. Gilbert Turner. *Ann. Soc. Entomol. Belg.* **39**: 417–428 [417]. Type data: syntypes, GMNH W, ANIC W, from Mackay, Qld.

Distribution: NE coastal, Qld. Ecology: terrestrial, noctidiurnal, omnivore, woodland, open forest; nest in ground layer.

Camponotus latrunculus Wheeler, 1915

Camponotus latrunculus latrunculus Wheeler, 1915

Camponotus (Myrmoturba) latrunculus Wheeler, W.M. (1915). Hymenoptera. *Trans. R. Soc. S. Aust.* **39**: 805–823 pls 64–66 [Dec. 1915] [814]. Type data: holotype, MCZ *W, from Todmorden, S.A.

Distribution: Lake Eyre basin, S.A. Ecology: terrestrial, noctidiurnal, omnivore, desert, woodland; nest in soil.

Camponotus latrunculus victoriensis Santschi, 1928

Camponotus (Myrmoturba) latrunculus victoriensis Santschi, F. (1928). Nouvelles fourmis d'Australie. *Bull. Soc. Vaud. Sci. Nat.* **56**: 465–483 [30 Aug. 1928] [479]. Type data: syntypes, BNHM W,M, from Elsternwick and Belgrave, Vic., see The Zoological Society of London (1929). *The Zoological Record*. Vol. 65 relating chiefly to the year 1928. London : Gurney & Jackson.

Distribution: SE coastal, Vic. Ecology: terrestrial, noctidiurnal, omnivore, woodland, open forest; nest in ground layer.

Camponotus leae Wheeler, 1915

Camponotus (Myrmosphincta) leae Wheeler, W.M. (1915). Hymenoptera. *Trans. R. Soc. S. Aust.* **39**: 805–823 pls 64–66 [Dec. 1915] [819]. Type data: syntypes, MCZ *W, from Flat Rock Hole in the Musgrave Ranges, S.A.

Distribution: W plateau, S.A. Ecology: terrestrial, noctidiurnal, omnivore, desert, woodland, open forest; nest in soil.

Camponotus lividicoxis Viehmeyer, 1925

Camponotus (Myrmophyma) lividicoxis Viehmeyer, H. (1925). Formiciden der australischen Faunenregion. *Entomol. Mitt.* **14**: 139–149 [142]. Type data: syntypes (probable), ZMB *W, from Trial Bay, N.S.W.

Distribution: SE coastal, N.S.W. Ecology: terrestrial, noctidiurnal, omnivore, woodland, open forest; nest in ground layer.

Camponotus lownei Forel, 1895

Formica nitida Lowne, B.T. (1865). Contributions to the natural history of Australian ants. *Entomologist* **2**: 275–280 [277] [*non Formica nitida* F. Smith, 1858]. Type data: holotype, BMNH (probable) *W, from Sidney (=Sydney), N.S.W.

Camponotus lownei Forel, A. (1895). Nouvelles fourmis de diverses provenances, surtout d'Australie. *Ann. Soc. Entomol. Belg.* **39**: 41–49 [43] [*nom. nov.* for *Formica nitida* Lowne, 1865].

Distribution: SE coastal, N.S.W. Ecology: terrestrial, noctidiurnal, omnivore, woodland, open forest; nest in ground layer.

Camponotus maculatus (Fabricius, 1781)

Formica maculata Fabricius, J.C. (1781). *Species Insectorum exhibentes eorum Differentias specificas, Synonyma auctorum, Loca Natalia, Metamorphosis adiectis observationibus, Descriptionibus*. Hamburgi et Kilonii : C.E. Bohnii Vol. 1 [491]. Type data: status unknown, ?BMNH, from " Africa Aequinoctiale".

Camponotus maculatus humilior Forel, 1902

Camponotus maculatus humilior Forel, A. (1902). Fourmis nouvelles d'Australie. *Rev. Suisse Zool.* **10**: 405–548 [497]. Type data: syntypes, GMNH W, ANIC W, from Cairns, Qld.

Distribution: NE coastal, Qld. Ecology: terrestrial, noctidiurnal, omnivore, open forest, closed forest; nest in ground layer.

Camponotus michaelseni Forel, 1907

Camponotus michaelseni Forel, A. (1907). Formicidae. pp. 263–310 *in* Michaelsen, W. & Hartmeyer, R. (eds.) *Die Fauna Südwest-Australiens*. Jena : G. Fischer Vol. 1

[303]. Type data: syntypes, GMNH W, from Mundaring Weir, Jarrahdale, Gooseberry Hill and Pickering Brook, W.A.

Distribution: SW coastal, W.A. Ecology: terrestrial, noctidiurnal, omnivore, woodland, open forest; nest in ground layer.

Camponotus midas Froggatt, 1896

Camponotus midas Froggatt, W.W. (1896). Honey ants. pp. 385-392 *in* Spencer, B. (ed.) *Report on the work of the Horn Scientific Expedition to Central Australia.* Melbourne : Melville, Mullen & Slade Pt. 2 Zoology [390 pl 27]. Type data: syntypes, AM W,F,M, from Illamurta in the James Range, N.T.

Distribution: Lake Eyre basin, N.T. Ecology: terrestrial, noctidiurnal, omnivore, desert, woodland, open forest; nest in soil.

Camponotus molossus Forel, 1907

Camponotus molossus Forel, A. (1907). Formicidae. pp. 263-310 *in* Michaelsen, W. & Hartmeyer, R. (eds.) *Die Fauna Südwest-Australiens.* Jena : G. Fischer Vol. 1 [306]. Type data: syntypes, GMNH W, ANIC W, from Buckland Hill and Serpentine, W.A.

Distribution: SW coastal, W.A. Ecology: terrestrial, noctidiurnal, omnivore, woodland, open forest; nest in ground layer.

Camponotus myoporus Clark, 1938

Camponotus (Tanaemyrmex) myoporus Clark, J. (1938). Reports of the McCoy Society for Field Investigation and Research. No. 2. Sir Joseph Bank Islands. Part I. Formicidae (Hymenoptera). *Proc. R. Soc. Vict.* **50**: 356-382 [379]. Type data: syntypes, NMV *W, from Reevesby Is., S.A.

Distribution: S Gulfs, S.A. Ecology: terrestrial, noctidiurnal, omnivore, woodland; nest in soil.

Camponotus nigriceps (F. Smith, 1858)

Camponotus nigriceps nigriceps (F. Smith, 1858)

Formica nigriceps Smith, F. (1858). *Catalogue of hymenopterous insects in the collection of the British Museum.* Part 6. Formicidae. London : British Museum 216 pp. 14 pls [27 Mar. 1858] [38]. Publication date established from Donisthorpe, H. (1932). On the identity of Smith's types of Formicidae (Hymenoptera) collected by Alfred Russell Wallace in the Malay Archipelago, with descriptions of two new species. *Ann. Mag. Nat. Hist. (10)* **10**: 441-476. Type data: syntypes (probable), BMNH *W, from Australia.

Distribution: SE coastal, N.S.W. Ecology: terrestrial, noctidiurnal, omnivore, woodland, open forest; nest in soil.

Camponotus nigriceps clarior Forel, 1902

Camponotus nigriceps clarior Forel, A. (1902). Fourmis nouvelles d'Australie. *Rev. Suisse Zool.* **10**: 405-548 [506]. Type data: syntypes, GMNH W, from Bendigo, Vic.

Distribution: Murray-Darling basin, Vic. Ecology: terrestrial, noctidiurnal, omnivore, woodland, open forest; nest in soil.

Camponotus nigriceps lividipes Emery, 1887

Camponotus nigriceps lividipes Emery, C. (1887). Catalogo delle formiche esistenti nelle collezioni del Museo Civico di Genova. Parte terza. Formiche della regione Indo-Malese e dell'Australia. *Ann. Mus. Civ. Stor. Nat. Giacomo Doria* **25**: 209-258 pls 3-4 [211]. Type data: syntypes, MCG *W, from Adelaide, S.A. and Qld.

Distribution: SW coastal, W.A. Ecology: terrestrial, noctidiurnal, omnivore, woodland, open forest; nest in soil.

Camponotus nigriceps obniger Forel, 1902

Camponotus nigriceps obniger Forel, A. (1902). Fourmis nouvelles d'Australie. *Rev. Suisse Zool.* **10**: 405-548 [506]. Type data: syntypes, GMNH W, ANIC W, from S.A.

Distribution: S Gulfs, W plateau, S.A. Ecology: terrestrial, noctidiurnal, omnivore, woodland, open forest; nest in soil.

Camponotus nigriceps pallidiceps Emery, 1887

Camponotus nigriceps pallidiceps Emery, C. (1887). Catalogo delle formiche esistenti nelle collezioni del Museo Civico di Genova. Parte terza. Formiche della regione Indo-Malese e dell'Australia. *Ann. Mus. Civ. Stor. Nat. Giacomo Doria* **25**: 209-258 pls 3-4 [211]. Type data: syntypes, MCG *W,F, from Mt. Victoria and Blue Mts., N.S.W.

Distribution: SE coastal, N.S.W. Ecology: terrestrial, noctidiurnal, omnivore, woodland, open forest; nest in soil.

Camponotus nigroaeneus (F. Smith, 1858)

Camponotus nigroaeneus nigroaeneus (F. Smith, 1858)

Formica nigroaenea Smith, F. (1858). *Catalogue of hymenopterous insects in the collection of the British Museum.* Part 6. Formicidae. London : British Museum 216 pp. 14 pls [27 Mar. 1858] [40]. Publication date established from Donisthorpe, H. (1932). On the identity of Smith's types of Formicidae (Hymenoptera) collected by Alfred Russell Wallace in the Malay Archipelago, with descriptions of two new species. *Ann. Mag. Nat. Hist. (10)* **10**: 441-476. Type data: syntypes (probable), BMNH *W, from Melbourne, Vic.

Distribution: SE coastal, Vic. Ecology: terrestrial, noctidiurnal, omnivore, woodland, open forest; nest in ground layer.

Camponotus nigroaeneus divus Forel, 1907

Camponotus nigroaeneus divus Forel, A. (1907). Formicides du Musée National Hongrois. *Ann. Hist.-Nat. Mus. Natl. Hung.* **5**: 1-42 [30 June 1907] [34]. Type data: syntypes (probable), probably in GMNH or MNH, from Mt. Victoria, Blue Mts., N.S.W.

Distribution: SE coastal, N.S.W. Ecology: terrestrial, noctidiurnal, omnivore, woodland, open forest; nest in ground layer.

Camponotus nitidiceps Viehmeyer, 1925

Camponotus (Myrmophyma) nitidiceps Viehmeyer, H. (1925). Formiciden der australischen Faunenregion. *Entomol. Mitt.* **14**: 139-149 [141]. Type data: syntypes, ZMB *W,F, from Liverpool and Trial Bay, N.S.W.

Distribution: SE coastal, N.S.W. Ecology: terrestrial, noctidiurnal, omnivore, woodland, open forest; nest in ground layer.

Camponotus novaehollandiae Mayr, 1870

Camponotus novaehollandiae Mayr, G.L. (1870). Neue Formiciden. *Verh. Zool.-Bot. Ges. Wien* **20**: Abhand. 939-996 [31 Dec. 1870] [939]. Type data: syntypes, NHMW W, from Cape York, Qld.

Distribution: N Gulf, Qld. Ecology: terrestrial, noctidiurnal, omnivore, woodland; nest in ground layer.

Camponotus oetkeri Forel, 1910

Camponotus oetkeri oetkeri Forel, 1910

Camponotus oetkeri Forel, A. (1910). Formicides australiens reçus de MM. Froggatt et Rowland Turner. *Rev. Suisse Zool.* **18**: 1-94 [75]. Type data: syntypes, GMNH W, ANIC W, from Tennant Creek, N.T.

Distribution: W plateau, N.T. Ecology: terrestrial, noctidiurnal, omnivore, desert, woodland; nest in soil.

Camponotus oetkeri voltai Forel, 1913

Camponotus (Myrmogonia) oetkeri voltai Forel, A. (1913). Fourmis de Tasmanie et d'Australie récoltées par MM. Lea, Froggatt etc. *Bull. Soc. Vaud. Sci. Nat.* **49**: 173-196 pl 2 [191]. Type data: syntypes, GMNH W, from Tas.

Distribution: Tas. Ecology: terrestrial, noctidiurnal, omnivore, woodland, open forest; nest in ground layer.

Camponotus oxleyi Forel, 1902

Camponotus oxleyi Forel, A. (1902). Fourmis nouvelles d'Australie. *Rev. Suisse Zool.* **10**: 405-548 [501]. Type data: syntypes, GMNH W, ANIC W, from Bong Bong, N.S.W.

Distribution: SE coastal, N.S.W. Ecology: terrestrial, noctidiurnal, omnivore, woodland, open forest; nest in ground layer.

Camponotus pellax Santschi, 1919

Camponotus (Myrmocamelus) pellax Santschi, F. (1919). Cinq notes myrmécologiques. *Bull. Soc. Vaud. Sci. Nat.* **52**: 325-350 [330]. Type data: syntypes, BNHM W, from Townsville, Qld.

Distribution: NE coastal, Qld. Ecology: terrestrial, noctidiurnal, omnivore, woodland, open forest; nest in ground layer.

Camponotus postcornutus Clark, 1930

Camponotus (Tanaemyrmex) postcornutus Clark, J. (1930). Some new Australian Formicidae. *Proc. R. Soc. Vict.* **42**: 116-128 [10 Mar. 1930] [121]. Type data: syntypes, NMV *W, from Bungulla, W.A.

Distribution: SW coastal, W.A. Ecology: terrestrial, noctidiurnal, omnivore, woodland; nest in soil.

Camponotus punctiventris Emery, 1920

Camponotus (Myrmogonia) punctiventris Emery, C. (1920). Studi sui *Camponotus*. *Boll. Soc. Entomol. Ital.* **52**: 1-49 [6 Dec. 1920] [31]. Type data: holotype, MCG *W, from Kamerunga, Qld.

Distribution: NE coastal, Qld. Ecology: terrestrial, noctidiurnal, omnivore, woodland, open forest; nest in ground layer.

Camponotus reticulatus Roger, 1863

Camponotus reticulatus Roger, J. (1863). Die neu aufgeführten Gattungen und Arten meines Formiaden-Verzeichnisses. *Berl. Entomol. Z.* **7**: 129-214 [139]. Type data: status unknown, ?ZMB, from Manilla (Philippines?).

Camponotus reticulatus mackayensis Forel, 1902

Camponotus reticulatus mackayensis Forel, A. (1902). Fourmis nouvelles d'Australie. *Rev. Suisse Zool.* **10**: 405-548 [506]. Type data: syntypes, GMNH W, ANIC W, from Mackay, Qld.

Distribution: NE coastal, Qld. Ecology: terrestrial, noctidiurnal, omnivore, woodland, open forest; nest in ground layer.

Camponotus rubiginosus Mayr, 1876

Camponotus rubiginosus Mayr, G.L. (1876). Die australischen Formiciden. *J. Mus. Godeffroy* **5**: 56-115 [66]. Type data: syntypes, whereabouts unknown, from Peak Downs, Qld.

Distribution: NE coastal, Qld. Ecology: terrestrial, noctidiurnal, omnivore, woodland; nest in ground layer.

Camponotus rufus Crawley, 1925

Camponotus (Dinomyrmex) rufus Crawley, W.C. (1925). New ants from Australia. II. *Ann. Mag. Nat. Hist. (9)* **16**: 577–598 [596]. Type data: syntypes, OUM *W,F, from W.A.

Distribution: SW coastal, W.A. Ecology: terrestrial, noctidiurnal, omnivore, woodland, open forest; nest in ground layer.

Camponotus sanguinea McAreavey, 1949

Camponotus (Myrmogonia) sanguinea McAreavey, J.J. (1949). Australian Formicidae. New genera and species. *Proc. Linn. Soc. N.S.W.* **74**: 1–25 [15 June 1949] [18]. Type data: holotype, ANIC W, from Broome, W.A.

Distribution: N coastal, W.A. Ecology: terrestrial, noctidiurnal, omnivore, desert, woodland; nest in ground layer.

Camponotus sanguinifrons Viehmeyer, 1925

Camponotus (Colobopsis) sanguinifrons Viehmeyer, H. (1925). Formiciden der australischen Faunenregion. *Entomol. Mitt.* **14**: 139–149 [143]. Type data: syntypes, ZMB *W, from Trial Bay, N.S.W.

Distribution: SE coastal, N.S.W. Ecology: terrestrial, noctidiurnal, omnivore, woodland, open forest; nest arboreal.

Camponotus scratius Forel, 1907

Camponotus scratius scratius Forel, 1907

Camponotus scratius Forel, A. (1907). Formicidae. pp. 263–310 *in* Michaelsen, W. & Hartmeyer, R. (eds.) *Die Fauna Südwest-Australiens.* Jena : G. Fischer Vol. 1 [304]. Type data: syntypes, GMNH W,F, ANIC W, from Buckland Hill and Fremantle, W.A.

Distribution: SW coastal, W.A. Ecology: terrestrial, noctidiurnal, omnivore, woodland, open forest; nest in ground layer.

Camponotus scratius nuntius Forel, 1907

Camponotus scratius nuntius Forel, A. (1907). Formicidae. pp. 263–310 *in* Michaelsen, W. & Hartmeyer, R. (eds.) *Die Fauna Südwest-Australiens.* Jena : G. Fischer Vol. 1 [306]. Type data: holotype, probably destroyed in ZMH in WW II, from Dirk Hartog Brown Station, W.A.

Distribution: NW coastal, W.A. Ecology: terrestrial, noctidiurnal, omnivore, desert, woodland; nest in soil.

Camponotus semicarinatus (Forel, 1895)

Colobopsis rufifrons semicarinata Forel, A. (1895). Nouvelles fourmis d'Australie, récoltée à The Ridge, Mackay, Queensland par M. Gilbert Turner. *Ann. Soc. Entomol. Belg.* **39**: 417–428 [418]. Type data: syntypes, GMNH W, ANIC W, from Mackay, Qld.

Distribution: NE coastal, Qld. Ecology: terrestrial, noctidiurnal, omnivore, woodland, open forest; nest in ground layer. Biological references: Emery, C. (1925). Hymenoptera Fam. Formicidae subfam. Formicinae. *in* Wytsman, P. (ed.) *Genera Insectorum.* Fasc. 183 302 pp. 4 pls (raised to species).

Camponotus simulator Forel, 1915

Camponotus (Dinomyrmex) simulator Forel, A. (1915). Results of Dr. E. Mjöbergs Swedish Scientific Expeditions to Australia 1910–1913. 2. Ameisen. *Ark. Zool.* **9**: 1–119 pls 1–3 [4 Dec. 1915] [96]. Type data: syntypes, GMNH W, other syntypes may exist, from Atherton and Herberton, Qld.

Distribution: NE coastal, Qld. Ecology: terrestrial, noctidiurnal, omnivore, woodland, open forest; nest in ground layer.

Camponotus spenceri Clark, 1930

Camponotus reticulatus Kirby, W.F. (1896). Hymenoptera. pp. 203–209 *in* Spencer, B. (ed.) *Report on the work of the Horn Scientific Expedition to Central Australia.* Melbourne : Melville, Mullen & Slade Pt. 1 supplement [204] [*non Camponotus reticulatus* Roger, 1863]. Type data: syntypes, BMNH (probable) *W, from Paisley Bluff, N.T.

Camponotus (Tanaemyrmex) spenceri Clark, J. (1930). New Formicidae, with notes on some little-known species. *Proc. R. Soc. Vict.* **43**: 2–25 [30 Aug. 1930] [18] [*nom. nov.* for *Camponotus reticulatus* W.F. Kirby, 1896].

Distribution: Lake Eyre basin, N.T. Ecology: terrestrial, noctidiurnal, omnivore, desert, woodland; nest in soil.

Camponotus spinitarsus Emery, 1920

Camponotus (Dinomyrmex) spinitarsus Emery, C. (1920). Studi sui *Camponotus. Boll. Soc. Entomol. Ital.* **52**: 1–49 [6 Dec. 1920] [22]. Type data: holotype, MCG *W, from Cooktown, Qld.

Distribution: NE coastal, Qld. Ecology: terrestrial, noctidiurnal, omnivore, woodland, open forest; nest in ground layer.

Camponotus sponsorum Forel, 1910

Camponotus sponsorum Forel, A. (1910). Formicides australiens reçus de MM. Froggatt et Rowland Turner. *Rev. Suisse Zool.* **18**: 1–94 [76]. Type data: syntypes, GMNH W,M, ANIC W, from Tennant Creek, N.T.

Distribution: W plateau, N.T. Ecology: terrestrial, noctidiurnal, omnivore, desert, woodland; nest in soil.

Camponotus subnitidus Mayr, 1876

Camponotus subnitidus subnitidus Mayr, 1876

Camponotus subnitidus Mayr, G.L. (1876). Die australischen Formiciden. *J. Mus. Godeffroy* **5**: 56–115 [65]. Type data: syntypes, NHMW W, from Peak Downs, Qld.

Distribution: NE coastal, Qld. Ecology: terrestrial, noctidiurnal, omnivore, woodland, open forest; nest in ground layer.

Camponotus subnitidus famelicus Emery, 1887

Camponotus subnitidus famelicus Emery, C. (1887). Catalogo delle formiche esistenti nelle collezioni del Museo Civico di Genova. Parte terza. Formiche della regione Indo-Malese e dell'Australia. *Ann. Mus. Civ. Stor. Nat. Giacomo Doria* **25**: 209–258 pls 3–4 [214]. Type data: syntypes, MCG *W, from Adelaide, S.A.

Distribution: S Gulfs, S.A. Ecology: terrestrial, noctidiurnal, omnivore, woodland, open forest; nest in ground layer.

Camponotus subnitidus longinodis Forel, 1915

Camponotus (Dinomyrmex) subnitidus longinodis Forel, A. (1915). Results of Dr. E. Mjöbergs Swedish Scientific Expeditions to Australia 1910–1913. 2. Ameisen. *Ark. Zool.* **9**: 1–119 pls 1–3 [4 Dec. 1915] [96]. Type data: syntypes, whereabouts unknown, from Cape York Peninsula, Qld.

Distribution: N Gulf, Qld. Ecology: terrestrial, noctidiurnal, omnivore, woodland; nest in ground layer.

Camponotus suffusus (F. Smith, 1858)

Camponotus suffusus suffusus (F. Smith, 1858)

Formica suffusa Smith, F. (1858). *Catalogue of hymenopterous insects in the collection of the British Museum.* Part 6. Formicidae. London : British Museum 216 pp. 14 pls [27 Mar. 1858] [38]. Publication date established from Donisthorpe, H. (1932). On the identity of Smith's types of Formicidae (Hymenoptera) collected by Alfred Russell Wallace in the Malay Archipelago, with descriptions of two new species. *Ann. Mag. Nat. Hist.* (10) **10**: 441–476. Type data: syntypes (probable), BMNH *F, from Australia.

Formica piliventris Smith, F. (1858). *Catalogue of hymenopterous insects in the collection of the British Museum.* Part 6. Formicidae. London : British Museum 216 pp. 14 pls [27 Mar. 1858] [39]. Publication date established from Donisthorpe, H. (1932). On the identity of Smith's types of Formicidae (Hymenoptera) collected by Alfred Russell Wallace in the Malay Archipelago, with descriptions of two new species. *Ann. Mag. Nat. Hist.* (10) **10**: 441–476. Type data: syntypes (probable), BMNH *W, from S.A.

Camponotus schencki Mayr, G.L. (1862). Myrmecologische Studien. *Verh. Zool.-Bot. Ges. Wien* **12**: Abhand. 649–776 [674 pl 19]. Type data: uncertain, whereabouts unknown, from Australia (as New Holland).

Synonymy that of Emery, C. (1925). Hymenoptera Fam. Formicidae subfam. Formicinae. *in* Wytsman, P. (ed.) *Genera Insectorum.* Fasc. 183 302 pp. 4 pls [114].

Distribution: S Gulfs, S.A. Ecology: terrestrial, noctidiurnal, omnivore, woodland, open forest; nest in ground layer.

Camponotus suffusus bendigensis Forel, 1902

Camponotus suffusus bendigensis Forel, A. (1902). Fourmis nouvelles d'Australie. *Rev. Suisse Zool.* **10**: 405–548 [493]. Type data: syntypes, GMNH W, from Bendigo, Vic.

Distribution: Murray-Darling basin, Vic. Ecology: terrestrial, noctidiurnal, omnivore, woodland, open forest; nest in ground layer.

Camponotus tasmani Forel, 1902

Camponotus tasmani Forel, A. (1902). Fourmis nouvelles d'Australie. *Rev. Suisse Zool.* **10**: 405–548 [503]. Type data: syntypes, GMNH W, ANIC W, from S.A.

Distribution: S Gulfs, S.A. Ecology: terrestrial, noctidiurnal, omnivore, woodland, open forest; nest in ground layer.

Camponotus testaceipes (F. Smith, 1858)

Formica testaceipes Smith, F. (1858). *Catalogue of hymenopterous insects in the collection of the British Museum.* Part 6. Formicidae. London : British Museum 216 pp. 14 pls [27 Mar. 1858] [39]. Publication date established from Donisthorpe, H. (1932). On the identity of Smith's types of Formicidae (Hymenoptera) collected by Alfred Russell Wallace in the Malay Archipelago, with descriptions of two new species. *Ann. Mag. Nat. Hist.* (10) **10**: 441–476. Type data: syntypes (probable), BMNH *W, from King George's Sound (=King George Sound), W.A.

Formica terebrans Lowne, B.T. (1865). Contributions to the natural history of Australian ants. *Entomologist* **2**: 275–280 [278]. Type data: syntypes, BMNH (probable) *W,F, from Sidney (=Sydney), N.S.W.

Camponotus (Myrmophyma) darlingtoni Wheeler, W.M. (1934). Contributions to the fauna of Rottnest Island, Western Australia No. IX. The ants. *J. R. Soc. West. Aust.* **20**: 137–163 [5 Oct. 1934] [160]. Type data: syntypes, MCZ *W,F, from Longreach Bay and Government House, Rottnest Is. and Kings Park, Perth and Margaret River, W.A.

Camponotus (Myrmophyma) rottnesti Donisthorpe, H. (1941). Synonymical notes, etc., on Formicidae (Hym.). *Entomol. Mon. Mag.* **77**: 237–240 [1 Oct. 1941] [239] [unnecessarily proposed *nom. nov.* for *Camponotus (Myrmophyma) darlingtoni* Wheeler, 1934].

Synonymy that of Emery, C. (1925). Hymenoptera Fam. Formicidae subfam. Formicinae. *in* Wytsman, P. (ed.) *Genera Insectorum.* Fasc. 183 302 pp. 4 pls [102]; Brown, W.L. jr. (1956). Some synonymies in the ant genus *Camponotus. Psyche Camb.* **63**: 38–40 [39].

Distribution: SE coastal, SW coastal, N.S.W., W.A. Ecology: terrestrial, noctidiurnal, omnivore, woodland; nest in soil.

Camponotus tricoloratus Clark, 1941

Camponotus (Tanaemyrmex) tricoloratus Clark, J. (1941). Australian Formicidae. Notes and new species. *Mem. Natl. Mus. Vict.* 12: 71–94 [90 pl 13]. Type data: syntypes, NMV *W, from near Mildura, Vic.

Distribution: Murray-Darling basin, Vic. Ecology: terrestrial, noctidiurnal, omnivore, desert, woodland, open forest; nest in ground layer.

Camponotus tristis Clark, 1930

Camponotus (Myrmophyma) tristis Clark, J. (1930). Some new Australian Formicidae. *Proc. R. Soc. Vict.* 42: 116–128 [10 Mar. 1930] [124]. Type data: syntypes, NMV *W, from Eradu, W.A.

Distribution: NW coastal, W.A. Ecology: terrestrial, noctidiurnal, omnivore, woodland, open forest; nest in ground layer.

Camponotus tumidus Crawley, 1922

Camponotus (Myrmogonia) tumidus Crawley, W.C. (1922). New ants from Australia. *Ann. Mag. Nat. Hist. (9)* 10: 16–36 [35]. Type data: syntypes, OUM *W, from Byford, W.A.

Distribution: SW coastal, W.A. Ecology: terrestrial, noctidiurnal, omnivore, woodland, open forest; nest in ground layer.

Camponotus versicolor Clark, 1930

Camponotus (Myrmosaulus) versicolor Clark, J. (1930). Some new Australian Formicidae. *Proc. R. Soc. Vict.* 42: 116–128 [10 Mar. 1930] [122]. Type data: syntypes, NMV *W, from Emu Rocks, W.A.

Distribution: SW coastal, W.A. Ecology: terrestrial, noctidiurnal, omnivore, woodland; nest in soil.

Camponotus villosus Crawley, 1915

Camponotus (Myrmoturba) villosa Crawley, W.C. (1915). Ants from north and central Australia, collected by G.F. Hill. Part I. *Ann. Mag. Nat. Hist. (8)* 15: 130–136 [135]. Type data: syntypes, BMNH *W, from Batchelor, N.T.

Distribution: N coastal, N.T. Ecology: terrestrial, noctidiurnal, omnivore, desert, woodland; nest in soil.

Camponotus vitreus (F. Smith, 1860)

Formica vitrea Smith, F. (1860). Catalogue of hymenopterous insects collected by Mr A.R. Wallace in the islands of Bachian, Kaisaa, Amboyne, Gilolo, and at Dory in New Guinea. *J. Linn. Soc. Zool.* 5: 93–143 pl 1 [18 July 1860] [94]. Publication date established from Donisthorpe, H. (1932). On the identity of Smith's types of Formicidae (Hymenoptera) collected by Alfred Russell Wallace in the Malay Archipelago, with descriptions of two new species. *Ann. Mag. Nat. Hist. (10)* 10: 441–476. Type data: syntypes (probable), BMNH *W, from Bachian, Indonesia.

Prenolepis adlerzii Forel, A. (1886). Études myrmécologiques en 1886. *Ann. Soc. Entomol. Belg.* 30: 131–215 [209]. Type data: syntypes (probable), GMNH *W, from Darnley Is., Qld.

Synonymy that of Emery, C. (1925). Hymenoptera Fam. Formicidae subfam. Formicinae. *in* Wytsman, P. (ed.) *Genera Insectorum*. Fasc. 183 302 pp. 4 pls [148].

Distribution: NE coastal, Qld. Ecology: terrestrial, noctidiurnal, omnivore, woodland, open forest; nest in ground layer.

Camponotus walkeri Forel, 1893

Camponotus walkeri walkeri Forel, 1893

Camponotus walkeri Forel, A. (1893). Nouvelles fourmis d'Australie et des Canaries. *Ann. Soc. Entomol. Belg.* 37: 454–466 [454]. Type data: syntypes, GMNH W, from Baudin Is., W.A.

Distribution: NW coastal, W.A. Ecology: terrestrial, noctidiurnal, omnivore, desert, woodland; nest in ground layer.

Camponotus walkeri bardus Forel, 1910

Camponotus walkeri bardus Forel, A. (1910). Formicides australiens reçus de MM. Froggatt et Rowland Turner. *Rev. Suisse Zool.* 18: 1–94 [73]. Type data: holotype, GMNH W, from Perth, W.A.

Distribution: SW coastal, W.A. Ecology: terrestrial, noctidiurnal, omnivore, woodland, open forest; nest in ground layer.

Camponotus whitei Wheeler, 1915

Camponotus (Myrmosphincta) whitei Wheeler, W.M. (1915). Hymenoptera. *Trans. R. Soc. S. Aust.* 39: 805–823 pls 64–66 [Dec. 1915] [818]. Type data: syntypes, MCZ *W, from Flat Rock Hole in the Musgrave Ranges, S.A.

Camponotus (Myrmosaulus) scutellus Clark, J. (1930). Some new Australian Formicidae. *Proc. R. Soc. Vict.* 42: 116–128 [10 Mar. 1930] [123]. Type data: syntypes, NMV *W, from Tammin, Emu Rocks, Bungulla and Merredin, W.A.

Synonymy that of Brown, W.L. jr. (1956). Some synonymies in the ant genus *Camponotus*. *Psyche Camb.* 63: 38–40 [40].

Distribution: W plateau, SW coastal, S.A., W.A. Ecology: terrestrial, noctidiurnal, omnivore, desert, woodland; nest in soil.

Camponotus wiederkehri Forel, 1894

Camponotus wiederkehri wiederkehri Forel, 1894

Camponotus wiederkehri Forel, A. (1894). Quelques fourmis de Madagascar (récoltées par M. le Dr. Völtzkow); de Nouvelle Zélande (récoltées par M. W.W. Smith); de Nouvelle Calédonie (récoltées par M.

Sommer); de Queensland (Australie) récoltées par M. Wiederkehr; et de Perth (Australie occidentale) récoltées par M. Chase. *Ann. Soc. Entomol. Belg.* **38**: 226-237 [232]. Type data: syntypes, GMNH W, ANIC W, from Charters Towers, Qld.

Distribution: NE coastal, Qld. Ecology: terrestrial, noctidiurnal, omnivore, woodland, open forest; nest in soil.

Camponotus wiederkehri lucidior Forel, 1910

Camponotus wiederkehri lucidior Forel, A. (1910). Formicides australiens reçus de MM. Froggatt et Rowland Turner. *Rev. Suisse Zool.* **18**: 1-94 [81]. Type data: syntypes, GMNH W,M, ANIC W, from Tennant Creek, N.T.

Distribution: W plateau, N.T. Ecology: terrestrial, noctidiurnal, omnivore, desert, woodland, open forest; nest in soil.

Echinopla F. Smith, 1857

Echinopla Smith, F. (1857). Catalogue of the hymenopterous insects collected at Sarawak, Borneo, Mount Ophir, Malacca; and at Singapore by A. R. Wallace. *J. Linn. Soc. Zool.* **2**: 42-130 [2 Nov. 1857] [79 pls 1-2]. Publication date established from Donisthorpe, H. (1932). On the identity of Smith's types of Formicidae (Hymenoptera) collected by Alfred Russell Wallace in the Malay Archipelago, with descriptions of two new species. *Ann. Mag. Nat. Hist. (10)* **10**: 441-476. Type species *Echinopla melanarctos* F. Smith, 1857 by subsequent designation, see Wheeler, W. M. (1911). A list of the type species of the genera and subgenera of Formicidae. *Ann. N.Y. Acad. Sci.* **21**: 157-175 [17 Oct. 1911].

This group is also found in the Oriental Region; New Guinea in the Australian Region, see Brown, W.L. jr. (1973). A comparison of the Hylean and Congo-West African rain forest ant faunas. pp. 161-185 *in* Meggers, B.J., Ayensu, E.S. & Duckworth, W.D. (eds.) *Tropical forest ecosystems in Africa and South America: a comparative review.* Washington : Smithsonian Institution Press.

Echinopla australis Forel, 1901

Echinopla australis Forel, A. (1901). Formiciden des Naturhistorischen Museums zu Hamburg. Neue *Calyptomyrmex-, Dacryon-, Podomyrma-,* und *Echinopla*-Arten. *Mitt. Naturh. Mus. Hamb.* **18**: 45-82 [75]. Type data: syntypes, GMNH W, ANIC W, from Mackay, Qld.

Distribution: NE coastal, Qld. Ecology: terrestrial, diurnal, omnivore, closed forest; nest arboreal.

Echinopla turneri Forel, 1901

Echinopla turneri turneri Forel, 1901

Echinopla turneri Forel, A. (1901). Formiciden des Naturhistorischen Museums zu Hamburg. Neue *Calyptomyrmex-, Dacryon-, Podomyrma-,* und *Echinopla*-Arten. *Mitt. Naturh. Mus. Hamb.* **18**: 45-82 [76]. Type data: syntypes, GMNH W, ANIC W, from Mackay, Qld.

Distribution: NE coastal, Qld. Ecology: terrestrial, diurnal, omnivore, closed forest; nest arboreal.

Echinopla turneri pictipes Forel, 1901

Echinopla turneri pictipes Forel, A. (1901). Formiciden des Naturhistorischen Museums zu Hamburg. Neue *Calyptomyrmex-, Dacryon-, Podomyrma-,* und *Echinopla*-Arten. *Mitt. Naturh. Mus. Hamb.* **18**: 45-82 [76]. Type data: syntypes, GMNH W,F, ANIC W, from Mackay, Qld.

Distribution: NE coastal, Qld. Ecology: terrestrial, diurnal, omnivore, closed forest; nest arboreal.

Melophorus Lubbock, 1883

Melophorus Lubbock, J. (1883). Observations on Ants, Bees and Wasps - Part X. With a Description of a New Genus of Honey-Ant. *J. Linn. Soc. Zool.* **17**: 41-52 [17 Apr. 1883] [51 pl 2]. Type species *Melophorus bagoti* Lubbock, 1883 by monotypy.
Erimelophorus Wheeler, W.M. (1935). Myrmecological notes. *Psyche Camb.* **42**: 68-72 [71] [proposed with subgeneric rank in *Melophorus* Lubbock, 1883]. Type species *Melophorus wheeleri* Forel, 1910 by original designation.
Trichomelophorus Wheeler, W.M. (1935). Myrmecological notes. *Psyche Camb.* **42**: 68-72 [71] [proposed with subgeneric rank in *Melophorus* Lubbock, 1883]. Type species *Melophorus hirsutus* Forel, 1902 by original designation.

Synonymy that of Brown, W.L. jr. (1955). A revision of the Australian ant genus *Notoncus* Emery, with notes on the other genera of Melophorini. *Bull. Mus. Comp. Zool.* **113**: 469-494 [474].

Melophorus aeneovirens (Lowne, 1865)

Formica aeneovirens Lowne, B.T. (1865). Contributions to the natural history of Australian ants. *Entomologist* **2**: 275-280 [276]. Type data: syntypes, BMNH (probable) *W, from Port Jackson, N.S.W.

Distribution: SE coastal, N.S.W. Ecology: terrestrial, noctidiurnal, omnivore, granivore, tussock grassland, woodland, open forest; nest in soil.

Melophorus bagoti Lubbock, 1883

Melophorus bagoti Lubbock, J. (1883). Observations on Ants, Bees and Wasps. - Part X. With a Description of a New Genus of Honey-Ant. *J. Linn. Soc. Zool.* **17**: 41-52 [17 Apr. 1883] [52 pl 2 figs 1-10]. Type data: syntypes (probable), BMNH (probable) *W, from Australia (lat. 21 S) [*sic*].

Distribution: W plateau, Lake Eyre basin, W.A., S.A., Qld., N.S.W. Ecology: terrestrial, noctidiurnal, omnivore, granivore, desert, hummock grassland, woodland; nest in soil.

Melophorus biroi Forel, 1907

Melophorus biroi Forel, A. (1907). Formicides du Musée National Hongrois. *Ann. Hist.- Nat. Mus. Natl. Hung.* **5**: 1–42 [30 June 1907] [29]. Type data: syntypes (probable), probably in GMNH or MNH, from Mt. Victoria, Blue Mts., N.S.W.

Distribution: SE coastal, N.S.W. Ecology: terrestrial, noctidiurnal, omnivore, granivore, woodland, open forest; nest in soil.

Melophorus bruneus McAreavey, 1949

Melophorus (Melophorus) brunea McAreavey, J.J. (1949). Australian Formicidae. New genera and species. *Proc. Linn. Soc. N.S.W.* **74**: 1–25 [15 June 1949] [20]. Type data: holotype, ANIC W, from Nyngan, N.S.W.

Distribution: Murray-Darling basin, N.S.W. Ecology: terrestrial, noctidiurnal, omnivore, granivore, desert, woodland; nest in soil.

Melophorus constans Santschi, 1928

Melophorus constans Santschi, F. (1928). Nouvelles fourmis d'Australie. *Bull. Soc. Vaud. Sci. Nat.* **56**: 465–483 [30 Aug. 1928] [475]. Type data: syntypes, BNHM W,F, from Idatlle Glen, Vic.

Distribution: SE coastal, N.S.W. Ecology: terrestrial, noctidiurnal, omnivore, granivore, tussock grassland, woodland, open forest; nest in soil.

Melophorus curtus Forel, 1902

Melophorus curtus Forel, A. (1902). Fourmis nouvelles d'Australie. *Rev. Suisse Zool.* **10**: 405–548 [485]. Type data: syntypes, GMNH W,F, ANIC W, from Mackay, Qld.

Distribution: NE coastal, Qld. Ecology: terrestrial, noctidiurnal, omnivore, granivore, tussock grassland, woodland, open forest; nest in soil.

Melophorus fieldi Forel, 1910

Melophorus fieldi fieldi Forel, 1910

Melophorus fieldi Forel, A. (1910). Formicides australiens reçus de MM. Froggatt et Rowland Turner. *Rev. Suisse Zool.* **18**: 1–94 [62]. Type data: holotype, GMNH W, from Tennant Creek, N.T.

Distribution: W plateau, N.T. Ecology: terrestrial, noctidiurnal, omnivore, granivore, desert, hummock grassland, woodland; nest in soil.

Melophorus fieldi major Forel, 1915

Melophorus fieldi major Forel, A. (1915). Results of Dr. E. Mjöbergs Swedish Scientific Expeditions to Australia 1910–1913. 2. Ameisen. *Ark. Zool.* **9**: 1–119 pls 1–3 [4 Dec. 1915] [87]. Type data: syntypes, GMNH W, other syntypes may exist, from Kimberley distr., W.A.

Distribution: N coastal, W.A. Ecology: terrestrial, noctidiurnal, omnivore, granivore, desert, woodland; nest in soil.

Melophorus fieldi propinqua Viehmeyer, 1925

Melophorus fieldi propinqua Viehmeyer, H. (1925). Formiciden der australischen Faunenregion. *Entomol. Mitt.* **14**: 25–39 [36]. Type data: syntypes, ZMB *W, from Liverpool, N.S.W.

Distribution: SE coastal, N.S.W. Ecology: terrestrial, noctidiurnal, omnivore, granivore, woodland, open forest; nest in soil.

Melophorus fulvihirtus Clark, 1941

Melophorus fulvihirtus Clark, J. (1941). Australian Formicidae. Notes and new species. *Mem. Natl. Mus. Vict.* **12**: 71–94 [88 pl 13]. Type data: syntypes, NMV *W, from Patho, Vic.

Distribution: Murray-Darling basin, Vic. Ecology: terrestrial, noctidiurnal, omnivore, granivore, woodland, open forest; nest in soil.

Melophorus hirsutus Forel, 1902

Melophorus hirsutus Forel, A. (1902). Fourmis nouvelles d'Australie. *Rev. Suisse Zool.* **10**: 405–548 [488]. Type data: syntypes, GMNH W, from Mackay, Qld.

Distribution: NE coastal, Qld. Ecology: terrestrial, noctidiurnal, omnivore, granivore, tussock grassland, woodland, open forest.

Melophorus insularis Wheeler, 1934

Melophorus insularis Wheeler, W.M. (1934). Contributions to the fauna of Rottnest Island, Western Australia No. IX. The ants. *J. R. Soc. West. Aust.* **20**: 137–163 [5 Oct. 1934] [151]. Type data: syntypes, MCZ *W, from White Hill and City of York Bay, Rottnest Is., W.A.

Distribution: SW coastal, W.A. Ecology: terrestrial, noctidiurnal, omnivore, granivore, woodland; nest in soil.

Melophorus iridescens (Emery, 1887)

Melophorus iridescens iridescens (Emery, 1887)

Myrmecocystus iridescens Emery, C. (1887). Catalogo delle formiche esistenti nelle collezioni del Museo Civico di Genova. Parte terza. Formiche della regione Indo-Malese e dell'Australia. *Ann. Mus. Civ. Stor. Nat. Giacomo Doria* **25**: 209–258 pls 3–4 [247]. Type data: syntypes, MCG *W, from Mt. Victoria, N.S.W.

Distribution: SE coastal, N.S.W. Ecology: terrestrial, noctidiurnal, omnivore, granivore, tussock grassland, woodland, open forest; nest in soil.

Melophorus iridescens fraudatrix Forel, 1915

Melophorus iridescens fraudatrix Forel, A. (1915). Results of Dr. E. Mjöbergs Swedish Scientific

Expeditions to Australia 1910–1913. 2. Ameisen. *Ark. Zool.* **9**: 1–119 pls 1–3 [4 Dec. 1915] [87]. Type data: syntypes, GMNH W, other syntypes may exist, from Healesville, Vic.

Distribution: SE coastal, Vic. Ecology: terrestrial, noctidiurnal, omnivore, granivore, tussock grassland, woodland, open forest; nest in soil.

Melophorus iridescens froggatti Forel, 1902

Melophorus iridescens froggatti Forel, A. (1902). Fourmis nouvelles d'Australie. *Rev. Suisse Zool.* **10**: 405–548 [487]. Type data: syntypes, GMNH W,F, ANIC W, from Sydney, N.S.W.

Distribution: SE coastal, N.S.W. Ecology: terrestrial, noctidiurnal, omnivore, granivore, tussock grassland, woodland, open forest; nest in soil.

Melophorus laticeps Wheeler, 1915

Melophorus laticeps Wheeler, W.M. (1915). Hymenoptera. *Trans. R. Soc. S. Aust.* **39**: 805–823 pls 64–66 [Dec. 1915] [813]. Type data: holotype, MCZ *F, from between Todmorden and Wantapella, S.A.

Distribution: Lake Eyre basin, S.A. Ecology: terrestrial, noctidiurnal, omnivore, granivore, desert, hummock grassland, woodland; nest in soil.

Melophorus ludius Forel, 1902

Melophorus ludius ludius Forel, 1902

Melophorus ludius Forel, A. (1902). Fourmis nouvelles d'Australie. *Rev. Suisse Zool.* **10**: 405–548 [484]. Type data: syntypes, GMNH W, from Mackay, Qld.

Distribution: NE coastal, Qld. Ecology: terrestrial, noctidiurnal, omnivore, granivore, tussock grassland, woodland, open forest; nest in soil.

Melophorus ludius sulla Forel, 1910

Melophorus ludius sulla Forel, A. (1910). Formicides australiens reçus de MM. Froggatt et Rowland Turner. *Rev. Suisse Zool.* **18**: 1–94 [66]. Type data: syntypes, GMNH W,F,M, ANIC W, from Tennant Creek, N.T.

Distribution: W plateau, N.T. Ecology: terrestrial, noctidiurnal, omnivore, granivore, hummock grassland, woodland, open forest; nest in soil.

Melophorus marius Forel, 1910

Melophorus marius Forel, A. (1910). Formicides australiens reçus de MM. Froggatt et Rowland Turner. *Rev. Suisse Zool.* **18**: 1–94 [66]. Type data: holotype, GMNH W, from Tennant Creek, N.T.

Distribution: W plateau, N.T. Ecology: terrestrial, noctidiurnal, omnivore, granivore, hummock grassland, woodland, open forest; nest in soil.

Melophorus mjobergi Forel, 1915

Melophorus mjobergi Forel, A. (1915). Results of Dr. E. Mjöbergs Swedish Scientific Expeditions to Australia 1910–1913. 2. Ameisen. *Ark. Zool.* **9**: 1–119 pls 1–3 [4 Dec. 1915] [88]. Type data: syntypes, GMNH W, ANIC W, other syntypes may exist, from Broome, W.A.

Distribution: N coastal, W.A. Ecology: terrestrial, noctidiurnal, omnivore, granivore, woodland, open forest; nest in soil.

Melophorus omniparens Forel, 1915

Melophorus omniparens Forel, A. (1915). Results of Dr. E. Mjöbergs Swedish Scientific Expeditions to Australia 1910–1913. 2. Ameisen. *Ark. Zool.* **9**: 1–119 pls 1–3 [4 Dec. 1915] [85]. Type data: syntypes, GMNH W, other syntypes may exist, from Alice River, Qld.

Distribution: NE coastal, Qld. Ecology: terrestrial, noctidiurnal, omnivore, granivore, tussock grassland, woodland, open forest; nest in soil.

Melophorus pillipes Santschi, 1919

Melophorus pillipes Santschi, F. (1919). Cinq notes myrmécologiques. *Bull. Soc. Vaud. Sci. Nat.* **52**: 325–350 [329]. Type data: syntypes, BNHM W, from Townsville, Qld.

Distribution: NE coastal, Qld. Ecology: terrestrial, noctidiurnal, omnivore, granivore, tussock grassland, woodland, open forest; nest in soil.

Melophorus potteri McAreavey, 1947

Melophorus potteri McAreavey, J.J. (1947). New species of the genera *Prolasius* Forel and *Melophorus* Lubbock (Hymenoptera : Formicidae). *Mem. Natl. Mus. Vict.* **15**: 7–27 [Oct. 1947] [25 pl 1]. Type data: syntypes, NMV *W,F, from Patho, Vic.

Distribution: Murray-Darling basin, Vic. Ecology: terrestrial, noctidiurnal, omnivore, granivore, desert, woodland; nest in soil.

Melophorus scipio Forel, 1915

Melophorus scipio Forel, A. (1915). Results of Dr. E. Mjöbergs Swedish Scientific Expeditions to Australia 1910–1913. 2. Ameisen. *Ark. Zool.* **9**: 1–119 pls 1–3 [4 Dec. 1915] [86]. Type data: holotype, whereabouts unknown, from Mt. Bellenden Ker, Qld.

Distribution: NE coastal, Qld. Ecology: terrestrial, noctidiurnal, omnivore, granivore, woodland, open forest; nest in soil.

Melophorus turneri Forel, 1910

Melophorus turneri turneri Forel, 1910

Melophorus turneri Forel, A. (1910). Formicides australiens reçus de MM. Froggatt et Rowland Turner. *Rev. Suisse Zool.* **18**: 1–94 [63]. Type data: syntypes, GMNH W, ANIC W, from Cape York, Qld.

Distribution: N Gulf, Qld. Ecology: terrestrial, noctidiurnal, omnivore, granivore, hummock grassland, woodland, open forest; nest in soil.

Melophorus turneri aesopus Forel, 1910

Melophorus turneri aesopus Forel, A. (1910). Formicides australiens reçus de MM. Froggatt et Rowland Turner. *Rev. Suisse Zool.* **18**: 1-94 [64]. Type data: syntypes, GMNH W,M,F, ANIC W, from Tennant Creek, N.T.

Distribution: W plateau, N.T. Ecology: terrestrial, noctidiurnal, omnivore, granivore, hummock grassland, woodland, open forest; nest in soil.

Melophorus turneri candidus Santschi, 1919

Melophorus turneri candida Santschi, F. (1919). Cinq notes myrmécologiques. *Bull. Soc. Vaud. Sci. Nat.* **52**: 325-350 [328]. Type data: syntypes, BNHM W, from Vic.

Distribution: (SE coastal), Vic.; type locality as Vic. only. Ecology: terrestrial, noctidiurnal, omnivore, granivore, woodland, open forest; nest in soil.

Melophorus turneri perthensis Wheeler, 1934

Melophorus turneri perthensis Wheeler, W.M. (1934). Contributions to the fauna of Rottnest Island, Western Australia No. IX. The ants. *J. R. Soc. West. Aust.* **20**: 137-163 [5 Oct. 1934] [152]. Type data: syntypes, MCZ *W, from Rottnest Is., W.A.

Distribution: SW coastal, W.A. Ecology: terrestrial, noctidiurnal, omnivore, granivore, hummock grassland, woodland, open forest; nest in soil.

Melophorus wheeleri Forel, 1910

Melophorus wheeleri Forel, A. (1910). Formicides australiens reçus de MM. Froggatt et Rowland Turner. *Rev. Suisse Zool.* **18**: 1-94 [60]. Type data: syntypes, GMNH W,M, ANIC W, from Tennant Creek, N.T.

Distribution: W plateau, N.T. Ecology: terrestrial, noctidiurnal, omnivore, granivore, desert, woodland, open forest; nest in soil.

Myrmecorhynchus E. André, 1896

Myrmecorhynchus André, E. (1896). Fourmis nouvelles d'Asie et d'Australie. *Rev. Entomol.* **15**: 251-265 [253] [redefined in Wheeler, W.M. (1917). The Australian ant-genus *Myrmecorhynchus* (Ern. André) and its position in the sub-family Camponotinae. *Trans. R. Soc. S. Aust.* **61**: 14-19 pl 1]. Type species *Myrmecorhynchus emeryi* E. André, 1896 by monotypy.

Myrmecorhynchus carteri Clark, 1934

Myrmecorhynchus carteri Clark, J. (1934). New Australian ants. *Mem. Natl. Mus. Vict.* **8**: 21-47 [43 pls 2-3]. Type data: syntypes, NMV *W, from Barrington Tops, N.S.W. and Kinglake, Vic.

Distribution: SE coastal, Murray-Darling basin, A.C.T., N.S.W., Vic. Ecology: terrestrial, noctidiurnal, omnivore, woodland, open forest; nest arboreal.

Myrmecorhynchus emeryi E. André, 1896

Myrmecorhynchus emeryi André, E. (1896). Fourmis nouvelles d'Asie et d'Australie. *Rev. Entomol.* **15**: 251-265 [254]. Type data: holotype, MNHP W, from Victorian Alps.

Distribution: Murray-Darling basin, A.C.T., Vic. Ecology: terrestrial, noctidiurnal, omnivore, woodland, open forest; nest arboreal.

Myrmecorhynchus musgravei Clark, 1934

Myrmecorhynchus musgravei Clark, J. (1934). New Australian ants. *Mem. Natl. Mus. Vict.* **8**: 21-47 [43 pls 2-3]. Type data: syntypes, AM *M, from National Park", Qld.

Distribution: NE coastal, Qld. Ecology: terrestrial, noctidiurnal, omnivore, woodland, open forest; nest arboreal.

Myrmecorhynchus nitidus Clark, 1934

Myrmecorhynchus nitidus Clark, J. (1934). New Australian ants. *Mem. Natl. Mus. Vict.* **8**: 21-47 [44 pls 2-3]. Type data: syntypes, NMV *W,F,M, from Cheltenham, Vic. and Canberra, A.C.T.

Distribution: SE coastal, Murray-Darling basin, Vic., A.C.T. Ecology: terrestrial, noctidiurnal, omnivore, woodland, open forest; nest arboreal.

Myrmecorhynchus rufithorax Clark, 1934

Myrmecorhynchus rufithorax Clark, J. (1934). New Australian ants. *Mem. Natl. Mus. Vict.* **8**: 21-47 [46 pls 2-3]. Type data: syntypes, NMV *W, from Warburton, Vic.

Distribution: SE coastal, Vic. Ecology: terrestrial, noctidiurnal, omnivore, woodland, open forest; nest arboreal.

Notoncus Emery, 1895

Notoncus Emery, C. (1895). Descriptions de quelques fourmis nouvelles d'Australie. *Ann. Soc. Entomol. Belg.* **39**: 345-358 [352]. Type species *Camponotus ectatommoides* Forel, 1892 by monotypy.

Diodontolepis Wheeler, W.M. (1920). The Subfamilies of Formicidae, and other taxonomic notes. *Psyche Camb.* **27**: 46-55 [53]. Type species *Melophorus spinisquamis* E. André, 1896 by original designation.

Synonymy that of Brown, W.L. jr. (1955). A revision of the Australian ant genus *Notoncus* Emery, with notes on the other genera of Melophorini. *Bull. Mus. Comp. Zool.* **113**: 469-494 [477].

This group is also found in south New Guinea, one species in *Eucalyptus* savanna.

Notoncus ectatommoides (Forel, 1892)

Camponotus ectatommoides Forel, A. (1892). Die Ameisen Neu-Seelands. *Mitt. Schweiz. Entomol. Ges.* **8**: 331-343 [333]. Type data: holotype, MCG (probable) *F, from probably (South) Australia, see Brown, W.L. jr.

(1955). A revision of the Australian ant genus *Notoncus* Emery, with notes on the other genera of Melophorini. *Bull. Mus. Comp. Zool.* **113**: 469–494 [480].

Notoncus foreli André, E. (1896). Fourmis nouvelles d'Asie et d'Australie. *Rev. Entomol.* **15**: 251–265 [256]. Type data: holotype, MNHP W, from W.A.

Notoncus foreli subdentata Forel, A. (1910). Formicides australiens reçus de MM. Froggatt et Rowland Turner. *Rev. Suisse Zool.* **18**: 1–94 [68]. Type data: syntypes, GMNH W, ANIC W, from Forset Reefs, N.S.W.

Notoncus foreli dentata Forel, A. (1910). Formicides australiens reçus de MM. Froggatt et Rowland Turner. *Rev. Suisse Zool.* **18**: 1–94 [68]. Type data: syntypes, GMNH W, ANIC W, from Gembrook, Vic.

Notoncus foreli acuminata Viehmeyer, H. (1925). Formiciden der australischen Faunenregion. *Entomol. Mitt.* **14**: 25–39 [37]. Type data: syntypes (probable), ZMB *W, from probably Liverpool or Trial Bay, N.S.W.

Notoncus rodwayi Donisthorpe, H. (1941). Descriptions of new ants (Hym., Formicidae) from various localities. *Ann. Mag. Nat. Hist. (11)* **8**: 199–210 [206]. Type data: holotype, BMNH *F, from Nowra, N.S.W.

Synonymy that of Brown, W.L. jr. (1955). A revision of the Australian ant genus *Notoncus* Emery, with notes on the other genera of Melophorini. *Bull. Mus. Comp. Zool.* **113**: 469–494 [485].

Distribution: SE coastal, NE coastal, Murray-Darling basin, S Gulfs, Qld., S.A., A.C.T., N.S.W., Vic. Ecology: terrestrial, noctidiurnal, omnivore, woodland, open forest; nest in ground layer.

Notoncus enormis Szabó, 1910

Notoncus enormis Szabó, J. (1910). Formicides nouveaux ou peu connus des collections du Musée National Hongrois. *Ann. Hist.-Nat. Mus. Natl. Hung.* **8**: 364–369 [368]. Type data: syntypes, NMH *W, from Mt. Victoria, N.S.W.

Notoncus capitatus Forel, A. (1915). Results of Dr. E. Mjöbergs Swedish Scientific Expeditions to Australia 1910–1913. 2. Ameisen. *Ark. Zool.* **9**: 1–119 pls 1–3 [4 Dec. 1915] [90]. Type data: syntypes, GMNH W, ANIC W, other syntypes may exist, from Mt. Tambourine (=Tamborine Mt.), Qld.

Notoncus mjobergi Forel, A. (1915). Results of Dr. E. Mjöbergs Swedish Scientific Expeditions to Australia 1910–1913. 2. Ameisen. *Ark. Zool.* **9**: 1–119 pls 1–3 [4 Dec. 1915] [91]. Type data: holotype (probable), whereabouts unknown, from Colosseum, Qld.

Notoncus capitatus minor Viehmeyer, H. (1925). Formiciden der australischen Faunenregion. *Entomol. Mitt.* **14**: 139–149 [139]. Type data: syntypes, ZMB *W, from probably Liverpool or Trial Bay, N.S.W.

Synonymy that of Brown, W.L. jr. (1955). A revision of the Australian ant genus *Notoncus* Emery, with notes on the other genera of Melophorini. *Bull. Mus. Comp. Zool.* **113**: 469–494 [489].

Distribution: NE coastal, SE coastal, Qld., N.S.W. Ecology: terrestrial, noctidiurnal, omnivore, woodland, open forest; nest in ground layer.

Notoncus gilberti Forel, 1895

Notoncus gilberti Forel, A. (1895). Nouvelles fourmis d'Australie, récoltée à The Ridge, Mackay, Queensland par M. Gilbert Turner. *Ann. Soc. Entomol. Belg.* **39**: 417–428 [418]. Type data: syntypes, GMNH W,F, ANIC W, from Mackay, Qld.

Notoncus gilberti gracilior Forel, A. (1907). Formicidae. pp. 263–310 *in* Michaelsen, W. & Hartmayer, R. (eds.) *Die Fauna Südwest-Australiens.* Jena : G. Fischer Vol. 1 [299]. Type data: holotype, probably destroyed in ZMH in WW II, from Fremantle, W.A.

Notoncus politus Viehmeyer, H. (1925). Formiciden der australischen Faunenregion. *Entomol. Mitt.* **14**: 25–39 [38]. Type data: syntypes, ZMB *W, ANIC W, from Liverpool, N.S.W.

Notoncus gilberti annectens Wheeler, W.M. (1934). Contributions to the fauna of Rottnest Island, Western Australia No. IX. The ants. *J. R. Soc. West. Aust.* **20**: 137–163 [5 Oct. 1934] [154]. Type data: syntypes, MCZ *W, from Enoggera, Qld., see Brown, W.L. jr. (1955). A revision of the Australian ant genus *Notoncus* Emery, with notes on the other genera of Melophorini. *Bull. Mus. Comp. Zool.* **113**: 469–494.

Synonymy that of Brown, W.L. jr. (1955). A revision of the Australian ant genus *Notoncus* Emery, with notes on the other genera of Melophorini. *Bull. Mus. Comp. Zool.* **113**: 469–494 [490].

Distribution: NE coastal, SE coastal, SW coastal, Qld., N.S.W., W.A. Ecology: terrestrial, noctidiurnal, omnivore, woodland, open forest; nest in ground layer.

Notoncus hickmani Clark, 1930

Notoncus hickmani Clark, J. (1930). Some new Australian Formicidae. *Proc. R. Soc. Vict.* **42**: 116–128 [10 Mar. 1930] [126]. Type data: syntypes, NMV *W,F, from Trevallyn, Tas.

Notoncus rotundiceps Clark, J. (1930). Some new Australian Formicidae. *Proc. R. Soc. Vict.* **42**: 116–128 [10 Mar. 1930] [127]. Type data: syntypes, NMV *W, from Albany, W.A.

Synonymy that of Brown, W.L. jr. (1955). A revision of the Australian ant genus *Notoncus* Emery, with notes on the other genera of Melophorini. *Bull. Mus. Comp. Zool.* **113**: 469–494 [492].

Distribution: SW coastal, SE coastal, Murray-Darling basin, S Gulfs, W plateau, S.A., Vic., N.S.W., W.A. Ecology: terrestrial, noctidiurnal, omnivore, woodland, open forest; nest in ground layer.

Notoncus spinisquamis (E. André, 1896)

Melophorus spinisquamis André, E. (1896). Fourmis nouvelles d'Asie et d'Australie. *Rev. Entomol.* **15**: 251–265 [254]. Type data: syntypes, MNHP W,F,M, ANIC W, from Victorian Alps.

Distribution: Murray-Darling basin, SE coastal, N.S.W., A.C.T., Tas., Vic. Ecology: terrestrial, noctidiurnal, omnivore, woodland, open forest, closed forest; nest in ground layer.

Notostigma Emery, 1920

Notostigma Emery, C. (1920). Le genre *Camponotus* Mayr. Nouvel essai de sa subdivision en sous-genres. *Rev. Zool. Afr.* **8**: 229–260 [252]. Type species *Camponotus carazzii* Emery, 1895 by original designation.

Notostigma carazzii (Emery, 1895)

Camponotus carazzii Emery, C. (1895). Descriptions de quelques fourmis nouvelles d'Australie. *Ann. Soc. Entomol. Belg.* **39**: 345–358 [354]. Type data: syntypes, MCG *W, from Mt. Bellenden Ker, Qld.

Distribution: NE coastal, Qld. Ecology: terrestrial, nocturnal, omnivore, closed forest; nest in ground layer.

Notostigma foreli Emery, 1920

Notostigma foreli Emery, C. (1920). Le genre *Camponotus* Mayr. Nouvel essai de sa subdivision en sous-genres. *Rev. Zool. Afr.* **8**: 229–260 [253]. Type data: syntypes, MCG *W,F,M, from N.S.W.

Distribution: NE coastal, SE coastal, Qld., N.S.W. Ecology: terrestrial, nocturnal, omnivore, closed forest; nest in soil.

Notostigma podenzanai (Emery, 1895)

Camponotus podenzanai Emery, C. (1895). Descriptions de quelques fourmis nouvelles d'Australie. *Ann. Soc. Entomol. Belg.* **39**: 345–358 [355]. Type data: syntypes, MCG *W,M, from Kamerunga, Qld.

Distribution: NE coastal, Qld. Ecology: terrestrial, nocturnal, omnivore, woodland, open forest; nest in ground layer.

Notostigma sanguinea Clark, 1930

Notostigma sanguinea Clark, J. (1930). Some new Australian Formicidae. *Proc. R. Soc. Vict.* **42**: 116–128 [10 Mar. 1930] [116]. Type data: syntypes, NMV *W, from Perth and Ludlow, W.A.

Distribution: SW coastal, W.A. Ecology: terrestrial, nocturnal, omnivore, woodland, open forest; nest in ground layer.

Oecophylla F. Smith, 1860

Oecophylla Smith, F. (1860). Catalogue of hymenopterous insects collected by Mr. A.R. Wallace in the islands of Bachian, Kaisaa, Amboyna, Gilolo, and at Dory in New Guinea. *J. Linn. Soc. Zool.* **5**: 93–143 [18 July 1860] [101 pl 1]. Publication date established from Donisthorpe, H. (1932). On the identity of Smith's types of Formicidae (Hymenoptera) collected by Alfred Russell Wallace in the Malay Archipelago, with descriptions of two new species. *Ann. Mag. Nat. Hist. (10)* **10**: 441–476.

Type species *Formica smaragdina* Fabricius, 1775 by monotypy.

This group is also found in the north Ethiopian and Oriental regions; New Guinea and east Melanesia in the Australian Region, see Brown, W.L. jr. (1973). A comparison of the Hylean and Congo-West African rain forest ant faunas. pp. 161–185 *in* Meggers, B.J., Ayensu, E.S. & Duckworth, W.D. (eds.) *Tropical forest ecosystems in Africa and South America: a comparative review*. Washington : Smithsonian Institution Press.

Oecophylla smaragdina (Fabricius, 1775)

Formica smaragdina Fabricius, J.C. (1775). *Systema Entomologiae*, sistens insectorum classes, ordines, genera, species, adiectis synonymis, locis, descriptionibus, observationibus. Flensburgi et Lipsiae [Appendix,828]. Type data: syntypes (probable), whereabouts uncertain, from India.

Formica virescens Fabricius, J.C. (1775). *Systema Entomologiae*, sistens insectorum classes, ordines, genera, species, adiectis synonymis, locis, descriptionibus, observationibus. Flensburgi et Lipsiae [392]. Type data: uncertain, BMNH W, from Australia (as New Holland).

Formica viridis Kirby, W. (1818). A description of several new species of insects collected in New Holland by Robert Brown, Esq., F.R.S., Lib. Linn. Soc. *Trans. Linn. Soc. Lond.* **12**: 454–482 pls 21–23 [478]. Type data: uncertain, BMNH *W, from northern Australia.

Synonymy that of Mayr, G.L. (1872). Formicidae Borneenses. *Ann. Mus. Civ. Stor. Nat. Giacomo Doria* **2**: 134–155 [143].

Distribution: NE coastal, Qld. Ecology: terrestrial, noctidiurnal, omnivore, open forest, closed forest; nest arboreal.

Opisthopsis Emery, 1893

Myrmecopsis Smith, F. (1865). Desriptions of new species of hymenopterous insects from the islands of Sumatra, Sula, Gilolo, Salwatty, and New Guinea, collected by Mr A. R. Wallace. *J. Linn. Soc. Zool.* **8**: 61–94 [13 Jan. 1865] [68 pl 4] [*non Myrmecopsis* Newman, 1850; proposed with subgeneric rank in *Formica* Linnaeus, 1758]. Publication date established from Donisthorpe, H. (1932). On the identity of Smith's types of Formicidae (Hymenoptera) collected by Alfred Russell Wallace in the Malay Archipelago, with descriptions of two new species. *Ann. Mag. Nat. Hist. (10)* **10**: 441–476. Type species *Formica (Myrmecopsis) respiciens* F. Smith, 1865 by monotypy.

Opisthopsis Emery, C. (1893). *in* Dalla Torre, C.G. de (1893). *Catalogus hymenopterorum hucusque descriptorum systematicus et synonymicus*. Formicidae (Heterogyna). Lipsiae : G. Engelmann Vol. 7 289 pp. [219] [*nom. nov.* for *Myrmecopsis* F. Smith, 1865].

This group is also found in New Guinea and east Melanesia in the Australian Region.

Opisthopsis diadematus Wheeler, 1918

Opisthopsis diadematus diadematus Wheeler, 1918

Opisthopsis diadematus Wheeler, W.M. (1918). The ants of the genus *Opisthopsis* Emery. *Bull. Mus. Comp. Zool.* **62**: 341–362 pls 1–3 [357]. Type data: syntypes, MCZ *W, from Townsville, Qld.

Distribution: NE coastal, N Gulf, N coastal, N.T., Qld. Ecology: terrestrial, diurnal, omnivore, woodland, open forest; nest in soil.

Opisthopsis diadematus dubius Wheeler, 1918

Opisthopsis diadematus dubius Wheeler, W.M. (1918). The ants of the genus *Opisthopsis* Emery. *Bull. Mus. Comp. Zool.* **62**: 341–362 pls 1–3 [358]. Type data: holotype, MCZ *W, from Longreach, Qld.

Distribution: Lake Eyre basin, Qld. Ecology: terrestrial, diurnal, omnivore, woodland; nest in soil.

Opisthopsis haddoni Emery, 1893

Opisthopsis haddoni haddoni Emery, 1893

Opisthopsis haddoni Emery, C. (1893). Formicides de l'Archipel Malais. *Rev. Suisse Zool.* **1**: 187–229 [226 pl 8]. Type data: syntypes, MCG *W, from Mer Is. of the Murray Group, Qld.

Distribution: N coastal, N Gulf, NE coastal, Lake Eyre basin, W plateau, Murray-Darling basin, N.T., Qld., N.S.W., S.A., W.A. Ecology: terrestrial, diurnal, omnivore, woodland, open forest; nest in soil.

Opisthopsis haddoni rufoniger Forel, 1910

Opisthopsis haddoni rufoniger Forel, A. (1910). Formicides australiens reçus de MM. Froggatt et Rowland Turner. *Rev. Suisse Zool.* **18**: 1–94 [70]. Type data: syntypes, GMNH W, ANIC W, from Tennant Creek, N.T.

Distribution: W plateau, N.T. Ecology: terrestrial, diurnal, omnivore, woodland, open forest; nest in soil.

Opisthopsis jocosus Wheeler, 1918

Opisthopsis jocosus Wheeler, W.M. (1918). The ants of the genus *Opisthopsis* Emery. *Bull. Mus. Comp. Zool.* **62**: 341–362 pls 1–3 [359]. Type data: syntypes, MCZ *W, from Baron Falls at Kuranda, Qld.

Distribution: NE coastal, Qld. Ecology: terrestrial, diurnal, omnivore, woodland, open forest; nest in soil.

Opisthopsis lienosus Wheeler, 1918

Opisthopsis lienosus Wheeler, W.M. (1918). The ants of the genus *Opisthopsis* Emery. *Bull. Mus. Comp. Zool.* **62**: 341–362 pls 1–3 [356]. Type data: syntypes, MCZ *W, from Koah, Qld.

Distribution: NE coastal, Qld. Ecology: terrestrial, diurnal, omnivore, woodland, open forest; nest in soil.

Opisthopsis major Forel, 1902

Opisthopsis major Forel, A. (1902). Fourmis nouvelles d'Australie. *Rev. Suisse Zool.* **10**: 405–548 [492]. Type data: syntypes, GMNH W,F, ANIC W, from Mackay, Qld.

Distribution: NE coastal, Qld. Ecology: terrestrial, diurnal, omnivore, woodland, open forest; nest in soil.

Opisthopsis maurus Wheeler, 1918

Opisthopsis maurus Wheeler, W.M. (1918). The ants of the genus *Opisthopsis* Emery. *Bull. Mus. Comp. Zool.* **62**: 341–362 pls 1–3 [350]. Type data: holotype, MCZ *W, from Koah, Qld.

Distribution: NE coastal, Qld. Ecology: terrestrial, diurnal, omnivore, woodland, open forest; nest in soil.

Opisthopsis pictus Emery, 1895

Opisthopsis pictus pictus Emery, 1895

Opisthopsis pictus Emery, C. (1895). Descriptions de quelques fourmis nouvelles d'Australie. *Ann. Soc. Entomol. Belg.* **39**: 345–358 [354]. Type data: syntypes, MCG *W, from Kamerunga, Qld.

Distribution: NE coastal, Qld. Ecology: terrestrial, diurnal, omnivore, woodland, open forest; nest in soil.

Opisthopsis pictus bimaculatus Wheeler, 1918

Opisthopsis pictus bimaculatus Wheeler, W.M. (1918). The ants of the genus *Opisthopsis* Emery. *Bull. Mus. Comp. Zool.* **62**: 341–362 pls 1–3 [352]. Type data: holotype, MCZ *W, from mountain west of Townsville, Qld.

Distribution: NE coastal, Qld. Ecology: terrestrial, diurnal, omnivore, woodland, open forest; nest in soil.

Opisthopsis pictus lepidus Wheeler, 1918

Opisthopsis pictus lepidus Wheeler, W.M. (1918). The ants of the genus *Opisthopsis* Emery. *Bull. Mus. Comp. Zool.* **62**: 341–362 pls 1–3 [352]. Type data: syntypes, MCZ *W, from Townsville, Qld.

Distribution: NE coastal, Qld. Ecology: terrestrial, diurnal, omnivore, woodland, open forest; nest in soil.

Opisthopsis pictus palliatus Wheeler, 1918

Opisthopsis pictus palliatus Wheeler, W.M. (1918). The ants of the genus *Opisthopsis* Emery. *Bull. Mus. Comp. Zool.* **62**: 341–362 pls 1–3 [352]. Type data: syntypes, MCZ *W, from Sunnybank, near Brisbane, Qld.

Distribution: NE coastal, Qld. Ecology: terrestrial, diurnal, omnivore, woodland, open forest; nest in soil.

Opisthopsis respiciens (F. Smith, 1865)

Opisthopsis respiciens respiciens (F. Smith, 1865)

Formica (Myrmecopsis) respiciens Smith, F. (1865). Descriptions of new species of hymenopterous insects from the islands of Sumatra, Sula, Gilolo, Salwatty, and New Guinea, collected by Mr A.R. Wallace. *J. Linn. Soc. Zool.* **8**: 61-94 pl 4 [13 Jan. 1865] [68]. Publication date established from Donisthorpe, H. (1932). On the identity of Smith's types of Formicidae (Hymenoptera) collected by Alfred Russell Wallace in the Malay Archipelago, with descriptions of two new species. *Ann. Mag. Nat. Hist. (10)* **10**: 441-476. Type data: holotype, BMNH *W, from New Guinea.

Distribution: N coastal, N Gulf, NE coastal, SE coastal, N.T., Qld., N.S.W. Ecology: terrestrial, diurnal, omnivore, open forest, closed forest; nest in soil.

Opisthopsis respiciens moestus Wheeler, 1918

Opisthopsis respiciens moestus Wheeler, W.M. (1918). The ants of the genus *Opisthopsis* Emery. *Bull. Mus. Comp. Zool.* **62**: 341-362 pls 1-3 [348]. Type data: syntypes, SAMA *W,F,M, from Townsville, Qld.

Distribution: NE coastal, Qld. Ecology: terrestrial, diurnal, omnivore, open forest, closed forest; nest in soil.

Opisthopsis rufithorax Emery, 1895

Opisthopsis rufithorax Emery, C. (1895). Descriptions de quelques fourmis nouvelles d'Australie. *Ann. Soc. Entomol. Belg.* **39**: 345-358 [354]. Type data: syntypes, MCG (probable) *W, from Peak Downs, Qld.

Distribution: NE coastal, N coastal, N Gulf, Murray-Darling basin, SE coastal, S Gulfs, W plateau, N.T., N.S.W., A.C.T., S.A., W.A., Qld. Ecology: terrestrial, diurnal, omnivore, woodland, open forest; nest in soil.

Paratrechina Motschoulsky, 1863

Paratrechina Motschoulsky, V. von. (1863). Essai d'un catalogue des insectes de l'île Ceylon. *Byull. Mosk. Obshch. Ispyt. Prir.* **26**: 1-153 [13]. Type species *Formica longicornis* Latreille, 1802 (as *Paratrechina currens* Motschoulsky, 1863) by monotypy. Compiled from secondary source: Wheeler, W.M. (1911). A list of the type species of the genera and subgenera of Formicidae. *Ann. N.Y. Acad. Sci.* **21**: 157-175 [17 Oct. 1911].

This group is also found in the Neotropical, Nearctic, south Palearctic, Ethiopian, Malagasy and Oriental regions; widespread in the Australian Region except New Zealand, see Brown, W.L. jr. (1973). A comparison of the Hylean and Congo-West African rain forest ant faunas. pp. 161-185 *in* Meggers, B.J., Ayensu, E.S. & Duckworth, W.D. (eds.) *Tropical forest ecosystems in Africa and South America: a comparative review.* Washington : Smithsonian Institution Press.

Paratrechina braueri (Mayr, 1868)

Paratrechina braueri braueri (Mayr, 1868)

Prenolepis braueri Mayr, G.L. *in* Brauer, F. (1868). Neuropteren. *in, Reise der österreichischen fregatte Novara um die Erde in der Jahren 1857, 1858, 1859.* Zool. 2 Abt. 1A4: 1-107 pl 1-2 [49]. Type data: syntypes, NHMW (probable) *W, from Sidney (=Sydney), N.S.W.

Distribution: SE coastal, N.S.W. Ecology: terrestrial, noctidiurnal, omnivore, woodland, open forest; nest in ground layer.

Paratrechina braueri glabrior (Forel, 1902)

Prenolepis braueri glabrior Forel, A. (1902). Fourmis nouvelles d'Australie. *Rev. Suisse Zool.* **10**: 405-548 [490]. Type data: syntypes, GMNH W,F,M, ANIC W, from Mackay, Qld.

Distribution: NE coastal, Qld. Ecology: terrestrial, noctidiurnal, omnivore, woodland, open forest; nest in ground layer.

Paratrechina minutula (Forel, 1901)

Prenolepis minutula Forel, A. (1901). Formiciden aus dem Bismarck-Archipel, auf Grundlage des von Prof. Dr. F. Dahl gesammelten Materials bearbeitet. *Mitt. Zool. Mus. Berl.* **2**: 1-37 [3 Apr. 1901] [25]. Type data: syntypes, GMNH W, ANIC W, from N.S.W.

Distribution: NE coastal, SE coastal, Murray-Darling basin, Qld., N.S.W. Vic. Ecology: terrestrial, noctidiurnal, omnivore, woodland, open forest; nest in ground layer.

Paratrechina nana Santschi, 1928

Paratrechina (Nylanderia) nana Santschi, F. (1928). Nouvelles fourmis d'Australie. *Bull. Soc. Vaud. Sci. Nat.* **56**: 465-483 [30 Aug. 1928] [478]. Type data: syntypes, whereabouts uncertain, from Ringwood, Vic., see The Zoological Society of London (1929). *The Zoological Record.* Vol. 65 relating chiefly to the year 1928. London : Gurney & Jackson.

Distribution: SE coastal, Vic. Ecology: terrestrial, noctidiurnal, omnivore, woodland, open forest; nest in ground layer.

Paratrechina obscura (Mayr, 1862)

Prenolepis obscura Mayr, G.L. (1862). Myrmecologische Studien. *Verh. Zool.-Bot. Ges. Wien* **12**: Abhand. 649-776 [698 pl 19]. Type data: syntypes, NHMW *W,F, from Sidney (=Sydney), N.S.W.

Distribution: SE coastal, N.S.W. Ecology: terrestrial, noctidiurnal, omnivore, woodland, open forest; nest in ground layer.

Paratrechina rosae (Forel, 1902)

Prenolepis rosae Forel, A. (1902). Fourmis nouvelles d'Australie. *Rev. Suisse Zool.* **10**: 405–548 [489]. Type data: syntypes, GMNH W,F,M, ANIC W, from Sydney, N.S.W.

Distribution: SE coastal, N.S.W. Ecology: terrestrial, noctidiurnal, omnivore, woodland, open forest; nest in ground layer.

Paratrechina tasmaniensis (Forel, 1913)

Prenolepis (Nylanderia) tasmaniensis Forel, A. (1913). Fourmis de Tasmanie et d'Australie récoltées par MM. Lea, Froggatt etc. *Bull. Soc. Vaud. Sci. Nat.* **49**: 173–196 pl 2 [190]. Type data: syntypes, GMNH W, from Tas.

Distribution: Tas. Ecology: terrestrial, noctidiurnal, omnivore, woodland, open forest; nest in ground layer.

Paratrechina vaga (Forel, 1901)

Prenolepis obscura vaga Forel, A. (1901). Formiciden aus dem Bismarck-Archipel, auf Grundlage des von Prof. Dr. F. Dahl gesammelten Materials bearbeitet. *Mitt. Zool. Mus. Berl.* **2**: 1–37 [3 Apr. 1901] [26]. Type data: syntypes, probably in GMNH, from Ralum, New Britain.

Distribution: NE coastal, N coastal, N.T., Qld.; introduced(?), found from Philippines to Juan Fernandez Is. Ecology: terrestrial, noctidiurnal, omnivore, woodland, open forest; nest in ground layer. Biological references: Emery, C. (1914). Les fourmis de la Nouvelle-Calédonie et des Îles Loyalty. pp. 393–435 *in* Sarasin, F. & Roux, J. (eds.) *Nova Caledonia, Zoologie.* Vol. 1 No. 11 Wiesbaden : C.W. Kreidels Verl. (raised to species)

Plagiolepis Mayr, 1861

Plagiolepis Mayr, G.L. (1861). *Die europëischen Formiciden. (Ameisen.) Nach der analytischen Methode bearbeitet.* Vienna : Carl Gerolds Sohn 80 pp. 1 pl [42]. Type species *Formica pygmaea* Latreille, 1798 by monotypy.

This group is also found in the Palearctic, Ethiopian, Malagasy and Oriental regions; New Guinea, east Melanesia and parts of Polynesia in the Australian Region, see Brown, W.L. jr. (1973). A comparison of the Hylean and Congo-West African rain forest ant faunas. pp. 161–185 *in* Meggers, B.J., Ayensu, E.S. & Duckworth, W.D. (eds.) *Tropical forest ecosystems in Africa and South America: a comparative review.* Washington : Smithsonian Institution Press.

Plagiolepis clarki Wheeler, 1934

Plagiolepis clarki clarki Wheeler, 1934

Plagiolepis clarki Wheeler, W.M. (1934). Contributions to the fauna of Rottnest Island, Western Australia No. IX. The ants. *J. R. Soc. West. Aust.* **20**: 137–163 [5 Oct. 1934] [157]. Type data: syntypes, MCZ *W,F,M, from Mundaring Weir, Margaret River and Pemberton, W.A.

Distribution: SW coastal, W.A. Ecology: terrestrial, noctidiurnal, omnivore, woodland, open forest; nest in ground layer.

Plagiolepis clarki impasta Wheeler, 1934

Plagiolepis clarki impasta Wheeler, W.M. (1934). Contributions to the fauna of Rottnest Island, Western Australia No. IX. The ants. *J. R. Soc. West. Aust.* **20**: 137–163 [5 Oct. 1934] [158]. Type data: syntypes, MCZ *W, from Jenolan Caves in the Blue Mts., N.S.W.

Distribution: SE coastal, N.S.W. Ecology: terrestrial, noctidiurnal, omnivore, woodland, open forest; nest in ground layer.

Plagiolepis exigua Forel, 1894

Plagiolepis exigua Forel, A. (1894). Les Formicides de l'Empire des Indes et de Ceylan. Part N. *J. Bombay Nat. Hist. Soc.* **8**: 396–420 [415]. Type data: status unknown, ?GMNH, from India.

Plagiolepis exigua quadrimaculata Forel, 1902

Plagiolepis exigua quadrimaculata Forel, A. (1902). Fourmis nouvelles d'Australie. *Rev. Suisse Zool.* **10**: 405–548 [483]. Type data: syntypes, GMNH W,M, from Mackay, Qld.

Distribution: NE coastal, Qld. Ecology: terrestrial, noctidiurnal, omnivore, woodland, open forest; nest in ground layer.

Plagiolepis lucidula Wheeler, 1934

Plagiolepis lucidula Wheeler, W.M. (1934). Contributions to the fauna of Rottnest Island, Western Australia No. IX. The ants. *J. R. Soc. West. Aust.* **20**: 137–163 [5 Oct. 1934] [155]. Type data: syntypes, MCZ *W, from Lady Edeline Beach, Rottnest Is., W.A.

Distribution: SW coastal, W.A. Ecology: terrestrial, noctidiurnal, omnivore, woodland, open forest; nest in ground layer.

Plagiolepis nynganensis McAreavey, 1949

Plagiolepis nynganensis McAreavey, J.J. (1949). Australian Formicidae. New genera and species. *Proc. Linn. Soc. N.S.W.* **74**: 1–25 [15 June 1949] [23]. Type data: holotype, ANIC W, from Nyngan, N.S.W.

Distribution: Murray-Darling basin, N.S.W. Ecology: terrestrial, noctidiurnal, omnivore, woodland, open forest; nest in ground layer.

Plagiolepis squamulosa Wheeler, 1934

Plagiolepis squamulosa Wheeler, W.M. (1934). Contributions to the fauna of Rottnest Island, Western Australia No. IX. The ants. *J. R. Soc. West. Aust.* **20**: 137-163 [5 Oct. 1934] [156]. Type data: syntypes, MCZ *W, from sand dunes S of Geraldton, W.A.

Distribution: NW coastal, W.A. Ecology: terrestrial, noctidiurnal, omnivore, woodland; nest in ground layer.

Polyrhachis F. Smith, 1857

Polyrhachis Smith, F. (1857). Catalogue of the hymenopterous insects collected at Sarawak, Borneo, Mount Ophir, Malacca; and at Singapore by A. R. Wallace. *J. Linn. Soc. Zool.* **2**: 42-130 [2 Nov. 1857] [58 pls 1-2]; Publication date established from Donisthorpe, H. (1932). On the identity of Smith's types of Formicidae (Hymenoptera) collected by Alfred Russell Wallace in the Malay Archipelago, with descriptions of two new species. *Ann. Mag. Nat. Hist. (10)* **10**: 441-476. Type species *Formica bihamata* Drury, 1773 by original designation.

Hagiomyrma Wheeler, W.M. (1911). Three Formicid names which have been overlooked. *Science (ns)* **33**: 858-860 [860] [proposed with subgeneric rank in *Polyrhachis* F. Smith, 1857]. Type species *Formica ammon* Fabricus, 1775 by original designation. Compiled from secondary source: Donisthorpe, H. (1934). A list of the type species of the genera and subgenera of the Formicidae. *Ann. Mag. Nat. Hist. (11)* **10**: 649-688.

Chariomyrma Forel, A. (1915). Results of Dr. E. Mjöbergs Swedish Scientific Expeditions to Australia. 1910-1913. 2. Ameisen. *Ark. Zool.* **9**: 1-119 [4 Dec. 1915] [107 pls 1-3] [proposed with subgeneric rank in *Polyrhachis* F. Smith, 1857]. Type species *Polyrhachis guerini* Roger, 1863 by original designation.

Hedomyrma Forel, A. (1915). Results of Dr. E. Mjöbergs Swedish Scientific Expeditions to Australia. 1910-1913. 2. Ameisen. *Ark. Zool.* **9**: 1-119 [4 Dec. 1915] [107 pls 1-3] [proposed with subgeneric rank in *Polyrhachis* F. Smith, 1857]. Type species *Polyrhachis ornata* Mayr, 1876 by original designation.

Synonymy that of Brown, W.L. jr. (1973). A comparison of the Hylean and Congo-West African rain forest ant faunas. pp. 161-185 *in* Meggers, B.J., Ayensu, E.S. & Duckworth, W.D. (eds.) *Tropical forest ecosystems in Africa and South America: a comparative review.* Washington : Smithsonian Institution Press [177]; the subgenera of *Polyrhachis* are discussed in Hung, A.C.F. (1967). A revision of the ant genus *Polyrhachis* at the subgeneric level (Hymenoptera : Formicidae). *Trans. Am. Entomol. Soc.* **93**: 395-422 [20 Dec. 1967].

This group is also found in the south Palearctic, Ethiopian and Oriental regions; widespread in the Australian Region except New Zealand, see Brown, W.L. jr. (1973). A comparison of the Hylean and Congo-West African rain forest ant faunas. pp. 161-185 *in* Meggers, B.J., Ayensu, E.S. & Duckworth, W.D. (eds.) *Tropical forest ecosystems in Africa and South America: a comparative review.* Washington : Smithsonian Institution Press.

Polyrhachis aeschyle Forel, 1915

Polyrhachis (Hedomyrma) aeschyle Forel, A. (1915). Results of Dr. E. Mjöbergs Swedish Scientific Expeditions to Australia 1910-1913. 2. Ameisen. *Ark. Zool.* **9**: 1-119 pls 1-3 [4 Dec. 1915] [111]. Type data: holotype, whereabouts unknown, from Cedar Creek, Qld.

Distribution: NE coastal, Qld. Ecology: terrestrial, noctidiurnal, omnivore, open forest; nest in soil.

Polyrhachis ammon (Fabricius, 1775)

Polyrhachis ammon ammon (Fabricius, 1775)

Formica ammon Fabricius, J.C. (1775). *Systema Entomologiae,* sistens insectorum classes, ordines, genera, species, adiectis synonymis, locis, descriptionibus, observationibus. Flensburgi et Lipsiae [394]. Type data: uncertain, BMNH W, from Australia (as New Holland).

Distribution: NE coastal, Qld. Ecology: terrestrial, noctidiurnal, omnivore, woodland, open forest; nest in soil.

Polyrhachis ammon angusta Forel, 1902

Polyrhachis ammon angusta Forel, A. (1902). Fourmis nouvelles d'Australie. *Rev. Suisse Zool.* **10**: 405-548 [524]. Type data: syntypes, GMNH W,F,M, ANIC W, from Mackay, Qld.

Distribution: NE coastal, Qld. Ecology: terrestrial, noctidiurnal, omnivore, woodland, open forest; nest in soil.

Polyrhachis ammon angustata Forel, 1902

Polyrhachis ammon angustata Forel, A. (1902). Fourmis nouvelles d'Australie. *Rev. Suisse Zool.* **10**: 405-548 [525]. Type data: holotype (probable), GMNH W, from Australia.

Distribution: NE coastal, Qld. Ecology: terrestrial, noctidiurnal, omnivore, woodland, open forest; nest in soil.

Polyrhachis ammonoeides Roger, 1863

Polyrhachis ammonoeides ammonoeides Roger, 1863

Polyrhachis ammonoeides Roger, J. (1863). Die neu aufgeführten Gattungen und Arten meines Formiciden-Verzeichnisses. *Berl. Entomol. Z.* **7**: 129-214 [June 1863] [157]. Type data: syntypes (probable), MNHP *W, from Port Jackson, N.S.W.

Distribution: SE coastal, N.S.W. Ecology: terrestrial, noctidiurnal, omnivore, woodland, open forest; nest in soil.

Polyrhachis ammonoeides crawleyi Forel, 1916

Polyrhachis (Hagiomyrma) ammonoeides crawleyi Forel, A. (1916). Fourmis du Congo et d'autres provenances récoltées par MM. Hermann, Kohl, Luja, Mayné, etc. *Rev. Suisse Zool.* **24**: 397–460 [447]. Type data: syntypes, GMNH (probable) *W, from N Australia.

Distribution: N coastal, N.T., W.A. Ecology: terrestrial, noctidiurnal, omnivore, woodland, open forest; nest in soil.

Polyrhachis anguliceps Viehmeyer, 1925

Polyrhachis (Hedomyrma) anguliceps Viehmeyer, H. (1925). Formiciden der australischen Faunenregion. *Entomol. Mitt.* **14**: 139–149 [148]. Type data: syntypes, ZMB *W, from Trial Bay, N.S.W.

Distribution: SE coastal, N.S.W. Ecology: terrestrial, noctidiurnal, omnivore, woodland, open forest; nest in soil.

Polyrhachis appendiculata Emery, 1893

Polyrhachis appendiculata appendiculata Emery, 1893

Polyrhachis appendiculata Emery, C. (1893). Formicides de l'Archipel Malais. *Rev. Suisse Zool.* **1**: 187–229 [227 pl 8]. Type data: syntypes, MCG *W, from Mer Is. of the Murray Group, Qld.

Distribution: Torres Strait. Ecology: terrestrial, noctidiurnal, omnivore, woodland, open forest; nest in soil.

Polyrhachis appendiculata schoopae Forel, 1902

Polyrhachis appendiculata schoopae Forel, A. (1902). Fourmis nouvelles d'Australie. *Rev. Suisse Zool.* **10**: 405–548 [520]. Type data: syntypes, GMNH W, ANIC W, from Mackay, Qld.

Distribution: NE coastal, Qld. Ecology: terrestrial, noctidiurnal, omnivore, woodland, open forest; nest in soil.

Polyrhachis arcuata (Le Guillou, 1841)

Formica arcuata Le Guillou, E.J.F. (1841). Catalogue raisonné des insectes hyménoptères recueillis dans le voyage de circumnavigation des corvettes l'*Astrolabe* et la *Zélée. Ann. Soc. Entomol. Fr.* **10**: 311–324 [315]. Type data: syntypes, MNHP (probable) *W,F, from Borneo and northern Australia.

Distribution: NE coastal, Qld. Ecology: terrestrial, noctidiurnal, omnivore, woodland, open forest; nest in soil.

Polyrhachis aurea Mayr, 1876

Polyrhachis aurea aurea Mayr, 1876

Polyrhachis guerini aurea Mayr, G.L. (1876). Die australischen Formiciden. *J. Mus. Godeffroy* **5**: 56–115 [74]. Type data: syntypes, NHMW *W,F, from Rockhampton and Gayndah, Qld.

Distribution: NE coastal, Qld. Ecology: terrestrial, noctidiurnal, omnivore, woodland, open forest; nest in soil. Biological references: Emery, C. (1897). Viaggio do Lamberto Loria nella Papuasia orientale 18. Formiche raccolte nelle Nuova Guinea. *Ann. Mus. Civ. Stor. Nat. Giacomo Doria* **38**: 546–594 pl 1 (raised to species).

Polyrhachis aurea depilis Emery, 1897

Polyrhachis aurea depilis Emery, C. (1897). Viaggio do Lamberto Loria nella Papuasia orientale 18. Formiche raccolte nelle Nuova Guinea. *Ann. Mus. Civ. Stor. Nat. Giacomo Doria* **38**: 546–594 [22 Nov. 1897] [589 pl 1]. Type data: syntypes (probable), MCG *W, from Qld.

Distribution: (NE coastal), Qld. Ecology: terrestrial, noctidiurnal, omnivore, woodland, open forest; nest in soil.

Polyrhachis barnardi Clark, 1928

Polyrhachis (Myrmhopla) barnardi Clark, J. (1928). Australian Formicidae. *J. R. Soc. West. Aust.* **14**: 29–41 pl 1 [24 Apr. 1928] [39]. Type data: syntypes, NMV *W, MCZ *W, from Cape York, Qld.

Distribution: N Gulf, Qld. Ecology: terrestrial, noctidiurnal, omnivore, closed forest; nest arboreal.

Polyrhachis barretti Clark, 1928

Polyrhachis (Hedomyrma) barretti Clark, J. (1928). Ants from North Queensland. *Vict. Nat.* **45**: 169–171 [10 Oct. 1928] [170]. Type data: syntypes, NMV *W, from Daintree River, Qld.

Distribution: NE coastal, Qld. Ecology: terrestrial, noctidiurnal, omnivore, closed forest; nest arboreal.

Polyrhachis bedoti Forel, 1902

Polyrhachis bedoti Forel, A. (1902). Fourmis nouvelles d'Australie. *Rev. Suisse Zool.* **10**: 405–548 [518]. Type data: holotype (probable), GMNH W, from probably Australia or New Guinea.

Distribution: distribution and ecology unknown.

Polyrhachis bellicosa F. Smith, 1859

Polyrhachis bellicosus Smith, F. (1859). Catalogue of hymenopterous insects collected by Mr A.R. Wallace at the islands of Aru and Key. *J. Linn. Soc. Zool.* **3**: 132–178 [1 Feb. 1859] [142]. Publication date established from Donisthorpe, H, (1932). On the identity of Smith's types of Formicidae (Hymenoptera) collected by Alfred Russell Wallace in the Malay Archipelago, with descriptions of two new species. *Ann. Mag. Nat. Hist. (10)* **10**: 441–476. Type data: syntypes (probable), BMNH *W, from Aru, Indonesia.

Distribution: NE coastal, Qld.; widespread in SE Asia. Ecology: terrestrial, noctidiurnal, omnivore, closed forest; nest arboreal.

Polyrhachis cataulacoidea Stitz, 1911

Polyrhachis cataulacoidea Stitz, H. (1911). Australische Ameisen (Neu-Guinea und Salomons-Inseln, Festland, Neu-Seeland). *Sber. Ges. Naturf. Freunde Berl.* **1911**: 351–381 [377]. Type data: holotype, ZMB *W, from Sidney (=Sydney), N.S.W.

Distribution: SE coastal, N.S.W. Ecology: terrestrial, noctidiurnal, omnivore, woodland, open forest; nest in soil.

Polyrhachis chalchas Forel, 1907

Polyrhachis chalchas Forel, A. (1907). Formicidae. pp. 263–310 *in* Michaelsen, W. & Hartmeyer, R. (eds.) *Die Fauna Südwest-Australiens.* Jena : G. Fischer Vol. 1 [307]. Type data: syntypes, GMNH W, ANIC W, from Denham, Geraldton and Dongarra, W.A.

Distribution: NW coastal, W.A. Ecology: terrestrial, noctidiurnal, omnivore, woodland, open forest; nest in soil.

Polyrhachis chrysothorax Viehmeyer, 1925

Polyrhachis (Hedomyrma) chrysothorax Viehmeyer, H. (1925). Formiciden der australischen Faunenregion. *Entomol. Mitt.* **14**: 139–149 [148]. Type data: syntypes, ZMB *W,F, from Trial Bay, N.S.W.

Distribution: SE coastal, N.S.W. Ecology: terrestrial, noctidiurnal, omnivore, woodland, open forest; nest arboreal.

Polyrhachis cleopatra Forel, 1902

Polyrhachis cleopatra Forel, A. (1902). Fourmis nouvelles d'Australie. *Rev. Suisse Zool.* **10**: 405–548 [513]. Type data: syntypes, GMNH W, ANIC W, from Mackay, Qld.

Distribution: NE coastal, Qld. Ecology: terrestrial, noctidiurnal, omnivore, woodland, open forest; nest arboreal.

Polyrhachis clio Forel, 1902

Polyrhachis clio Forel, A. (1902). Fourmis nouvelles d'Australie. *Rev. Suisse Zool.* **10**: 405–548 [515]. Type data: syntypes, GMNH W, from Mackay, Qld.

Distribution: NE coastal, Qld. Ecology: terrestrial, noctidiurnal, omnivore, woodland, open forest; nest arboreal.

Polyrhachis clotho Forel, 1902

Polyrhachis clotho Forel, A. (1902). Fourmis nouvelles d'Australie. *Rev. Suisse Zool.* **10**: 405–548 [525]. Type data: syntypes, GMNH W, ANIC W, from Mackay, Qld.

Distribution: NE coastal, Qld. Ecology: terrestrial, noctidiurnal, arboreal, omnivore, woodland, open forest; nest arboreal.

Polyrhachis constricta Emery, 1897

Polyrhachis constricta Emery, C. (1897). Viaggio do Lamberto Loria nella Papuasia orientale 18. Formiche raccolte nelle Nuova Guinea. *Ann. Mus. Civ. Stor. Nat. Giacomo Doria* **38**: 546–594 [22 Nov. 1897] [584 pl 1]. Type data: holotype, MCG *W, from Qld.

Distribution: NE coastal, Qld. Ecology: terrestrial, noctidiurnal, omnivore, woodland, open forest; nest in soil.

Polyrhachis contemta Mayr, 1876

Polyrhachis contemta Mayr, G.L. (1876). Die australischen Formiciden. *J. Mus. Godeffroy* **5**: 56–115 [74]. Type data: syntypes, NHMW *W, from Gayndah, Qld.

Distribution: NE coastal, Qld. Ecology: terrestrial, noctidiurnal, omnivore, woodland, open forest; nest in soil.

Polyrhachis daemeli Mayr, 1876

Polyrhachis daemeli daemeli Mayr, 1876

Polyrhachis daemeli Mayr, G.L. (1876). Die australischen Formiciden. *J. Mus. Godeffroy* **5**: 56–115 [72]. Type data: syntypes, NHMW *W, from Rockhampton and Peak Downs, Qld.

Distribution: NE coastal, Qld. Ecology: terrestrial, noctidiurnal, omnivore, open forest; nest arboreal.

Polyrhachis daemeli argentosa Forel, 1902

Polyrhachis daemeli argentosa Forel, A. (1902). Fourmis nouvelles d'Australie. *Rev. Suisse Zool.* **10**: 405–548 [515]. Type data: syntypes, GMNH W, ANIC W, from Mackay, Qld.

Distribution: NE coastal, Qld. Ecology: terrestrial, noctidiurnal, omnivore, open forest; nest arboreal.

Polyrhachis daemeli exlex Forel, 1915

Polyrhachis (Hedomyrma) daemeli exlex Forel, A. (1915). Results of Dr. E. Mjöbergs Swedish Scientific Expeditions to Australia 1910–1913. 2. Ameisen. *Ark. Zool.* **9**: 1–119 pls 1–3 [4 Dec. 1915] [110]. Type data: holotype, SMNH ?* W, from Yarrabah, Qld.

Distribution: NE coastal, Qld. Ecology: terrestrial, noctidiurnal, omnivore, open forest; nest arboreal.

Polyrhachis doddi Donisthorpe, 1938

Polyrhachis (Cyrtomyrma) doddi Donisthorpe, H. (1938). The subgenus *Cyrtomyrma* Forel of *Polyrhachis* Smith, with descriptions of new species, etc. *Ann. Mag. Nat. Hist. (11)* **1**: 246–267 [263]. Type data: syntypes, BMNH *W,F, from Qld.

Distribution: NE coastal, Qld. Ecology: terrestrial, noctidiurnal, omnivore, open forest, closed forest; nest arboreal.

Polyrhachis erato Forel, 1902

Polyrhachis erato Forel, A. (1902). Fourmis nouvelles d'Australie. *Rev. Suisse Zool.* **10**: 405–548 [512]. Type data: syntypes, GMNH W, ANIC W, from Mackay, Qld.

Distribution: NE coastal, Qld. Ecology: terrestrial, noctidiurnal, omnivore, open forest; nest arboreal.

Polyrhachis euterpe Forel, 1902

Polyrhachis euterpe Forel, A. (1902). Fourmis nouvelles d'Australie. *Rev. Suisse Zool.* **10**: 405-548 [511]. Type data: holotype (probable), GMNH W, from Mackay, Qld.

Distribution: NE coastal, Qld. Ecology: terrestrial, noctidiurnal, omnivore, open forest; nest arboreal.

Polyrhachis exulans Clark, 1941

Polyrhachis (Myrmhopla) exulans Clark, J. (1941). Australian Formicidae. Notes and new species. *Mem. Natl. Mus. Vict.* **12**: 71-94 [91 pl 13]. Type data: syntypes, NMV *W, from Koolpinyah, N.T.

Distribution: N coastal, N.T. Ecology: terrestrial, noctidiurnal, omnivore, closed forest; nest arboreal.

Polyrhachis femorata F. Smith, 1858

Polyrhachis femoratus Smith, F. (1858). *Catalogue of hymenopterous insects in the collection of the British Museum. Part 6. Formicidae.* London : British Museum 216 pp. 14 pls [27 Mar. 1858] [73]. Publication date established from Donisthorpe, H. (1932). On the identity of Smith's types of Formicidae (Hymenoptera) collected by Alfred Russell Wallace in the Malay Archipelago, with descriptions of two new species. *Ann. Mag. Nat. Hist. (10)* **10**: 441-476. Type data: syntypes (probable), BMNH *W, from Melbourne, Vic.

Camponotus emeryi Forel, A. (1880). Études myrmécologiques en 1879. *Bull. Soc. Vaud. Sci. Nat.* **16**: 53-128 [113 pl 1]. Type data: holotype, possibly in GMNH, from Australia.

Synonymy that of Emery, C. (1925). Hymenoptera Fam. Formicidae subfam. Formicinae *in* Wytsman, P. (ed.) *Genera Insectorum.* Fasc. 183 302 pp. 4 pls [179].

Distribution: SE coastal, Vic. Ecology: terrestrial, noctidiurnal, omnivore, woodland, open forest; nest in soil.

Polyrhachis flavibasis Clark, 1930

Polyrhachis (Campomyrma) flavibasis Clark, J. (1930). New Formicidae, with notes on some little-known species. *Proc. R. Soc. Vict.* **43**: 2-25 [30 Aug. 1930] [16]. Type data: syntypes, NMV *W,F, from Brooklana and Dorrigo, N.S.W.

Distribution: SE coastal, N.S.W. Ecology: terrestrial, noctidiurnal, omnivore, woodland, open forest; nest in soil.

Polyrhachis froggatti Forel, 1910

Polyrhachis froggatti Forel, A. (1910). Formicides australiens reçus de MM. Froggatt et Rowland Turner. *Rev. Suisse Zool.* **18**: 1-94 [89]. Type data: syntypes, GMNH W, ANIC W, from Bombala, N.S.W.

Distribution: SE coastal, N.S.W. Ecology: terrestrial, noctidiurnal, omnivore, woodland, open forest; nest in soil.

Polyrhachis fuscipes Mayr, 1862

Polyrhachis fuscipes Mayr, G.L. (1862). Myrmecologische Studien. *Verh. Zool.-Bot. Ges. Wien* **12**: Abhand. 649-776 [679 pl 19]. Type data: syntypes (probable), NHMW *W, from Tas.

Distribution: Tas. Ecology: terrestrial, noctidiurnal, omnivore, woodland, open forest; nest in soil.

Polyrhachis gab Forel, 1880

Polyrhachis gab gab Forel, 1880

Polyrhachis guerini gab Forel, A. (1880). Études myrmécologiques en 1879. *Bull. Soc. Vaud. Sci. Nat.* **16**: 53-128 [116 pl 1]. Type data: syntypes, possibly in GMNH, from Australia.

Polyrhachis (Chariomyrma) gab tripellis Forel, A. (1915). Results of Dr. E. Mjöbergs Swedish Scientific Expeditions to Australia 1910-1913. 2. Ameisen. *Ark. Zool.* **9**: 1-119 pls 1-3 [4 Dec. 1915] [108]. Type data: syntypes, GMNH W,F, ANIC W, other syntypes may exist, from Kimberley distr., Derby and Noonkanbah, W.A.

Polyrhachis comata Crawley, W.C. (1915). Ants from north and south-west Australia (G.F. Hill, Rowland Turner and Christmas Island, Straits Settlements. Part II. *Ann. Mag. Nat. Hist. (8)* **15**: 232-239 [237] [*non Polyrhachis bicolor comata* Emery, 1911]. Type data: syntypes (probable), BMNH *W, from Stapleton, N.T.

Polyrhachis crawleyella Santschi, F. (1916). Rectifications à la nomenclature de quelques formicides [Hym.]. *Bull. Soc. Entomol. Fr.* **1916**: 242-243 [243] [*nom. nov.* for *Polyrhachis comata* Crawley, 1915].

Synonymy that of Bolton, B. (1974). New synonymy and a new name in the ant genus *Polyrhachis* F. Smith (Hym., Formicidae). *Entomol. Mon. Mag.* **109**: 172-180 [173].

Distribution: N coastal, W.A., N.T. Ecology: terrestrial, noctidiurnal, omnivore, woodland, open forest; nest in soil.

Polyrhachis gab aegra Forel, 1915

Polyrhachis (Chariomyrma) gab aegra Forel, A. (1915). Results of Dr. E. Mjöbergs Swedish Scientific Expeditions to Australia 1910-1913. 2. Ameisen. *Ark. Zool.* **9**: 1-119 pls 1-3 [4 Dec. 1915] [109]. Type data: syntypes, GMNH W, ANIC W, other syntypes may exist, from Atherton, Qld.

Distribution: NE coastal, Qld. Ecology: terrestrial, noctidiurnal, omnivore, open forest; nest in soil.

Polyrhachis gab senilis Forel, 1902

Polyrhachis gab senilis Forel, A. (1902). Fourmis nouvelles d'Australie. *Rev. Suisse Zool.* **10**: 405-548 [520]. Type data: syntypes, GMNH W, ANIC W, from Townsville, Qld.

Distribution: NE coastal, Qld. Ecology: terrestrial, noctidiurnal, omnivore, woodland, open forest; nest in soil.

Polyrhachis glabrinota Clark, 1930

Polyrhachis (Myrmhopla) glabrinotum Clark, J. (1930). New Formicidae, with notes on some little-known species. *Proc. R. Soc. Vict.* **43**: 2–25 [30 Aug. 1930] [13]. Type data: syntypes, NMV *W, from Cape York, Qld.

Distribution: N Gulf, Qld. Ecology: terrestrial, noctidiurnal, omnivore, open forest, closed forest; nest arboreal.

Polyrhachis gravis Clark, 1930

Polyrhachis (Campomyrma) gravis Clark, J. (1930). New Formicidae, with notes on some little-known species. *Proc. R. Soc. Vict.* **43**: 2–25 [30 Aug. 1930] [15]. Type data: syntypes, NMV *W, from Burt Plains, N.T.

Distribution: Lake Eyre basin, N.T. Ecology: terrestrial, noctidiurnal, omnivore, desert, woodland; nest in soil.

Polyrhachis guerini Roger, 1863

Polyrhachis guerini guerini Roger, 1863

Polyrhachis guerini Roger, J. (1863). Die neu aufgeführten Gattungen und Arten meines Formiciden-Verzeichnisses. *Berl. Entomol. Z.* **7**: 129–214 [June 1863] [157]. Type data: holotype, MNHP *W, from Australia (as New Holland).

Distribution: SE coastal, N.S.W. Ecology: terrestrial, noctidiurnal, omnivore, woodland, open forest; nest in soil.

Polyrhachis guerini lata Emery, 1895

Polyrhachis guerini lata Emery, C. (1895). Descriptions de quelques fourmis nouvelles d'Australie. *Ann. Soc. Entomol. Belg.* **39**: 345–358 [357]. Type data: syntypes, MCG *W, from Somerset, Qld.

Distribution: NE coastal, Qld. Ecology: terrestrial, noctidiurnal, omnivore, woodland, open forest; nest in soil.

Polyrhachis guerini pallescens Mayr, 1876

Polyrhachis guerini pallescens Mayr, G.L. (1876). Die australischen Formiciden. *J. Mus. Godeffroy* **5**: 56–115 [74]. Type data: syntypes (probable), NHMW *W, from Rockhampton, Qld.

Distribution: NE coastal, Qld. Ecology: terrestrial, noctidiurnal, omnivore, woodland, open forest; nest in soil.

Polyrhachis guerini vermiculosa Mayr, 1876

Polyrhachis guerini vermiculosa Mayr, G.L. (1876). Die australischen Formiciden. *J. Mus. Godeffroy* **5**: 56–115 [74]. Type data: syntypes, NHMW *W,F,M, from Rockhampton and Peak Downs, Qld. and Sidney (=Sydney), N.S.W.

Distribution: NE coastal, SE coastal, Qld., N.S.W. Ecology: terrestrial, noctidiurnal, omnivore, woodland, open forest; nest in soil.

Polyrhachis hecuba Forel, 1902

Polyrhachis hecuba Forel, A. (1902). Fourmis nouvelles d'Australie. *Rev. Suisse Zool.* **10**: 405–548 [527]. Type data: syntypes, GMNH W,F,M, ANIC W, from Mackay, Qld.

Distribution: NE coastal, Qld. Ecology: terrestrial, noctidiurnal, omnivore, woodland, open forest; nest in soil.

Polyrhachis heinlethii Forel, 1895

Polyrhachis heinlethii heinlethii Forel, 1895

Polyrhachis heinlethii Forel, A. (1895). Nouvelles fourmis de diverses provenances, surtout d'Australie. *Ann. Soc. Entomol. Belg.* **39**: 41–49 [47]. Type data: syntypes, ANIC W, GMNH W, from Mackay, Qld.

Distribution: NE coastal, Qld. Ecology: terrestrial, noctidiurnal, omnivore, woodland, open forest; nest in soil.

Polyrhachis heinlethii sophiae Forel, 1902

Polyrhachis heinlethii sophiae Forel, A. (1902). Fourmis nouvelles d'Australie. *Rev. Suisse Zool.* **10**: 405–548 [521]. Type data: syntypes, GMNH W, ANIC W, from Mackay, Qld.

Distribution: NE coastal, Qld. Ecology: terrestrial, noctidiurnal, omnivore, woodland, open forest; nest in soil.

Polyrhachis hermione Emery, 1895

Polyrhachis hermione hermione Emery, 1895

Polyrhachis hermione Emery, C. (1895). Descriptions de quelques fourmis nouvelles d'Australie. *Ann. Soc. Entomol. Belg.* **39**: 345–358 [357]. Type data: syntypes, MCG *W, from Mt. Bellenden Ker, Qld.

Distribution: NE coastal, Qld. Ecology: terrestrial, noctidiurnal, omnivore, woodland, open forest; nest in soil.

Polyrhachis hermione cupreata Emery, 1895

Polyrhachis hermione cupreata Emery, C. (1895). Descriptions de quelques fourmis nouvelles d'Australie. *Ann. Soc. Entomol. Belg.* **39**: 345–358 [357]. Type data: holotype, MCG *W, from Cairus (=Cairns), Qld.

Distribution: NE coastal, Qld. Ecology: terrestrial, noctidiurnal, omnivore, woodland, open forest; nest in soil.

Polyrhachis hexacantha (Erichson, 1842)

Formica hexacantha Erichson, W.F. (1842). Beitrag zur Fauna von Vandiemansland mit besonderer rucksicht auf die geographische Verbreitung der Insecten. *Arch. Naturg.* **8**: 83–287 [260]. Type data: holotype (probable), ZMB *W, from Tas.

Distribution: Tas. Ecology: terrestrial, noctidiurnal, omnivore, woodland, open forest; nest in soil.

Polyrhachis hirsuta Mayr, 1876

Polyrhachis hirsuta hirsuta Mayr, 1876

Polyrhachis hirsuta Mayr, G.L. (1876). Die australischen Formiciden. *J. Mus. Godeffroy* **5**: 56-115 [75]. Type data: syntypes (probable), NHMW *W, from Rockhampton, Qld.

Distribution: NE coastal, Qld. Ecology: terrestrial, noctidiurnal, omnivore, woodland, open forest; nest in soil.

Polyrhachis hirsuta quinquedentata Viehmeyer, 1925

Polyrhachis (Campomyrma) hirsuta quinquedentata Viehmeyer, H. (1925). Formiciden der australischen Faunenregion. *Entomol. Mitt.* **14**: 139-149 [147]. Type data: syntypes (probable), ZMB *W, from Liverpool, N.S.W.

Distribution: SE coastal, N.S.W. Ecology: terrestrial, noctidiurnal, omnivore, woodland, open forest; nest in soil.

Polyrhachis hookeri Lowne, 1865

Polyrhachis hookeri hookeri Lowne, 1865

Polyrhachis hookeri Lowne, B.T. (1865). Contributions to the natural history of Australian ants. *Entomologist* **2**: 331-336 [334]. Type data: syntypes (probable), BMNH (probable) *W, from Sidney (=Sydney), N.S.W.

Distribution: SE coastal, N.S.W. Ecology: terrestrial, noctidiurnal, omnivore, woodland, open forest; nest in soil.

Polyrhachis hookeri aerea Forel, 1902

Polyrhachis hookeri aerea Forel, A. (1902). Fourmis nouvelles d'Australie. *Rev. Suisse Zool.* **10**: 405-548 [521]. Type data: syntypes, GMNH W,F,M, from Mackay, Qld.

Distribution: NE coastal, Qld. Ecology: terrestrial, noctidiurnal, omnivore, woodland, open forest; nest in soil.

Polyrhachis hookeri lownei Forel, 1895

Polyrhachis hookeri lownei Forel, A. (1895). Nouvelles fourmis de diverses provenances, surtout d'Australie. *Ann. Soc. Entomol. Belg.* **39**: 41-49 [44]. Type data: syntypes, GMNH W, ANIC W, from Mackay, Qld.

Distribution: NE coastal, Qld. Ecology: terrestrial, noctidiurnal, omnivore, woodland, open forest; nest in soil.

Polyrhachis hookeri obscura Forel, 1895

Polyrhachis hookeri obscura Forel, A. (1895). Nouvelles fourmis de diverses provenances, surtout d'Australie. *Ann. Soc. Entomol. Belg.* **39**: 41-49 [44]. Type data: syntypes, GMNH W, ANIC W, from Mackay, Qld.

Distribution: NE coastal, Qld. Ecology: terrestrial, noctidiurnal, omnivore, woodland, open forest; nest in soil.

Polyrhachis humerosa Emery, 1921

Polyrhachis (Hedomyrma) humerosa Emery, C. (1921). Le genre *Polyrhachis*. Classification; espèces nouvelles ou critiques. *Bull. Soc. Vaud. Sci. Nat.* **54**: 17-25 [18]. Type data: syntypes, GMNH (probable) W, from Adelaide, S.A.

Distribution: S Gulfs, S.A. Ecology: terrestrial, noctidiurnal, omnivore, open forest; nest arboreal.

Polyrhachis inconspicua Emery, 1887

Polyrhachis inconspicua inconspicua Emery, 1887

Polyrhachis inconspicua Emery, C. (1887). Catalogo delle formiche esistenti nelle collezioni del Museo Civico di Genova. Parte terza. Formiche della regione Indo-Malese e dell'Australia. *Ann. Mus. Civ. Stor. Nat. Giacomo Doria* **25**: 209-258 pls 3-4 [225]. Type data: syntypes, MCG *W, from Somerset, Qld.

Distribution: NE coastal, Qld. Ecology: terrestrial, noctidiurnal, omnivore, woodland, closed forest; nest in soil.

Polyrhachis inconspicua subnitens Emery, 1895

Polyrhachis inconspicua subnitens Emery, C. (1895). Descriptions de quelques fourmis nouvelles d'Australie. *Ann. Soc. Entomol. Belg.* **39**: 345-358 [357]. Type data: holotype, MCG *W, from Kamerunga, Qld.

Distribution: NE coastal, Qld. Ecology: terrestrial, noctidiurnal, omnivore, woodland, open forest; nest in soil.

Polyrhachis ithona F. Smith, 1860

Polyrhachis hector Smith, F. (1859). Catalogue of hymenopterous insects collected by Mr A.R. Wallace at the islands of Aru and Key. *J. Linn. Soc. Zool.* **3**: 132-178 [1 Feb. 1859] [142] [*non Polyrhachis hector* F. Smith, 1857]. Publication date established from Donisthorpe, H. (1932). On the identity of Smith's types of Formicidae (Hymenoptera) collected by Alfred Russell Wallace in the Malay Archipelago, with descriptions of two new species. *Ann. Mag. Nat. Hist. (10)* **10**: 441-476. Type data: holotype, OUM *W, from Aru Ils., Indonesia.

Polyrhachis ithonus Smith, F. (1860). Catalogue of hymenopterous insects collected by Mr A.R. Wallace in the islands of Bachian, Kaisaa, Amboyna, Gilolo, and Dory in New Guinea. *J. Linn. Soc. Zool.* **5**: 93-143 pl 1 [18 July 1860] [99]. Type data: syntypes, OUM *W,F, from Bachian, Indonesia.

Polyrhachis andromache Roger, J. (1863). Verzeichniss der Formiciden-Gattungen und Arten. *Berl. Entomol. Z.* **7** appendix to vol.: 1-65 [8] [*nom. nov.* for *Polyrhachis hector* F. Smith, 1859].

Synonymy that of Bolton, B. (1974). New synonymy and a new name in the ant genus *Polyrhachis* F. Smith (Hym., Formicidae). *Entomol. Mon. Mag.* **109**: 172-180 [177].

Distribution: NE coastal, Qld.; widespread in SE Asia. Ecology: terrestrial, noctidiurnal, omnivore, woodland, open forest; nest in soil.

Polyrhachis jacksoniana Roger, 1863

Polyrhachis jacksoniana Roger, J. (1863). Die neu aufgeführten Gattungen und Arten meines Formiciden-Verzeichnisses. *Berl. Entomol. Z.* **7**: 129–214 [June 1863] [158]. Type data: holotype, MNHP *W, from Port Jackson, N.S.W.

Distribution: SE coastal, N.S.W. Ecology: terrestrial, noctidiurnal, omnivore, woodland, open forest; nest in soil.

Polyrhachis kershawi Clark, 1930

Polyrhachis (Hedomyrma) kershawi Clark, J. (1930). New Formicidae, with notes on some little-known species. *Proc. R. Soc. Vict.* **43**: 2–25 [30 Aug. 1930] [12]. Type data: syntypes, NMV *W, from Claudie River, Qld.

Distribution: NE coastal, Qld. Ecology: terrestrial, noctidiurnal, omnivore, open forest, closed forest; nest arboreal.

Polyrhachis lachesis Forel, 1897

Polyrhachis lachesis Forel, A. *in* Emery, C. (1897). Viaggio do Lamberto Loria nella Papuasia orientale 18. Formiche raccolte nelle Nuova Guinea. *Ann. Mus. Civ. Stor. Nat. Giacomo Doria* **38**: 546–594 [22 Nov. 1897] [582 pl 1]. Type data: syntypes, GMNH W, from Mackay, Qld.

Distribution: NE coastal, Qld. Ecology: terrestrial, noctidiurnal, omnivore, open forest, closed forest; nest arboreal.

Polyrhachis latreillii (Guérin, 1831)

Formica latreillii Guérin-Meneville, F.E. (1831). Chapter 12, Insectes. *in* Duperrey, M.L.I. (1838). *Voyage autour du monde, exécuté par ordre du roi, sur la corvette de La Majesté, La Coquille, pendant les années 1822, 1823, 1824 et 1825.* Vol. 2 part 2, division 1: 57–302 Atlas (1830–1832), Ins pls 1–21 [203 pl 8 fig 4]. Publication date established from Bequaret, J. (1926). The date of publication of the Hymenoptera and Diptera described by Guérin in Duperrey's "Voyage de la *Coquille*". *Entomol. Mitt.* **15**: 186–195 [20 Mar. 1926]. Type data: holotype, MNHP (probable) *W, from Australia (as New Holland).

Distribution: (SE coastal), (N.S.W.). Ecology: terrestrial, noctidiurnal, omnivore, woodland, open forest; nest in soil.

Polyrhachis leae Forel, 1913

Polyrhachis leae leae Forel, 1913

Polyrhachis leae Forel, A. (1913). Fourmis de Tasmanie et d'Australie récoltées par MM. Lea, Froggatt etc. *Bull. Soc. Vaud. Sci. Nat.* **49**: 173–196 pl 2 [193]. Type data: syntypes, GMNH W, from Hobart, Tas.

Distribution: Tas. Ecology: terrestrial, noctidiurnal, omnivore, woodland, open forest; nest in soil.

Polyrhachis leae cedarensis Forel, 1915

Polyrhachis (Campomyrma) leae cedarensis Forel, A. (1915). Results of Dr. E. Mjöbergs Swedish Scientific Expeditions to Australia 1910–1913. 2. Ameisen. *Ark. Zool.* **9**: 1–119 pls 1–3 [4 Dec. 1915] [114]. Type data: syntypes, GMNH W,F, other syntypes may exist, from Cedar Creek, Qld.

Distribution: NE coastal, Qld. Ecology: terrestrial, noctidiurnal, omnivore, woodland, open forest; nest in soil.

Polyrhachis levior Roger, 1863

Polyrhachis laevissima Smith, F. (1859). Catalogue of hymenopterous insects collected by Mr A.R. Wallace at the islands of Aru and Key. *J. Linn. Soc. Zool.* **3**: 132–178 [1 Feb. 1859] [141] [*non Polyrhachis laevissima* Smith, 1858]. Publication date established from Donisthorpe, H. (1932). On the identity of Smith's types of Formicidae (Hymenoptera) collected by Alfred Russell Wallace in the Malay Archipelago, with descriptions of two new species. *Ann. Mag. Nat. Hist. (10)* **10**: 441–476. Type data: syntypes (probable), BMNH *W, from Aru Ils., Indonesia.

Polyrhachis levior Roger, J. (1863). Verzeichniss der Formiciden-Gattungen und Arten. *Berl. Entomol. Z.* **7** appendix to vol.: 1–65 [8] [*nom. nov.* for *Polyrhachis laevissima* F. Smith, 1859].

Polyrhachis australis Mayr, G.L. (1870). Neue Formiciden. *Verh. Zool.-Bot. Ges. Wien* **20**: Abhand. 939–996 [31 Dec. 1870] [945]. Type data: syntypes (probable), NHMW *W, from Port Mackay, Qld.

Synonymy that of Emery, C. (1925). Fam. Formicidae subfam. Formicinae. *in* Wytsman, P. (ed.) *Genera Insectorum.* Fasc. 183 302 pp. 4 pls [208].

Distribution: NE coastal, Qld; also in E Indonesia and Papua New Guinea. Ecology: omnivore, arboreal, closed forest; nest arboreal.

Polyrhachis lombokensis Emery, 1898

Polyrhachis lombokensis Emery, C. (1898). Descrizioni di formiche nuove Malesi e Australiane. Note sinonimiche. *Rec. Sess. Accad. Sci. Ist. Bologna (ns)* **2**: 231–245 [239]. Type data: status unknown, ?MCG, from Lombok, Indonesia.

Polyrhachis lombokensis yarrabahensis Forel, 1915

Polyrhachis (Myrmatopa) lombokensis yarrabahensis Forel, A. (1915). Results of Dr. E. Mjöbergs Swedish Scientific Expeditions to Australia 1910–1913. 2. Ameisen. *Ark. Zool.* **9**: 1–119 pls 1–3 [4 Dec. 1915] [115]. Type data: syntypes, GMNH W, ANIC W, other syntypes may exist, from Malanda and Yarrabah, Qld.

Distribution: NE coastal, Qld. Ecology: terrestrial, noctidiurnal, omnivore, closed forest; nest arboreal.

Polyrhachis lysistrata Santschi, 1920

Polyrhachis (Myrmothrinax) lysistrata Santschi, F. (1920). Quelques nouveaux Camponotinae d'Indochine et

Australie. *Bull. Soc. Vaud. Sci. Nat.* **52**: 565–569 [569] [introduced as *Polyrhachys*]. Type data: syntypes, BNHM W, from Townsville, Qld.

Distribution: NE coastal, Qld. Ecology: terrestrial, noctidiurnal, omnivore, closed forest; nest arboreal.

Polyrhachis machaon Santschi, 1920

Polyrhachis (Hedomyrma) machaon Santschi, F. (1920). Quelques nouveaux Camponotinae d'Indochine et Australie. *Bull. Soc. Vaud. Sci. Nat.* **52**: 565–569 [568] [introduced as *Polyrhachys*]. Type data: holotype, BNHM W, from Townsville, Qld., see The Zoological Society of London (1922). *The Zoological Record.* Vol. 57, relating chiefly to the year 1920. London : Gurney & Jackson.

Distribution: NE coastal, Qld. Ecology: terrestrial, noctidiurnal, omnivore, woodland, open forest; nest arboreal.

Polyrhachis mackayi Donisthorpe, 1938

Polyrhachis (Cyrtomyrma) mackayi Donisthorpe, H. (1938). The subgenus *Cyrtomyrma* Forel of *Polyrhachis* Smith, with descriptions of new species, etc. *Ann. Mag. Nat. Hist. (11)* **1**: 246–267 [258]. Type data: syntypes, BMNH *W,F, from Mackay, Qld.

Distribution: NE coastal, Qld. Ecology: terrestrial, noctidiurnal, omnivore, closed forest; nest arboreal.

Polyrhachis macropus Wheeler, 1916

Polyrhachis (Campomyrma) longipes Wheeler, W.M. (1915). Hymenoptera. *Trans. R. Soc. S. Aust.* **39**: 805–823 pls 64–66 [Dec. 1915] [821] [*non Polyrhachis longipes* F. Smith, 1858]. Type data: syntypes, MCZ *W, from Everard Range S.A.

Polyrhachis macropus Wheeler, W.M. (1916). *Prodiscothyrea*, a new genus of ponerine ants from Queensland. *Trans. R. Soc. S. Aust.* **40**: 33–37 [23 Dec. 1916] [37 pl 4] [*nom. nov.* for *Polyrhachis longipes* Wheeler, 1915].

Distribution: W plateau, S.A. Ecology: terrestrial, noctidiurnal, omnivore, woodland; nest in soil.

Polyrhachis micans Mayr, 1876

Polyrhachis micans micans Mayr, 1876

Polyrhachis micans Mayr, G.L. (1876). Die australischen Formiciden. *J. Mus. Godeffroy* **5**: 56–115 [76]. Type data: syntypes, NHMW *W,F, from Rockhampton and Peak Downs, Qld.

Distribution: NE coastal, Qld. Ecology: terrestrial, noctidiurnal, omnivore, woodland, open forest; nest in soil.

Polyrhachis micans ops Forel, 1907

Polyrhachis micans ops Forel, A. (1907). Formicidae. pp. 263–310 *in* Michaelsen, W. & Hartmeyer, R. (eds.) *Die Fauna Südwest-Australiens.* Jena : G. Fischer Vol. 1 [308]. Type data: holotype, GMNH W, from Albany, W.A.

Distribution: SW coastal, W.A. Ecology: terrestrial, noctidiurnal, omnivore, woodland, open forest; nest in soil.

Polyrhachis mjobergi Forel, 1915

Polyrhachis (Hedomyrma) mjobergi Forel, A. (1915). Results of Dr. E. Mjöbergs Swedish Scientific Expeditions to Australia 1910–1913. 2. Ameisen. *Ark. Zool.* **9**: 1–119 pls 1–3 [4 Dec. 1915] [112]. Type data: syntypes, GMNH W, other syntypes may exist, from Glen Lamington, Qld.

Distribution: NE coastal, Qld. Ecology: terrestrial, noctidiurnal, omnivore, open forest, closed forest; nest arboreal.

Polyrhachis nox Donisthorpe, 1938

Polyrhachis (Cyrtomyrma) nox Donisthorpe, H. (1938). The subgenus *Cyrtomyrma* Forel of *Polyrhachis* Smith, with descriptions of new species, etc. *Ann. Mag. Nat. Hist. (11)* **1**: 246–267 [249]. Type data: syntypes, BMNH *W, from Mackay, Qld.

Distribution: NE coastal, Qld. Ecology: terrestrial, noctidiurnal, omnivore, closed forest; nest arboreal.

Polyrhachis ornata Mayr, 1876

Polyrhachis ornata Mayr, G.L. (1876). Die australischen Formiciden. *J. Mus. Godeffroy* **5**: 56–115 [73]. Type data: syntypes, NHMW *W, from Rockhampton, Qld.

Distribution: NE coastal, Qld. Ecology: terrestrial, noctidiurnal, omnivore, woodland, open forest; nest arboreal.

Polyrhachis patiens Santschi, 1920

Polyrhachis (Campomyrma) patiens Santschi, F. (1920). Cinq nouvelles notes sur les fourmis. *Bull. Soc. Vaud. Sci. Nat.* **53**: 163–186 [185]. Type data: holotype, BNHM W, from Kabrinville, Vic.

Distribution: SE coastal, Vic. Ecology: terrestrial, noctidiurnal, omnivore, woodland, open forest; nest in soil.

Polyrhachis penelope Forel, 1895

Polyrhachis penelope Forel, A. (1895). Nouvelles fourmis de diverses provenances, surtout d'Australie. *Ann. Soc. Entomol. Belg.* **39**: 41–49 [46]. Type data: syntypes, GMNH W, ANIC W, from Mackay, Qld.

Distribution: NE coastal, Qld. Ecology: terrestrial, noctidiurnal, omnivore, woodland, open forest; nest in soil.

Polyrhachis phryne Forel, 1907

Polyrhachis phryne Forel, A. (1907). Formicides du Musée National Hongrois. *Ann. Hist.- Nat. Mus. Natl. Hung.* **5**: 1–42 [30 June 1907] [41]. Type data: syntypes (probable), probably in GMNH or MNH, from Mt. Victoria, Blue Mts., N.S.W.

Distribution: SE coastal, N.S.W. Ecology: terrestrial, noctidiurnal, omnivore, woodland, open forest; nest in soil.

Polyrhachis polymnia Forel, 1902

Polyrhachis polymnia polymnia Forel, 1902

Polyrhachis polymnia Forel, A. (1902). Fourmis nouvelles d'Australie. *Rev. Suisse Zool.* **10**: 405-548 [532]. Type data: syntypes, GMNH W,F, from Mackay, Qld.

Distribution: NE coastal, Qld. Ecology: terrestrial, noctidiurnal, omnivore, woodland, open forest; nest in soil.

Polyrhachis polymnia maculata Forel, 1915

Polyrhachis (Campomyrma) polymnia maculata Forel, A. (1915). Results of Dr. E. Mjöbergs Swedish Scientific Expeditions to Australia 1910-1913. 2. Ameisen. *Ark. Zool.* **9**: 1-119 pls 1-3 [4 Dec. 1915] [115]. Type data: syntypes, GMNH W, other syntypes may exist, from Malanda, Cedar Creek and Atherton, Qld.

Distribution: NE coastal, Qld. Ecology: terrestrial, noctidiurnal, omnivore, woodland, open forest; nest in soil.

Polyrhachis prometheus Santschi, 1920

Polyrhachis (Campomyrma) prometheus Santschi, F. (1920). Quelques nouveaux Camponotinae d'Indochine et Australie. *Bull. Soc. Vaud. Sci. Nat.* **52**: 565-569 [566]. Type data: syntypes, BNHM W, from Townsville, Qld.

Distribution: NE coastal, Qld. Ecology: terrestrial, noctidiurnal, omnivore, woodland, open forest; nest in soil.

Polyrhachis pseudothrinax Hung, 1967

Polyrhachis pseudothrinax Hung, A.C.F. (1967). A new species and two new names of the *Polyrhachis* ants (Hymenoptera : Formicidae). *Mushi* **40**: 199-202 [24 Mar. 1967] [199]. Type data: holotype, AMNH *W, from Daly River, N.T.

Distribution: N coastal, N.T. Ecology: terrestrial, noctidiurnal, omnivore, woodland, open forest; nest in soil.

Polyrhachis punctiventris Mayr, 1876

Polyrhachis punctiventris Mayr, G.L. (1876). Die australischen Formiciden. *J. Mus. Godeffroy* **5**: 56-115 [73]. Type data: syntypes, NHMW *W,F, from Rockhampton, Qld.

Distribution: NE coastal, Qld. Ecology: terrestrial, noctidiurnal, omnivore, woodland, open forest; nest in soil.

Polyrhachis pyrrhus Forel, 1910

Polyrhachis pyrrhus Forel, A. (1910). Formicides australiens reçus de MM. Froggatt et Rowland Turner. *Rev. Suisse Zool.* **18**: 1-94 [90]. Type data: syntypes, GMNH W, ANIC W, from Tennant Creek, N.T.

Distribution: W plateau, N.T. Ecology: terrestrial, noctidiurnal, omnivore, woodland, open forest; nest in soil.

Polyrhachis quadricuspis Mayr, 1870

Polyrhachis quadricuspis Mayr, G.L. (1870). Neue Formiciden. *Verh. Zool.-Bot. Ges. Wien* **20**: Abhand. 939-996 [31 Dec. 1870] [946]. Type data: syntypes (probable), NHMW (probable) *W, from N.S.W.

Distribution: SE coastal, N.S.W. Ecology: terrestrial, noctidiurnal, omnivore, woodland, open forest; nest in soil.

Polyrhachis queenslandica Emery, 1895

Polyrhachis queenslandica Emery, C. (1895). Descriptions de quelques fourmis nouvelles d'Australie. *Ann. Soc. Entomol. Belg.* **39**: 345-358 [356]. Type data: syntypes (probable), MCG *W, from Kamerunga, Qld.

Polyrhachis delicata Crawley, W.C. (1915). Ants from north and south-west Australia (G.F. Hill, Rowland Turner) and Christmas Island, Straits Settlements. Part II. *Ann. Mag. Nat. Hist. (8)* **15**: 232-239 [238]. Type data: syntypes, BMNH *W, from Darwin, N.T.

Synonymy that of Emery, C. (1925). Hymenoptera Fam. Formicidae subfam. Formicinae. *in* Wytsman, P. (ed.) *Genera Insectorum.* Fasc. 183 302 pp. 4 pls [184].

Distribution: NE coastal, N coastal, Qld., N.T. Ecology: terrestrial, noctidiurnal, omnivore, open forest, closed forest; nest arboreal.

Polyrhachis rastellata (Latreille, 1802)

Formica rastellata Latreille, P.A. (1802). *Histoire naturelle des fourmis,* et recueil de mémoires et d'observations sur les abeilles, les araignées, les faucheurs, et autre insects. Paris : Crapelet 445 pp. 12 pls [130]. Type data: status unknown, ?MNHP, from Indes Orientales.

Polyrhachis rastellata yorkana Forel, 1915

Polyrhachis (Cyrtomyrma) rastellata yorkana Forel, A. (1915). Results of Dr. E. Mjöbergs Swedish Scientific Expeditions to Australia 1910-1913. 2. Ameisen. *Ark. Zool.* **9**: 1-119 pls 1-3 [4 Dec. 1915] [110]. Type data: syntypes, GMNH W, ANIC W, other syntypes may exist, from Cape York Peninsula, Qld.

Distribution: N Gulf, Qld. Ecology: terrestrial, noctidiurnal, omnivore, closed forest; nest arboreal.

Polyrhachis relucens (Latreille, 1802)

Formica relucens Latreille, P.A. (1802). *Histoire naturelle des fourmis,* et recueil de mémoires et d'observations sur les abeilles, les araignées, les faucheurs, et autres insectes. Paris : Crapelet 445 pp. 12 pls [131]. Type data: uncertain, MNHP (probable) *W, from East Indies.

Polyrhachis relucens australiae Emery, 1887

Polyrhachis connectens australiae Emery, C. (1887). Catalogo delle formiche esistenti nelle collezioni del Museo Civico di Genova. Parte terza. Formiche della regione Indo-Malese e dell'Australia. *Ann. Mus. Civ. Stor. Nat. Giacomo Doria* **25**: 209–258 pls 3–4 [231]. Type data: syntypes (probable), MCG *W, from Somerset, Qld.

Distribution: NE coastal, Qld. Ecology: terrestrial, noctidiurnal, omnivore, open forest; nest in soil.

Polyrhachis rowlandi Forel, 1910

Polyrhachis rowlandi Forel, A. (1910). Formicides australiens reçus de MM. Froggatt et Rowland Turner. *Rev. Suisse Zool.* **18**: 1–94 [85]. Type data: syntypes, GMNH W, ANIC W, from Cape York, Qld.

Distribution: N Gulf, Qld. Ecology: terrestrial, noctidiurnal, omnivore, woodland, open forest; nest in soil.

Polyrhachis schenkii Forel, 1886

Polyrhachis schenkii schenkii Forel, 1886

Polyrhachis schenkii Forel, A. (1886). Études myrmécologiques en 1886. *Ann. Soc. Entomol. Belg.* **30**: 131–215 [198]. Type data: syntypes, GMNH W, from Darnley Is., Qld. and New Guinea.

Distribution: Qld.; Torres Strait. Ecology: terrestrial, noctidiurnal, omnivore, open forest, closed forest; nest arboreal.

Polyrhachis schenkii lydiae Forel, 1902

Polyrhachis schenkii lydiae Forel, A. (1902). Fourmis nouvelles d'Australie. *Rev. Suisse Zool.* **10**: 405–548 [523]. Type data: syntypes, GMNH W,F, ANIC W, from Mackay, Qld.

Distribution: NE coastal, Qld. Ecology: terrestrial, noctidiurnal, omnivore, open forest, closed forest; nest arboreal.

Polyrhachis schwiedlandi Forel, 1902

Polyrhachis schwiedlandi Forel, A. (1902). Fourmis nouvelles d'Australie. *Rev. Suisse Zool.* **10**: 405–548 [529]. Type data: syntypes, GMNH W,F, ANIC W, from Sydney, N.S.W.

Distribution: SE coastal, N.S.W. Ecology: terrestrial, noctidiurnal, omnivore, woodland, open forest; nest in soil.

Polyrhachis semiaurata Mayr, 1876

Polyrhachis semiaurata Mayr, G.L. (1876). Die australischen Formiciden. *J. Mus. Godeffroy* **5**: 56–115 [71]. Type data: syntypes (probable), NHMW *W, from Sidney (=Sydney), N.S.W.

Distribution: SE coastal, N.S.W. Ecology: terrestrial, noctidiurnal, omnivore, woodland, open forest; nest in soil.

Polyrhachis semipolita E. André, 1896

Polyrhachis semipolita semipolita E. André, 1896

Polyrhachis semipolita André, E. (1896). Fourmis nouvelles d'Asie et d'Australie. *Rev. Entomol.* **15**: 251–265 [251]. Type data: syntypes, MNHP W, from Victorian Alps.

Distribution: Murray-Darling basin, Vic. Ecology: terrestrial, noctidiurnal, omnivore, woodland, open forest; nest in soil.

Polyrhachis semipolita hestia Forel, 1911

Polyrhachis semipolita hestia Forel, A. (1911). Die Ameisen des K. zoologischen Museums in München. *Sber. Beyer Akad. Wiss., Nat.-Hist. Klasse* **41**: Abhand. 249–303 [295]. Type data: holotype, ZSM W, from Australia.

Distribution: Murray-Darling basin, Vic., N.S.W. Ecology: terrestrial, noctidiurnal, omnivore, woodland, open forest; nest in soil.

Polyrhachis sempronia Forel, 1907

Polyrhachis sempronia Forel, A. (1907). Formicides du Musée National Hongrois. *Ann. Hist.- Nat. Mus. Natl. Hung.* **5**: 1–42 [30 June 1907] [39]. Type data: syntypes (probable), probably in GMNH or MNH, from Mt. Victoria, Blue Mts., N.S.W.

Distribution: SE coastal, N.S.W. Ecology: terrestrial, noctidiurnal, omnivore, woodland, open forest; nest in soil.

Polyrhachis sexspinosa (Latreille, 1802)

Formica sexspinosa Latreille, P.A. (1802). *Histoire naturelle des fourmis,* et recueil de mémoires et d'observations sur les abeilles, les araignées, les faucheurs, et autres insectes. Paris 445 pp. pls 12 [126]. Type data: holotype (probable), lost, from East Indies.

Distribution: NE coastal, Qld.; widespread on New Guinea. Ecology: terrestrial, noctidiurnal, omnivore, closed forest; nest arboreal.

Polyrhachis sidnica Mayr, 1866

Polyrhachis sidnica sidnica Mayr, 1866

Polyrhachis sidnica Mayr, G.L. (1866). Diagnosen neuer and wenig gekannter Formiciden. *Verh. Zool.-Bot. Ges. Wien* **16**: Abhand. 885–908 [886 pl 20]. Type data: syntypes (probable), NHMW (probable) *W, from Sidney (=Sydney), N.S.W.

Distribution: SE coastal, N.S.W. Ecology: terrestrial, noctidiurnal, omnivore, woodland, open forest; nest in soil.

Polyrhachis sidnica perthensis Crawley, 1922

Polyrhachis (Campomyrma) sidnica perthensis Crawley, W.C. (1922). New ants from Australia. *Ann. Mag. Nat. Hist.* (9) **10**: 16-36 [36]. Type data: syntypes, OUM *W, BMNH *W, from Perth, W.A.

Distribution: SW coastal, W.A. Ecology: terrestrial, noctidiurnal, omnivore, woodland, open forest; nest in soil.

Polyrhachis sidnica tambourinensis Forel, 1915

Polyrhachis (Campomyrma) sidnica tambourinensis Forel, A. (1915). Results of Dr. E. Mjöbergs Swedish Scientific Expeditions to Australia 1910-1913. 2. Ameisen. *Ark. Zool.* **9**: 1-119 pls 1-3 [4 Dec. 1915] [113]. Type data: holotype, SMNH ?* W, from Mt. Tambourine (=Tamborine Mt.), Qld.

Distribution: NE coastal, Qld. Ecology: terrestrial, noctidiurnal, omnivore, woodland, open forest; nest in soil.

Polyrhachis sokolova Forel, 1902

Polyrhachis sokolova sokolova Forel, 1902

Polyrhachis sokolova Forel, A. (1902). Fourmis nouvelles d'Australie. *Rev. Suisse Zool.* **10**: 405-548 [522]. Type data: syntypes, GMNH W, from Mackay, Qld.

Distribution: NE coastal, Qld. Ecology: terrestrial, noctidiurnal, omnivore, woodland, open forest; nest in soil.

Polyrhachis sokolova degener Forel, 1910

Polyrhachis sokolova degener Forel, A. (1910). Formicides australiens reçus de MM. Froggatt et Rowland Turner. *Rev. Suisse Zool.* **18**: 1-94 [84]. Type data: holotype, GMNH W, from Mackay, Qld.

Distribution: NE coastal, Qld. Ecology: terrestrial, noctidiurnal, omnivore, woodland, open forest; nest in soil.

Polyrhachis templi Forel, 1902

Polyrhachis templi Forel, A. (1902). Fourmis nouvelles d'Australie. *Rev. Suisse Zool.* **10**: 405-548 [531]. Type data: syntypes, GMNH W, ANIC W, from Mackay, Qld.

Distribution: NE coastal, Qld. Ecology: terrestrial, noctidiurnal, omnivore, woodland, open forest; nest in soil.

Polyrhachis terpsichore Forel, 1893

Polyrhachis terpsichore terpsichore Forel, 1893

Polyrhachis terpsichore Forel, A. (1893). Nouvelles fourmis d'Australie et des Canaries. *Ann. Soc. Entomol. Belg.* **37**: 454-466 [455]. Type data: syntypes, GMNH W, from Adelaide River, N.T.

Distribution: N coastal, N.T. Ecology: terrestrial, noctidiurnal, omnivore, open forest, closed forest; nest arboreal.

Polyrhachis terpischore elegans Forel, 1910

Polyrhachis terpsichore elegans Forel, A. (1910). Formicides australiens reçus de MM. Froggatt et Rowland Turner. *Rev. Suisse Zool.* **18**: 1-94 [84]. Type data: syntypes, GMNH W, from Kuranda near Cairns, Qld.

Distribution: NE coastal, Qld. Ecology: terrestrial, noctidiurnal, omnivore, open forest, closed forest; nest arboreal.

Polyrhachis terpsichore rufifemur Forel, 1907

Polyrhachis terpsichore rufifemur Forel, A. (1907). Formicides du Musée National Hongrois. *Ann. Hist.- Nat. Mus. Natl. Hung.* **5**: 1-42 [30 June 1907] [41]. Type data: syntypes (probable), probably in GMNH or MNH, from Springwood, N.S.W.

Distribution: SE coastal, N.S.W. Ecology: terrestrial, noctidiurnal, omnivore, open forest; nest arboreal.

Polyrhachis thais Forel, 1910

Polyrhachis thais Forel, A. (1910). Formicides australiens reçus de MM. Froggatt et Rowland Turner. *Rev. Suisse Zool.* **18**: 1-94 [86]. Type data: syntypes, GMNH W, from Kuranda near Cairns, Qld.

Distribution: NE coastal, Qld. Ecology: terrestrial, noctidiurnal, omnivore, woodland, open forest; nest in soil.

Polyrhachis thalia Forel, 1902

Polyrhachis thalia thalia Forel, 1902

Polyrhachis thalia Forel, A. (1902). Fourmis nouvelles d'Australie. *Rev. Suisse Zool.* **10**: 405-548 [530]. Type data: syntypes, GMNH W, from Charters Towers, Qld.

Distribution: NE coastal, Qld. Ecology: terrestrial, noctidiurnal, omnivore, woodland, open forest; nest in soil.

Polyrhachis thalia io Forel, 1915

Polyrhachis (Campomyrma) thalia io Forel, A. (1915). Results of Dr. E. Mjöbergs Swedish Scientific Expeditions to Australia 1910-1913. 2. Ameisen. *Ark. Zool.* **9**: 1-119 pls 1-3 [4 Dec. 1915] [114]. Type data: syntypes, GMNH W, other syntypes may exist, from Derby, W.A.

Distribution: N coastal, W.A. Ecology: terrestrial, noctidiurnal, omnivore, woodland, open forest; nest in soil.

Polyrhachis thusnelda Forel, 1902

Polyrhachis thusnelda Forel, A. (1902). Fourmis nouvelles d'Australie. *Rev. Suisse Zool.* **10**: 405-548 [509]. Type data: syntypes, GMNH W,F,M, ANIC W, from Mackay, Qld.

Distribution: NE coastal, Qld. Ecology: terrestrial, noctidiurnal, omnivore, woodland, open forest; nest in soil.

Polyrhachis townsvillei Donisthorpe, 1938

Polyrhachis (Cyrtomyrma) townsvillei Donisthorpe, H. (1938). The subgenus *Cyrtomyrma* Forel of *Polyrhachis* Smith, with descriptions of new species, etc. *Ann. Mag. Nat. Hist. (11)* **1**: 246–267 [251]. Type data: syntypes, BMNH *W,F, from Townsville, Qld.

Distribution: NE coastal, Qld. Ecology: terrestrial, noctidiurnal, omnivore, closed forest; nest arboreal.

Polyrhachis trapezoidea Mayr, 1876

Polyrhachis trapezoidea Mayr, G.L. (1876). Die australischen Formiciden. *J. Mus. Godeffroy* **5**: 56–115 [72]. Type data: syntypes, NHMW *W,F,M, from Rockhampton and Peak Downs, Qld.

Distribution: NE coastal, Qld. Ecology: terrestrial, noctidiurnal, omnivore, woodland, open forest; nest in soil.

Polyrhachis tubifera Forel, 1902

Polyrhachis tubifera Forel, A. (1902). Fourmis nouvelles d'Australie. *Rev. Suisse Zool.* **10**: 405–548 [517]. Type data: syntypes, GMNH W,M, from Mackay, Qld.

Distribution: NE coastal, Qld. Ecology: terrestrial, noctidiurnal, omnivore, woodland, open forest; nest in soil.

Polyrhachis turneri Forel, 1895

Polyrhachis turneri Forel, A. (1895). Nouvelles fourmis de diverses provenances, surtout d'Australie. *Ann. Soc. Entomol. Belg.* **39**: 41–49 [45]. Type data: syntypes, GMNH W, ANIC W, from Mackay, Qld.

Distribution: NE coastal, Qld. Ecology: terrestrial, noctidiurnal, omnivore, open forest, closed forest; nest arboreal.

Polyrhachis urania Forel, 1902

Polyrhachis urania Forel, A. (1902). Fourmis nouvelles d'Australie. *Rev. Suisse Zool.* **10**: 405–548 [516]. Type data: syntypes, GMNH W, from Mackay, Qld.

Distribution: NE coastal, Qld. Ecology: terrestrial, noctidiurnal, omnivore, woodland, open forest; nest in soil.

Polyrhachis zimmerae Clark, 1941

Polyrhachis (Campomyrma) zimmerae Clark, J. (1941). Australian Formicidae. Notes and new species. *Mem. Natl. Mus. Vict.* **12**: 71–94 [92 pl 13]. Type data: syntypes, NMV *W, from Mt. Manfred, N.S.W.

Distribution: Murray-Darling basin, N.S.W. Ecology: terrestrial, noctidiurnal, omnivore, woodland, open forest; nest in soil.

Prolasius Forel, 1892

Prolasius Forel, A. (1892). Die Ameisen Neu-Seelands. *Mitt. Schweiz. Entomol. Ges.* **8**: 331–343 [331] [proposed with subgeneric rank in *Melophorus* Lubbock, 1883]. Type species *Formica advena* F. Smith, 1862 by monotypy.

This group is also found in New Guinea and New Zealand.

Prolasius abruptus Clark, 1934

Prolasius abruptus Clark, J. (1934). Ants from the Otway Ranges. *Mem. Natl. Mus. Vict.* **8**: 48–73 [66 pl 4]. Type data: syntypes (probable), NMV *W, from Gellibrand, Vic.

Distribution: SE coastal, N.S.W., Vic. Ecology: terrestrial, noctidiurnal, omnivore, open forest, closed forest; nest in ground layer.

Prolasius antennatus McAreavey, 1947

Prolasius antennata McAreavey, J.J. (1947). New species of the genera *Prolasius* Forel and *Melophorus* Lubbock (Hymenoptera : Formicidae). *Mem. Natl. Mus. Vict.* **15**: 7–27 [Oct. 1947] [13 pl 1]. Type data: syntypes, NMV *W, from Ludlow, W.A.

Distribution: SW coastal, W.A. Ecology: terrestrial, noctidiurnal, omnivore, woodland, open forest; nest in ground layer.

Prolasius bruneus McAreavey, 1947

Prolasius brunea McAreavey, J.J. (1947). New species of the genera *Prolasius* Forel and *Melophorus* Lubbock (Hymenoptera : Formicidae). *Mem. Natl. Mus. Vict.* **15**: 7–27 [Oct. 1947] [16 pl 1]. Type data: syntypes (probable), NMV *W, from Millgrove, Vic.

Distribution: SE coastal, N.S.W., Vic. Ecology: terrestrial, noctidiurnal, omnivore, granivore, woodland, open forest; nest in ground layer.

Prolasius clarki McAreavey, 1947

Prolasius clarki McAreavey, J.J. (1947). New species of the genera *Prolasius* Forel and *Melophorus* Lubbock (Hymenoptera : Formicidae). *Mem. Natl. Mus. Vict.* **15**: 7–27 [Oct. 1947] [15 pl 1]. Type data: syntypes, NMV *W,F, from Barrington Tops, N.S.W.

Distribution: SE coastal, N.S.W. Ecology: terrestrial, noctidiurnal, omnivore, woodland, open forest; nest in ground layer.

Prolasius convexus McAreavey, 1947

Prolasius convexus McAreavey, J.J. (1947). New species of the genera *Prolasius* Forel and *Melophorus* Lubbock (Hymenoptera : Formicidae). *Mem. Natl. Mus. Vict.* **15**: 7–27 [Oct. 1947] [15 pl 1]. Type data: syntypes, NMV *W, from Dorrigo, N.S.W.

Distribution: SE coastal, N.S.W. Ecology: terrestrial, noctidiurnal, omnivore, open forest, closed forest; nest in ground layer.

Prolasius depressiceps (Emery, 1914)

Prolasius depressiceps depressiceps (Emery, 1914)

Melophorus depressiceps Emery, C. (1914). Formiche d'Australia e di Samoa raccolte dal Prof. Silvestri nel 1913. *Boll. Lab. Zool. Gen. Agr. R. Scuola Agric. Portici* **8**: 179–186 [30 Jan. 1914] [186]. Type data: syntypes, MCG *W, from Katoomba, N.S.W.

Distribution: SE coastal, NE coastal, Vic., Qld., N.S.W. Ecology: terrestrial, noctidiurnal, omnivore, woodland, open forest; nest in ground layer.

Prolasius depressiceps similis McAreavey, 1947

Prolasius depressiceps similis McAreavey, J.J. (1947). New species of the genera *Prolasius* Forel and *Melophorus* Lubbock (Hymenoptera : Formicidae). *Mem. Natl. Mus. Vict.* **15**: 7–27 [Oct. 1947] [23 pl 1]. Type data: syntypes (probable), NMV *W, from Mt. Kosciusko, N.S.W.

Distribution: Murray-Darling basin, N.S.W. Ecology: terrestrial, noctidiurnal, omnivore, alpine, woodland, open forest; nest in ground layer.

Prolasius flavicornis Clark, 1934

Prolasius flavicornis flavicornis Clark, 1934

Prolasius flavicornis Clark, J. (1934). Ants from the Otway Ranges. *Mem. Natl. Mus. Vict.* **8**: 48–73 [69 pl 4]. Type data: syntypes, NMV *W,F, from Beech Forest, Vic.

Distribution: SE coastal, Vic., N.S.W., Tas. Ecology: terrestrial, noctidiurnal, omnivore, granivore, woodland, open forest; nest in ground layer.

Prolasius flavicornis minor McAreavey, 1947

Prolasius flavicornis minor McAreavey, J.J. (1947). New species of the genera *Prolasius* Forel and *Melophorus* Lubbock (Hymenoptera : Formicidae). *Mem. Natl. Mus. Vict.* **15**: 7–27 [Oct. 1947] [21 pl 1]. Type data: syntypes (probable), NMV *W, from Sherbrooke Forest, Vic.

Distribution: SE coastal, Vic. Ecology: terrestrial, noctidiurnal, omnivore, woodland, open forest; nest in ground layer.

Prolasius flavidiscus McAreavey, 1947

Prolasius flavidiscus McAreavey, J.J. (1947). New species of the genera *Prolasius* Forel and *Melophorus* Lubbock (Hymenoptera : Formicidae). *Mem. Natl. Mus. Vict.* **15**: 7–27 [Oct. 1947] [21 pl 1]. Type data: syntypes, NMV *W,F, from Mt. Ben Cairn, Vic.

Distribution: SE coastal, Vic. Ecology: terrestrial, noctidiurnal, omnivore, woodland, open forest; nest in ground layer.

Prolasius hellenae McAreavey, 1947

Prolasius hellenae McAreavey, J.J. (1947). New species of the genera *Prolasius* Forel and *Melophorus* Lubbock (Hymenoptera : Formicidae). *Mem. Natl. Mus. Vict.* **15**: 7–27 [Oct. 1947] [13 pl 1]. Type data: syntypes (probable), NMV *W, from Katoomba, N.S.W.

Distribution: SE coastal, Vic., N.S.W. Ecology: terrestrial, noctidiurnal, omnivore, granivore, woodland, open forest; nest in ground layer.

Prolasius hemiflavus Clark, 1934

Prolasius hemiflavus hemiflavus Clark, 1934

Prolasius hemiflavus Clark, J. (1934). Ants from the Otway Ranges. *Mem. Natl. Mus. Vict.* **8**: 48–73 [68 pl 4]. Type data: syntypes, NMV *W,F, from Beech Forest, Vic.

Distribution: SE coastal, Murray-Darling basin, N.S.W., Tas., Vic. Ecology: terrestrial, noctidiurnal, omnivore, woodland, open forest; nest in ground layer.

Prolasius hemiflavus wilsoni McAreavey, 1947

Prolasius hemiflavus wilsoni McAreavey, J.J. (1947). New species of the genera *Prolasius* Forel and *Melophorus* Lubbock (Hymenoptera : Formicidae). *Mem. Natl. Mus. Vict.* **15**: 7–27 [Oct. 1947] [18 pl 1]. Type data: syntypes (probable), NMV *W, from Bogong Plains, Vic.

Distribution: Murray-Darling basin, Vic. Ecology: terrestrial, noctidiurnal, omnivore, alpine, woodland, open forest; nest in ground layer.

Prolasius mjoebergella (Forel, 1916)

Prenolepis mjobergi Forel, A. (1915). Results of Dr. E. Mjöbergs Swedish Scientific Expeditions to Australia 1910–1913. 2. Ameisen. *Ark. Zool.* **9**: 1–119 pls 1–3 [4 Dec. 1915] [93] [*non Prenolepis vividula mjobergi* Forel, 1908]. Type data: syntypes, GMNH W, ANIC W, other syntypes may exist, from Malanda, Qld.
Prenolepis mjoebergella Forel, A. *in* Santschi, F. (1916). Rectifications à la nomenclature de quelques formicides [Hym.]. *Bull. Soc. Entomol. Fr.* **1916**: 242–243 [242] [*nom. nov.* for *Prenolepis mjobergi* Forel, 1915].

Distribution: NE coastal, Qld. Ecology: terrestrial, noctidiurnal, omnivore, closed forest; nest in ground layer.

Prolasius niger Clark, 1934

Prolasius niger Clark, J. (1934). Ants from the Otway Ranges. *Mem. Natl. Mus. Vict.* **8**: 48–73 [68 pl 4]. Type data: syntypes, NMV *W, from Beech Forest, Vic.

Distribution: SE coastal, Murray-Darling basin, N.S.W., Vic. Ecology: terrestrial, noctidiurnal, omnivore, woodland, open forest; nest in ground layer.

Prolasius nigriventris McAreavey, 1947

Prolasius nigriventris McAreavey, J.J. (1947). New species of the genera *Prolasius* Forel and *Melophorus* Lubbock (Hymenoptera : Formicidae). *Mem. Natl. Mus. Vict.* **15**: 7–27 [Oct. 1947] [17 pl 1]. Type data: syntypes, NMV *W,M, from Deal Is., Vic.

Distribution: SE coastal, Tas., Vic.; Bass Strait. Ecology: terrestrial, noctidiurnal, omnivore, woodland, open forest; nest in ground layer.

Prolasius nitidissimus (E. André, 1896)

Prolasius nitidissimus nitidissimus (E. André, 1896)

Formica nitidissima André, E. (1896). Fourmis nouvelles d'Asie et d'Australie. *Rev. Entomol.* **15**: 251–265 [255]. Type data: syntypes, MNHP W, ANIC W, from Victorian Alps.

Distribution: Murray-Darling basin, Vic. Ecology: terrestrial, noctidiurnal, omnivore, alpine, woodland, open forest; nest in ground layer.

Prolasius nitidissimus formicoides (Forel, 1902)

Melophorus formicoides Forel, A. (1902). Fourmis nouvelles d'Australie. *Rev. Suisse Zool.* **10**: 405–548 [483]. Type data: syntypes, GMNH W,F, ANIC W, from Mackay, Qld.

Distribution: NE coastal, Qld. Ecology: terrestrial, noctidiurnal, omnivore, woodland, open forest; nest in ground layer.

Prolasius pallidus Clark, 1934

Prolasius pallidus Clark, J. (1934). Ants from the Otway Ranges. *Mem. Natl. Mus. Vict.* **8**: 48–73 [67 pl 4]. Type data: syntypes, NMV *W,F, from Beech Forest, Vic.

Distribution: SE coastal, Murray-Darling basin, N.S.W., Tas., Vic. Ecology: terrestrial, noctidiurnal, omnivore, granivore, woodland, open forest; nest in ground layer.

Prolasius quadratus McAreavey, 1947

Prolasius quadrata McAreavey, J.J. (1947). New species of the genera *Prolasius* Forel and *Melophorus* Lubbock (Hymenoptera : Formicidae). *Mem. Natl. Mus. Vict.* **15**: 7–27 [Oct. 1947] [19 pl 1]. Type data: syntypes, NMV *W, from Mt. Kosciusko, N.S.W.

Distribution: Murray-Darling basin, N.S.W. Ecology: terrestrial, noctidiurnal, omnivore, granivore, alpine, woodland, open forest; nest in ground layer.

Prolasius reticulatus McAreavey, 1947

Prolasius reticulata McAreavey, J.J. (1947). New species of the genera *Prolasius* Forel and *Melophorus* Lubbock (Hymenoptera : Formicidae). *Mem. Natl. Mus. Vict.* **15**: 7–27 [Oct. 1947] [22 pl 1]. Type data: syntypes, NMV *W, from Mundaring, W.A.

Distribution: SW coastal, W.A. Ecology: terrestrial, noctidiurnal, omnivore, woodland, open forest; nest in ground layer.

Prolasius robustus McAreavey, 1947

Prolasius robustus McAreavey, J.J. (1947). New species of the genera *Prolasius* Forel and *Melophorus* Lubbock (Hymenoptera : Formicidae). *Mem. Natl. Mus. Vict.* **15**: 7–27 [Oct. 1947] [20 pl 1]. Type data: syntypes, NMV *W, from Fern Tree Gully, Vic.

Distribution: SE coastal, Murray-Darling basin, N.S.W., Tas., Vic. Ecology: terrestrial, noctidiurnal, omnivore, open forest, closed forest; nest in ground layer.

Prolasius wheeleri McAreavey, 1947

Prolasius wheeleri McAreavey, J.J. (1947). New species of the genera *Prolasius* Forel and *Melophorus* Lubbock (Hymenoptera : Formicidae). *Mem. Natl. Mus. Vict.* **15**: 7–27 [Oct. 1947] [22 pl 1]. Type data: syntypes, NMV *W, from King's Park, Perth, W.A.

Distribution: SW coastal, W.A. Ecology: terrestrial, noctidiurnal, omnivore, woodland, open forest; nest in ground layer.

Pseudolasius Emery, 1887

Pseudolasius Emery, C. (1887). Catalogo delle Formiche esistenti nelle collezioni del Museo Civico di Genova. Parte terza. Formiche della regione Indo-Malese e dell'Australia. *Ann. Mus. Civ. Stor. Nat. Giacomo Doria* **25**: 209–258 [244 pls 3–4]. Type species *Formica familiaris* F. Smith, 1859 by subsequent designation, see Bingham, C.T. (1903). *The Fauna of British India, including Ceylon and Burma.* Hymenoptera. Vol. 2 Ants and cuckoo-wasps. London : Taylor & Francis [337].

This group is also found in the Ethiopian and Oriental regions; New Guinea and east Melanesia in Australian Region, see Brown, W.L. jr. (1973). A comparison of the Hylean and Congo-West African rain forest ant faunas. pp. 161–185 *in* Meggers, B.J., Ayensu, E.S. & Duckworth, W.D. (eds.) *Tropical forest ecosystems in Africa and South America: a comparative review.* Washington : Smithsonian Institution Press.

Pseudolasius australis Forel, 1915

Pseudolasius australis Forel, A. (1915). Results of Dr. E. Mjöbergs Swedish Scientific Expeditions to Australia 1910–1913. 2. Ameisen. *Ark. Zool.* **9**: 1–119 pls 1–3 [4 Dec. 1915] [94]. Type data: syntypes, GMNH W, other syntypes may exist, from Australia.

Distribution: NE coastal, Qld. Ecology: terrestrial, noctidiurnal, omnivore, closed forest; nest in ground layer.

Pseudonotoncus Clark, 1934

Pseudonotoncus Clark, J. (1934). Ants from the Otway Ranges. *Mem. Natl. Mus. Vict.* **8**: 48–73 [64 pl 4]. Type species *Pseudonotoncus hirsutus* Clark, 1934 by original designation.

Pseudonotoncus hirsutus Clark, 1934

Pseudonotoncus hirsutus Clark, J. (1934). Ants from the Otway Ranges. *Mem. Natl. Mus. Vict.* **8**: 48-73 [65 pl 4]. Type data: syntypes, NMV *W,F, from Gellibrand, Vic.

Distribution: SE coastal, Vic. Ecology: terrestrial, noctidiurnal, omnivore, open forest, closed forest; nest in ground layer.

Pseudonotoncus turneri Donisthorpe, 1937

Pseudonotoncus turneri Donisthorpe, H. (1937). Some new forms of Formicidae and a correction. *Ann. Mag. Nat. Hist. (10)* **19**: 619-628 [619]. Type data: holotype, BMNH *W, from Tambourin (=Tamborine) Mt., Qld.

Distribution: NE coastal, Qld. Ecology: terrestrial, noctidiurnal, omnivore, open forest, closed forest; nest in ground layer.

Stigmacros Forel, 1905

Acrostigma Forel, A. (1902). Fourmis nouvelles d'Australie. *Rev. Suisse Zool.* **10**: 405-548 [477] [*non Acrostigma* Emery, 1890; described with subgeneric rank in *Acantholepis* Mayr, 1861]. Type species *Acantholepis (Acrostigma) froggatti* Forel, 1902 by subsequent designation, see Wheeler, W.M. (1911). A list of the type species of the genera and subgenera of Formicidae. *Ann. N.Y. Acad. Sci.* **21**: 157-175 [17 Oct. 1911].
Stigmacros Forel, A. (1905). Miscellanea myrmécolgiques 2 (1905). *Ann. Soc. Entomol. Belg.* **49**: 155-185 [179] [*nom. nov.* for *Acrostigma* Forel, 1902].
Hagiostigmacros McAreavey, J.J. (1957). Revision of the genus *Stigmacros* Forel. *Mem. Natl. Mus. Vict.* **21**: 7-64 [6 Aug. 1957] [19] [proposed with subgeneric rank in *Stigmacros* Forel, 1905]. Type species *Stigmacros barretti* Santschi, 1928 by original designation.
Chariostigmacros McAreavey, J.J. (1957). Revision of the genus *Stigmacros* Forel. *Mem. Natl. Mus. Vict.* **21**: 7-64 [6 Aug. 1957] [23] [proposed with subgeneric rank in *Stigmacros* Forel, 1905]. Type species *Stigmacros (Chariostigmacros) hirsuta* McAreavey, 1957 by original designation.
Pseudostigmacros McAreavey, J.J. (1957). Revision of the genus *Stigmacros* Forel. *Mem. Natl. Mus. Vict.* **21**: 7-64 [6 Aug. 1957] [24] [proposed with subgeneric rank in *Stigmacros* Forel, 1905]. Type species *Stigmacros (Pseudostigmacros) inermis* McAreavey, 1957 by original designation.
Campostigmacros McAreavey, J.J. (1957). Revision of the genus *Stigmacros* Forel. *Mem. Natl. Mus. Vict.* **21**: 7-64 [6 Aug. 1957] [25] [proposed with subgeneric rank in *Stigmacros* Forel, 1905]. Type species *Acantholepis (Stigmacros) aemula* Forel, 1907 by original designation.
Cyrtostigmacros McAreavey, J.J. (1957). Revision of the genus *Stigmacros* Forel. *Mem. Natl. Mus. Vict.* **21**: 7-64 [6 Aug. 1957] [35] [proposed with subgeneric rank in *Stigmacros* Forel, 1905]. Type species *Acantholepis (Acrostigma) australis* Forel, 1902 by original designation.

Synonymy that of Brown, W.L. jr. (1973). A comparison of the Hylean and Congo-West African rain forest ant faunas. pp. 161-185 *in* Meggers, B.J., Ayensu, E.S. & Duckworth, W.D. (eds.) *Tropical forest ecosystems in Africa and South America: a comparative review.* Washington : Smithsonian Institution Press.

Stigmacros aciculata McAreavey, 1957

Stigmacros (Cyrtostigmacros) aciculata McAreavey, J.J. (1957). Revision of the genus *Stigmacros* Forel. *Mem. Natl. Mus. Vict.* **21**: 7-64 [6 Aug. 1957] [50]. Type data: syntypes (probable), NMV *W, from Brisbane, Qld.

Distribution: NE coastal, Qld. Ecology: terrestrial, noctidiurnal, omnivore, woodland, open forest; nest in ground layer.

Stigmacros acuta McAreavey, 1957

Stigmacros (Stigmacros) acuta McAreavey, J.J. (1957). Revision of the genus *Stigmacros* Forel. *Mem. Natl. Mus. Vict.* **21**: 7-64 [6 Aug. 1957] [12]. Type data: syntypes, NMV *W, from Mt. Lofty, S.A.

Distribution: S Gulfs, S.A. Ecology: terrestrial, noctidiurnal, omnivore, woodland, open forest; nest in ground layer.

Stigmacros aemula (Forel, 1907)

Acantholepis (Stigmacros) aemula Forel, A. (1907). Formicidae. pp. 263-310 *in* Michaelsen, W. & Hartmeyer, R. (eds.) *Die Fauna Südwest-Australiens.* Jena : G. Fischer Vol.1 [298]. Type data: holotype, probably destroyed in ZMH in WW II, from Fremantle, W.A.

Distribution: SW coastal, W.A. Ecology: terrestrial, noctidiurnal, omnivore, woodland, open forest; nest in ground layer.

Stigmacros anthracina McAreavey, 1957

Stigmacros (Campostigmacros) anthracina McAreavey, J.J. (1957). Revision of the genus *Stigmacros* Forel. *Mem. Natl. Mus. Vict.* **21**: 7-64 [6 Aug. 1957] [29]. Type data: syntypes (probable), NMV *W, from Mt. Lofty, S.A.

Distribution: S Gulfs, S.A. Ecology: terrestrial, noctidiurnal, omnivore, woodland, open forest; nest in ground layer.

Stigmacros armstrongi McAreavey, 1957

Stigmacros (Cyrtostigmacros) armstrongi McAreavey, J.J. (1957). Revision of the genus *Stigmacros* Forel. *Mem. Natl. Mus. Vict.* **21**: 7-64 [6 Aug. 1957] [52]. Type data: syntypes, NMV *W, from Nyngan, N.S.W.

Distribution: Murray-Darling basin, N.S.W. Ecology: terrestrial, noctidiurnal, omnivore, woodland; nest in ground layer.

Stigmacros australis (Forel, 1902)

Acantholepis (Acrostigma) australis Forel, A. (1902). Fourmis nouvelles d'Australie. *Rev. Suisse Zool.* **10**: 405–548 [479]. Type data: syntypes, GMNH W, ANIC W, from Wollongbar, Richmond River, N.S.W.

Distribution: SE coastal, N.S.W. Ecology: terrestrial, noctidiurnal, omnivore, woodland, open forest; nest in ground layer.

Stigmacros barretti Santschi, 1928

Stigmacros barretti Santschi, F. (1928). Nouvelles fourmis d'Australie. *Bull. Soc. Vaud. Sci. Nat.* **56**: 465–483 [30 Aug. 1928] [477]. Type data: syntypes, BNHM W, from Ringwood, Vic.

Distribution: SE coastal, Murray-Darling basin, N.S.W, Vic. Ecology: terrestrial, noctidiurnal, omnivore, woodland, open forest; nest in ground layer.

Stigmacros bosii (Forel, 1902)

Acantholepis (Acrostigma) bosii Forel, A. (1902). Fourmis nouvelles d'Australie. *Rev. Suisse Zool.* **10**: 405–548 [481]. Type data: syntypes, GMNH W,F, ANIC W, from Queanbeyan, N.S.W.

Distribution: Murray-Darling basin, N.S.W. Ecology: terrestrial, noctidiurnal, omnivore, woodland, open forest; nest in ground layer.

Stigmacros brachytera McAreavey, 1957

Stigmacros (Campostigmacros) brachytera McAreavey, J.J. (1957). Revision of the genus *Stigmacros* Forel. *Mem. Natl. Mus. Vict.* **21**: 7–64 [6 Aug. 1957] [27]. Type data: syntypes, NMV *W,F, from Margaret River, W.A.

Distribution: SW coastal, W.A. Ecology: terrestrial, noctidiurnal, omnivore, woodland, open forest; nest in ground layer.

Stigmacros brevispina McAreavey, 1957

Stigmacros (Stigmacros) brevispina McAreavey, J.J. (1957). Revision of the genus *Stigmacros* Forel. *Mem. Natl. Mus. Vict.* **21**: 7–64 [6 Aug. 1957] [14]. Type data: syntypes, NMV *W, from Bogong Plains, Vic.

Distribution: Murray-Darling basin, Vic. Ecology: terrestrial, noctidiurnal, omnivore, woodland, open forest; nest in ground layer.

Stigmacros brooksi McAreavey, 1957

Stigmacros (Cyrtostigmacros) brooksi McAreavey, J.J. (1957). Revision of the genus *Stigmacros* Forel. *Mem. Natl. Mus. Vict.* **21**: 7–64 [6 Aug. 1957] [42]. Type data: syntypes, NMV *W,F,M, from Manjimup, W.A.

Distribution: SW coastal, W.A. Ecology: terrestrial, noctidiurnal, omnivore, woodland, open forest; nest in ground layer.

Stigmacros castanea McAreavey, 1957

Stigmacros (Cyrtostigmacros) castanea McAreavey, J.J. (1957). Revision of the genus *Stigmacros* Forel. *Mem. Natl. Mus. Vict.* **21**: 7–64 [6 Aug. 1957] [49]. Type data: syntypes, NMV *W,F,M, from Canberra, A.C.T.

Distribution: Murray-Darling basin, N.S.W., A.C.T. Ecology: terrestrial, noctidiurnal, omnivore, woodland, open forest; nest in ground layer.

Stigmacros clarki McAreavey, 1957

Stigmacros (Cyrtostigmacros) clarki McAreavey, J.J. (1957). Revision of the genus *Stigmacros* Forel. *Mem. Natl. Mus. Vict.* **21**: 7–64 [6 Aug. 1957] [41]. Type data: syntypes (probable), NMV *W, from Ludlow, W.A.

Distribution: SW coastal, W.A. Ecology: terrestrial, noctidiurnal, omnivore, woodland, open forest; nest in ground layer.

Stigmacros clivispina (Forel, 1902)

Acantholepis (Acrostigma) clivispina Forel, A. (1902). Fourmis nouvelles d'Australie. *Rev. Suisse Zool.* **10**: 405–548 [482]. Type data: syntypes, GMNH W, ANIC W, from Cooma, N.S.W.

Distribution: Murray-Darling basin, N.S.W. Ecology: terrestrial, noctidiurnal, omnivore, woodland, open forest; nest in ground layer.

Stigmacros elegans McAreavey, 1949

Stigmacros elegans McAreavey, J.J. (1949). Australian Formicidae. New genera and species. *Proc. Linn. Soc. N.S.W.* **74**: 1–25 [15 June 1949] [24]. Type data: holotype, ANIC W, from Nyngan, N.S.W.

Distribution: Murray-Darling basin, N.S.W. Ecology: terrestrial, noctidiurnal, omnivore, woodland, open forest; nest in ground layer.

Stigmacros epinotalis McAreavey, 1957

Stigmacros (Campostigmacros) epinotalis McAreavey, J.J. (1957). Revision of the genus *Stigmacros* Forel. *Mem. Natl. Mus. Vict.* **21**: 7–64 [6 Aug. 1957] [28] [introduced as *Compostigmacros*]. Type data: syntypes, NMV *W, from Booang, W.A.

Distribution: NW coastal, SW coastal, W.A. Ecology: terrestrial, noctidiurnal, omnivore, woodland, open forest; nest in ground layer.

Stigmacros extreminigra McAreavey, 1957

Stigmacros (Cyrtostigmacros) extreminigra McAreavey, J.J. (1957). Revision of the genus *Stigmacros* Forel. *Mem. Natl. Mus. Vict.* **21**: 7–64 [6 Aug. 1957] [48]. Type data: syntypes, NMV *W, from Wyperfeld, Vic.

Distribution: Murray-Darling basin, Vic. Ecology: terrestrial, noctidiurnal, omnivore, woodland, open forest; nest in ground layer.

Stigmacros ferruginea McAreavey, 1957

Stigmacros (Cyrtostigmacros) ferruginea McAreavey, J.J. (1957). Revision of the genus *Stigmacros* Forel. *Mem. Natl. Mus. Vict.* **21**: 7-64 [6 Aug. 1957] [46]. Type data: syntypes, NMV *W, from Mt. Lofty, S.A.

Distribution: S Gulfs, SE coastal, Vic., N.S.W., S.A. Ecology: terrestrial, noctidiurnal, omnivore, woodland, open forest; nest in ground layer.

Stigmacros flava McAreavey, 1957

Stigmacros (Cyrtostigmacros) flava McAreavey, J.J. (1957). Revision of the genus *Stigmacros* Forel. *Mem. Natl. Mus. Vict.* **21**: 7-64 [6 Aug. 1957] [40] [introduced as *Crytostigmacros*]. Type data: syntypes (probable), NMV *W, from Mundaring, W.A.

Distribution: SW coastal, W.A. Ecology: terrestrial, noctidiurnal, omnivore, woodland, open forest; nest in ground layer.

Stigmacros flavinodis Clark, 1938

Stigmacros flavinodis Clark, J. (1938). Reports of the McCoy Society for Field Investigation and Research. No. 2. Sir Joseph Bank Islands. Part I. Formicidae (Hymenoptera). *Proc. R. Soc. Vict.* **50**: 356-382 [375]. Type data: syntypes, NMV *W, from Reevesby Is., S.A.

Distribution: S Gulfs, S.A. Ecology: terrestrial, noctidiurnal, omnivore, woodland, open forest; nest in ground layer.

Stigmacros froggatti (Forel, 1902)

Acantholepis (Acrostigma) froggatti Forel, A. (1902). Fourmis nouvelles d'Australie. *Rev. Suisse Zool.* **10**: 405-548 [478]. Type data: syntypes, GMNH W,F,M, ANIC W, from Bong Bong, N.S.W.

Acantholepis (Stigmacros) fossulata Viehmeyer, H. (1925). Formiciden der australischen Faunenregion. *Entomol. Mitt.* **14**: 25-39 [34]. Type data: syntypes (probable), ZMB *W, from Trial Bay, N.S.W.

Acantholepis (Stigmacros) foreli Viehmeyer, H. (1925). Formiciden der australischen Faunenregion. *Entomol. Mitt.* **14**: 25-39 [34]. Type data: syntypes, ZMB *W,M,F, ANIC W, from Trial Bay, N.S.W.

Synonymy that of McAreavey, J.J. (1957). Revision of the genus *Stigmacros* Forel. *Mem. Natl. Mus. Vict.* **21**: 7-64 [10].

Distribution: SE coastal, N.S.W. Ecology: terrestrial, noctidiurnal, omnivore, woodland, open forest; nest in ground layer.

Stigmacros glauerti McAreavey, 1957

Stigmacros (Cyrtostigmacros) glauerti McAreavey, J.J. (1957). Revision of the genus *Stigmacros* Forel. *Mem. Natl. Mus. Vict.* **21**: 7-64 [6 Aug. 1957] [41]. Type data: syntypes (probable), NMV *W, from Darlington, W.A.

Distribution: SW coastal, W.A. Ecology: terrestrial, noctidiurnal, omnivore, woodland, open forest; nest in ground layer.

Stigmacros hirsuta McAreavey, 1957

Stigmacros (Chariostigmacros) hirsuta McAreavey, J.J. (1957). Revision of the genus *Stigmacros* Forel. *Mem. Natl. Mus. Vict.* **21**: 7-64 [6 Aug. 1957] [23]. Type data: syntypes, NMV *W, from Kuranda, Qld.

Distribution: NE coastal, Qld. Ecology: terrestrial, noctidiurnal, omnivore, open forest, closed forest; nest in ground layer.

Stigmacros impressa McAreavey, 1957

Stigmacros (Stigmacros) impressa McAreavey, J.J. (1957). Revision of the genus *Stigmacros* Forel. *Mem. Natl. Mus. Vict.* **21**: 7-64 [6 Aug. 1957] [14]. Type data: syntypes, NMV *W, from Taggerty, Vic.

Distribution: Murray-Darling basin, Vic. Ecology: terrestrial, noctidiurnal, omnivore, woodland, open forest; nest in ground layer.

Stigmacros inermis McAreavey, 1957

Stigmacros (Pseudostigmacros) inermis McAreavey, J.J. (1957). Revision of the genus *Stigmacros* Forel. *Mem. Natl. Mus. Vict.* **21**: 7-64 [6 Aug. 1957] [24]. Type data: syntypes, NMV *W, from Nyngan, N.S.W.

Distribution: Murray-Darling basin, N.S.W. Ecology: terrestrial, noctidiurnal, omnivore, woodland, open forest; nest in ground layer.

Stigmacros intacta (Viehmeyer, 1925)

Acantholepis (Stigmacros) aemula intacta Viehmeyer, H. (1925). Formiciden der australischen Faunenregion. *Entomol. Mitt.* **14**: 25-39 [34]. Type data: syntypes, ZMB *W, from Trial Bay, N.S.W.

Distribution: SE coastal, N.S.W. Ecology: terrestrial, noctidiurnal, omnivore, woodland, open forest; nest in ground layer. Biological references: McAreavey, J.J. (1957). Revision of the genus *Stigmacros* Forel. *Mem. Natl. Mus. Vict.* **21**: 7-64 (raised to species).

Stigmacros lanaris McAreavey, 1957

Stigmacros (Cyrtostigmacros) lanaris McAreavey, J.J. (1957). Revision of the genus *Stigmacros* Forel. *Mem. Natl. Mus. Vict.* **21**: 7-64 [6 Aug. 1957] [43]. Type data: syntypes, NMV *W,F, from Pymble, N.S.W.

Distribution: SE coastal, N.S.W. Ecology: terrestrial, noctidiurnal, omnivore, woodland, open forest; nest in ground layer.

Stigmacros major McAreavey, 1957

Stigmacros (Cyrtostigmacros) major McAreavey, J.J. (1957). Revision of the genus *Stigmacros* Forel. *Mem. Natl. Mus. Vict.* **21**: 7-64 [6 Aug. 1957] [39]. Type data: syntypes (probable), NMV *W, from National Park", Qld.

Distribution: NE coastal, Qld. Ecology: terrestrial, noctidiurnal, omnivore, woodland, open forest; nest in ground layer.

Stigmacros marginata McAreavey, 1957

Stigmacros (Campostigmacros) marginata McAreavey, J.J. (1957). Revision of the genus *Stigmacros* Forel. *Mem. Natl. Mus. Vict.* **21**: 7–64 [6 Aug. 1957] [27]. Type data: syntypes (probable), NMV *W, from Gosford, N.S.W.

Distribution: SE coastal, Vic., N.S.W. Ecology: terrestrial, noctidiurnal, omnivore, woodland, open forest; nest in ground layer.

Stigmacros medioreticulata (Viehmeyer, 1925)

Acantholepis (Stigmacros) medioreticulata Viehmeyer, H. (1925). Formiciden der australischen Faunenregion. *Entomol. Mitt.* **14**: 25–39 [32]. Type data: holotype, ZMB *W, from Trial Bay, N.S.W.

Distribution: SE coastal, N.S.W. Ecology: terrestrial, noctidiurnal, omnivore, woodland, open forest; nest in ground layer.

Stigmacros minor McAreavey, 1957

Stigmacros (Stigmacros) minor McAreavey, J.J. (1957). Revision of the genus *Stigmacros* Forel. *Mem. Natl. Mus. Vict.* **21**: 7–64 [6 Aug. 1957] [17]. Type data: syntypes, NMV *W, from Brisbane, Qld.

Distribution: NE coastal, Qld. Ecology: terrestrial, noctidiurnal, omnivore, woodland, open forest; nest in ground layer.

Stigmacros nitida McAreavey, 1957

Stigmacros (Campostigmacros) nitida McAreavey, J.J. (1957). Revision of the genus *Stigmacros* Forel. *Mem. Natl. Mus. Vict.* **21**: 7–64 [6 Aug. 1957] [30]. Type data: syntypes, NMV *W, from Fern Tree Gully, Vic.

Distribution: SE coastal, Murray-Darling basin, N.S.W., Vic. Ecology: terrestrial, noctidiurnal, omnivore, woodland, open forest; nest in ground layer.

Stigmacros occidentalis (Crawley, 1922)

Acantholepis (Stigmacros) occidentalis Crawley, W.C. (1922). New ants from Australia. *Ann. Mag. Nat. Hist. (9)* **10**: 16–36 [30]. Type data: syntypes (probable), OUM *W, from Murray River, W.A.

Distribution: SW coastal, W.A. Ecology: terrestrial, noctidiurnal, omnivore, woodland, open forest; nest in ground layer.

Stigmacros pilosella (Viehmeyer, 1925)

Acantholepis (Stigmacros) pilosella Viehmeyer, H. (1925). Formiciden der australischen Faunenregion. *Entomol. Mitt.* **14**: 25–39 [33]. Type data: holotype, ZMB *W, from Liverpool, N.S.W.

Distribution: SE coastal, N.S.W. Ecology: terrestrial, noctidiurnal, omnivore, woodland, open forest; nest in ground layer.

Stigmacros proxima McAreavey, 1957

Stigmacros (Cyrtostigmacros) proxima McAreavey, J.J. (1957). Revision of the genus *Stigmacros* Forel. *Mem. Natl. Mus. Vict.* **21**: 7–64 [6 Aug. 1957] [51]. Type data: syntypes, NMV *W, from Athol, N.S.W.

Distribution: Murray-Darling basin, N.S.W. Ecology: terrestrial, noctidiurnal, omnivore, woodland, open forest; nest in ground layer.

Stigmacros punctatissima McAreavey, 1957

Stigmacros (Hagiostigmacros) punctatissima McAreavey, J.J. (1957). Revision of the genus *Stigmacros* Forel. *Mem. Natl. Mus. Vict.* **21**: 7–64 [6 Aug. 1957] [22]. Type data: syntypes, NMV *W, from Leura, N.S.W.

Distribution: SE coastal, N.S.W. Ecology: terrestrial, noctidiurnal, omnivore, woodland, open forest; nest in ground layer.

Stigmacros pusilla McAreavey, 1957

Stigmacros (Stigmacros) pusilla McAreavey, J.J. (1957). Revision of the genus *Stigmacros* Forel. *Mem. Natl. Mus. Vict.* **21**: 7–64 [6 Aug. 1957] [16]. Type data: syntypes, NMV *W, from Canberra, A.C.T.

Distribution: Murray-Darling basin, N.S.W., A.C.T. Ecology: terrestrial, noctidiurnal, omnivore, woodland, open forest; nest in ground layer.

Stigmacros rectangularis McAreavey, 1957

Stigmacros (Stigmacros) rectangularis McAreavey, J.J. (1957). Revision of the genus *Stigmacros* Forel. *Mem. Natl. Mus. Vict.* **21**: 7–64 [6 Aug. 1957] [15]. Type data: syntypes, NMV *W,M, from Mundaring, W.A.

Distribution: SW coastal, W.A. Ecology: terrestrial, noctidiurnal, omnivore, woodland, open forest; nest in ground layer.

Stigmacros reticulata Clark, 1930

Stigmacros reticulata Clark, J. (1930). Some new Australian Formicidae. *Proc. R. Soc. Vict.* **42**: 116–128 [10 Mar. 1930] [127]. Type data: syntypes, NMV *W,F, from Perth, W.A.

Distribution: SW coastal, W.A. Ecology: terrestrial, noctidiurnal, omnivore, woodland, open forest; nest in ground layer.

Stigmacros rufa McAreavey, 1957

Stigmacros (Stigmacros) rufa McAreavey, J.J. (1957). Revision of the genus *Stigmacros* Forel. *Mem. Natl. Mus. Vict.* **21**: 7–64 [6 Aug. 1957] [13]. Type data: syntypes (probable), NMV *W, from Kallista, Vic.

Distribution: SE coastal, Vic. Ecology: terrestrial, noctidiurnal, omnivore, woodland, open forest; nest in ground layer.

Stigmacros sordida McAreavey, 1957

Stigmacros (Cyrtostigmacros) sordida McAreavey, J.J. (1957). Revision of the genus *Stigmacros* Forel. *Mem. Natl. Mus. Vict.* **21**: 7–64 [6 Aug. 1957] [52]. Type data: syntypes (probable), NMV *W, from Adelaide, S.A.

Distribution: S Gulfs, S.A. Ecology: terrestrial, noctidiurnal, omnivore, woodland, open forest; nest in ground layer.

Stigmacros spinosa McAreavey, 1957

Stigmacros (Hagiostigmacros) spinosa McAreavey, J.J. (1957). Revision of the genus *Stigmacros* Forel. *Mem. Natl. Mus. Vict.* **21**: 7–64 [6 Aug. 1957] [19]. Type data: syntypes, NMV *W,F, from Nyngan, N.S.W.

Distribution: Murray-Darling basin, A.C.T., N.S.W. Ecology: terrestrial, noctidiurnal, omnivore, woodland, open forest; nest in ground layer.

Stigmacros stanleyi McAreavey, 1957

Stigmacros (Campostigmacros) stanleyi McAreavey, J.J. (1957). Revision of the genus *Stigmacros* Forel. *Mem. Natl. Mus. Vict.* **21**: 7–64 [6 Aug. 1957] [34]. Type data: syntypes (probable), NMV *W, from Greensborough, Vic.

Distribution: SE coastal, N.S.W., Vic. Ecology: terrestrial, noctidiurnal, omnivore, woodland, open forest; nest in ground layer.

Stigmacros striata McAreavey, 1957

Stigmacros (Cyrtostigmacros) striata McAreavey, J.J. (1957). Revision of the genus *Stigmacros* Forel. *Mem. Natl. Mus. Vict.* **21**: 7–64 [6 Aug. 1957] [38] [introduced as *Crytostigmacros*]. Type data: syntypes, NMV *W,F,M, from Hornsby, N.S.W.

Distribution: SE coastal, N.S.W. Ecology: terrestrial, noctidiurnal, omnivore, woodland, open forest; nest in ground layer.

Stigmacros termitoxenus Wheeler, 1936

Stigmacros termitoxenus Wheeler, W.M. (1936). Ecological relations of ponerine and other ants to termites. *Proc. Am. Acad. Arts Sci.* **71**: 159–243 [215]. Type data: syntypes, MCZ *W,F, from Mullewa, W.A.

Distribution: NW coastal, W.A. Ecology: terrestrial, noctidiurnal, omnivore, woodland, open forest; nest in ground layer.

Stigmacros wilsoni McAreavey, 1957

Stigmacros (Stigmacros) wilsoni McAreavey, J.J. (1957). Revision of the genus *Stigmacros* Forel. *Mem. Natl. Mus. Vict.* **21**: 7–64 [6 Aug. 1957] [11]. Type data: syntypes, NMV *W, from Cobunga (=Cobungra), Vic.

Distribution: Murray-Darling basin, N.S.W., Vic. Ecology: terrestrial, noctidiurnal, omnivore, woodland, open forest; nest in ground layer.

Teratomyrmex McAreavey, 1957

Teratomyrmex McAreavey, J.J. (1957). Revision of the genus *Stigmacros* Forel. *Mem. Natl. Mus. Vict.* **21**: 7–64 [6 Aug. 1957] [54]. Type species *Teratomyrmex greavesi* McAreavey, 1957 by original designation.

Teratomyrmex greavesi McAreavey, 1957

Teratomyrmex greavesi McAreavey, J.J. (1957). Revision of the genus *Stigmacros* Forel. *Mem. Natl. Mus. Vict.* **21**: 7–64 [6 Aug. 1957] [55]. Type data: syntypes, NMV *W, from Blackall Range, Qld.

Distribution: NE coastal, Qld. Ecology: terrestrial, noctidiurnal, omnivore, closed forest; nest in ground layer.

Incertae sedis

Formica amyoti Le Guillou, E.J.F. (1841). Catalogue raisonné des insectes hyménoptères recueillis dans le voyage de circumnavigation des corvettes l'*Astrolabe* et la *Zélée*. *Ann. Soc. Entomol. Fr.* **10**: 311–324 [315]. Type data: syntypes (probable), MNHP (probable) *W, from northern Australia.

Ponera oculata Smith, F. (1858). *Catalogue of hymenopterous insects in the collection of the British Museum. Part 6. Formicidae.* London : British Museum 216 pp. 14 pls [27 Mar. 1858] [93]. Publication date established from Donisthorpe, H. (1932). On the identity of Smith's types of Formicidae (Hymenoptera) collected by Alfred Russell Wallace in the Malay Archipelago, with descriptions of two new species. *Ann. Mag. Nat. Hist. (10)* **10**: 441–476. Type data: syntypes (probable), BMNH *M, from Macintyre, N.S.W.

Formica inequalis Lowne, B.T. (1865). Contributions to the natural history of Australian ants. *Entomologist* **2**: 331–336 [331]. Type data: syntypes, BMNH (probable) *M,F, from Sidney (=Sydney), N.S.W.

Formica minuta Lowne, B.T. (1865). Contributions to the natural history of Australian ants. *Entomologist* **2**: 331–336 [331]. Type data: syntypes (probable), BMNH (probable) *W, from Sidney (=Sydney), N.S.W.

Formica purpurescens Lowne, B.T. (1865). Contributions to the natural history of Australian ants. *Entomologist* **2**: 331–336 [331]. Type data: syntypes, BMNH (probable) *W,F, from Sidney (=Sydney), N.S.W.

VESPOIDEA AND SPHECOIDEA

Josephine C. Cardale

INTRODUCTION

The Sphecoidea and Vespoidea are among the largest and most conspicuous aculeate Hymenoptera. The habits of the Vespidae (papernest wasps, hornets) and of some of the mud-nest builders bring them into direct conflict with man, but they are also useful, as biological control agents (preying on other insects, especially larval Lepidoptera) and as potential pollination agents.

As predatory wasps, the females collect insects or other arthropods to feed their larvae, except for the Masaridae where it appears that most species provision their nest cells with pollen and nectar. Adults of some species feed on the body fluids of their prey, but in most species the adults require carbohydrates, usually taken as nectar, but sometimes as honeydew or plant sap. Except for the family Vespidae, these wasps are solitary. In general, after mating, each female constructs a cell (in a burrow in the soil, a previously existing cavity or specially-built nest), lays an egg before or after provisioning the cell, seals the cell, and commences another cell. Large nesting aggregations may be formed, especially in soil-nesting species, but these aggregations are not social. The social wasps (Vespidae) show cooperation and at least some division of labour occurs between females (mothers and daughters, or sisters) in the construction and provisioning of their "paper" nests. Larvae are fed progressively and the cell is not sealed until the larva is ready to pupate. Some species of Sphecidae also show "subsocial" behaviour: communal nesting or progressive feeding.

Study of the diversity and complexity of behaviour during nest construction and provisioning among these wasps has been undertaken both in the field and laboratory. Studies of the interactions between individuals and division of labour among subsocial and social wasps have contributed to the understanding of the organisation of insect societies and the development of social behaviour. There are, however, comparatively few Australian species whose biology is known, and behavioural research here is hindered by problems in identifying species. Although these two superfamilies are among the best known of the Australian wasps, the identity of many species is uncertain. The presence of much type material, often single specimens, in museums outside Australia, the need for redescription of species associated with some of the early nomenclature and description of the unnamed material found in virtually every museum collection has hindered the work essential to a better knowledge of these wasps. A particularly striking example of the problems facing students of Australian wasps is shown by *Bembix*, a genus of comparatively large, conspicuous species. Although Evans and Matthews (1973) were interested in comparative behaviour, they found that it was first necessary to study the systematics of the genus. Prior to their work it was believed that there were about 35 species of *Bembix* in Australia; their revision recognised 80 species, 55 of which were described as new. Since then, Evans (1982) has described two new species.

In many species adults emerge, mate and build nests in a few weeks. This short period of activity, which is related to the availability of flowers for nectar, water for nest building, and suitable prey for their larvae, makes systematic collection of the species of any given area quite difficult. The climatic extremes in Australia of drought or flood may be the most significant factor controlling reproduction among these wasps.

The first description of Hymenoptera from Australia was by Fabricius (1775). The only published catalogue of Australian Hymenoptera is that of Froggatt (1891, 1892); Australian species have been included in various world catalogues, such as those of Dalla Torre (1894, 1897) and the *Genera Insectorum* series (Dalla Torre, 1904). The modern *Hymenopterorum Catalogus* series covers comparatively few families as yet though the Palaearctic Eumenidae has been catalogued by van der Vecht and Fischer (1972). Thus, for anyone starting research on Australian wasps, extensive library work is needed to discover which species have been described or recorded from Australia, what revisionary work, if any, has been done and whether any biological studies have been made. The Australian National Insect Collection card catalogue containing all references to Australian Hymenoptera, the revision of Masaridae by Richards (1962), the revision of Australian Vespidae by Richards (1978) and the world generic revision of the Sphecidae by Bohart and Menke (1976) were used as a framework for the compilation of this portion of the *Zoological Catalogue of Australia*. The absence of a generic revision of the Eumenidae has meant that this section lacks the authoritative structuring of the remainder of the catalogue.

CLASSIFICATION

A conservative classification, as used by Riek (1970) and Riek & Cardale (1974), has been followed here; the Vespoidea is divided into the families Masaridae, Vespidae and Eumenidae. Only one family, the Sphecidae, is included in the Sphecoidea, following Bohart and Menke (1976) who treated the Ampulicinae as a subfamily of the Sphecidae. Although the Australian genus *Sericogaster* Westwood, 1835 was believed by early workers to belong in the Sphecidae, it was transferred to the Colletidae (Apoidea) by Menke & Michener (1973) and is not included here.

Carpenter (1982) has proposed a reclassification of the Vespoidea, with the single family Vespidae divided into six subfamilies. Under this scheme, Australian species fall into the subfamilies Masarinae, Eumeninae and Polistinae, with *Vespula* in the Vespinae. Brothers (1975) reviewed the phylogeny and classification of aculeate Hymenoptera and divided them into three superfamilies: the Bethyloidea, the Vespoidea including the Scolioidea, Pompiloidea and Formicoidea of Riek (1970) and the Sphecoidea including Apoidea of Riek (1970).

NOTES ON THE CATALOGUE

Family Introductions
A brief account is given of the size and biology of each family, concluding with the major Australian taxonomic reference and general biological references.

Generic Arrangement
The arrangement within each family, except the Eumenidae, follows that of the most recent revision. In the Eumenidae the genera are listed in alphabetical order.

Name Combinations
Species names are usually listed in the latest published combination; previously used combinations are given after the reference in which they first occur.

Synonymy
Full details are given for all Australian genera and species for which synonymies occur and a reference is given in which these details occur.

Type Information
This is given almost exclusively for holotypes; details of "allotypes" and paratypes will be found by consulting the references. The listed information is based on the original publication, notes in revisionary works and, for specimens in Australian museums, from labels or information supplied by curators (see Acknowledgements). It is possible that a few of the holotypes listed here may not have that status. Types in non-Australian collections were not checked though many of the holotypes in the British Museum (Natural History) were seen by L.F. Graham in 1929 and this information is noted where appropriate in this catalogue.

Date of Publication

The date, or order of publication, listed by Musgrave (1932) is used.

Distribution

The species distributions are from published data; where five or fewer localities were published, they are listed in full. It should be noted that the A.C.T. is seldom mentioned among the States because there are few published distribution records; many of the species recorded from the Murray-Darling basin, N.S.W., occur in the A.C.T. On older labels, "Tasmania" is often erroneous, the species being found only on mainland Australia. "New Guinea is used in the zoogeographic sense; many of the references to distribution pre-date the formation of Papua New Guinea and West Irian. Some non-Australian localities have been given their modern names (or spelling) but it was not possible to check all of them.

Ecological Information

Comparatively little is known of the ecology of these wasps except for a few well studied species. All available biological references, including redescriptions and taxonomic decisions, are listed for most species.

ACKNOWLEDGEMENTS

Preparation of this catalogue was undertaken as part of my work in the Australian National Insect Collection (A.N.I.C.), Division of Entomology, C.S.I.R.O., Canberra, with full use of their resources and facilities, and several members of the staff have assisted me. Dr E.F. Riek, formerly of the A.N.I.C., instigated my work on the card catalogue; Dr I.D. Naumann has supported and encouraged publication of this compilation; Dr K.H.L. Key gave advice on a nomenclatural problem; Dr M.J. Dallwitz developed a computer prompts program (ENCAT) especially for use in this work; Mrs J.E. Pyke has been of great assistance in solving problems connected with data entry.

The Australian Biological Resources Study provided funds for assistance during the preparation of the catalogue, and I am grateful to Dr D.W. Walton and Dr B. Richardson (Bureau of Flora and Fauna) for their editorial advice.

The assistance of the following is also acknowledged with gratitude: Dr J. van der Vecht (Putten, The Netherlands) provided advice on taxonomic problems in the Eumenidae; Dr M.C. Day (BMNH) and Dr H.E. Evans (Colorado State University) assisted in problems in the Vespidae and Sphecidae; information on the types of Vespoidea and Sphecoidea in their collections was provided by Mr E.C. Dahms and Ms G. Sarnes (QM), Dr E.G. Matthews (SAM), Dr T.F. Houston (WAM), Mr K.L. Walker (NMV), and Mr G.R. Brown (DARI); Mr G.A. Holloway (AM) provided assistance during my visits to that institution.

J.C.C.

References

Bohart, R.M. & Menke, A.S. (1976). *Sphecid Wasps of the World : a Generic Revision.* Berkeley : Univ. California Press ix 695 pp.

Brothers, D.J. (1975). Phylogeny and classification of the aculeate Hymenoptera, with special reference to the Mutillidae. *Univ. Kansas Sci. Bull.* **50**: 483–648

Carpenter, J.M. (1982). The phylogenetic relationships and natural classification of the Vespoidea (Hymenoptera). *Syst. Entomol.* **7**: 11–38

Dalla Torre, K.W. (1894). *Catalogus Hymenopterorum Hucusque Descriptorum Systematicus et Synonymicus.* Vol. 9 Vespidae (Diploptera). Leipzig : G. Engelmann 181 pp.

Dalla Torre, K.W. (1897). *Catalogus Hymenopterorum Hucusque Descriptorum Systematicus et Synonymicus.* Vol. 8 Fossores (Sphegidae). Leipzig : G. Engelmann viii 749 pp.

Dalla Torre, K.W. (1904). Hymenoptera. Fam. Vespidae. *Genera Insectorum* **19**: 1–108

Evans, H.E. (1982). Two new species of Australian *Bembix* sand wasps, with notes on other species of the genus (Hymenoptera, Sphecidae). *Aust. Entomol. Mag.* **9**: 7–12

Evans, H.E. & Matthews, R.W. (1973). Systematics and nesting behavior of Australian *Bembix* sand wasps (Hymenoptera, Sphecidae). *Mem. Am. Entomol. Inst.* **20**: iv 387 pp.

Fabricius, J.C. (1775). *Systema Entomologiae,* sistens insectorum classes, ordines, genera, species, adiectis synonymis, locis, descriptionibus, observationibus. Flensburgi et Lipsiae : Kortii xxvii 832 pp.

Froggatt, W.W. (1891). Catalogue of the described Hymenoptera of Australia. Part I. *Proc. Linn. Soc. N.S.W. (2)* **5**: 689–762

Froggatt, W.W. (1892). Catalogue of the described Hymenoptera of Australia. Part II. *Proc. Linn. Soc. N.S.W. (2)* **7**: 205–248

Menke, A.S. & Michener, C.D. (1973). *Sericogaster* Westwood, a senior synonym of *Holohesma* Michener. *J. Aust. Entomol. Soc.* **12**: 173–174

Musgrave, A. (1932). *Bibliography of Australian Entomology 1775–1930* with biographical notes on authors and collectors. Sydney : Royal Zoological Society of New South Wales viii 380 pp.

Richards, O.W. (1962). *A Revisional Study of the Masarid Wasps (Hymenoptera, Vespoidea).* London : British Museum 294 pp.

Richards, O.W. (1978). The Australian social wasps (Hymenoptera : Vespidae). *Aust. J. Zool. Suppl. Ser.* **61**: 1–132

Riek, E.F. (1970). Hymenoptera. pp. 867–943 *in* CSIRO (1970). *The Insects of Australia.* A textbook for students and research workers. Melbourne : Melbourne Univ. Press

Riek, E.F. & Cardale, J.C. (1974). Hymenoptera. pp. 107–109 *in* CSIRO. (1974). *The Insects of Australia.* A textbook for students and research workers. Supplement 1974. Melbourne : Melbourne Univ. Press

van der Vecht, J. & Fischer, F.C.J. (1972). Palaearctic Eumenidae. *Hymenopterorum Catalogus* **8**: 1–199

MASARIDAE

INTRODUCTION

This family contains medium to large sized, mostly non-predatory, solitary wasps, found on all continents, though the species are usually rare. Their distribution within Australia seems to have been reduced as settlement spread. There are four endemic genera with 28 included species and subspecies, with greatest diversity (and most of the recently collected specimens) found in arid areas. Adults are usually collected on flowers or at water. It appears that the Australian species provision their cells with nectar and pollen. Probably all nest in burrows in the ground though some non-Australian species construct mud-cells attached to trees and shrubs. Very little is known about their biology in Australia, apart from the species studied by Houston (1984).

References

Richards, O.W. (1962). *A Revisional Study of the Masarid Wasps (Hymenoptera, Vespoidea)*. London : British Museum 294 pp.

Houston, T.F. (1984). Bionomics of a pollen-collecting wasp, *Paragia tricolor* Smith (Hymenoptera : Masaridae), in Western Australia. *Rec. W. Aust. Mus.* **11**: 141–151.

Riekia Richards, 1962

Riekia Richards, O.W. (1962). *A Revisional Study of the Masarid Wasps (Hymenoptera, Vespoidea)*. London : British Museum 294 pp. [54]. Type species *Riekia nocatunga* Richards, 1962 by monotypy and original designation.

Riekia angulata Richards, 1968

Riekia angulata Richards, O.W. (1968). New records and new species of Australian Masaridae (Hymenoptera : Vespoidea). *J. Aust. Entomol. Soc.* **7**: 101–104 [101]. Type data: holotype, ANIC F. adult, from Cunnamulla, Qld.

Distribution: Murray-Darling basin, Qld., N.S.W.; only published localities Cunnamulla and 90 km W of Cobar. Ecology: larva - sedentary, soil, mellivore : adult - volant, burrowing.

Riekia nocatunga Richards, 1962

Riekia nocatunga Richards, O.W. (1962). *A Revisional Study of the Masarid Wasps (Hymenoptera, Vespoidea)*. London : British Museum 294 pp. [55]. Type data: holotype, ANIC F. adult, from 7 miles N of Nocatunga, N.S.W." in original description, should be Nockatunga, Qld.

Distribution: Bulloo River basin, Murray-Darling basin, Qld., N.S.W.; only published localities Nockatunga, Bourke and Cobar area. Ecology: larva - sedentary, soil, mellivore : adult - volant, burrowing. Biological references: Richards, O.W. (1968). New records and new species of Australian Masaridae (Hymenoptera : Vespoidea). *J. Aust. Entomol. Soc.* **7**: 101–104 (flower record, locality).

Rolandia Richards, 1962

Rolandia Richards, O.W. (1962). *A Revisional Study of the Masarid Wasps (Hymenoptera, Vespoidea)*. London : British Museum 294 pp. [57]. Type species *Paragia maculata* Meade-Waldo, 1910 by monotypy and original designation.

Rolandia maculata (Meade-Waldo, 1910)

Paragia maculata Meade-Waldo, G. (1910). New species of Diploptera in the collection of the British Museum. *Ann. Mag. Nat. Hist.* (8) **5**: 30–51 [32]. Type data: holotype, BMNH Hym.18.23 *F. adult (seen 1929 by L.F. Graham), from W.A.

Distribution: SW coastal, W.A.; only published localities Yallingup, Perth and Yanchep. Ecology: larva - sedentary, soil, mellivore : adult - volant, burrowing. Biological references: Bequaert, J.C. (1928). A study of certain types of diplopterous wasps in the collection of the British Museum. *Ann. Mag. Nat. Hist. (10)* **2**: 138–176 (note on type); Richards, O.W. (1962). *A Revisional Study of the Masarid Wasps (Hymenoptera, Vespoidea).* London : British Museum 294 pp. (redescription).

Metaparagia Meade-Waldo, 1911

Metaparagia Meade-Waldo, G. (1911). Notes on the family Masaridae (Hymenoptera), with descriptions of a new genus and three new species. *Ann. Mag. Nat. Hist. (8)* **8**: 747–750 [748]. Type species *Paragia pictifrons* Smith, 1857 by original designation.

Metaparagia doddi Meade-Waldo, 1911

Metaparagia doddi Meade-Waldo, G. (1911). Notes on the family Masaridae (Hymenoptera), with descriptions of a new genus and three new species. *Ann. Mag. Nat. Hist. (8)* **8**: 747–750 [748]. Type data: holotype, BMNH Hym.18.22 *F. adult (seen 1929 by L.F. Graham), from Cairns, Qld.

Distribution: NE coastal, Qld.; type locality only. Ecology: larva - sedentary, soil, mellivore : adult - volant, burrowing. Biological references: Richards, O.W. (1962). *A Revisional Study of the Masarid Wasps (Hymenoptera, Vespoidea).* London : British Museum 294 pp. (redescription).

Metaparagia pictifrons (Smith, 1857)

Paragia pictifrons Smith, F. (1857). *Catalogue of Hymenopterous Insects in the Collection of the British Museum. Part V. Vespidae* pp. 1–147 London : British Museum [2 pl I fig 1]. Type data: holotype, BMNH Hym.18.21 *F. adult (seen 1929 by L.F. Graham), from Swan River, W.A.

Distribution: SW coastal, W.A.; type locality only. Ecology: larva - sedentary, soil, mellivore : adult - volant, burrowing. Biological references: Richards, O.W. (1962). *A Revisional Study of the Masarid Wasps (Hymenoptera, Vespoidea).* London : British Museum 294 pp. (redescription).

Paragia Shuckard, 1838

Taxonomic decision of Richards, O.W. (1962). *A Revisional Study of the Masarid Wasps (Hymenoptera, Vespoidea).* London : British Museum 294 pp.

Paragia (Cygnaea) Richards, 1962

Cygnaea Richards, O.W. (1962). *A Revisional Study of the Masarid Wasps (Hymenoptera, Vespoidea).* London : British Museum 294 pp. [53] [proposed with subgeneric rank in *Paragia* Shuckard, 1838]. Type species *Paragia vespiformis* Smith, 1865 by monotypy and original designation.

Paragia (Cygnaea) vespiformis Smith, 1865

Paragia vespiformis Smith, F. (1865). Descriptions of some new species of hymenopterous insects belonging to the families Thynnidae, Masaridae, and Apidae. *Trans. Entomol. Soc. Lond. (3)* **2**: 389–399 [393]. Type data: holotype, BMNH Hym.18.5 *F. adult (seen 1929 by L.F. Graham), from Swan River, W.A.

Distribution: SW coastal, NW coastal, W.A.; only published localities Swan River, Wurrarga (presumably Wurarga is meant) and Champion Bay. Ecology: larva - sedentary, soil, mellivore : adult - volant, burrowing. Biological references: Smith, F. (1868). Descriptions of aculeate Hymenoptera from Australia. *Trans. Entomol. Soc. Lond.* **1868**: 231–258 (description of male); Smith, F. (1869). Descriptions of new genera and species of exotic Hymenoptera. *Trans. Entomol. Soc. Lond.* **1869**: 301–311 pl vi (illustration); Richards, O.W. (1962). *A Revisional Study of the Masarid Wasps (Hymenoptera, Vespoidea).* London : British Museum 294 pp. (redescription).

Paragia (Paragia) Shuckard, 1838

Paragia Shuckard, W.E. (1838). Descriptions of new exotic aculeate Hymenoptera. *Trans. Entomol. Soc. Lond.* **2**(1): 68–82 pl viii [81]. Type species *Paragia decipiens* Shuckard, 1838 by monotypy.

Paragia (Paragia) decipiens Shuckard, 1838

Taxonomic decision of Richards, O.W. (1962). *A Revisional Study of the Masarid Wasps (Hymenoptera, Vespoidea).* London : British Museum 294 pp. [63].

Paragia (Paragia) decipiens decipiens Shuckard, 1838

Paragia decipiens Shuckard, W.E. (1838). Descriptions of new exotic aculeate Hymenoptera. *Trans. Entomol. Soc. Lond.* **2**(1): 68–82 pl viii [82 pl viii fig 3]. Type data: holotype, BMNH *F. adult (seen 1929 by L.F. Graham), from N.S.W.

Distribution: Murray-Darling basin, S Gulfs, Qld., N.S.W., Vic., S.A.; only published localities Qld., Lachlan River (Coopers Bridge), Lake Hattah and Adelaide. Ecology: larva - sedentary, soil, mellivore : adult - volant, burrowing. Biological references: Smith, F. (1865). Descriptions of some new species of hymenopterous insects belonging to the families Thynnidae, Masaridae, and Apidae. *Trans. Entomol. Soc. Lond. (3)* **2**: 389–399 (description of male); Richards, O.W. (1968). New records and new species of Australian Masaridae (Hymenoptera : Vespoidea). *J. Aust. Entomol. Soc.* **7**: 101–104 (locality, behaviour).

Paragia (Paragia) decipiens aliciae Richards, 1962

Paragia (Paragia) decipiens aliciae Richards, O.W. (1962). *A Revisional Study of the Masarid Wasps (Hymenoptera, Vespoidea).* London : British Museum 294 pp. [64]. Type data: holotype, ANIC F. adult, from Alice Springs, N.T.

Distribution: Murray-Darling basin, Lake Eyre basin, Qld., N.T.; only published localities Cunnamulla and Alice Springs. Ecology: larva - sedentary, soil, mellivore : adult - volant, burrowing.

Paragia (Paragia) schulthessi Turner, 1936

Paragia schulthessi Turner, R.E. (1936). A new masarid wasp from Australia. *Ann. Mag. Nat. Hist. (10)* **18**: 352. Type data: holotype, BMNH Hym.18.4 *F. adult, from Dedari, W.A.

Distribution: W plateau, W.A.; type locality only. Ecology: larva - sedentary, soil, mellivore : adult - volant, burrowing. Biological references: Richards, O.W. (1962). *A Revisional Study of the Masarid Wasps (Hymenoptera, Vespoidea)*. London : British Museum 294 pp. (redescription).

Paragia (Paragia) smithii Saussure, 1854

Paragia smithii Saussure, H. de (1854). *Études sur la Famille des Vespides*. Troisième Partie comprenant la Monographie des Masariens et un Supplement à la Monographie des Eumeniens. Paris : Masson 352 pp. pls i-xvi (1854-1856) [55 pl II figs 1,1a]. Type data: holotype, BMNH *M. adult (seen 1929 by L.F. Graham), from Adelaide, S.A.

Distribution: SE coastal, Murray-Darling basin, Lake Eyre basin, S Gulfs, Vic., S.A., (W.A.), N.T.; only published localities Melbourne, Lake Hattah, Adelaide, Hermannsburg, records from "W. Australia" and "Africa" are probably incorrect. Ecology: larva - sedentary, soil, mellivore : adult - volant, burrowing. Biological references: Bequaert, J.C. (1928). A study of certain types of diplopterous wasps in the collection of the British Museum. *Ann. Mag. Nat. Hist. (10)* **2**: 138-176 (taxonomy); Richards, O.W. (1962). *A Revisional Study of the Masarid Wasps (Hymenoptera, Vespoidea)*. London : British Museum 294 pp. (nest, redescription, distribution).

Paragia (Paragia) tricolor Smith, 1850

Paragia tricolor Smith, F. (1850). Descriptions of two new species of exotic Hymenoptera. *Trans. Entomol. Soc. Lond. (2)* **1**: 41-42 pl v [41 pl v figs 1e-k]. Type data: lectotype, BMNH Hym.18.3 *F. adult (seen 1929 by L.F. Graham), from Perth, W.A., designation by Saussure, H. de (1854-1856). *Études sur la Famille des Vespides*. Troisième Partie comprenant la Monographie des Masariens et un Supplement à la Monographie des Eumeniens. Paris : Masson 352 pp. pls i-xvi.

Paragia saussurii Smith, F. (1857). *Catalogue of Hymenopterous Insects in the Collection of the British Museum*. Part V. Vespidae pp. 1-147 London : British Museum [2] [unnecessary *nom. nov.* for *Paragia tricolor* Smith, 1850].

Taxonomic decision of Bequaert, J.C. (1928). A study of certain types of diplopterous wasps in the collection of the British Museum. *Ann. Mag. Nat. Hist. (10)* **2**: 138-176 [141].

Distribution: SW coastal, NW coastal, W plateau, W.A. Ecology: larva - sedentary, soil, mellivore : adult - volant, burrowing. Biological references: Bradley, J.C. (1922). The taxonomy of the masarid wasps, including a monograph on the North American species. *Univ. Calif. Publ. Entomol. Tech. Bull.* **1**: 369-464 (figs); Richards, O.W. (1962). *A Revisional Study of the Masarid Wasps (Hymenoptera, Vespoidea)*. London : British Museum 294 pp. (redescription, distribution); Houston, T.F. (1984). Bionomics of a pollen-collecting wasp, *Paragia tricolor* (Hymenoptera : Vespidae : Masarinae), in Western Australia. *Rec. West. Aust. Mus.* **11**: 141-151 (biology).

Paragia (Paragiella) Richards, 1962

Paragiella Richards, O.W. (1962). *A Revisional Study of the Masarid Wasps (Hymenoptera, Vespoidea)*. London : British Museum 294 pp. [53] [proposed with subgeneric rank in *Paragia* Shuckard, 1838]. Type species *Paragia odyneroides* Smith, 1850 by original designation.

Paragia (Paragiella) australis Saussure, 1853

Taxonomic decision of Richards, O.W. (1962). *A Revisional Study of the Masarid Wasps (Hymenoptera, Vespoidea)*. London : British Museum 294 pp. [67].

Paragia (Paragiella) australis australis Saussure, 1853

Paragia australis Saussure, H. de (1853). Note sur la tribu des Masariens et principalement sur le *Masaris vespiformis*. *Ann. Soc. Entomol. Fr. Bull. (3)* **1**: xvii-xxi [xxi]. Type data: holotype, MNHP *M. adult, from Australia (as New Holland).

Distribution: N.S.W.; type locality only (specimen labelled Australia, described from New Holland but later said to be from Tas. which is regarded as unlikely). Ecology: larva - sedentary, soil, mellivore : adult - volant, burrowing.

Paragia (Paragiella) australis borealis Richards, 1962

Paragia (Paragiella) australis borealis Richards, O.W. (1962). *A Revisional Study of the Masarid Wasps (Hymenoptera, Vespoidea)*. London : British Museum 294 pp. [71]. Type data: holotype, ANIC F. adult, from 17 miles W of Morven, Qld." in original description; specimen labelled female type is from 30 (miles) SE Charleville, Qld.

Distribution: Murray-Darling basin, Qld. Ecology: larva - sedentary, soil, mellivore : adult - volant, burrowing.

Paragia (Paragiella) bicolor Saussure, 1853

Paragia bicolor Saussure, H. de (1853). Note sur la tribu des Masariens et principalement sur le *Masaris vespiformis*. *Ann. Soc. Entomol. Fr. Bull.* (3) **1**: xvii–xxi [xxi]. Type data: holotype, MNHP *M. adult, from Australia (as New Holland).

Distribution: N.S.W.; type locality only. Ecology: larva - sedentary, soil, mellivore : adult - volant, burrowing. Biological references: Saussure, H. de (1854–1856). *Études sur la Famille des Vespides*. Troisième Partie comprenant la Monographie des Masariens et un Supplement à la Monographie des Eumeniens. Paris : Masson 352 pp. pls i–xvi (redescription, illustration); Richards, O.W. (1962). *A Revisional Study of the Masarid Wasps (Hymenoptera, Vespoidea)*. London : British Museum 294 pp. (redescription).

Paragia (Paragiella) calida Smith, 1865

Paragia calida Smith, F. (1865). Descriptions of some new species of hymenopterous insects belonging to the families Thynnidae, Masaridae, and Apidae. *Trans. Entomol. Soc. Lond.* (3) **2**: 389–399 [392]. Type data: holotype, BMNH Hym.18.8 *M. adult (seen 1929 by L.F. Graham), from Champion Bay, W.A.

Distribution: (Murray-Darling basin), S Gulfs, NW coastal, (Vic.), S.A., W.A.; only published localities (Lake Hattah), Adelaide and Champion Bay. Ecology: larva - sedentary, soil, mellivore : adult - volant, burrowing. Biological references: Bequaert, J.C. (1928). A study of certain types of diplopterous wasps in the collection of the British Museum. *Ann. Mag. Nat. Hist.* (10) **2**: 138–176 (note on type); Richards, O.W. (1962). *A Revisional Study of the Masarid Wasps (Hymenoptera, Vespoidea)*. London : British Museum 294 pp. (redescription); Richards, O.W. (1968). New records and new species of Australian Masaridae (Hymenoptera : Vespoidea). *J. Aust. Entomol. Soc.* **7**: 101–104 (description of female, possibly this species, from Lake Hattah).

Paragia (Paragiella) deceptrix Smith, 1862

Paragia deceptor Smith, F. (1862). Descriptions of new species of Australian Hymenoptera, and of a species of *Formica* from New Zealand. *Trans. Entomol. Soc. Lond.* (3) **1**: 53–62 [56]. Type data: holotype, BMNH Hym.18.14 *F. adult (seen 1929 by L.F. Graham), from Australia.

Distribution: NE coastal, SE coastal, Qld., N.S.W.; only published localities Brisbane, Rockhampton and Kenthurst. Ecology: larva - sedentary, soil, mellivore : adult - volant, burrowing. Biological references: Schulz, W.A. (1906). *Spolia Hymenopterologica*. Paderborn : Pape iii 356 pp. 1 pl (emendation of name); Richards, O.W. (1962). *A Revisional Study of the Masarid Wasps (Hymenoptera, Vespoidea)*. London : British Museum 294 pp. (redescription); Richards, O.W. (1968). New records and new species of Australian Masaridae (Hymenoptera : Vespoidea). *J. Aust. Entomol. Soc.* **7**: 101–104 (locality).

Paragia (Paragiella) generosa Richards, 1962

Paragia (Paragiella) generosa Richards, O.W. (1962). *A Revisional Study of the Masarid Wasps (Hymenoptera, Vespoidea)*. London : British Museum 294 pp. [69]. Type data: holotype, ANIC F. adult, from 30 (miles) SSW Ayr, Qld.

Distribution: NE coastal, Qld.; type locality only. Ecology: larva - sedentary, soil, mellivore : adult - volant, burrowing.

Paragia (Paragiella) hirsuta Meade-Waldo, 1911

Paragia hirsuta Meade-Waldo, G. (1911). Notes on the family Masaridae (Hymenoptera), with descriptions of a new genus and three new species. *Ann. Mag. Nat. Hist.* (8) **8**: 747–750 [749]. Type data: holotype, BMNH Hym.18.18 *M. adult (seen 1929 by L.F. Graham), from Cairns, Qld.

Distribution: NE coastal, Qld.; only published localities Cairns and Brisbane. Ecology: larva - sedentary, soil, mellivore : adult - volant, burrowing. Biological references: Hacker, H. (1915). Notes on the genus *Megachile* and some rare insects collected during 1913-14. *Mem. Qd. Mus.* **3**: 137–141 (locality); Richards, O.W. (1962). *A Revisional Study of the Masarid Wasps (Hymenoptera, Vespoidea)*. London : British Museum 294 pp. (redescription).

Paragia (Paragiella) magdalena Turner, 1908

Paragia magdalena Turner, R.E. (1908). Two new diplopterous Hymenoptera from Queensland. *Trans. Entomol. Soc. Lond.* **1908**: 89–91 [89]. Type data: holotype, BMNH Hym.18.16 *F. adult (seen 1929 by L.F. Graham), from Mackay, Qld.

Distribution: NE coastal, Qld.; type locality only. Ecology: larva - sedentary, soil, mellivore : adult - volant, burrowing. Biological references: Bequaert, J.C. (1928). A study of certain types of diplopterous wasps in the collection of the British Museum. *Ann. Mag. Nat. Hist.* (10) **2**: 138–176 (states type is male); Richards, O.W. (1962). *A Revisional Study of the Masarid Wasps (Hymenoptera, Vespoidea)*. London : British Museum 294 pp. (redescription).

Paragia (Paragiella) mimetica Richards, 1968

Paragia (Paragiella) mimetica Richards, O.W. (1968). New records and new species of Australian Masaridae (Hymenoptera : Vespoidea). *J. Aust. Entomol. Soc.* **7**: 101–104 [102]. Type data: holotype, ANIC M. adult, from 10 mi W Watheroo, W.A.

Distribution: SW coastal, W.A.; type locality only. Ecology: larva - sedentary, soil, mellivore : adult - volant, burrowing.

Paragia (Paragiella) morosa Smith, 1868

Paragia morosa Smith, F. (1868). Descriptions of aculeate Hymenoptera from Australia. *Trans. Entomol. Soc. Lond.* **1868**: 231–258 [251]. Type data: holotype, BMNH Hym.18.19 *F. adult (seen 1929 by L.F. Graham), from Champion Bay, W.A.

Distribution: SW coastal, NW coastal, W plateau, W.A.; only published localities Champion Bay, Southern Cross and Dedari. Ecology: larva - sedentary, soil, mellivore : adult - volant, burrowing. Biological references: Richards, O.W. (1962). *A Revisional Study of the Masarid Wasps (Hymenoptera, Vespoidea).* London : British Museum 294 pp. (redescription).

Paragia (Paragiella) nasuta Smith, 1868

Paragia nasuta Smith, F. (1868). Descriptions of aculeate Hymenoptera from Australia. *Trans. Entomol. Soc. Lond.* **1868**: 231–258 [252]. Type data: holotype, BMNH Hym.18.12 *F. adult (seen 1929 by L.F. Graham), from Champion Bay, W.A.

Distribution: SW coastal, NW coastal, W plateau, W.A., (Qld.); only published localities Champion Bay, Swan River, Perenjori, Coolgardie and a questionable Qld. record. Ecology: larva - sedentary, soil, mellivore : adult - volant, burrowing. Biological references: Richards, O.W. (1962). *A Revisional Study of the Masarid Wasps (Hymenoptera, Vespoidea).* London : British Museum 294 pp. (redescription, distribution).

Paragia (Paragiella) odyneroides Smith, 1850

Paragia odyneroides Smith, F. (1850). Descriptions of two new species of exotic Hymenoptera. *Trans. Entomol. Soc. Lond.* (2) **1**: 41–42 pl v [42 pl v fig 2]. Type data: holotype, BMNH Hym.18.20 *M. adult (seen 1929 by L.F. Graham), from Hunter River, N.S.W.

Paragia bidens Saussure, H. de (1855). *Études sur la Famille des Vespides.* Troisième Partie comprenant la Monographie des Masariens et un Supplement à la Monographie des Eumeniens. Paris : Masson 352 pp. pls i–xvi (1854–1856) [59]. Type data: holotype, BMNH Hym.18.6 *M. adult (seen 1929 by L.F. Graham), from Adelaide, S.A.

Paragia praedator Saussure, H. de (1855). *Études sur la Famille des Vespides.* Troisième Partie comprenant la Monographie des Masariens et un Supplement à la Monographie des Eumeniens. Paris : Masson 352 pp. pls i–xvi (1854–1856) [59]. Type data: holotype, BMNH Hym.18.7 *F. adult (seen 1929 by L.F. Graham), from Australia.

Taxonomic decision of Richards, O.W. (1962). *A Revisional Study of the Masarid Wasps (Hymenoptera, Vespoidea).* London : British Museum 294 pp. [75].

Distribution: NE coastal, SE coastal, Murray-Darling basin, S Gulfs, Qld., N.S.W., Vic., S.A. Ecology: larva - sedentary, soil, mellivore : adult - volant, burrowing. Biological references: Bequaert, J.C. (1928). A study of certain types of diplopterous wasps in the collection of the British Museum. *Ann. Mag. Nat. Hist.* (10) **2**: 138–176 (states type is female); Richards, O.W. (1968). New records and new species of Australian Masaridae (Hymenoptera : Vespoidea). *J. Aust. Entomol. Soc.* **7**: 101–104 (localities).

Paragia (Paragiella) perkinsi Meade-Waldo, 1911

Paragia perkinsi Meade-Waldo, G. (1911). Notes on the family Masaridae (Hymenoptera), with descriptions of a new genus and three new species. *Ann. Mag. Nat. Hist.* (8) **8**: 747–750 [750]. Type data: holotype, BMNH Hym.18.17 *F. adult (seen 1929 by L.F. Graham), from Cairns, Qld.

Distribution: NE coastal, Qld.; only published localities Cairns, near Springsure. Ecology: larva - sedentary, soil, mellivore : adult - volant, burrowing. Biological references: Richards, O.W. (1962). *A Revisional Study of the Masarid Wasps (Hymenoptera, Vespoidea).* London : British Museum 294 pp. (redescription).

Paragia (Paragiella) propodealis Richards, 1968

Paragia (Paragiella) propodealis Richards, O.W. (1968). New records and new species of Australian Masaridae (Hymenoptera : Vespoidea). *J. Aust. Entomol. Soc.* **7**: 101–104 [104]. Type data: holotype, ANIC F. adult, from Caldwell, N.S.W.

Distribution: Murray-Darling basin, N.S.W.; type locality only. Ecology: larva - sedentary, soil, mellivore : adult - volant, burrowing.

Paragia (Paragiella) sobrina Smith, 1869

Paragia sobrina Smith, F. (1869). Descriptions of new genera and species of exotic Hymenoptera. *Trans. Entomol. Soc. Lond.* **1869**: 301–311 pl vi [309]. Type data: holotype, BMNH Hym.18.10 *F. adult (seen 1929 by L.F. Graham), from Champion Bay, W.A.

Paragia excellens Smith, F. (1869). Descriptions of new genera and species of exotic Hymenoptera. *Trans. Entomol. Soc. Lond.* **1869**: 301–311 pl vi [309]. Type data: holotype, BMNH Hym.18.11 *F. adult (seen 1929 by L.F. Graham), from Swan River, W.A. (not Champion Bay as given in original description), see Bequaert, J.C. (1928). A study of certain types of diplopterous wasps in the collection of the British Museum. *Ann. Mag. Nat. Hist.* (10) **2**: 138–176.

Taxonomic decision of Richards, O.W. (1962). *A Revisional Study of the Masarid Wasps (Hymenoptera, Vespoidea).* London : British Museum 294 pp. [75].

Distribution: SW coastal, NW coastal, W.A.; only published localities Champion Bay, Southern Cross, Swan River and Galena. Ecology: larva - sedentary, soil, mellivore : adult - volant, burrowing.

Paragia (Paragiella) venusta Smith, 1865

Paragia venusta Smith, F. (1865). Descriptions of some new species of hymenopterous insects belonging to the families Thynnidae, Masaridae, and Apidae. *Trans. Entomol. Soc. Lond. (3)* **2**: 389–399 [393]. Type data: holotype, BMNH Hym.18.9 *F. adult (seen 1929 by L.F. Graham), from Swan River, W.A.

Paragia concinna Smith, F. (1868). Descriptions of aculeate Hymenoptera from Australia. *Trans. Entomol. Soc. Lond.* **1868**: 231–258 [251]. Type data: holotype, BMNH Hym.18.15 *F. adult (seen 1929 by L.F. Graham), from Champion Bay, W.A.

Taxonomic decision of Richards, O.W. (1962). *A Revisional Study of the Masarid Wasps (Hymenoptera, Vespoidea)*. London : British Museum 294 pp. [69].

Distribution: SW coastal, NW coastal, W.A.; only published localities Swan River and Champion Bay. Ecology: larva - sedentary, soil, mellivore : adult - volant, burrowing.

Paragia (Paragiella) walkeri Meade-Waldo, 1910

Paragia walkeri Meade-Waldo, G. (1910). New species of Diploptera in the collection of the British Museum. *Ann. Mag. Nat. Hist. (8)* **5**: 30–51 [33]. Type data: holotype, BMNH Hym.18.13 *M. adult (seen 1929 by L.F. Graham), from Adelaide River, N.T.

Distribution: Murray-Darling basin, N coastal, Qld., N.T.; only published localities Toowoomba, Stanthorpe and Adelaide River. Ecology: larva - sedentary, soil, mellivore : adult - volant, burrowing. Biological references: Richards, O.W. (1962). *A Revisional Study of the Masarid Wasps (Hymenoptera, Vespoidea)*. London : British Museum 294 pp. (redescription).

EUMENIDAE

INTRODUCTION

This cosmopolitan family, the mud, potter or mason wasps, contains small to large solitary wasps, with over 300 described Australian species and subspecies in 35 genera, many of them endemic. Some species are very common and conspicuous, building large mud nests on walls of houses. Adults are often collected on flowers or at water. Nests may be of mud cells built in the open (on house walls, rocks or vegetation), in burrows in the ground, or in wood. Some species use a paste of vegetable fibres to make their cells, and some species use abandoned mud nests of other species (Eumenidae or Sphecidae), partitioning large cells to the correct size. All Australian species appear to use larval Lepidoptera to provision their cells, though larval Coleoptera and Symphyta are also used by non-Australian eumenids.

Recent studies on the behaviour of Australian eumenids, not mentioned in the checklist because the species were not identified, include: Smith (1978) on *Paralastor* sp.; Smith & Alcock (1980) on *Epsilon* sp. (no Australian species named in this small tropical genus yet); Naumann (1983) on *Odynerus* sp., *Paralastor* sp., and *Abispa* sp.

References

Riek, E.F. (1970). Hymenoptera. pp. 867–943 *in* CSIRO (1970). *The Insects of Australia*. A textbook for students and research workers. Melbourne : Melbourne Univ. Press

Evans, H.E. & West Eberhard, M.J. (1970). *The Wasps*. Ann Arbor : Univ. Michigan Press 265 pp.

Smith, A.P. (1978). An investigation of the mechanisms underlying nest construction in the mud wasp *Paralastor* sp. (Hymenoptera : Eumenidae). *Anim. Behav.* **26**: 232–240

Smith, A.P. & Alcock, J. (1980). A comparative study of the mating systems of Australian eumenid wasps (Hymenoptera). *Z. Tierpsychol.* **53**: 41–60

Naumann, I.D. (1983). The biology of mud nesting Hymenoptera (and their associates) and Isoptera in rock shelters of the Kakadu Region, Northern Territory. *Aust. Natl. Parks & Wldlf. Serv. Spec. Publ. 10* pp. 127–189

Abispa Mitchell, 1838

Taxonomic decision of van der Vecht, J. (1960). On *Abispa* and some other Eumenidae from the Australian region (Hymenoptera, Vespoidea). *Nova Guinea Zool.* **6**: 91–115 [94].

Abispa (Abispa) Mitchell, 1838

Abispa Mitchell, T.L. (1838). *Three Expeditions into the Interior of Eastern Australia*, with descriptions of the recently explored region of Australia Felix, and of the present colony of New South Wales. Vol. i Journey in search of the Kindur in 1831–2. Expedition sent to explore the course of the River Darling in 1835. London : T. & W. Boone [104] [proposed with subgeneric rank in *Vespa* Linnaeus, 1758]. Type species *Vespa (Abispa) australiana* Mitchell, 1838 by monotypy.

Monerebia Saussure, H. de (1852). *Études sur la Famille des Vespides*. 1. Monographie des guêpes solitaires, ou de la tribu des Euméniens, comprenant la classification et la description de toutes les espèces connues jusqu'à ce jour, et servant de complément au Manuel de Lepeletier de Saint Fargeau. Paris : Masson 286 pp. pls i–xx (1852–1853) [98]. Publication date established from Griffin, F.J. (1939). On the dates of publication of Saussure (H. de) : Études sur la famille des Vespides 1–3,

1852–1858. *J. Soc. Bibliogr. Nat. Hist.* **1**: 211–212. Type species *Odynerus splendidus* Guérin, 1838 by subsequent designation, see van der Vecht, J. (1960). On *Abispa* and some other Eumenidae from the Australian region (Hymenoptera, Vespoidea). *Nova Guinea Zool.* **6**: 91–115.

This group is also found on New Guinea, see van der Vecht, J. (1960). On *Abispa* and some other Eumenidae from the Australian region (Hymenoptera, Vespoidea). *Nova Guinea Zool.* **6**: 91–115.

Abispa (Abispa) australiana Mitchell, 1838

Vespa (Abispa) australiana Mitchell, T.L. (1838). *Three Expeditions into the Interior of Eastern Australia, with descriptions of the recently explored region of Australia Felix, and of the present colony of New South Wales.* Vol. i Journey in search of the Kindur in 1831–2. Expedition sent to explore the course of the River Darling in 1835. London : T. & W. Boone [104]. Type data: syntypes (probable), adult whereabouts unknown, from Karaula River, N.S.W.

Distribution: Murray-Darling basin, N.S.W.; type locality only. Ecology: larva - sedentary, predator : adult - volant; mud-nest, prey larval Lepidoptera. Biological references: Bridwell, J.C. (1919). Miscellaneous notes on Hymenoptera. With descriptions of new genera and species. *Proc. Hawaii. Entomol. Soc.* **4**: 109–165 (taxonomy); van der Vecht, J. (1960). On *Abispa* and some other Eumenidae from the Australian region (Hymenoptera, Vespoidea). *Nova Guinea Zool.* **6**: 91–115 (taxonomy).

Abispa (Abispa) ephippium (Fabricius, 1775)

Vespa ephippium Fabricius, J.C. (1775). *Systema Entomologiae, sistens insectorum classes, ordines, genera, species, adiectis synonymis, locis, descriptionibus, observationibus.* Flensburgi et Lipsiae : Kortii xxvii 832 pp. [362]. Type data: holotype, BMNH *F. adult, from Australia (as New Holland).

Abispa meadewaldoensis Perkins, R.C.L. (1914). On the species of *Alastor (Paralastor)* Sauss. and some other Hymenoptera of the family Eumenidae. *Proc. Zool. Soc. Lond.* **1914**: 563–624 pl I [623]. Type data: holotype, BMNH 18.238 *adult, from Darwin, N.T.

Taxonomic decision of van der Vecht, J. (1960). On *Abispa* and some other Eumenidae from the Australian region (Hymenoptera, Vespoidea). *Nova Guinea Zool.* **6**: 91–115 [96].

Distribution: NE coastal, SE coastal, Murray-Darling basin, Lake Eyre basin, SW coastal, NW coastal, N coastal, Qld., N.S.W., Tas., W.A., N.T. Ecology: larva - sedentary, predator : adult - volant; mud-nest, prey larval Lepidoptera (Psychidae). Biological references: Hacker, H. (1918). Entomological contributions. *Mem. Qd. Mus.* **6**: 106–111 (biology, prey, as *Monerebia ephippium*); Raff, J.W. (1940). Notes on the mason-wasp and its nest. *Vict. Nat.* **56**: 139–141 (biology); Smith, A.P. & Alcock, J. (1980). A comparative study of the mating systems of Australian eumenid wasps (Hymenoptera). *Z. Tierpsychol.* **53**: 41–60.

Abispa (Abispa) laticincta van der Vecht, 1960

Abispa (Abispa) laticincta van der Vecht, J. (1960). On *Abispa* and some other Eumenidae from the Australian region (Hymenoptera, Vespoidea). *Nova Guinea Zool.* **6**: 91–115 [97]. Type data: holotype, RMNH *F. adult, from Cooktown, Qld.

Distribution: NE coastal, Qld.; only published localities Cooktown and Somerset. Ecology: larva - (sedentary), (predator) : adult - (volant); (mud-nest), (prey larval Lepidoptera).

Abispa (Abispa) splendida (Guérin, 1838)

Taxonomic decision of van der Vecht, J. (1960). On *Abispa* and some other Eumenidae from the Australian region (Hymenoptera, Vespoidea). *Nova Guinea Zool.* **6**: 91–115 [96].

Abispa (Abispa) splendida splendida (Guérin, 1838)

Odynerus splendidus Guérin-Méneville, F.E. (1838). Crustacés, Arachnides et Insectes. in Duperrey, L.J. (1838). *Voyage Autour du Monde, Exécuté par Ordre du Roi, sur la Corvette de la Majesté, La Coquille, Pendant les Années 1822, 1823, 1824, et 1825, ...* Zool. Vol. ii Pt. 2 Div. 1 Chap. xiii pp. 57–302 Paris : Bertrand [265]. Type data: holotype, RMNH *M. adult, from Australia (as New Holland).

Distribution: NE coastal, SE coastal, Murray-Darling basin, Qld., N.S.W., Tas. Ecology: larva - sedentary, predator : adult - volant; mud-nest, prey larval Lepidoptera. Biological references: Roth, H.L. (1885). Notes on the habits of some Australian Hymenoptera Aculeata. (With description of a new species by W.F. Kirby). *J. Linn. Soc. Zool.* **18**: 318–328 (nest); Froggatt, W.W. (1894). On the nests and habits of Australian Vespidae and Larridae. *Proc. Linn. Soc. N.S.W. (2)* **9**: 27–34 (biology); Brewster, M.N., Brewster, A.A. & Crouch, N. (1946). *Life Stories of Australian Insects.* 2nd edn. Sydney : Dymock's viii 332 pp. (biology); Smith, A.P. & Alcock, J. (1980). A comparative study of the mating systems of Australian eumenid wasps (Hymenoptera). *Z. Tierpsychol.* **53**: 41–60.

Abispa (Abispa) splendida australis Smith, 1857

Abispa australis Smith, F. (1857). *Catalogue of Hymenopterous Insects in the Collection of the British Museum.* Pt. V. Vespidae. pp. 1–147 London : British Museum [42]. Type data: holotype, BMNH 18.237 *F. adult (seen 1929 by L.F. Graham), from Port Essington, N.T.

Distribution: N coastal, N.T.; only published localities Port Essington, Darwin, Litchfield and "Daly". Ecology: larva - sedentary, predator : adult - volant; mud-nest, prey larval Lepidoptera. Biological references: Bridwell, J.C. (1919). Miscellaneous notes on Hymenoptera. With descriptions of new genera and species. *Proc. Hawaii. Entomol. Soc.* **4**: 109-165 (taxonomy).

Acarodynerus Giordani Soika, 1962

Acarodynerus Giordani Soika, A. (1962). Gli *Odynerus* sensu antiquo del continente australiano e della Tasmania. *Boll. Mus. Civ. Stor. Nat. Venezia* **14**: 57-202 [146]. Type species *Odynerus clypeatus* Saussure, 1853 by original designation.

Acarodynerus acarophilus Giordani Soika, 1962

Acarodynerus acarophilus Giordani Soika, A. (1962). Gli *Odynerus* sensu antiquo del continente australiano e della Tasmania. *Boll. Mus. Civ. Stor. Nat. Venezia* **14**: 57-202 [157]. Type data: holotype, BMNH *F. adult, from Stradbroke Is., Qld.

Distribution: NE coastal, Qld., N.S.W.; only published localities Stradbroke Is., Sunnybank, Nambour and N.S.W. Ecology: larva - sedentary, predator : adult - volant; prey larval Lepidoptera. Biological references: Giordani Soika, A. (1977). Contributo alla conoscenza degli Eumenidi australiani (Hymenoptera). *Mem. Soc. Entomol. Ital.* **55**: 109-138 (distribution).

Acarodynerus clypeatus (Saussure, 1853)

Odynerus (Leionotus) clypeatus Saussure, H. de (1853). Études sur la Famille des Vespides. 1. Monographie des guêpes solitaires, ou de la tribu des Euméniens, comprenant la classification et la description de toutes les espèces connues jusqu'à ce jour, et servant de complément au Manuel de Lepeletier de Saint Fargeau. Paris : Masson 286 pp. pls i-xx (1852-1853) [200]. Type data: syntypes (probable), MNHP *adult, from Tas.; this locality is doubtful, see Giordani Soika, A. (1962). Gli *Odynerus* sensu antiquo del continente australiano e della Tasmania. *Boll. Mus. Civ. Stor. Nat. Venezia* **14**: 57-202.

Distribution: NE coastal, Qld.; only published localities Brisbane and "Tas." Ecology: larva - sedentary, predator : adult - volant; prey larval Lepidoptera. Biological references: Giordani Soika, A. (1962). Gli *Odynerus* sensu antiquo del continente australiano e della Tasmania. *Boll. Mus. Civ. Stor. Nat. Venezia* **14**: 57-202 (redescription, locality); Giordani Soika, A. (1977). Contributo alla conoscenza degli Eumenidi australiani (Hymenoptera). *Mem. Soc. Entomol. Ital.* **55**: 109-138 (taxonomy).

Acarodynerus denticulatus Giordani Soika, 1962

Acarodynerus denticulatus Giordani Soika, A. (1962). Gli *Odynerus* sensu antiquo del continente australiano e della Tasmania. *Boll. Mus. Civ. Stor. Nat. Venezia* **14**: 57-202 [160]. Type data: syntypes, ZMB *5F.,1M. adult, from Adelaide, S.A.

Distribution: S Gulfs, S.A.; type locality only. Ecology: larva - sedentary, predator : adult - volant; prey larval Lepidoptera. Biological references: Giordani Soika, A. (1977). Contributo alla conoscenza degli Eumenidi australiani (Hymenoptera). *Mem. Soc. Entomol. Ital.* **55**: 109-138 (taxonomy).

Acarodynerus dietrichianus (Saussure, 1869)

Taxonomic decision of Giordani Soika, A. (1962). Gli *Odynerus* sensu antiquo del continente australiano e della Tasmania. *Boll. Mus. Civ. Stor. Nat. Venezia* **14**: 57-202 [153].

Acarodynerus dietrichianus dietrichianus (Saussure, 1869)

Odynerus (Odynerus) dietrichianus Saussure, H. de (1869). Hyménoptères divers du Musée Godeffroy. *Stettin. Entomol. Ztg.* **30**: 53-64 [54]. Type data: syntypes (probable), MGH *F. adult, from Rockhampton, Qld.

Distribution: NE coastal, N coastal, Qld., N.T. Ecology: larva - sedentary, predator : adult - volant; prey larval Lepidoptera. Biological references: Giordani Soika, A. (1973). Designazione di lectotipi ed elenco dei tipi di Eumenidi, Vespidi e Masaridi da me descritti negli anni 1934-1960. *Boll. Mus. Civ. Stor. Nat. Venezia* **24**: 7-53 (taxonomy); Giordani Soika, A. (1977). Contributo alla conoscenza degli Eumenidi australiani (Hymenoptera). *Mem. Soc. Entomol. Ital.* **55**: 109-138 (taxonomy).

Acarodynerus dietrichianus rufocaudatus Giordani Soika, 1962

Acarodynerus dietrichianus rufocaudatus Giordani Soika, A. (1962). Gli *Odynerus* sensu antiquo del continente australiano e della Tasmania. *Boll. Mus. Civ. Stor. Nat. Venezia* **14**: 57-202 [155]. Type data: holotype, SAMA *F. adult, from Stewart River, Qld.

Distribution: NE coastal, Qld., S.A.; only published localities Stewart River and S.A. Ecology: larva - sedentary, predator : adult - volant; prey larval Lepidoptera. Biological references: Giordani Soika, A. (1977). Contributo alla conoscenza degli Eumenidi australiani (Hymenoptera). *Mem. Soc. Entomol. Ital.* **55**: 109-138 (description of male).

Acarodynerus drewsenianus Giordani Soika, 1962

Acarodynerus drewsenianus Giordani Soika, A. (1962). Gli *Odynerus* sensu antiquo del continente australiano e della Tasmania. *Boll. Mus. Civ. Stor. Nat. Venezia* **14**: 57-202 [156]. Type data: holotype, UZM *F. adult, from Australia (as New Holland).

Distribution: no locality specified. Ecology: larva - sedentary, predator : adult - volant; prey larval Lepidoptera.

Acarodynerus exarmatus (Giordani Soika, 1937)

Odynerus (Stenodynerus) exarmatus Giordani Soika, A. (1937). Description of three new *Stenodynerus* recently collected by R.E. Turner in W. Australia. *Ann. Mag. Nat. Hist. (10)* **20**: 356-360 [359]. Type data: lectotype, BMNH *M adult, from Merredin, W.A., designation by Giordani Soika, A. (1973). Designazione di lectotipi ed elenco dei tipi di Eumenidi, Vespidi e Masaridi da me descritti negli anni 1934-1960. *Boll. Mus. Civ. Stor. Nat. Venezia* **24**: 7-53.

Distribution: SW coastal, W.A.; type locality only. Ecology: larva - sedentary, predator : adult - volant; prey larval Lepidoptera. Biological references: Giordani Soika, A. (1962). Gli *Odynerus* sensu antiquo del continente australiano e della Tasmania. *Boll. Mus. Civ. Stor. Nat. Venezia* **14**: 57-202 (redescription); Giordani Soika, A. (1977). Contributo alla conoscenza degli Eumenidi australiani (Hymenoptera). *Mem. Soc. Entomol. Ital.* **55**: 109-138 (taxonomy).

Acarodynerus legatus Giordani Soika, 1977

Acarodynerus legatus Giordani Soika, A. (1977). Contributo alla conoscenza degli Eumenidi australiani (Hymenoptera). *Mem. Soc. Entomol. Ital.* **55**: 109-138 [127]. Type data: holotype, ZMB *M. adult, from Marloo Station, Wurarga, W.A.

Distribution: SW coastal, NW coastal, W.A.; only published localities Marloo Station Wurarga and Wubin. Ecology: larva - sedentary, predator : adult - volant; prey larval Lepidoptera.

Acarodynerus lunaris Giordani Soika, 1977

Acarodynerus lunaris Giordani Soika, A. (1977). Contributo alla conoscenza degli Eumenidi australiani (Hymenoptera). *Mem. Soc. Entomol. Ital.* **55**: 109-138 [127]. Type data: holotype, WAM adult not found, from Canning, W.A.

Distribution: (W plateau), (SW coastal), W.A.; type locality only. Ecology: larva - sedentary, predator : adult - volant; prey larval Lepidoptera.

Acarodynerus paleovariatus Giordani Soika, 1962

Acarodynerus paleovariatus Giordani Soika, A. (1962). Gli *Odynerus* sensu antiquo del continente australiano e della Tasmania. *Boll. Mus. Civ. Stor. Nat. Venezia* **14**: 57-202 [162]. Type data: holotype, BMNH *F. adult, from Champion Bay, W.A.

Distribution: NW coastal, W.A.; type locality only. Ecology: larva - sedentary, predator : adult - volant; prey larval Lepidoptera. Biological references: Giordani Soika, A. (1977). Contributo alla conoscenza degli Eumenidi australiani (Hymenoptera). *Mem. Soc. Entomol. Ital.* **55**: 109-138 (taxonomy).

Acarodynerus posttegulatus (Giordani Soika, 1937)

Odynerus (Stenodynerus) posttegulatus Giordani Soika, A. (1937). Description of three new *Stenodynerus* recently collected by R.E. Turner in W. Australia. *Ann. Mag. Nat. Hist. (10)* **20**: 356-360 [356]. Type data: lectotype, BMNH *M. adult, from Dedari, W.A., designation by Giordani Soika, A. (1973). Designazione di lectotipi ed elenco dei tipi di Eumenidi, Vespidi e Masaridi da me descritti negli anni 1934-1960. *Boll. Mus. Civ. Stor. Nat. Venezia* **24**: 7-53.

Distribution: SW coastal, W plateau, W.A.; only published localities Dedari, Southern Cross and Salmon Gums. Ecology: larva - sedentary, predator : adult - volant; prey larval Lepidoptera. Biological references: Giordani Soika, A. (1962). Gli *Odynerus* sensu antiquo del continente australiano e della Tasmania. *Boll. Mus. Civ. Stor. Nat. Venezia* **14**: 57-202 (redescription); Giordani Soika, A. (1977). Contributo alla conoscenza degli Eumenidi australiani (Hymenoptera). *Mem. Soc. Entomol. Ital.* **55**: 109-138 (description of female).

Acarodynerus propodalaris Giordani Soika, 1962

Acarodynerus propodalaris Giordani Soika, A. (1962). Gli *Odynerus* sensu antiquo del continente australiano e della Tasmania. *Boll. Mus. Civ. Stor. Nat. Venezia* **14**: 57-202 [165]. Type data: holotype, BMNH *F. adult, from Dedari, W.A.

Distribution: W plateau, W.A.; type locality only. Ecology: larva - sedentary, predator : adult - volant; prey larval Lepidoptera.

Acarodynerus quadrangolum Giordani Soika, 1977

Acarodynerus quadrangolum Giordani Soika, A. (1977). Contributo alla conoscenza degli Eumenidi australiani (Hymenoptera). *Mem. Soc. Entomol. Ital.* **55**: 109-138 [126]. Type data: holotype, NMV *F. adult, from Caldwell, N.S.W.

Distribution: Murray-Darling basin, N.S.W., (Vic.); only published localities Caldwell and Kangaroo". Ecology: larva - sedentary, predator : adult - volant; prey larval Lepidoptera.

Acarodynerus queenslandicus Giordani Soika, 1962

Acarodynerus queenslandicus Giordani Soika, A. (1962). Gli *Odynerus* sensu antiquo del continente australiano e della Tasmania. *Boll. Mus. Civ. Stor. Nat. Venezia* **14**: 57-202 [156]. Type data: holotype, BMNH *F. adult, from Mackay, Qld.

Distribution: NE coastal, Qld.; only published localities Mackay, Townsville and Claudie River. Ecology: larva - sedentary, predator : adult - volant; prey larval Lepidoptera. Biological references: Giordani Soika, A. (1977). Contributo alla conoscenza degli Eumenidi australiani (Hymenoptera). *Mem. Soc. Entomol. Ital.* **55**: 109–138 (description of male).

Acarodynerus spargovillensis (Giordani Soika, 1937)

Odynerus (Stenodynerus) spargovillensis Giordani Soika, A. (1937). Description of three new *Stenodynerus* recently collected by R.E. Turner in W. Australia. *Ann. Mag. Nat. Hist. (10)* **20**: 356–360 [359]. Type data: lectotype, BMNH *M. adult, from Spargoville (28 mi W of Coolgardie), W.A., designation by Giordani Soika, A. (1973). Designazione di lectotipi ed elenco dei tipi di Eumenidi, Vespidi e Masaridi da me descritti negli anni 1934–1960. *Boll. Mus. Civ. Stor. Nat. Venezia* **24**: 7–53.

Distribution: W plateau, W.A.; type locality only. Ecology: larva - sedentary, predator : adult - volant; prey larval Lepidoptera. Biological references: Giordani Soika, A. (1962). Gli *Odynerus* sensu antiquo del continente australiano e della Tasmania. *Boll. Mus. Civ. Stor. Nat. Venezia* **14**: 57–202 (generic placement, redescription); Giordani Soika, A. (1973). Designazione di lectotipi ed elenco dei tipi di Eumenidi, Vespidi e Masaridi da me descritti negli anni 1934–1960. *Boll. Mus. Civ. Stor. Nat. Venezia* **24**: 7–53 (locality); Giordani Soika, A. (1977). Contributo alla conoscenza degli Eumenidi australiani (Hymenoptera). *Mem. Soc. Entomol. Ital.* **55**: 109–138 (taxonomy).

Acarodynerus spectrum Giordani Soika, 1962

Acarodynerus spectrum Giordani Soika, A. (1962). Gli *Odynerus* sensu antiquo del continente australiano e della Tasmania. *Boll. Mus. Civ. Stor. Nat. Venezia* **14**: 57–202 [155]. Type data: holotype, ZMB *F. adult, from Carshalton, W.A. (this locality could not be found in the gazetteer; the collector, E. Clement, did collect in the Sherlock River area, W.A.).

Distribution: (NW coastal), W.A.; type locality only. Ecology: larva - sedentary, predator : adult - volant; prey larval Lepidoptera.

Acarodynerus triangulum (Saussure, 1855)

Odynerus triangulum Saussure, H. de (1855). *Études sur la Famille des Vespides*. Troisième Partie comprenant la Monographie des Masariens et un Supplement à la Monographie des Eumeniens. Paris : Masson 352 pp. pls i–xvi (1854–1856) [285 pl xiv fig 8]. Type data: syntypes (probable), M. adult whereabouts unknown, from Australia.

Distribution: NE coastal, SE coastal, Qld., N.S.W.; only published localities Brisbane and Hornsby. Ecology: larva - sedentary, predator : adult - volant; prey larval Lepidoptera. Biological references: Giordani Soika, A. (1962). Gli *Odynerus* sensu antiquo del continente australiano e della Tasmania. *Boll. Mus. Civ. Stor. Nat. Venezia* **14**: 57–202 (redescription, distribution).

Acarozumia Bequaert, 1921

Acarozumia Bequaert, J. (1921). Description d'une espèce congolaise du genre *Montezumia*" (Hyménoptères, Vespides) suivie de remarques taxonomiques sur ce groupe. *Rev. Zool. Bot. Afr.* **9**: 235–251 [249] [proposed with subgeneric rank in *Montezumia* Saussure, 1852]. Type species *Nortonia amaliae* Saussure, 1869 by monotypy and original designation.

Acarozumia amaliae (Saussure, 1869)

Nortonia amaliae Saussure, H. de (1869). Hyménoptères divers du Musée Godeffroy. *Stettin. Entomol. Ztg.* **30**: 53–64 [53]. Type data: holotype, MGH *M. adult, from Rockhampton, Qld.

Montezumia australensis Perkins, R.C.L. (1908). Some remarkable Australian Hymenoptera. *Proc. Hawaii. Entomol. Soc.* **2**: 27–35 [33]. Type data: holotype, BMNH (probable) *adult, from middle Qld.

Taxonomic decision of Meade-Waldo, G. (1914). Notes on the Hymenoptera in the collection of the British Museum, with descriptions of new species. V. *Ann. Mag. Nat. Hist. (8)* **14**: 450–464 [461].

Distribution: NE coastal, Qld., N.S.W.; only published localities Rockhampton, Brisbane and N.S.W. Ecology: larva - sedentary, predator : adult - volant; prey larval Lepidoptera. Biological references: Schulthess-Rechberg, A. von (1904). Beiträge zur Kenntnis der *Nortonia*-Arten. *Z. Syst. Hymenopterol. Dipterol.* **4**: 270–283 (redescription); Bequaert, J. (1921). Description d'une espèce congolaise du genre *Montezumia*" (Hyménoptères, Vespides) suivie de remarques taxonomiques sur ce groupe. *Rev. Zool. Bot. Afr.* **9**: 235–251 (taxonomy as *Montezumia (Acarozumia) amaliae*); Riek, E.F. *in* CSIRO (1970). *The Insects of Australia. A textbook for students and research workers.* Melbourne : Melbourne Univ. Press 1029 pp. (generic placement, symbiotic mites); Giordani Soika, A. (1977). Contributo alla conoscenza degli Eumenidi australiani (Hymenoptera). *Mem. Soc. Entomol. Ital.* **55**: 109–138 (distribution as *Nortozumia amaliae*).

Alastoroides Saussure, 1856

Alastoroides Saussure, H. de (1856). *Études sur la Famille des Vespides*. Troisième Partie comprenant la Monographie des Masariens et un Supplement à la Monographie des Eumeniens. Paris : Masson 352 pp. pls i–xvi (1854–1856) [327] [proposed with subgeneric rank in *Alastor* Lepeletier, 1841; placed on Official List of Generic Names in Zoology, see International Commission on Zoological Nomenclature (1970). Opinion 893. Eumenidae names of Saussure (Hymenoptera) : grant of availability to certain names proposed for secondary divisions of genera. *Bull. Zool. Nomen.* **26**: 187–191]. Publication date established from Griffin, F.J. (1939). On the dates of publication of Saussure (H. de) : Études sur la famille des Vespides 1–3, 1852–1858. *J. Soc. Bibliogr. Nat. Hist.* **1**: 211–212. Type species *Alastor clotho* Lepeletier, 1841 by subsequent designation, see Ashmead, W.H. (1902). Classification of the fossorial, predaceous and parasitic wasps, or the superfamily Vespoidea. *Can. Entomol.* **34**: 203–210.

Paralastoroides Saussure, H. de (1856). *Études sur la Famille des Vespides*. Troisième Partie comprenant la Monographie des Masariens et un Supplement à la Monographie des Eumeniens. Paris : Masson 352 pp. pls i–xvi (1854–1856) [328] [unavailable name; proposed for secondary division of *Alastoroides* Saussure, 1856; placed on Official Index of Rejected and Invalid Generic Names in Zoology, see International Commission on Zoological Nomenclature (1970). Opinion 893. Eumenidae names of Saussure (Hymenoptera) : grant of availability to certain names proposed for secondary divisions of genera. *Bull. Zool. Nomen.* **26**: 187–191]. Publication date established from Griffin, F.J. (1939). On the dates of publication of Saussure (H. de) : Études sur la famille des Vespides 1–3, 1852–1858. *J. Soc. Bibliogr. Nat. Hist.* **1**: 211–212. Type species *Alastor clotho* Lepeletier, 1841 by monotypy.

Alastoroides clotho (Lepeletier, 1841)

Alastor clotho Lepeletier, A.L.M. (1841). *Histoire Naturelle des Insectes. Hyménoptères*. Paris : Roret Vol. ii 680 pp. [668]. Type data: syntypes (probable), MNHP *F. adult, from Australia (as New Holland).

Distribution: no locality specified. Ecology: larva - sedentary, predator : adult - volant; prey larval Lepidoptera. Biological references: Perkins, R.C.L. (1914). On the species of *Alastor (Paralastor)* Sauss. and some other Hymenoptera of the family Eumenidae. *Proc. Zool. Soc. Lond.* **1914**: 563–624 pl I (taxonomy as *Paralastor clotho*); International Commission on Zoological Nomenclature (1970). Opinion 893. Eumenidae names of Saussure (Hymenoptera) : grant of availability to certain names proposed for secondary divisions of genera. *Bull. Zool. Nomen.* **26**: 187–191 (placed on Official List of Specific Names).

Allorhynchium van der Vecht, 1963

Allorhynchium van der Vecht, J. (1963). Studies on Indo-Australian and east-Asiatic Eumenidae (Hymenoptera, Vespoidea). *Zool. Verh.* **60**: 1–116 [58]. Type species *Vespa argentata* Fabricius, 1804 by original designation.

This group is also found in the Oriental Region, see van der Vecht, J. (1963). Studies on Indo-Australian and east-Asiatic Eumenidae (Hymenoptera, Vespoidea). *Zool. Verh.* **60**: 1–116.

Allorhynchium iridipenne (Smith, 1861)

Rhynchium iridipenne Smith, F. (1861). Catalogue of hymenopterous insects collected by Mr. A.R. Wallace in the islands of Bachian, Kaisaa, Amboyna, Gilolo, and at Dory in New Guinea. *J. Proc. Linn. Soc. Lond. Zool.* **5**: 93–143 [128]. Type data: holotype, OUM *F. adult, from Amboina (as Amboyna).

Distribution: N coastal, N.T.; only published locality Marrakai, also found in Amboina, Philippines, Celebes, India and Indochina. Ecology: larva - sedentary, predator : adult - volant; prey larval Lepidoptera. Biological references: Schulthess-Rechberg, A. von (1935). Hymenoptera aus den Sundainseln und Nordaustralien (mit Ausschluss der Blattwespen, Schlupfwespen und Ameisen). *Rev. Suisse Zool.* **42**: 293–323 (Australian record); van der Vecht, J. (1963). Studies on Indo-Australian and east-Asiatic Eumenidae (Hymenoptera, Vespoidea). *Zool. Verh.* **60**: 1–116 (generic placement); Baltazar, C.R. (1966). A catalogue of Philippine Hymenoptera (with a bibliography, 1758–1963). *Pac. Insects Monogr.* **8**: 1–488 (distribution).

Ancistrocerus Wesmael, 1836

Ancistrocerus Wesmael, C. (1836). Supplément à la Monographie des Odynères. *Bull. Acad. Brux.* **3**: 44–54 [45] [proposed as subgenus of *Odynerus* Latreille, 1802]. Type species *Vespa parietum* Linnaeus, 1758 by subsequent designation, see Girard, M. (1879). *Traité élémentaire d'Entomologie*. Tom. ii Fasc. 2 Hymenoptères porte-aiguillon. pp. 577–1028 7 pls Paris : Baillière. Compiled from secondary source: van der Vecht, J. & Fischer, F.C.J. (1972). Palearctic Eumenidae. *Hymenopterorum Catalogus* **8**: 1–199.

This group is mainly Holarctic, also in Ethiopian and Neotropical Regions, see Richards, O.W. (1980). Scolioidea, Vespoidea and Sphecoidea. Hymenoptera, Aculeata. *Handbk. Ident. Br. Insects* **6**(3b): 1–118.

Ancistrocerus fluvialis (Saussure, 1855)

Odynerus (Ancistrocerus) fluvialis Saussure, H. de (1855). *Études sur la Famille des Vespides*. Troisième Partie comprenant la Monographie des Masariens et un Supplement à la Monographie des Eumeniens. Paris : Masson 352 pp. pls i–xvi (1854–1856) [215]. Type data: syntypes (probable), MZUT *adult, from Swan River, W.A. (as New Holland).

Distribution: SW coastal, W.A.; type locality only. Ecology: larva - sedentary, predator : adult - volant; prey larval Lepidoptera. Biological references: Giordani Soika, A. (1962). Gli *Odynerus* sensu antiquo del continente australiano e della Tasmania. *Boll. Mus. Civ. Stor. Nat. Venezia* **14**: 57–202 (taxonomy).

Antamenes Giordani Soika, 1958

Taxonomic decision of Giordani Soika, A. (1962). Gli *Odynerus* sensu antiquo del continente australiano e della Tasmania. *Boll. Mus. Civ. Stor. Nat. Venezia* **14**: 57–202 [185].

Antamenes (Antamenes) Giordani Soika, 1958

Antamenes Giordani Soika, A. (1958). Biogeografia, evoluzione e sistematica dei Vespidi solitari della Polinesia meridionale. *Boll. Mus. Civ. Stor. Nat. Venezia* **10**: 183–221 [214]. Type species *Odynerus flavocinctus* Smith, 1857 (= *Odynerus vernalis* Saussure, 1853) by original designation.

Antamenes (Antamenes) pseudoneotropicus (Giordani Soika, 1943)

Pachymenes pseudoneotropicus Giordani Soika, A. (1943). Le specie indo-australiane del genere *Pachymenes* (Hym. Vespidae). *Mem. Soc. Entomol. Ital.* **22**: 102–117 [113]. Type data: lectotype, A. Giordani Soika pers. coll. *M. adult, from N.S.W., designation by Giordani Soika, A. (1973). Designazione di lectotipi ed elenco dei tipi di Eumenidi, Vespidi e Masaridi da me descritti negli anni 1934–1960. *Boll. Mus. Civ. Stor. Nat. Venezia* **24**: 7–53.

Distribution: N.S.W.; no locality specified. Ecology: larva - sedentary, predator : adult - volant; prey larval Lepidoptera. Biological references: Giordani Soika, A. (1962). Gli *Odynerus* sensu antiquo del continente australiano e della Tasmania. *Boll. Mus. Civ. Stor. Nat. Venezia* **14**: 57–202 (generic placement, redescription); Giordani Soika, A. (1974). Prime ricerche sugli Eumenidi ipsobionti. I. Caratteristiche generali degli Eumenidi ipsobionti del globo. *Redia* **55**: 287–302 (colouration).

Antamenes (Antamenes) tasmaniae (Giordani Soika, 1943)

Pachymenes tasmaniae Giordani Soika, A. (1943). Le specie indo-australiane del genere *Pachymenes* (Hym. Vespidae). *Mem. Soc. Entomol. Ital.* **22**: 102–117 [113]. Type data: lectotype, A. Giordani Soika pers. coll. *M. adult, from Tas., designation by Giordani Soika, A. (1973). Designazione di lectotipi ed elenco dei tipi di Eumenidi, Vespidi e Masaridi da me descritti negli anni 1934–1960. *Boll. Mus. Civ. Stor. Nat. Venezia* **24**: 7–53.

Distribution: NE coastal, Qld., Tas.; only published localities Mackay and Tas. Ecology: larva - sedentary, predator : adult - volant; prey larval Lepidoptera. Biological references: Giordani Soika, A. (1962). Gli *Odynerus* sensu antiquo del continente australiano e della Tasmania. *Boll. Mus. Civ. Stor. Nat. Venezia* **14**: 57–202 (generic placement, redescription); Giordani Soika, A. (1974). Prime ricerche sugli Eumenidi ipsobionti. I. Caratteristiche generali degli Eumenidi ipsobionti del globo. *Redia* **55**: 287–302 (colouration).

Antamenes (Antamenes) vernalis (Saussure, 1853)

Odynerus (Ancistrocerus) vernalis Saussure, H. de (1853). *Études sur la Famille des Vespides*. 1. Monographie des guêpes solitaires, ou de la tribu des Euméniens, comprenant la classification et la description de toutes les espèces connues jusqu'à ce jour, et servant de complément au Manuel de Lepeletier de Saint Fargeau. Paris : Masson 286 pp. pls i–xx (1852–1853) [148]. Type data: syntypes (probable), MNHP *F. adult, from Tas.

Odynerus flavocinctus Smith, F. (1857). *Catalogue of Hymenopterous Insects in the Collection of the British Museum*. Pt. V. Vespidae pp. 1–147 London : British Museum [64]. Type data: holotype, BMNH *M. adult (seen 1929 by L.F. Graham), from Australia (as New Holland).

Odynerus (Leionotus) bisulcatus Cameron, P. (1906). Description of a new species of *Odynerus (Leionotus)* from Australia. *Entomologist* **39**: 78–79 [78]. Type data: syntypes probable, adult whereabouts unknown, from Australia.

Pachymenes rectispina Giordani Soika, A. (1943). Le specie indo-australiane del genere *Pachymenes* (Hym. Vespidae). *Mem. Soc. Entomol. Ital.* **22**: 102–117 [112]. Type data: holotype, A. Giordani Soika pers. coll. *M. adult, from Vic.

Taxonomic decision of Giordani Soika, A. (1962). Gli *Odynerus* sensu antiquo del continente australiano e della Tasmania. *Boll. Mus. Civ. Stor. Nat. Venezia* **14**: 57–202 [188].

Distribution: NE coastal, Murray-Darling basin, SE coastal, S Gulfs, Qld., N.S.W., Vic., Tas., S.A. Ecology: larva - sedentary, predator : adult - volant; prey larval Lepidoptera. Biological references: Giordani Soika, A. (1977). Contributo alla conoscenza degli Eumenidi australiani (Hymenoptera). *Mem. Soc. Entomol. Ital.* **55**: 109–138 (distribution).

Antamenes (Antamenes) vorticosus (Giordani Soika, 1943)

Pachymenes vorticosus Giordani Soika, A. (1943). Le specie indo-australiane del genere *Pachymenes* (Hym. Vespidae). *Mem. Soc. Entomol. Ital.* **22**: 102–117 [110]. Type data: lectotype, A. Giordani Soika pers. coll. *adult, from Emerald, Vic., designation by Giordani Soika, A. (1973). Designazione di lectotipi ed elenco dei tipi di Eumenidi, Vespidi e Masaridi da me descritti negli anni 1934–1960. *Boll. Mus. Civ. Stor. Nat. Venezia* **24**: 7–53.

Distribution: SE coastal, N.S.W., Vic. Ecology: larva - sedentary, predator : adult - volant; prey larval Lepidoptera. Biological references: Giordani Soika, A. (1962). Gli *Odynerus* sensu antiquo del continente australiano e della Tasmania. *Boll. Mus. Civ. Stor. Nat. Venezia* **14**: 57–202 (generic placement, redescription); Giordani Soika, A. (1977). Contributo alla conoscenza degli Eumenidi australiani (Hymenoptera). *Mem. Soc. Entomol. Ital.* **55**: 109–138 (distribution).

Antamenes (Australochilus) Giordani Soika, 1962

Australochilus Giordani Soika, A. (1962). Gli *Odynerus* sensu antiquo del continente australiano e della Tasmania. *Boll. Mus. Civ. Stor. Nat. Venezia* **14**: 57–202 [184] [proposed with subgeneric rank in *Antamenes* Giordani Soika, 1958]. Type species *Odynerus citreocinctus* Saussure, 1867 by subsequent designation, see Giordani Soika, A. (1973). Designazione di lectotipi ed elenco dei tipi di Eumenidi, Vespidi e Masaridi da me descritti negli anni 1934–1960. *Boll. Mus. Civ. Stor. Nat. Venezia* **24**: 7–53.

Antamenes (Australochilus) amicus (Giordani Soika, 1943)

Pachymenes amicus Giordani Soika, A. (1943). Le specie indo-australiane del genere *Pachymenes* (Hym. Vespidae). *Mem. Soc. Entomol. Ital.* **22**: 102–117 [114]. Type data: lectotype, BMNH *M. adult, from Mackay, Qld., designation by Giordani Soika, A. (1973). Designazione di lectotipi ed elenco dei tipi di Eumenidi, Vespidi e Masaridi da me descritti negli anni 1934–1960. *Boll. Mus. Civ. Stor. Nat. Venezia* **24**: 7–53.

Distribution: NE coastal, Qld.; only published localities Mackay and Kuranda. Ecology: larva - sedentary, predator : adult - volant; prey larval Lepidoptera. Biological references: Giordani Soika, A. (1962). Gli *Odynerus* sensu antiquo del continente australiano e della Tasmania. *Boll. Mus. Civ. Stor. Nat. Venezia* **14**: 57–202 (redescription).

Antamenes (Australochilus) citreocinctus (Saussure, 1867)

Odynerus (Leionotus) citreocinctus Saussure, H. de (1867). Hymenoptera. in, Reise der österreichischen Fregatte Novara um die Erde in den Jahren 1857, 1858, 1859 unter den Befehlen des Commodore B. von Wüllerstorf-Urbair. Zoologischer Theil. Wien : K-K Hof- und Staatsdrückerei Vol. 2(1a) 138 pp. [10 pl I fig 5]. Type data: syntypes (probable), MNHP (probable) *adult, from Sydney, N.S.W.

Distribution: NE coastal, SE coastal, Qld., N.S.W.; only published localities Stradbroke Is. and Sydney. Ecology: larva - sedentary, predator : adult - volant; prey larval Lepidoptera. Biological references: Giordani Soika, A. (1943). Le specie indo-australiane del genere *Pachymenes* (Hym. Vespidae). *Mem. Soc. Entomol. Ital.* **22**: 102–117 (redescription as *Pachymenes citreocinctus*); Giordani Soika, A. (1962). Gli *Odynerus* sensu antiquo del continente australiano e della Tasmania. *Boll. Mus. Civ. Stor. Nat. Venezia* **14**: 57–202 (generic placement, redescription).

Antamenes (Australochilus) ferrugineus Giordani Soika, 1977

Antamenes (Australochilus) ferrugineus Giordani Soika, A. (1977). Contributo alla conoscenza degli Eumenidi australiani (Hymenoptera). *Mem. Soc. Entomol. Ital.* **55**: 109–138 [133]. Type data: holotype, BMNH *F. adult, from Port Darwin, N.T.

Distribution: N coastal, N.T.; type locality only. Ecology: larva - sedentary, predator : adult - volant; prey larval Lepidoptera.

Antamenes (Australochilus) flavoniger Giordani Soika, 1977

Antamenes (Australochilus) flavoniger Giordani Soika, A. (1977). Contributo alla conoscenza degli Eumenidi australiani (Hymenoptera). *Mem. Soc. Entomol. Ital.* **55**: 109–138 [132]. Type data: holotype, BMNH *adult, from Herbert, Qld. (presumably Herbert River or Herberton is meant).

Distribution: NE coastal, Qld.; only published localities Herbert and Cairns. Ecology: larva - sedentary, predator : adult - volant; prey larval Lepidoptera.

Antamenes (Australochilus) hackeri Giordani Soika, 1962

Antamenes (Australochilus) hackeri Giordani Soika, A. (1962). Gli *Odynerus* sensu antiquo del continente australiano e della Tasmania. *Boll. Mus. Civ. Stor. Nat. Venezia* **14**: 57–202 [198]. Type data: holotype, QM T8538 *adult, from Brisbane, Qld.

Distribution: NE coastal, Qld.; only published localities Brisbane and Stradbroke Is. Ecology: larva - sedentary, predator : adult - volant; prey larval Lepidoptera.

Antamenes (Australochilus) hostilis Giordani Soika, 1962

Antamenes (Australochilus) hostilis Giordani Soika, A. (1962). Gli *Odynerus* sensu antiquo del continente australiano e della Tasmania. *Boll. Mus. Civ. Stor. Nat. Venezia* **14**: 57–202 [195]. Type data: holotype, QM T5983 *F. adult, from Brisbane, Qld.

Distribution: NE coastal, Qld.; only published localities Brisbane and Burleigh. Ecology: larva - sedentary, predator : adult - volant; prey larval Lepidoptera. Biological references: Giordani Soika, A. (1977). Contributo alla conoscenza degli Eumenidi australiani (Hymenoptera). *Mem. Soc. Entomol. Ital.* **55**: 109–138 (description of male).

Antamenes (Australochilus) jugulatus Giordani Soika, 1977

Antamenes (Australochilus) jugulatus Giordani Soika, A. (1977). Contributo alla conoscenza degli Eumenidi australiani (Hymenoptera). *Mem. Soc. Entomol. Ital.* **55**: 109–138 [131]. Type data: holotype, NMV *M. adult, from Melton, Vic.

Distribution: SE coastal, Vic.; type locality only. Ecology: larva - sedentary, predator : adult - volant; prey larval Lepidoptera.

Australodynerus Giordani Soika, 1962

Australodynerus Giordani Soika, A. (1962). Gli *Odynerus* sensu antiquo del continente australiano e della Tasmania. *Boll. Mus. Civ. Stor. Nat. Venezia* **14**: 57–202 [114]. Type species *Odynerus pusillus* Saussure, 1855 by original designation.

Australodynerus convexus Giordani Soika, 1977

Australodynerus convexus Giordani Soika, A. (1977). Contributo alla conoscenza degli Eumenidi australiani (Hymenoptera). *Mem. Soc. Entomol. Ital.* **55**: 109–138 [119]. Type data: holotype, USNM *adult, from Brisbane, Qld.

Distribution: NE coastal, Qld.; only published localities Brisbane and Burleigh. Ecology: larva - sedentary, predator : adult - volant; prey larval Lepidoptera.

Australodynerus merredinensis Giordani Soika, 1962

Taxonomic decision of Giordani Soika, A. (1977). Contributo alla conoscenza degli Eumenidi australiani (Hymenoptera). *Mem. Soc. Entomol. Ital.* **55**: 109–138 [119].

Australodynerus merredinensis merredinensis Giordani Soika, 1962

Australodynerus merredinensis Giordani Soika, A. (1962). Gli *Odynerus* sensu antiquo del continente australiano e della Tasmania. *Boll. Mus. Civ. Stor. Nat. Venezia* **14**: 57–202 [122]. Type data: holotype, BMNH *F. adult, from Merredin, W.A.

Distribution: SW coastal, W plateau, NW coastal, W.A.; only published localities Merredin, Marloo Station and Dedari. Ecology: larva - sedentary, predator : adult - volant; prey larval Lepidoptera.

Australodynerus merredinensis everardensis Giordani Soika, 1977

Australodynerus merredinensis everardensis Giordani Soika, A. (1977). Contributo alla conoscenza degli Eumenidi australiani (Hymenoptera). *Mem. Soc. Entomol. Ital.* **55**: 109–138 [119]. Type data: holotype, SAMA *F. adult, from Everard Park Station near Victory Well, S.A.

Distribution: Lake Eyre basin, S.A.; type locality only. Ecology: larva - sedentary, predator : adult - volant; prey larval Lepidoptera.

Australodynerus merredinensis victoriensis Giordani Soika, 1977

Australodynerus merredinensis victoriensis Giordani Soika, A. (1977). Contributo alla conoscenza degli Eumenidi australiani (Hymenoptera). *Mem. Soc. Entomol. Ital.* **55**: 109–138 [120]. Type data: holotype, NMV *M. adult, from Purnong, S.A. (as Pomong, Vic.); Mr. K. Walker checked the label data on the holotype.

Distribution: Murray-Darling basin, Vic., S.A.; only published localities Purnong, Mooroopna and Mallee. Ecology: larva - sedentary, predator : adult - volant; prey larval Lepidoptera.

Australodynerus punctiventris Giordani Soika, 1977

Australodynerus punctiventris Giordani Soika, A. (1977). Contributo alla conoscenza degli Eumenidi australiani (Hymenoptera). *Mem. Soc. Entomol. Ital.* **55**: 109–138 [120]. Type data: holotype, NMV *F. adult, from Caldwell, N.S.W.

Distribution: Murray-Darling basin, N.S.W.; type locality only. Ecology: larva - sedentary, predator : adult - volant; prey larval Lepidoptera.

Australodynerus pusilloides Giordani Soika, 1962

Taxonomic decision of Giordani Soika, A. (1962). Gli *Odynerus* sensu antiquo del continente australiano e della Tasmania. *Boll. Mus. Civ. Stor. Nat. Venezia* **14**: 57–202 [119].

Australodynerus pusilloides pusilloides Giordani Soika, 1962

Australodynerus pusilloides Giordani Soika, A. (1962). Gli *Odynerus* sensu antiquo del continente australiano e della Tasmania. *Boll. Mus. Civ. Stor. Nat. Venezia* **14**: 57–202 [119]. Type data: holotype, BMNH *F. adult, from Yanchep, W.A.

Distribution: SW coastal, NW coastal, W.A.; only published localities Yanchep, Dongarra, Maya, Merredin and Midland. Ecology: larva - sedentary, predator : adult - volant; prey larval Lepidoptera. Biological references: Giordani Soika, A. (1977). Contributo alla conoscenza degli Eumenidi australiani (Hymenoptera). *Mem. Soc. Entomol. Ital.* **55**: 109–138 (distribution).

Australodynerus pusilloides impudicus Giordani Soika, 1962

Australodynerus pusilloides impudicus Giordani Soika, A. (1962). Gli *Odynerus* sensu antiquo del continente australiano e della Tasmania. *Boll. Mus. Civ. Stor. Nat. Venezia* **14**: 57–202 [120]. Type data: holotype, BMNH *F. adult, from Yallingup, W.A.

Distribution: SW coastal, W.A.; type locality only. Ecology: larva - sedentary, predator : adult - volant; prey larval Lepidoptera. Biological references: Giordani Soika, A. (1977). Contributo alla conoscenza degli Eumenidi australiani (Hymenoptera). *Mem. Soc. Entomol. Ital.* **55**: 109–138 (taxonomy).

Australodynerus pusillus (Saussure, 1855)

Odynerus pusillus Saussure, H. de (1855). *Études sur la Famille des Vespides*. Troisième Partie comprenant la Monographie des Masariens et un Supplement à la Monographie des Eumeniens. Paris : Masson 352 pp. pls i–xvi (1854–1856) [287]. Type data: syntypes (probable), F. adult whereabouts unknown, from Australia (as New Holland).

Odynerus (Leionotus) macilentus Saussure, H. de (1867). Hymenoptera. *in, Reise der österreichischen Fregatte Novara um die Erde in den Jahren 1857, 1858, 1859 unter den Befehlen des Commodore B. von Wüllerstorf-Urbair*. Zoologischer Theil. Wien : K-K Hof- und Staatsdrückerei Vol. 2(1a) 138 pp. [16 pl I fig 10]. Type data: syntypes (probable), BMNH *adult, from Sydney, N.S.W.

Taxonomic decision of Giordani Soika, A. (1962). Gli *Odynerus* sensu antiquo del continente australiano e della Tasmania. *Boll. Mus. Civ. Stor. Nat. Venezia* **14**: 57–202 [116].

Distribution: SE coastal, S Gulfs, N.S.W., Vic., S.A.; only published localities Sydney, Kewell and Adelaide. Ecology: larva - sedentary, predator : adult - volant; prey larval Lepidoptera. Biological references: Giordani Soika, A. (1973). Designazione di lectotipi ed elenco dei tipi di Eumenidi, Vespidi e Masaridi da me descritti negli anni 1934–1960. *Boll. Mus. Civ. Stor. Nat. Venezia* **24**: 7–53 (taxonomy); Giordani Soika, A. (1977). Contributo alla conoscenza degli Eumenidi australiani (Hymenoptera). *Mem. Soc. Entomol. Ital.* **55**: 109–138 (distribution).

Australozethus Giordani Soika, 1969

Australozethus Giordani Soika, A. (1969). Revisione dei Discoeliinae Australiani. *Boll. Mus. Civ. Stor. Nat. Venezia* **19**: 25–100 [29]. Type species *Australozethus tasmaniensis* Giordani Soika, 1969 by original designation.

Australozethus continentalis Giordani Soika, 1969

Australozethus continentalis Giordani Soika, A. (1969). Revisione dei Discoeliinae Australiani. *Boll. Mus. Civ. Stor. Nat. Venezia* **19**: 25–100 [36]. Type data: holotype, AM K60161 F. adult, from Barrington Tops, N.S.W.

Distribution: SE coastal, N.S.W.; type locality only. Ecology: larva - sedentary, predator : adult - volant; prey larval Lepidoptera.

Australozethus occidentalis Giordani Soika, 1969

Australozethus occidentalis Giordani Soika, A. (1969). Revisione dei Discoeliinae Australiani. *Boll. Mus. Civ. Stor. Nat. Venezia* **19**: 25–100 [37]. Type data: holotype, QM T7235 *F. adult, from Mundaring, W.A.

Distribution: SW coastal, W.A.; type locality only. Ecology: larva - sedentary, predator : adult - volant; prey larval Lepidoptera.

Australozethus tasmaniensis Giordani Soika, 1969

Taxonomic decision of Giordani Soika, A. (1969). Revisione dei Discoeliinae Australiani. *Boll. Mus. Civ. Stor. Nat. Venezia* **19**: 25–100 [35].

Australozethus tasmaniensis tasmaniensis Giordani Soika, 1969

Australozethus tasmaniensis Giordani Soika, A. (1969). Revisione dei Discoeliinae Australiani. *Boll. Mus. Civ. Stor. Nat. Venezia* **19**: 25–100 [31]. Type data: holotype, BMNH *F. adult, from Eaglehawk Neck, Tas.

Distribution: Murray-Darling basin, Vic., Tas.; only published localities Mt. Buffalo, Eaglehawk Neck, Hobart and Launceston. Ecology: larva - sedentary, predator : adult - volant; prey larval Lepidoptera. Biological references: Giordani Soika, A. (1974). Prime ricerche sugli Eumenidi ipsobionti. I. Caratteristiche generali degli Eumenidi ipsobionti del globo. *Redia* **55**: 287–302 (colouration).

Australozethus tasmaniensis montanus Giordani Soika, 1969

Pseudozethus tasmaniensis montanus Giordani Soika, A. (1969). Revisione dei Discoeliinae Australiani. *Boll. Mus. Civ. Stor. Nat. Venezia* **19**: 25–100 [35] [*Pseudozethus* was printed in error for *Australozethus*]. Type data: holotype, AM F. adult, from Bunya Mts., Qld.

Distribution: NE coastal, Qld.; only published localities Bunya Mts. and Mt. Glorious. Ecology: larva - sedentary, predator : adult - volant; prey larval Lepidoptera.

Bidentodynerus Giordani Soika, 1977

Bidentodynerus Giordani Soika, A. (1977). Contributo alla conoscenza degli Eumenidi australiani (Hymenoptera). *Mem. Soc. Entomol. Ital.* **55**: 109–138 [117]. Type species ' *Odynerus bicolor* Saussure, 1855 by original designation.

Bidentodynerus bicolor (Saussure, 1855)

Taxonomic decision of Giordani Soika, A. (1962). Gli *Odynerus* sensu antiquo del continente australiano e della Tasmania. *Boll. Mus. Civ. Stor. Nat. Venezia* **14**: 57–202 [89].

Bidentodynerus bicolor bicolor (Saussure, 1855)

Odynerus bicolor Saussure, H. de (1855). *Études sur la Famille des Vespides*. Troisième Partie comprenant la Monographie des Masariens et un Supplement à la Monographie des Eumeniens. Paris : Masson 352 pp. pls i–xvi (1854–1856) [284]. Type data: holotype, BMNH *F. adult (seen 1929 by L.F. Graham), from MacIntyre River, N.S.W.

Distribution: NE coastal, Murray-Darling basin, SE coastal, N Gulf, N coastal, Qld., N.S.W., Vic., W.A., N.T. Ecology: larva - sedentary, predator : adult - volant; nest in abandoned mud-nests of *Sceliphron laetum* Smith, prey larval Lepidoptera (Pyralidae). Biological references: Roth, H.L. (1885). Notes on the habits of some Australian Hymenoptera Aculeata. (With description of a new species by W.F. Kirby). *J. Linn. Soc. Zool.* **18**:

318–328 (biology as *Odyncrus bicolor*); Giordani Soika, A. (1941). Studi sui Vespidi solitari. *Boll. Soc. Veneziana Stor. Nat.* **2**: 130–279 (taxonomy as *Odynerus (Rhynchium) bicolor*); Smith, A. (1972). The Michelangelo of mud wasps. *Animals* **14**: 496–497 (nest, biology as *Pseudepipona bicolor*); Giordani Soika, A. (1977). Contributo alla conoscenza degli Eumenidi australiani (Hymenoptera). *Mem. Soc. Entomol. Ital.* **55**: 109–138 (generic placement, distribution); Naumann, I.D. (1983).The biology of mud nesting Hymenoptera (and their associates) and Isoptera in rock shelters of the Kakadu Region, Northern Territory. *Aust. Natl. Parks & Wildlf. Ser. Spec. Publ. 10* pp. 127–189 (biology as *Odynerus bicolor*).

Bidentodynerus bicolor aurantiopicta (Giordani Soika, 1941)

Odynerus (Rhynchium) bicolor aurantiopictus Giordani Soika, A. (1941). Studi sui Vespidi solitari. *Boll. Soc. Veneziana Stor. Nat.* **2**: 130–279 [258] [described as variety]. Type data: lectotype, A. Giordani Soika pers. coll. *F. adult, from Australia, designation by Giordani Soika, A. (1973). Designazione di lectotipi ed elenco dei tipi di Eumenidi, Vespidi e Masaridi da me descritti negli anni 1934-1960. *Boll. Mus. Civ. Stor. Nat. Venezia* **24**: 7–53.

Distribution: W plateau, N coastal, W.A., N.T.; only published localities Meekatharra, Billiluna and Darwin. Ecology: larva - sedentary, predator : adult - volant; prey larval Lepidoptera.

Bidentodynerus bicolor flavescentulus (Giordani Soika, 1941)

Odynerus (Rhynchium) bicolor flavescentulus Giordani Soika, A. (1941). Studi sui Vespidi solitari. *Boll. Soc. Veneziana Stor. Nat.* **2**: 130–279 [258] [described as variety]. Type data: lectotype, BMNH *M. adult, from Alexandria, N.T., designation by Giordani Soika, A. (1973). Designazione di lectotipi ed elenco dei tipi di Eumenidi, Vespidi e Masaridi da me descritti negli anni 1934-1960. *Boll. Mus. Civ. Stor. Nat. Venezia* **24**: 7–53.

Distribution: N Gulf, W plateau, W.A., N.T.; only published localities Meekatharra and Alexandria. Ecology: larva - sedentary, predator : adult - volant; prey larval Lepidoptera.

Bidentodynerus bicolor nigrocinctoides (Giordani Soika, 1941)

Odynerus (Rhynchium) bicolor nigrocinctoides Giordani Soika, A. (1941). Studi sui Vespidi solitari. *Boll. Soc. Veneziana Stor. Nat.* **2**: 130–279 [258] [described as variety]. Type data: lectotype, BMNH *F. adult, from Kalamunda, W.A., designation by Giordani Soika, A. (1973). Designazione di lectotipi ed elenco dei tipi di Eumenidi, Vespidi e Masaridi da me descritti negli anni 1934-1960. *Boll. Mus. Civ. Stor. Nat. Venezia* **24**: 7–53. from Kalamunda, W.A.

Distribution: SW coastal, NW coastal, Lake Eyre basin, W plateau, W.A., N.T. Ecology: larva - sedentary, predator : adult - volant; prey larval Lepidoptera. Biological references: Giordani Soika, A. (1977). Contributo alla conoscenza degli Eumenidi australiani (Hymenoptera). *Mem. Soc. Entomol. Ital.* **55**: 109–138 (generic placement, distribution).

Delta Saussure, 1855

Delta Saussure, H. de (1855). *Études sur la Famille des Vespides*. Troisième Partie comprenant la Monographie des Masariens et un Supplement à la Monographie des Eumeniens. Paris : Masson 352 pp. pls i–xvi (1854–1856) [130,132,143] [proposed as a subgenus of *Eumenes* Latreille, 1802]. Publication date established from Griffin, F.J. (1939). On the dates of publication of Saussure (H. de) : Études sur la famille des Vespides 1–3, 1852–1858. *J. Soc. Bibliogr. Nat. Hist.* **1**: 211–212. Type species *Vespa maxillosa* DeGeer, 1775 (= *Vespa emarginata* Linnaeus, 1758) by subsequent designation, see Bequaert, J. (1926). The genus *Eumenes* Latreille, in South Africa, with a revision of the Ethiopian species (Hymenoptera). *Ann. S. Afr. Mus.* **23**: 483–577. Compiled from secondary source: van der Vecht, J. & Fischer, F.C.J. (1972). Palearctic Eumenidae. *Hymenopterorum Catalogus* **8**: 1–199.

This group is also found in the Palaearctic, Ethiopian, Neotropical and Oriental Regions, see Giordani Soika, A. (1961). Les lignées philétiques des *Eumenes* s.l. du globe (Hym. Vesp.). *Trans. 11th Int. Congr. Entomol.* (Vienna, 1960) **1**: 240–245.

Delta arcuata (Fabricius, 1775)

Taxonomic decision of van der Vecht, J. (1959). On *Eumenes arcuatus* (Fabricius) and some allied Indo-Australian wasps (Hymenoptera, Vespidae). *Zool. Verh.* **41**: 1–71 [52].

Delta arcuata arcuata (Fabricius, 1775)

Vespa arcuata Fabricius, J.C. (1775). *Systema Entomologiae*, sistens insectorum classes, ordines, genera, species, adiectis synonymis, locis, descriptionibus, observationibus. Flensburgi et Lipsiae : Kortii xxvii 832 pp. [371]. Type data: holotype, BMNH *F. adult, from Australia (as New Holland).

Distribution: NE coastal, N coastal, Qld., W.A.; also in New Guinea, D'Entrecasteaux Is., Trobriand Is., Aru Ils. and Taiwan. Ecology: larva - sedentary, predator : adult - volant; (mud-nest), (prey larval Lepidoptera). Biological references: van der Vecht, J. (1960). On *Abispa* and some other Eumenidae from the Australian region (Hymenoptera, Vespoidea). *Nova Guinea Zool.* **6**: 91–115 (distribution); Giordani Soika, A. (1961). Les lignées philétiques des *Eumenes* s.l. du globe (Hym. Vesp.). *Trans. 11th Int. Congr. Entomol.* (Vienna, 1960) **1**: 240–245 (taxonomy as *Delta* *arcuata*); van der Vecht, J. (1961). Evolution in a group of Indo-Australian *Eumenes* (Hymenoptera, Eumenidae). *Evolution* **15**: 468–477 (evolution).

Delta bicinctus (Saussure, 1852)

Eumenes bicincta Saussure, H. de (1852). *Études sur la Famille des Vespides*. 1. Monographie des guêpes solitaires, ou de la tribu des Euméniens, comprenant la classification et la description de toutes les espèces connues jusqu'à ce jour, et servant de complément au Manuel de Lepeletier de Saint Fargeau. Paris : Masson 286 pp. pls i–xx (1852–1853) [44]. Type data: syntypes (probable), GMNH *adult, from Australia (as New Holland).

Distribution: NE coastal, Murray-Darling basin, SE coastal, Lake Eyre basin, SW coastal, NW coastal, N coastal, W plateau, Qld., N.S.W., Vic., S.A., W.A., N.T. Ecology: larva - sedentary, predator : adult - volant; mud-nest, prey larval Lepidoptera. Biological references: Giordani Soika, A. (1941). Studi sui Vespidi solitari. *Boll. Soc. Veneziana Stor. Nat.* **2**: 130–279 (distribution as *Eumenes (Delta) bicinctus*); James, C.T. (1956). Some nesting habits of *Eumenes bicincta*. *S. Aust. Nat.* **31**: 24 (nest); Giordani Soika, A. (1961). Les lignées philétiques des *Eumenes* s.l. du globe (Hym. Vesp.). *Trans. 11th Int. Congr. Entomol.* (Vienna, 1960) **1**: 240–245 (taxonomy as *Delta bicinctus*); Callan, E.M. (1981). Further records of *Macrosiagon* (Coleoptera : Rhipiphoridae) reared from eumenid and sphecid wasps in Australia. *Aust. Entomol. Mag.* **7**: 81–83 (host, as *Eumenes bicinctus*).

Delta campaniformis (Fabricius, 1775)

Taxonomic decision of Giordani Soika, A. (1934). *Labus* ed *Eumenes* nuovi o poci noti (Hym. Vespidae). *Mem. Soc. Entomol. Ital.* **12**: 215–228 [224].

Delta campaniformis campaniformis (Fabricius, 1775)

Vespa campaniformis Fabricius, J.C. (1775). *Systema Entomologiae*, sistens insectorum classes, ordines, genera, species, adiectis synonymis, locis, descriptionibus, observationibus. Flensburgi et Lipsiae : Kortii xxvii 832 pp. [371]. Type data: holotype, BMNH *F. adult, from Australia (as New Holland).

Distribution: NE coastal, N coastal, NW coastal, Qld., W.A.; also in New Guinea, Philippines, Java and Yule Is. Ecology: larva - sedentary, predator : adult - volant; mud-nest, prey larval Lepidoptera. Biological references: Giordani Soika, A. (1961). Les lignées philétiques des *Eumenes* s.l. du globe (Hym. Vesp.). *Trans. 11th Int. Congr. Entomol.* (Vienna, 1960) **1**: 240–245 (taxonomy, generic placement); Baltazar, C.R. (1966). A catalogue of Philippine Hymenoptera (with a bibliography, 1758–1963). *Pac. Insects Monogr.* **8**: 1–488 (distribution); Anderson, D.L., Sedgley, M., Short, J.R.T. & Allwood, A.J. (1982). Insect pollination of mango in Northern Australia. *Aust. J. Agric. Res.* **33**: 541–548 (as pollinator).

Delta campaniformis assatus (Giordani Soika, 1934)

Eumenes campaniformis assatus Giordani Soika, A. (1934). *Labus* ed *Eumenes* nuovi o poci noti (Hym. Vespidae). *Mem. Soc. Entomol. Ital.* **12**: 215–228 [224] [described as variety]. Type data: lectotype, A. Giordani Soika pers. coll. *F. adult, from Qld., designation by Giordani Soika, A. (1973). Designazione di lectotipi ed elenco dei tipi di Eumenidi, Vespidi e Masaridi da me descritti negli anni 1934–1960. *Boll. Mus. Civ. Stor. Nat. Venezia* **24**: 7–53.

Distribution: Qld.; no locality specified. Ecology: larva - sedentary, predator : adult - volant; mud-nest, prey larval Lepidoptera.

Delta incola (Giordani Soika, 1935)

Taxonomic decision of van der Vecht, J. (1959). On *Eumenes arcuatus* (Fabricius) and some allied Indo-Australian wasps (Hymenoptera, Vespidae). *Zool. Verh.* **41**: 1–71 [28].

Delta incola aruensis (Giordani Soika, 1935)

Eumenes (Delta) incola aruensis Giordani Soika, A. (1935). Ricerche sistematiche sugli *Eumenes* e *Pareumenes* dell'Archipelago Malese e della Nuova Guinea. *Ann. Mus. Civ. Stor. Nat. Giacomo Doria* **57**: 114–151 [137] [described as variety]. Type data: lectotype, MCG *F. adult, from Wokan, Aru, designation by Giordani Soika, A. (1973). Designazione di lectotipi ed elenco dei tipi di Eumenidi, Vespidi e Masaridi da me descritti negli anni 1934–1960. *Boll. Mus. Civ. Stor. Nat. Venezia* **24**: 7–53.

Distribution: NE coastal, Qld.; only published localities Mackay and Cairns, also Aru Ils. Ecology: larva - sedentary, predator : adult - volant; mud-nest, prey larval Lepidoptera. Biological references: Giordani Soika, A. (1941). Studi sui Vespidi solitari. *Boll. Soc. Veneziana Stor. Nat.* **2**: 130–279 (distribution).

Delta incola teleporus (van der Vecht, 1959)

Eumenes incola teleporus van der Vecht, J. (1959). On *Eumenes arcuatus* (Fabricius) and some allied Indo-Australian wasps (Hymenoptera, Vespidae). *Zool. Verh.* **41**: 1–71 [30]. Type data: holotype, BMNH *F. adult, from Mackay, Qld.

Distribution: NE coastal, Qld.; only published localities Mackay, Cairns and Kuranda. Ecology: larva - sedentary, predator : adult - volant; mud-nest, prey larval Lepidoptera.

Delta latreillei (Saussure, 1852)

Taxonomic decision of van der Vecht, J. (1960). On *Abispa* and some other Eumenidae from the Australian region (Hymenoptera, Vespoidea). *Nova Guinea Zool.* **6**: 91–115 [111].

Delta latreillei latreillei (Saussure, 1852)

Eumenes latreillei Saussure, H. de (1852). *Études sur la Famille des Vespides*. 1. Monographie des guêpes solitaires, ou de la tribu des Euméniens, comprenant la classification et la description de toutes les espèces connues jusqu'à ce jour, et servant de complément au Manuel de Lepeletier de Saint Fargeau. Paris : Masson 286 pp. pls i–xx (1852–1853) [51 pl x figs 5a–b]. Type data: holotype, MCG *F. adult, from Australia (as New Holland).

Distribution: NE coastal, Lake Eyre basin, SE coastal, N coastal, Qld., Vic., W.A., N.T.; also in New Guinea. Ecology: larva - sedentary, predator : adult - volant; mud-nest, prey larval Lepidoptera. Biological references: Roth, H.L. (1885). Notes on the habits of some Australian Hymenoptera Aculeata. (With description of a new species by W.F. Kirby). *J. Linn. Soc. Zool.* **18**: 318–328 (biology); Girault, A.A. (1914). Observations on an Australian mud dauber which uses in part its own saliva in nest construction. *Z. Wiss. InsektBiol.* **10**: 28–32 (biology); Giordani Soika, A. (1941). Studi sui Vespidi solitari. *Boll. Soc. Veneziana Stor. Nat.* **2**: 130–279 (taxonomy as *Eumenes (Delta) pyriformis* var. *latreillei*); Donnell, F.O. (1944). Cell-building by a mason wasp. *Vict. Nat.* **61**: 67–68 (biology); Giordani Soika, A. (1958). Biogeografia, evoluzione e sistematica dei Vespidi solitari della Polinesia meridionale. *Boll. Mus. Civ. Stor. Nat. Venezia* **10**: 183–221 (taxonomy, distribution); Giordani Soika, A. (1961). Les lignées philétiques des *Eumenes* s.l. du globe (Hym. Vesp.). *Trans. 11th Int. Congr. Entomol.* (Vienna, 1960) **1**: 240–245 (taxonomy, generic placement); Callan, E.M. (1981). Further records of *Macrosiagon* (Coleoptera : Rhipiphoridae) reared from eumenid and sphecid wasps in Australia. *Aust. Entomol. Mag.* **7**: 81–83 (as host).

Delta nigritarsis (Meade-Waldo, 1910)

Eumenes nigritarsis Meade-Waldo, G. (1910). New species of Diploptera in the collection of the British Museum. *Ann. Mag. Nat. Hist. (8)* **5**: 30–51 [43]. Type data: holotype, BMNH *M. adult (seen 1929 by L.F. Graham), from Darwin, N.T.

Distribution: N coastal, N.T.; type locality only. Ecology: larva - sedentary, predator : adult - volant; mud-nest, prey larval Lepidoptera. Biological references: Bequaert, J.C. (1928). A study of certain types of diplopterous wasps in the collection of the British Museum. *Ann. Mag. Nat. Hist. (10)* **2**: 138–176 (taxonomy as *Eumenes (Delta) pyriformis* var. *nigritarsis*); Giordani Soika, A. (1941). Studi sui Vespidi solitari. *Boll. Soc. Veneziana Stor. Nat.* **2**: 130–279 (taxonomy as *Eumenes (Delta) pyriformis* var. *nigritarsis*).

Delta philantes (Saussure, 1852)

Eumenes philantes Saussure, H. de (1852). *Études sur la Famille des Vespides*. 1. Monographie des guêpes solitaires, ou de la tribu des Euméniens, comprenant la classification et la description de toutes les espèces connues jusqu'à ce jour, et servant de complément au Manuel de Lepeletier de Saint Fargeau. Paris : Masson 286 pp. pls i–xx (1852–1853) [54]. Type data: holotype, BMNH *F. adult (seen 1929 by L.F. Graham), from Australia (as New Holland).

Distribution: NE coastal, N coastal, Lake Eyre basin, Qld., W.A., S.A.; only published localities Cairns, Ord River and Killalpanima. Ecology: larva - sedentary, predator : adult - volant; mud-nest, prey larval Lepidoptera. Biological references: Giordani Soika, A. (1941). Studi sui Vespidi solitari. *Boll. Soc. Veneziana Stor. Nat.* **2**: 130–279 (taxonomy as *Eumenes (Delta) philantes*); Giordani Soika, A. (1958). Biogeografia, evoluzione e sistematica dei Vespidi solitari della Polinesia meridionale. *Boll. Mus. Civ. Stor. Nat. Venezia* **10**: 183–221 (taxonomy); Giordani Soika, A. (1973). Designazione di lectotipi ed elenco dei tipi di Eumenidi, Vespidi e Masaridi da me descritti negli anni 1934–1960. *Boll. Mus. Civ. Stor. Nat. Venezia* **24**: 7–53 (taxonomy).

Delta transmarinum (van der Vecht, 1959)

Eumenes transmarinus van der Vecht, J. (1959). On *Eumenes arcuatus* (Fabricius) and some allied Indo-Australian wasps (Hymenoptera, Vespidae). *Zool. Verh.* **41**: 1–71 [71]. Type data: holotype, BMNH *F. adult, from De Freycinet Is., W.A.

Distribution: N coastal, W.A.; only published localities De Freycinet Is. and in Kimberley area. Ecology: larva - sedentary, predator : adult - volant; mud-nest, prey larval Lepidoptera. Biological references: van der Vecht, J. (1981). Indo-Australian solitary wasps. *Proc. K. Ned. Akad. Wet. (c)* **84**: 443–464 (taxonomy, distribution).

Delta xanthurum (Saussure, 1852)

Taxonomic decision of Giordani Soika, A. (1958). Biogeografia, evoluzione e sistematica dei Vespidi solitari della Polinesia meridionale. *Boll. Mus. Civ. Stor. Nat. Venezia* **10**: 183–221 [201].

Delta xanthurum xanthurum (Saussure, 1852)

Eumenes xanthura Saussure, H. de (1852). *Études sur la Famille des Vespides*. 1. Monographie des guêpes solitaires, ou de la tribu des Euméniens, comprenant la classification et la description de toutes les espèces connues jusqu'à ce jour, et servant de complément au

Manuel de Lepeletier de Saint Fargeau. Paris : Masson 286 pp. pls i–xx (1852–1853) [46 pl x fig 4]. Type data: holotype, MNHP *F. adult, from East Indies (as Indes or.).

Distribution: NE coastal, Qld.; only published localities Murray Is., also Loyalty Ils., India, Indo-China, Malay Archipelago, New Caledonia and New Hebrides. Ecology: larva - sedentary, predator : adult - volant; mud-nest, prey larval Lepidoptera. Biological references: Giordani Soika, A. (1974). Prime ricerche sugli Eumenidi ipsobionti. I. Caratteristiche generali degli Eumenidi ipsobionti del globo. *Redia* **55**: 287–302 (generic placement, Australian record).

Diemodynerus Giordani Soika, 1962

Diemodynerus Giordani Soika, A. (1962). Gli *Odynerus* sensu antiquo del continente australiano e della Tasmania. *Boll. Mus. Civ. Stor. Nat. Venezia* **14**: 57–202 [141]. Type species *Odynerus diemensis* Saussure, 1853 by original designation.

Diemodynerus decipiens (Saussure, 1867)

Taxonomic decision of Giordani Soika, A. (1977). Contributo alla conoscenza degli Eumenidi australiani (Hymenoptera). *Mem. Soc. Entomol. Ital.* **55**: 109–138 [125].

Diemodynerus decipiens decipiens (Saussure, 1867)

Odynerus (Leionotus) decipiens Saussure, H. de (1867). Hymenoptera. in, *Reise der österreichischen Fregatte Novara um die Erde in den Jahren 1857, 1858, 1859 unter den Befehlen des Commodore B. von Wüllerstorf-Urbair.* Zoologischer Theil. Wien : K-K Hof- und Staatsdrückerei Vol. 2(1a) 138 pp. [11 pl I fig 6]. Type data: syntypes (probable), NHMW *.F adult, from Sydney, N.S.W.

Distribution: SE coastal, Murray-Darling basin, S Gulfs, N.S.W., Vic., S.A. Ecology: larva - sedentary, predator : adult - volant; prey larval Lepidoptera. Biological references: Giordani Soika, A. (1941). Studi sui Vespidi solitari. *Boll. Soc. Veneziana Stor. Nat.* **2**: 130–279 (description of male as *Odynerus (Rhynchium) decipiens*); Giordani Soika, A. (1962). Gli *Odynerus* sensu antiquo del continente australiano e della Tasmania. *Boll. Mus. Civ. Stor. Nat. Venezia* **14**: 57–202 (redescription, generic placement); Giordani Soika, A. (1973). Designazione di lectotipi ed elenco dei tipi di Eumenidi, Vespidi e Masaridi da me descritti negli anni 1934–1960. *Boll. Mus. Civ. Stor. Nat. Venezia* **24**: 7–53 (taxonomy).

Diemodynerus decipiens positus Giordani Soika, 1977

Diemodynerus decipiens positus Giordani Soika, A. (1977). Contributo alla conoscenza degli Eumenidi australiani (Hymenoptera). *Mem. Soc. Entomol. Ital.* **55**: 109–138 [125]. Type data: holotype, NMV *M. adult not found, from Port Lincoln, S.A.

Distribution: S Gulfs, S.A.; type locality only. Ecology: larva - sedentary, predator : adult - volant; prey larval Lepidoptera.

Diemodynerus diemensis (Saussure, 1853)

Odynerus (Leionotus) diemensis Saussure, H. de (1853). *Études sur la Famille des Vespides.* 1. Monographie des guêpes solitaires, ou de la tribu des Euméniens, comprenant la classification et la description de toutes les espèces connues jusqu'à ce jour, et servant de complément au Manuel de Lepeletier de Saint Fargeau. Paris : Masson 286 pp. pls i–xx (1852–1853) [201]. Type data: holotype, MNHP *F. adult, from Tas.

Distribution: NE coastal, SE coastal, Qld., N.S.W., Tas.; only published localities Brisbane, Sydney and Tas. Ecology: larva - sedentary, predator : adult - volant; prey larval Lepidoptera. Biological references: Giordani Soika, A. (1962). Gli *Odynerus* sensu antiquo del continente australiano e della Tasmania. *Boll. Mus. Civ. Stor. Nat. Venezia* **14**: 57–202 (generic placement).

Diemodynerus pseudacarodynerus Giordani Soika, 1962

Diemodynerus pseudacarodynerus Giordani Soika, A. (1962). Gli *Odynerus* sensu antiquo del continente australiano e della Tasmania. *Boll. Mus. Civ. Stor. Nat. Venezia* **14**: 57–202 [142]. Type data: holotype, AM K68217 F. adult, from King George Sound, W.A.

Distribution: SW coastal, W.A.; only published localities King George Sound, Bunbury and Gnowangerup (as Ghovargi). Ecology: larva - sedentary, predator : adult - volant; prey larval Lepidoptera. Biological references: Giordani Soika, A. (1977). Contributo alla conoscenza degli Eumenidi australiani (Hymenoptera). *Mem. Soc. Entomol. Ital.* **55**: 109–138 (description of male).

Diemodynerus saucius (Saussure, 1856)

Odynerus saucius Saussure, H. de (1856). *Études sur la Famille des Vespides.* Troisième Partie comprenant la Monographie des Masariens et un Supplement à la Monographie des Eumeniens. Paris : Masson 352 pp. pls i–xvi (1854–1856) [280]. Type data: holotype, BMNH *F. adult (seen 1929 by L.F. Graham), from Australia (as New Holland).

Distribution: NE coastal, SE coastal, Murray-Darling basin, S Gulfs, Qld., N.S.W., Vic., S.A. Ecology: larva - sedentary, predator : adult - volant; prey larval Lepidoptera. Biological references: Giordani Soika, A. (1962). Gli *Odynerus* sensu antiquo del continente australiano e della Tasmania. *Boll. Mus. Civ. Stor. Nat. Venezia* **14**: 57–202 (redescription, generic placement); Giordani Soika, A. (1977). Contributo alla conoscenza degli Eumenidi australiani (Hymenoptera). *Mem. Soc. Entomol. Ital.* **55**: 109–138 (distribution).

Ectopioglossa Perkins, 1912

Ectopioglossa Perkins, R.C.L. (1912). Notes, with descriptions of new species, on aculeate Hymenoptera of the Australian Region. *Ann. Mag. Nat. Hist. (8)* **9**: 96–121 [118]. Type species *Ectopioglossa australensis* Perkins, 1912 (=*Eumenes australensis* Meade-Waldo, 1910) by monotypy.

This group is also found in the Oriental Region, see van der Vecht, J. (1963). Studies on Indo-Australian and east-Asiatic Eumenidae (Hymenoptera, Vespoidea). *Zool. Verh.* **60**: 1–116.

Ectopioglossa polita (Smith, 1861)

Taxonomic decision of van der Vecht, J. (1981). Indo-Australian solitary wasps. *Proc. K. Ned. Akad. Wet. (c)* **84**: 443–464 [456].

Ectopioglossa polita australensis (Meade-Waldo, 1910)

Eumenes (Pareumenes) australensis Meade-Waldo, G. (1910). New species of Diploptera in the collection of the British Museum. *Ann. Mag. Nat. Hist. (8)* **5**: 30–51 [44]. Type data: holotype, BMNH *M. adult (seen 1929 by L.F. Graham), from Kuranda, Qld.

Ectopioglossa australensis Perkins, R.C.L. (1912). Notes, with descriptions of new species, on aculeate Hymenoptera of the Australian Region. *Ann. Mag. Nat. Hist. (8)* **9**: 96–121 [119]. Type data: holotype, BMNH *adult, from Cairns, Qld.

Distribution: NE coastal, Qld.; only published localities Kuranda and Cairns. Ecology: larva - sedentary, predator : adult - volant; prey larval Lepidoptera. Biological references: Meade-Waldo, G. (1914). Notes on the Hymenoptera in the collection of the British Museum, with descriptions of new species. V. *Ann. Mag. Nat. Hist. (8)* **14**: 450–464 (redescription as *Pareumenes australensis*); van der Vecht, J. (1963). Studies on Indo-Australian and east-Asiatic Eumenidae (Hymenoptera, Vespoidea). *Zool. Verh.* **60**: 1–116 (taxonomy).

Elimus Saussure, 1852

Elimus Saussure, H. de (1852). *Études sur la Famille des Vespides.* 1. Monographie des guêpes solitaires, ou de la tribu des Euméniens, comprenant la classification et la description de toutes les espèces connues jusqu'à ce jour, et servant de complément au Manuel de Lepeletier de Saint Fargeau. Paris : Masson 286 pp. pls i-xx (1852–1853) [7]. Publication date established from Griffin, F.J. (1939). On the dates of publication of Saussure (H. de) : Études sur la famille des Vespides 1–3, 1852-1858. *J. Soc. Bibliogr. Nat. Hist.* **1**: 211–212. Type species *Elimus australis* Saussure, 1852 by monotypy.

Elimus australis Saussure, 1852

Elimus australis Saussure, H. de (1852). *Études sur la Famille des Vespides.* 1. Monographie des guêpes solitaires, ou de la tribu des Euméniens, comprenant la classification et la description de toutes les espèces connues jusqu'à ce jour, et servant de complément au Manuel de Lepeletier de Saint Fargeau. Paris : Masson 286 pp. pls i-xx (1852–1853) [8 pl iii fig 1a–c]. Type data: holotype, MNHP *M. adult, from S.A.

Distribution: NE coastal, SE coastal, Qld., N.S.W., S.A.; only published localities Stradbroke Is., Sydney and S.A. (New Guinea record is considered doubtful). Ecology: larva - sedentary, predator : adult - volant; prey larval Lepidoptera. Biological references: Giordani Soika, A. (1969). Revisione dei Discoeliinae Australiani. *Boll. Mus. Civ. Stor. Nat. Venezia* **19**: 25–100 (redescription).

Elimus mackayensis Meade-Waldo, 1910

Elimus mackayensis Meade-Waldo, G. (1910). New species of Diploptera in the collection of the British Museum. *Ann. Mag. Nat. Hist. (8)* **5**: 30–51 [39]. Type data: holotype, BMNH *F. adult (seen 1929 by L.F. Graham), from Mackay, Qld.

Distribution: NE coastal, Qld.; only published localities Mackay and Brisbane. Ecology: larva - sedentary, predator : adult - volant; prey larval Lepidoptera. Biological references: Bequaert, J.C. (1928). A study of certain types of diplopterous wasps in the collection of the British Museum. *Ann. Mag. Nat. Hist. (10)* **2**: 138–176 (taxonomy); Giordani Soika, A. (1969). Revisione dei Discoeliinae Australiani. *Boll. Mus. Civ. Stor. Nat. Venezia* **19**: 25–100 (redescription).

Epiodynerus Giordani Soika, 1958

Epiodynerus Giordani Soika, A. (1958). Biogeografia, evoluzione e sistematica dei Vespidi solitari della Polinesia meridionale. *Boll. Mus. Civ. Stor. Nat. Venezia* **10**: 183–221 [195] [proposed with subgeneric rank in *Pseudepipona* Saussure, 1856]. Type species *Odynerus alecto* Lepeletier, 1841 by original designation.

This group is also found in the Oriental Region and Pacific islands, see van der Vecht, J. (1963). Studies on Indo-Australian and east-Asiatic Eumenidae (Hymenoptera, Vespoidea). *Zool. Verh.* **60**: 1–116.

Epiodynerus decoratus (Saussure, 1855)

Rhynchium decoratum Saussure, H. de (1855). *Études sur la Famille des Vespides.* Troisième Partie comprenant la Monographie des Masariens et un Supplement à la Monographie des Eumeniens. Paris : Masson 352 pp. pls i-xvi (1854–1856) [180 pl ix fig 6]. Type data: holotype, MNHP *F. adult, from Australia (as New Holland).

Distribution: no locality specified. Ecology: larva - sedentary, predator : adult - volant; prey larval Lepidoptera. Biological references: Giordani Soika, A. (1962). Gli *Odynerus* sensu antiquo del continente australiano e della Tasmania. *Boll. Mus. Civ. Stor. Nat. Venezia* **14**: 57-202 (redescription as *Pseudepipona (Epiodynerus) decorata*).

Epiodynerus nigrocinctus (Saussure, 1853)

Taxonomic decision of van der Vecht, J. (1963). Studies on Indo-Australian and east-Asiatic Eumenidae (Hymenoptera, Vespoidea). *Zool. Verh.* **60**: 1-116 [93].

Epiodynerus nigrocinctus nigrocinctus (Saussure, 1853)

Odynerus (Leionotus) nigrocinctus Saussure, H. de (1853). *Études sur la Famille des Vespides.* 1. Monographie des guêpes solitaires, ou de la tribu des Euméniens, comprenant la classification et la description de toutes les espèces connues jusqu'à ce jour, et servant de complément au Manuel de Lepeletier de Saint Fargeau. Paris : Masson 286 pp. pls i-xx (1852-1853) [201]. Type data: holotype, MNHP *M. adult, from Tas.

Distribution: NE coastal, Murray-Darling basin, Lake Eyre basin, SW coastal, N coastal, NW coastal, Qld., N.S.W., Vic., W.A., N.T., Tas.; also in New Guinea and Solomon Ils. (probably introduced). Ecology: larva - sedentary, predator : adult - volant; nest in hole in mud brick wall, prey larval Lepidoptera. Biological references: Giordani Soika, A. (1941). Studi sui Vespidi solitari. *Boll. Soc. Veneziana Stor. Nat.* **2**: 130-279 (redescription as *Odynerus (Rhynchium) nigrocinctus*); Giordani Soika, A. (1962). Gli *Odynerus* sensu antiquo del continente australiano e della Tasmania. *Boll. Mus. Civ. Stor. Nat. Venezia* **14**: 57-202 (redescription, distribution, as *Pseudepipona (Epiodynerus) nigrocinctus*); McKenzie, J. (1975). Studies of the behaviour of some insects nesting in a mud brick wall. *Qd. Nat.* **21**: 63-64 (biology as *Pseudepipona nigrocincta*); Giordani Soika, A. (1977). Contributo alla conoscenza degli Eumenidi australiani (Hymenoptera). *Mem. Soc. Entomol. Ital.* **55**: 109-138 (distribution); Wilson, A.G.L. & Greenup, L.R. (1977). The relative injuriousness of insect pests of cotton in the Namoi Valley, New South Wales. *Aust. J. Ecol.* **2**: 319-328 (prey).

Epiodynerus tamarinus (Saussure, 1853)

Taxonomic decision of Giordani Soika, A. (1977). Contributo alla conoscenza degli Eumenidi australiani (Hymenoptera). *Mem. Soc. Entomol. Ital.* **55**: 109-138 [118].

Epiodynerus tamarinus tamarinus (Saussure, 1853)

Odynerus (Leionotus) tamarinus Saussure, H. de (1853). *Études sur la Famille des Vespides.* 1. Monographie des guêpes solitaires, ou de la tribu des Euméniens, comprenant la classification et la description de toutes les espèces connues jusqu'à ce jour, et servant de complément au Manuel de Lepeletier de Saint Fargeau. Paris : Masson 286 pp. pls i-xx (1852-1853) [203]. Type data: holotype, MNHP *adult, from Tas. (locality is probably incorrect), see Giordani Soika, A. (1962). Gli *Odynerus* sensu antiquo del continente australiano e della Tasmania. *Boll. Mus. Civ. Stor. Nat. Venezia* **14**: 57-202.

Distribution: NE coastal, SE coastal, SW coastal, Qld., N.S.W., W.A., (Tas.). Ecology: larva - sedentary, predator : adult - volant; prey larval Lepidoptera. Biological references: Giordani Soika, A. (1962). Gli *Odynerus* sensu antiquo del continente australiano e della Tasmania. *Boll. Mus. Civ. Stor. Nat. Venezia* **14**: 57-202 (redescription as *Pseudepipona (Epiodynerus) tamarina*).

Epiodynerus tamarinus inviolatus Giordani Soika, 1977

Epiodynerus tamarinus inviolatus Giordani Soika, A. (1977). Contributo alla conoscenza degli Eumenidi australiani (Hymenoptera). *Mem. Soc. Entomol. Ital.* **55**: 109-138 [118]. Type data: holotype, MNH *adult, from N.S.W.

Distribution: N.S.W.; no locality specified. Ecology: larva - sedentary, predator : adult - volant; prey larval Lepidoptera.

Epiodynerus tasmaniensis (Saussure, 1853)

Odynerus (Leionotus) tasmaniensis Saussure, H. de (1853). *Études sur la Famille des Vespides.* 1. Monographie des guêpes solitaires, ou de la tribu des Euméniens, comprenant la classification et la description de toutes les espèces connues jusqu'à ce jour, et servant de complément au Manuel de Lepeletier de Saint Fargeau. Paris : Masson 286 pp. pls i-xx (1852-1853) [199 pl xviii fig 4]. Type data: holotype, MNHP *M. adult, from Tas.

Rhynchium abispoides Meade-Waldo, G. (1910). New species of Diploptera in the collection of the British Museum. *Ann. Mag. Nat. Hist.* (8) **5**: 30-51 [50]. Type data: holotype, BMNH *F. adult (seen 1929 by L.F. Graham), from Townsville, Qld., see Cheesman, L.E. (1954). A new species of *Odynerus*, subgen. *Rhygchium* (Eumeninae), from the Loyalty Islands. *Ann. Mag. Nat. Hist.* (12) **7**: 385-390.

Taxonomic decision of Giordani Soika, A. (1962). Gli *Odynerus* sensu antiquo del continente australiano e della Tasmania. *Boll. Mus. Civ. Stor. Nat. Venezia* **14**: 57-202 [92].

Distribution: NE coastal, Murray-Darling basin, S Gulfs, W plateau, SW coastal, NW coastal, Qld., N.S.W., Vic., S.A., W.A., Tas. Ecology: larva - sedentary, predator : adult - volant; prey larval Lepidoptera. Biological references: CSIRO (1970). *The Insects of Australia.* A textbook for students and research workers. Melbourne : Melbourne

Univ. Press 1029 pp. (plate 6 K - model for mimicry); Giordani Soika, A. (1977). Contributo alla conoscenza degli Eumenidi australiani (Hymenoptera). *Mem. Soc. Entomol. Ital.* **55**: 109-138 (distribution).

Eudiscoelius Friese, 1904

Eudiscoelius Friese, H. (1904). Eine metallisch gefärbte Vespide. *Z. Syst. Hymenopt. Dipterol.* **4**: 16. Type species *Eudiscoelius metallicus* Friese, 1904 by monotypy.

Euchalcomenes Turner, R.E. (1908). Two new diplopterous Hymenoptera from Queensland. *Trans. Entomol. Soc. Lond.* **1908**: 89-91 [90]. Type species *Euchalcomenes gilberti* Turner, 1908 by original designation.

Taxonomic decision of Giordani Soika, A. (1943). Le specie indo-australiane del genere *Pachymenes* (Hym. Vespidae). *Mem. Soc. Entomol. Ital.* **22**: 102-117 [102].

This group is also found in the Oriental Region, see Giordani Soika, A. (1943). Le specie indo-australiane del genere *Pachymenes* (Hym. Vespidae). *Mem. Soc. Entomol. Ital.* **22**: 102-117.

Eudiscoelius gilberti (Turner, 1908)

Euchalcomenes gilberti Turner, R.E. (1908). Two new diplopterous Hymenoptera from Queensland. *Trans. Entomol. Soc. Lond.* **1908**: 89-91 [90]. Type data: holotype, BMNH *F. adult (seen 1929 by L.F. Graham), from Kuranda, Qld.

Distribution: NE coastal, Qld.; only published localities Kuranda and Cairns. Ecology: larva - sedentary, predator : adult - volant; prey larval Lepidoptera. Biological references: Meade-Waldo, G. (1910). New species of Diploptera in the collection of the British Museum. *Ann. Mag. Nat. Hist. (8)* **5**: 30-51 (taxonomy as *Nortonia gilberti*); Giordani Soika, A. (1943). Le specie indo-australiane del genere *Pachymenes* (Hym. Vespidae). *Mem. Soc. Entomol. Ital.* **22**: 102-117 (taxonomy as *Pachymenes gilberti*); Riek, E.F. *in* CSIRO (1970). *The Insects of Australia*. A textbook for students and research workers. Melbourne : Melbourne Univ. Press 1029 pp. (generic placement).

Eumenes Latreille, 1802

Eumenes Latreille, P.A. (1802). *Histoire Naturelle, Générale et Particulière des Crustacés et des Insectes.* Paris : F. Dufart Vol. 3 xii 13+467 pp. (1802-1803) [360]. Type species *Vespa coarctata* Linnaeus, 1758 by subsequent designation, see Latreille, P.A. (1810). *Considérations Générales sur l'Ordre Naturel des Animaux Composant les Classes des Crustacès, des Arachnides, et des Insectes;* avec un tableau méthodique de leurs genres, disposés en familles. Paris : F. Schoell 444 pp.

This group is found worldwide, see Richards, O.W. (1980). Scolioidea, Vespoidea and Sphecoidea. Hymenoptera, Aculeata. *Handbk. Ident. Br. Insects* **6**(3b): 1-118. Species now known not to occur in Australia: *Eumenes fluctuans* Saussure, 1852 (now placed in *Katamenes* Meade-Waldo, 1910), described from Australia (but without locality label on the holotype), is from the Palaearctic Region teste van der Vecht, J. & Fischer, F.C.J. (1972). Palearctic Eumenidae. *Hymenopterorum Catalogus* **8**: 1-199.

Eumenes apicalis Macleay, 1826

Eumenes apicalis Macleay, W.S. (1826). Annulosa. Catalogue of insects, collected by Captain King, R.N. pp. 438-469 *in* King, P.P. *Narrative of a Survey of the Intertropical and Western Coasts of Australia Performed between the Years 1818 and 1822.* London : John Murray Vol. 2 [457]. Type data: syntypes (probable), adult whereabouts unknown, from Australia.

Distribution: no locality specified. Ecology: larva - sedentary, predator : adult - volant; prey larval Lepidoptera.

Eumenes simplicilamellatus Giordani Soika, 1935

Eumenes (Eumenes) simplicilamellatus Giordani Soika, A. (1935). Ricerche sistematiche sugli *Eumenes* e *Pareumenes* dell'Archipelago Malese e della Nuova Guinea. *Ann. Mus. Civ. Stor. Nat. Giacomo Doria* **57**: 114-151 [124 pl 2 fig 4]. Type data: lectotype, MCG *F. adult, from Kapakapa, New Guinea, designation by Giordani Soika, A. (1973). Designazione di lectotipi ed elenco dei tipi di Eumenidi, Vespidi e Masaridi da me descritti negli anni 1934-1960. *Boll. Mus. Civ. Stor. Nat. Venezia* **24**: 7-53. from Kapakapa, New Guinea.

Distribution: NE coastal, Qld.; only published localities Cairns, Cooktown and Kuranda, also in New Guinea. Ecology: larva - sedentary, predator : adult - volant; prey larval Lepidoptera. Biological references: Giordani Soika, A. (1941). Studi sui Vespidi solitari. *Boll. Soc. Veneziana Stor. Nat.* **2**: 130-279 (distribution); Giordani Soika, A. (1961). Les lignées philétiques des *Eumenes* s.l. du globe (Hym. Vesp.). *Trans. 11th Int. Congr. Entomol.* (Vienna, 1960) **1**: 240-245 (taxonomy).

Euodynerus Dalla Torre, 1904

Euodynerus Dalla Torre, K.W. von (1904). Hymenoptera. Fam. Vespidae. *Genera Insectorum* **19**: 1-108 [38] [proposed for secondary division of *Odynerus* Latreille, 1802; placed on Official List of Generic Names in Zoology, see International Commission on Zoological Nomenclature (1970). Opinion 893. Eumenidae names of Saussure (Hymenoptera) : grant of availability to certain names proposed for secondary divisions of genera. *Bull. Zool. Nomen.* **26**: 187-191]. Type species *Vespa dantici* Rossi, 1790 by subsequent designation, see Blüthgen, P. (1938). Systematisches Verzeichnis der Faltenwespen Mittel-europas, Skandinaviens und Englands. *Konowia* **16**: 270-295.

This group is also found in the Holarctic and Ethiopian Regions, see Richards, O.W. (1980). Scolioidea, Vespoidea and Sphecoidea. Hymenoptera, Aculeata. *Handbk. Ident. Br. Insects* **6**(3b): 1–118.

Euodynerus polyphemus (Kirby, 1888)

Odynerus polyphemus Kirby, W.F. (1888). On the insects (exclusive of Coleoptera and Lepidoptera) of Christmas Island. *Proc. Zool. Soc. Lond.* **1888**: 546–555 [551]. Type data: syntypes, BMNH *5M.,4F. adult, from Christmas Is.

Distribution: Christmas Is.; type locality only. Ecology: larva - sedentary, predator : adult - volant; prey larval Lepidoptera. Biological references: Kirby, W.F. (1900). Hymenoptera. pp. 81–88 *in* Andrews, C.W. *A Monograph of Christmas Island (Indian Ocean): physical features and geology; with descriptions of the fauna and flora by numerous contributors.* London : British Museum 337+20 pp. (repeats description); Giordani Soika, A. (1958). Biogeografia, evoluzione e sistematica dei Vespidi solitari della Polinesia meridionale. *Boll. Mus. Civ. Stor. Nat. Venezia* **10**: 183–221 (taxonomy).

Flammodynerus Giordani Soika, 1962

Flammodynerus Giordani Soika, A. (1962). Gli *Odynerus* sensu antiquo del continente australiano e della Tasmania. *Boll. Mus. Civ. Stor. Nat. Venezia* **14**: 57–202 [124]. Type species *Odynerus subalaris* Saussure, 1855 by original designation.

Flammodynerus flammiger (Saussure, 1856)

Taxonomic decision of Giordani Soika, A. (1962). Gli *Odynerus* sensu antiquo del continente australiano e della Tasmania. *Boll. Mus. Civ. Stor. Nat. Venezia* **14**: 57–202 [131].

Flammodynerus flammiger flammiger (Saussure, 1856)

Odynerus flammiger Saussure, H. de (1856). *Études sur la Famille des Vespides. Troisième Partie comprenant la Monographie des Masariens et un Supplement à la Monographie des Eumeniens.* Paris : Masson 352 pp. pls i–xvi (1854–1856) [282]. Type data: holotype, BMNH *M. adult (seen 1929 by L.F. Graham), from N.S.W.

Distribution: SE coastal, Murray-Darling basin, N.S.W., Vic. Ecology: larva - sedentary, predator : adult - volant; prey larval Lepidoptera. Biological references: Giordani Soika, A. (1962). Gli *Odynerus* sensu antiquo del continente australiano e della Tasmania. *Boll. Mus. Civ. Stor. Nat. Venezia* **14**: 57–202 (redescription, generic placement); Giordani Soika, A. (1977). Contributo alla conoscenza degli Eumenidi australiani (Hymenoptera). *Mem. Soc. Entomol. Ital.* **55**: 109–138 (taxonomy, distribution).

Flammodynerus flammiger nigroflammeus Giordani Soika, 1962

Flammodynerus flammiger nigroflammeus Giordani Soika, A. (1962). Gli *Odynerus* sensu antiquo del continente australiano e della Tasmania. *Boll. Mus. Civ. Stor. Nat. Venezia* **14**: 57–202 [131]. Type data: holotype, ZMB *F. adult, from Melbourne, Vic.

Distribution: SE coastal, Vic.; type locality only. Ecology: larva - sedentary, predator : adult - volant; prey larval Lepidoptera.

Flammodynerus pseudoloris Giordani Soika, 1962

Flammodynerus pseudoloris Giordani Soika, A. (1962). Gli *Odynerus* sensu antiquo del continente australiano e della Tasmania. *Boll. Mus. Civ. Stor. Nat. Venezia* **14**: 57–202 [125]. Type data: holotype, ZMB *F. adult, from Marloo Station, Wurarga, W.A.

Distribution: NW coastal, W.A.; type locality only. Ecology: larva - sedentary, predator : adult - volant; prey larval Lepidoptera.

Flammodynerus subalaris (Saussure, 1855)

Odynerus subalaris Saussure, H. de (1855). *Études sur la Famille des Vespides. Troisième Partie comprenant la Monographie des Masariens et un Supplement à la Monographie des Eumeniens.* Paris : Masson 352 pp. pls i–xvi (1854–1856) [240 pl xiv fig 5]. Type data: holotype, MNHP *F. adult, from Australia (as New Holland).

Distribution: NE coastal, SE coastal, Qld., N.S.W.; only published localities Caloundra, Brisbane, Sydney, South West Rocks and Trial Bay. Ecology: larva - sedentary, predator : adult - volant; prey larval Lepidoptera. Biological references: Giordani Soika, A. (1962). Gli *Odynerus* sensu antiquo del continente australiano e della Tasmania. *Boll. Mus. Civ. Stor. Nat. Venezia* **14**: 57–202 (redescription, generic placement); Giordani Soika, A. (1973). Designazione di lectotipi ed elenco dei tipi di Eumenidi, Vespidi e Masaridi da me descritti negli anni 1934–1960. *Boll. Mus. Civ. Stor. Nat. Venezia* **24**: 7–53 (distribution, taxonomy).

Ischnocoelia Perkins, 1908

Ischnocoelia Perkins, R.C.L. (1908). Some remarkable Australian Hymenoptera. *Proc. Hawaii. Entomol. Soc.* **2**: 27–35 [32]. Type species *Ischnocoelia xanthochroma* Perkins, 1908 by monotypy.

Ischnocoelia ecclesiastica (Rayment, 1954)

Discoelius ecclesiasticus Rayment, T. (1954). The trail of the running postman. *Proc. R. Zool. Soc. N.S.W.* **1952-53**: 18–22 [19]. Type data: holotype, ANIC F. adult, from Watsonia, Vic.

Distribution: SE coastal, Vic.; type locality only. Ecology: larva - sedentary, soil, predator : adult - volant; nest in soil, prey larval Lepidoptera (Geometridae). Biological references: Rayment, T.

(1953). Pictorial biology of a leafcutter bee. *Megachile chrysopyga* Smith. *Vict. Nat.* **70**: 50–51 (biology); van der Vecht, J. (1981). Indo-Australian solitary wasps. *Proc. K. Ned. Akad. Wet. (c)* **84**: 443–464 (generic placement).

Ischnocoelia elongata (Saussure, 1855)

Discoelius elongatus Saussure, H. de (1855). *Études sur la Famille des Vespides*. Troisième Partie comprenant la Monographie des Masariens et un Supplement à la Monographie des Eumeniens. Paris : Masson 352 pp. pls i–xvi (1854–1856) [124 pl vi fig 7]. Type data: syntypes (probable), OUM or BMNH *F. adult, from S.A. (as Australie méridionale).

Distribution: SE coastal, N.S.W., Vic., S.A.; only published localities Gordon, Kealesville (presumably Healesville) and S.A. Ecology: larva - sedentary, predator : adult - volant; prey larval Lepidoptera. Biological references: Giordani Soika, A. (1962). Gli *Odynerus* sensu antiquo del continente australiano e della Tasmania. *Boll. Mus. Civ. Stor. Nat. Venezia* **14**: 57–202 (taxonomy as *Pseudozethus elongatus*); Giordani Soika, A. (1969). Revisione dei Discoeliinae Australiani. *Boll. Mus. Civ. Stor. Nat. Venezia* **19**: 25–100 (redescription, generic placement).

Ischnocoelia ferruginea (Meade-Waldo, 1910)

Elimus ferrugineus Meade-Waldo, G. (1910). New species of Diploptera in the collection of the British Museum. *Ann. Mag. Nat. Hist. (8)* **5**: 30–51 [38]. Type data: holotype, BMNH *M. adult (seen 1929 by L.F. Graham), from S.A.

Distribution: SE coastal, Vic., S.A.; only published localities Melbourne and S.A. Ecology: larva - sedentary, predator : adult - volant; prey larval Lepidoptera. Biological references: Meade-Waldo, G. (1913). New species of Diploptera in the collection of the British Museum. Part IV. *Ann. Mag. Nat. Hist. (8)* **11**: 44–54 (generic placement); Giordani Soika, A. (1969). Revisione dei Discoeliinae Australiani. *Boll. Mus. Civ. Stor. Nat. Venezia* **19**: 25–100 (redescription).

Ischnocoelia fulva (Schulthess-Rechberg, 1910)

Taxonomic decision of Giordani Soika, A. (1969). Revisione dei Discoeliinae Australiani. *Boll. Mus. Civ. Stor. Nat. Venezia* **19**: 25–100 [81].

Ischnocoelia fulva fulva (Schulthess-Rechberg, 1910)

Stenolabus fulvus Schulthess-Rechberg, A. von (1910). Über einige neue und weniger bekannte Eumeniden (Vespiden, Hymenoptera). *Dt. Entomol. Z.* **1910**: 187–192 [190]. Type data: holotype, GMNH *M. adult, from Adelaide, S.A.

Distribution: S Gulfs, SW coastal, S.A., W.A.; only published localities Adelaide and Yanchep. Ecology: larva - sedentary, predator : adult - volant; prey larval Lepidoptera.

Ischnocoelia fulva major Meade-Waldo, 1914

Ischnocoelia integra major Meade-Waldo, G. (1914). Notes on the Hymenoptera in the collection of the British Museum, with descriptions of new species. V. *Ann. Mag. Nat. Hist. (8)* **14**: 450–464 [459] [described as variety]. Type data: holotype, BMNH *F. adult (seen 1929 by L.F. Graham), from Yallingup, W.A.

Distribution: SW coastal, W.A.; only published localities Yallingup and Bunbury. Ecology: larva - sedentary, predator : adult - volant; prey larval Lepidoptera.

Ischnocoelia integra (Schulthess-Rechberg, 1910)

Taxonomic decision of Giordani Soika, A. (1969). Revisione dei Discoeliinae Australiani. *Boll. Mus. Civ. Stor. Nat. Venezia* **19**: 25–100 [87].

Ischnocoelia integra integra (Schulthess-Rechberg, 1910)

Stenolabus integer Schulthess-Rechberg, A. von (1910). Über einige neue und weniger bekannte Eumeniden (Vespiden, Hymenoptera). *Dt. Entomol. Z.* **1910**: 187–192 [191]. Type data: holotype, ETHZ *M. adult, from N.S.W.

Distribution: NE coastal, Qld., N.S.W.; only published localities Brisbane, Stradbroke Is. and N.S.W. Ecology: larva - sedentary, predator : adult - volant; prey larval Lepidoptera. Biological references: Meade-Waldo, G. (1910). New species of Diploptera in the collection of the British Museum. *Ann. Mag. Nat. Hist. (8)* **5**: 30–51 (generic placement); Bohart, R.M. & Stange, L.A. (1965). A revision of the genus *Zethus* Fabricius in the western hemisphere (Hymenoptera : Eumenidae). *Univ. Calif. Publ. Entomol.* **40**: 1–208 (taxonomy).

Ischnocoelia integra carnowi Giordani Soika, 1969

Ischnocoelia integra carnowi Giordani Soika, A. (1969). Revisione dei Discoeliinae Australiani. *Boll. Mus. Civ. Stor. Nat. Venezia* **19**: 25–100 [87]. Type data: holotype, SAMA *F. adult, from Mt. Lofty Ranges, S.A.

Distribution: S Gulfs, S.A.; type locality only. Ecology: larva - sedentary, predator : adult - volant; prey larval Lepidoptera.

Ischnocoelia occidentalis Giordani Soika, 1969

Taxonomic decision of Giordani Soika, A. (1969). Revisione dei Discoeliinae Australiani. *Boll. Mus. Civ. Stor. Nat. Venezia* **19**: 25–100 [82].

Ischnocoelia occidentalis occidentalis Giordani Soika, 1969

Ischnocoelia occidentalis Giordani Soika, A. (1969). Revisione dei Discoeliinae Australiani. *Boll. Mus. Civ. Stor. Nat. Venezia* **19**: 25–100 [82]. Type data: holotype, BMNH *M. adult, from Dedari, W.A.

Distribution: SW coastal, W plateau, W.A.; only published localities Merredin and Dedari. Ecology: larva - sedentary, predator : adult - volant; prey larval Lepidoptera.

Ischnocoelia occidentalis blumburyensis Giordani Soika, 1969

Ischnocoelia occidentalis blumburyensis Giordani Soika, A. (1969). Revisione dei Discoeliinae Australiani. *Boll. Mus. Civ. Stor. Nat. Venezia* **19**: 25–100 [84]. Type data: holotype, AM M. adult, from Bunbury (as Blumbury), W.A.

Distribution: SW coastal, W.A.; type locality only. Ecology: larva - sedentary, predator : adult - volant; prey larval Lepidoptera.

Ischnocoelia polychroma Giordani Soika, 1969

Ischnocoelia polychroma Giordani Soika, A. (1969). Revisione dei Discoeliinae Australiani. *Boll. Mus. Civ. Stor. Nat. Venezia* **19**: 25–100 [92]. Type data: holotype, AM M. adult, from Parkes, N.S.W.

Distribution: Murray-Darling basin, Qld., N.S.W.; only published localities Condamine and Parkes. Ecology: larva - sedentary, predator : adult - volant; prey larval Lepidoptera.

Ischnocoelia robusta (Meade-Waldo, 1910)

Taxonomic decision of Giordani Soika, A. (1969). Revisione dei Discoeliinae Australiani. *Boll. Mus. Civ. Stor. Nat. Venezia* **19**: 25–100 [91].

Ischnocoelia robusta robusta (Meade-Waldo, 1910)

Elimus robustus Meade-Waldo, G. (1910). New species of Diploptera in the collection of the British Museum. *Ann. Mag. Nat. Hist.* (8) **5**: 30–51 [Jan. 1910] [40]. Type data: holotype, BMNH *F. adult (seen 1929 by L.F. Graham), from S.A.
Stenolabus vulneratus Schulthess-Rechberg, A. von (1910). Über einige neue und weniger bekannte Eumeniden (Vespiden, Hymenoptera). *Dt. Entomol. Z.* **1910**: 187–192 [Mar. 1910] [191]. Type data: lectotype, GMNH *F. adult, from Adelaide, S.A., designation by Giordani Soika, A. (1969). Revisione dei Discoeliinae Australiani. *Boll. Mus. Civ. Stor. Nat. Venezia* **19**: 25–100.
Taxonomic decision of Meade-Waldo, G. (1914). Notes on the Hymenoptera in the collection of the British Museum, with descriptions of new species. V. *Ann. Mag. Nat. Hist.* (8) **14**: 450–464 [459].

Distribution: S Gulfs, NW coastal, SW coastal, S.A., W.A.; only published localities Adelaide, Marloo Station, Wurarga and King George Sound. Ecology: larva - sedentary, predator : adult - volant; prey larval Lepidoptera. Biological references: Meade-Waldo, G. (1913). New species of Diploptera in the collection of the British Museum. Part IV. *Ann. Mag. Nat. Hist.* (8) **11**: 44–54 (generic placement).

Ischnocoelia robusta analis Giordani Soika, 1969

Ischnocoelia robusta analis Giordani Soika, A. (1969). Revisione dei Discoeliinae Australiani. *Boll. Mus. Civ. Stor. Nat. Venezia* **19**: 25–100 [91]. Type data: holotype, ZMB *M. adult, from Marloo Station, Wurarga, W.A.

Distribution: NW coastal, W plateau, W.A.; only published localities Marloo Station, Wurarga and Dedari. Ecology: larva - sedentary, predator : adult - volant; prey larval Lepidoptera.

Ischnocoelia robusta aurantiaca Giordani Soika, 1969

Ischnocoelia robusta aurantiaca Giordani Soika, A. (1969). Revisione dei Discoeliinae Australiani. *Boll. Mus. Civ. Stor. Nat. Venezia* **19**: 25–100 [91]. Type data: holotype, SAMA *F. adult, from Balladonia, W.A.

Distribution: W plateau, W.A.; type locality only. Ecology: larva - sedentary, predator : adult - volant; prey larval Lepidoptera.

Ischnocoelia robusta unicolor Giordani Soika, 1969

Ischnocoelia robusta unicolor Giordani Soika, A. (1969). Revisione dei Discoeliinae Australiani. *Boll. Mus. Civ. Stor. Nat. Venezia* **19**: 25–100 [92]. Type data: holotype, A. Giordani Soika pers. coll. *F. adult, from Central Australia.

Distribution: no locality specified. Ecology: larva - sedentary, predator : adult - volant; prey larval Lepidoptera.

Ischnocoelia xanthochroma Perkins, 1908

Ischnocoelia xanthochroma Perkins, R.C.L. (1908). Some remarkable Australian Hymenoptera. *Proc. Hawaii. Entomol. Soc.* **2**: 27–35 [32]. Type data: syntypes (probable), BMNH *adult, from middle Qld.

Distribution: NE coastal, SE coastal, N coastal, Qld., Vic., N.T. Ecology: larva - sedentary, predator : adult - volant; prey larval Lepidoptera. Biological references: Giordani Soika, A. (1969). Revisione dei Discoeliinae Australiani. *Boll. Mus. Civ. Stor. Nat. Venezia* **19**: 25–100 (redescription); Giordani Soika, A. (1977). Contributo alla conoscenza degli Eumenidi australiani (Hymenoptera). *Mem. Soc. Entomol. Ital.* **55**: 109–138 (distribution).

Leptomenoides Giordani Soika, 1962

Leptomenoides Giordani Soika, A. (1962). Gli *Odynerus* sensu antiquo del continente australiano e della Tasmania. *Boll. Mus. Civ. Stor. Nat. Venezia* **14**: 57–202 [171]. Type species *Leptomenoides placidior* Giordani Soika, 1962 by original designation.

Leptomenoides cairnensis Giordani Soika, 1962

Leptomenoides cairnensis Giordani Soika, A. (1962). Gli *Odynerus* sensu antiquo del continente australiano e della Tasmania. *Boll. Mus. Civ. Stor. Nat. Venezia* **14**: 57–202 [177]. Type data: holotype, BMNH (probable) *F. adult, from Cairns, Kuranda", Qld.

Distribution: NE coastal, Qld.; type locality only. Ecology: larva - sedentary, predator : adult - volant; prey larval Lepidoptera.

Leptomenoides extraneus (Saussure, 1855)

Odynerus (Leionotus) exilis Saussure, H. de (1853). *Études sur la Famille des Vespides*. 1. Monographie des guêpes solitaires, ou de la tribu des Euméniens, comprenant la classification et la description de toutes les espèces connues jusqu'à ce jour, et servant de complément au Manuel de Lepeletier de Saint Fargeau. Paris : Masson 286 pp. pls i–xx (1852–1853) [157 pl xvii fig 2] [*non Odynerus exilis* Schaeffer, 1841]. Type data: holotype, MNHP *F. adult, from Tas.

Odynerus (Leionotus) extraneus Saussure, H. de (1855). *Études sur la Famille des Vespides*. Troisième Partie comprenant la Monographie des Masariens et un Supplement à la Monographie des Eumeniens. Paris : Masson 352 pp. pls i–xvi (1854–1856) [224] [*nom. nov.* for *Odynerus (Leionotus) exilis* Saussure, 1853].

Distribution: NE coastal, Qld., Tas.; only published localities Mackay and Tas. Ecology: larva - sedentary, predator : adult - volant; prey larval Lepidoptera. Biological references: Giordani Soika, A. (1941). Studi sui Vespidi solitari. *Boll. Soc. Veneziana Stor. Nat.* **2**: 130–279 (description male as *Pachymenes extraneus*); Giordani Soika, A. (1943). Le specie indo-australiane del genere *Pachymenes* (Hym. Vespidae). *Mem. Soc. Entomol. Ital.* **22**: 102–117 (taxonomy); Giordani Soika, A. (1962). Gli *Odynerus* sensu antiquo del continente australiano e della Tasmania. *Boll. Mus. Civ. Stor. Nat. Venezia* **14**: 57–202 (redescription, generic placement).

Leptomenoides histrio Giordani Soika, 1962

Leptomenoides histrio Giordani Soika, A. (1962). Gli *Odynerus* sensu antiquo del continente australiano e della Tasmania. *Boll. Mus. Civ. Stor. Nat. Venezia* **14**: 57–202 [180]. Type data: holotype, USNM *F. adult, from Mosman, N.S.W.

Distribution: SE coastal, N.S.W.; type locality only. Ecology: larva - sedentary, predator : adult - volant; prey larval Lepidoptera.

Leptomenoides mackayensis Giordani Soika, 1962

Leptomenoides mackayensis Giordani Soika, A. (1962). Gli *Odynerus* sensu antiquo del continente australiano e della Tasmania. *Boll. Mus. Civ. Stor. Nat. Venezia* **14**: 57–202 [175]. Type data: holotype, BMNH *F. adult, from Mackay, Qld.

Distribution: NE coastal, Qld.; type locality only. Ecology: larva - sedentary, predator : adult - volant; prey larval Lepidoptera.

Leptomenoides pachymeniformis Giordani Soika, 1962

Leptomenoides pachymeniformis Giordani Soika, A. (1962). Gli *Odynerus* sensu antiquo del continente australiano e della Tasmania. *Boll. Mus. Civ. Stor. Nat. Venezia* **14**: 57–202 [178]. Type data: holotype, BMNH *F. adult, from N.S.W.

Distribution: NE coastal, Qld., N.S.W.; only published localities Tamborine Mt. and N.S.W. Ecology: larva - sedentary, predator : adult - volant; prey larval Lepidoptera.

Leptomenoides placidior Giordani Soika, 1962

Leptomenoides placidior Giordani Soika, A. (1962). Gli *Odynerus* sensu antiquo del continente australiano e della Tasmania. *Boll. Mus. Civ. Stor. Nat. Venezia* **14**: 57–202 [180]. Type data: holotype, BMNH *F. adult, from Mackay, Qld.

Distribution: NE coastal, Qld.; only published localities Mackay and Somerset. Ecology: larva - sedentary, predator : adult - volant; prey larval Lepidoptera.

Leptomenoides pronotalis Giordani Soika, 1962

Leptomenoides pronotalis Giordani Soika, A. (1962). Gli *Odynerus* sensu antiquo del continente australiano e della Tasmania. *Boll. Mus. Civ. Stor. Nat. Venezia* **14**: 57–202 [173]. Type data: holotype, BMNH *F. adult, from Mackay, Qld.

Distribution: NE coastal, Qld.; only published localities Mackay, Kuranda, Cairns district and Brisbane. Ecology: larva - sedentary, predator : adult - volant; prey larval Lepidoptera.

Macrocalymma Perkins, 1908

Macrocalymma Perkins, R.C.L. (1908). Some remarkable Australian Hymenoptera. *Proc. Hawaii. Entomol. Soc.* **2**: 27–35 [31]. Type species *Macrocalymma smithianum* Perkins, 1908 by monotypy.

Macrocalymma aliciae Meade-Waldo, 1914

Macrocalymma aliciae Meade-Waldo, G. (1914). Notes on the Hymenoptera in the collection of the British Museum, with descriptions of new species. V. *Ann. Mag. Nat. Hist. (8)* **14**: 450–464 [459]. Type data: holotype, BMNH *F. adult (seen 1929 by L.F. Graham), from Yallingup, W.A.

Distribution: SW coastal, W.A.; type locality only. Ecology: larva - sedentary, predator : adult - volant; prey larval Lepidoptera. Biological references: Giordani Soika, A. (1969). Revisione dei Discoeliinae Australiani. *Boll. Mus. Civ. Stor. Nat. Venezia* **19**: 25-100 (redescription).

Macrocalymma smithianum Perkins, 1908

Macrocalymma smithianum Perkins, R.C.L. (1908). Some remarkable Australian Hymenoptera. *Proc. Hawaii. Entomol. Soc.* **2**: 27-35 [31]. Type data: holotype, BMNH *M. adult (seen 1929 by L.F. Graham), from Qld. ("common in middle Qld.").

Distribution: NE coastal, Qld.; only published localities Mackay and Brisbane. Ecology: larva - sedentary, predator : adult - volant; prey larval Lepidoptera. Biological references: Giordani Soika, A. (1941). Studi sui Vespidi solitari. *Boll. Soc. Veneziana Stor. Nat.* **2**: 130-279 (description of female); Giordani Soika, A. (1969). Revisione dei Discoeliinae Australiani. *Boll. Mus. Civ. Stor. Nat. Venezia* **19**: 25-100 (redescription); Giordani Soika, A. (1973). Designazione di lectotipi ed elenco dei tipi di Eumenidi, Vespidi e Masaridi da me descritti negli anni 1934-1960. *Boll. Mus. Civ. Stor. Nat. Venezia* **24**: 7-53 (taxonomy).

Odynerus Latreille, 1802

Odynerus Latreille, P.A. (1802). *Histoire Naturelle, Générale et Particulière des Crustacés et des Insectes.* Paris : F. Dufart Vol. 3 xii 13+467 pp. (1802-1803) [362] [for discussion of type for genus, see Richards, O.W. & van der Vecht, J. (1968). The type of the genus *Odynerus* Latreille (Hymenoptera, Vespoidea). *Entomol. Ber. (Amst.)* **28**: 196-197]. Type species *Vespa spinipes* Linnaeus, 1758 by subsequent designation, see Shuckard, W.E. (1837). Description of a new British wasp, with an account of its development. *Mag. Nat. Hist. (ns)* **1**: 490-496.

This group is also found in the Palaearctic Region, see van der Vecht, J. & Fischer, F.C.J. (1972). Palearctic Eumenidae. *Hymenopterorum Catalogus* **8**: 1-199. Species now known not to occur in Australia: *Odynerus alecto* Lepeletier, 1841 (now placed in *Epiodynerus* Giordani Soika, 1958), described from "N. Holl.", is from New Caledonia *teste* Giordani Soika, A. (1958). Biogeografia, evoluzione e sistematica dei Vespidi solitari della Polinesia meridionale. *Boll. Mus. Civ. Stor. Nat. Venezia* **10**: 183-221 and *Odynerus drewseni* Saussure, 1857 (now placed in *Orancistrocerus* van der Vecht, 1963) described from "La Nouvelle Hollande", is from China *teste* van der Vecht, J. (1963). Studies on Indo-Australian and east-Asiatic Eumenidae (Hymenoptera, Vespoidea). *Zool. Verh.* **60**: 1-116.

Odynerus indecoratus Giordani Soika, 1962

Odynerus (Leionotus) decoratus Bingham, C.T. (1912). South African and Australian Aculeate Hymenoptera in the Oxford Museum. *Trans. Entomol. Soc. Lond.* **1912**: 375-383 [379] [as *Lionotus*; non Odynerus decoratus Saussure, 1855]. Type data: syntypes (probable), OUM *adult, from Towranna Plains between Yule River and Sherlock River, W.A.

Odynerus indecoratus Giordani Soika, A. (1962). Gli *Odynerus* sensu antiquo del continente australiano e della Tasmania. *Boll. Mus. Civ. Stor. Nat. Venezia* **14**: 57-202 [201] [*nom. nov.* for *Odynerus (Leionotus) decoratus* Bingham, 1912].

Distribution: NW coastal, W.A.; type locality only. Ecology: larva - sedentary, predator : adult - volant; prey larval Lepidoptera.

Pachycoelius Giordani Soika, 1969

Pachycoelius Giordani Soika, A. (1969). Revisione dei Discoeliinae Australiani. *Boll. Mus. Civ. Stor. Nat. Venezia* **19**: 25-100 [54]. Type species *Pachycoelius brevicornis* Giordani Soika, 1969 by original designation.

Pachycoelius brevicornis Giordani Soika, 1969

Pachycoelius brevicornis Giordani Soika, A. (1969). Revisione dei Discoeliinae Australiani. *Boll. Mus. Civ. Stor. Nat. Venezia* **19**: 25-100 [55]. Type data: holotype, WAM *M. adult lost, from Dudinin (as Dudiniu), W.A.

Distribution: SW coastal, W.A.; type locality only. Ecology: larva - sedentary, predator : adult - volant; prey larval Lepidoptera.

Pachycoelius carinatus (Meade-Waldo, 1910)

Discoelius carinatus Meade-Waldo, G. (1910). New species of Diploptera in the collection of the British Museum. *Ann. Mag. Nat. Hist. (8)* **5**: 30-51 [38]. Type data: holotype, BMNH *F. adult (seen 1929 by L.F. Graham), from Vic.

Distribution: Murray-Darling basin, S Gulfs, Vic., S.A.; only published localities Kiata and Rostrevor. Ecology: larva - sedentary, predator : adult - volant; prey larval Lepidoptera. Biological references: Bequaert, J.C. (1928). A study of certain types of diplopterous wasps in the collection of the British Museum. *Ann. Mag. Nat. Hist. (10)* **2**: 138-176 (type locality); Giordani Soika, A. (1969). Revisione dei Discoeliinae Australiani. *Boll. Mus. Civ. Stor. Nat. Venezia* **19**: 25-100 (redescription, generic placement, distribution).

Pachycoelius mediocris Giordani Soika, 1969

Pachycoelius mediocris Giordani Soika, A. (1969). Revisione dei Discoeliinae Australiani. *Boll. Mus. Civ. Stor. Nat. Venezia* **19**: 25-100 [60]. Type data: holotype, UZM *F. adult, from Australia (as New Holland).

Distribution: no locality specified. Ecology: larva - sedentary, predator : adult - volant; prey larval Lepidoptera.

Paralastor Saussure, 1856

Paralastor Saussure, H. de (1856). *Études sur la Famille des Vespides.* Troisième Partie comprenant la Monographie des Masariens et un Supplement à la Monographie des Eumeniens. Paris : Masson 352 pp. pls i–xvi (1854–1856) [328] [proposed with subgeneric rank in *Alastor* Lepeletier, 1841; placed on Official List of Generic Names in Zoology, see International Commission on Zoological Nomenclature (1970). Opinion 893. Eumenidae names of Saussure (Hymenoptera) : grant of availability to certain names proposed for secondary divisions of genera. *Bull. Zool. Nomen.* **26**: 187–191; van der Vecht, J. (1983). *Ancistroceroides* Saussure, 1855 : proposed change of type species in order to preserve the well-established name *Paralastor* Saussure, 1856 (Hymenoptera, Vespoidea, Eumenidae). Z.N.(S.)2280. *Bull. Zool. Nomen.* **40**: 111–113 states that *Paralastor* is undoubtedly a junior subjective synonym of *Ancistroceroides* Saussure, 1855, but in order to avoid the confusion which would be caused by implementing this synonymy (affecting the generic assignment of approximately 130 species), he has requested that the International Commission on Zoological Nomenclature use its plenary powers to designate an alternative type species for *Ancistroceroides*; *Ancistroceroides* would then be used for a small group of South American Eumenidae, and *Paralastor* would remain the valid name for the large Australian group; on the assumption that the case to the International Commission will be successful, the species *Odynerus cruentus* Saussure, 1855, designated as type species of *Ancistroceroides* in 1925, is transferred to *Paralastor*]. Publication date established from Griffin, F.J. (1939). On the dates of publication of Saussure (H. de) : Études sur la famille des Vespides 1–3, 1852–1858. *J. Soc. Bibliogr. Nat. Hist.* **1**: 211–212. Type species *Alastor tuberculatus* Saussure, 1853 by subsequent designation, see van der Vecht, J. (1967). The status of certain genus-group names in the Eumenidae (Hymenoptera, Vespoidea). Z.N.(S.) 1689. *Bull. Zool. Nomen.* **24**: 27–33.

Paralastor abnormis (Bingham, 1912)

Alastor abnormis Bingham, C.T. (1912). South African and Australian Aculeate Hymenoptera in the Oxford Museum. *Trans. Entomol. Soc. Lond.* **1912**: 375–383 [380]. Type data: syntypes (probable), OUM *M. adult, from Towranna Plains, W.A.

Distribution: N coastal, NW coastal, W plateau, W.A. Ecology: larva - sedentary, predator : adult - volant; prey larval Lepidoptera. Biological references: Giordani Soika, A. (1962). Gli *Odynerus* sensu antiquo del continente australiano e della Tasmania. *Boll. Mus. Civ. Stor. Nat. Venezia* **14**: 57–202 (taxonomy); Giordani Soika, A. (1973). Designazione di lectotipi ed elenco dei tipi di Eumenidi, Vespidi e Masaridi da me descritti negli anni 1934–1960. *Boll. Mus. Civ. Stor. Nat. Venezia* **24**: 7–53 (distribution).

Paralastor aequifasciatus Perkins, 1914

Paralastor aequifasciatus Perkins, R.C.L. (1914). New species of *Paralastor*, Sauss. (Hymenoptera, Fam. Eumenidae) collected by Mr. R.E. Turner in S.W. Australia. *Ann. Mag. Nat. Hist.* (8) **14**: 235–240 [239]. Type data: holotype, BMNH *F. adult (seen 1929 by L.F. Graham), from Yallingup, W.A.

Distribution: SW coastal, W.A.; type locality only. Ecology: larva - sedentary, predator : adult - volant; prey larval Lepidoptera.

Paralastor alastoripennis (Saussure, 1853)

Odynerus (Ancistrocerus) alastoripennis Saussure, H. de (1853). *Études sur la Famille des Vespides.* 1. Monographie des guêpes solitaires, ou de la tribu des Euméniens, comprenant la classification et la description de toutes les espèces connues jusqu'à ce jour, et servant de complément au Manuel de Lepeletier de Saint Fargeau. Paris : Masson 286 pp. pls i–xx (1852–1853) [147 pl xvi figs 5,5a]. Type data: holotype, MNHP *F. adult, from Tas.

Distribution: Tas.; type locality only. Ecology: larva - sedentary, predator : adult - volant; prey larval Lepidoptera. Biological references: Giordani Soika, A. (1962). Gli *Odynerus* sensu antiquo del continente australiano e della Tasmania. *Boll. Mus. Civ. Stor. Nat. Venezia* **14**: 57–202 (generic placement).

Paralastor albifrons (Fabricius, 1775)

Vespa albifrons Fabricius, J.C. (1775). *Systema Entomologiae,* sistens insectorum classes, ordines, genera, species, adiectis synonymis, locis, descriptionibus, observationibus. Flensburgi et Lipsiae : Kortii xxvii 832 pp. [366]. Type data: holotype, BMNH *M. adult (seen 1929 by L.F. Graham), from Australia (as New Holland).

Distribution: no locality specified. Ecology: larva - sedentary, predator : adult - volant; prey larval Lepidoptera. Biological references: Perkins, R.C.L. (1914). On the species of *Alastor (Paralastor)* Sauss. and some other Hymenoptera of the family Eumenidae. *Proc. Zool. Soc. Lond.* **1914**: 563–624 pl I (taxonomy); Bequaert, J. (1928). The diplopterous wasps of Fabricius, in the Banksian collection at the British Museum. *Bull. Brooklyn Entomol. Soc.* **23**: 53–63 (taxonomy).

Paralastor alexandriae Perkins, 1914

Paralastor alexandriae Perkins, R.C.L. (1914). On the species of *Alastor (Paralastor)* Sauss. and some other Hymenoptera of the family Eumenidae. *Proc. Zool. Soc. Lond.* **1914**: 563–624 pl I [618 pl I figs 7, 22]. Type data: holotype, BMNH *F. adult (seen 1929 by L.F. Graham), from Alexandria, N.T.

Distribution: N coastal, N Gulf, N.T.; only published localities Alexandria, Adelaide River and Kakadu Natl. Park. Ecology: larva - sedentary, predator : adult - volant; nest in abandoned mud-nests of *Sceliphron* spp., prey larval Lepidoptera (Gelechiidae). Biological references: Naumann, I.D. (1983). The biology of mud nesting Hymenoptera (and their associates) and Isoptera in rock shelters of the Kakadu Region, Northern Territory. *Aust. Natl. Parks & Wildlf. Ser. Spec. Publ. 10* pp. 127-189 (nest, biology).

Paralastor anostreptus Perkins, 1914

Paralastor anostreptus Perkins, R.C.L. (1914). On the species of *Alastor (Paralastor)* Sauss. and some other Hymenoptera of the family Eumenidae. *Proc. Zool. Soc. Lond.* **1914**: 563-624 pl I [613]. Type data: holotype, BMNH *F. adult (seen 1929 by L.F. Graham), from S Heywood Is., W.A.

Distribution: N coastal, W.A.; type locality only. Ecology: larva - sedentary, predator : adult - volant; prey larval Lepidoptera.

Paralastor arenicola Perkins, 1914

Paralastor arenicola Perkins, R.C.L. (1914). On the species of *Alastor (Paralastor)* Sauss. and some other Hymenoptera of the family Eumenidae. *Proc. Zool. Soc. Lond.* **1914**: 563-624 pl I [618]. Type data: holotype, BMNH *F. adult (seen 1929 by L.F. Graham), from Hermannsburg, N.T.

Distribution: Lake Eyre basin, N.T.; type locality only. Ecology: larva - sedentary, predator : adult - volant; prey larval Lepidoptera.

Paralastor argentifrons (Smith, 1857)

Alastor argentifrons Smith, F. (1857). *Catalogue of Hymenopterous Insects in the Collection of the British Museum*. Pt. V. Vespidae pp. 1-147 London : British Museum [90]. Type data: holotype, BMNH *F. adult (seen 1929 by L.F. Graham), from Australia.

Distribution: SE coastal, S Gulfs, Vic., S.A.; only published localities Melbourne and Adelaide. Ecology: larva - sedentary, predator : adult - volant; prey larval Lepidoptera. Biological references: Perkins, R.C.L. (1914). On the species of *Alastor (Paralastor)* Sauss. and some other Hymenoptera of the family Eumenidae. *Proc. Zool. Soc. Lond.* **1914**: 563-624 pl I (redescription).

Paralastor argyrias Perkins, 1914

Paralastor argyrias Perkins, R.C.L. (1914). On the species of *Alastor (Paralastor)* Sauss. and some other Hymenoptera of the family Eumenidae. *Proc. Zool. Soc. Lond.* **1914**: 563-624 pl I [597]. Type data: holotype, BMNH *M. adult (seen 1929 by L.F. Graham), from Wagga Wagga, N.S.W. (as Wagga).

Distribution: Murray-Darling basin, N.S.W.; type locality only. Ecology: larva - sedentary, predator : adult - volant; prey larval Lepidoptera.

Paralastor aterrimus Turner, 1919

Paralastor aterrimus Turner, R.E. (1919). New Australian diplopterous Hymenoptera. *Ann. Mag. Nat. Hist.* (9) **3**: 398-399 [398]. Type data: holotype, BMNH *M. adult (seen 1929 by L.F. Graham), from Townsville, Qld.

Distribution: NE coastal, Qld.; type locality only. Ecology: larva - sedentary, predator : adult - volant; prey larval Lepidoptera.

Paralastor atripennis Perkins, 1914

Paralastor atripennis Perkins, R.C.L. (1914). On the species of *Alastor (Paralastor)* Sauss. and some other Hymenoptera of the family Eumenidae. *Proc. Zool. Soc. Lond.* **1914**: 563-624 pl I [602 pl I fig 13]. Type data: holotype, OUM *F. adult, from Adelaide, S.A.

Distribution: S Gulfs, S.A.; type locality only. Ecology: larva - sedentary, predator : adult - volant; prey larval Lepidoptera.

Paralastor aurocinctus (Guérin, 1831)

Odynerus aurocinctus Guérin-Méneville, F.E. (1831). Insectes. *in* Duperrey, L.J. (1830-1832). *Voyage Autour du Monde, Exécuté par Ordre du Roi, sur la Corvette de la Majesté, La Coquille, Pendant les Années 1822, 1823, 1824, et 1825, ... Atlas. Histoire naturelle. Zoologie.* pls 1-21 Paris : Bertrand [pl 9 fig 4] [first published as an illustration, written description (page 266) in Guérin-Méneville, F.E. (1838). Crustacés, Arachnides et Insectes. *in* Duperrey, L.J. (1838). *Voyage Autour du Monde, Exécuté par Ordre du Roi, sur la Corvette de la Majesté, La Coquille, Pendant les Années 1822, 1823, 1824, et 1825, ... Zool.* Vol. ii Pt. 2 Div. 1 Chap. xiii pp. 57-302 Paris : Bertrand]. Type data: syntypes (probable), MCG *F. adult, from Australia.

Distribution: no locality specified. Ecology: larva - sedentary, predator : adult - volant; prey larval Lepidoptera. Biological references: Perkins, R.C.L. (1914). On the species of *Alastor (Paralastor)* Sauss. and some other Hymenoptera of the family Eumenidae. *Proc. Zool. Soc. Lond.* **1914**: 563-624 pl I (redescription as *Paralastor aureocinctus*); Guiglia, D. & Pasteels, J. (1961). Aggiunte ed osservazioni all'elenco delle specie di imenotteri descritte da Guérin-Méneville che si trovano nelle collezioni del Museo di Genova. *Ann. Mus. Civ. Stor. Nat. Giacomo Doria* **72**: 17-20 (note on type, as *aureocinctus*); Giordani Soika, A. (1962). Gli *Odynerus* sensu antiquo del continente australiano e della Tasmania. *Boll. Mus. Civ. Stor. Nat. Venezia* **14**: 57-202 (taxonomy, as *Paralastor aureocinctus* Perkins).

Paralastor auster Perkins, 1914

Paralastor auster Perkins, R.C.L. (1914). New species of *Paralastor*, Sauss. (Hymenoptera, Fam. Eumenidae) collected by Mr. R.E. Turner in S.W. Australia. *Ann. Mag. Nat. Hist. (8)* **14**: 235–240 [237]. Type data: holotype, BMNH *F. adult (seen 1929 by L.F. Graham), from Yallingup, W.A.

Distribution: SW coastal, W.A.; type locality only. Ecology: larva - sedentary, predator : adult - volant; prey larval Lepidoptera.

Paralastor australis (Saussure, 1853)

Alastor australis Saussure, H. de (1853). *Études sur la Famille des Vespides*. 1. Monographie des guêpes solitaires, ou de la tribu des Euméniens, comprenant la classification et la description de toutes les espèces connues jusqu'à ce jour, et servant de complément au Manuel de Lepeletier de Saint Fargeau. Paris : Masson 286 pp. pls i–xx (1852–1853) [250]. Type data: syntypes (probable), GMNH *F. adult, from Australia (as New Holland).

Distribution: no locality specified. Ecology: larva - sedentary, predator : adult - volant; prey larval Lepidoptera. Biological references: Perkins, R.C.L. (1914). On the species of *Alastor (Paralastor)* Sauss. and some other Hymenoptera of the family Eumenidae. *Proc. Zool. Soc. Lond.* **1914**: 563–624 pl I (taxonomy).

Paralastor bicarinatus Perkins, 1914

Paralastor bicarinatus Perkins, R.C.L. (1914). On the species of *Alastor (Paralastor)* Sauss. and some other Hymenoptera of the family Eumenidae. *Proc. Zool. Soc. Lond.* **1914**: 563–624 pl I [601]. Type data: holotype, BMNH *F. adult (seen 1929 by L.F. Graham), from Mackay, Qld.

Distribution: NE coastal, Qld.; type locality only. Ecology: larva - sedentary, predator : adult - volant; prey larval Lepidoptera.

Paralastor bischoffi Giordani Soika, 1961

Paralastor bischoffi Giordani Soika, A. (1961). Notulae Vespidologicae. XIII - *Paralastor* nuovi o poco noti. *Boll. Soc. Entomol. Ital.* **91**: 12–15 [12]. Type data: holotype, ZMB *F. adult, from Australia "Merinaids" (locality not found in gazetteer).

Distribution: no locality specified. Ecology: larva - sedentary, predator : adult - volant; prey larval Lepidoptera. Biological references: Giordani Soika, A. (1962). Gli *Odynerus* sensu antiquo del continente australiano e della Tasmania. *Boll. Mus. Civ. Stor. Nat. Venezia* **14**: 57–202 (taxonomy).

Paralastor brisbanensis Perkins, 1914

Paralastor brisbanensis Perkins, R.C.L. (1914). On the species of *Alastor (Paralastor)* Sauss. and some other Hymenoptera of the family Eumenidae. *Proc. Zool. Soc. Lond.* **1914**: 563–624 pl I [607]. Type data: holotype, BMNH *F. adult, from Brisbane, Qld.

Distribution: NE coastal, Qld.; type locality only. Ecology: larva - sedentary, predator : adult - volant; prey larval Lepidoptera. Biological references: Giordani Soika, A. (1962). Gli *Odynerus* sensu antiquo del continente australiano e della Tasmania. *Boll. Mus. Civ. Stor. Nat. Venezia* **14**: 57–202 (taxonomy).

Paralastor brunneus (Saussure, 1856)

Alastor (Paralastor) brunneus Saussure, H. de (1856). *Études sur la Famille des Vespides*. Troisième Partie comprenant la Monographie des Masariens et un Supplement à la Monographie des Eumeniens. Paris : Masson 352 pp. pls i–xvi (1854–1856) [337]. Type data: holotype, BMNH *F. adult (seen 1929 by L.F. Graham), from Australia (as New Holland).

Distribution: no locality specified. Ecology: larva - sedentary, predator : adult - volant; prey larval Lepidoptera. Biological references: Perkins, R.C.L. (1914). On the species of *Alastor (Paralastor)* Sauss. and some other Hymenoptera of the family Eumenidae. *Proc. Zool. Soc. Lond.* **1914**: 563–624 pl I (redescription).

Paralastor caprai Giordani Soika, 1977

Paralastor caprai Giordani Soika, A. (1977). Contributo alla conoscenza degli Eumenidi australiani (Hymenoptera). *Mem. Soc. Entomol. Ital.* **55**: 109–138 [135]. Type data: holotype, BMNH *F. adult, from Adelaide River, N.T.

Distribution: N coastal, N.T.; type locality only. Ecology: larva - sedentary, predator : adult - volant; prey larval Lepidoptera.

Paralastor carinatus (Smith, 1857)

Alastor carinatus Smith, F. (1857). *Catalogue of Hymenopterous Insects in the Collection of the British Museum. Pt. V. Vespidae* pp. 1–147 London : British Museum [90]. Type data: holotype, BMNH *F. adult (seen 1929 by L.F. Graham), from Adelaide, S.A.

Distribution: S Gulfs, S.A.; type locality only. Ecology: larva - sedentary, predator : adult - volant; prey larval Lepidoptera. Biological references: Perkins, R.C.L. (1914). On the species of *Alastor (Paralastor)* Sauss. and some other Hymenoptera of the family Eumenidae. *Proc. Zool. Soc. Lond.* **1914**: 563–624 pl I (redescription).

Paralastor clypeopunctatus Schulthess-Rechberg, 1925

Paralastor clypeopunctatus Schulthess-Rechberg, A. von (1925). Beitrag zur Kenntnis der Gattung *Alastor* Lep. (Hym. Vesp.). *Konowia* **4**: 57–65, 195–209, 257–263 [261]. Type data: holotype, NHRM *F. adult, from Herberton, Qld.

Distribution: NE coastal, Qld.; type locality only. Ecology: larva - sedentary, predator : adult - volant; prey larval Lepidoptera.

Paralastor commutatus Perkins, 1914

Paralastor commutatus Perkins, R.C.L. (1914). On the species of *Alastor (Paralastor)* Sauss. and some other Hymenoptera of the family Eumenidae. *Proc. Zool. Soc. Lond.* **1914**: 563–624 pl I [608]. Type data: holotype, BMNH *F. adult (seen 1929 by L.F. Graham), from Champion Bay, W.A.

Distribution: NW coastal, W.A.; type locality only. Ecology: larva - sedentary, predator : adult - volant; prey larval Lepidoptera.

Paralastor comptus Perkins, 1914

Taxonomic decision of Perkins, R.C.L. (1914). On the species of *Alastor (Paralastor)* Sauss. and some other Hymenoptera of the family Eumenidae. *Proc. Zool. Soc. Lond.* **1914**: 563–624 pl I [617].

Paralastor comptus comptus Perkins, 1914

Paralastor comptus Perkins, R.C.L. (1914). On the species of *Alastor (Paralastor)* Sauss. and some other Hymenoptera of the family Eumenidae. *Proc. Zool. Soc. Lond.* **1914**: 563–624 pl I [617]. Type data: holotype, BMNH *M. adult, from Herberton, Qld.

Distribution: NE coastal, Qld.; type locality only. Ecology: larva - sedentary, predator : adult - volant; prey larval Lepidoptera.

Paralastor comptus rubescens Perkins, 1914

Paralastor comptus rubescens Perkins, R.C.L. (1914). On the species of *Alastor (Paralastor)* Sauss. and some other Hymenoptera of the family Eumenidae. *Proc. Zool. Soc. Lond.* **1914**: 563–624 pl I [617] [described as variety]. Type data: holotype, BMNH *M. adult, from Herberton, Qld.

Distribution: NE coastal, Qld.; type locality only. Ecology: larva - sedentary, predator : adult - volant; prey larval Lepidoptera.

Paralastor conspiciendus Perkins, 1914

Paralastor conspiciendus Perkins, R.C.L. (1914). On the species of *Alastor (Paralastor)* Sauss. and some other Hymenoptera of the family Eumenidae. *Proc. Zool. Soc. Lond.* **1914**: 563–624 pl I [581]. Type data: holotype, BMNH *F. adult (seen 1929 by L.F. Graham), from Inkerman, near Townsville, Qld.

Distribution: NE coastal, Qld.; only published localities Townsville and Inkerman. Ecology: larva - sedentary, predator : adult - volant; prey larval Lepidoptera.

Paralastor conspicuus Perkins, 1914

Paralastor conspicuus Perkins, R.C.L. (1914). On the species of *Alastor (Paralastor)* Sauss. and some other Hymenoptera of the family Eumenidae. *Proc. Zool. Soc. Lond.* **1914**: 563–624 pl I [580]. Type data: holotype, BMNH *M. adult, from Cairns district, Qld.

Distribution: NE coastal, Qld.; type locality only. Ecology: larva - sedentary, predator : adult - volant; prey larval Lepidoptera.

Paralastor constrictus Perkins, 1914

Paralastor constrictus Perkins, R.C.L. (1914). On the species of *Alastor (Paralastor)* Sauss. and some other Hymenoptera of the family Eumenidae. *Proc. Zool. Soc. Lond.* **1914**: 563–624 pl I [615]. Type data: holotype, BMNH *F. adult (seen 1929 by L.F. Graham), from Mackay, Qld.

Distribution: NE coastal, Qld.; only published localities Mackay and Bundaberg. Ecology: larva - sedentary, predator : adult - volant; prey larval Lepidoptera. Biological references: Giordani Soika, A. (1962). Gli *Odynerus* sensu antiquo del continente australiano e della Tasmania. *Boll. Mus. Civ. Stor. Nat. Venezia* **14**: 57–202 (taxonomy); CSIRO (1970). *The Insects of Australia*. A textbook for students and research workers. Melbourne : Melbourne Univ. Press 1029 pp. (plate 6T - model for mimicry).

Paralastor cruentatus (Saussure, 1867)

Alastor (Paralastor) cruentatus Saussure, H. de (1867). Hymenoptera. in, Reise der österreichischen Fregatte Novara um die Erde in den Jahren 1857, 1858, 1859 unter den Befehlen des Commodore B. von Wüllerstorf-Urbair. Zoologischer Theil. Wien : K-K Hof- und Staatsdrückerei Vol. 2(1a) 138 pp. [18 pl I fig 12]. Type data: syntypes (probable), adult whereabouts unknown, from Sydney, N.S.W.

Distribution: SE coastal, N.S.W.; type locality only. Ecology: larva - sedentary, predator : adult - volant; prey larval Lepidoptera.

Paralastor cruentus (Saussure, 1855)

Odynerus (Ancistroceroides) cruentus Saussure, H. de (1855). *Études sur la Famille des Vespides.* Troisième Partie comprenant la Monographie des Masariens et un Supplement à la Monographie des Eumeniens. Paris : Masson 352 pp. pls i–xvi (1854–1856) [221]. Type data: holotype, BMNH *F. adult lost, from Australia (as New Holland).

Distribution: S.A.; no locality specified. Ecology: larva - sedentary, predator : adult - volant; prey larval Lepidoptera. Biological references: Bequaert, J. (1925). The genus *Ancistrocerus* in North America, with a partial key to the species. *Trans. Am. Entomol. Soc.* **51**: 57–117 (designation as type species of *Ancistroceroides*); International Commission on Zoological Nomenclature (1970). Opinion 893. Eumenidae names of Saussure (Hymenoptera) : grant of availability to certain names proposed for secondary divisions of genera.

Bull. Zool. Nomen. **26**: 187–191 (placed on Official List of Specific Names); van der Vecht, J. (1983). *Ancistroceroides* Saussure, 1855 : proposed change of type species in order to preserve the well-established name *Paralastor* Saussure, 1856 (Hymenoptera, Vespoidea, Eumenidae). Z.N.(S.)2280. *Bull. Zool. Nomen.* **40**: 111–113 (type lost, generic placement - see note under *Paralastor*).

Paralastor darwinianus Perkins, 1914

Paralastor darwinianus Perkins, R.C.L. (1914). On the species of *Alastor (Paralastor)* Sauss. and some other Hymenoptera of the family Eumenidae. *Proc. Zool. Soc. Lond.* **1914**: 563–624 pl I [617]. Type data: holotype, BMNH *F. adult, from Darwin, N.T.

Distribution: N coastal, N.T.; type locality only. Ecology: larva - sedentary, predator : adult - volant; prey larval Lepidoptera.

Paralastor debilis Perkins, 1914

Paralastor debilis Perkins, R.C.L. (1914). On the species of *Alastor (Paralastor)* Sauss. and some other Hymenoptera of the family Eumenidae. *Proc. Zool. Soc. Lond.* **1914**: 563–624 pl I [595 pl I fig 20]. Type data: holotype, OUM *M. adult, from Swan River, W.A.

Distribution: SW coastal, W.A.; type locality only. Ecology: larva - sedentary, predator : adult - volant; prey larval Lepidoptera.

Paralastor debilitatus Perkins, 1914

Paralastor debilitatus Perkins, R.C.L. (1914). On the species of *Alastor (Paralastor)* Sauss. and some other Hymenoptera of the family Eumenidae. *Proc. Zool. Soc. Lond.* **1914**: 563–624 pl I [611]. Type data: holotype, BMNH *F. adult (seen 1929 by L.F. Graham), from Vic.

Distribution: S Gulfs, S.A., N.S.W., Vic.; only published localities Adelaide, N.S.W. and Vic. Ecology: larva - sedentary, predator : adult - volant; prey larval Lepidoptera. Biological references: Perkins, R.C.L. (1914). New species of *Paralastor*, Sauss. (Hymenoptera, Fam. Eumenidae) collected by Mr. R.E. Turner in S.W. Australia. *Ann. Mag. Nat. Hist. (8)* **14**: 235–240 (taxonomy).

Paralastor dentiger Perkins, 1914

Paralastor dentiger Perkins, R.C.L. (1914). On the species of *Alastor (Paralastor)* Sauss. and some other Hymenoptera of the family Eumenidae. *Proc. Zool. Soc. Lond.* **1914**: 563–624 pl I [603 pl I fig 11]. Type data: holotype, BMNH *M. adult (seen 1929 by L.F. Graham), from Champion Bay, W.A.

Distribution: NW coastal, SW coastal, W.A.; only published localities Champion Bay and Swan River. Ecology: larva - sedentary, predator : adult - volant; prey larval Lepidoptera. Biological references: Giordani Soika, A. (1962). Gli *Odynerus* sensu antiquo del continente australiano e della Tasmania. *Boll. Mus. Civ. Stor. Nat. Venezia* **14**: 57–202 (taxonomy).

Paralastor despectus Perkins, 1914

Paralastor despectus Perkins, R.C.L. (1914). On the species of *Alastor (Paralastor)* Sauss. and some other Hymenoptera of the family Eumenidae. *Proc. Zool. Soc. Lond.* **1914**: 563–624 pl I [589]. Type data: holotype, BMNH *F. adult (seen 1929 by L.F. Graham), from W.A.

Distribution: Qld., W.A.; localities not specified. Ecology: larva - sedentary, predator : adult - volant; prey larval Lepidoptera. Biological references: Giordani Soika, A. (1961). Notulae Vespidologicae. XIII - *Paralastor* nuovi o poco noti. *Boll. Soc. Entomol. Ital.* **91**: 12–15 (description of male).

Paralastor diabolicus Turner, 1919

Paralastor diabolicus Turner, R.E. (1919). New Australian diplopterous Hymenoptera. *Ann. Mag. Nat. Hist. (9)* **3**: 398–399 [398]. Type data: holotype, BMNH *M. adult (seen 1929 by L.F. Graham), from Townsville, Qld.

Distribution: NE coastal, Qld.; type locality only. Ecology: larva - sedentary, predator : adult - volant; prey larval Lepidoptera.

Paralastor diadema Rayment, 1954

Paralastor diadema Rayment, T. (1954). She stands on the waters. *Proc. R. Zool. Soc. N.S.W.* **1952–53**: 15–18 [15]. Type data: syntypes (probable), adult whereabouts unknown, from Gunbower Is. and Sandringham, Vic.

Distribution: SE coastal, Murray-Darling basin, Vic.; only published localities Gunbower Is. and Sandringham. Ecology: larva - sedentary, predator : adult - volant; prey larval Lepidoptera.

Paralastor donatus Perkins, 1914

Paralastor donatus Perkins, R.C.L. (1914). On the species of *Alastor (Paralastor)* Sauss. and some other Hymenoptera of the family Eumenidae. *Proc. Zool. Soc. Lond.* **1914**: 563–624 pl I [588]. Type data: holotype, BMNH *adult, from Bacchus Marsh, Vic.

Distribution: SE coastal, Vic.; type locality only. Ecology: larva - sedentary, predator : adult - volant; prey larval Lepidoptera.

Paralastor dubiosus Perkins, 1914

Paralastor dubiosus Perkins, R.C.L. (1914). On the species of *Alastor (Paralastor)* Sauss. and some other Hymenoptera of the family Eumenidae. *Proc. Zool. Soc. Lond.* **1914**: 563–624 pl I [580]. Type data: holotype, BMNH *M. adult (seen 1929 by L.F. Graham), from Mackay, Qld.

Distribution: NE coastal, Qld.; type locality only. Ecology: larva - sedentary, predator : adult - volant; prey larval Lepidoptera.

Paralastor dyscritias Perkins, 1914

Paralastor dyscritias Perkins, R.C.L. (1914). On the species of *Alastor (Paralastor)* Sauss. and some other Hymenoptera of the family Eumenidae. *Proc. Zool. Soc. Lond.* **1914**: 563-624 pl I [614]. Type data: holotype, BMNH *F. adult (seen 1929 by L.F. Graham), from Australia.

Distribution: no locality specified. Ecology: larva - sedentary, predator : adult - volant; prey larval Lepidoptera.

Paralastor elegans Perkins, 1914

Paralastor elegans Perkins, R.C.L. (1914). On the species of *Alastor (Paralastor)* Sauss. and some other Hymenoptera of the family Eumenidae. *Proc. Zool. Soc. Lond.* **1914**: 563-624 pl I [581]. Type data: holotype, BMNH *F. adult, from N Qld.

Distribution: no locality specified. Ecology: larva - sedentary, predator : adult - volant; prey larval Lepidoptera.

Paralastor emarginatus (Saussure, 1853)

Alastor emarginatus Saussure, H. de (1853). *Études sur la Famille des Vespides*. 1. Monographie des guêpes solitaires, ou de la tribu des Euméniens, comprenant la classification et la description de toutes les espèces connues jusqu'à ce jour, et servant de complément au Manuel de Lepeletier de Saint Fargeau. Paris : Masson 286 pp. pls i-xx (1852-1853) [254]. Type data: syntypes (probable), MNHP *F. adult, from Tas.

Distribution: Tas.; only published locality Eaglehawk Neck. Ecology: larva - sedentary, predator : adult - volant; prey larval Lepidoptera. Biological references: Perkins, R.C.L. (1914). On the species of *Alastor (Paralastor)* Sauss. and some other Hymenoptera of the family Eumenidae. *Proc. Zool. Soc. Lond.* **1914**: 563-624 pl I (distribution); Giordani Soika, A. (1962). Gli *Odynerus* sensu antiquo del continente australiano e della Tasmania. *Boll. Mus. Civ. Stor. Nat. Venezia* **14**: 57-202 (doubts type locality).

Paralastor eriurgus (Saussure, 1853)

Alastor eriurgus Saussure, H. de (1853). *Études sur la Famille des Vespides*. 1. Monographie des guêpes solitaires, ou de la tribu des Euméniens, comprenant la classification et la description de toutes les espèces connues jusqu'à ce jour, et servant de complément au Manuel de Lepeletier de Saint Fargeau. Paris : Masson 286 pp. pls i-xx (1852-1853) [251 pl xxi fig 4]. Type data: syntypes (probable), GMNH *F. adult, from Australia (as New Holland).

Distribution: NE coastal, SE coastal, Qld., N.S.W.; only published localities Brisbane and Sydney. Ecology: larva - sedentary, predator : adult - volant; nest of "gum" from mango tree, prey larval Lepidoptera. Biological references: Froggatt, W.W. (1894). On the nests and habits of Australian Vespidae and Larridae. *Proc. Linn. Soc. N.S.W.* (2) **9**: 27-34 (biology); Steel, T. (1909). Notes and exhibits. *Proc. Linn. Soc. N.S.W.* **34**: 117 (nest); Perkins, R.C.L. (1914). On the species of *Alastor (Paralastor)* Sauss. and some other Hymenoptera of the family Eumenidae. *Proc. Zool. Soc. Lond.* **1914**: 563-624 pl I (redescription); Giordani Soika, A. (1962). Gli *Odynerus* sensu antiquo del continente australiano e della Tasmania. *Boll. Mus. Civ. Stor. Nat. Venezia* **14**: 57-202 (taxonomy).

Paralastor euclidias Perkins, 1914

Paralastor euclidias Perkins, R.C.L. (1914). On the species of *Alastor (Paralastor)* Sauss. and some other Hymenoptera of the family Eumenidae. *Proc. Zool. Soc. Lond.* **1914**: 563-624 pl I [598]. Type data: holotype, BMNH *M. adult (seen 1929 by L.F. Graham), from Gippsland, Vic.

Distribution: SE coastal, Vic.; type locality only. Ecology: larva - sedentary, predator : adult - volant; prey larval Lepidoptera.

Paralastor eugonias Perkins, 1914

Paralastor eugonias Perkins, R.C.L. (1914). On the species of *Alastor (Paralastor)* Sauss. and some other Hymenoptera of the family Eumenidae. *Proc. Zool. Soc. Lond.* **1914**: 563-624 pl I [596]. Type data: holotype, BMNH *M. adult (seen 1929 by L.F. Graham), from Adelaide, S.A.

Distribution: S Gulfs, S.A.; type locality only. Ecology: larva - sedentary, predator : adult - volant; prey larval Lepidoptera.

Paralastor eustomus Perkins, 1914

Paralastor eustomus Perkins, R.C.L. (1914). On the species of *Alastor (Paralastor)* Sauss. and some other Hymenoptera of the family Eumenidae. *Proc. Zool. Soc. Lond.* **1914**: 563-624 pl I [604]. Type data: holotype, BMNH *F. adult (seen 1929 by L.F. Graham), from Australia.

Distribution: no locality specified. Ecology: larva - sedentary, predator : adult - volant; prey larval Lepidoptera. Biological references: Giordani Soika, A. (1962). Gli *Odynerus* sensu antiquo del continente australiano e della Tasmania. *Boll. Mus. Civ. Stor. Nat. Venezia* **14**: 57-202 (taxonomy); Giordani Soika, A. (1977). Contributo alla conoscenza degli Eumenidi australiani (Hymenoptera). *Mem. Soc. Entomol. Ital.* **55**: 109-138 (taxonomy).

Paralastor eutretus Perkins, 1914

Paralastor eutretus Perkins, R.C.L. (1914). New species of *Paralastor*, Sauss. (Hymenoptera, Fam. Eumenidae) collected by Mr. R.E. Turner in S.W. Australia. *Ann. Mag. Nat. Hist. (8)* **14**: 235-240 [240]. Type data: holotype, BMNH *F. adult (seen 1929 by L.F. Graham), from Yallingup, W.A.

Distribution: SW coastal, W.A.; type locality only. Ecology: larva - sedentary, predator : adult - volant; prey larval Lepidoptera. Biological references: Giordani Soika, A. (1962). Gli *Odynerus* sensu antiquo del continente australiano e della Tasmania. *Boll. Mus. Civ. Stor. Nat. Venezia* **14**: 57-202 (taxonomy).

Paralastor fallax Perkins, 1914

Paralastor fallax Perkins, R.C.L. (1914). On the species of *Alastor (Paralastor)* Sauss. and some other Hymenoptera of the family Eumenidae. *Proc. Zool. Soc. Lond.* **1914**: 563-624 pl I [606]. Type data: holotype, BMNH *F. adult (seen 1929 by L.F. Graham), from W.A.

Distribution: W.A.; no locality specified. Ecology: larva - sedentary, predator : adult - volant; prey larval Lepidoptera. Biological references: Giordani Soika, A. (1962). Gli *Odynerus* sensu antiquo del continente australiano e della Tasmania. *Boll. Mus. Civ. Stor. Nat. Venezia* **14**: 57-202 (taxonomy).

Paralastor flaviceps (Saussure, 1856)

Alastor (Paralastor) flaviceps Saussure, H. de (1856). *Études sur la Famille des Vespides.* Troisième Partie comprenant la Monographie des Masariens et un Supplement à la Monographie des Eumeniens. Paris : Masson 352 pp. pls i-xvi (1854-1856) [336]. Type data: holotype, BMNH *F. adult (seen 1929 by L.F. Graham), from Australia (as New Holland).

Distribution: no locality specified. Ecology: larva - sedentary, predator : adult - volant; prey larval Lepidoptera. Biological references: Perkins, R.C.L. (1914). On the species of *Alastor (Paralastor)* Sauss. and some other Hymenoptera of the family Eumenidae. *Proc. Zool. Soc. Lond.* **1914**: 563-624 pl I (taxonomy).

Paralastor frater Perkins, 1914

Paralastor frater Perkins, R.C.L. (1914). On the species of *Alastor (Paralastor)* Sauss. and some other Hymenoptera of the family Eumenidae. *Proc. Zool. Soc. Lond.* **1914**: 563-624 pl I [585]. Type data: holotype, OUM *M. adult, from Albany, W.A.

Distribution: SW coastal, W.A.; type locality only. Ecology: larva - sedentary, predator : adult - volant; prey larval Lepidoptera.

Paralastor fraternus (Saussure, 1856)

Alastor (Paralastor) fraternus Saussure, H. de (1856). *Études sur la Famille des Vespides.* Troisième Partie comprenant la Monographie des Masariens et un Supplement à la Monographie des Eumeniens. Paris : Masson 352 pp. pls i-xvi (1854-1856) [330]. Type data: holotype, BMNH *F. adult (seen 1929 by L.F. Graham), from Australia.

Distribution: N.S.W.; no locality specified. Ecology: larva - sedentary, predator : adult - volant; prey larval Lepidoptera. Biological references: Perkins, R.C.L. (1914). On the species of *Alastor (Paralastor)* Sauss. and some other Hymenoptera of the family Eumenidae. *Proc. Zool. Soc. Lond.* **1914**: 563-624 pl I (description of male); Giordani Soika, A. (1962). Gli *Odynerus* sensu antiquo del continente australiano e della Tasmania. *Boll. Mus. Civ. Stor. Nat. Venezia* **14**: 57-202 (taxonomy).

Paralastor habilis Perkins, 1914

Paralastor habilis Perkins, R.C.L. (1914). On the species of *Alastor (Paralastor)* Sauss. and some other Hymenoptera of the family Eumenidae. *Proc. Zool. Soc. Lond.* **1914**: 563-624 pl I [583]. Type data: syntypes (probable), BMNH *adult, from N Qld.

Distribution: Qld; no locality specified. Ecology: larva - sedentary, predator : adult - volant; prey larval Lepidoptera.

Paralastor hilaris Perkins, 1914

Paralastor hilaris Perkins, R.C.L. (1914). On the species of *Alastor (Paralastor)* Sauss. and some other Hymenoptera of the family Eumenidae. *Proc. Zool. Soc. Lond.* **1914**: 563-624 pl I [600 pl I fig 18]. Type data: syntype, BMNH *M. adult (seen 1929 by L.F. Graham), from N Qld.

Distribution: N coastal, N.T., Qld.; only published localities Darwin and N Qld. Ecology: larva - sedentary, predator : adult - volant; prey larval Lepidoptera.

Paralastor icarioides Perkins, 1914

Paralastor icarioides Perkins, R.C.L. (1914). On the species of *Alastor (Paralastor)* Sauss. and some other Hymenoptera of the family Eumenidae. *Proc. Zool. Soc. Lond.* **1914**: 563-624 pl I [621 pl I fig 2]. Type data: holotype, BMNH *M. adult (seen 1929 by L.F. Graham), from Townsville, Qld.

Distribution: NE coastal, Qld.; only published localities Townsville, Kuranda and Cairns. Ecology: larva - sedentary, predator : adult - volant; prey larval Lepidoptera. Biological references: Giordani Soika, A. (1962). Gli *Odynerus* sensu antiquo del continente australiano e della Tasmania. *Boll. Mus. Civ. Stor. Nat. Venezia* **14**: 57-202 (taxonomy).

Paralastor ignotus Perkins, 1914

Paralastor ignotus Perkins, R.C.L. (1914). On the species of *Alastor (Paralastor)* Sauss. and some other

Hymenoptera of the family Eumenidae. *Proc. Zool. Soc. Lond.* **1914**: 563–624 pl I [621]. Type data: holotype, BMNH *M. adult (seen 1929 by L.F. Graham), from Swan River, W.A.

Distribution: SW coastal, W.A.; only published localities Swan River and Kalamunda. Ecology: larva - sedentary, predator : adult - volant; prey larval Lepidoptera. Biological references: Giordani Soika, A. (1962). Gli *Odynerus* sensu antiquo del continente australiano e della Tasmania. *Boll. Mus. Civ. Stor. Nat. Venezia* **14**: 57–202 (taxonomy).

Paralastor imitator Perkins, 1914

Paralastor imitator Perkins, R.C.L. (1914). On the species of *Alastor (Paralastor)* Sauss. and some other Hymenoptera of the family Eumenidae. *Proc. Zool. Soc. Lond.* **1914**: 563–624 pl I [595 pl I fig 10]. Type data: holotype, BMNH *F. adult (seen 1929 by L.F. Graham), from Champion Bay, W.A.

Distribution: NW coastal, W.A.; type locality only. Ecology: larva - sedentary, predator : adult - volant; prey larval Lepidoptera. Biological references: Giordani Soika, A. (1962). Gli *Odynerus* sensu antiquo del continente australiano e della Tasmania. *Boll. Mus. Civ. Stor. Nat. Venezia* **14**: 57–202 (taxonomy).

Paralastor infernalis (Saussure, 1856)

Alastor (Paralastor) infernalis Saussure, H. de (1856). *Études sur la Famille des Vespides*. Troisième Partie comprenant la Monographie des Masariens et un Supplement à la Monographie des Eumeniens. Paris : Masson 352 pp. pls i–xvi (1854–1856) [332]. Type data: holotype, BMNH *F. adult (seen 1929 by L.F. Graham), from Australia (as New Holland).

Distribution: NE coastal, N coastal, Qld., N.T., W.A.; only published localities Cairns, Burnside, Darwin and W.A. Ecology: larva - sedentary, predator : adult - volant; prey larval Lepidoptera. Biological references: Perkins, R.C.L. (1914). On the species of *Alastor (Paralastor)* Sauss. and some other Hymenoptera of the family Eumenidae. *Proc. Zool. Soc. Lond.* **1914**: 563–624 pl I (redescription); Schulthess-Rechberg, A. von (1935). Hymenoptera aus den Sundainseln und Nordaustralien (mit Ausschluss der Blattwespen, Schlupfwespen und Ameisen). *Rev. Suisse Zool.* **42**: 293–323 (distribution).

Paralastor infimus Perkins, 1914

Paralastor infimus Perkins, R.C.L. (1914). On the species of *Alastor (Paralastor)* Sauss. and some other Hymenoptera of the family Eumenidae. *Proc. Zool. Soc. Lond.* **1914**: 563–624 pl I [603]. Type data: holotype, BMNH *M. adult (seen 1929 by L.F. Graham), from Brisbane, Qld.

Distribution: NE coastal, Qld.; only published localities Brisbane and Stradbroke Is. Ecology: larva - sedentary, predator : adult - volant; prey larval Lepidoptera. Biological references: Giordani Soika, A. (1962). Gli *Odynerus* sensu antiquo del continente australiano e della Tasmania. *Boll. Mus. Civ. Stor. Nat. Venezia* **14**: 57–202 (taxonomy).

Paralastor insularis (Saussure, 1856)

Alastor (Paralastor) insularis Saussure, H. de (1856). *Études sur la Famille des Vespides*. Troisième Partie comprenant la Monographie des Masariens et un Supplement à la Monographie des Eumeniens. Paris : Masson 352 pp. pls i–xvi (1854–1856) [334 pl xvi fig 3]. Type data: syntypes, BMNH *2F. adult (seen 1929 by L.F. Graham), from Australia.

Distribution: SW coastal, W.A.; only published locality Swan River. Ecology: larva - sedentary, predator : adult - volant; prey larval Lepidoptera. Biological references: Perkins, R.C.L. (1914). On the species of *Alastor (Paralastor)* Sauss. and some other Hymenoptera of the family Eumenidae. *Proc. Zool. Soc. Lond.* **1914**: 563–624 pl I (redescription).

Paralastor lachesis (Saussure, 1853)

Alastor lachesis Saussure, H. de (1853). *Études sur la Famille des Vespides*. 1. Monographie des guêpes solitaires, ou de la tribu des Euméniens, comprenant la classification et la description de toutes les espèces connues jusqu'à ce jour, et servant de complément au Manuel de Lepeletier de Saint Fargeau. Paris : Masson 286 pp. pls i–xx (1852–1853) [251 pl xxi figs 5, 5a]. Type data: syntypes (probable), MNHP *F. adult, from Tas.; locality probably incorrect, see Perkins, R.C.L. (1914). On the species of *Alastor (Paralastor)* Sauss. and some other Hymenoptera of the family Eumenidae. *Proc. Zool. Soc. Lond.* **1914**: 563–624 pl I.

Distribution: Tas.; no locality specified. Ecology: larva - sedentary, predator : adult - volant; prey larval Lepidoptera. Biological references: Perkins, R.C.L. (1914). On the species of *Alastor (Paralastor)* Sauss. and some other Hymenoptera of the family Eumenidae. *Proc. Zool. Soc. Lond.* **1914**: 563–624 pl I (taxonomy).

Paralastor laetus Perkins, 1914

Paralastor laetus Perkins, R.C.L. (1914). On the species of *Alastor (Paralastor)* Sauss. and some other Hymenoptera of the family Eumenidae. *Proc. Zool. Soc. Lond.* **1914**: 563–624 pl I [585]. Type data: holotype, BMNH *M. adult (seen 1929 by L.F. Graham), from Fremantle, W.A.

Distribution: SW coastal, W.A.; only published localities Fremantle and Yallingup. Ecology: larva - sedentary, predator : adult - volant; prey larval Lepidoptera. Biological references: Perkins, R.C.L. (1914). New species of *Paralastor*, Sauss.

(Hymenoptera, Fam. Eumenidae) collected by Mr. R.E. Turner in S.W. Australia. *Ann. Mag. Nat. Hist. (8)* **14**: 235–240 (redescription).

Paralastor lateritius (Saussure, 1867)

Alastor (Paralastor) lateritius Saussure, H. de (1867). Hymenoptera. in, *Reise der österreichischen Fregatte Novara um die Erde in den Jahren 1857, 1858, 1859 unter den Befehlen des Commodore B. von Wüllerstorf-Urbair*. Zoologischer Theil. Wien : K-K Hof- und Staatsdrückerei Vol. 2(1a) 138 pp. [17]. Type data: syntypes (probable), adult whereabouts unknown, from Australia (as New Holland).

Distribution: no locality specified. Ecology: larva - sedentary, predator : adult - volant; prey larval Lepidoptera. Biological references: Perkins, R.C.L. (1914). On the species of *Alastor (Paralastor)* Sauss. and some other Hymenoptera of the family Eumenidae. *Proc. Zool. Soc. Lond.* **1914**: 563–624 pl I (taxonomy).

Paralastor leptias Perkins, 1914

Paralastor leptias Perkins, R.C.L. (1914). On the species of *Alastor (Paralastor)* Sauss. and some other Hymenoptera of the family Eumenidae. *Proc. Zool. Soc. Lond.* **1914**: 563–624 pl I [620]. Type data: syntype, BMNH *F. adult (seen 1929 by L.F. Graham), from Adelaide, S.A.

Distribution: S Gulfs, S.A.; type locality only. Ecology: larva - sedentary, predator : adult - volant; prey larval Lepidoptera.

Paralastor mackayensis Perkins, 1914

Paralastor mackayensis Perkins, R.C.L. (1914). On the species of *Alastor (Paralastor)* Sauss. and some other Hymenoptera of the family Eumenidae. *Proc. Zool. Soc. Lond.* **1914**: 563–624 pl I [607]. Type data: holotype, BMNH *F. adult (seen 1929 by L.F. Graham), from Mackay, Qld.

Distribution: NE coastal, Qld.; type locality only. Ecology: larva - sedentary, predator : adult - volant; prey larval Lepidoptera.

Paralastor maculiventris (Saussure, 1856)

Alastor (Paralastor) maculiventris Saussure, H. de (1856). *Études sur la Famille des Vespides*. Troisième Partie comprenant la Monographie des Masariens et un Supplement à la Monographie des Eumeniens. Paris : Masson 352 pp. pls i–xvi (1854–1856) [337 pl xvi fig 2]. Type data: holotype, BMNH *F. adult (seen 1929 by L.F. Graham), from Australia.

Distribution: no locality specified. Ecology: larva - sedentary, predator : adult - volant; prey larval Lepidoptera. Biological references: Perkins, R.C.L. (1914). On the species of *Alastor (Paralastor)* Sauss. and some other Hymenoptera of the family Eumenidae. *Proc. Zool. Soc. Lond.* **1914**: 563–624 pl I (redescription).

Paralastor medius Perkins, 1914

Paralastor medius Perkins, R.C.L. (1914). On the species of *Alastor (Paralastor)* Sauss. and some other Hymenoptera of the family Eumenidae. *Proc. Zool. Soc. Lond.* **1914**: 563–624 pl I [604]. Type data: holotype, BMNH *F. adult (seen 1929 by L.F. Graham), from Mackay, Qld.

Distribution: NE coastal, Qld.; type locality only. Ecology: larva - sedentary, predator : adult - volant; prey larval Lepidoptera.

Paralastor mesochlorus Perkins, 1914

Taxonomic decision of Perkins, R.C.L. (1914). On the species of *Alastor (Paralastor)* Sauss. and some other Hymenoptera of the family Eumenidae. *Proc. Zool. Soc. Lond.* **1914**: 563–624 pl I [616].

Paralastor mesochlorus mesochlorus Perkins, 1914

Paralastor mesochlorus Perkins, R.C.L. (1914). On the species of *Alastor (Paralastor)* Sauss. and some other Hymenoptera of the family Eumenidae. *Proc. Zool. Soc. Lond.* **1914**: 563–624 pl I [616]. Type data: holotype, BMNH *F. adult (seen 1929 by L.F. Graham), from Mackay, Qld.

Distribution: NE coastal, Qld.; type locality only. Ecology: larva - sedentary, predator : adult - volant; prey larval Lepidoptera.

Paralastor mesochlorus mesochloroides Perkins, 1914

Paralastor mesochlorus mesochloroides Perkins, R.C.L. (1914). On the species of *Alastor (Paralastor)* Sauss. and some other Hymenoptera of the family Eumenidae. *Proc. Zool. Soc. Lond.* **1914**: 563–624 pl I [616] [described as race]. Type data: holotype, BMNH *F. adult (seen 1929 by L.F. Graham), from Kuranda, Qld.

Distribution: NE coastal, Qld.; type locality only. Ecology: larva - sedentary, predator : adult - volant; prey larval Lepidoptera.

Paralastor microgonias Perkins, 1914

Paralastor microgonias Perkins, R.C.L. (1914). On the species of *Alastor (Paralastor)* Sauss. and some other Hymenoptera of the family Eumenidae. *Proc. Zool. Soc. Lond.* **1914**: 563–624 pl I [596]. Type data: holotype, BMNH *M. adult (seen 1929 by L.F. Graham), from Adelaide, S.A.

Distribution: S Gulfs, S.A.; type locality only. Ecology: larva - sedentary, predator : adult - volant; prey larval Lepidoptera.

Paralastor mimus Perkins, 1914

Paralastor mimus Perkins, R.C.L. (1914). On the species of *Alastor (Paralastor)* Sauss. and some other Hymenoptera of the family Eumenidae. *Proc. Zool. Soc. Lond.* **1914**: 563–624 pl I [594]. Type data: holotype, BMNH *F. adult (seen 1929 by L.F. Graham), from Swan River, W.A.

Distribution: SW coastal, W.A.; type locality only (1 female "N.S.W." probably label error). Ecology: larva - sedentary, predator : adult - volant; prey larval Lepidoptera. Biological references: Giordani Soika, A. (1962). Gli *Odynerus* sensu antiquo del continente australiano e della Tasmania. *Boll. Mus. Civ. Stor. Nat. Venezia* **14**: 57-202 (taxonomy).

Paralastor multicolor Perkins, 1914

Paralastor multicolor Perkins, R.C.L. (1914). On the species of *Alastor (Paralastor)* Sauss. and some other Hymenoptera of the family Eumenidae. *Proc. Zool. Soc. Lond.* **1914**: 563-624 pl I [612 pl I figs 9, 19]. Type data: syntype, BMNH *M. adult (seen 1929 by L.F. Graham), from Port Darwin, N.T.

Distribution: N coastal, N.T.; type locality only. Ecology: larva - sedentary, predator : adult - volant; prey larval Lepidoptera.

Paralastor mutabilis Perkins, 1914

Paralastor mutabilis Perkins, R.C.L. (1914). On the species of *Alastor (Paralastor)* Sauss. and some other Hymenoptera of the family Eumenidae. *Proc. Zool. Soc. Lond.* **1914**: 563-624 pl I [610 pl I fig 8]. Type data: holotype, BMNH *F. adult (seen 1929 by L.F. Graham), from Vic.

Distribution: Vic.; no locality specified. Ecology: larva - sedentary, predator : adult - volant; prey larval Lepidoptera.

Paralastor nautarum (Saussure, 1856)

Alastor (Paralastor) nautarum Saussure, H. de (1856). *Études sur la Famille des Vespides. Troisième Partie comprenant la Monographie des Masariens et un Supplement à la Monographie des Eumeniens.* Paris : Masson 352 pp. pls i-xvi (1854-1856) [330]. Type data: holotype, BMNH *M. adult (seen 1929 by L.F. Graham), from Australia.

Distribution: no locality specified. Ecology: larva - sedentary, predator : adult - volant; prey larval Lepidoptera. Biological references: Perkins, R.C.L. (1914). On the species of *Alastor (Paralastor)* Sauss. and some other Hymenoptera of the family Eumenidae. *Proc. Zool. Soc. Lond.* **1914**: 563-624 pl I (redescription).

Paralastor neglectus (Saussure, 1855)

Odynerus (Parodynerus) neglectus Saussure, H. de (1855). *Études sur la Famille des Vespides. Troisième Partie comprenant la Monographie des Masariens et un Supplement à la Monographie des Eumeniens.* Paris : Masson 352 pp. pls i-xvi (1854-1856) [245]. Type data: holotype, BMNH *F. adult (seen 1929 by L.F. Graham), from Australia (as New Holland).

Distribution: no locality specified. Ecology: larva - sedentary, predator : adult - volant; prey larval Lepidoptera. Biological references: Giordani Soika, A. (1941). Studi sui Vespidi solitari. *Boll. Soc. Veneziana Stor. Nat.* **2**: 130-279 (taxonomy).

Paralastor neochromus Perkins, 1914

Paralastor neochromus Perkins, R.C.L. (1914). New species of *Paralastor*, Sauss. (Hymenoptera, Fam. Eumenidae) collected by Mr. R.E. Turner in S.W. Australia. *Ann. Mag. Nat. Hist. (8)* **14**: 235-240 [238]. Type data: holotype, BMNH *F. adult (seen 1929 by L.F. Graham), from Yallingup, W.A.

Distribution: SW coastal, W.A.; type locality only. Ecology: larva - sedentary, predator : adult - volant; prey larval Lepidoptera.

Paralastor occidentalis Perkins, 1914

Paralastor occidentalis Perkins, R.C.L. (1914). On the species of *Alastor (Paralastor)* Sauss. and some other Hymenoptera of the family Eumenidae. *Proc. Zool. Soc. Lond.* **1914**: 563-624 pl I [598]. Type data: holotype, BMNH *M. adult (seen 1929 by L.F. Graham), from Swan River, W.A.

Distribution: SW coastal, W.A.; type locality only. Ecology: larva - sedentary, predator : adult - volant; prey larval Lepidoptera. Biological references: Giordani Soika, A. (1962). Gli *Odynerus* sensu antiquo del continente australiano e della Tasmania. *Boll. Mus. Civ. Stor. Nat. Venezia* **14**: 57-202 (taxonomy).

Paralastor odynericornis Giordani Soika, 1961

Paralastor odynericornis Giordani Soika, A. (1961). Notulae Vespidologicae. XIII - *Paralastor* nuovi o poco noti. *Boll. Soc. Entomol. Ital.* **91**: 12-15 [14]. Type data: holotype, ZMB *M. adult, from Marloo Station, Wurarga, W.A.

Distribution: NW coastal, W.A.; type locality only. Ecology: larva - sedentary, predator : adult - volant; prey larval Lepidoptera.

Paralastor odyneripennis Perkins, 1914

Paralastor odyneripennis Perkins, R.C.L. (1914). On the species of *Alastor (Paralastor)* Sauss. and some other Hymenoptera of the family Eumenidae. *Proc. Zool. Soc. Lond.* **1914**: 563-624 pl I [610]. Type data: holotype, BMNH *F. adult (seen 1929 by L.F. Graham), from Vic.

Distribution: Vic.; no locality specified. Ecology: larva - sedentary, predator : adult - volant; prey larval Lepidoptera.

Paralastor odyneroides Perkins, 1914

Paralastor odyneroides Perkins, R.C.L. (1914). On the species of *Alastor (Paralastor)* Sauss. and some other Hymenoptera of the family Eumenidae. *Proc. Zool. Soc. Lond.* **1914**: 563-624 pl I [610]. Type data: holotype, OUM *M. adult, from Australia.

Distribution: no locality specified. Ecology: larva - sedentary, predator : adult - volant; prey larval Lepidoptera.

Paralastor oloris Perkins, 1914

Paralastor oloris Perkins, R.C.L. (1914). On the species of *Alastor (Paralastor)* Sauss. and some other Hymenoptera of the family Eumenidae. *Proc. Zool. Soc. Lond.* **1914**: 563-624 pl I [608 pl I figs 4,16]. Type data: holotype, BMNH *M. adult (seen 1929 by L.F. Graham), from Swan River, W.A.

Distribution: SW coastal, NW coastal, W.A.; only published localities Swan River and Marloo Station, Wurarga. Ecology: larva - sedentary, predator : adult - volant; prey larval Lepidoptera. Biological references: Giordani Soika, A. (1961). Notulae Vespidologicae. XIII - *Paralastor* nuovi o poco noti. *Boll. Soc. Entomol. Ital.* **91**: 12-15 (description of female); Giordani Soika, A. (1962). Gli *Odynerus* sensu antiquo del continente australiano e della Tasmania. *Boll. Mus. Civ. Stor. Nat. Venezia* **14**: 57-202 (taxonomy).

Paralastor optabilis Perkins, 1914

Paralastor optabilis Perkins, R.C.L. (1914). On the species of *Alastor (Paralastor)* Sauss. and some other Hymenoptera of the family Eumenidae. *Proc. Zool. Soc. Lond.* **1914**: 563-624 pl I [587 pl I fig 12]. Type data: syntype, BMNH *F. adult (seen 1929 by L.F. Graham), from N Qld.

Distribution: Qld.; no locality specified. Ecology: larva - sedentary, predator : adult - volant; prey larval Lepidoptera.

Paralastor ordinarius Perkins, 1914

Paralastor ordinarius Perkins, R.C.L. (1914). On the species of *Alastor (Paralastor)* Sauss. and some other Hymenoptera of the family Eumenidae. *Proc. Zool. Soc. Lond.* **1914**: 563-624 pl I [586]. Type data: holotype, BMNH *F. adult (seen 1929 by L.F. Graham), from Wimmera, Vic.

Distribution: Murray-Darling basin, Vic.; type locality only. Ecology: larva - sedentary, predator : adult - volant; prey larval Lepidoptera.

Paralastor orientalis Perkins, 1914

Paralastor orientalis Perkins, R.C.L. (1914). On the species of *Alastor (Paralastor)* Sauss. and some other Hymenoptera of the family Eumenidae. *Proc. Zool. Soc. Lond.* **1914**: 563-624 pl I [599]. Type data: holotype, BMNH *F. adult (seen 1929 by L.F. Graham), from N.S.W.

Distribution: NE coastal, N.S.W., Qld.; only published localities N.S.W., Brisbane and Bundaberg. Ecology: larva - sedentary, predator : adult - volant; prey larval Lepidoptera. Biological references: Giordani Soika, A. (1962). Gli *Odynerus* sensu antiquo del continente australiano e della Tasmania. *Boll. Mus. Civ. Stor. Nat. Venezia* **14**: 57-202 (taxonomy).

Paralastor pallidus Perkins, 1914

Paralastor pallidus Perkins, R.C.L. (1914). On the species of *Alastor (Paralastor)* Sauss. and some other Hymenoptera of the family Eumenidae. *Proc. Zool. Soc. Lond.* **1914**: 563-624 pl I [584]. Type data: holotype, BMNH *F. adult (seen 1929 by L.F. Graham), from Mackay, Qld.

Distribution: NE coastal, Qld.; only published locality Mackay. Ecology: larva - sedentary, predator : adult - volant; prey larval Lepidoptera.

Paralastor parca (Saussure, 1853)

Alastor parca Saussure, H. de (1853). *Études sur la Famille des Vespides*. 1. Monographie des guêpes solitaires, ou de la tribu des Euméniens, comprenant la classification et la description de toutes les espèces connues jusqu'à ce jour, et servant de complément au Manuel de Lepeletier de Saint Fargeau. Paris : Masson 286 pp. pls i-xx (1852-1853) [254]. Type data: holotype, MNHP *adult, from Australia (as New Holland).

Distribution: Vic., Tas.; only published localities Vic., Mt. Wellington, Eaglehawk Neck and Franklin. Ecology: larva - sedentary, predator : adult - volant; prey larval Lepidoptera. Biological references: Perkins, R.C.L. (1914). On the species of *Alastor (Paralastor)* Sauss. and some other Hymenoptera of the family Eumenidae. *Proc. Zool. Soc. Lond.* **1914**: 563-624 pl I (taxonomy, distribution); Giordani Soika, A. (1962). Gli *Odynerus* sensu antiquo del continente australiano e della Tasmania. *Boll. Mus. Civ. Stor. Nat. Venezia* **14**: 57-202 (taxonomy); Giordani Soika, A. (1974). Prime ricerche sugli Eumenidi ipsobionti. I. Caratteristiche generali degli Eumenidi ipsobionti del globo. *Redia* **55**: 287-302 (colouration).

Paralastor petiolatus Schulthess-Rechberg, 1925

Paralastor petiolatus Schulthess-Rechberg, A. von (1925). Beitrag zur Kenntnis der Gattung *Alastor* Lep. (Hym. Vesp.). *Konowia* **4**: 57-65, 195-209, 257-263 [258]. Type data: holotype, GMNH *adult, from Parramatta, N.S.W.

Distribution: SE coastal, N.S.W.; type locality only. Ecology: larva - sedentary, predator : adult - volant; prey larval Lepidoptera.

Paralastor picteti (Saussure, 1853)

Alastor picteti Saussure, H. de (1853). *Études sur la Famille des Vespides*. 1. Monographie des guêpes solitaires, ou de la tribu des Euméniens, comprenant la classification et la description de toutes les espèces connues jusqu'à ce jour, et servant de complément au Manuel de Lepeletier de Saint Fargeau. Paris : Masson 286 pp. pls i-xx (1852-1853) [256]. Type data: syntypes (probable), GMNH *M. adult, from Tas.; locality probably incorrect, see Perkins, R.C.L. (1914). On the species of *Alastor (Paralastor)* Sauss. and some other Hymenoptera of the family Eumenidae. *Proc. Zool. Soc. Lond.* **1914**: 563–624 pl I.

Distribution: NE coastal, Qld., (Tas.); only published localities Cairns, Kuranda, Mackay, Bundaberg and "Tas." Ecology: larva - sedentary, predator : adult - volant; prey larval Lepidoptera. Biological references: Saussure, H. de (1854–1856). *Études sur la Famille des Vespides.* Troisième Partie comprenant la Monographie des Masariens et un Supplement à la Monographie des Eumeniens. Paris : Masson 352 pp. pls i–xvi (description of male); Perkins, R.C.L. (1914). On the species of *Alastor (Paralastor)* Sauss. and some other Hymenoptera of the family Eumenidae. *Proc. Zool. Soc. Lond.* **1914**: 563–624 pl I (taxonomy, distribution).

Paralastor placens Perkins, 1914

Paralastor placens Perkins, R.C.L. (1914). On the species of *Alastor (Paralastor)* Sauss. and some other Hymenoptera of the family Eumenidae. *Proc. Zool. Soc. Lond.* **1914**: 563–624 pl I [592]. Type data: holotype, BMNH *F. adult (seen 1929 by L.F. Graham), from Swan River, W.A.

Distribution: SW coastal, W.A.; type locality only. Ecology: larva - sedentary, predator : adult - volant; prey larval Lepidoptera. Biological references: Giordani Soika, A. (1962). Gli *Odynerus* sensu antiquo del continente australiano e della Tasmania. *Boll. Mus. Civ. Stor. Nat. Venezia* **14**: 57–202 (taxonomy).

Paralastor plebeius Perkins, 1914

Paralastor plebeius Perkins, R.C.L. (1914). On the species of *Alastor (Paralastor)* Sauss. and some other Hymenoptera of the family Eumenidae. *Proc. Zool. Soc. Lond.* **1914**: 563–624 pl I [611]. Type data: holotype, BMNH *F. adult (seen 1929 by L.F. Graham), from Australia.

Distribution: S Gulfs, Vic., S.A.; only published localities Vic. and Adelaide. Ecology: larva - sedentary, predator : adult - volant; prey larval Lepidoptera. Biological references: Giordani Soika, A. (1962). Gli *Odynerus* sensu antiquo del continente australiano e della Tasmania. *Boll. Mus. Civ. Stor. Nat. Venezia* **14**: 57–202 (taxonomy).

Paralastor princeps Perkins, 1914

Paralastor princeps Perkins, R.C.L. (1914). On the species of *Alastor (Paralastor)* Sauss. and some other Hymenoptera of the family Eumenidae. *Proc. Zool. Soc. Lond.* **1914**: 563–624 pl I [607]. Type data: holotype, BMNH *F. adult (seen 1929 by L.F. Graham), from W.A.

Distribution: W.A.; no locality specified. Ecology: larva - sedentary, predator : adult - volant; prey larval Lepidoptera.

Paralastor pseudochromus Perkins, 1914

Paralastor pseudochromus Perkins, R.C.L. (1914). On the species of *Alastor (Paralastor)* Sauss. and some other Hymenoptera of the family Eumenidae. *Proc. Zool. Soc. Lond.* **1914**: 563–624 pl I [605 pl I fig 14]. Type data: holotype, BMNH *F. adult (seen 1929 by L.F. Graham), from Vic.

Distribution: SE coastal, Vic.; only published locality Melbourne. Ecology: larva - sedentary, predator : adult - volant; prey larval Lepidoptera.

Paralastor punctulatus (Saussure, 1853)

Alastor punctulatus Saussure, H. de (1853). *Études sur la Famille des Vespides.* 1. Monographie des guêpes solitaires, ou de la tribu des Euméniens, comprenant la classification et la description de toutes les espèces connues jusqu'à ce jour, et servant de complément au Manuel de Lepeletier de Saint Fargeau. Paris : Masson 286 pp. pls i–xx (1852–1853) [255]. Type data: holotype, MNHP *F. adult, from Tas.

Alastor similis Saussure, H. de (1853). *Études sur la Famille des Vespides.* 1. Monographie des guêpes solitaires, ou de la tribu des Euméniens, comprenant la classification et la description de toutes les espèces connues jusqu'à ce jour, et servant de complément au Manuel de Lepeletier de Saint Fargeau. Paris : Masson 286 pp. pls i–xx (1852–1853) [256]. Type data: holotype, MNHP *F. adult, from Australia (as New Holland).

Alastor albocinctus Smith, F. (1857). *Catalogue of Hymenopterous Insects in the Collection of the British Museum.* Pt. V. Vespidae pp. 1–147 London : British Museum [91]. Type data: holotype, BMNH *M. adult (seen 1929 by L.F. Graham), from Tas.

Taxonomic decision of Perkins, R.C.L. (1914). On the species of *Alastor (Paralastor)* Sauss. and some other Hymenoptera of the family Eumenidae. *Proc. Zool. Soc. Lond.* **1914**: 563–624 pl I [586].

Distribution: Murray-Darling basin, N.S.W., Tas.; only published localities Mt. Kosciusko, Mt. Wellington, Eaglehawk Neck and Hobart. Ecology: larva - sedentary, predator : adult - volant; prey larval Lepidoptera. Biological references: Giordani Soika, A. (1962). Gli *Odynerus* sensu antiquo del continente australiano e della Tasmania. *Boll. Mus. Civ. Stor. Nat. Venezia* **14**: 57–202 (taxonomy); Giordani Soika, A. (1974). Prime ricerche sugli Eumenidi ipsobionti. I. Caratteristiche generali degli Eumenidi ipsobionti del globo. *Redia* **55**: 287–302 (colouration).

Paralastor pusillus (Saussure, 1856)

Alastor (Paralastor) pusillus Saussure, H. de (1856). *Études sur la Famille des Vespides.* Troisième Partie comprenant la Monographie des Masariens et un Supplement à la Monographie des Eumeniens. Paris : Masson 352 pp. pls i–xvi (1854–1856) [332 pl xvi fig 5]. Type data: holotype, BMNH *M. adult (seen 1929 by L.F. Graham), from N.S.W.

Distribution: N.S.W.; no locality specified. Ecology: larva - sedentary, predator : adult - volant; prey larval Lepidoptera. Biological references: Perkins, R.C.L. (1914). On the species of *Alastor (Paralastor)* Sauss. and some other Hymenoptera of the family Eumenidae. *Proc. Zool. Soc. Lond.* **1914**: 563–624 pl I (taxonomy).

Paralastor roseotinctus Perkins, 1914

Paralastor roseotinctus Perkins, R.C.L. (1914). On the species of *Alastor (Paralastor)* Sauss. and some other Hymenoptera of the family Eumenidae. *Proc. Zool. Soc. Lond.* **1914**: 563–624 pl I [591]. Type data: holotype, BMNH *F. adult (seen 1929 by L.F. Graham), from W.A.

Distribution: SW coastal, W.A.; only published locality Swan River. Ecology: larva - sedentary, predator : adult - volant; prey larval Lepidoptera.

Paralastor rubroviolaceus (Giordani Soika, 1941)

Odynerus (Rhynchium) rubroviolaceus Giordani Soika, A. (1941). Studi sui Vespidi solitari. *Boll. Soc. Veneziana Stor. Nat.* **2**: 130–279 [255]. Type data: lectotype, A. Giordani Soika pers. coll. *F. adult, from Mt. Tamborine, Qld., designation by Giordani Soika, A. (1973). Designazione di lectotipi ed elenco dei tipi di Eumenidi, Vespidi e Masaridi da me descritti negli anni 1934–1960. *Boll. Mus. Civ. Stor. Nat. Venezia* **24**: 7–53.

Distribution: NE coastal, Qld.; type locality only. Ecology: larva - sedentary, predator : adult - volant; prey larval Lepidoptera. Biological references: Giordani Soika, A. (1961). Notulae Vespidologicae. XIII - *Paralastor* nuovi o poco noti. *Boll. Soc. Entomol. Ital.* **91**: 12–15 (generic placement).

Paralastor rufipes Perkins, 1914

Paralastor rufipes Perkins, R.C.L. (1914). On the species of *Alastor (Paralastor)* Sauss. and some other Hymenoptera of the family Eumenidae. *Proc. Zool. Soc. Lond.* **1914**: 563–624 pl I [579]. Type data: holotype, BMNH *F. adult, from N Qld.

Distribution: Qld.; no locality specified. Ecology: larva - sedentary, predator : adult - volant; prey larval Lepidoptera.

Paralastor sanguineus (Saussure, 1856)

Alastor (Paralastor) sanguineus Saussure, H. de (1856). Études sur la Famille des Vespides. Troisième Partie comprenant la Monographie des Masariens et un Supplement à la Monographie des Eumeniens. Paris : Masson 352 pp. pls i–xvi (1854–1856) [331]. Type data: holotype, BMNH *M. adult (seen 1929 by L.F. Graham), from Australia (as New Holland).

Distribution: no locality specified. Ecology: larva - sedentary, predator : adult - volant; prey larval Lepidoptera. Biological references: Perkins, R.C.L. (1914). On the species of *Alastor (Paralastor)* Sauss. and some other Hymenoptera of the family Eumenidae. *Proc. Zool. Soc. Lond.* **1914**: 563–624 pl I (redescription).

Paralastor saussurei Perkins, 1914

Paralastor saussurei Perkins, R.C.L. (1914). On the species of *Alastor (Paralastor)* Sauss. and some other Hymenoptera of the family Eumenidae. *Proc. Zool. Soc. Lond.* **1914**: 563–624 pl I [579]. Type data: holotype, BMNH *F. adult, from N Qld.

Distribution: Qld.; no locality specified. Ecology: larva - sedentary, predator : adult - volant; prey larval Lepidoptera.

Paralastor semirufus Schulthess-Rechberg, 1925

Paralastor semirufus Schulthess-Rechberg, A. von (1925). Beitrag zur Kenntnis der Gattung *Alastor* Lep. (Hym. Vesp.). *Konowia* **4**: 57–65, 195–209, 257–263 [257]. Type data: holotype, NHRM *F. adult, from Kimberley district, W.A.

Distribution: N coastal, W.A.; type locality only. Ecology: larva - sedentary, predator : adult - volant; prey larval Lepidoptera.

Paralastor simillimus Perkins, 1914

Paralastor simillimus Perkins, R.C.L. (1914). On the species of *Alastor (Paralastor)* Sauss. and some other Hymenoptera of the family Eumenidae. *Proc. Zool. Soc. Lond.* **1914**: 563–624 pl I [619]. Type data: holotype, BMNH *M. adult (seen 1929 by L.F. Graham), from Mackay, Qld.

Distribution: NE coastal, Qld.; only published localities Mackay and Bundaberg. Ecology: larva - sedentary, predator : adult - volant; prey larval Lepidoptera.

Paralastor simplex Perkins, 1914

Paralastor simplex Perkins, R.C.L. (1914). On the species of *Alastor (Paralastor)* Sauss. and some other Hymenoptera of the family Eumenidae. *Proc. Zool. Soc. Lond.* **1914**: 563–624 pl I [594]. Type data: holotype, BMNH *F. adult, from Albany, W.A.

Distribution: SW coastal, W.A.; type locality only. Ecology: larva - sedentary, predator : adult - volant; prey larval Lepidoptera.

Paralastor simulator Perkins, 1914

Paralastor simulator Perkins, R.C.L. (1914). On the species of *Alastor (Paralastor)* Sauss. and some other Hymenoptera of the family Eumenidae. *Proc. Zool. Soc. Lond.* **1914**: 563–624 pl I [588]. Type data: holotype, BMNH *M. adult (seen 1929 by L.F. Graham), from Melbourne, Vic.

Distribution: SE coastal, S Gulfs, Vic., S.A.; only published localities Melbourne and Adelaide. Ecology: larva - sedentary, predator : adult - volant; prey larval Lepidoptera. Biological references: Giordani Soika, A. (1962). Gli

Odynerus sensu antiquo del continente australiano e della Tasmania. *Boll. Mus. Civ. Stor. Nat. Venezia* **14**: 57–202 (taxonomy).

Paralastor smithii (Saussure, 1856)

Alastor (Paralastor) smithii Saussure, H. de (1856). *Études sur la Famille des Vespides*. Troisième Partie comprenant la Monographie des Masariens et un Supplement à la Monographie des Eumeniens. Paris : Masson 352 pp. pls i–xvi (1854–1856) [333 pl xvi fig 4]. Type data: holotype, BMNH *F. adult (seen 1929 by L.F. Graham), from Australia.

Distribution: no locality specified. Ecology: larva - sedentary, predator : adult - volant; prey larval Lepidoptera. Biological references: Perkins, R.C.L. (1914). On the species of *Alastor (Paralastor)* Sauss. and some other Hymenoptera of the family Eumenidae. *Proc. Zool. Soc. Lond.* **1914**: 563–624 pl I (generic placement).

Paralastor solitarius Perkins, 1914

Paralastor solitarius Perkins, R.C.L. (1914). On the species of *Alastor (Paralastor)* Sauss. and some other Hymenoptera of the family Eumenidae. *Proc. Zool. Soc. Lond.* **1914**: 563–624 pl I [600 pl I fig 6]. Type data: holotype, BMNH Hym 18.644 *M. adult, from Bundaberg, Qld.

Distribution: NE coastal, Qld.; type locality only. Ecology: larva - sedentary, predator : adult - volant; prey larval Lepidoptera.

Paralastor subhabilis Perkins, 1914

Paralastor subhabilis Perkins, R.C.L. (1914). On the species of *Alastor (Paralastor)* Sauss. and some other Hymenoptera of the family Eumenidae. *Proc. Zool. Soc. Lond.* **1914**: 563–624 pl I [583]. Type data: holotype, BMNH *F. adult (seen 1929 by L.F. Graham), from Mackay, Qld.

Distribution: NE coastal, Qld.; type locality only. Ecology: larva - sedentary, predator : adult - volant; prey larval Lepidoptera.

Paralastor submersus Turner, 1919

Paralastor submersus Turner, R.E. (1919). New Australian diplopterous Hymenoptera. *Ann. Mag. Nat. Hist. (9)* **3**: 398–399 [399]. Type data: holotype, BMNH *F. adult (seen 1929 by L.F. Graham), from Lolworth Station, Qld.

Distribution: (Murray-Darling basin), (Lake Eyre basin), Qld.; type locality only. Ecology: larva - sedentary, predator : adult - volant; prey larval Lepidoptera.

Paralastor subobscurus Perkins, 1914

Paralastor subobscurus Perkins, R.C.L. (1914). On the species of *Alastor (Paralastor)* Sauss. and some other Hymenoptera of the family Eumenidae. *Proc. Zool. Soc. Lond.* **1914**: 563–624 pl I [593]. Type data: holotype, BMNH *M. adult, from N Qld.

Distribution: Qld; no locality specified. Ecology: larva - sedentary, predator : adult - volant; prey larval Lepidoptera.

Paralastor suboloris Perkins, 1914

Paralastor suboloris Perkins, R.C.L. (1914). On the species of *Alastor (Paralastor)* Sauss. and some other Hymenoptera of the family Eumenidae. *Proc. Zool. Soc. Lond.* **1914**: 563–624 pl I [608]. Type data: holotype, OUM *M. adult, from W.A.

Distribution: W.A.; no locality specified. Ecology: larva - sedentary, predator : adult - volant; prey larval Lepidoptera.

Paralastor subplebeius Perkins, 1914

Paralastor subplebeius Perkins, R.C.L. (1914). On the species of *Alastor (Paralastor)* Sauss. and some other Hymenoptera of the family Eumenidae. *Proc. Zool. Soc. Lond.* **1914**: 563–624 pl I [611 pl I fig 15]. Type data: syntype, BMNH *F. adult (seen 1929 by L.F. Graham), from N Qld.

Distribution: Qld.; no locality specified. Ecology: larva - sedentary, predator : adult - volant; prey larval Lepidoptera.

Paralastor subpunctulatus Perkins, 1914

Paralastor subpunctulatus Perkins, R.C.L. (1914). New species of *Paralastor*, Sauss. (Hymenoptera, Fam. Eumenidae) collected by Mr. R.E. Turner in S.W. Australia. *Ann. Mag. Nat. Hist. (8)* **14**: 235–240 [238]. Type data: holotype, BMNH *M. adult (seen 1929 by L.F. Graham), from Yallingup, W.A.

Distribution: SW coastal, W.A.; type locality only. Ecology: larva - sedentary, predator : adult - volant; prey larval Lepidoptera. Biological references: Giordani Soika, A. (1962). Gli *Odynerus* sensu antiquo del continente australiano e della Tasmania. *Boll. Mus. Civ. Stor. Nat. Venezia* **14**: 57–202 (taxonomy); Giordani Soika, A. (1974). Prime ricerche sugli Eumenidi ipsobionti. I. Caratteristiche generali degli Eumenidi ipsobionti del globo. *Redia* **55**: 287–302 (colouration).

Paralastor summus Perkins, 1914

Paralastor summus Perkins, R.C.L. (1914). On the species of *Alastor (Paralastor)* Sauss. and some other Hymenoptera of the family Eumenidae. *Proc. Zool. Soc. Lond.* **1914**: 563–624 pl I [604]. Type data: holotype, BMNH *F. adult, from Cairns, Qld.

Distribution: NE coastal, Qld.; type locality only. Ecology: larva - sedentary, predator : adult - volant; prey larval Lepidoptera.

Paralastor synchromus Perkins, 1914

Paralastor synchromus Perkins, R.C.L. (1914). On the species of *Alastor (Paralastor)* Sauss. and some other Hymenoptera of the family Eumenidae. *Proc. Zool. Soc.*

Lond. **1914**: 563–624 pl I [619]. Type data: holotype, BMNH *M. adult (seen 1929 by L.F. Graham), from Mackay, Qld.

Distribution: NE coastal, Qld.; only published localities Mackay, Bundaberg and Cairns. Ecology: larva - sedentary, predator : adult - volant; prey larval Lepidoptera.

Paralastor tasmaniensis (Saussure, 1853)

Alastor tasmaniensis Saussure, H. de (1853). *Études sur la Famille des Vespides.* 1. Monographie des guêpes solitaires, ou de la tribu des Euméniens, comprenant la classification et la description de toutes les espèces connues jusqu'à ce jour, et servant de complément au Manuel de Lepeletier de Saint Fargeau. Paris : Masson 286 pp. pls i–xx (1852–1853) [253]. Type data: holotype, MNHP *M. adult, from Tas.; locality probably incorrect, see Perkins, R.C.L. (1914). On the species of *Alastor (Paralastor)* Sauss. and some other Hymenoptera of the family Eumenidae. *Proc. Zool. Soc. Lond.* **1914**: 563–624 pl I.

Distribution: NE coastal, (Tas.), Qld.; only published localities "Tas." and Brisbane. Ecology: larva - sedentary, predator : adult - volant; prey larval Lepidoptera. Biological references: Perkins, R.C.L. (1914). On the species of *Alastor (Paralastor)* Sauss. and some other Hymenoptera of the family Eumenidae. *Proc. Zool. Soc. Lond.* **1914**: 563–624 pl I (redescription); Giordani Soika, A. (1962). Gli *Odynerus* sensu antiquo del continente australiano e della Tasmania. *Boll. Mus. Civ. Stor. Nat. Venezia* **14**: 57–202 (taxonomy).

Paralastor tricarinulatus Perkins, 1914

Paralastor tricarinulatus Perkins, R.C.L. (1914). On the species of *Alastor (Paralastor)* Sauss. and some other Hymenoptera of the family Eumenidae. *Proc. Zool. Soc. Lond.* **1914**: 563–624 pl I [582]. Type data: holotype, BMNH *M. adult (seen 1929 by L.F. Graham), from Vic.

Distribution: SE coastal, Murray-Darling basin, N.S.W., Vic.; only published localities Sydney, Binnaway, Warrumbungle and Vic. Ecology: larva - sedentary, predator : adult - volant; nest in insect-borer holes in dead trees, prey larval Lepidoptera. Biological references: Schulthess-Rechberg, A. von (1925). Beitrag zur Kenntnis der Gattung *Alastor* Lep. (Hym. Vesp.). *Konowia* **4**: 57–65, 195–209, 257–263 (redescription); Giordani Soika, A. (1962). Gli *Odynerus* sensu antiquo del continente australiano e della Tasmania. *Boll. Mus. Civ. Stor. Nat. Venezia* **14**: 57–202 (taxonomy); Smith, A.P. & Alcock, J. (1980). A comparative study of the mating systems of Australian eumenid wasp (Hymenoptera). *Z. Tierpsychol.* **53**: 41–60.

Paralastor tricolor Perkins, 1914

Paralastor tricolor Perkins, R.C.L. (1914). On the species of *Alastor (Paralastor)* Sauss. and some other Hymenoptera of the family Eumenidae. *Proc. Zool. Soc. Lond.* **1914**: 563–624 pl I [590 pl I fig 5]. Type data: holotype, BMNH *F. adult (seen 1929 by L.F. Graham), from Cairns, Qld.

Distribution: NE coastal, Qld.; only published localities Cairns, Kuranda and Mackay. Ecology: larva - sedentary, predator : adult - volant; prey larval Lepidoptera. Biological references: Giordani Soika, A. (1962). Gli *Odynerus* sensu antiquo del continente australiano e della Tasmania. *Boll. Mus. Civ. Stor. Nat. Venezia* **14**: 57–202 (taxonomy).

Paralastor tuberculatus (Saussure, 1853)

Alastor tuberculatus Saussure, H. de (1853). *Études sur la Famille des Vespides.* 1. Monographie des guêpes solitaires, ou de la tribu des Euméniens, comprenant la classification et la description de toutes les espèces connues jusqu'à ce jour, et servant de complément au Manuel de Lepeletier de Saint Fargeau. Paris : Masson 286 pp. pls i–xx (1852–1853) [253 pl xxi figs 6,6a–6b]. Type data: holotype, MNHP *adult, from Tas.

Distribution: S Gulfs, S.A., Vic., Tas.; only published localities Adelaide, Vic. and Tas. Ecology: larva - sedentary, predator : adult - volant; prey larval Lepidoptera. Biological references: Perkins, R.C.L. (1914). On the species of *Alastor (Paralastor)* Sauss. and some other Hymenoptera of the family Eumenidae. *Proc. Zool. Soc. Lond.* **1914**: 563–624 pl I (distribution); Giordani Soika, A. (1962). Gli *Odynerus* sensu antiquo del continente australiano e della Tasmania. *Boll. Mus. Civ. Stor. Nat. Venezia* **14**: 57–202 (taxonomy); International Commission on Zoological Nomenclature (1970). Opinion 893. Eumenidae names of Saussure (Hymenoptera) : grant of availability to certain names proposed for secondary divisions of genera. *Bull. Zool. Nomen.* **26**: 187–191 (placed on Official List of Specific Names).

Paralastor victor Giordani Soika, 1977

Paralastor victor Giordani Soika, A. (1977). Contributo alla conoscenza degli Eumenidi australiani (Hymenoptera). *Mem. Soc. Entomol. Ital.* **55**: 109–138 [135]. Type data: holotype, SAMA *M. adult, from Everard Park near Victory Well, S.A.

Distribution: Lake Eyre basin, S.A.; type locality only. Ecology: larva - sedentary, predator : adult - volant; prey larval Lepidoptera.

Paralastor viduus Perkins, 1914

Paralastor viduus Perkins, R.C.L. (1914). On the species of *Alastor (Paralastor)* Sauss. and some other Hymenoptera of the family Eumenidae. *Proc. Zool. Soc.*

Lond. **1914**: 563–624 pl I [609]. Type data: holotype, BMNH *M. adult (seen 1929 by L.F. Graham), from Melbourne, Vic.

Distribution: SE coastal, Vic.; only published locality Melbourne. Ecology: larva - sedentary, predator : adult - volant; prey larval Lepidoptera.

Paralastor vulneratus (Saussure, 1856)

Alastor (Paralastor) vulneratus Saussure, H. de (1856). *Études sur la Famille des Vespides*. Troisième Partie comprenant la Monographie des Masariens et un Supplement à la Monographie des Eumeniens. Paris : Masson 352 pp. pls i–xvi (1854–1856) [334 pl xvi fig 7]. Type data: syntypes (probable), F. adult whereabouts unknown, from Australia (as New Holland).

Distribution: S Gulfs, S.A., Vic.; only published localities Adelaide and Vic. Ecology: larva - sedentary, predator : adult - volant; prey larval Lepidoptera. Biological references: Perkins, R.C.L. (1914). On the species of *Alastor (Paralastor)* Sauss. and some other Hymenoptera of the family Eumenidae. *Proc. Zool. Soc. Lond.* **1914**: 563–624 pl I (redescription, distribution).

Paralastor vulpinus (Saussure, 1856)

Taxonomic decision of Perkins, R.C.L. (1914). On the species of *Alastor (Paralastor)* Sauss. and some other Hymenoptera of the family Eumenidae. *Proc. Zool. Soc. Lond.* **1914**: 563–624 pl I [587].

Paralastor vulpinus vulpinus (Saussure, 1856)

Alastor (Paralastor) vulpinus Saussure, H. de (1856). *Études sur la Famille des Vespides*. Troisième Partie comprenant la Monographie des Masariens et un Supplement à la Monographie des Eumeniens. Paris : Masson 352 pp. pls i–xvi (1854–1856) [335 pl xvi fig 6]. Type data: syntypes (probable), F. adult whereabouts unknown, from Australia.

Distribution: SE coastal, S Gulfs, Vic., S.A.; only published localities Croydon and Adelaide. Ecology: larva - sedentary, predator : adult - volant; prey larval Lepidoptera. Biological references: Perkins, R.C.L. (1914). On the species of *Alastor (Paralastor)* Sauss. and some other Hymenoptera of the family Eumenidae. *Proc. Zool. Soc. Lond.* **1914**: 563–624 pl I (redescription, distribution).

Paralastor vulpinus excisus Perkins, 1914

Paralastor vulpinus excisus Perkins, R.C.L. (1914). On the species of *Alastor (Paralastor)* Sauss. and some other Hymenoptera of the family Eumenidae. *Proc. Zool. Soc. Lond.* **1914**: 563–624 pl I [587] [described as race]. Type data: holotype, BMNH *F. adult (seen 1929 by L.F. Graham), from Australia (Vic. or N.S.W.).

Distribution: SE coastal, N.S.W., Vic.; only published localities Mittagong, Port Stephen, Melbourne, Cumberland and Woodford. Ecology: larva - sedentary, predator : adult - volant; prey larval Lepidoptera.

Paralastor xanthochromus Perkins, 1914

Paralastor xanthochromus Perkins, R.C.L. (1914). On the species of *Alastor (Paralastor)* Sauss. and some other Hymenoptera of the family Eumenidae. *Proc. Zool. Soc. Lond.* **1914**: 563–624 pl I [614]. Type data: holotype, BMNH *F. adult, from Cairns, Qld.

Distribution: NE coastal, Qld.; only published localities Cairns, (Townsville). Ecology: larva - sedentary, predator : adult - volant; prey larval Lepidoptera.

Paralastor xanthus Giordani Soika, 1977

Paralastor xanthus Giordani Soika, A. (1977). Contributo alla conoscenza degli Eumenidi australiani (Hymenoptera). *Mem. Soc. Entomol. Ital.* **55**: 109–138 [134]. Type data: holotype, RMNH *F. adult, from Sturt Creek (Kimberley), W.A.

Distribution: N coastal, W.A.; type locality only. Ecology: larva - sedentary, predator : adult - volant; prey larval Lepidoptera.

Paralastor xerophilus Perkins, 1914

Taxonomic decision of Giordani Soika, A. (1977). Contributo alla conoscenza degli Eumenidi australiani (Hymenoptera). *Mem. Soc. Entomol. Ital.* **55**: 109–138 [134].

Paralastor xerophilus xerophilus Perkins, 1914

Paralastor xerophilus Perkins, R.C.L. (1914). On the species of *Alastor (Paralastor)* Sauss. and some other Hymenoptera of the family Eumenidae. *Proc. Zool. Soc. Lond.* **1914**: 563–624 pl I [591]. Type data: holotype, BMNH *M. adult (seen 1929 by L.F. Graham), from Hermannsburg, N.T.

Distribution: Lake Eyre basin, N.T.; type locality only. Ecology: larva - sedentary, predator : adult - volant; prey larval Lepidoptera.

Paralastor xerophilus meesi Giordani Soika, 1977

Paralastor xerophilus meesi Giordani Soika, A. (1977). Contributo alla conoscenza degli Eumenidi australiani (Hymenoptera). *Mem. Soc. Entomol. Ital.* **55**: 109–138 [134]. Type data: holotype, RMNH *F. adult, from Carranya, Kimberley, W.A.

Distribution: N coastal, W.A.; type locality only. Ecology: larva - sedentary, predator : adult - volant; prey larval Lepidoptera.

Parifodynerus Giordani Soika, 1962

Parifodynerus Giordani Soika, A. (1962). Gli *Odynerus* sensu antiquo del continente australiano e della Tasmania. *Boll. Mus. Civ. Stor. Nat. Venezia* **14**: 57–202

[167]. Type species *Parifodynerus parificus* Giordani Soika, 1962 by original designation.

Parifodynerus alariformis (Saussure, 1856)

Odynerus alariformis Saussure, H. de (1856). *Études sur la Famille des Vespides.* Troisième Partie comprenant la Monographie des Masariens et un Supplement à la Monographie des Eumeniens. Paris : Masson 352 pp. pls i–xvi (1854–1856) [282 pl 14 fig 6]. Type data: holotype, MNHP *F. adult, from Australia (as New Holland).

Odynerus (Stenodyneroides) subalaris Giordani Soika, A. (1941). Studi sui Vespidi solitari. *Boll. Soc. Veneziana Stor. Nat.* **2**: 130–279 [248]. Type data: holotype, BMNH *F. adult, from Brisbane, Qld.

Taxonomic decision of Giordani Soika, A. (1962). Gli *Odynerus* sensu antiquo del continente australiano e della Tasmania. *Boll. Mus. Civ. Stor. Nat. Venezia* **14**: 57–202 [169].

Distribution: NE coastal, Murray-Darling basin, Qld., N.S.W. Ecology: larva - sedentary, predator : adult - volant; prey larval Lepidoptera. Biological references: Giordani Soika, A. (1941). Studi sui Vespidi solitari. *Boll. Soc. Veneziana Stor. Nat.* **2**: 130–279 (description of male, distribution, as *Odynerus (Stenodyneroides) alariformis*); Giordani Soika, A. (1973). Designazione di lectotipi ed elenco dei tipi di Eumenidi, Vespidi e Masaridi da me descritti negli anni 1934–1960. *Boll. Mus. Civ. Stor. Nat. Venezia* **24**: 7–53 (taxonomy); Giordani Soika, A. (1977). Contributo alla conoscenza degli Eumenidi australiani (Hymenoptera). *Mem. Soc. Entomol. Ital.* **55**: 109–138 (distribution).

Parifodynerus parificus Giordani Soika, 1962

Parifodynerus parificus Giordani Soika, A. (1962). Gli *Odynerus* sensu antiquo del continente australiano e della Tasmania. *Boll. Mus. Civ. Stor. Nat. Venezia* **14**: 57–202 [168]. Type data: holotype, BMNH *F. adult, from Australia.

Distribution: SE coastal, Vic.; only published locality Melbourne. Ecology: larva - sedentary, predator : adult - volant; prey larval Lepidoptera.

Parodynerus Saussure, 1855

Parodynerus Saussure, H. de (1855). *Études sur la Famille des Vespides.* Troisième Partie comprenant la Monographie des Masariens et un Supplement à la Monographie des Eumeniens. Paris : Masson 352 pp. pls i–xvi (1854–1855) [245] [proposed for secondary division of *Odynerus* Latreille, 1802; placed on Official List of Generic Names in Zoology, see International Commission on Zoological Nomenclature (1970). Opinion 893. Eumenidae names of Saussure (Hymenoptera) : grant of availability to certain names proposed for secondary divisions of genera. *Bull. Zool. Nomen.* **26**: 187–191]. Publication date established from Griffin, F.J. (1939). On the dates of publication of Saussure (H. de) : Études sur la famille des Vespides 1–3, 1852–1858. *J. Soc. Bibliogr. Nat. Hist.* **1**: 211–212. Type species *Odynerus bizonatus* Boisduval, 1835 (= *Vespa bicincta* Fabricius, 1781) by subsequent designation, see Giordani Soika, A. (1958). Biogeografia, evoluzione e sistematica dei Vespidi solitari della Polinesia meridionale. *Boll. Mus. Civ. Stor. Nat. Venezia* **10**: 183–221.

This group is also found in the Pacific islands, see Giordani Soika, A. (1958). Biogeografia, evoluzione e sistematica dei Vespidi solitari della Polinesia meridionale. *Boll. Mus. Civ. Stor. Nat. Venezia* **10**: 183–221.

Parodynerus bicincta (Fabricius, 1781)

Vespa bicincta Fabricius, J.C. (1781). *Species insectorum exhibentes eorum differentias specificas, synonyma auctorum, loca natalia, metamorphosin adiectis observationibus, descriptionibus.* Hamburgi et Kilonii Vols i–ii [465]. Type data: holotype, BMNH *F. adult, from Cap. bon. sp."; this locality is erroneous, the species is found on Pacific islands, see Bequaert, J. (1918). A revision of the Vespidae of the Belgian Congo based on the collection of the American Museum Congo expedition, with a list of Ethiopian diplopterous wasps. *Bull. Am. Mus. Nat. Hist.* **39**: 1–384.

Odynerus bizonatus Boisduval, J.B.A.D. de (1835). *Faune Entomologique de l'Ocean Pacifique, Découvertes de l'Astrolabe exécuté pendant les années 1826–29, sous le commandement de M. J. Dumont d'Urville.* 2 vols vii 716 pp. Paris : Tastu [658 pl 12 fig 5]. Type data: syntypes (probable), MNHP *adult, from Tonga Ils., see Bequaert, J. (1928). The diplopterous wasps of Fabricius, in the Banksian collection at the British Museum. *Bull. Brooklyn Entomol. Soc.* **23**: 53–63.

Taxonomic decision of Smith, F. (1857). *Catalogue of Hymenopterous Insects in the Collection of the British Museum.* Pt. V. Vespidae pp. 1–147 London : British Museum [63].

Distribution: Qld.; record doubtful, also Pacific islands of Samoa, Tonga, Marquesas, Fiji, New Caledonia, Society Ils. Ecology: larva - sedentary, predator : adult - volant; prey larval Lepidoptera. Biological references: Saussure, H. de (1852–1853). *Études sur la Famille des Vespides.* 1. Monographie des guêpes solitaires, ou de la tribu des Euméniens, comprenant la classification et la description de toutes les espèces connues jusqu'à ce jour, et servant de complément au Manuel de Lepeletier de Saint Fargeau. Paris : Masson 286 pp. pls i–xx (Pacific islands and New Holland as *Odynerus bizonatus*); Bequaert, J. (1928). The diplopterous wasps of Fabricius, in the Banksian collection at the British Museum. *Bull. Brooklyn Entomol. Soc.* **23**: 53–63 (taxonomy); Giordani Soika, A. (1943). Le specie indo-australiane del genere *Pachymenes* (Hym. Vespidae). *Mem. Soc. Entomol. Ital.* **22**: 102–117 (taxonomy, Australian record); Giordani Soika, A. (1958). Biogeografia, evoluzione e sistematica dei Vespidi solitari della Polinesia meridionale. *Boll. Mus. Civ. Stor. Nat. Venezia* **10**: 183–221 (taxonomy, generic placement); Baltazar, C.R. (1966). A catalogue of Philippine Hymenoptera (with a bibliography, 1758–1963). *Pac. Insects Monogr.* **8**: 1–488 (distribution); International Commission on Zoological Nomenclature (1970). Opinion 893. Eumenidae names of Saussure (Hymenoptera) : grant of availability to certain names proposed for secondary divisions of genera. *Bull. Zool. Nomen.* **26**: 187–191 (*Odynerus bizonatus* placed on Official List of Specific Names).

Pseudabispa van der Vecht, 1960

Pseudabispa van der Vecht, J. (1960). On *Abispa* and some other Eumenidae from the Australian region (Hymenoptera, Vespoidea). *Nova Guinea Zool.* **6**: 91–115 [102]. Type species *Odynerus abispoides* Perkins, 1912 by original designation.

This group is also found in New Guinea, see van der Vecht, J. (1960). On *Abispa* and some other Eumenidae from the Australian region (Hymenoptera, Vespoidea). *Nova Guinea Zool.* **6**: 91–115.

Pseudabispa confusa van der Vecht, 1960

Pseudabispa confusa van der Vecht, J. (1960). On *Abispa* and some other Eumenidae from the Australian region (Hymenoptera, Vespoidea). *Nova Guinea Zool.* **6**: 91–115 [104]. Type data: holotype, RIB *F. adult, from N.S.W.

Distribution: NE coastal, Qld., N.S.W.; only published localities Mackay, (Cape York), N.S.W., also New Guinea. Ecology: larva - sedentary, predator : adult - volant; prey larval Lepidoptera.

Pseudabispa ephippioides van der Vecht, 1960

Pseudabispa ephippioides van der Vecht, J. (1960). On *Abispa* and some other Eumenidae from the Australian region (Hymenoptera, Vespoidea). *Nova Guinea Zool.* **6**: 91–115 [103]. Type data: holotype, MCG *M. adult, from Australia.

Distribution: no locality specified. Ecology: larva - sedentary, predator : adult - volant; prey larval Lepidoptera. Biological references: Giordani Soika, A. (1962). Gli *Odynerus* sensu antiquo del continente australiano e della Tasmania. *Boll. Mus. Civ. Stor. Nat. Venezia* **14**: 57–202 (taxonomy).

Pseudabispa paragioides (Meade-Waldo, 1910)

Abispa paragioides Meade-Waldo, G. (1910). New species of Diploptera in the collection of the British Museum. *Ann. Mag. Nat. Hist. (8)* **5**: 30–51 [49]. Type data: holotype, BMNH *M. adult (seen 1929 by L.F. Graham), from Port Darwin, N.T.

Distribution: NE coastal, N coastal, Qld., N.T.; only published localities Cooktown and Port Darwin. Ecology: larva - sedentary, predator : adult - volant; prey larval Lepidoptera. Biological references: van der Vecht, J. (1960). On *Abispa* and some other Eumenidae from the Australian region (Hymenoptera, Vespoidea). *Nova Guinea Zool.* **6**: 91–115 (generic placement).

Pseudalastor Giordani Soika, 1962

Pseudalastor Giordani Soika, A. (1962). Gli *Odynerus* sensu antiquo del continente australiano e della Tasmania. *Boll. Mus. Civ. Stor. Nat. Venezia* **14**: 57–202 [131]. Type species *Odynerus concolor* Saussure, 1853 by original designation.

This group is also found in New Guinea, see van der Vecht, J. (1981). Indo-Australian solitary wasps. *Proc. K. Ned. Akad. Wet. (c)* **84**: 443–464.

Pseudalastor anguloides Giordani Soika, 1962

Pseudalastor anguloides Giordani Soika, A. (1962). Gli *Odynerus* sensu antiquo del continente australiano e della Tasmania. *Boll. Mus. Civ. Stor. Nat. Venezia* **14**: 57–202 [138]. Type data: syntypes, ZMB *2F.,3M. adult, from Adelaide, S.A.

Distribution: S Gulfs, S.A.; type locality only. Ecology: larva - sedentary, predator : adult - volant; prey larval Lepidoptera.

Pseudalastor cavifemur Giordani Soika, 1962

Pseudalastor cavifemur Giordani Soika, A. (1962). Gli *Odynerus* sensu antiquo del continente australiano e della Tasmania. *Boll. Mus. Civ. Stor. Nat. Venezia* **14**: 57–202 [139]. Type data: holotype, BMNH *M. adult, from Merredin, W.A.

Distribution: SW coastal, W.A.; type locality only. Ecology: larva - sedentary, predator : adult - volant; prey larval Lepidoptera.

Pseudalastor concolor (Saussure, 1853)

Taxonomic decision of Giordani Soika, A. (1977). Contributo alla conoscenza degli Eumenidi australiani (Hymenoptera). *Mem. Soc. Entomol. Ital.* **55**: 109–138 [122].

Pseudalastor concolor concolor (Saussure, 1853)

Odynerus (Leionotus) concolor Saussure, H. de (1853). *Études sur la Famille des Vespides*. 1. Monographie des guêpes solitaires, ou de la tribu des Euméniens, comprenant la classification et la description de toutes les espèces connues jusqu'à ce jour, et servant de complément au Manuel de Lepeletier de Saint Fargeau. Paris : Masson 286 pp. pls i–xx (1852–1853) [202 pl xviii fig 7]. Type data: holotype, MNHP *F. adult, from Tas.

Distribution: NE coastal, Murray-Darling basin, Qld., N.S.W., Tas., W.A.; only published localities Brisbane, Narrabri, Biniguy, Tas. and W.A. Ecology: larva - sedentary, predator : adult - volant; prey larval Lepidoptera. Biological references: Giordani Soika, A. (1962). Gli *Odynerus* sensu antiquo del continente australiano e della Tasmania. *Boll. Mus. Civ. Stor. Nat. Venezia* **14**: 57–202 (redescription, generic placement); Giordani Soika, A. (1977). Contributo alla conoscenza degli Eumenidi australiani (Hymenoptera). *Mem. Soc. Entomol. Ital.* **55**: 109-138 (distribution).

Pseudalastor concolor rapax Giordani Soika, 1977

Pseudalastor concolor rapax Giordani Soika, A. (1977). Contributo alla conoscenza degli Eumenidi australiani (Hymenoptera). *Mem. Soc. Entomol. Ital.* **55**: 109-138 [122]. Type data: holotype, NMV *F. adult, from Caldwell, N.S.W.

Distribution: Murray-Darling basin, N.S.W., Vic.; only published localities Caldwell and Gunbower. Ecology: larva - sedentary, predator : adult - volant; prey larval Lepidoptera.

Pseudalastor metathoracicus (Saussure, 1855)

Odynerus metathoracicus Saussure, H. de (1855). *Études sur la Famille des Vespides*. Troisième Partie comprenant la Monographie des Masariens et un Supplement à la Monographie des Eumeniens. Paris : Masson 352 pp. pls i–xvi (1854–1856) [286]. Type data: holotype, BMNH *F. adult (seen 1929 by L.F. Graham), from Australia (as New Holland).

Odynerus (Ancistroceroides) sanguinolentus Saussure, H. de (1856). *Études sur la Famille des Vespides*. Troisième Partie comprenant la Monographie des Masariens et un Supplement à la Monographie des Eumeniens. Paris : Masson 352 pp. pls i–xvi (1854–1856) [221]. Type data: holotype, BMNH *M. adult (seen 1929 by L.F. Graham), from Australia.

Taxonomic decision of Giordani Soika, A. (1962). Gli *Odynerus* sensu antiquo del continente australiano e della Tasmania. *Boll. Mus. Civ. Stor. Nat. Venezia* **14**: 57–202 [134].

Distribution: Murray-Darling basin, S Gulfs, N coastal, N.S.W., Vic., S.A., N.T.; only published localities Caldwell, Hattah, Adelaide and Darwent Creek. Ecology: larva - sedentary, predator : adult - volant; prey larval Lepidoptera. Biological references: Giordani Soika, A. (1977). Contributo alla conoscenza degli Eumenidi australiani (Hymenoptera). *Mem. Soc. Entomol. Ital.* **55**: 109-138 (distribution).

Pseudalastor superbus Giordani Soika, 1977

Pseudalastor superbus Giordani Soika, A. (1977). Contributo alla conoscenza degli Eumenidi australiani (Hymenoptera). *Mem. Soc. Entomol. Ital.* **55**: 109-138 [124]. Type data: holotype, NMV *F. adult, from Kuranda, Qld.

Distribution: NE coastal, N Gulf, Qld., N.T.; only published localities Kuranda and Borroloola. Ecology: larva - sedentary, predator : adult - volant; prey larval Lepidoptera.

Pseudalastor tridentatus (Schulthess-Rechberg, 1935)

Taxonomic decision of Giordani Soika, A. (1977). Contributo alla conoscenza degli Eumenidi australiani (Hymenoptera). *Mem. Soc. Entomol. Ital.* **55**: 109-138 [123].

Pseudalastor tridentatus tridentatus (Schulthess-Rechberg, 1935)

Alastor (Paralastor) tridentatus Schulthess-Rechberg, A. von (1935). Hymenoptera aus den Sundainseln und Nordaustralien (mit Ausschluss der Blattwespen, Schlupfwespen und Ameisen). *Rev. Suisse Zool.* **42**: 293–323 [302]. Type data: syntypes, NHMB *F. adult, from Burnside, N.T.

Distribution: NE coastal, N coastal, Qld., N.T. Ecology: larva - sedentary, predator : adult - volant; prey larval Lepidoptera. Biological references: Giordani Soika, A. (1962). Gli *Odynerus* sensu antiquo del continente australiano e della Tasmania. *Boll. Mus. Civ. Stor. Nat. Venezia* **14**: 57–202 (redescription, generic placement); Giordani Soika, A. (1973). Designazione di lectotipi ed elenco dei tipi di Eumenidi, Vespidi e Masaridi da me descritti negli anni 1934–1960. *Boll. Mus. Civ. Stor. Nat. Venezia* **24**: 7–53 (description of male); Giordani Soika, A. (1977). Contributo alla conoscenza degli Eumenidi australiani (Hymenoptera). *Mem. Soc. Entomol. Ital.* **55**: 109 138 (distribution).

Pseudalastor tridentatus paganus Giordani Soika, 1977

Pseudalastor tridentatus paganus Giordani Soika, A. (1977). Contributo alla conoscenza degli Eumenidi australiani (Hymenoptera). *Mem. Soc. Entomol. Ital.* **55**: 109-138 [123]. Type data: holotype, BMNH *F. adult, from Mackay, Qld.

Distribution: NE coastal, Qld.; type locality only. Ecology: larva - sedentary, predator : adult - volant; prey larval Lepidoptera.

Pseudalastor tridentatus septentrionalis Giordani Soika, 1977

Pseudalastor tridentatus septentrionalis Giordani Soika, A. (1977). Contributo alla conoscenza degli Eumenidi australiani (Hymenoptera). *Mem. Soc. Entomol. Ital.* **55**: 109-138 [123]. Type data: holotype, BMNH *F. adult, from Port Darwin, N.T.

Distribution: N coastal, N.T.; type locality only. Ecology: larva - sedentary, predator : adult - volant; prey larval Lepidoptera.

Pseudalastor tridentatus transgrediens Giordani Soika, 1977

Pseudalastor tridentatus transgrediens Giordani Soika, A. (1977). Contributo alla conoscenza degli Eumenidi australiani (Hymenoptera). *Mem. Soc. Entomol. Ital.* **55**: 109–138 [123]. Type data: holotype, SAMA *F. adult, from Mornington Is., Qld.

Distribution: NE coastal, Qld.; type locality only. Ecology: larva - sedentary, predator : adult - volant; prey larval Lepidoptera.

Pseudepipona Saussure, 1856

Pseudepipona Saussure, H. de (1856). *Études sur la Famille des Vespides.* Troisième Partie comprenant la Monographie des Masariens et un Supplement à la Monographie des Eumeniens. Paris : Masson 352 pp. pls i–xvi (1854–1856) [309] [proposed for secondary division of *Odynerus* Latreille, 1802; placed on Official List of Generic Names in Zoology, see International Commission on Zoological Nomenclature (1970). Opinion 893. Eumenidae names of Saussure (Hymenoptera) : grant of availability to certain names proposed for secondary divisions of genera. *Bull. Zool. Nomen.* **26**: 187–191]. Publication date established from Griffin, F.J. (1939). On the dates of publication of Saussure (H. de) : Études sur la famille des Vespides 1–3, 1852–1858. *J. Soc. Bibliogr. Nat. Hist.* **1**: 211–212. Type species *Odynerus herrichii* Saussure, 1856 by monotypy.

This group is also found in the Palaearctic Region, see Richards, O.W. (1980). Scolioidea, Vespoidea and Sphecoidea. Hymenoptera, Aculeata. *Handbk. Ident. Br. Insects* **6**(3b): 1–118.

Pseudepipona alaris (Saussure, 1853)

Odynerus (Leionotus) alaris Saussure, H. de (1853). *Études sur la Famille des Vespides.* 1. Monographie des guêpes solitaires, ou de la tribu des Euméniens, comprenant la classification et la description de toutes les espèces connues jusqu'à ce jour, et servant de complément au Manuel de Lepeletier de Saint Fargeau. Paris : Masson 286 pp. pls i–xx (1852–1853) [203 pl xviii fig 5]. Type data: lectotype, MNHP *F. adult, from Tas., designation by Giordani Soika, A. (1962). Gli *Odynerus* sensu antiquo del continente australiano e della Tasmania. *Boll. Mus. Civ. Stor. Nat. Venezia* **14**: 57–202.

Distribution: NE coastal, SE coastal, Qld., N.S.W., Vic., Tas. Ecology: larva - sedentary, predator : adult - volant; prey larval Lepidoptera. Biological references: Giordani Soika, A. (1962). Gli *Odynerus* sensu antiquo del continente australiano e della Tasmania. *Boll. Mus. Civ. Stor. Nat. Venezia* **14**: 57–202 (redescription, distribution, generic placement).

Pseudepipona angulata (Saussure, 1856)

Taxonomic decision of Giordani Soika, A. (1962). Gli *Odynerus* sensu antiquo del continente australiano e della Tasmania. *Boll. Mus. Civ. Stor. Nat. Venezia* **14**: 57–202 [107].

Pseudepipona angulata angulata (Saussure, 1856)

Odynerus angulatus Saussure, H. de (1856). *Études sur la Famille des Vespides.* Troisième Partie comprenant la Monographie des Masariens et un Supplement à la Monographie des Eumeniens. Paris : Masson 352 pp. pls i–xvi (1854–1856) [284 pl xiv fig 7]. Type data: holotype, BMNH *M. adult (seen 1929 by L.F. Graham), from Australia (as New Holland).

Distribution: NE coastal, SE coastal, Murray-Darling basin, S Gulfs, Lake Eyre basin, SW coastal, W plateau, NW coastal, Qld., N.S.W., Vic., S.A., W.A., N.T. Ecology: larva - sedentary, predator : adult - volant; prey larval Lepidoptera. Biological references: Giordani Soika, A. (1941). Studi sui Vespidi solitari. *Boll. Soc. Veneziana Stor. Nat.* **2**: 130–279 (distribution as *Odynerus (Rhynchium) angulatus*); Giordani Soika, A. (1962). Gli *Odynerus* sensu antiquo del continente australiano e della Tasmania. *Boll. Mus. Civ. Stor. Nat. Venezia* **14**: 57–202 (redescription, generic placement).

Pseudepipona angulata alexandriae (Giordani Soika, 1941)

Odynerus (Rhynchium) angulatus alexandriae Giordani Soika, A. (1941). Studi sui Vespidi solitari. *Boll. Soc. Veneziana Stor. Nat.* **2**: 130–279 [255] [described as variety]. Type data: lectotype, BMNH *F. adult, from Alexandria, N.T., designation by Giordani Soika, A. (1973). Designazione di lectotipi ed elenco dei tipi di Eumenidi, Vespidi e Masaridi da me descritti negli anni 1934–1960. *Boll. Mus. Civ. Stor. Nat. Venezia* **24**: 7–53.

Distribution: N coastal, W plateau, N Gulf, (Murray-Darling basin), W.A., N.T., (Vic.); only published localities Ord River, Meekatharra, NW coast, Alexandria and (Bamawn). Ecology: larva - sedentary, predator : adult - volant; prey larval Lepidoptera. Biological references: Giordani Soika, A. (1962). Gli *Odynerus* sensu antiquo del continente australiano e della Tasmania. *Boll. Mus. Civ. Stor. Nat. Venezia* **14**: 57–202 (redescription, generic placement).

Pseudepipona aspra Giordani Soika, 1962

Pseudepipona (Pseudepipona) aspra Giordani Soika, A. (1962). Gli *Odynerus* sensu antiquo del continente australiano e della Tasmania. *Boll. Mus. Civ. Stor. Nat. Venezia* **14**: 57–202 [85]. Type data: holotype, QM T5978 *adult, from Brisbane, Qld.

Distribution: NE coastal, Qld., N.S.W., Vic.; only published localities Brisbane, Mackay, N.S.W. and Vic. Ecology: larva - sedentary, predator : adult - volant; prey larval Lepidoptera. Biological references: Giordani Soika, A. (1977). Contributo alla conoscenza degli Eumenidi australiani (Hymenoptera). *Mem. Soc. Entomol. Ital.* **55**: 109–138 (distribution).

Pseudepipona chartergiformis Giordani Soika, 1962

Pseudepipona chartergiformis Giordani Soika, A. (1962). Gli *Odynerus* sensu antiquo del continente australiano e della Tasmania. *Boll. Mus. Civ. Stor. Nat. Venezia* **14**: 57–202 [112]. Type data: holotype, BMNH (probable) *F. adult, from Australia.

Distribution: NE coastal, Qld.; only published localities Cairns and Kuranda. Ecology: larva - sedentary, predator : adult - volant; prey larval Lepidoptera.

Pseudepipona clypalaris Giordani Soika, 1962

Pseudepipona clypalaris Giordani Soika, A. (1962). Gli *Odynerus* sensu antiquo del continente australiano e della Tasmania. *Boll. Mus. Civ. Stor. Nat. Venezia* **14**: 57–202 [111]. Type data: holotype, ANIC F. adult, from Yass, N.S.W. (wrongly published as S.A.).

Distribution: Murray-Darling basin, N.S.W.; type locality only. Ecology: larva - sedentary, predator : adult - volant; prey larval Lepidoptera.

Pseudepipona pallida Giordani Soika, 1977

Pseudepipona pallida Giordani Soika, A. (1977). Contributo alla conoscenza degli Eumenidi australiani (Hymenoptera). *Mem. Soc. Entomol. Ital.* **55**: 109–138 [115]. Type data: holotype, BMNH *F. adult, from 56 mi W Barnato, N.S.W.

Distribution: Murray-Darling basin, Lake Eyre basin, NW coastal, N.S.W., S.A., W.A.; only published localities near Barnato, Everard Park Station and Carnarvon area. Ecology: larva - sedentary, predator : adult - volant; prey larval Lepidoptera.

Pseudepipona succincta (Saussure, 1853)

Taxonomic decision of Giordani Soika, A. (1977). Contributo alla conoscenza degli Eumenidi australiani (Hymenoptera). *Mem. Soc. Entomol. Ital.* **55**: 109–138 [115].

Pseudepipona succincta succincta (Saussure, 1853)

Odynerus (Leionotus) succinctus Saussure, H. de (1853). *Études sur la Famille des Vespides*. 1. Monographie des guêpes solitaires, ou de la tribu des Euméniens, comprenant la classification et la description de toutes les espèces connues jusqu'à ce jour, et servant de complément au Manuel de Lepeletier de Saint Fargeau. Paris : Masson 286 pp. pls i–xx (1852–1853) [204]. Type data: holotype, probably MNHP or MCG *F. adult, from Australia (as New Holland or Tas.).

Odynerus balyi Saussure, H. de (1855). *Études sur la Famille des Vespides*. Troisième Partie comprenant la Monographie des Masariens et un Supplement à la Monographie des Eumeniens. Paris : Masson 352 pp. pls i–xvi (1854–1856) [283 pl xiv fig 6]. Type data: syntypes (probable), F. adult whereabouts unknown, from Australia (as New Holland).

Distribution: NE coastal, SE coastal, Murray-Darling basin, N coastal, Qld., N.S.W., N.T., (Tas.). Ecology: larva - sedentary, predator : adult - volant; prey larval Lepidoptera. Biological references: Giordani Soika, A. (1941). Studi sui Vespidi solitari. *Boll. Soc. Veneziana Stor. Nat.* **2**: 130–279 (description of male as *Odynerus (Rhynchium) succinctus*); Giordani Soika, A. (1962). Gli *Odynerus* sensu antiquo del continente australiano e della Tasmania. *Boll. Mus. Civ. Stor. Nat. Venezia* **14**: 57–202 (redescription, generic placement); Giordani Soika, A. (1973). Designazione di lectotipi ed elenco dei tipi di Eumenidi, Vespidi e Masaridi da me descritti negli anni 1934–1960. *Boll. Mus. Civ. Stor. Nat. Venezia* **24**: 7–53 (taxonomy).

Pseudepipona succincta purgata Giordani Soika, 1977

Pseudepipona succincta purgata Giordani Soika, A. (1977). Contributo alla conoscenza degli Eumenidi australiani (Hymenoptera). *Mem. Soc. Entomol. Ital.* **55**: 109–138 [115]. Type data: holotype, WAM *adult not found, from Landor Station, W.A.

Distribution: NW coastal, N coastal, W.A., N.T.; only published localities Landor Station and Port Darwin. Ecology: larva - sedentary, predator : adult - volant; prey larval Lepidoptera.

Pseudozethus Perkins, 1914

Pseudozethus Perkins, R.C.L. (1914). On the species of *Alastor (Paralastor)* Sauss. and some other Hymenoptera of the family Eumenidae. *Proc. Zool. Soc. Lond.* **1914**: 563–624 pl I [622]. Type species *Pseudozethus australensis* Perkins, 1914 by monotypy.

Pseudozethus australensis Perkins, 1914

Pseudozethus australensis Perkins, R.C.L. (1914). On the species of *Alastor (Paralastor)* Sauss. and some other Hymenoptera of the family Eumenidae. *Proc. Zool. Soc. Lond.* **1914**: 563–624 pl I [623]. Type data: holotype, BMNH *M. adult, from N Qld.

Distribution: Qld.; no locality specified. Ecology: larva - sedentary, predator : adult - volant; prey larval Lepidoptera. Biological references: Giordani Soika, A. (1969). Revisione dei Discoeliinae Australiani. *Boll. Mus. Civ. Stor. Nat. Venezia* **19**: 25–100 (taxonomy).

Pseudozethus confusus Giordani Soika, 1969

Pseudozethus confusus Giordani Soika, A. (1969). Revisione dei Discoeliinae Australiani. *Boll. Mus. Civ. Stor. Nat. Venezia* **19**: 25–100 [46]. Type data: holotype, A. Giordani Soika pers. coll. *F. adult, from S.A.

Distribution: S Gulfs, S.A.; only published locality Orroroo. Ecology: larva - sedentary, predator : adult - volant; prey larval Lepidoptera.

Pseudozethus ephippium (Saussure, 1855)

Discoelius ephippium Saussure, H. de (1855). *Études sur la Famille des Vespides*. Troisième Partie comprenant la Monographie des Masariens et un Supplement à la

Monographie des Eumeniens. Paris : Masson 352 pp. pls i–xvi (1854–1856) [125 pl vi fig 8]. Type data: holotype, MNHP *M. adult, from Australia (as New Holland).

Distribution: NE coastal, SE coastal, Qld., N.S.W.; only published localities Wide Bay and Dorrigo. Ecology: larva - sedentary, predator : adult - volant; prey larval Lepidoptera. Biological references: Giordani Soika, A. (1969). Revisione dei Discoeliinae Australiani. *Boll. Mus. Civ. Stor. Nat. Venezia* **19**: 25–100 (redescription, distribution).

Pseudozethus insignis (Saussure, 1856)

Discoelius insignis Saussure, H. de (1856). *Études sur la Famille des Vespides*. Troisième Partie comprenant la Monographie des Masariens et un Supplement à la Monographie des Eumeniens. Paris : Masson 352 pp. pls i–xvi (1854–1856) [126]. Type data: holotype, BMNH *F. adult (seen 1929 by L.F. Graham), from Australia (as New Holland).

Distribution: NE coastal, Qld., N.S.W.; only published localities Peak Downs and N.S.W. Ecology: larva - sedentary, predator : adult - volant; prey larval Lepidoptera. Biological references: Smith, F. (1873). Natural history notices. Insects, Hymenoptera Aculeata. pp. 456–463 pls xliii–xlv *in* Brenchley, J.B. *Jottings During the Cruise of H.M.S. Curaçoa among the South Sea Islands in 1865*. London : Longmans, Green & Co. (illustration); Giordani Soika, A. (1941). Studi sui Vespidi solitari. *Boll. Soc. Veneziana Stor. Nat.* **2**: 130–279 (taxonomy as *Macrocalymma insignis*); Giordani Soika, A. (1969). Revisione dei Discoeliinae Australiani. *Boll. Mus. Civ. Stor. Nat. Venezia* **19**: 25–100 (redescription, generic placement).

Pseudozethus pseudospinosus Giordani Soika, 1969

Pseudozethus pseudospinosus Giordani Soika, A. (1969). Revisione dei Discoeliinae Australiani. *Boll. Mus. Civ. Stor. Nat. Venezia* **19**: 25–100 [51]. Type data: holotype, AM F. adult, from Sydney, N.S.W.

Distribution: SE coastal, Murray-Darling basin, N.S.W., A.C.T., Vic., S.A.; only published localities Sydney, Hornsby, Canberra, Melton and S.A. Ecology: larva - sedentary, predator : adult - volant; prey larval Lepidoptera. Biological references: Giordani Soika, A. (1977). Contributo alla conoscenza degli Eumenidi australiani (Hymenoptera). *Mem. Soc. Entomol. Ital.* **55**: 109–138 (distribution as *Deuterodiscoelius pseudospinosus*).

Pseudozethus spinosus (Saussure, 1855)

Discoelius spinosus Saussure, H. de (1855). *Études sur la Famille des Vespides*. Troisième Partie comprenant la Monographie des Masariens et un Supplement à la Monographie des Eumeniens. Paris : Masson 352 pp. pls i–xvi (1854–1856) [125]. Type data: holotype, F. adult whereabouts unknown, from N.S.W.

Distribution: N.S.W.; no locality specified. Ecology: larva - sedentary, predator : adult - volant; prey larval Lepidoptera. Biological references: Giordani Soika, A. (1969). Revisione dei Discoeliinae Australiani. *Boll. Mus. Civ. Stor. Nat. Venezia* **19**: 25–100 (generic placement).

Pseudozethus verreauxii (Saussure, 1852)

Discoelius verreauxii Saussure, H. de (1852). *Études sur la Famille des Vespides*. 1. Monographie des guêpes solitaires, ou de la tribu des Euméniens, comprenant la classification et la description de toutes les espèces connues jusqu'à ce jour, et servant de complément au Manuel de Lepeletier de Saint Fargeau. Paris : Masson 286 pp. pls i–xx (1852–1853) [26 pl ix fig 4a–c]. Type data: holotype, MNHP *F. adult, from Tas.

Distribution: NE coastal, SE coastal, Murray-Darling basin, Qld., N.S.W., Vic., S.A., Tas. Ecology: larva - sedentary, predator : adult - volant; prey larval Lepidoptera. Biological references: Giordani Soika, A. (1969). Revisione dei Discoeliinae Australiani. *Boll. Mus. Civ. Stor. Nat. Venezia* **19**: 25–100 (redescription, generic placement); Giordani Soika, A. (1977). Contributo alla conoscenza degli Eumenidi australiani (Hymenoptera). *Mem. Soc. Entomol. Ital.* **55**: 109–138 (distribution as *Deuterodiscoelius verreauxi*).

Rhynchium Spinola, 1806

Rhynchium Spinola, M. (1806). *Insectorum Liguriae species novae aut rariores, quae in agro ligustico nuper detexit, descripsit et iconibus illustravit*. Genoa : Gravier Vol. 1 159+17 pp. [84] [Spinola originally spelled the genus as *Rygchium*; this was suppressed by International Commission on Zoological Nomenclature (1965). Opinion 747. *Rygchium* Spinola, 1806 (Insecta, Hymenoptera): validation of emendation to *Rhynchium*. *Bull. Zool. Nomen.* **22**: 186–187 which validated the emendation *Rhynchium* of Billberg, G.J. (1820). *Enumeratio Insectorum in Museo Billberg*. Stockholm : Gadel 138 pp.]. Type species *Rygchium europaeum* Spinola, 1806 (= *Vespa oculata* Fabricius, 1781) by monotypy.

This group is also found in the Palaearctic and Oriental Regions (tropics and subtropics of Old World), see van der Vecht, J. (1963). Studies on Indo-Australian and east-Asiatic Eumenidae (Hymenoptera, Vespoidea). *Zool. Verh.* **60**: 1–116.

Rhynchium atrum Saussure, 1852

Rhynchium atrum Saussure, H. de (1852). *Études sur la Famille des Vespides*. 1. Monographie des guêpes solitaires, ou de la tribu des Euméniens, comprenant la classification et la description de toutes les espèces connues jusqu'à ce jour, et servant de complément au Manuel de Lepeletier de Saint Fargeau. Paris : Masson

286 pp. pls i–xx (1852–1853) [109]. Type data: neotype, USNM *F. adult, from Manila, Philippines, designation by van der Vecht, J. (1968). The *Rhynchium* species of the Philippine Islands (Hymenoptera, Eumenidae). *Zool. Meded.* **42**: 255–259,

Distribution: doubtfully Australian, in Philippines and other parts of the Oriental Region. Ecology: larva - sedentary, predator : adult - volant; prey larval Lepidoptera. Biological references: Baltazar, C.R. (1966). A catalogue of Philippine Hymenoptera (with a bibliography, 1758–1963). *Pac. Insects Monogr.* **8**: 1–488 (distribution includes Australia); van der Vecht, J. (1968). The *Rhynchium* species of the Philippine Islands (Hymenoptera, Eumenidae). *Zool. Meded.* **42**: 255–259 (taxonomy).

Rhynchium australense Perkins, 1914

Rhynchium australense Perkins, R.C.L. (1914). On the species of *Alastor (Paralastor)* Sauss. and some other Hymenoptera of the family Eumenidae. *Proc. Zool. Soc. Lond.* **1914**: 563–624 pl I [623]. Type data: holotype, BMNH *M. adult, from N Qld.

Distribution: no locality specified. Ecology: larva - sedentary, predator : adult - volant; prey larval Lepidoptera.

Rhynchium magnificum Smith, 1869

Rhynchium magnificum Smith, F. (1869). Descriptions of new genera and species of exotic Hymenoptera. *Trans. Entomol. Soc. Lond.* **1869**: 301–311 pl vi [310]. Type data: holotype, BMNH *F. adult (seen 1929 by L.F. Graham), from NW Australia.

Distribution: NE coastal, NW coastal, Qld., W.A.; only published localities Townsville and Nicol Bay. Ecology: larva - sedentary, predator : adult - volant; prey larval Lepidoptera. Biological references: Smith, F. (1873). Natural history notices. Insects, Hymenoptera Aculeata. pp. 456–463 pls xliii–xlv in Brenchley, J.B. *Jottings During the Cruise of H.M.S. Curaçoa among the South Sea Islands in 1865*. London : Longmans, Green & Co. (illustration); Froggatt, W.W. (1892). Catalogue of the described Hymenoptera of Australia. *Proc. Linn. Soc. N.S.W. (2)* **7**: 205–248 (locality); Giordani Soika, A. (1941). Studi sui Vespidi solitari. *Boll. Soc. Veneziana Stor. Nat.* **2**: 130–279 (taxonomy as *Odynerus (Rhynchium) haemorrhoidalis* var. *magnificus*).

Rhynchium mirabile Saussure, 1852

Rygchium mirabile Saussure, H. de (1852). *Études sur la Famille des Vespides*. 1. Monographie des guêpes solitaires, ou de la tribu des Euméniens, comprenant la classification et la description de toutes les espèces connues jusqu'à ce jour, et servant de complément au Manuel de Lepeletier de Saint Fargeau. Paris : Masson 286 pp. pls i–xx (1852–1853) [106 pl xiv fig 5]. Type data: holotype, MNHP *F. adult, from Tas. (locality probably incorrect).

Rhynchium rothi Kirby, W.F. (1885). in Roth, H.L. Notes on the habits of some Australian Hymenoptera Aculeata. (With description of a new species by W.F. Kirby). *J. Linn. Soc. Zool.* **18**: 324–326 [324]. Type data: holotype, BMNH *M. adult (seen 1929 by L.F. Graham), from Mackay, Qld.

Taxonomic decision of Giordani Soika, A. (1962). Gli *Odynerus* sensu antiquo del continente australiano e della Tasmania. *Boll. Mus. Civ. Stor. Nat. Venezia* **14**: 57–202 [200].

Distribution: NE coastal, N coastal, Qld., W.A., (Tas.); only published localities Mackay, Torres Strait, Moreton Bay, Ord River, also on Aru and Key Ils. and New Pomerania. Ecology: larva - sedentary, predator : adult - volant; mud-nest, prey larval Lepidoptera. Biological references: Smith, F. (1864). Notes on the geographical distribution of the aculeate Hymenoptera collected by Mr. A.R. Wallace in the eastern archipelago. *J. Linn. Soc. Lond. Zool.* **7**: 109–145 (distribution); Roth, H.L. (1885). Notes on the habits of some Australian Hymenoptera Aculeata. (With description of a new species by W.F. Kirby). *J. Linn. Soc. Zool.* **18**: 318–328 (biology as *Rhynchium rothi*); Tillyard, R.J. (1926). *The Insects of Australia and New Zealand*. Sydney : Angus & Robertson 560 pp. (plate 21); Cheesman, L.E. (1954). A new species of *Odynerus*, subgen. *Rhygchium* (Eumeninae), from the Loyalty Islands. *Ann. Mag. Nat. Hist. (12)* **7**: 385–390 (redescription as *Odynerus (Rhygchium) rothi*).

Rhynchium nigrolimbatum Bingham, 1912

Rhynchium nigrolimbatum Bingham, C.T. (1912). South African and Australian Aculeate Hymenoptera in the Oxford Museum. *Trans. Entomol. Soc. Lond.* **1912**: 375–383 [380]. Type data: holotype, OUM *F. adult, from Towranna Plains, W.A.

Distribution: NW coastal, W.A.; type locality only. Ecology: larva - sedentary, predator : adult - volant; prey larval Lepidoptera.

Rhynchium rufipes (Fabricius, 1775)

Vespa rufipes Fabricius, J.C. (1775). *Systema Entomologiae, sistens insectorum classes, ordines, genera, species, adiectis synonymis, locis, descriptionibus, observationibus.* Flensburgi et Lipsiae : Kortii xxvii 832 pp. [367]. Type data: holotype, BMNH *F. adult, from islands of the Pacific Ocean.

Distribution: no precise locality, islands of Pacific Ocean, Loo-choo, Samoa, Fiji and Rarotonga. Ecology: larva - sedentary, predator : adult - volant; prey larval Lepidoptera. Biological references: Cheesman, L.E. (1928). A contribution towards the insect fauna of French Oceania. Part II. *Ann. Mag. Nat. Hist. (10)* **1**: 169–194 (distribution includes Australia).

Rhynchium superbum Saussure, 1852

Rygchium superbum Saussure, H. de (1852). *Études sur la Famille des Vespides*. 1. Monographie des guêpes solitaires, ou de la tribu des Euméniens, comprenant la classification et la description de toutes les espèces connues jusqu'à ce jour, et servant de complément au Manuel de Lepeletier de Saint Fargeau. Paris : Masson 286 pp. pls i–xx (1852–1853) [113]. Type data: holotype, MNHP *F. adult, from Australia (as New Holland).

Distribution: NE coastal, Qld.; only published locality Thursday Is., also on New Guinea, Aru and Key Ils. and East Indies. Ecology: larva - sedentary, predator : adult - volant; prey larval Lepidoptera. Biological references: Smith, F. (1864). Notes on the geographical distribution of the aculeate Hymenoptera collected by Mr. A.R. Wallace in the eastern archipelago. *J. Linn. Soc. Lond. Zool.* **7**: 109–145 (distribution); Schulz, W.A. (1904). Ein Beitrag zur Kenntnis der papuanischen Hymenopteren-Fauna. *Berl. Entomol. Z.* **49**: 209–239 (taxonomy as *Rhynchium mirabile superbum*); Cameron, P. (1906). On the Malay fossorial Hymenoptera and Vespidae of the Museum of the R. Zool. Soc. "Natura artis magistra" at Amsterdam. *Tijdschr. Entomol.* **48**: 48–78 (taxonomy, distribution); Cockerell, T.D.A. (1930). The bees of Australia. *Aust. Zool.* **6**: 137–156 (distribution).

Stenodyneriellus Giordani Soika, 1962

Stenodyneriellus Giordani Soika, A. (1962). Gli *Odynerus* sensu antiquo del continente australiano e della Tasmania. *Boll. Mus. Civ. Stor. Nat. Venezia* **14**: 57–202 [71]. Type species *Stenodyneriellus turneriellus* Giordani Soika, 1962 by original designation.

Stenodyneriellus bicoloratus (Saussure, 1856)

Odynerus bicoloratus Saussure, H. de (1856). *Études sur la Famille des Vespides*. Troisième Partie comprenant la Monographie des Masariens et un Supplement à la Monographie des Eumeniens. Paris : Masson 352 pp. pls i–xvi (1854–1856) [281]. Type data: holotype, BMNH Hym 18.339 *F. adult (seen 1929 by L.F. Graham), from Australia (as New Holland).

Distribution: NE coastal, SE coastal, Murray-Darling basin, S Gulfs, Qld., N.S.W., Vic., S.A. Ecology: larva - sedentary, predator : adult - volant; prey larval Lepidoptera. Biological references: Giordani Soika, A. (1962). Gli *Odynerus* sensu antiquo del continente australiano e della Tasmania. *Boll. Mus. Civ. Stor. Nat. Venezia* **14**: 57–202 (redescription, generic placement); Giordani Soika, A. (1977). Contributo alla conoscenza degli Eumenidi australiani (Hymenoptera). *Mem. Soc. Entomol. Ital.* **55**: 109–138 (distribution).

Stenodyneriellus brisbanensis Giordani Soika, 1962

Stenodyneriellus brisbanensis Giordani Soika, A. (1962). Gli *Odynerus* sensu antiquo del continente australiano e della Tasmania. *Boll. Mus. Civ. Stor. Nat. Venezia* **14**: 57–202 [75]. Type data: holotype, BMNH *adult, from Brisbane, Qld.

Distribution: NE coastal, SE coastal, Murray-Darling basin, Lake Eyre basin, S Gulfs, Qld., N.S.W., Vic., S.A., N.T. Ecology: larva - sedentary, predator : adult - volant; prey larval Lepidoptera. Biological references: Giordani Soika, A. (1977). Contributo alla conoscenza degli Eumenidi australiani (Hymenoptera). *Mem. Soc. Entomol. Ital.* **55**: 109–138 (distribution).

Stenodyneriellus carnarvonensis Giordani Soika, 1977

Stenodyneriellus carnarvonensis Giordani Soika, A. (1977). Contributo alla conoscenza degli Eumenidi australiani (Hymenoptera). *Mem. Soc. Entomol. Ital.* **55**: 109–138 [113]. Type data: holotype, NMV *M. adult, from Carnarvon, W.A.

Distribution: NW coastal, W.A.; type locality only. Ecology: larva - sedentary, predator : adult - volant; prey larval Lepidoptera.

Stenodyneriellus darnleyensis Giordani Soika, 1977

Stenodyneriellus darnleyensis Giordani Soika, A. (1977). Contributo alla conoscenza degli Eumenidi australiani (Hymenoptera). *Mem. Soc. Entomol. Ital.* **55**: 109–138 [113]. Type data: holotype, BMNH *M. adult, from Darnley Is., Qld.

Distribution: NE coastal, Qld.; type locality only. Ecology: larva - sedentary, predator : adult - volant; prey larval Lepidoptera.

Stenodyneriellus novempunctatus Giordani Soika, 1977

Stenodyneriellus novempunctatus Giordani Soika, A. (1977). Contributo alla conoscenza degli Eumenidi australiani (Hymenoptera). *Mem. Soc. Entomol. Ital.* **55**: 109–138 [110]. Type data: holotype, NMV *F. adult, from Cheshunt, Vic.

Distribution: SE coastal, Murray-Darling basin, Vic.; only published localities Cheshunt, Croydon, Gippsland, Healesville and Woodend. Ecology: larva - sedentary, predator : adult - volant; prey larval Lepidoptera.

Stenodyneriellus pseudancistrocerus (Giordani Soika, 1962)

Pseudonortonia pseudancistrocerus Giordani Soika, A. (1962). Gli *Odynerus* sensu antiquo del continente australiano e della Tasmania. *Boll. Mus. Civ. Stor. Nat. Venezia* **14**: 57–202 [69]. Type data: holotype, BMNH *F. adult, from Mackay, Qld.

Distribution: NE coastal, SE coastal, S Gulfs, Qld., N.S.W., S.A. Ecology: larva - sedentary, predator : adult - volant; prey larval Lepidoptera. Biological references: Giordani Soika, A. (1977). Contributo alla conoscenza degli Eumenidi australiani (Hymenoptera). *Mem. Soc. Entomol. Ital.* **55**: 109–138 (distribution, generic placement).

Stenodyneriellus punctatissimus Giordani Soika, 1977

Stenodyneriellus punctatissimus Giordani Soika, A. (1977). Contributo alla conoscenza degli Eumenidi australiani (Hymenoptera). *Mem. Soc. Entomol. Ital.* **55**: 109–138 [111]. Type data: holotype, NMV *F. adult, from Westwood, Qld.

Distribution: NE coastal, Murray-Darling basin, Qld.; only published localities Westwood and Mt. Emlyn. Ecology: larva - sedentary, predator : adult - volant; prey larval Lepidoptera.

Stenodyneriellus spinosiusculus Giordani Soika, 1962

Stenodyneriellus spinosiusculus Giordani Soika, A. (1962). Gli *Odynerus* sensu antiquo del continente australiano e della Tasmania. *Boll. Mus. Civ. Stor. Nat. Venezia* **14**: 57–202 [74]. Type data: holotype, QM T5970 *M. adult, from Brisbane, Qld.

Distribution: NE coastal, Qld.; type locality only. Ecology: larva - sedentary, predator : adult - volant; prey larval Lepidoptera. Biological references: Giordani Soika, A. (1977). Contributo alla conoscenza degli Eumenidi australiani (Hymenoptera). *Mem. Soc. Entomol. Ital.* **55**: 109–138 (description of female).

Stenodyneriellus tricoloratus Giordani Soika, 1962

Stenodyneriellus tricoloratus Giordani Soika, A. (1962). Gli *Odynerus* sensu antiquo del continente australiano e della Tasmania. *Boll. Mus. Civ. Stor. Nat. Venezia* **14**: 57–202 [78]. Type data: holotype, SAMA *F. adult, from Port Douglas, Qld.

Distribution: NE coastal, Qld. Ecology: larva - sedentary, predator : adult - volant; prey larval Lepidoptera. Biological references: Giordani Soika, A. (1977). Contributo alla conoscenza degli Eumenidi australiani (Hymenoptera). *Mem. Soc. Entomol. Ital.* **55**: 109–138 (description of male, distribution).

Stenodyneriellus turneriellus Giordani Soika, 1962

Stenodyneriellus turneriellus Giordani Soika, A. (1962). Gli *Odynerus* sensu antiquo del continente australiano e della Tasmania. *Boll. Mus. Civ. Stor. Nat. Venezia* **14**: 57–202 [73]. Type data: holotype, BMNH *adult, from Mackay, Qld.

Distribution: NE coastal, SE coastal, N coastal, Vic., Qld., N.T.; only published localities Mackay, Ayr, Warburton and Darwin. Ecology: larva - sedentary, predator : adult - volant; prey larval Lepidoptera. Biological references: Giordani Soika, A. (1977). Contributo alla conoscenza degli Eumenidi australiani (Hymenoptera). *Mem. Soc. Entomol. Ital.* **55**: 109–138 (distribution).

Stenodyneriellus yanchepensis (Giordani Soika, 1962)

Taxonomic decision of Giordani Soika, A. (1962). Gli *Odynerus* sensu antiquo del continente australiano e della Tasmania. *Boll. Mus. Civ. Stor. Nat. Venezia* **14**: 57–202 [115].

Stenodyneriellus yanchepensis yanchepensis (Giordani Soika, 1962)

Australodynerus yanchepensis Giordani Soika, A. (1962). Gli *Odynerus* sensu antiquo del continente australiano e della Tasmania. *Boll. Mus. Civ. Stor. Nat. Venezia* **14**: 57–202 [121]. Type data: holotype, BMNH *F. adult, from Yanchep, W.A.

Distribution: SW coastal, W.A.; only published localities Yanchep, Perth and Kalamunda. Ecology: larva - sedentary, predator : adult - volant; prey larval Lepidoptera. Biological references: Giordani Soika, A. (1977). Contributo alla conoscenza degli Eumenidi australiani (Hymenoptera). *Mem. Soc. Entomol. Ital.* **55**: 109–138 (taxonomy).

Stenodyneriellus yanchepensis nigrithorax (Giordani Soika, 1962)

Australodynerus yanchepensis nigrithorax Giordani Soika, A. (1962). Gli *Odynerus* sensu antiquo del continente australiano e della Tasmania. *Boll. Mus. Civ. Stor. Nat. Venezia* **14**: 57–202 [122]. Type data: syntypes, ZMB *4M. adult, from Adelaide, S.A.

Distribution: S Gulfs, S.A.; type locality only. Ecology: larva - sedentary, predator : adult - volant; prey larval Lepidoptera. Biological references: Giordani Soika, A. (1977). Contributo alla conoscenza degli Eumenidi australiani (Hymenoptera). *Mem. Soc. Entomol. Ital.* **55**: 109–138 (taxonomy).

Subancistrocerus Saussure, 1855

Subancistrocerus Saussure, H. de (1855). *Études sur la Famille des Vespides*. Troisième Partie comprenant la Monographie des Masariens et un Supplement à la Monographie des Eumeniens. Paris : Masson 352 pp. pls i–xvi (1854–1856) [206] [proposed for secondary division of *Ancistrocerus* Wesmael, 1836; placed on Official List of Generic Names in Zoology, see International Commission on Zoological Nomenclature (1970). Opinion 893. Eumenidae names of Saussure (Hymenoptera) : grant of availability to certain names proposed for secondary divisions of genera. *Bull. Zool. Nomen.* **26**: 187–191]. Publication date established from Griffin, F.J. (1939). On the dates of publication of Saussure (H. de) : Études sur la famille des Vespides 1–3, 1852–1858. *J. Soc. Bibliogr. Nat. Hist.* **1**: 211–212. Type species *Odynerus sichelii* Saussure, 1855 by subsequent designation, see Bequaert, J. (1925). The genus *Ancistrocerus* in North America, with a partial key to the species. *Trans. Am. Entomol. Soc.* **51**: 57–117.

This group is also found in the Oriental and Neotropical Regions, see Giordani Soika, A. (1941). Studi sui Vespidi solitari. *Boll. Soc. Veneziana Stor. Nat.* **2**: 130–279.

Subancistrocerus monstricornis (Giordani Soika, 1941)

Ancistrocerus (Subancistrocerus) monstricornis Giordani Soika, A. (1941). Studi sui Vespidi solitari. *Boll. Soc. Veneziana Storr Nat.* **2**: 130–279 [241]. Type data: lectotype, BMNH *M. adult, from Mackay, Qld., designation by Giordani Soika, A. (1973). Designazione di lectotipi ed elenco dei tipi di Eumenidi, Vespidi e Masaridi da me descritti negli anni 1934–1960. *Boll. Mus. Civ. Stor. Nat. Venezia* **24**: 7–53.

Distribution: NE coastal, N coastal, Qld., N.T. Ecology: larva - sedentary, predator : adult - volant; prey larval Lepidoptera. Biological references: Giordani Soika, A. (1962). Gli *Odynerus* sensu antiquo del continente australiano e della Tasmania. *Boll. Mus. Civ. Stor. Nat. Venezia* **14**: 57–202 (redescription, generic placement); Giordani Soika, A. (1977). Contributo alla conoscenza degli Eumenidi australiani (Hymenoptera). *Mem. Soc. Entomol. Ital.* **55**: 109–138 (distribution).

Syneuodynerus Blüthgen, 1951

Syneuodynerus Blüthgen, P. (1951). Die *Euodynerus*-Arten des Balkans (Hym. Vespidae Eumeninae). *Boll. Soc. Entomol. Ital.* **81**: 66–76 [75] [proposed with subgeneric rank in *Euodynerus* Dalla Torre, 1904]. Type species *Odynerus egregius* Herrich-Schaeffer, 1839 by original designation.

This group is also found in the Palaearctic Region, see van der Vecht, J. & Fischer, F.C.J. (1972). Palearctic Eumenidae. *Hymenopterorum Catalogus* **8**: 1–199.

Syneuodynerus aurantiopilosellus (Giordani Soika, 1962)

Pseudepipona (Syneuodynerus) aurantiopilosella Giordani Soika, A. (1962). Gli *Odynerus* sensu antiquo del continente australiano e della Tasmania. *Boll. Mus. Civ. Stor. Nat. Venezia* **14**: 57–202 [102]. Type data: holotype, SAMA *F. adult, from Birthday Well, Cariewerloo (as Birthday Carlewerloo), S.A.

Distribution: S Gulfs, S.A.; type locality only. Ecology: larva - sedentary, predator : adult - volant; prey larval Lepidoptera.

Syneuodynerus longebispinosus (Giordani Soika, 1962)

Pseudepipona (Syneuodynerus) longebispinosa Giordani Soika, A. (1962). Gli *Odynerus* sensu antiquo del continente australiano e della Tasmania. *Boll. Mus. Civ. Stor. Nat. Venezia* **14**: 57–202 [101]. Type data: holotype, BMNH *F. adult, from Adelaide, S.A.

Distribution: SE coastal, Murray-Darling basin, S Gulfs, N.S.W., Vic., S.A. Ecology: larva - sedentary, predator : adult - volant; prey larval Lepidoptera. Biological references: Giordani Soika, A. (1977). Contributo alla conoscenza degli Eumenidi australiani (Hymenoptera). *Mem. Soc. Entomol. Ital.* **55**: 109–138 (description of male, distribution).

Syneuodynerus occidentatus (Giordani Soika, 1962)

Pseudepipona (Syneuodynerus) occidentata Giordani Soika, A. (1962). Gli *Odynerus* sensu antiquo del continente australiano e della Tasmania. *Boll. Mus. Civ. Stor. Nat. Venezia* **14**: 57–202 [102]. Type data: holotype, BMNH (probable) *F. adult, from Yanchep, W.A.

Distribution: SW coastal, W.A.; type locality only. Ecology: larva - sedentary, predator : adult - volant; prey larval Lepidoptera.

VESPIDAE

INTRODUCTION

This family, the papernest wasps, contains small to quite large, social wasps. There are some 50 species and subspecies from two genera native to Australia. The greatest diversity is in north Queensland; no species occur naturally in Tasmania or in the south of Western Australia. Two introduced species of a third genus, *Vespula*, have become established and are spreading within eastern Australia. They cause concern on medical grounds (there is a risk of severe allergic reaction to the sting) and on economic grounds (damage to hives of honey bees and to fruit). One species of *Vespula* is established in Tasmania and one eastern Australian and one European species of *Polistes* in Western Australia.

The "paper" nests, constructed from masticated plant fibres, are founded by one or several females. As adults remain after emergence, the number of individuals in each nest increases with age, up to several hundreds in *Ropalidia* and *Polistes*, to many thousands for *Vespula* in Tasmania. Adults are often collected on flowers, and larvae are fed progressively on masticated insects, mainly larval Lepidoptera.

References

Richards, O.W. (1978). The Australian social wasps (Hymenoptera : Vespidae). *Aust. J. Zool. Suppl. Ser.* **61**: 1–132

Evans, H.E. & West Eberhard, M.J. (1970). *The Wasps*. Ann Arbor : Univ. Michigan Press 265 pp.

Breed, M.D., Michener, C.D. & Evans, H.E. (eds.) (1982). *The Biology of Social Insects*. Proceedings of the Ninth Congress of the International Union for the study of social Insects, Boulder, Colorado, August 1982. Boulder : Westview Press 419 pp.

Vespula Thomson, 1869

Taxonomic decision of Richards, O.W. (1978). The Australian social wasps (Hymenoptera : Vespidae). *Aust. J. Zool. Suppl. Ser.* **61**: 1–132 [3].

Vespula (Paravespula) Blüthgen, 1938

Paravespula Blüthgen, P. (1938). Systematisches Verzeichnis der Faltenwespen Mittel-europas, Skandinaviens und Englands. *Konowia* **16**: 270–295 [271] [proposed with subgeneric rank in *Dolichovespula* Rohwer, 1916]. Type species *Vespa vulgaris* Linnaeus, 1758 by original designation.

This group has a natural distribution in the Holarctic Region only, see Richards, O.W. (1978). The Australian social wasps (Hymenoptera : Vespidae). *Aust. J. Zool. Suppl. Ser.* **61**: 1– 132.

Vespula (Paravespula) germanica (Fabricius, 1793)

Vespa germanica Fabricius, J.C. (1793). *Entomologia Systematica Emendata et Aucta*. Secundum classes, ordines, genera, species. Adjectis synonimis, locis, observationibus, descriptionibus. Hafniae : C.G. Profit Vol. 2 viii 519 pp. [256]. Type data: syntypes (probable), adult whereabouts unknown, from Kiel (as Kiliae), West Germany.

Distribution: SE coastal, N.S.W., Vic., Tas., S.A., W.A.; native to Europe, accidentally introduced into Australia (N.S.W. 1975, Vic. 1977, Tas. 1959, S.A. 1978, W.A. 1977), still spreading in N.S.W. and Vic., eradication attempted and apparently successful in S.A. and W.A., also introduced into New Zealand, North and South America, and South Africa. Ecology: larva - sedentary, predator : adult - volant; social, nests of "paper". Biological references: Thomas, C.R. (1960). The European wasp (*Vespula germanica* Fab.) in New Zealand. *Inf. Ser. Dept. Sci. Ind. Res. N.Z.* **27**: 1–74 (biology in New Zealand); Spradbery, J.P. (1973). The European social wasp, *Paravespula germanica* (F.) (Hymenoptera : Vespidae) in Tasmania, Australia. *Proc. VII Congr. I.U.S.S.I.* (London, 1973) pp. 375–380 (biology); Richards, O.W. (1978). The Australian social wasps (Hymenoptera : Vespidae). *Aust. J. Zool. Suppl. Ser.* **61**: 1–132 (distribution, illustration); Smithers, C.N. & Holloway, G.A. (1978). Establishment of *Vespula germanica* (Fabricius) (Hymenoptera : Vespidae) in New South Wales. *Aust. Entomol. Mag.* **5**: 55–59 (biology in N.S.W.); Edwards, R. (1980). *Social Wasps : Their Biology and Control.* East Grinstead : Rentokil Ltd. 398 pp. (biology); Madden, J.L. (1981). Factors influencing the abundance of the European wasp (*Paravespula germanica* (F.)). *J. Aust. Entomol. Soc.* **20**: 59–65 (biology); Anon. (1981). New and unusual insect records in Victoria. *Rep. Vict. Plant Res. Inst.* **10**: 94–96 (discovery in Vic.); White, B.R. (1983). Field day report. European wasps. *Australas. Beekpr.* **85**: 17 (spread in N.S.W.).

Vespula (Paravespula) vulgaris (Linnaeus, 1758)

Vespa vulgaris Linnaeus, C. von (1758). *Systema Naturae per regna tria naturae, secundum classes, ordines, genera, species, cum characteribus, differentiis, synonymis, locis.* 10th edn. Stockholm : Laurentii Salvii Vol. 1 823 pp. [572]. Type data: lectotype, LS *F. adult, from Europa, designation by Day, M.C. (1979). The species of Hymenoptera described by Linnaeus in the genera *Sphex, Chrysis, Vespa, Apis* and *Mutilla. Biol. J. Linn. Soc.* **12**: 45–84.

Distribution: SE coastal, Vic.; native to Europe, accidentally introduced into Vic. (1958) and spreading slowly from Melbourne area in spite of attempted eradication, also introduced in North America and Hawaii, found in Asia Minor and Palaearctic Asia. Ecology: larva - sedentary, predator : adult - volant; social, nests of "paper". Biological references: Richards, O.W. (1978). The Australian social wasps (Hymenoptera : Vespidae). *Aust. J. Zool. Suppl. Ser.* **61**: 1–132 (distribution); Edwards, R. (1980). *Social Wasps: Their Biology and Control.* East Grinstead : Rentokil 398 pp. (biology).

Polistes Latreille, 1802

Taxonomic decision of Richards, O.W. (1973). The subgenera of *Polistes* Latreille (Hymenoptera, Vespidae). *Rev. Bras. Entomol.* **17**: 85–104 [85].

Polistes (Polistes) Latreille, 1802

Polistes Latreille, P.A. (1802). *Histoire Naturelle, Générale et Particulière des Crustacés et des Insectes.* Paris : F. Dufart Vol. 3 xii 13+467 pp. (1802–1803) [363]. Type species *Vespa gallica* Linnaeus, 1767 by subsequent designation, see Blanchard, E. (1840). Histoire naturelle des insectes; orthoptères, névroptères, hémiptères, hyménoptères, lépidoptères et diptères. *in* Castelnau, F.L. (1840). *Histoire Naturelle des Animaux Articulés.* Paris : P. Dumenil Vol. 3 672 pp.

This group is also found in the Oriental, Palaearctic and Ethiopian Regions, see Richards, O.W. (1973). The subgenera of *Polistes* Latreille (Hymenoptera, Vespidae). *Rev. Bras. Entomol.* **17**: 85–104.

Polistes (Polistes) dominulus (Christ, 1791)

Vespa dominula Christ, J.L. (1791). *Naturgeschichte, Classification und Nomenclatur der Insecten vom Bienen, Wespen und Ameisengeschlecht.* Frankfurt-am-Main : Hermann 535 pp. [229] [this is a new record for Australia, based on O.W. Richards' identification of specimens as *Polistes gallicus* (Linnaeus, 1767)]. Type data: syntypes, adult whereabouts unknown, from Kronberg, West Germany. Compiled from secondary source: Blüthgen, P. (1961). Die Faltenwespen Mitteleuropas (Hymenoptera, Diploptera). *Abh. Dt. Akad. Wiss. Berl. Kl. Chem. Geol. Biol.* **1961**(2): 1–251.

Polistes gallicus Auctorum [this is not *Polistes gallicus* (Linnaeus, 1767), originally described in *Vespa*].

Taxonomic decision of Day, M.C. (1979). The species of Hymenoptera described by Linnaeus in the genera *Sphex, Chrysis, Vespa, Apis* and *Mutilla. Biol. J. Linn. Soc.* **12**: 45–84 [63].

Distribution: SW coastal, W.A.; accidentally introduced into the Perth area, established by 1977, native to the Palearctic Region and introduced into Massachusetts (U.S.A.). Ecology: larva - sedentary, predator : adult - volant; social, nests of "paper". Biological references: Guiglia, D. (1972). Les guêpes sociales (Hymenoptera Vespidae) d'Europe occidentale et septtentrionale. *Faune de l'Europe et du Bassin Méditerranéen* **6**: 1–181 (biology as *Polistes gallicus*); Hathaway, M. (1982). *Polistes gallicus* in Massachusetts (Hymenoptera : Vespidae). *Psyche Camb.* **88**: 169–173 (U.S.A. record as *Polistes gallicus*).

Polistes (Megapolistes) van der Vecht, 1968

Megapolistes van der Vecht, J. (1968). The geographic variation of *Polistes* (*Megapolistes* subgen. n.) *rothneyi* Cameron. *Bijdr. Dierkd.* **38**: 97–109 [97] [proposed with subgeneric rank in *Polistes* Latreille, 1802]. Type species *Vespa olivacea* Degeer, 1773 by original designation.

This group is also found in the Oriental Region and Pacific islands, see Richards, O.W. (1978). *The Social Wasps of the Americas: excluding the Vespinae.* London : British Museum 580 pp.

Polistes (Megapolistes) balder Kirby, 1888

Polistes balder Kirby, W.F. (1888). On the insects (exclusive of Coleoptera and Lepidoptera) of Christmas

Island. *Proc. Zool. Soc. Lond.* **1888**: 546–555 [552]. Type data: holotype, BMNH *F. adult, from Christmas Is.

Distribution: Christmas Is., Cocos (Keeling) Is. Ecology: larva - sedentary, predator : adult - volant; social, nests of "paper". Biological references: Richards, O.W. (1978). The Australian social wasps (Hymenoptera : Vespidae). *Aust. J. Zool. Suppl. Ser.* **61**: 1–132 (redescription).

Polistes (Megapolistes) erythrinus Holmgren, 1868

Polistes erythrinus Holmgren, A.E. (1868). Hymenoptera, species novas descripsit. *Kongliga Svenska Fregatten Eugenies resa omkring Jorden under befäl af C.A. Virgen Aren 1851-53.* II Zoologi 1 Insecta pp. 391–442 pl viii [440]. Type data: holotype, NHRM *F. adult, from Sydney, N.S.W.

Distribution: SE coastal, Murray-Darling basin, Qld., N.S.W., A.C.T., Vic., S.A. Ecology: larva - sedentary, predator : adult - volant; social, nests of "paper". Biological references: Richards, O.W. (1978). The Australian social wasps (Hymenoptera : Vespidae). *Aust. J. Zool. Suppl. Ser.* **61**: 1–132 (redescription, nest, biology).

Polistes (Megapolistes) facilis Saussure, 1853

Polistes facilis Saussure, H. de (1853). *Études sur la Famille des Vespides.* 2. Monographie des Guêpes Sociales, ou de la Tribu des Vespiens, ouvrage faisant suite à la Monographie des Guêpes Solitaires. Paris : Masson cxcix 256 pp. pls i–xxxvii (1853–1858) [53]. Type data: holotype, MZUT (probable) *adult, from Australia (as New Holland).

Distribution: NE coastal, Qld.; only published localities Gladstone, Rockhampton and Yeppoon. Ecology: larva - sedentary, predator : adult - volant; social, nests of "paper". Biological references: Richards, O.W. (1978). The Australian social wasps (Hymenoptera : Vespidae). *Aust. J. Zool. Suppl. Ser.* **61**: 1–132 (redescription).

Polistes (Megapolistes) olivaceus (Degeer, 1773)

Vespa olivacea Degeer, C. (1773). *Mémoires pour Servir à l'Histoire des Insectes.* Stockholm Vol. 3 [582 pl 24 fig 9]. Type data: holotype, adult whereabouts unknown, from America. Compiled from secondary source: Richards, O.W. (1978). *The Social Wasps of the Americas: excluding the Vespinae.* London : British Museum 580 pp.

Distribution: NE coastal, Qld.; accidentally introduced into Brisbane (1946, 1953), possibly temporarily established, normal distribution is Indian area to S China, Pacific islands, and accidentally introduced into New Zealand but not established. Ecology: larva - sedentary, predator : adult - volant; social, nests of "paper". Biological references: Richards, O.W. (1978). *The Social Wasps of the Americas: excluding the Vespinae.* London : British Museum 580 pp. (redescription, distribution); Richards, O.W. (1978). The Australian social wasps (Hymenoptera : Vespidae). *Aust. J. Zool. Suppl. Ser.* **61**: 1–132 (larva, redescription, nest).

Polistes (Megapolistes) schach (Fabricius, 1781)

Vespa schach Fabricius, J.C. (1781). *Species Insectorum exhibentes eorum differentias specificas, synonyma auctorum, loca natalia, metamorphosin adiectis observationibus, descriptionibus.* Hamburgi et Kilonii Vols i–ii [461]. Type data: holotype, BMNH *F. adult, from Australia (as New Holland).

Distribution: NE coastal, Murray-Darling basin, N coastal, Qld., N.S.W., W.A., N.T. Ecology: larva - sedentary, predator : adult - volant; social, nests of "paper". Biological references: Richards, O.W. (1978). The Australian social wasps (Hymenoptera : Vespidae). *Aust. J. Zool. Suppl. Ser.* **61**: 1–132 (redescription).

Polistes (Megapolistes) tepidus (Fabricius, 1775)

Taxonomic decision of Richards, O.W. (1978). The Australian social wasps (Hymenoptera : Vespidae). *Aust. J. Zool. Suppl. Ser.* **61**: 1–132 [19]. (with a list of non-Australian subspecies, including *Polistes tepidus picteti* Saussure, 1853, described from New Holland but actually occurring only in Ambon, Seram, and West Irian)

Polistes (Megapolistes) tepidus tepidus (Fabricius, 1775)

Vespa tepida Fabricius, J.C. (1775). *Systema Entomologiae, sistens insectorum classes, ordines, genera, species, adiectis synonymis, locis, descriptionibus, observationibus.* Flensburgi et Lipsiae : Kortii xxvii 832 pp. [366]. Type data: holotype, BMNH *F. adult, from Australia (as New Holland).

Polistes malayanus Cameron, P. (1906). Hymenoptera I (all families except Apidae and Formicidae). pp 41–65 *in* Wichmann, A. *Nova Guinea, Résultats de l'Expédition Scientifique Néerlandaise à la Nouvelle- Guinée en 1903, sous les Auspices de Arthur Wichmann, Chef de l'Expédition.* Leiden : E.J. Brill Vol. 5 [60]. Type data: holotype, BMNH *F. adult, from Manokwari, New Guinea.

Distribution: NE coastal, SE coastal, Qld., N.S.W., Vic.; also New Guinea, Bougainville Is., Aru Is., Ki Is., Russell Is. and Solomon Ils. Ecology: larva - sedentary, predator : adult - volant; social, nests of "paper". Biological references: Hook, A. (1982). Observations on a declining nest of *Polistes tepidus* (F.) (Hymenoptera : Vespidae). *J. Aust. Entomol. Soc.* **21**: 277–278 (biology).

Polistes (Stenopolistes) van der Vecht, 1972

Stenopolistes van der Vecht, J. (1972). The subgenera *Megapolistes* and *Stenopolistes* in the Solomon Islands (Hymenoptera, Vespidae, *Polistes* Latreille). pp. 87–106 *in, Entomological Essays to Commemorate the Retirement of Professor K. Yasumatsu 1971*. Tokyo : Hokuryukan Publishing Co. Ltd. [101] [proposed with subgeneric rank in *Polistes* Latreille, 1802]. Type species *Polistes lateritius* Smith, 1857 by original designation.

This group is also found in the Oriental Region, see Richards, O.W. (1973). The subgenera of *Polistes* Latreille (Hymenoptera, Vespidae). *Rev. Bras. Entomol.* **17**: 85–104.

Polistes (Stenopolistes) laevigatissimus Giordani Soika, 1975

Polistes laevigatissimus Giordani Soika, A. (1975). Notulae vespidologicae XXXVII. Nuovi *Polistes* del continente australiano (Hymenoptera Vespidae). *Boll. Soc. Entomol. Ital.* **107**: 20–25 [25]. Type data: holotype, AM F. adult not found, from Lombardia, Broome, W.A.

Distribution: N coastal, W.A.; type locality only. Ecology: larva - sedentary, predator : adult - volant; social, nests of "paper". Biological references: Richards, O.W. (1978). The Australian social wasps (Hymenoptera : Vespidae). *Aust. J. Zool. Suppl. Ser.* **61**: 1–132 (redescription).

Polistes (Stenopolistes) riekii Richards, 1978

Polistes (Stenopolistes) riekii Richards, O.W. (1978). The Australian social wasps (Hymenoptera : Vespidae). *Aust. J. Zool. Suppl. Ser.* **61**: 1–132 [23]. Type data: holotype, ANIC F. adult, from Iron Range, Qld.

Distribution: NE coastal, N Gulf, Qld.; only published localities Iron Range, Lockerbie, Claudie River and Mt. Tozer. Ecology: larva - sedentary, predator : adult - volant; social, nests of "paper".

Polistes (Polistella) Ashmead, 1904

Polistella Ashmead, W.H. (1904). Descriptions of new genera and species of Hymenoptera from the Philippine Islands. *Proc. U.S. Natl. Mus.* **28**: 127–158 [133]. Type species *Polistes manillensis* Saussure, 1853 by monotypy and original designation.

This group is also found in the Oriental and Ethiopian Regions, see Richards, O.W. (1973). The subgenera of *Polistes* Latreille (Hymenoptera, Vespidae). *Rev. Bras. Entomol.* **17**: 85–104.

Polistes (Polistella) bernardii Le Guillou, 1841

Taxonomic decision of Richards, O.W. (1978). The Australian social wasps (Hymenoptera : Vespidae). *Aust. J. Zool. Suppl. Ser.* **61**: 1–132 [25].

Polistes (Polistella) bernardii bernardii Le Guillou, 1841

Polistes bernardii Le Guillou, E.J.F. (1841). Description de 20 espèces nouvelles appartenant à diverses familles d'Hyménoptères. *Rev. Zool.* **4**: 322–325 [325]. Type data: holotype, F. adult whereabouts unknown, from N Australia (as Aust. sept.).

Distribution: N Gulf, N coastal, N.T.; only published localities Darwin and Groote Eylandt. Ecology: larva - sedentary, predator : adult - volant; social, nests of "paper".

Polistes (Polistella) bernardii duplicinctus Richards, 1978

Polistes (Polistella) bernardii duplicinctus Richards, O.W. (1978). The Australian social wasps (Hymenoptera : Vespidae). *Aust. J. Zool. Suppl. Ser.* **61**: 1–132 [28]. Type data: holotype, ANIC F. adult, from Millstream, W.A.

[Polistes townsvillensis] Giordani Soika, A. (1975). Notulae vespidologicae XXXVII. Nuovi *Polistes* del continente australiano (Hymenoptera Vespidae). *Boll. Soc. Entomol. Ital.* **107**: 20–25 [22] [for species assignment of the two paratypes see Richards, O.W. (1978). The Australian social wasps (Hymenoptera : Vespidae). *Aust. J. Zool. Suppl. Ser.* **61**: 1–132]. Type data: paratypes, SAMA *1M., 1F., from Hamersley Range, Fortescue River, W.A.

Distribution: N coastal, NW coastal, W.A.; only published localities Millstream, Hamersley Range, W Kimberleys and Wotjulum. Ecology: larva - sedentary, predator : adult - volant; social, nests of "paper".

Polistes (Polistella) bernardii insulae Richards, 1978

Polistes (Polistella) bernardii insulae Richards, O.W. (1978). The Australian social wasps (Hymenoptera : Vespidae). *Aust. J. Zool. Suppl. Ser.* **61**: 1–132 [30]. Type data: holotype, ANIC F. adult, from Dauan Is., Qld.

Distribution: NE coastal, Qld.; only published localities Dauan Is., Banks Is., Murray Is., Thursday Is., also in New Guinea. Ecology: larva - sedentary, predator : adult - volant; social, nests of "paper".

Polistes (Polistella) bernardii richardsi Giordani Soika, 1975

Polistes richardsi Giordani Soika, A. (1975). Notulae vespidologicae XXXVII. Nuovi *Polistes* del continente australiano (Hymenoptera Vespidae). *Boll. Soc. Entomol. Ital.* **107**: 20–25 [23]. Type data: holotype, SAMA *F. adult, from Roper River, N.T.

Distribution: N Gulf, N coastal, NE coastal, Qld., N.T. Ecology: larva - sedentary, predator : adult - volant; social, nests of "paper".

Polistes (Polistella) humilis (Fabricius, 1781)

Taxonomic decision of Richards, O.W. (1978). The Australian social wasps (Hymenoptera : Vespidae). *Aust. J. Zool. Suppl. Ser.* **61**: 1–132 [43].

Polistes* (*Polistella*) *humilis humilis (Fabricius, 1781)

Vespa humilis Fabricius, J.C. (1781). *Species Insectorum exhibentes eorum differentias specificas, synonyma auctorum, loca natalia, metamorphosin adiectis observationibus, descriptionibus.* Hamburgi et Kilonii Vol. i [461]. Type data: lectotype, BMNH *M. adult, from Australia (as New Holland), designation by Richards, O.W. (1978). The Australian social wasps (Hymenoptera : Vespidae). *Aust. J. Zool. Suppl. Ser.* **61**: 1–132.

Polistes tasmaniensis Saussure, H. de (1853). *Études sur la Famille des Vespides.* 2. Monographie des Guêpes Sociales, ou de la Tribu des Vespiens, ouvrage faisant suite à la Monographie des Guêpes Solitaires. Paris : Masson cxcix 256 pp. pls i–xxxvii (1853–1858) [66 pl vi fig 6, pl viii fig 3]. Type data: lectotype, MNHP *F. adult, from Australia (as New Holland), designation by Richards, O.W. (1978). The Australian social wasps (Hymenoptera : Vespidae). *Aust. J. Zool. Suppl. Ser.* **61**: 1–132.

Polistes tricolor Saussure, H. de (1853). *Études sur la Famille des Vespides.* 2. Monographie des Guêpes Sociales, ou de la Tribu des Vespiens, ouvrage faisant suite à la Monographie des Guêpes Solitaires. Paris : Masson cxcix 256 pp. pls i–xxxvii (1853–1858) [67]. Type data: holotype, BMNH *F. adult (seen 1929 by L.F. Graham), from Australia (as New Holland).

Polistes humilis pseudoscach Giordani Soika, A. (1975). Notulae vespidologicae XXXVII. Nuovi *Polistes* del continente australiano (Hymenoptera Vespidae). *Boll. Soc. Entomol. Ital.* **107**: 20–25 [21]. Type data: holotype, AM F. adult not found, from Dubbo, N.S.W.

Distribution: SE coastal, Murray-Darling basin, Qld., N.S.W., A.C.T., Vic., S.A.; accidentally introduced to New Zealand. Ecology: larva - sedentary, predator : adult - volant; social, nests of "paper". Biological references: Robertson, P.L. (1968). A morphological and functional study of the venom apparatus in representatives of some major groups of Hymenoptera. *Aust. J. Zool.* **16**: 133–166 (venom apparatus); Owen, M.D. (1979). Chemical components in the venoms of *Ropalidia revolutionalis* and *Polistes humilis* (Hymenoptera, Vespidae). *Toxicon* **17**: 519–523 (venom analysis).

Polistes* (*Polistella*) *humilis centrocontinentalis Giordani Soika, 1975

Polistes humilis centrocontinentalis Giordani Soika, A. (1975). Notulae vespidologicae XXXVII. Nuovi *Polistes* del continente australiano (Hymenoptera Vespidae). *Boll. Soc. Entomol. Ital.* **107**: 20–25 [20]. Type data: holotype, A. Giordani Soika pers. coll. *F. adult, from Central Australia.

Distribution: type locality only. Ecology: larva - sedentary, predator : adult - volant; social, nests of "paper".

Polistes* (*Polistella*) *humilis synoecus Saussure, 1853

Polistes synoecus Saussure, H. de (1853). *Études sur la Famille des Vespides.* 2. Monographie des Guêpes Sociales, ou de la Tribu des Vespiens, ouvrage faisant suite à la Monographie des Guêpes Solitaires. Paris : Masson cxcix 256 pp. pls i–xxxvii (1853–1858) [65 pl vi fig 5]. Type data: lectotype, MNHP *M. adult, from Australia (as New Holland), designation by Richards, O.W. (1978). The Australian social wasps (Hymenoptera : Vespidae). *Aust. J. Zool. Suppl. Ser.* **61**: 1–132.

Polistes variabilis reginae Meade-Waldo, G. (1911). New species of Diploptera in the collection of the British Museum. iii. *Ann. Mag. Nat. Hist.* (8) **7**: 98–113 [101] [described as variety]. Type data: holotype, BMNH *F. adult (seen 1929 by L.F. Graham), from Cooktown, Qld.

Polistes humilis xanthorrhoicus Giordani Soika, A. (1975). Notulae vespidologicae XXXVII. Nuovi *Polistes* del continente australiano (Hymenoptera Vespidae). *Boll. Soc. Entomol. Ital.* **107**: 20–25 [21]. Type data: holotype, BMNH *F. adult, from N Qld.

Polistes humilis clarior Giordani Soika, A. (1975). Notulae vespidologicae XXXVII. Nuovi *Polistes* del continente australiano (Hymenoptera Vespidae). *Boll. Soc. Entomol. Ital.* **107**: 20–25 [22]. Type data: holotype, BMNH *F. adult, from N Qld.

Distribution: NE coastal, SE coastal, Murray-Darling basin, SW coastal, (N coastal), Qld., N.S.W., W.A., (N.T.); accidentally introduced to Perth area in about 1950, also introduced into the Society Ils. Ecology: larva - sedentary, predator : adult - volant; social, nests of "paper".

Polistes* (*Polistella*) *sgarambus Giordani Soika, 1975

Polistes sgarambus Giordani Soika, A. (1975). Notulae vespidologicae XXXVII. Nuovi *Polistes* del continente australiano (Hymenoptera Vespidae). *Boll. Soc. Entomol. Ital.* **107**: 20–25 [24]. Type data: holotype, AM F. adult not found, from East Alligator River, Oenpelli, N.T.

Distribution: N coastal, N.T.; only published localities East Alligator River, Oenpelli, Mt. Cahill and Darwin. Ecology: larva - sedentary, predator : adult - volant; social, nests of "paper". Biological references: Richards, O.W. (1978). The Australian social wasps (Hymenoptera : Vespidae). *Aust. J. Zool. Suppl. Ser.* **61**: 1–132 (redescription).

Polistes* (*Polistella*) *townsvillensis Giordani Soika, 1975

Taxonomic decision of Richards, O.W. (1978). The Australian social wasps (Hymenoptera : Vespidae). *Aust. J. Zool. Suppl. Ser.* **61**: 1–132 [36].

Polistes* (*Polistella*) *townsvillensis townsvillensis Giordani Soika, 1975

Polistes townsvillensis Giordani Soika, A. (1975). Notulae vespidologicae XXXVII. Nuovi *Polistes* del continente australiano (Hymenoptera Vespidae). *Boll. Soc. Entomol. Ital.* **107**: 20–25 [22] [two paratypes transferred to *Polistes bernardii duplicinctus* Richards, 1978]. Type data: holotype, SAMA *F. adult, from Townsville, Qld.

Distribution: NE coastal, N Gulf, N coastal, Qld., W.A., N.T. Ecology: larva - sedentary, predator : adult - volant; social, nests of "paper".

Polistes (Polistella) townsvillensis austrinus Richards, 1978

Polistes (Polistella) townsvillensis austrinus Richards, O.W. (1978). The Australian social wasps (Hymenoptera : Vespidae). *Aust. J. Zool. Suppl. Ser.* **61**: 1–132 [37]. Type data: holotype, ANIC F. adult, from Bundaberg, Qld.

Polistes synoecus Auctorum [this is not *Polistes synoecus* Saussure, 1853, which has been placed as a subspecies of *Polistes humilis* (Fabricius, 1781)].

Distribution: NE coastal, SE coastal, Murray-Darling basin, Qld., N.S.W., Vic. Ecology: larva - sedentary, predator : adult - volant; social, nests of "paper".

Polistes (Polistella) variabilis (Fabricius, 1781)

Vespa variabilis Fabricius, J.C. (1781). *Species Insectorum exhibentes eorum differentias specificas, synonyma auctorum, loca natalia, metamorphosin adiectis observationibus, descriptionibus.* Hamburgi et Kilonii Vol. i [466]. Type data: holotype, BMNH *M. adult (seen 1929 by L.F. Graham), from Australia (as New Holland).

Distribution: NE coastal, Qld. Ecology: larva - sedentary, predator : adult - volant; social, nests of "paper". Biological references: Richards, O.W. (1978). The Australian social wasps (Hymenoptera : Vespidae). *Aust. J. Zool. Suppl. Ser.* **61**: 1–132 (redescription).

Ropalidia Guérin, 1831

Taxonomic decision of Richards, O.W. (1978). The Australian social wasps (Hymenoptera : Vespidae). *Aust. J. Zool. Suppl. Ser.* **61**: 1–132 [56].

Species now known not to occur in Australia: *Ropalidia punctum* (Fabricius, 1804), proposed in *Polistes* Latreille, 1802, from "Nova Cambria", is from the Ethiopian Region *teste* van der Vecht, J. (1958). On some Fabrician types of Indo-Australian Vespidae (Hymenoptera). *Arch. Néerl. Zool.* **13** 1 Suppl.: 234–247.

Ropalidia (Ropalidia) Guérin, 1831

Ropalidia Guérin-Méneville, F.E. (1831). Insectes. *in* Duperrey, L.J. (1830–1832). *Voyage Autour du Monde, Exécuté par Ordre du Roi, sur la Corvette de la Majesté, La Coquille, Pendant les Années 1822, 1823, 1824, et 1825, ...* Atlas. Histoire naturelle. Zoologie. pls 1–21 Paris : Bertrand [pl 9 fig 8]. Type species *Ropalidia maculiventris* Guérin, 1831 by monotypy.

Rhopalidia Guérin-Méneville, F.E. (1838). Crustacés, Arachnides et Insectes. pp. 57–302 *in* Duperrey, L.J. (1838). *Voyage Autour du Monde, Exécuté par Ordre du roi, sur la Corvette de la Majesté,* La Coquille, *Pendant les Années 1822, 1823, 1824, et 1825, ...* Zool. Vol. ii Pt. 2 Div. 1 Chap. xiii Paris : Bertrand [266] [emend. and description of *Ropalidia* Guérin, 1831; non *Rhopalidia* Lepeletier, 1836 suppressed by International Commission on Zoological Nomenclature (1976). Opinion 1051. *Rhopalidia* Lepeletier, 1836 (Insecta : Hymenoptera): suppressed under the plenary powers. *Bull. Zool. Nomen.* **32**: 240–241].

Icaria Saussure, H. de (1853). *Études sur la Famille des Vespides.* 2. Monographie des Guêpes Sociales, ou de la Tribu des Vespiens, ouvrage faisant suite à la Monographie des Guêpes Solitaires. Paris : Masson cxcix 256 pp. pls i–xxxvii (1853–1858) [22]. Type species *Ropalidia maculiventris* Guérin, 1831 by subsequent designation, see Bingham, C.T. (1897). *The Fauna of British India, including Ceylon and Burma.* Hymenoptera. Vol. I Wasps and bees. London : Taylor & Francis xxx 579 pp.

This group is also found in the Oriental Region, see Richards, O.W. (1978). The Australian social wasps (Hymenoptera : Vespidae). *Aust. J. Zool. Suppl. Ser.* **61**: 1–132.

Ropalidia (Ropalidia) eboraca Richards, 1978

Ropalidia (Ropalidia) eboraca Richards, O.W. (1978). The Australian social wasps (Hymenoptera : Vespidae). *Aust. J. Zool. Suppl. Ser.* **61**: 1–132 [66]. Type data: holotype, ANIC F. adult, from Lockerbie, Cape York, Qld.

Distribution: NE coastal, N Gulf, Qld. Ecology: larva - sedentary, predator : adult - volant; social, nests of "paper".

Ropalidia (Ropalidia) fulvopruinosa (Cameron, 1906)

Odynerus (Leionotus) fulvopruinosus Cameron, P. (1906). Hymenoptera of the Dutch expedition to New Guinea in 1904 and 1905. Part 1, Thynnidae, Scoliidae, Pompilidae, Sphegidae and Vespidae. *Tijdschr. Entomol.* **49**: 215–233 [225]. Type data: holotype, ZMA *F. adult, from Etna Bay, New Guinea.

Distribution: NE coastal, N Gulf, Qld.; also New Guinea. Ecology: larva - sedentary, predator : adult - volant; social, nests of "paper". Biological references: Richards, O.W. (1978). The Australian social wasps (Hymenoptera : Vespidae). *Aust. J. Zool. Suppl. Ser.* **61**: 1–132 (redescription, nest).

Ropalidia (Polistratus) Cameron, 1906

Polistratus Cameron, P. (1906). Hymenoptera I (all families except Apidae and Formicidae). pp 41–65 *in* Wichmann, A. *Nova Guinea, Résultats de l'Expédition Scientifique Néerlandaise à la Nouvelle- Guinée en 1903, sous les Auspices de Arthur Wichmann, Chef de l'Expédition.* Leiden : E.J. Brill Vol. 5 [59]. Type species

Polistratus cariniscutis Cameron, 1906 (= *Icaria brunnea* Smith, 1858) by monotypy.

This group is also found in the Oriental Region and Madagascar, see Richards, O.W. (1978). The Australian social wasps (Hymenoptera : Vespidae). *Aust. J. Zool. Suppl. Ser.* **61**: 1-132.

Ropalidia (Polistratus) latetergum Richards, 1978

Ropalidia (Polistratus) latetergum Richards, O.W. (1978). The Australian social wasps (Hymenoptera : Vespidae). *Aust. J. Zool. Suppl. Ser.* **61**: 1-132 [69]. Type data: holotype, NMV T7676 *F. adult, from Kuranda, Qld.

Distribution: NE coastal, Qld.; type locality only. Ecology: larva - sedentary, predator : adult - volant; social, nests of "paper".

Ropalidia (Icariola) Dalla Torre, 1904

Icariola Dalla Torre, K.W. von (1904). Hymenoptera. Fam. Vespidae. *Genera Insectorum* **19**: 1-108 [72] [proposed with subgeneric rank in *Icaria* Saussure, 1853]. Type species *Icaria gregaria* Saussure, 1853 by subsequent designation, see Meade-Waldo, G. (1913). New species of Diploptera in the collection of the British Museum. Part IV. *Ann. Mag. Nat. Hist.* (8) **11**: 44-54.

Zuba Cheesman, L.E. (1952). *Ropalidia* of Papuasia. *Ann. Mag. Nat. Hist.* (12) **5**: 1-26 4 pls [15] [proposed with subgeneric rank in *Ropalidia* Guérin, 1831]. Type species *Icaria gregaria* Saussure, 1853 by subsequent designation, see Richards, O.W. (1978). The Australian social wasps (Hymenoptera : Vespidae). *Aust. J. Zool. Suppl. Ser.* **61**: 1-132.

This group is found in all tropical regions except America, see Richards, O.W. (1978). The Australian social wasps (Hymenoptera : Vespidae). *Aust. J. Zool. Suppl. Ser.* **61**: 1-132.

Ropalidia (Icariola) darwini Richards, 1978

Ropalidia (Icariola) darwini Richards, O.W. (1978). The Australian social wasps (Hymenoptera : Vespidae). *Aust. J. Zool. Suppl. Ser.* **61**: 1-132 [72]. Type data: holotype, AM K69358 F. adult, from Milners Swamp, N.T.

Distribution: N coastal, N.T.; only published localities Milners Swamp, Cahills Crossing (East Alligator River) and Darwin. Ecology: larva - sedentary, predator : adult - volant; social, nests of "paper".

Ropalidia (Icariola) deceptor (Smith, 1864)

Icaria deceptor Smith, F. (1864). Catalogue of hymenopterous insects collected by Mr. A.R. Wallace in the Islands of Mysol, Ceram, Waigiou, Bouru and Timor. *J. Linn. Soc. Lond. Zool.* **7**: 6-48 [42]. Type data: holotype, OUM *F. adult, from Misool Is. (as Mysol), New Guinea.

Distribution: NE coastal, N Gulf, Qld.; also West Irian (Misool Is.). Ecology: larva - sedentary, predator : adult - volant; social, nests of "paper". Biological references: Richards, O.W. (1978). The Australian social wasps (Hymenoptera : Vespidae). *Aust. J. Zool. Suppl. Ser.* **61**: 1-132 (redescription, nest).

Ropalidia (Icariola) elegantula Richards, 1978

Ropalidia (Icariola) elegantula Richards, O.W. (1978). The Australian social wasps (Hymenoptera : Vespidae). *Aust. J. Zool. Suppl. Ser.* **61**: 1-132 [70]. Type data: holotype, ANIC F. adult, from Bamaga, Cape York Peninsula, Qld.

Distribution: NE coastal, N Gulf, Qld. Ecology: larva - sedentary, predator : adult - volant; social, nests of "paper". Biological references: Richards, O.W. (1978). The Australian social wasps (Hymenoptera : Vespidae). *Aust. J. Zool. Suppl. Ser.* **61**: 1-132 (nest).

Ropalidia (Icariola) eurostoma Richards, 1978

Ropalidia (Icariola) eurostoma Richards, O.W. (1978). The Australian social wasps (Hymenoptera : Vespidae). *Aust. J. Zool. Suppl. Ser.* **61**: 1-132 [93]. Type data: holotype, SAMA F. adult not found, paratypes SAMA 2F., from Mornington Is., Qld.

Distribution: NE coastal, Qld.; only published localities Mornington Is. and Australia (as New Holland). Ecology: larva - sedentary, predator : adult - volant; social, nests of "paper".

Ropalidia (Icariola) gracilenta Richards, 1978

Ropalidia (Icariola) gracilenta Richards, O.W. (1978). The Australian social wasps (Hymenoptera : Vespidae). *Aust. J. Zool. Suppl. Ser.* **61**: 1-132 [98]. Type data: holotype, AM K69363 F. adult, from North Creek near Ballina, N.S.W.

Distribution: NE coastal, (N Gulf), SE coastal, Qld., N.S.W.; only published localities Stradbroke Is., Fraser Is., Mackay, ?Coen and near Ballina. Ecology: larva - sedentary, predator : adult - volant; social, nests of "paper".

Ropalidia (Icariola) gregaria (Saussure, 1854)

Taxonomic decision of (including extralimital synonymy) Richards, O.W. (1978). The Australian social wasps (Hymenoptera : Vespidae). *Aust. J. Zool. Suppl. Ser.* **61**: 1-132 [83].

Ropalidia (Icariola) gregaria gregaria (Saussure, 1854)

Polistes bioculata Fabricius, J.C. (1804). *Systema Piezatorum*. Brunsvigae : C. Reichard xiv 439 pp. [278] [Richards, O.W. (1978). The Australian social wasps (Hymenoptera : Vespidae). *Aust. J. Zool. Suppl. Ser.* **61**: 1-132 examined the holotype and stated that the condition of the specimen was such that accurate assessment was not possible but tentatively placed it with *gregaria*]. Type data: holotype, UZM *F. adult, from Nova Cambria".

Icaria gregaria Saussure, H. de (1853). *Études sur la Famille des Vespides*. 2. Monographie des Guêpes Sociales, ou de la Tribu des Vespiens, ouvrage faisant suite à la Monographie des Guêpes Solitaires. Paris : Masson cxcix 256 pp. pls i–xxxvii (1853–1858) [236]. Type data: syntypes (probable), F. adult whereabouts unknown, from Australia (as New Holland).

Distribution: N coastal, N.T. Ecology: larva - sedentary, predator : adult - volant; social, nests of "paper".

Ropalidia (Icariola) gregaria spilocephala (Cameron, 1906)

Icaria spilocephala Cameron, P. (1906). Hymenoptera of the Dutch expedition to New Guinea in 1904 and 1905. Part 1, Thynnidae, Scoliidae, Pompilidae, Sphegidae and Vespidae. *Tijdschr. Entomol.* **49**: 215–233 [230]. Type data: holotype, ZMA *M. adult, from Etna Bay, New Guinea.

Distribution: NE coastal, Murray-Darling basin, Qld., N.S.W., Vic.; also New Guinea, New Ireland and Solomon Ils. Ecology: larva - sedentary, predator : adult - volant; social, nests of "paper".

Ropalidia (Icariola) hirsuta Richards, 1978

Ropalidia (Icariola) hirsuta Richards, O.W. (1978). The Australian social wasps (Hymenoptera : Vespidae). *Aust. J. Zool. Suppl. Ser.* **61**: 1–132 [81]. Type data: holotype, BMNH *F. adult, from Port Darwin, N.T.

Distribution: N coastal, N.T.; type locality only. Ecology: larva - sedentary, predator : adult - volant; social, nests of "paper".

Ropalidia (Icariola) interrupta van der Vecht, 1941

Taxonomic decision of Richards, O.W. (1978). The Australian social wasps (Hymenoptera : Vespidae). *Aust. J. Zool. Suppl. Ser.* **61**: 1–132 [89].

Ropalidia (Icariola) interrupta interrupta van der Vecht, 1941

Ropalidia variegata interrupta van der Vecht, J. (1941). The Indo-Australian species of the genus *Ropalidia* (=*Icaria*) (Hym., Vespidae) (First Part). *Treubia* **18**: 103–190 [158]. Type data: holotype, MCZ F. adult lost, from Thursday Is., Qld.

Distribution: NE coastal, Qld.; type locality only. Ecology: larva - sedentary, predator : adult - volant; social, nests of "paper".

Ropalidia (Icariola) interrupta flavinoda van der Vecht, 1941

Ropalidia variegata flavinoda van der Vecht, J. (1941). The Indo-Australian species of the genus *Ropalidia* (=*Icaria*) (Hym., Vespidae) (First Part). *Treubia* **18**: 103–190 [158]. Type data: holotype, MCZ F. adult lost, from Cape York, Qld.

Distribution: NE coastal, Qld.; type locality only. Ecology: larva - sedentary, predator : adult - volant; social, nests of "paper".

Ropalidia (Icariola) kurandae Richards, 1978

Ropalidia (Icariola) kurandae Richards, O.W. (1978). The Australian social wasps (Hymenoptera : Vespidae). *Aust. J. Zool. Suppl. Ser.* **61**: 1–132 [106]. Type data: holotype, ANIC F. adult, from Ellis Beach N of Cairns, Qld.

Distribution: NE coastal, N Gulf, Qld.; one N.S.W. record possibly a stray. Ecology: larva - sedentary, predator : adult - volant; social, nests of "paper".

Ropalidia (Icariola) mackayensis Richards, 1978

Ropalidia (Icariola) mackayensis Richards, O.W. (1978). The Australian social wasps (Hymenoptera : Vespidae). *Aust. J. Zool. Suppl. Ser.* **61**: 1–132 [105]. Type data: holotype, ANIC F. adult, from 35 mi SE Ayr, Qld.

Distribution: NE coastal, N Gulf, Qld. Ecology: larva - sedentary, predator : adult - volant; social, nests of "paper".

Ropalidia (Icariola) marginata (Lepeletier, 1836)

Taxonomic decision of van der Vecht, J. (1941). The Indo-Australian species of the genus *Ropalidia* (=*Icaria*) (Hym., Vespidae) (First Part). *Treubia* **18**: 103–190 [120] (including non-Australian subspecies).

Ropalidia (Icariola) marginata jucunda (Cameron, 1898)

Icaria jucunda Cameron, P. (1898). Hymenoptera orientalia, or contributions to a knowledge of the Hymenoptera of the oriental zoological region. Part VII. *Mem. Proc. Manchr. Lit. Phil. Soc.* **42**(11): 1–84 pl iv [46]. Type data: holotype, OUM *F. adult, from New Guinea.

Distribution: NE coastal, N Gulf, Qld.; also in New Guinea, New Britain, and the Philippines. Ecology: larva - sedentary, predator : adult - volant; social, nests of "paper". Biological references: Richards, O.W. (1978). The Australian social wasps (Hymenoptera : Vespidae). *Aust. J. Zool. Suppl. Ser.* **61**: 1–132 (redescription); Crosskey, R.W. (1973). A conspectus of the Tachinidae (Diptera) of Australia, including keys to the supraspecific taxa and taxonomic and host catalogues. *Bull. Br. Mus. Nat. Hist. (Entomol. Suppl.)* **21**: 1–221 (dipterous parasite).

Ropalidia (Icariola) mutabilis Richards, 1978

Taxonomic decision of Richards, O.W. (1978). The Australian social wasps (Hymenoptera : Vespidae). *Aust. J. Zool. Suppl. Ser.* **61**: 1–132.

Ropalidia (Icariola) mutabilis mutabilis Richards, 1978

Ropalidia (Icariola) mutabilis mutabilis Richards, O.W. (1978). The Australian social wasps (Hymenoptera : Vespidae). *Aust. J. Zool. Suppl. Ser.* **61**: 1–132 [94]. Type data: holotype, ANIC *F. adult, from 2 mi ENE Victoria River Downs Homestead, N.T.

Distribution: N Gulf, N coastal, N.T., W.A. Ecology: larva - sedentary, predator : adult - volant; social, nests of "paper".

Ropalidia (Icariola) mutabilis torresiana Richards, 1978

Ropalidia (Icariola) mutabilis torresiana Richards, O.W. (1978). The Australian social wasps (Hymenoptera : Vespidae). *Aust. J. Zool. Suppl. Ser.* **61**: 1–132 [96]. Type data: holotype, ANIC *F. adult, from Iron Range, Qld.

Distribution: NE coastal, N Gulf, Qld.; only published localities Iron Range, Somerset and Prince of Wales Is. Ecology: larva - sedentary, predator : adult - volant; social, nests of "paper".

Ropalidia (Icariola) plebiana Richards, 1978

Icaria plebeja Saussure, H. de (1863). Mélanges hyménoptérologiques. *Mém. Soc. Phys. Hist. Nat. Genève* **17**: 171–244 [235] [*non Icaria plebeja* Saussure, 1862, described from Celebes; no new name proposed by O.W. Richards as he believed this species to be the same as *Ropalidia (Icariola) plebiana* Richards, 1978]. Type data: syntypes (probable), adult whereabouts unknown, from Australia (as New Holland).

Ropalidia (Icariola) plebiana Richards, O.W. (1978). The Australian social wasps (Hymenoptera : Vespidae). *Aust. J. Zool. Suppl. Ser.* **61**: 1–132 [75]. Type data: holotype, ANIC *F. adult, from Nelligen, N.S.W.

Taxonomic decision of Richards, O.W. (1978). The Australian social wasps (Hymenoptera : Vespidae). *Aust. J. Zool. Suppl. Ser.* **61**: 1–132 [75].

Distribution: NE coastal, SE coastal, Murray-Darling basin, Qld., N.S.W., A.C.T., Vic. Ecology: larva - sedentary, predator : adult - volant; social, nests of "paper". Biological references: Hook, A. & Evans, H.E. (1982). Observations on the nesting behaviour of three species of *Ropalidia* Guérin-Méneville (Hymenoptera : Vespidae). *J. Aust. Entomol. Soc.* **21**: 271–275 (biology).

Ropalidia (Icariola) proletaria Richards, 1978

Ropalidia (Icariola) proletaria Richards, O.W. (1978). The Australian social wasps (Hymenoptera : Vespidae). *Aust. J. Zool. Suppl. Ser.* **61**: 1–132 [96]. Type data: holotype, BMNH *F. adult, from N Qld.

Distribution: (NE coastal), (N coastal), (N Gulf), Qld., (N.T.); only published localities North Queensland, Queensland and ?Port Darwin. Ecology: larva - sedentary, predator : adult - volant; social, nests of "paper".

Ropalidia (Icariola) revolutionalis (Saussure, 1853)

Icaria revolutionalis Saussure, H. de (1853). *Études sur la Famille des Vespides.* 2. Monographie des Guêpes Sociales, ou de la Tribu des Vespiens, ouvrage faisant suite à la Monographie des Guêpes Solitaires. Paris : Masson cxcix 256 pp. pls i–xxxvii (1853–1858) [29 pl v fig 7]. Type data: holotype, MNHP *F. adult, from Australia ("La Nouvelle Hollande ou la Tasmanie", Australie" on label).

Distribution: NE coastal, Murray-Darling basin, Qld.; possibly also New Britain. Ecology: larva - sedentary, predator : adult - volant; social, nests of "paper". Biological references: Richards, O.W. (1978). The Australian social wasps (Hymenoptera : Vespidae). *Aust. J. Zool. Suppl. Ser.* **61**: 1–132 (redescription, nest); Owen, M.D. (1979). Chemical components in the venoms of *Ropalidia revolutionalis* and *Polistes humilis* (Hymenoptera, Vespidae). *Toxicon* **17**: 519–523 (venom analysis - species identification is doubtful, locality outside normal range); Hook, A. & Evans, H.E. (1982). Observations on the nesting behaviour of three species of *Ropalidia* Guérin-Méneville (Hymenoptera : Vespidae). *J. Aust. Entomol. Soc.* **21**: 271–275 (biology).

Ropalidia (Icariola) socialistica (Saussure, 1853)

Icaria socialistica Saussure, H. de (1853). *Études sur la Famille des Vespides.* 2. Monographie des Guêpes Sociales, ou de la Tribu des Vespiens, ouvrage faisant suite à la Monographie des Guêpes Solitaires. Paris : Masson cxcix 256 pp. pls i–xxxvii (1853–1858) [27 pl iv fig 6]. Type data: lectotype, MNHP *F. adult, from "Tasmanie" (probably erroneous), designation by Richards, O.W. (1978). The Australian social wasps (Hymenoptera : Vespidae). *Aust. J. Zool. Suppl. Ser.* **61**: 1–132.

Distribution: NE coastal, N Gulf, Murray-Darling basin, SE coastal, (N coastal), Qld., N.S.W., (N.T.); N.T. specimens may be incorrectly labelled. Ecology: larva - sedentary, predator : adult - volant; social, nests of "paper". Biological references: Richards, O.W. (1978). The Australian social wasps (Hymenoptera : Vespidae). *Aust. J. Zool. Suppl. Ser.* **61**: 1–132 (redescription); Hook, A. & Evans, H.E. (1982). Observations on the nesting behaviour of three species of *Ropalidia* Guérin-Méneville (Hymenoptera : Vespidae). *J. Aust. Entomol. Soc.* **21**: 271–275 (biology).

Ropalidia (Icariola) trichophthalma Richards, 1978

Ropalidia (Icariola) trichophthalma Richards, O.W. (1978). The Australian social wasps (Hymenoptera : Vespidae). *Aust. J. Zool. Suppl. Ser.* **61**: 1–132 [102]. Type data: holotype, ANIC F. adult, from Forest Road near Ingham, Qld.

Distribution: NE coastal, N Gulf, (N coastal), Qld., (N.T.), (W.A.); specimens other than from Qld. believed to be incorrectly labelled. Ecology: larva - sedentary, predator : adult - volant; social, nests of "paper".

Ropalidia (Icariola) turneri Richards, 1978

Ropalidia (Icariola) turneri Richards, O.W. (1978). The Australian social wasps (Hymenoptera : Vespidae). *Aust. J. Zool. Suppl. Ser.* **61**: 1–132 [73]. Type data: holotype, ANIC F. adult, from Kuranda, Qld.

Distribution: NE coastal, Qld. Ecology: larva - sedentary, predator : adult - volant; social, nests of "paper".

Ropalidia (Icarielia) Dalla Torre, 1904

Icarielia Dalla Torre, K.W. von (1904). Hymenoptera. Fam. Vespidae. *Genera Insectorum* **19**: 1–108 [72] [proposed with subgeneric rank in *Icaria* Saussure, 1853]. Type species *Icaria flavopicta* Smith, 1857 by subsequent designation, see Meade-Waldo, G. (1913). New species of Diploptera in the collection of the British Museum. Part IV. *Ann. Mag. Nat. Hist. (8)* **11**: 44–54.

This group is also found in the Oriental Region, see Richards, O.W. (1978). The Australian social wasps (Hymenoptera : Vespidae). *Aust. J. Zool. Suppl. Ser.* **61**: 1–132.

Ropalidia (Icarielia) nigrior Richards, 1978

Ropalidia (Icarielia) nigrior Richards, O.W. (1978). The Australian social wasps (Hymenoptera : Vespidae). *Aust. J. Zool. Suppl. Ser.* **61**: 1–132 [113]. Type data: holotype, ANIC F. adult, from Iron Range, Qld.

Distribution: NE coastal, N Gulf, Qld.; possibly also in New Guinea. Ecology: larva - sedentary, predator : adult - volant; social, nests of "paper".

Ropalidia (Icarielia) romandi (Le Guillou, 1841)

Taxonomic decision of Richards, O.W. (1978). The Australian social wasps (Hymenoptera : Vespidae). *Aust. J. Zool. Suppl. Ser.* **61**: 1–132 [109].

Ropalidia (Icarielia) romandi romandi (Le Guillou, 1841)

Polistes romandi Le Guillou, E.J.F. (1841). Description de 20 espèces nouvelles appartenant à diverses familles d'Hyménoptères. *Rev. Zool.* **4**: 322–325 [322]. Type data: holotype, MNHP *F. adult, from N Australia (as "Aust. sept.").

Distribution: N coastal, N.T. Ecology: larva - sedentary, predator : adult - volant; social, nests of "paper".

Ropalidia (Icarielia) romandi cabeti (Saussure, 1853)

Icaria cabeti Saussure, H. de (1853). *Études sur la Famille des Vespides*. 2. Monographie des Guêpes Sociales, ou de la Tribu des Vespiens, ouvrage faisant suite à la Monographie des Guêpes Solitaires. Paris : Masson cxcix 256 pp. pls i–xxxvii (1853–1858) [26 pl iv fig 2, pl v fig 2]. Type data: lectotype, MNHP *F. adult, from Tas. (doubtful), designation by Richards, O.W. (1978). The Australian social wasps (Hymenoptera : Vespidae). *Aust. J. Zool. Suppl. Ser.* **61**: 1–132.

Distribution: NE coastal, N Gulf, SE coastal, Qld., N.S.W.; possibly the N.S.W. specimens are strays, but common in Qld. Ecology: larva - sedentary, predator : adult - volant; social, nests of "paper".

SPHECIDAE

INTRODUCTION

This cosmopolitan family contains very small to very large solitary wasps. Australia has about 600 described species and subspecies in over 50 genera of which about a quarter are endemic. Adults are often collected on flowers or at nesting sites. Nests are made by burrowing in the ground, by using existing cavities in the ground, in dead wood or the pith of plants, by constructing mud cells in the open, on house walls or rocks or tree trunks, and by using abandoned mud nests. One genus (*Acanthostethus*) is cleptoparasitic. Adults of other genera provision their cells with insects - there are records from almost all the orders - or spiders or Collembola. Most genera exhibit some degree of prey specificity. *Bembix* is unusual in this respect, for while most northern hemisphere species studied prey on Diptera, about one third of the Australian species whose prey is known use other orders (Hymenoptera, Odonata and Neuroptera) and two species have been found to prey on more than one order of insects. Recent work on the biology of *Arpactophilus* sp., *Spilomena* sp. and *Pison* sp., not mentioned as a biological reference because the species were not identified, was published by Naumann (1983) and on *Lyroda* sp. by Evans & Hook (1984).

References

Bohart, R.M. & Menke, A.S. (1976). *Sphecid Wasps of the World : a Generic Revision.* Berkeley : Univ. California Press ix 695 pp.

Evans, H.E. & Hook, A.W. (1984). Nesting behaviour of a *Lyroda* predator (Hymenoptera : Sphecidae) on *Tridactylus* (Orthoptera : Tridactylidae). *Aust. Entomol. Mag.* **11**: 16–18

Evans, H.E. & West Eberhard, M.J. (1970). *The Wasps.* Ann Arbor : Univ. Michigan Press 265 pp.

Naumann, I.D. (1983). The biology of mud nesting Hymenoptera (and their associates) and Isoptera in rock shelters of the Kakadu Region, Northern Territory. *Aust. Natl. Parks & Wldlf. Serv. Spec. Publ.* 10 pp. 127–189

Dolichurus Latreille, 1809

Dolichurus Latreille, P.A. (1809). *Genera Crustaceorum et Insectorum* secundem ordinem naturalem in familias disposita, iconibus exemplisque plurimis explicata. Paris : A. Koenig Vol. 4 397 pp. [387]. Type species *Pompilus corniculus* Spinola, 1808 by subsequent designation, see Latreille, P.A. (1810). *Considérations Générales sur l'Ordre Naturel des Animaux Composant les Classes des Crustacés, des Arachnides, et des Insectes;* avec un Tableau Méthodique de leurs Genres, Disposés en Familles. Paris : F. Schoell 444 pp. Compiled from secondary source: Bohart, R.M. & Menke, A.S. (1976). *Sphecid Wasps of the World : a Generic Revision.* Berkeley : Univ. California Press ix 695 pp.

This group is found worldwide, see Bohart, R.M. & Menke, A.S. (1976). *Sphecid Wasps of the World : a Generic Revision.* Berkeley : Univ. California Press ix 695 pp. [66].

Dolichurus carbonarius Smith, 1869

Dolichurus carbonarius Smith, F. (1869). Descriptions of new genera and species of exotic Hymenoptera. *Trans. Entomol. Soc. Lond.* **1869**: 301–311 [303]. Type data: holotype, BMNH *F. adult (seen 1929 by L.F. Graham), from Champion Bay, W.A.

Distribution: NE coastal, NW coastal, Qld., W.A.; only published localities Mackay, Kuranda, Dunk Is., Brisbane and Champion Bay. Ecology: larva - sedentary, predator : adult - volant; prey Blattodea, nest in pre-existing cavity. Biological references: Turner, R.E. (1915). Notes on fossorial Hymenoptera. XV. New Australian Crabronidae. *Ann. Mag. Nat. Hist. (8)* **15**: 62–96 (behaviour); Riek, E.F. (1955). Australian Ampulicidae (Hymenoptera : Sphecoidea). *Aust. J. Zool.* **3**: 131–144 (redescription).

Aphelotoma Westwood, 1841

Aphelotoma Westwood, J.O. (1841). *in,* Proceedings of the Entomological Society of London. (Descriptions of the following exotic hymenopterous insects belonging to the family Sphegidae). *Ann. Mag. Nat. Hist. (1)* **7**: 151–152 [152]. Type species *Aphelotoma tasmanica* Westwood, 1841 by monotypy.

Aphelotoma affinis Turner, 1910

Aphelotoma affinis Turner, R.E. (1910). Additions to our knowledge of the fossorial wasps of Australia. *Proc. Zool. Soc. Lond.* **1910**: 253–356 [341]. Type data: holotype, BMNH *F. adult (seen 1929 by L.F. Graham), from Townsville, Qld.

Distribution: NE coastal, Qld.; type locality only. Ecology: larva - sedentary, predator : adult - volant; prey Blattodea.

Aphelotoma auricula Riek, 1955

Aphelotoma auricula Riek, E.F. (1955). Australian Ampulicidae (Hymenoptera : Sphecoidea). *Aust. J. Zool.* **3**: 131–144 [139 pl 1 fig 8]. Type data: holotype, ANIC M. adult, from 10 mi S of Bowen, Qld.

Distribution: NE coastal, Qld.; only published localities near Bowen and Caloundra. Ecology: larva - sedentary, predator : adult - volant; prey Blattodea.

Aphelotoma fuscata Riek, 1955

Aphelotoma fuscata Riek, E.F. (1955). Australian Ampulicidae (Hymenoptera : Sphecoidea). *Aust. J. Zool.* **3**: 131–144 [139 pl 1 fig 7]. Type data: holotype, ANIC F. adult, from Catherine Hill, N.S.W.

Distribution: SE coastal, N.S.W.; type locality only. Ecology: larva - sedentary, predator : adult - volant; prey Blattodea.

Aphelotoma melanogaster Riek, 1955

Aphelotoma melanogaster Riek, E.F. (1955). Australian Ampulicidae (Hymenoptera : Sphecoidea). *Aust. J. Zool.* **3**: 131–144 [135 pl 1 figs 2–3]. Type data: holotype, ANIC M. adult, from Blundells, A.C.T.

Distribution: NE coastal, SE coastal, Murray-Darling basin, Qld., N.S.W., A.C.T. Ecology: larva - sedentary, predator : adult - volant; prey Blattodea.

Aphelotoma nigricula Riek, 1955

Aphelotoma nigricula Riek, E.F. (1955). Australian Ampulicidae (Hymenoptera : Sphecoidea). *Aust. J. Zool.* **3**: 131–144 [138 pl 1 fig 10]. Type data: holotype, ANIC M. adult, from Blundells, A.C.T.

Distribution: NE coastal, Murray-Darling basin, SE coastal, Qld., N.S.W., A.C.T.; only published localities Stanthorpe, Barrington Tops, Goulburn and Blundells. Ecology: larva - sedentary, predator : adult - volant; prey Blattodea.

Aphelotoma rufiventris Turner, 1914

Aphelotoma rufiventris Turner, R.E. (1914). New fossorial Hymenoptera from Australia and Tasmania. *Proc. Linn. Soc. N.S.W.* **38**: 608–623 [618]. Type data: holotype, BMNH *M. adult (seen 1929 by L.F. Graham), from Kuranda, Qld.

Distribution: NE coastal, Qld.; only published localities Kuranda, Bowen, Stradbroke Is., Caloundra and Stanthorpe. Ecology: larva - sedentary, predator : adult - volant; prey Blattodea. Biological references: Riek, E.F. (1955). Australian Ampulicidae (Hymenoptera : Sphecoidea). *Aust. J. Zool.* **3**: 131–144 (redescription).

Aphelotoma striaticollis Turner, 1910

Aphelotoma striaticollis Turner, R.E. (1910). Additions to our knowledge of the fossorial wasps of Australia. *Proc. Zool. Soc. Lond.* **1910**: 253–356 [341]. Type data: holotype, BMNH *F. adult (seen 1929 by L.F. Graham), from Townsville, Qld.

Distribution: NE coastal, Qld.; only published localities Townsville, Kuranda. Ecology: larva - sedentary, predator : adult - volant; prey Blattodea.

Aphelotoma tasmanica Westwood, 1841

Taxonomic decision of Riek, E.F. (1955). Australian Ampulicidae (Hymenoptera : Sphecoidea). *Aust. J. Zool.* **3**: 131–144 [136–137].

Aphelotoma tasmanica tasmanica Westwood, 1841

Aphelotoma tasmanica Westwood, J.O. (1841). *in* Proceedings of the Entomological Society of London. (Descriptions of the following exotic hymenopterous insects belonging to the family Sphegidae). *Ann. Mag. Nat. Hist. (1)* **7**: 151–152 [152]. Type data: syntypes (probable), OUM or BMNH *F. adult, from Tas.

Distribution: SE coastal, Vic., Tas. Ecology: larva - sedentary, predator : adult - volant; prey Blattodea.

Aphelotoma tasmanica auriventris Turner, 1907

Aphelotoma auriventris Turner, R.E. (1907). New species of Sphegidae from Australia. *Ann. Mag. Nat. Hist. (7)* **19**: 268–276 [269]. Type data: holotype, BMNH *M. adult, from Vic.

Distribution: Murray-Darling basin, SE coastal, S Gulfs, SW coastal, A.C.T., Vic., S.A., W.A. Ecology: larva - sedentary, predator : adult - volant; prey Blattodea.

Austrotoma Riek, 1955

Austrotoma Riek, E.F. (1955). Australian Ampulicidae (Hymenoptera : Sphecoidea). *Aust. J. Zool.* **3**: 131-144 [141]. Type species *Aphelotoma aterrima* Turner, 1907 by monotypy and original designation.

Austrotoma aterrima (Turner, 1907)

Aphelotoma aterrima Turner, R.E. (1907). New species of Sphegidae from Australia. *Ann. Mag. Nat. Hist. (7)* **19**: 268-276 [268]. Type data: holotype, BMNH *M. adult (seen 1929 by L.F. Graham), from Mackay, Qld.

Distribution: NE coastal, Qld.; only published localities Kuranda and Mackay. Ecology: larva - sedentary, predator: adult - volant. Biological references: Riek, E.F. (1955). Australian Ampulicidae (Hymenoptera : Sphecoidea). *Aust. J. Zool.* **3**: 131-144 (redescription, generic placement).

Ampulex Jurine, 1807

Ampulex Jurine, L. (1807). *Nouvelle Méthode de Classer les Hyménoptères et les Diptères.* Hyménoptères. Genève : J.J. Paschoud Vol. 1 319+4 pp. [132]. Type species *Ampulex fasciata* Jurine, 1807 by monotypy.

This group is widespread, mostly in the Ethiopian, Oriental, and Neotropical Regions, see Bohart, R.M. & Menke, A.S. (1976). *Sphecid Wasps of the World : a Generic Revision.* Berkeley : Univ. California Press ix 695 pp. [74]. Species now known not to occur in Australia: *Ampulex micans* Kohl, 1893 originally attributed to Australia or Mexico" vide Bohart, R.M. & Menke, A.S. (1976). *Sphecid Wasps of the World : a Generic Revision.* Berkeley : Univ. California Press ix 695 pp. [78].

Ampulex compressa (Fabricius, 1781)

Sphex compressa Fabricius, J.C. (1781). *Species insectorum exhibentes eorum differentias specificas, synonyma auctorum, loca natalia, metamorphosin adiectis observationibus, descriptionibus.* Hamburgi et Kilonii Vols i-ii [445]. Type data: holotype, BMNH *F. adult, from Malabar.

Distribution: NE coastal, Qld.; also in the Oriental and Ethiopian Regions and Venezuala. Ecology: larva - sedentary, predator : adult - volant; nest in stems or crevices, prey Blattodea. Biological references: Smith, F. (1856). *Catalogue of Hymenopterous Insects in the Collection of the British Museum.* Part IV, Sphegidae, Larridae and Crabronidae. pp. 207-497 London : British Museum (generic placement); Krombein, K.V. (1979). Biosystematic studies of Ceylonese wasps, V : A monograph of the Ampulicidae (Hymenoptera : Sphecoidea). *Smithson. Contrib. Zool.* **298**: 1-29 (redescription, biology); Menke, A.S. & Yustiz, E. (1983). *Ampulex compressa* (F.) in Venezuala (Hymenoptera : Sphecidae). *Proc. Entomol. Soc. Wash.* **85**: 180.

Chalybion Dahlbom, 1843

Chalybion Dahlbom, A.G. (1843). *Hymenoptera Europaea Praecipue Borealia;* formis typicis nonnullis specierum generumve exoticorum aut extraneorum propter nexum systematicum associatis; per familias, genera, species et varietates disposita atque descripta. I. *Sphex* in sensu Linneano. Lund : Lundberg, Berolini Vol. 1 528 pp. (1843-1845) [21]. Type species *Sphex caeruleus* Linnaeus, 1767 [*non Sphex caeruleus* Linnaeus, 1758 (= *Pelopeus californicus* Saussure, 1867)] by subsequent designation, see Patton, W.H. (1881). Some characters useful in the study of the Sphecidae. *Proc. Bost. Soc. Nat. Hist.* **20**: 378-385. Compiled from secondary source: Bohart, R.M. & Menke, A.S. (1976). *Sphecid Wasps of the World : a Generic Revision.* Berkeley : Univ. California Press ix 695 pp.

This group is mostly Old World, two species in North and Central America, see Bohart, R.M. & Menke, A.S. (1976). *Sphecid Wasps of the World : a Generic Revision.* Berkeley : Univ. California Press ix 695 pp. [98].

Chalybion bengalense (Dahlbom, 1845)

Pelopoeus bengalensis Dahlbom, A.G. (1845). *Hymenoptera Europaea Praecipue Borealia;* formis tipicis nonnullis specierum generumve exoticorum aut extraneorum propter nexum systematicum associatis; per familias, genera, species et varietates disposita atque descripta. I. *Sphex* in sensu Linneano. Lund : Lundberg, Berolini Vol. 1 528 pp. (1843-1845) [433]. Type data: lectotype, UZM *F. adult, from India, designation by van der Vecht, J. (1961). Hymenoptera Sphecoidea Fabriciana. *Zool. Verh. Leiden* **48**: 1-85.

Sphex violacea Fabricius, J.C. (1775). *Systema Entomologiae,* sistens insectorum classes, ordines, genera, species, adiectis synonymis, locis, descriptionibus, observationibus. Flensburgi et Lipsiae : Kortii xxvii 832 pp. [346]. Type data: lectotype, UZM *adult, from Cape of Good Hope, designation by van der Vecht, J. (1961). Hymenoptera Sphecoidea Fabriciana. *Zool. Verh. Leiden* **48**: 1-85.

Taxonomic decision of Kohl, F.F. (1918). Die Hautflüglergruppe "Sphecinae". iv. Teil. Die natürliche Gattung *Sceliphron* Klug (*Pelopoeus* Kirby). *Ann. Naturh. Hofmus. Wien* **32**: 1-171 [54].

Distribution: NE coastal, Qld.; Cape York, also the Ethiopian, Oriental, and Palearctic Regions. Ecology: larva - sedentary, arboreal, predator : adult - volant; prey spiders, nest in pre-existing cavity. Biological references: Williams, F.X. (1928). The natural history of a Philippine nipa house with descriptions of new wasps. *Philipp. J. Sci.* **35**: 53-118 (biology: as *violacea*).

Sceliphron Klug, 1801

Taxonomic decision of Bohart, R. M. & Menke, A. S. (1976). *Sphecid Wasps of the World : a Generic Revision.* Berkeley : Univ. California Press ix 695 pp. [39].

Sceliphron (Sceliphron) Klug, 1801

Sceliphron Klug, J.C.F. (1801). Absonderung einiger Raupentödter und Vereinigung derselben zu einer neuen Gattung *Sceliphron*. Neue Schrift. *Ges. Naturf. Freunde Berl.* **3**: 555–566 [561]. Type species *Sphex spirifex* Linnaeus, 1758 by subsequent designation, see Bingham, C.T. (1897). *The Fauna of British India, Including Ceylon and Burma.* Hymenoptera. Vol. I Wasps and Bees. London : Taylor & Francis xxx 579 pp. [235]. Compiled from secondary source: Bohart, R.M. & Menke, A.S. (1976). *Sphecid Wasps of the World : a Generic Revision.* Berkeley : Univ. California Press ix 695 pp. [39].

This group is found worldwide, see Bohart, R.M. & Menke, A.S. (1976). *Sphecid Wasps of the World : a Generic Revision.* Berkeley : Univ. California Press ix 695 pp. [103].

Sceliphron (Sceliphron) caementarium (Drury, 1770)

Sphex caementarium Drury, D. (1770). *Illustrations of Natural History.* London : White Vol. I 30+130 pp. pls 51 [105 pl 44 fig 6]. Type data: syntypes (probable), lost, from Jamaica.

Distribution: NE coastal, (Lake Eyre basin), Qld.; accidentally established in Brisbane, one specimen found in crate from North America in Alice Springs, also found in Europe, the Americas and many Pacific islands. Ecology: larva - sedentary, arboreal, predator : adult - volant; prey spiders, nest built of mud in sheltered place. Biological references: Bohart, R.M. & Menke, A.S. (1976). *Sphecid Wasps of the World : a Generic Revision.* Berkeley : Univ. California Press ix 695 pp. (list of synonyms, nest, biology); Evans, H.E. & Lin, C.S. (1956). Studies on the larvae of digger wasps (Hymenoptera, Sphecidae) Part I Sphecinae. *Trans. Am. Entomol. Soc.* **81**: 131–153 8 pls (larva); van der Vecht, J. & van Breugel, F.M.A. (1968). Revision of the nominate subgenus *Sceliphron* Latreille (Hymenoptera, Sphecidae) (Studies on the Sceliphronini, Part I). *Tijdschr. Entomol.* **111**: 185–255 (redescription); Naumann, I.D. (1983). The biology of mud nesting Hymenoptera (and their associates) and Isoptera in rock shelters of the Kakadu Region, Northern Territory. *Aust. Nat. Parks & Wildlife Service Spec. Publ.* **10** pp. 127–189 (Alice Springs record).

Sceliphron (Sceliphron) laetum (Smith, 1856)

Taxonomic decision of van der Vecht, J. & van Breugel, F.M.A. (1968). Revision of the nominate subgenus *Sceliphron* Latreille (Hymenoptera, Sphecidae) (studies on the Sceliphronini, Part I). *Tijdschr. Entomol.* **111**: 185–255 [250].

Sceliphron (Sceliphron) laetum laetum (Smith, 1856)

Pelopoeus laetus Smith, F. (1856). *Catalogue of Hymenopterous Insects in the Collection of the British Museum.* Part IV, Sphegidae, Larridae and Crabronidae. pp. 207–497 London : British Museum [229 pl 7 fig 1]. Type data: lectotype, BMNH *F. adult, from MacIntyre River, Australia, designation by van der Vecht, J. & van Breugel, F.M.A. (1968). Revision of the nominate subgenus *Sceliphron* Latreille (Hymenoptera, Sphecidae) (Studies on the Sceliphronini, Part I). *Tijdschr. Entomol.* **111**: 185–255 [234].

Sceliphron laetum cygnorum Turner, R.E. (1910). Additions to our knowledge of the fossorial wasps of Australia. *Proc. Zool. Soc. Lond.* **1910**: 253–356 [343]. Type data: holotype, BMNH *F. adult (seen 1929 by L.F. Graham), from SW Australia.

Distribution: NE coastal, Murray-Darling basin, SE coastal, Lake Eyre basin, S Gulfs, W plateau, SW coastal, N coastal, Qld., N.S.W., Vic., S.A., W.A., N.T.; also New Guinea, Indonesia and Pacific islands. Ecology: larva - sedentary, arboreal, predator : adult - volant; prey spiders, nest built of mud in sheltered place. Biological references: Maindron, M. (1878). Notes pour servir à l'histoire des Hyménoptères de l'Archipel Indien et de la Nouvelle-Guinée, i. Observations sur quelques Sphégiens (g. *Pelopaeus*) de l'Archipel Indien. (Métamorphoses. - Descriptions d'espèces). *Ann. Soc. Entomol. Fr. (5)* **8**: 385–398 (larva); Whittell, H.R. (1884). On some habits of *Pelopoeus laetus* and a species of *Larrada*. *Proc. Linn. Soc. N.S.W.* **8**: 29–33 (biology); Roth, H.L. (1885). Notes on the habits of some Australian Hymenoptera Aculeata. (With description of a new species by W.F. Kirby). *J. Linn. Soc. Zool.* **18**: 318–328 (biology); McCarthy, T. (1917). Some observations on solitary wasps at Hay, N.S.W.. *Aust. Nat.* **3**: 195–200 (biology); Brewster, M.N., Brewster, A.A. & Crouch, N. (1920). *Life Stories of Australian Insects.* Sydney : Dymock's Book Arcade 424 pp. (biology); Smith, A. (1974). Mud wasps. A study of their nests reveals an interesting series of builders and renters. *Wildlife, Lond.* **16**(6) : 300–301 (biology); Smith, A. (1979). Life strategy and mortality factors of *Sceliphron laetum* (Smith) (Hymenoptera : Sphecidae) in Australia. *Aust. J. Ecol.* **4**: 181–186; Bohart, R.M. & Menke, A.S. (1976). *Sphecid Wasps of the World : a Generic Revision.* Berkeley : Univ. California Press ix 695 pp.; Naumann, I.D. (1983). The biology of mud nesting Hymenoptera (and their associates) and Isoptera in rock shelters of the

Kakadu Region, Northern Territory. *Aust. Nat. Parks & Wildlife Service Spec. Publ. 10* pp. 127–189 (biology, parasites).

Sceliphron (Prosceliphron) van der Vecht, 1968

Prosceliphron van der Vecht, J. & van Breugel, F.M.A. (1968). Revision of the nominate subgenus *Sceliphron* Latreille (Hymenoptera, Sphecidae) (Studies on the Sceliphronini, Part I). *Tijdschr. Entomol.* **111**: 185–255 [192] [described with subgeneric rank in *Sceliphron* Klug, 1801]. Type species *Pelopaeus coromandelicus* Lepeletier, 1845 by original designation.

This group is also found in the Palearctic and Oriental Regions, see Bohart, R.M. & Menke, A.S. (1976). *Sphecid Wasps of the World : a Generic Revision.* Berkeley : Univ. California Press ix 695 pp. [103].

Sceliphron (Prosceliphron) formosum (Smith, 1856)

Pelopoeus formosus Smith, F. (1856). *Catalogue of Hymenopterous Insects in the Collection of the British Museum*. Part IV, Sphegidae, Larridae and Crabronidae. pp. 207–497 London : British Museum [230]. Type data: holotype, BMNH *F. adult (seen 1929 by L.F. Graham), from Australia.

Distribution: NE coastal, SE coastal, SW coastal, N coastal, Qld., N.S.W., W.A., N.T.; also New Guinea and Indonesia. Ecology: larva - sedentary, arboreal, predator : adult - volant; prey spiders, nest built of mud in sheltered place. Biological references: Bohart, R.M. & Menke, A.S. (1976). *Sphecid Wasps of the World : a Generic Revision.* Berkeley : Univ. California Press ix 695 pp. (generic placement); Callan, E.M. (1981). Further records of *Macrosiagon* (Coleoptera : Rhipiphoridae) reared from eumenid and sphecid wasps in Australia. *Aust. Entomol. Mag.* **7**: 81–83 (parasite); Naumann, I.D. (1983). The biology of mud nesting Hymenoptera (and their associates) and Isoptera in rock shelters of the Kakadu Region, Northern Territory. *Aust. Nat. Parks & Wildlife Service Spec. Publ. 10* pp. 127–189.

Sphex Linnaeus, 1758

Taxonomic decision of Bohart, R. M. & Menke, A. S. (1976). *Sphecid Wasps of the World : a Generic Revision.* Berkeley : Univ. California Press ix 695 pp. [39].

Sphex (Sphex) Linnaeus, 1758

Sphex Linnaeus, C. von (1758). *Systema Naturae per Regna Tria Naturae, secundum classes, ordines, genera, species, cum characteribus, differentiis, synonymis, locis.* Holmiae : Laurentii Salvii 10th edn. Vol.1 823 pp. [569]. Type species *Pepsis flavipennis* Fabricius, 1793 by subsequent designation, see International Commission on Zoological Nomenclature (1946). Opinion 180. On the status of the names *Sphex* Linnaeus, 1758, and *Ammophila* Kirby, 1798 (Class Insecta, Order Hymenoptera). *Opin. Decl. Int. Comm. Zool. Nomen.* **2**: 569–585 [571]. Compiled from secondary source: Bohart, R.M. & Menke, A.S. (1976). *Sphecid Wasps of the World : a Generic Revision.* Berkeley : Univ. California Press ix 695 pp. [39].

This group is found worldwide, see Bohart, R.M. & Menke, A.S. (1976). *Sphecid Wasps of the World : a Generic Revision.* Berkeley : Univ. California Press ix 695 pp. [109]. Species now known not to occur in Australia: *Sphex clavigera* Smith, 1856 a Holarctic species *vide* van der Vecht, J. (1973). Contribution to the taxonomy of the Oriental and Australian Sphecini (Hymenoptera, Sphecoidea). *Proc. K. Ned. Akad. Wet. (c)* **76**: 341–353 [351] [as *Isodontia clavigera* (Smith)]; *Sphex princeps* Kohl, 1890 now in synonymy with a North American species *vide* Bohart, R.M. & Menke, A.S. (1976). *Sphecid Wasps of the World : a Generic Revision.* Berkeley : Univ. California Press ix 695 pp. [115].

Sphex (Sphex) ahasverus Kohl, 1890

Sphex (Sphex) ahasverus Kohl, F.F. (1890). Die Hymenopterengruppe der Sphecinen. I Monographie der natürlichen Gattung *Sphex* Linné (*sens.lat.*). Abt. I–II *Ann. Naturh. Hofmus. Wien* **5**: 77–194, 317–462 [397]. Type data: syntypes (probable), whereabouts unknown, from S.A.

Distribution: S.A.; type locality only as S.A. Ecology: larva - sedentary, soil, predator : adult - volant, burrowing; prey Orthoptera.

Sphex (Sphex) argentatus Fabricius, 1787

Sphex argentatus Fabricius, J.C. (1787). *Mantissa insectorum sistens eorum species nuper detectas adiectis characteribus genericus, differentiis specificis, emendationibus, observationibus.* Hafniae vols i–ii [274]. Type data: lectotype, UZM *F. adult, from Coromandel, designation by van der Vecht, J. (1961). Hymenoptera Sphecoidea Fabriciana. *Zool. Verh. Leiden* **48**: 1–85.

Sphex umbrosus Christ, J.L. (1791). *Naturgeschichte, Classification und Nomenclatur der Insecten vom Bienen, Wespen und Ameisengeschlecht.* Frankfurt-am-Main : Hermann 535 pp. col pls 60 [293]. Type data: syntypes (probable), lost, from type locality unknown.

Taxonomic decision of van der Vecht, J. (1973). Contribution to the taxonomy of the Oriental and Australian Sphecini (Hymenoptera, Sphecoidea). *Proc. K. Ned. Akad. Wet. (c)* **76**: 341–353 [341].

Distribution: NE coastal, N coastal, Qld., N.T.; also in New Guinea, Africa and Oriental Region. Ecology: larva - sedentary, soil, predator : adult - volant, burrowing; prey Orthoptera. Biological references: Evans, H.E. & Lin, C.S. (1956). Studies on the larvae of digger wasps (Hymenoptera, Sphecidae) Part I Sphecinae. *Trans. Am. Entomol. Soc.* **81**: 131–153 8 pls (larva).

Sphex (Sphex) basilicus (Turner, 1915)

Chlorion (Proterosphex) basilicus Turner, R.E. (1915). Notes on fossorial Hymenoptera. XV. New Australian Crabronidae. *Ann. Mag. Nat. Hist. (8)* **15**: 62–96 [65]. Type data: holotype, BMNH *F. adult (seen 1929 by L.F. Graham), from N Qld. (probably Cape York Peninsula).

Distribution: NE coastal, Qld.; type locality only. Ecology: larva - sedentary, soil, predator : adult - volant, burrowing; prey Orthoptera. Biological references: Bohart, R.M. & Menke, A.S. (1976). *Sphecid Wasps of the World : a Generic Revision.* Berkeley : Univ. California Press ix 695 pp. (generic placement).

Sphex (Sphex) bilobatus Kohl, 1895

Sphex canescens Smith, F. (1856). *Catalogue of Hymenopterous Insects in the Collection of the British Museum.* Part IV, Sphegidae, Larridae and Crabronidae. pp. 207–497 London : British Museum [246] [*non Sphex canescens* Dahlbom, 1843]. Type data: holotype, BMNH *M. adult (seen 1929 by L.F. Graham), from Australia.

Sphex bilobatus Kohl, F.F. (1895). Zur Monographie der natürlichen Gattung *Sphex* Linné. *Ann. Naturh. Hofmus. Wien* **10**: 42–74 [59 pl 4 figs 10, 24]. Type data: syntypes (probable), ZMB *adult, from Adelaide, S.A.

Taxonomic decision of Turner, R.E. (1910). Additions to our knowledge of the fossorial wasps of Australia. *Proc. Zool. Soc. Lond.* **1910**: 253–356 [346].

Distribution: SE coastal, Murray-Darling basin, S Gulfs, Lake Eyre basin, N.S.W., A.C.T., S.A., N.T.; only published localities Como, Cumberland, Adelaide and Crown Point. Ecology: larva - sedentary, soil, predator : adult - volant, burrowing; prey Orthoptera (Tettigoniidae). Biological references: Rayment, T. (1946). Habits of a sphegid wasp (*Sphex canescens*). *Vict. Nat.* **63**: 185 (biology); Evans, H.E., Hook, A.W. & Matthews, R.W. (1982). Nesting behaviour of Australian wasps of the genus *Sphex* (Hymenoptera, Sphecidae). *J. Nat. Hist.* **16**: 219–225 (biology).

Sphex (Sphex) carbonicolor van der Vecht, 1973

Sphex carbonaria Smith, F. (1856). *Catalogue of Hymenopterous Insects in the Collection of the British Museum.* Part IV, Sphegidae, Larridae and Crabronidae. pp. 207–497 London : British Museum [247] [*non Sphex carbonaria* Scopoli, 1763]. Type data: holotype, BMNH *F. adult, from Sydney, N.S.W.

Sphex carbonicolor van der Vecht, J. (1973). Contribution to the taxonomy of the Oriental and Australian Sphecini (Hymenoptera, Sphecoidea). *Proc. K. Ned. Akad. Wet. (c)* **76**: 341–353 [342] [*nom. nov.* for *Sphex carbonaria* Smith, 1856].

Distribution: SE coastal, Qld., N.S.W., S.A., W.A., N.T. Ecology: larva - sedentary, soil, predator : adult - volant, burrowing; prey Orthoptera.

Sphex (Sphex) cognatus Smith, 1856

Sphex cognata Smith, F. (1856). *Catalogue of Hymenopterous Insects in the Collection of the British Museum.* Part IV, Sphegidae, Larridae and Crabronidae. pp. 207–497 London : British Museum [248]. Type data: holotype, BMNH 21.692 *F. adult (seen 1929 by L.F. Graham), from Australia.

Sphex amator Smith, F. (1856). *Catalogue of Hymenopterous Insects in the Collection of the British Museum.* Part IV, Sphegidae, Larridae and Crabronidae. pp. 207–497 London : British Museum [246]. Type data: holotype, OUM *M. adult, from Australia.

Sphex opulenta Smith, F. (1856). *Catalogue of Hymenopterous Insects in the Collection of the British Museum.* Part IV, Sphegidae, Larridae and Crabronidae. pp. 207–497 London : British Museum [250]. Type data: holotype, BMNH 21.693 *M. adult (seen 1929 by L.F. Graham), from Richmond R., N.S.W.

Sphex formosa Smith, F. (1856). *Catalogue of Hymenopterous Insects in the Collection of the British Museum.* Part IV, Sphegidae, Larridae and Crabronidae. pp. 207–497 London : British Museum [254]. Type data: holotype, BMNH 21.694 *F. adult (seen 1929 by L.F. Graham), from Ceram.

Taxonomic decision of van der Vecht, J. (1973). Contribution to the taxonomy of the Oriental and Australian Sphecini (Hymenoptera, Sphecoidea). *Proc. K. Ned. Akad. Wet. (c)* **76**: 341–353 [342].

Distribution: NE coastal, SE coastal, Murray-Darling basin, S Gulfs, Qld., N.S.W., A.C.T., S.A.; also in New Guinea and Oriental Region. Ecology: larva - sedentary, soil, predator : adult - volant, burrowing; prey Orthoptera (Tettigoniidae). Biological references: Ribi, W.A. (1978). A unique hymenopteran compound eye, the retina fine structure of the digger wasp *Sphex cognatus* Smith (Hymenoptera, Sphecidae). *Zool. Jb. Anat.* **100**: 299–342 (eye structure); Ribi, W.A. & Ribi, L. (1979). Natural history of the Australian digger wasp *Sphex cognatus* Smith (Hymenoptera, Sphecidae). *J. Nat. Hist.* **13**: 693–701; Evans, H.E., Hook, A.W. & Matthews, R.W. (1982). Nesting behaviour of Australian wasps of the genus *Sphex* (Hymenoptera, Sphecidae). *J. Nat. Hist.* **16**: 219–225 (biology).

Sphex (Sphex) darwiniensis Turner, 1912

Sphex darwiniensis Turner, R.E. (1912). Notes on fossorial Hymenoptera. IX. On some new species from the Australian and Austro-Malayan regions. *Ann. Mag. Nat. Hist. (8)* **10**: 48–63 [56]. Type data: holotype, BMNH *F. adult (seen 1929 by L.F. Graham), from Port Darwin, N.T.

Distribution: N coastal, N.T.; type locality only. Ecology: larva - sedentary, soil, predator : adult - volant, burrowing; prey Orthoptera.

Sphex (Sphex) decoratus Smith, 1873

Sphex decorata Smith, F. (1873). Natural history notices. Insects, Hymenoptera Aculeata. pp. 456–463 pls xliii–xlv *in* Brenchley, J.B. *Jottings During the Cruise of H.M.S. Curaçoa among the South Sea Islands in 1865*. London : Longmans, Green & Co. [461 pl 44 fig 4]. Type data: holotype, BMNH *F. adult (seen 1929 by L.F. Graham), from NW coast of Australia.

Distribution: NE coastal, NW coastal, Qld., W.A.; only published localities N Qld. and NW coast of Australia. Ecology: larva - sedentary, soil, predator : adult - volant, burrowing; prey Orthoptera.

Sphex (Sphex) ephippium Smith, 1856

Sphex ephippium Smith, F. (1856). *Catalogue of Hymenopterous Insects in the Collection of the British Museum*. Part IV, Sphegidae, Larridae and Crabronidae. pp. 207–497 London : British Museum [249 pl 6 fig 2]. Type data: holotype, BMNH 21.706 *F. adult (seen 1929 by L.F. Graham), from Port Essington, N.T.

Distribution: NE coastal, N coastal, Qld., W.A., N.T. Ecology: larva - sedentary, soil, predator : adult - volant, burrowing; prey Orthoptera (Tettigoniidae). Biological references: Roth, H.L. (1885). Notes on the habits of some Australian Hymenoptera Aculeata. (With description of a new species by W.F. Kirby). *J. Linn. Soc. (Zool.)* **18**: 318–328 (nest, ?prey); Evans, H.E., Hook, A.W. & Matthews, R.W. (1982). Nesting behaviour of Australian wasps of the genus *Sphex* (Hymenoptera, Sphecidae). *J. Nat. Hist.* **16**: 219–225 (biology).

Sphex (Sphex) ermineus Kohl, 1890

Sphex (Sphex) ermineus Kohl, F.F. (1890). Die Hymenopterengruppe der Sphecinen. I Monographie der natürlichen Gattung *Sphex* Linné (*sens.lat.*). Abt. I–II *Ann. Naturh. Hofmus. Wien* **5**: 77–194, 317–462 [412]. Type data: syntypes (probable), whereabouts unknown, from Swan River, W.A.

Distribution: SW coastal, N coastal, Qld., N.S.W., Vic., W.A., N.T. Ecology: larva - sedentary, soil, predator : adult - volant, burrowing; prey Orthoptera.

Sphex (Sphex) finschii Kohl, 1890

Sphex finschii Kohl, F.F. (1890). Die Hymenopterengruppe der Sphecinen. I Monographie der natürlichen Gattung *Sphex* Linné (*sens.lat.*). Abt. I–II *Ann. Naturh. Hofmus. Wien* **5**: 77–194, 317–462 [412]. Type data: syntypes (probable), ZMB *adult, from New Britain.

Distribution: Australia but no specific localities available, also in New Guinea, Solomon Ils. and Indonesia. Ecology: larva - sedentary, soil, predator : adult - volant, burrowing; prey Orthoptera. Biological references: Bohart, R.M. & Menke, A.S. (1976). *Sphecid Wasps of the World : a Generic Revision*. Berkeley : Univ. California Press ix 695 pp. (Australian record).

Sphex (Sphex) formosellus van der Vecht, 1957

Sphex formosellus van der Vecht, J. (1957). The Sphecoidea of the Lesser Sunda Islands (Hym.). I. Sphecinae. *Verh. Naturf. Ges. Basel* **68**: 358–372 [366]. Type data: syntypes (probable), RMNH *adult, from Timor.

Distribution: SW coastal, W.A.; only published locality SW Australia, also in Indonesia. Ecology: larva - sedentary, soil, predator : adult - volant, burrowing; prey Orthoptera.

Sphex (Sphex) fumipennis Smith, 1856

Sphex fumipennis Smith, F. (1856). *Catalogue of Hymenopterous Insects in the Collection of the British Museum*. Part IV, Sphegidae, Larridae and Crabronidae. pp. 207–497 London : British Museum [249]. Type data: holotype, BMNH *F. adult (seen 1929 by L.F. Graham), from Adelaide, S.A.

Distribution: NE coastal, SE coastal, Murray-Darling basin, S Gulfs, Qld., N.S.W., A.C.T., S.A., W.A.; also on New Caledonia, New Hebrides and Loyalty Ils. Ecology: larva - sedentary, soil, predator : adult - volant, burrowing; prey Orthoptera. Biological references: Bohart, R.M. & Menke, A.S. (1976). *Sphecid Wasps of the World : a Generic Revision*. Berkeley : Univ. California Press ix 695 pp. (subspecies listed).

Sphex (Sphex) gilberti Turner, 1908

Sphex gilberti Turner, R.E. (1908). Notes on the Australian fossorial wasps of the family Sphegidae, with descriptions of new species. *Proc. Zool. Soc. Lond.* **1908**: 457–535 pl xxvi [468]. Type data: holotype, BMNH *F. adult (seen 1929 by L.F. Graham), from Mackay, Qld.

Distribution: NE coastal, Qld.; only published localities Mackay and North-West Islet, Capricorn Group. Ecology: larva - sedentary, soil, predator : adult - volant, burrowing; prey Orthoptera.

Sphex (Sphex) luctuosus Smith, 1856

Sphex luctuosa Smith, F. (1856). *Catalogue of Hymenopterous Insects in the Collection of the British Museum*. Part IV, Sphegidae, Larridae and Crabronidae. pp. 207–497 London : British Museum [250]. Type data: holotype, BMNH *F. adult (seen 1929 by L.F. Graham), from Swan River, W.A.

Distribution: Lake Eyre basin, SW coastal, Vic., S.A., W.A., N.T.; only published localities Vic., Dalhousie, Swan River and Alice Springs. Ecology: larva - sedentary, soil, predator : adult - volant, burrowing; prey Orthoptera.

Sphex (Sphex) mimulus Turner, 1910

Sphex mimulus Turner, R.E. (1910). New fossorial Hymenoptera from Australia. *Trans. Entomol. Soc. Lond.* **1910**: 407–429 pl 50 [419]. Type data: holotype, BMNH *F. adult (seen 1929 by L.F. Graham), from Cairns, Qld.

Distribution: NE coastal, Qld.; type locality only. Ecology: larva - sedentary, soil, predator : adult - volant, burrowing; prey Orthoptera.

Sphex (Sphex) modestus Smith, 1856

Sphex modesta Smith, F. (1856). *Catalogue of Hymenopterous Insects in the Collection of the British Museum. Part IV, Sphegidae, Larridae and Crabronidae.* pp. 207–497 London : British Museum [248]. Type data: holotype, BMNH *F. adult (seen 1929 by L.F. Graham), from Australia.

Sphex (Sphex) dolichocerus Kohl, F.F. (1890). Die Hymenopterengruppe der Sphecinen. I Monographie der natürlichen Gattung *Sphex* Linné (*sens.lat.*). Abt. I–II *Ann. Naturh. Hofmus. Wien* **5**: 77–194, 317–462 [390]. Type data: syntypes (probable), MNH *M. adult, from Australia.

Sphex bannitus Kohl, F.F. (1895). Zur Monographie der natürlichen Gattung *Sphex* Linné. *Ann. Naturh. Hofmus. Wien* **10**: 42–74 [61 pl 4 fig 21]. Type data: syntypes (probable), ZMB *F. adult, from Australia (as New Holland).

Taxonomic decision of Turner, R.E. (1910). Additions to our knowledge of the fossorial wasps of Australia. *Proc. Zool. Soc. Lond.* **1910**: 253–356 [346].

Distribution: SE coastal, Lake Eyre basin, SW coastal, N Gulf, N.S.W., W.A., N.T.; only published localities Como, Perth, Alexandria and Urinilla Springs to Deep Well. Ecology: larva - sedentary, soil, predator : adult - volant, burrowing; prey Orthoptera.

Sphex (Sphex) resplendens Kohl, 1885

Sphex nitidiventris Smith, F. (1859). Catalogue of hymenopterous insects collected by Mr. A.R. Wallace at the islands of Aru and Key. *J. Linn. Soc. Lond. Zool.* **3**: 132–178 [158] [*non Sphex nitidiventris* Spinola, 1851]. Type data: holotype, OUM *F. adult.

Sphex gratiosa Smith, F. (1859). Catalogue of hymenopterous insects collected by Mr. A.R. Wallace at the islands of Aru and Key. *J. Linn. Soc. Lond. Zool.* **3**: 132–178 [158] [*non Sphex gratiosa* Smith, 1856]. Type data: holotype, OUM *M. adult, from Aru.

Sphex resplendens Kohl, F.F. (1885). Die Gattungen der Sphecinen und die palearktischen *Sphex*-Arten. *Természetr. Füz.* **9**: 154–207 pls 7–8 [200] [*nom. nov.* for *Sphex nitidiventris* Smith, 1859].

Sphex gratiosissimus Dalla Torre, C.G. de (1897). *Catalogus Hymenopterorum Hucusque Descriptorum Systematicus et Synonymicus.* Fossores (Sphegidae). Leipzig : G. Engelmann Vol. 8 viii 749 pp. [424] [*nom. nov.* for *Sphex gratiosa* Smith, 1859].

Sphex wallacei Turner, R.E. (1908). Notes on the Australian fossorial wasps of the family Sphegidae, with descriptions of new species. *Proc. Zool. Soc. Lond.* **1908**: 457–535 pl xxvi [467] [*nom. nov.* for *Sphex nitidiventris* Smith, 1859].

Taxonomic decision of van der Vecht, J. (1973). Contribution to the taxonomy of the Oriental and Australian Sphecini (Hymenoptera, Sphecoidea). *Proc. K. Ned. Akad. Wet. (c)* **76**: 341–353 [348].

Distribution: NE coastal, Qld.; Indonesian species, N Qld. records may be in error. Ecology: larva - sedentary, soil, predator : adult - volant, burrowing; prey Orthoptera.

Sphex (Sphex) rhodosoma (Turner, 1915)

Chlorion (Proterosphex) rhodosoma Turner, R.E. (1915). Notes on fossorial Hymenoptera. XV. New Australian Crabronidae. *Ann. Mag. Nat. Hist. (8)* **15**: 62–96 [65]. Type data: holotype, BMNH *F. adult (seen 1929 by L.F. Graham), from Cue, Cunderdin, W.A.".

Distribution: SW coastal, W plateau, W.A.; only published localities Cue and Cunderdin. Ecology: larva - sedentary, soil, predator : adult - volant, burrowing; prey Orthoptera. Biological references: Bohart, R.M. & Menke, A.S. (1976). *Sphecid Wasps of the World : a Generic Revision.* Berkeley : Univ. California Press ix 695 pp. (generic placement).

Sphex (Sphex) rugifer Kohl, 1890

Sphex (Sphex) rugifer Kohl, F.F. (1890). Die Hymenopterengruppe der Sphecinen. I Monographie der natürlichen Gattung *Sphex* Linné (*sens.lat.*). Abt. I–II *Ann. Naturh. Hofmus. Wien* **5**: 77–194, 317–462 [393]. Type data: syntypes, ZMB *F. adult, NHMW *F. adult, from Swan River, W.A.

Distribution: SW coastal, NW coastal, W.A.; only published localities Swan River, Perth and Geraldton. Ecology: larva - sedentary, soil, predator : adult - volant, burrowing; prey Orthoptera.

Sphex (Sphex) semifossulatus van der Vecht, 1973

Sphex argentifrons Smith, F. (1868). Descriptions of aculeate Hymenoptera from Australia. *Trans. Entomol. Soc. Lond.* **1868**: 231–258 [248] [*non Sphex argentifrons* Lepeletier, 1845]. Type data: holotype, BMNH 21.709 *M. adult (seen 1929 by L.F. Graham), from Champion Bay, W.A.

Sphex semifossulatus van der Vecht, J. (1973). Contribution to the taxonomy of the Oriental and Australian Sphecini (Hymenoptera, Sphecoidea). *Proc. K. Ned. Akad. Wet. (c)* **76**: 341–353 [349] [*nom. nov.* for *Sphex argentifrons* Smith, 1868].

Distribution: NW coastal, W.A.; type locality only. Ecology: larva - sedentary, soil, predator : adult - volant, burrowing; prey Orthoptera.

Sphex (Sphex) sericeus (Fabricius, 1804)

Taxonomic decision of van der Vecht, J. & Krombein, K.V. (1955). The subspecies of *Sphex sericeus* (Fabr.) (= *S. aurulentus* auct. *nec* Fabr. 1787) (Hymenoptera, Sphecidae). *Idea* **10**: 33-43 [43].

Sphex (Sphex) sericeus godeffroyi Saussure, 1869

Sphex godeffroyi Saussure, H. de (1869). Hyménoptères divers du Musée Godeffroy. *Stettin. Entomol. Ztg.* **30**: 53-64 [57]. Type data: syntypes (probable), MGH *F. adult, from Cape York, Qld.

Sphex aurifex Smith, F. (1873). Natural history notices. Insects, Hymenoptera Aculeata. pp. 456-463 pls xliii-xlv *in* Brenchley, J.B. *Jottings During the Cruise of H.M.S. Curaçoa among the South Sea Islands in 1865*. London : Longmans, Green & Co. [460 pl 44 fig 3]. Type data: syntypes (probable), BMNH *adult, from NW coast of Australia.

Distribution: NE coastal, NW coastal, Qld. W.A.; only published localities Cape York, Duaringa, Babinda and NW coast, also on New Guinea. Ecology: larva - sedentary, soil, predator : adult - volant, burrowing; prey Orthoptera. Biological references: Bohart, R.M. & Menke, A.S. (1976). *Sphecid Wasps of the World : a generic revision.* Berkeley : Univ. California Press ix 695 pp. (Australian record).

Sphex (Sphex) staudingeri Gribodo, 1894

Sphex staudingeri Gribodo, G. (1894). Hymenopterorum novorum diagnoses praecursoriae. *Miscnea Entomol.* **2**: 2-3, 22-23 [2]. Type data: syntypes (probable), whereabouts unknown, from New Guinea.

Distribution: Australia but no available specific localities, also on New Guinea. Ecology: larva - sedentary, soil, predator : adult - volant, burrowing; prey Orthoptera.

Sphex (Sphex) vestitus Smith, 1856

Sphex vestita Smith, F. (1856). *Catalogue of Hymenopterous Insects in the Collection of the British Museum.* Part IV, Sphegidae, Larridae and Crabronidae. pp. 207-497 London : British Museum [248]. Type data: holotype, BMNH *F. adult (seen 1929 by L.F. Graham), from Australia.

Sphex praetexta Smith, F. (1873). Natural history notices. Insects, Hymenoptera Aculeata. pp. 456-463 pls xliii-xlv *in* Brenchley, J.B. *Jottings During the Cruise of H.M.S. Curaçoa among the South Sea Islands in 1865*. London : Longmans, Green & Co. [461 pl 44 fig 5]. Type data: holotype, BMNH *M. adult (seen 1929 by L.F. Graham), from Moreton Bay, Qld.

Sphex (Sphex) imperialis Kohl, F.F. (1890). Die Hymenopterengruppe der Sphecinen. I Monographie der natürlichen Gattung *Sphex* Linné (*sens.lat.*). Abt. I–II *Ann. Naturh. Hofmus. Wien* **5**: 77-194, 317-462 [398]. Type data: syntypes (probable), whereabouts unknown, from Gayndah, Qld.

Taxonomic decision of Turner, R.E. (1910). Additions to our knowledge of the fossorial wasps of Australia. *Proc. Zool. Soc. Lond.* **1910**: 253-356 [345].

Distribution: NE coastal, Qld.; only published localities Gayndah, Mackay, Cairns and Moreton Bay. Ecology: larva - sedentary, soil, predator : adult - volant, burrowing; prey Orthoptera.

Isodontia Patton, 1881

Isodontia Patton, W.H. (1881). Some characters useful in the study of the Sphecidae. *Proc. Bost. Soc. Nat. Hist.* **20**: 378-385 [380]. Type species *Sphex philadelphicus* Lepeletier, 1845 by original designation.

This group is found worldwide, see Bohart, R.M. & Menke, A.S. (1976). *Sphecid Wasps of the World : a Generic Revision.* Berkeley : Univ. California Press ix 695 pp. [119].

Isodontia abdita (Kohl, 1895)

Taxonomic decision of Turner, R.E. (1910). Additions to our knowledge of the fossorial wasps of Australia. *Proc. Zool. Soc. Lond.* **1910**: 253-356.

Isodontia abdita nugenti (Turner, 1910)

Sphex (Isodontia) abditus nugenti Turner, R.E. (1910). Additions to our knowledge of the fossorial wasps of Australia. *Proc. Zool. Soc. Lond.* **1910**: 253-356 [343]. Type data: holotype, BMNH *F. adult (seen 1929 by L.F. Graham), from Cairns, Qld.

Distribution: NE coastal, Qld.; type locality only. Ecology: larva - sedentary, predator : adult - volant; prey Orthoptera, nest in pre-existing cavity.

Isodontia albohirta (Turner, 1908)

Sphex (Isodontia) albohirtus Turner, R.E. (1908). Notes on the Australian fossorial wasps of the family Sphegidae, with descriptions of new species. *Proc. Zool. Soc. Lond.* **1908**: 457-535 pl xxvi [466]. Type data: holotype, BMNH *F. adult (seen 1929 by L.F. Graham), from Mackay and Cairns, Qld.

Distribution: NE coastal, Qld.; type locality only. Ecology: larva - sedentary, predator : adult - volant; prey Orthoptera, nest in pre-existing cavity.

Isodontia aurifrons (Smith, 1859)

Sphex aurifrons Smith, F. (1859). Catalogue of hymenopterous insects collected by Mr. A.R. Wallace at the islands of Aru and Key. *J. Linn. Soc. Lond. Zool.* **3**: 132-178 [157]. Type data: holotype, BMNH *F. adult (seen 1929 by L.F. Graham), from Aru.

Distribution: NE coastal, Qld.; described from Aru, ranges from India to N Qld. Ecology: larva - sedentary, predator : adult - volant; prey Orthoptera, nest in pre-existing cavity. Biological references: Bohart, R.M. & Menke, A.S. (1976). *Sphecid Wasps of the World : a Generic Revision.* Berkeley : Univ. California Press ix 695 pp. (generic placement).

Isodontia nigella (Smith, 1856)

Sphex nigella Smith, F. (1856). *Catalogue of Hymenopterous Insects in the Collection of the British Museum. Part IV, Sphegidae, Larridae and Crabronidae.* pp. 207–497 London : British Museum [255]. Type data: holotype, BMNH *F. adult (seen 1929 by L.F. Graham), from Shanghai.

Distribution: NE coastal, SE coastal, SW coastal, Qld., N.S.W., Tas., W.A.; wide ranging in Australia and eastern Asia. Ecology: larva - sedentary, predator : adult - volant; prey Orthoptera, nest in pre-existing cavity, nest closed with grass. Biological references: Patton, W.H. (1881). Some characters useful in the study of the Sphecidae. *Proc. Bost. Soc. Nat. Hist.* **20**: 378–385 (generic placement); Hacker, H. (1913). Some field notes on Queensland insects. *Mem. Qd. Mus.* **2**: 96–100 (behaviour); Tsuneki, K. (1963). Comparative studies on the nesting biology of the genus *Sphex* (*s.l.*) in East Asia (Hymenoptera, Sphecidae). *Mem. Fac. Lib. Arts Fukui Univ.* (2, Nat. Sci.) **13**: 13–78 (biology).

Isodontia obscurella (Smith, 1856)

Sphex obscurella Smith, F. (1856). *Catalogue of Hymenopterous Insects in the Collection of the British Museum. Part IV, Sphegidae, Larridae and Crabronidae.* pp. 207–497 London : British Museum [251]. Type data: holotype, BMNH *F. adult (seen 1929 by L.F. Graham), from Tas. (as Van Diemen's Land).

Distribution: Tas.; only published locality Hobart. Ecology: larva - sedentary, predator : adult - volant; prey Orthoptera, nest in pre-existing cavity. Biological references: van der Vecht, J. (1973). Contribution to the taxonomy of the Oriental and Australian Sphecini (Hymenoptera, Sphecoidea). *Proc. K. Ned. Akad. Wet. (c)* **76**: 341–353 (generic placement).

Isodontia vidua (Smith, 1856)

Sphex vidua Smith, F. (1856). *Catalogue of Hymenopterous Insects in the Collection of the British Museum. Part IV, Sphegidae, Larridae and Crabronidae.* pp. 207–497 London : British Museum [249]. Type data: holotype, OUM *F. adult, from N.S.W., see van der Vecht, J. (1973). Contribution to the taxonomy of the Oriental and Australian Sphecini (Hymenoptera, Sphecoidea). *Proc. K. Ned. Akad. Wet. (c)* **76**: 341–353 [351].

Distribution: S Gulfs, N.S.W., S.A.; only published localities N.S.W. and Adelaide. Ecology: larva - sedentary, predator : adult - volant; prey Orthoptera, nest in pre-existing cavity. Biological references: van der Vecht, J. (1973). Contribution to the taxonomy of the Oriental and Australian Sphecini (Hymenoptera, Sphecoidea). *Proc. K. Ned. Akad. Wet. (c)* **76**: 341–353 (generic placement).

Palmodes Kohl, 1890

Palmodes Kohl, F.F. (1890). Die Hymenopterengruppe der Sphecinen. I Monographie der natürlichen Gattung *Sphex* Linné (*sens. lat.*). Abt. I–II *Ann. Naturh. Hofmus. Wien* **5**: 77–194, 317–462 [112] [described with subgeneric rank in *Sphex* Linnaeus, 1758]. Type species *Sphex occitanicus* Lepeletier and Serville, 1828 by subsequent designation, see Fernald, H.T. (1906). The digger wasps of North America and the West Indies belonging to the subfamily Chlorioninae. *Proc. U.S. Natl. Mus.* **31**: 291–423. Compiled from secondary source: Bohart, R.M. & Menke, A.S. (1976). *Sphecid Wasps of the World : a Generic Revision.* Berkeley : Univ. California Press ix 695 pp.

This group is non Australian, found only in the Holarctic Region, see Bohart, R.M. & Menke, A.S. (1976). *Sphecid Wasps of the World : a Generic Revision.* Berkeley : Univ. California Press ix 695 pp. [124]. Species now known not to occur in Australia: *Palmodes australis* (Saussure, 1867) a Palaearctic species (described in *Harpactopus*) and *Palmodes sagax* (Kohl, 1890) probably a Palaearctic species (described in *Sphex* (*Palmodes*)) vide van der Vecht, J. (1973). Contribution to the taxonomy of the Oriental and Australian Sphecini (Hymenoptera, Sphecoidea). *Proc. K. Ned. Akad. Wet. (c)* **76**: 341–353 [352].

Prionyx Van der Linden, 1827

Prionyx Van der Linden, P.L. (1827). Observations sur les Hyménoptères d'Europe de la famille des fouisseurs. Part 1. *Mém. Acad. R. Sci. Lettr. Belg.* **4**: 271–367 [362]. Type species *Ammophila kirbii* Van der Linden, 1827 by monotypy. Compiled from secondary source: Bohart, R.M. & Menke, A.S. (1976). *Sphecid Wasps of the World : a Generic Revision.* Berkeley : Univ. California Press ix 695 pp. [39].

This group is found worldwide, see Bohart, R.M. & Menke, A.S. (1976). *Sphecid Wasps of the World : a Generic Revision.* Berkeley : Univ. California Press ix 695 pp. [128].

Prionyx globosus (Smith, 1856)

Sphex globosa Smith, F. (1856). *Catalogue of Hymenopterous Insects in the Collection of the British Museum. Part IV, Sphegidae, Larridae and Crabronidae.* pp. 207–497 London : British Museum [251]. Type data: holotype, BMNH *F. adult (seen 1929 by L.F. Graham), from Tas. (as Van Diemen's Land).

Distribution: NE coastal, Murray-Darling basin, SE coastal, SW coastal, Qld., N.S.W., A.C.T., Vic., Tas., W.A. Ecology: larva - sedentary, soil, predator : adult - volant, burrowing; prey Orthoptera (Acrididae). Biological references: Bohart, R.M. & Menke, A.S. (1963). A reclassification of the Sphecinae with a revision of the Nearctic species of the tribes Sceliphronini and Sphecini (Hymenoptera, Sphecidae). *Univ. Calif. Publs. Entomol.* **30**: 91–182 (generic placement);

Chandler, L.G. (1928). Notes on two grasshopper-wasps. *Vict. Nat.* **45**: 176–181 (biology); Evans, H.E., Hook, A.W. & Matthews, R.W. (1982). Nesting behaviour of Australian wasps of the genus *Sphex* (Hymenoptera, Sphecidae). *J. Nat. Hist.* **16**: 219–225 (biology).

Prionyx saevus (Smith, 1856)

Harpactopus saevus Smith, F. (1856). *Catalogue of Hymenopterous Insects in the Collection of the British Museum.* Part IV, Sphegidae, Larridae and Crabronidae. pp. 207–497 London : British Museum [265]. Type data: holotype, BMNH *F. adult (seen 1929 by L.F. Graham), from Swan River, Cape Upstart, W.A.

Distribution: Murray-Darling basin, SE coastal, Lake Eyre basin, SW coastal, N coastal, Qld., N.S.W., A.C.T., W.A.; central Australia, also in Amboina. Ecology: larva - sedentary, soil, predator : adult - volant, burrowing; prey Orthoptera (Acrididae). Biological references: Bohart, R.M. & Menke, A.S. (1976). *Sphecid Wasps of the World : a Generic Revision.* Berkeley : Univ. California Press ix 695 pp. (generic placement); Common, I.F.B. (1948). The yellow-winged locust, *Gastrimargus musicus* Fabr., in central Queensland. *Qd. J. Agric. Sci.* **5**: 153–219 (prey); Evans, H.E., Hook, A.W. & Matthews, R.W. (1982). Nesting behaviour of Australian wasps of the genus *Sphex* (Hymenoptera, Sphecidae). *J. Nat. Hist.* **16**: 219–225 (biology); Baker, G.L. & Pigott, R. (1983). Parasitism of the Australian plague locust *Chortoicetes terminifera* (Walker) (Orthoptera : Acrididae) by *Prionyx saevus* (Smith)(Hymenoptera : Sphecidae). *Aust. Entomol. Mag.* **10**: 67–74.

Parapsammophila Taschenberg, 1869

Parapsammophila Taschenberg, E. (1869). Die Sphegidae des zoologischen Museums der Universität in Halle. *Z. Naturw.* **34**: 407–435 [429]. Type species *Parapsammophila miles* Taschenberg, 1869 by subsequent designation, see Pate, V.S.L. (1937). The generic names of the sphecoid wasps and their type species. *Mem. Am. Entomol. Soc.* **9**: 1–103 [48].

This group is also found in the Old World, see Bohart, R.M. & Menke, A.S. (1976). *Sphecid Wasps of the World : a Generic Revision.* Berkeley : Univ. California Press ix 695 pp. [137].

Parapsammophila eremophila (Turner, 1910)

Ammophila (Parapsammophila) eremophila Turner, R.E. (1910). Additions to our knowledge of the fossorial wasps of Australia. *Proc. Zool. Soc. Lond.* **1910**: 253–356 [342 pl 32 fig 12]. Type data: syntypes (probable), BMNH *M. adult, from Hermannsburg, N.T.

Distribution: Lake Eyre basin, N.T.; type locality only. Ecology: larva - sedentary, predator : adult - volant.

Podalonia Fernald, 1927

Podalonia Fernald, H.T. (1927). The digger wasps of North America of the genus *Podalonia (Psammophila). Proc. U.S. Natl. Mus.* **71**(2681) : 1–42 [11]. Type species *Ammophila violaceipennis* Lepeletier, 1845 by subsequent designation, see International Commission on Zoological Nomenclature (1968). Opinion 857. *Podalonia* Fernald, 1927 (Insecta, Hymenoptera) : validation and designation of a type-species under the plenary powers. *Bull. Zool. Nomen.* **25**: 88–89. Compiled from secondary source: Bohart, R.M. & Menke, A.S. (1976). *Sphecid Wasps of the World : a Generic Revision.* Berkeley : Univ. California Press ix 695 pp.

This group is found worldwide except in South America, see Bohart, R.M. & Menke, A.S. (1976). *Sphecid Wasps of the World : a Generic Revision.* Berkeley : Univ. California Press ix 695 pp. [141].

Podalonia tydei (Le Guillou, 1841)

Taxonomic decision of Bohart, R.M. & Menke, A.S. (1976). *Sphecid Wasps of the World : a Generic Revision.* Berkeley : Univ. California Press ix 695 pp. [145].

Podalonia tydei suspiciosa (Smith, 1856)

Ammophila suspiciosa Smith, F. (1856). *Catalogue of Hymenopterous Insects in the Collection of the British Museum.* Part IV, Sphegidae, Larridae and Crabronidae. pp. 207–497 London : British Museum [214]. Type data: holotype, BMNH *F. adult (seen 1929 by L.F. Graham), from NW Coast, Swan River, Hunter River," W.A.

Distribution: NE coastal, Murray-Darling basin, SE coastal, S Gulfs, Lake Eyre basin, SW coastal, N coastal, Qld., N.S.W., A.C.T., Vic., Tas., S.A., W.A.; also in New Zealand. Ecology: larva - sedentary, soil, predator : adult - volant, burrowing; prey Lepidoptera larvae. Biological references: Riek, E.F. (1970). Hymenoptera. pp. 867–959 *in* C.S.I.R.O. *The Insects of Australia.* Carlton : Melbourne Univ. Press xiv 1029 pp. (generic placement); McCarthy, T. (1917). Some observations on solitary wasps at Hay, N.S.W. *Aust. Nat.* **3**: 195–200 (behaviour); Chandler, L.G. (1925). Habits of the sand-wasp. *Vict. Nat.* **42**: 107–114 (behaviour); Bristowe, W.S. (1971). The habits of a West Australian sphecid wasp. *Entomologist* **104**: 42–44 (behaviour); Faulds, W. (1978). Notes on an Australian sphecid wasp, *Podalonia suspiciosa* (Hymenoptera : Sphecidae), now established in New Zealand. *N.Z. Entomol.* **6**: 312–313.

Ammophila Kirby, 1798

Ammophila Kirby, W. (1798). *Ammophila*, a new genus of insects in the class Hymenoptera, including the *Sphex sabulosa* of Linnaeus. *Trans. Linn. Soc. Lond.* **4**: 195–212 [199]. Type species *Sphex sabulosa* Linnaeus, 1758 by subsequent designation, see International Commission on Zoological Nomenclature (1946).

Opinion 180. On the status of the names *Sphex* Linnaeus, 1758, and *Ammophila* Kirby, 1798 (Class Insecta, Order Hymenoptera). *Opin. Decl. Int. Comm. Zool. Nomen.* **2**: 569–585 [571]. Compiled from secondary source: Bohart, R.M. & Menke, A.S. (1976). *Sphecid Wasps of the World : a Generic Revision.* Berkeley : Univ. California Press ix 695 pp. [39].

This group is found worldwide, see Bohart, R.M. & Menke, A.S. (1976). *Sphecid Wasps of the World : a Generic Revision.* Berkeley : Univ. California Press ix 695 pp. [147].

Ammophila ardens Smith, 1868

Ammophila ardens Smith, F. (1868). Descriptions of aculeate Hymenoptera from Australia. *Trans. Entomol. Soc. Lond.* **1868**: 231–258 [247]. Type data: holotype, BMNH *F. adult (seen 1929 by L.F. Graham), from Swan River, W.A.

Distribution: NE coastal, SW coastal, Qld., W.A.; only published localities Mackay and Swan River. Ecology: larva - sedentary, soil, predator : adult - volant, burrowing; prey Lepidoptera larvae.

Ammophila atripes Smith, 1852

Ammophila atripes Smith, F. (1852). Descriptions of new species of hymenopterous insects, with notes on their economy by Ezra T. Downes Esq.. *Ann. Mag. Nat. Hist.* (2) **9**: 44–50 [46]. Type data: syntypes (probable), whereabouts unknown, from India.

Distribution: N coastal, N.T.; N.T. records (from Katherine, Darwin, Burnside) doubtful, the species is found in India, S and SE Asia. Ecology: larva - sedentary, soil, predator : adult - volant, burrowing; prey larval Lepidoptera. Biological references: Tsuneki, K. (1967). Studies on the Formosan Sphecidae (III). The subfamily Sphecinae with special reference to the genus *Ammophila* in eastern Asia (Hymenoptera). *Etizenia* **26**: 1–824 (redescription); Bohart, R.M. & Menke, A.S. (1976). *Sphecid Wasps of the World : a Generic Revision.* Berkeley : Univ. California Press ix 695 pp. (species from Oriental Region).

Ammophila aurifera Turner, 1908

Ammophila aurifera Turner, R.E. (1908). Notes on the Australian fossorial wasps of the family Sphegidae, with descriptions of new species. *Proc. Zool. Soc. Lond.* **1908**: 457–535 [464 pl xxvi fig 3]. Type data: syntypes (probable), BMNH *M., F. adult (seen 1929 by L.F. Graham), from Port Darwin, N.T.

Distribution: N coastal, N.T.; type locality only. Ecology: larva - sedentary, soil, predator : adult - volant, burrowing; prey Lepidoptera larvae.

Ammophila clavus (Fabricius, 1775)

Sphex clavus Fabricius, J.C. (1775). *Systema Entomologiae*, sistens insectorum classes, ordines, genera, species, adiectis synonymis, locis, descriptionibus, observationibus. Flensburgi et Lipsiae : Kortii xxvii 832 pp. [348]. Type data: holotype, BMNH *F. adult (seen 1929 by L.F. Graham), from Australia (as New Holland).

Distribution: NE coastal, N coastal, Qld., W.A. Ecology: larva - sedentary, soil, predator : adult - volant, burrowing; prey Lepidoptera larvae. Biological references: Smith, F. (1856). *Catalogue of Hymenopterous Insects in the Collection of the British Museum.* Part IV, Sphegidae, Larridae and Crabronidae. pp. 207–497 London : British Museum (generic placement); Turner, R.E. (1908). Notes on the Australian fossorial wasps of the family Sphegidae, with descriptions of new species. *Proc. Zool. Soc. Lond.* **1908**: 457–535 pl xxvi (redescription); Tsuneki, K. (1967). Studies on the Formosan Sphecidae (III). The subfamily Sphecinae with special reference to the genus *Ammophila* in eastern Asia (Hymenoptera). *Etizenia* **26**: 1–24 (non-Australian subspecies).

Ammophila eyrensis Turner, 1908

Ammophila eyrensis Turner, R.E. (1908). Notes on the Australian fossorial wasps of the family Sphegidae, with descriptions of new species. *Proc. Zool. Soc. Lond.* **1908**: 457–535 pl xxvi [465]. Type data: holotype, BMNH *F. adult (seen 1929 by L.F. Graham), from Killalpanima, S.A.

Distribution: Lake Eyre basin, S.A.; type locality only. Ecology: larva - sedentary, soil, predator : adult - volant, burrowing; prey Lepidoptera larvae.

Ammophila instabilis Smith, 1856

Ammophila instabilis Smith, F. (1856). *Catalogue of Hymenopterous Insects in the Collection of the British Museum.* Part IV, Sphegidae, Larridae and Crabronidae. pp. 207–497 London : British Museum [214]. Type data: holotype, BMNH *F. adult, from Swan River, Port Essington, Australia".

Ammophila impatiens Smith, F. (1868). Descriptions of aculeate Hymenoptera from Australia. *Trans. Entomol. Soc. Lond.* **1868**: 231–258 [247]. Type data: syntypes (probable), BMNH *F. adult, from Champion Bay, W.A.

Taxonomic decision of Turner, R.E. (1908). Notes on the Australian fossorial wasps of the family Sphegidae, with descriptions of new species. *Proc. Zool. Soc. Lond.* **1908**: 457–535 pl xxvi [466].

Distribution: NE coastal, Lake Eyre basin, SW coastal, N coastal, Qld., W.A., N.T. Ecology: larva - sedentary, soil, predator : adult - volant, burrowing; prey Lepidoptera larvae.

Psenulus Kohl, 1896

Psenulus Kohl, F.F. (1896). Die Gattungen der Sphegiden. *Ann. Naturh. Hofmus. Wien* **11**: 233–516 [293]. Type species *Psen fuscipennis* Dahlbom, 1843 by subsequent designation, see Ashmead, W.H. (1899). Classification of the entomophilous wasps, or the superfamily Sphegoidea. *Can. Entomol.* **31**: 212–225 [224].

This group is found worldwide, mainly in the Oriental Region, see Bohart, R.M. & Menke, A.S. (1976). *Sphecid Wasps of the World : a Generic Revision*. Berkeley : Univ. California Press ix 695 pp. [171].

Psenulus carinifrons (Cameron, 1902)

Taxonomic decision of van Lith, J.P. (1966). The group of *Psenulus pulcherrimus* (Bingham) (Hymenoptera, Sphecidae). *Tijdschr. Entomol.* **109**: 35–48 [43].

Psenulus carinifrons scutellatus Turner, 1912

Psenulus scutellatus Turner, R.E. (1912). Notes on fossorial Hymenoptera. IX. On some new species from the Australian and Austro-Malayan regions. *Ann. Mag. Nat. Hist. (8)* **10**: 48–63 [54]. Type data: holotype, BMNH *F. adult (seen 1929 by L.F. Graham), from Cairns, Qld.

Distribution: NE coastal, Qld.; only published localities Cairns and Halifax, also in Papua New Guinea, Indonesia and Philippines. Ecology: larva - sedentary, predator : adult - volant; prey small Hemiptera, nest in cavities in plants. Biological references: van Lith, J.P. (1969). Descriptions of some Indo-Australian *Psenulus* and revision of the group of *Psenulus pulcherrimus* (Bingham) (Hymenoptera, Sphecidae, Psenini). *Tijdschr. Entomol.* **112**: 197–212 (redescription).

Psenulus interstitialis Cameron, 1906

Psenulus interstitialis Cameron, P. (1906). Hymenoptera of the Dutch expedition to New Guinea in 1904 and 1905. Part 1, Thynnidae, Scoliidae, Pompilidae, Sphegidae and Vespidae. *Tijdschr. Entomol.* **49**: 215–233 [222]. Type data: syntypes (probable), whereabouts unknown, from Etna Bay, New Guinea.

Psen lutescens Turner, R.E. (1907). New species of Sphegidae from Australia. *Ann. Mag. Nat. Hist. (7)* **19**: 268–276 [273]. Type data: holotype, BMNH 21.838 *F. adult, from Mackay, Qld.

Taxonomic decision of Turner, R.E. (1908). Notes on the Australian fossorial wasps of the family Sphegidae, with descriptions of new species. *Proc. Zool. Soc. Lond.* **1908**: 457–535 pl xxvi [463].

Distribution: NE coastal, Qld.; only published localities Mackay, Cairns and Kuranda, also New Guinea. Ecology: larva - sedentary, predator : adult - volant; prey small Hemiptera, nest in cavities in plants. Biological references: van Lith, J.P. (1972). Contribution to the knowledge of Oriental *Psenulus* (Hymenoptera, Sphecidae, Psenini). *Tijdschr. Entomol.* **115**: 153–203 (redescription).

Polemistus Saussure, 1892

Polemistus Saussure, H. de (1892). Histoire naturelle des Hyménoptères. Vol. 20 xxi 590 pp. *in* Grandidier, A. (1890–1892) *Histoire Physique, Naturelle et Politique de Madagascar*. Paris : Imprimerie Nationale [565]. Type species *Polemistus macilentus* Saussure, 1892 by subsequent designation, see Pate, V.S.L. (1937). The generic names of the sphecoid wasps and their type species. *Mem. Am. Entomol. Soc.* **9**: 1–103. Compiled from secondary source: Bohart, R.M. & Menke, A.S. (1976). *Sphecid Wasps of the World : a Generic Revision*. Berkeley : Univ. California Press ix 695 pp.

This group is found worldwide, see Bohart, R.M. & Menke, A.S. (1976). *Sphecid Wasps of the World : a Generic Revision*. Berkeley : Univ. California Press ix 695 pp. [184].

Polemistus exul Turner, 1907

Polemistus exul Turner, R.E. (1907). New species of Sphegidae from Australia. *Ann. Mag. Nat. Hist. (7)* **19**: 268–276 [274]. Type data: holotype, BMNH *F. adult (seen 1929 by L.F. Graham), from Mackay, Qld.

Distribution: NE coastal, Qld.; only published localities Mackay and whole eastern coast. Ecology: larva - sedentary, predator : adult - volant; nest in pre-existing holes.

Arpactophilus Smith, 1864

Arpactophilus Smith, F. (1864). Catalogue of hymenopterous insects collected by Mr. A.R. Wallace in the Islands of Mysol, Ceram, Waigiou, Bouru and Timor. *J. Linn. Soc. Lond. Zool.* **7**: 6–48 [36]. Type species *Arpactophilus bicolor* Smith, 1864 by monotypy.

Harpactophilus Kohl, F.F. (1896). Die Gattungen der Sphegiden. *Ann. Naturh. Hofmus. Wien* **11**: 233–516 [276] [emend. of *Arpactophilus* Smith, 1864].

Austrostigmus Turner, R.E. (1912). Notes on fossorial Hymenoptera. IX. On some new species from the Australian and Austro-Malayan regions. *Ann. Mag. Nat. Hist. (8)* **10**: 48–63 [55]. Type species *Stigmus queenslandensis* Turner, 1908 by original designation.

Taxonomic decision of Bohart, R.M. & Menke, A.S. (1976). *Sphecid Wasps of the World : a Generic Revision*. Berkeley : Univ. California Press ix 695 pp. [41].

This group is also found on Misool Is., Indonesia, see Bohart, R.M. & Menke, A.S. (1976). *Sphecid Wasps of the World : a Generic Revision*. Berkeley : Univ. California Press ix 695 pp. [186].

Arpactophilus approximatus (Turner, 1916)

Austrostigmus approximatus Turner, R.E. (1916). Notes on fossorial Hymenoptera. XIX. On new species from Australia. *Ann. Mag. Nat. Hist. (8)* **17**: 116–136 [131]. Type data: holotype, BMNH *F. adult (seen 1929 by L.F. Graham), from Kuranda, Qld.

Distribution: NE coastal, Qld.; type locality only. Ecology: larva - sedentary, predator : adult - volant.

Arpactophilus arator (Turner, 1908)

Harpactophilus arator Turner, R.E. (1908). Notes on the Australian fossorial wasps of the family Sphegidae, with descriptions of new species. *Proc. Zool. Soc. Lond.* **1908**: 457–535 [461 pl xxvi fig 1]. Type data: holotype, BMNH *F. adult (seen 1929 by L.F. Graham), from Cairns, Qld.

Distribution: NE coastal, Qld.; type locality only.
Ecology: larva - sedentary, predator : adult - volant.

Arpactophilus dubius (Turner, 1916)

Austrostigmus dubius Turner, R.E. (1916). Notes on fossorial Hymenoptera. XIX. On new species from Australia. *Ann. Mag. Nat. Hist. (8)* **17**: 116–136 [131]. Type data: holotype, BMNH *F. adult (seen 1929 by L.F. Graham), from Kuranda, Qld.

Distribution: NE coastal, Qld.; type locality only.
Ecology: larva - sedentary, predator : adult - volant.

Arpactophilus glabrellus (Turner, 1916)

Austrostigmus glabrellus Turner, R.E. (1916). Notes on fossorial Hymenoptera. XIX. On new species from Australia. *Ann. Mag. Nat. Hist. (8)* **17**: 116–136 [132]. Type data: holotype, BMNH *F. adult (seen 1929 by L.F. Graham), from Kalamunda, Darling Ranges, W.A.

Distribution: SW coastal, W.A.; type locality only.
Ecology: larva - sedentary, predator : adult - volant.

Arpactophilus kohlii (Turner, 1908)

Harpactophilus kohlii Turner, R.E. (1908). Notes on the Australian fossorial wasps of the family Sphegidae, with descriptions of new species. *Proc. Zool. Soc. Lond.* **1908**: 457–535 pl xxvi [459]. Type data: holotype, BMNH *F. adult (seen 1929 by L.F. Graham), from Mackay, Qld.

Distribution: NE coastal, Qld.; type locality only.
Ecology: larva - sedentary, predator : adult - volant.

Arpactophilus queenslandensis (Turner, 1908)

Stigmus queenslandensis Turner, R.E. (1908). Notes on the Australian fossorial wasps of the family Sphegidae, with descriptions of new species. *Proc. Zool. Soc. Lond.* **1908**: 457–535 pl xxvi [457]. Type data: holotype, BMNH *F. adult (seen 1929 by L.F. Graham), from Mackay, Qld.

Distribution: NE coastal, Qld.; type locality only.
Ecology: larva - sedentary, predator : adult - volant. Biological references: Bohart, R.M. & Menke, A.S. (1976). *Sphecid Wasps of the World : a Generic Revision*. Berkeley : Univ. California Press ix 695 pp. (generic placement).

Arpactophilus reticulatus (Turner, 1912)

Austrostigmus reticulatus Turner, R.E. (1912). Notes on fossorial Hymenoptera. IX. On some new species from the Australian and Austro-Malayan regions. *Ann. Mag. Nat. Hist. (8)* **10**: 48–63 [55]. Type data: holotype, BMNH *M. adult (seen 1929 by L.F. Graham), from Cairns district, Qld.

Distribution: NE coastal, N coastal, Qld., N.T.; only published localities Cairns district, Darwin.
Ecology: larva - sedentary, predator : adult - volant.

Arpactophilus ruficollis (Turner, 1916)

Austrostigmus ruficollis Turner, R.E. (1916). Notes on fossorial Hymenoptera. XIX. On new species from Australia. *Ann. Mag. Nat. Hist. (8)* **17**: 116–136 [133]. Type data: holotype, BMNH *F. adult (seen 1929 by L.F. Graham), from Kuranda, Qld.

Distribution: NE coastal, Qld.; type locality only.
Ecology: larva - sedentary, predator : adult - volant.

Arpactophilus steindachneri Kohl, 1884

Taxonomic decision of Turner, R.E. (1936). Notes on fossorial Hymenoptera. XLV. On new sphegid wasps from Australia. *Ann. Mag. Nat. Hist. (10)* **18**: 533–545 [534].

Arpactophilus steindachneri steindachneri Kohl, 1884

Arpactophilus steindachneri Kohl, F.F. (1884). Neue Hymenopteren in den Sammlungen des k.k. zoologischen Hof-Cabinetes zu Wien. ii. *Verh. Zool.-Bot. Ges. Wien* **33**: 331–386 [334 pl 18 figs 1–2]. Type data: syntypes (probable), NHMW *F. adult, from Australia.

Distribution: NE coastal, Qld.; only published localities Cooktown, Cairns, Mackay and Yaamba.
Ecology: larva - sedentary, predator : adult - volant. Biological references: Evans, H.E. (1964). Further studies on the larvae of digger wasps (Hymenoptera : Sphecidae). *Trans. Am. Entomol. Soc.* **90**: 235–299 pls 12 (description of larva).

Arpactophilus steindachneri deserticolus Turner, 1936

Harpactophilus steindachneri deserticolus Turner, R.E. (1936). Notes on fossorial Hymenoptera. XLV. On new sphegid wasps from Australia. *Ann. Mag. Nat. Hist. (10)* **18**: 533–545 [534]. Type data: holotype, BMNH *F. adult, from Dedari, W.A.

Distribution: W plateau, W.A.; type locality only.
Ecology: larva - sedentary, predator : adult - volant.

Arpactophilus sulcatus (Turner, 1908)

Harpactophilus sulcatus Turner, R.E. (1908). Notes on the Australian fossorial wasps of the family Sphegidae, with descriptions of new species. *Proc. Zool. Soc. Lond.* **1908**: 457–535 [460]. Type data: holotype, BMNH *F. adult (seen 1929 by L.F. Graham), from Kuranda, Qld.

Distribution: NE coastal, Qld.; type locality only.
Ecology: larva - sedentary, predator : adult - volant.

Arpactophilus tricolor (Turner, 1908)

Harpactophilus tricolor Turner, R.E. (1908). Notes on the Australian fossorial wasps of the family Sphegidae, with descriptions of new species. *Proc. Zool. Soc. Lond.* **1908**: 457–535 pl xxvi [462 pl xxvi fig 2]. Type data: holotype, BMNH *F. adult (seen 1929 by L.F. Graham), from Mackay, Qld.

Distribution: NE coastal, Qld.; type locality only. Ecology: larva - sedentary, predator : adult - volant.

Paracrabro Turner, 1907

Paracrabro Turner, R.E. (1907). New species of Sphegidae from Australia. *Ann. Mag. Nat. Hist.* (7) **19**: 268–276 [275]. Type species *Paracrabro froggatti* Turner, 1907 by monotypy and original designation.

Paracrabro froggatti Turner, 1907

Paracrabro froggatti Turner, R.E. (1907). New species of Sphegidae from Australia. *Ann. Mag. Nat. Hist.* (7) **19**: 268–276 [275]. Type data: holotype, BMNH *F. adult (seen 1929 by L.F. Graham), from Vic.

Distribution: N.S.W., Vic.; only published localities as states. Ecology: larva - sedentary, predator : adult - volant. Biological references: Turner, R.E. (1910). Additions to our knowledge of the fossorial wasps of Australia. *Proc. Zool. Soc. Lond.* **1910**: 253–356 (redescription).

Spilomena Shuckard, 1838

Celia Shuckard, W.E. (1837). *Essay on the Indigenous Fossorial Hymenoptera;* comprising a description of all the British species of burrowing sand wasps contained in the metropolitan collections; with their habits as far as they have been observed. London : Shuckard xii 259 pp. [182] [*non Celia* Zimmermann, 1832]. Type species *Stigmus troglodytes* Van der Linden, 1829 by monotypy and original designation.

Spilomena Shuckard, W.E. (1838). Descriptions of new exotic aculeate Hymenoptera. *Trans. Entomol. Soc. Lond.* **2**: 68–82 [79] [*nom. nov.* for *Celia* Shuckard, 1837].

Microglossa Rayment, T. (1930). *Microglossa* and *Melitribus*, new genera of Australian bees. *Proc. R. Soc. Vict. (ns)* **42**: 211–220 pl 21 [212] [*non Microglossa* Voight, 1831]. Type species *Microglossa longifrons* Rayment, 1930 by original designation.

Microglossella Rayment, T. (1935). *A Cluster of Bees*. Sydney : Endeavour Press 750 pp. [634] [*nom. nov.* for *Microglossa* Rayment, 1930].

Taxonomic decision of Bohart, R.M. & Menke, A.S. (1976). *Sphecid Wasps of the World : a Generic Revision*. Berkeley : Univ. California Press ix 695 pp. [41].

This group is found worldwide, see Bohart, R.M. & Menke, A.S. (1976). *Sphecid Wasps of the World : a Generic Revision*. Berkeley : Univ. California Press ix 695 pp. [192].

Spilomena australis Turner, 1910

Spilomena australis Turner, R.E. (1910). New fossorial Hymenoptera from Australia. *Trans. Entomol. Soc. Lond.* **1910**: 407–429 pl 50 [418 pl 50 fig 9]. Type data: holotype, BMNH *F. adult (seen 1929 by L.F. Graham), from Kuranda, Qld.

Distribution: NE coastal, Qld.; only published localities Kuranda and Cairns. Ecology: larva - sedentary, predator : adult - volant.

Spilomena bimaculata (Rayment, 1930)

Microglossa bimaculata Rayment, T. (1930). *Microglossa* and *Melitribus*, new genera of Australian bees. *Proc. R. Soc. Vict. (ns)* **42**: 211–220 pl 21 [216]. Type data: holotype, ANIC F. adult, from Sandringham, Vic.

Distribution: SE coastal, Vic.; type locality only. Ecology: larva - sedentary, predator : adult - volant.

Spilomena elegantula Turner, 1916

Spilomena elegantula Turner, R.E. (1916). Notes on fossorial Hymenoptera. XIX. On new species from Australia. *Ann. Mag. Nat. Hist.* (8) **17**: 116–136 [135]. Type data: holotype, BMNH *F. adult (seen 1929 by L.F. Graham), from Kuranda, Qld.

Distribution: NE coastal, Qld.; type locality only. Ecology: larva - sedentary, predator : adult - volant.

Spilomena hobartia Turner, 1914

Spilomena hobartia Turner, R.E. (1914). New fossorial Hymenoptera from Australia and Tasmania. *Proc. Linn. Soc. N.S.W.* **38**: 608–623 [622]. Type data: holotype, BMNH *F. adult (seen 1929 by L.F. Graham), from Eaglehawk Neck, Tas.

Distribution: Tas.; only published localities Eaglehawk Neck and Hobart. Ecology: larva - sedentary, predator : adult - volant.

Spilomena iridescens Turner, 1916

Spilomena iridescens Turner, R.E. (1916). Notes on fossorial Hymenoptera. XIX. On new species from Australia. *Ann. Mag. Nat. Hist.* (8) **17**: 116–136 [135]. Type data: holotype, BMNH *F. adult (seen 1929 by L.F. Graham), from Yallingup, W.A.

Distribution: SW coastal, W.A.; type locality only. Ecology: larva - sedentary, predator : adult - volant.

Spilomena longiceps Turner, 1916

Spilomena longiceps Turner, R.E. (1916). Notes on fossorial Hymenoptera. XIX. On new species from Australia. *Ann. Mag. Nat. Hist.* (8) **17**: 116–136 [134]. Type data: holotype, BMNH *F. adult (seen 1929 by L.F. Graham), from Kuranda, Qld.

Distribution: NE coastal, Qld.; type locality only. Ecology: larva - sedentary, predator : adult - volant.

Spilomena longifrons (Rayment, 1930)

Microglossa longifrons Rayment, T. (1930). *Microglossa* and *Melitribus*, new genera of Australian bees. *Proc. R. Soc. Vict. (ns)* **42**: 211-220 pl 21 [213]. Type data: holotype, ANIC M. adult (unnumbered), from Sandringham, Vic.

Distribution: SE coastal, Murray-Darling basin, Vic.; only published localities Sandringham, Gunbower Is. and Port Phillip Bay. Ecology: larva - sedentary, predator : adult - volant. Biological references: Rayment, T. (1935). *A Cluster of Bees.* Sydney : Endeavour Press 750 pp. (redescription).

Spilomena luteiventris Turner, 1936

Spilomena luteiventris Turner, R.E. (1936). Notes on fossorial Hymenoptera. XLV. On new sphegid wasps from Australia. *Ann. Mag. Nat. Hist. (10)* **18**: 533-545 [533]. Type data: holotype, BMNH *F. adult, from Tambourine Mt., Qld.

Distribution: NE coastal, Qld.; type locality only. Ecology: larva - sedentary, predator : adult - volant.

Spilomena rufitarsus (Rayment, 1930)

Microglossa rufitarsus Rayment, T. (1930). *Microglossa* and *Melitribus*, new genera of Australian bees. *Proc. R. Soc. Vict. (ns)* **42**: 211-220 pl 21 [215]. Type data: syntypes (probable), ANIC *F. adult, from Sandringham, Vic.

Distribution: SE coastal, Vic.; only published localities Sandringham and Malvern. Ecology: larva - sedentary, predator : adult - volant.

Astata Latreille, 1796

Astata Latreille, P.A. (1796). *Précis des Caractères Génériques des Insectes,* disposés dans un ordre naturel. Bordeaux : Brive xiv 208 pp. [xiii]. Type species *Tiphia abdominalis* Panzer, 1798 by subsequent designation, see Latreille, P.A. (1802-1803). *Histoire Naturelle, Générale et Particulière des Crustacés et des Insectes.* Paris : F. Dufart Vol. 3 xii 13+467 pp. [337]. Compiled from secondary source: Bohart, R.M. & Menke, A.S. (1976). *Sphecid Wasps of the World : a Generic Revision.* Berkeley : Univ. California Press ix 695 pp.

This group is found on all continents except Australia, see Bohart, R.M. & Menke, A.S. (1976). *Sphecid Wasps of the World : a Generic Revision.* Berkeley : Univ. California Press ix 695 pp. [211]. Species now known not to occur in Australia: *Astata australasiae* Shuckard, 1838 a South American species *vide* Pulawski, W. (1975). Synonymical notes on Larrinae and Astatinae (Hymenoptera : Sphecidae). *J. Wash. Acad. Sci.* **64**: 308-323 [320].

Larra Fabricius, 1793

Taxonomic decision of Bohart, R.M. & Menke, A.S. (1976). *Sphecid Wasps of the World : a Generic Revision.* Berkeley : Univ. California Press ix 695 pp. [42].

Larra (Larra) Fabricius, 1793

Larra Fabricius, J.C. (1793). *Entomologia Systematica Emendata et Aucta.* Secundum classes, ordines, genera, species. Adjectis synonimis, locis, observationibus, descriptionibus. Hafniae : C.G. Profit Vol. 2 viii 519 pp. [220]. Type species *Larra ichneumoniformis* Fabricius, 1793 by subsequent designation, see Latreille, P.A. (1810). *Considérations Générales sur l'Ordre Naturel des Animaux Composant les Classes des Crustacès, des Arachnides, et des Insectes;* avec un Tableau Méthodique de leurs Genres, Disposés en Familles. Paris : F. Schoell 444 pp.

This group is found worldwide, see Bohart, R.M. & Menke, A.S. (1976). *Sphecid Wasps of the World : a Generic Revision.* Berkeley : Univ. California Press ix 695 pp. [233]. Species now known not to occur in Australia: *Larra psilocera* Kohl, 1884 an Austro-Malayan or Melanesian species *vide* Turner, R.E. (1916). Notes on fossorial Hymenoptera. XX. On some Larrinae in the British Museum. *Ann. Mag. Nat. Hist. (8)* **17**: 248-259 [250].

Larra (Larra) melanocnemis Turner, 1916

Larra melanocnemis Turner, R.E. (1916). Notes on fossorial Hymenoptera. XX. On some Larrinae in the British Museum. *Ann. Mag. Nat. Hist. (8)* **17**: 248-259 [249]. Type data: holotype, BMNH *F. adult (seen 1929 by L.F. Graham), from Mackay, Qld.

Distribution: NE coastal, S Gulfs, N coastal, Qld., S.A., N.T.; only published localities Mackay, Adelaide and Adelaide River. Ecology: larva - ectoparasitic : adult - volant; prey Orthoptera (Gryllotalpidae), no nest constructed.

Larra (Cratolarra) Cameron, 1900

Cratolarra Cameron, P. (1900). Descriptions of new genera and species of aculeate Hymenoptera from the Oriental zoological region. *Ann. Mag. Nat. Hist. (7)* **8**: 116-122 [34] Type species *Cratolarra femorata* Cameron, 1900 by monotypy. [type species is a secondary homonym of *Tachytes femorata* Saussure, 1855 *vide* Bohart, R.M. & Menke, A.S. (1976). *Sphecid Wasps of the World : a Generic Revision.* Berkeley : Univ. California Press ix 695 pp.].

This group is also found in the Old World tropics, see Bohart, R.M. & Menke, A.S. (1976). *Sphecid Wasps of the World : a Generic Revision.* Berkeley : Univ. California Press ix 695 pp. [233].

Larra (Cratolarra) femorata (Saussure, 1855)

Tachytes femoratus Saussure, H. de (1855). Mélanges hyménoptérologiques. *Mém. Soc. Phys. Hist. Nat.*

Genève **14**: 1–67 [20 pl 8 fig 6]. Type data: syntypes (probable), whereabouts unknown, from Australia (as New Holland).

Larra scelesta Turner, R.E. (1908). Notes on the Australian fossorial wasps of the family Sphegidae, with descriptions of new species. *Proc. Zool. Soc. Lond.* **1908**: 457–535 pl xxvi [474]. Type data: holotype, BMNH *F. adult (seen 1929 by L.F. Graham), from Mackay, Qld.

Taxonomic decision of Turner, R.E. (1916). Notes on fossorial Hymenoptera. XX. On some Larrinae in the British Museum. *Ann. Mag. Nat. Hist. (8)* **17**: 248–259 [251].

Distribution: NE coastal, SE coastal, N coastal, Qld., N.S.W., Tas., N.T. Ecology: larva - ectoparasitic : adult - volant; prey Orthoptera (Gryllotalpidae), no nest constructed. Biological references: Bohart, R.M. & Menke, A.S. (1976). *Sphecid Wasps of the World : a Generic Revision*. Berkeley : Univ. California Press ix 695 pp. (generic placement); Williams, F.X. (1928). Studies in tropical wasps - their hosts and associates (with descriptions of new species). *Bull. Hawaiian Sug. Pltrs'. Ass. Exp. Stn. Entomol. Ser.* **19**: 1–179 (biology).

Larra (Unplaced)

Larra alecto (Smith, 1858)

Larrada alecto Smith, F. (1858). Catalogue of the hymenopterous insects collected at Sarawak, Borneo; Mount Ophir, Malacca; and Singapore, by A.R. Wallace. *J. Linn. Soc. Lond. Zool.* **2**: 42–130 [103]. Type data: syntypes (probable), OUM *F. adult, from Singapore.

Distribution: Christmas Is.; also Singapore and Celebes. Ecology: larva - ectoparasitic : adult - volant; prey Orthoptera Gryllotalpidae, no nest constructed. Biological references: Bohart, R.M. & Menke, A.S. (1976). *Sphecid Wasps of the World : a Generic Revision*. Berkeley : Univ. California Press ix 695 pp. (generic placement).

Liris Fabricius, 1804

Taxonomic decision of Bohart, R.M. & Menke, A.S. (1976). *Sphecid Wasps of the World : a Generic Revision*. Berkeley : Univ. California Press ix 695 pp. [43].

Liris (Liris) Fabricius, 1804

Liris Fabricius, J.C. (1804). *Systema Piezatorum*. Brunsvigae : C. Reichard xiv 439 pp. [227]. Type species *Sphex auratus* Fabricius, 1787 by subsequent designation, see Patton, W.H. (1881). List of the North American Larradae. *Proc. Bost. Soc. Nat. Hist.* **20**: 385–397 [386].

This group is also found in the Palearctic, Ethiopian and Oriental Regions, see Bohart, R.M. & Menke, A.S. (1976). *Sphecid Wasps of the World : a Generic Revision*. Berkeley : Univ. California Press ix 695 pp. [238].

Liris (Liris) melania Turner, 1916

Liris melania Turner, R.E. (1916). Notes on fossorial Hymenoptera. XX. On some Larrinae in the British Museum. *Ann. Mag. Nat. Hist. (8)* **17**: 248–259 [248]. Type data: holotype, BMNH *F. adult (seen 1929 by L.F. Graham), from Cairns district, Qld.

Distribution: NE coastal, Qld.; only published localities Cairns district and Halifax, also Solomon Ils. Ecology: larva - sedentary, soil, predator : adult - volant, burrowing; prey Orthoptera (Gryllidae), pre-existing holes preferred. Biological references: Williams, F.X. (1936). Notes on some larrid wasps from the British Solomon Islands Protectorate, with the description of one new species. *Ann. Mag. Nat. Hist. (10)* **18**: 124–130 (description of male, distribution).

Liris (Leptolarra) Cameron, 1900

Leptolarra Cameron, P. (1900). Descriptions of new genera and species of aculeate Hymenoptera from the Oriental zoological region. *Ann. Mag. Nat. Hist. (7)* **8**: 116–122 [29]. Type species *Leptolarra reticulata* Cameron, 1900 by subsequent designation, see Richards, O.W. (1935). Notes on the nomenclature of the aculeate Hymenoptera, with special reference to British genera and species. *Trans. R. Entomol. Soc. Lond.* **83**: 143–176.

This group is found worldwide, see Bohart, R.M. & Menke, A.S. (1976). *Sphecid Wasps of the World : a Generic Revision*. Berkeley : Univ. California Press ix 695 pp. [238].

Liris (Leptolarra) abbreviata (Turner, 1908)

Notogonia abbreviata Turner, R.E. (1908). Notes on the Australian fossorial wasps of the family Sphegidae, with descriptions of new species. *Proc. Zool. Soc. Lond.* **1908**: 457–535 pl xxvi [481]. Type data: holotype, BMNH *F. adult (seen 1929 by L.F. Graham), from Cairns, Qld.

Distribution: NE coastal, Qld.; only published localities Cairns, Kuranda and Mackay. Ecology: larva - sedentary, soil, predator : adult - volant, burrowing; prey Orthoptera (Gryllidae), pre-existing holes preferred. Biological references: Bohart, R.M. & Menke, A.S. (1976). *Sphecid Wasps of the World : a Generic Revision*. Berkeley : Univ. California Press ix 695 pp. (generic placement).

Liris (Leptolarra) agitata (Turner, 1908)

Notogonia agitata Turner, R.E. (1908). Notes on the Australian fossorial wasps of the family Sphegidae, with descriptions of new species. *Proc. Zool. Soc. Lond.* **1908**: 457–535 pl xxvi [477]. Type data: holotype, BMNH *F. adult (seen 1929 by L.F. Graham), from Mackay, Qld.

Distribution: NE coastal, Qld.; only published localities Mackay, Cairns and Brisbane. Ecology: larva - sedentary, soil, predator : adult - volant, burrowing; prey Orthoptera (Gryllidae), pre-existing holes preferred. Biological references:

Bohart, R.M. & Menke, A.S. (1976). *Sphecid Wasps of the World : a Generic Revision.* Berkeley : Univ. California Press ix 695 pp. (generic placement).

Liris (Leptolarra) basilissa (Turner, 1908)

Notogonia basilissa Turner, R.E. (1908). Notes on the Australian fossorial wasps of the family Sphegidae, with descriptions of new species. *Proc. Zool. Soc. Lond.* **1908**: 457–535 pl xxvi [476]. Type data: holotype, BMNH *F. adult (seen 1929 by L.F. Graham), from Mackay, Qld.

Distribution: NE coastal, Qld.; only published localities Mackay and Cairns. Ecology: larva - sedentary, soil, predator : adult - volant, burrowing; prey Orthoptera (Gryllidae), pre-existing holes preferred. Biological references: Bohart, R.M. & Menke, A.S. (1976). *Sphecid Wasps of the World : a Generic Revision.* Berkeley : Univ. California Press ix 695 pp. (generic placement).

Liris (Leptolarra) chrysonota (Smith, 1869)

Larrada chrysonota Smith, F. (1869). Descriptions of new genera and species of exotic Hymenoptera. *Trans. Entomol. Soc. Lond.* **1869**: 301–311 [304]. Type data: holotype, BMNH *F. adult (seen 1929 by L.F. Graham), from Champion Bay, W.A.

Larrada crassipes Smith, F. (1873). Descriptions of new species of fossorial Hymenoptera in the collection of the British Museum. *Ann. Mag. Nat. Hist. (4)* **12**: 291–300, 402–415 [294]. Type data: syntypes (probable), BMNH *F,M adult, from S.A.

Taxonomic decision of Turner, R.E. (1908). Notes on the Australian fossorial wasps of the family Sphegidae, with descriptions of new species. *Proc. Zool. Soc. Lond.* **1908**: 457–535 pl xxvi [475].

Distribution: S Gulfs, SW coastal, NW coastal, N coastal, S.A., W.A.; only published localities Adelaide, Swan River, Coolgardie, Champion Bay and Ord River. Ecology: larva - sedentary, soil, predator : adult - volant, burrowing; prey Orthoptera (Gryllidae), pre-existing holes preferred. Biological references: Bohart, R.M. & Menke, A.S. (1976). *Sphecid Wasps of the World : a Generic Revision.* Berkeley : Univ. California Press ix 695 pp. (generic placement).

Liris (Leptolarra) commixta (Turner, 1908)

Notogonia commixta Turner, R.E. (1908). Notes on the Australian fossorial wasps of the family Sphegidae, with descriptions of new species. *Proc. Zool. Soc. Lond.* **1908**: 457–535 pl xxvi [480]. Type data: holotype, BMNH *F. adult (seen 1929 by L.F. Graham), from Cairns, Qld.

Distribution: NE coastal, Qld.; only published localities Cairns and Kuranda. Ecology: larva - sedentary, soil, predator : adult - volant, burrowing; prey Orthoptera (Gryllidae), pre-existing holes preferred. Biological references:

Bohart, R.M. & Menke, A.S. (1976). *Sphecid Wasps of the World : a Generic Revision.* Berkeley : Univ. California Press ix 695 pp. (generic placement).

Liris (Leptolarra) festinans (Smith, 1859)

Larrada festinans Smith, F. (1859). Catalogue of hymenopterous insects collected at Celebes by Mr. A.R. Wallace. *J. Linn. Soc. Lond. Zool* **3**: 4–27 [17]. Type data: syntypes (probable), ?OUM * adult, from Celebes.

Notogonia manilae Ashmead, W.H. (1904). Descriptions of new genera and species of Hymenoptera from the Philippine Islands. *Proc. U.S. Natl. Mus.* **28**: 127–158 [130]. Type data: holotype, USNM 7996 *adult, from Philippines.

Notogonia retiaria Turner, R.E. (1908). Notes on the Australian fossorial wasps of the family Sphegidae, with descriptions of new species. *Proc. Zool. Soc. Lond.* **1908**: 457–535 pl xxvi [479]. Type data: holotype, BMNH *F. adult (seen 1929 by L.F. Graham), from Perth, W.A.

Taxonomic decision of Bohart, R.M. & Menke, A.S. (1976). *Sphecid Wasps of the World : a Generic Revision.* Berkeley : Univ. California Press ix 695 pp. [245].

Distribution: NE coastal, SW coastal, Qld., W.A.; only published localities Mackay, Kuranda, Perth and Kalamunda, also in Oriental Region and on many Pacific islands. Ecology: larva - sedentary, soil, predator : adult - volant, burrowing; prey Orthoptera (Gryllidae), pre-existing holes preferred. Biological references: Tsuneki, K. (1983). Larrinae of New Guinea in the collection of the Hungarian National Museum of Natural History Budapest (Hymenoptera, Sphecidae). *Spec. Publ. Japan Hymenopterists Assoc.* **25**: 6–53 (redescription).

Liris (Leptolarra) haemorrhoidalis (Fabricius, 1804)

Taxonomic decision of Turner, R.E. (1916). Notes on fossorial Hymenoptera. XX. On some Larrinae in the British Museum. *Ann. Mag. Nat. Hist. (8)* **17**: 248–259 [248].

Liris (Leptolarra) haemorrhoidalis magnifica Kohl, 1884

Liris magnifica Kohl, F.F. (1884). Neue Hymenopteren in den Sammlungen des k.k. zoologischen Hof-Cabinetes zu Wien. ii. *Verh. Zool.-Bot. Ges. Wien* **33**: 331–386 [356]. Type data: syntypes (probable), NHMW *F. adult, from N. Australia. [248].

Distribution: NE coastal, Qld.; only published localities Mackay to Cape York. Ecology: larva - sedentary, soil, predator : adult - volant, burrowing; prey Orthoptera (Gryllidae), pre-existing holes preferred. Biological references: Williams, F.X. (1928). Studies in tropical wasps - their hosts and associates (with descriptions of new species). *Bull. Hawaiian Sug. Pltrs'. Ass. Exp. Stn.*

Entomol. Ser. **19**: 1–179 (biology); Evans, H.E. (1958). Studies on the larvae of digger wasps (Hymenoptera, Sphecidae). Part IV Astatinae, Larrinae and Pemphredoninae. *Trans. Am. Entomol. Soc.* **84**: 109–139 (larva).

Liris (Leptolarra) obliquetruncata (Turner, 1908)

Notogonia obliquetruncata Turner, R.E. (1908). Notes on the Australian fossorial wasps of the family Sphegidae, with descriptions of new species. *Proc. Zool. Soc. Lond.* **1908**: 457–535 pl xxvi [479]. Type data: holotype, BMNH *F. adult (seen 1929 by L.F. Graham), from Port Darwin, N.T.

Distribution: SW coastal, N coastal, W.A., N.T.; only published localities Yallingup, S Perth and Darwin. Ecology: larva - sedentary, soil, predator : adult - volant, burrowing; prey Orthoptera (Gryllidae), pre-existing holes preferred. Biological references: Bohart, R.M. & Menke, A.S. (1976). *Sphecid Wasps of the World : a Generic Revision*. Berkeley : Univ. California Press ix 695 pp. (generic placement).

Liris (Leptolarra) recondita (Turner, 1916)

Notogonia recondita Turner, R.E. (1916). Notes on fossorial Hymenoptera. XXIII. *Ann. Mag. Nat. Hist. (8)* **18**: 277–288 [284]. Type data: holotype, BMNH *F. adult (seen 1929 by L.F. Graham), from Mackay, Qld.

Distribution: NE coastal, Qld.; only published localities Mackay and Kuranda. Ecology: larva - sedentary, soil, predator : adult - volant, burrowing; prey Orthoptera (Gryllidae), pre-existing holes preferred. Biological references: Bohart, R.M. & Menke, A.S. (1976). *Sphecid Wasps of the World : a Generic Revision*. Berkeley : Univ. California Press ix 695 pp. (generic placement).

Liris (Leptolarra) regina (Turner, 1908)

Notogonia regina Turner, R.E. (1908). Notes on the Australian fossorial wasps of the family Sphegidae, with descriptions of new species. *Proc. Zool. Soc. Lond.* **1908**: 457–535 [475 pl xxvi fig 7]. Type data: holotype, BMNH *F. adult (seen 1929 by L.F. Graham), from Cairns, Qld.

Distribution: NE coastal, Qld.; only published localities Cairns, Mackay, Cape York and Kuranda. Ecology: larva - sedentary, soil, predator : adult - volant, burrowing; prey Orthoptera (Gryllidae), pre-existing holes preferred. Biological references: Bohart, R.M. & Menke, A.S. (1976). *Sphecid Wasps of the World : a Generic Revision*. Berkeley : Univ. California Press ix 695 pp. (generic placement).

Liris (Leptolarra) serena (Turner, 1908)

Notogonia serena Turner, R.E. (1908). Notes on the Australian fossorial wasps of the family Sphegidae, with descriptions of new species. *Proc. Zool. Soc. Lond.* **1908**: 457–535 pl xxvi [478]. Type data: holotype, BMNH *F. adult (seen 1929 by L.F. Graham), from Mackay, Qld.

Distribution: NE coastal, N coastal, Qld., W.A.; only published localities Mackay, Kuranda, Cairns and Ord River. Ecology: larva - sedentary, soil, predator : adult - volant, burrowing; prey Orthoptera (Gryllidae), pre-existing holes preferred. Biological references: Bohart, R.M. & Menke, A.S. (1976). *Sphecid Wasps of the World : a Generic Revision*. Berkeley : Univ. California Press ix 695 pp. (generic placement).

Liris (Leptolarra) spathulifera (Turner, 1916)

Notogonia spathulifera Turner, R.E. (1916). Notes on fossorial Hymenoptera. XXIII. *Ann. Mag. Nat. Hist. (8)* **18**: 277–288 [282]. Type data: holotype, BMNH *F. adult (seen 1929 by L.F. Graham), from Port Darwin, N.T.

Distribution: N coastal, N.T.; only published localities Darwin and Bathurst Is. Ecology: larva - sedentary, soil, predator : adult - volant, burrowing; prey Orthoptera (Gryllidae), pre-existing holes preferred. Biological references: Bohart, R.M. & Menke, A.S. (1976). *Sphecid Wasps of the World : a Generic Revision*. Berkeley : Univ. California Press ix 695 pp. (generic placement).

Liris (Unplaced)

Liris australis (Saussure, 1855)

Tachytes australis Saussure, H. de (1855). Mélanges hyménoptérologiques. *Mém. Soc. Phys. Hist. Nat. Genève* **14**: 1–67 [19]. Type data: syntypes (probable), whereabouts unknown, from Australia (as New Holland).

Distribution: NE coastal, S Gulfs, SW coastal, Qld., Tas., S.A., W.A.; only published localities Cape York, Eaglehawk Neck, Adelaide and Yallingup. Ecology: larva - sedentary, soil, predator : adult - volant, burrowing; prey Orthoptera (Gryllidae), pre-existing holes preferred. Biological references: Bohart, R.M. & Menke, A.S. (1976). *Sphecid Wasps of the World : a Generic Revision*. Berkeley : Univ. California Press ix 695 pp. (generic placement); Turner, R.E. (1916). Notes on fossorial Hymenoptera. XX. On some Larrinae in the British Museum. *Ann. Mag. Nat. Hist. (8)* **17**: 248–259 (as *Notogonia australis*, redescription).

Liris nigripes (Saussure, 1867)

Larrada nigripes Saussure, H. de (1867). Hymenoptera. in, *Reise der österreichischen Fregatte Novara um die Erde in den Jahren 1857, 1858, 1859 unter den Befehlen des Commodore B. von Wüllerstorf-Urbair*. Zoologischer theil. Vol. 2(2) 138 pp. Wien : K-K Hof- und Staatsdrückerei [74]. Type data: holotype, GMNH *M. adult, from Tas.

Distribution: Tas.; type locality only. Ecology: larva - sedentary, soil, predator : adult - volant, burrowing; prey Orthoptera (Gryllidae), pre-existing holes preferred. Biological references: Schulz, W.A. (1911). Zweihundert alte Hymenopteren. *Zool. Ann. Würzburg* **4**: 1–220 (redescription); Bohart, R.M. & Menke, A.S. (1976). *Sphecid Wasps of the World : a Generic Revision.* Berkeley : Univ. California Press ix 695 pp. (generic placement).

Dicranorhina Shuckard, 1840

Dicranorhina Shuckard, W.E. (1840). *in* Swainson, W. & Shuckard, W. *On the History and Natural Arrangement of Insects.* Vol. 129 of the *Cabinet Cyclopedia* of D. Lardner. London : Longman iv 406 pp. [181]. Type species *Piagetia intaminata* Turner, 1910 by subsequent designation, see Pate, V.S.L. (1937). The generic names of the sphecoid wasps and their type species. *Mem. Am. Entomol. Soc.* **9**: 1–103. Compiled from secondary source: Bohart, R.M. & Menke, A.S. (1976). *Sphecid Wasps of the World : a Generic Revision.* Berkeley : Univ. California Press ix 695 pp.

This group is also found in Africa and the Oriental Region, see Bohart, R.M. & Menke, A.S. (1976). *Sphecid Wasps of the World : a Generic Revision.* Berkeley : Univ. California Press ix 695 pp. [250].

Dicranorhina intaminata (Turner, 1910)

Piagetia intaminata Turner, R.E. (1910). New fossorial Hymenoptera from Australia. *Trans. Entomol. Soc. Lond.* **1910**: 407–429 pl 50 [426 pl 50 fig 14]. Type data: holotype, BMNH *F. adult (seen 1929 by L.F. Graham), from Cairns, Qld.

Distribution: NE coastal, Qld.; type locality only. Ecology: larva - sedentary, soil, predator : adult - volant, burrowing; prey Orthoptera (crickets).

Tachytes Panzer, 1806

Tachytes Panzer, G.W.F. (1806). *Kritische Revision der Insektenfaune Deutschlands nach dem System bearbeitet.* Nürnberg : Felssecker Vol. 2 xii 271 pp. [129]. Type species *Pompilus tricolor* Fabricius, 1798 by monotypy.

This group is found worldwide, see Bohart, R.M. & Menke, A.S. (1976). *Sphecid Wasps of the World : a Generic Revision.* Berkeley : Univ. California Press ix 695 pp. [260].

Tachytes aestuans Turner, 1916

Tachytes aestuans Turner, R.E. (1916). Notes on fossorial Hymenoptera. XXI. On the Australian Larrinae of the genus *Tachytes. Ann. Mag. Nat. Hist. (8)* **17**: 299–306 [302]. Type data: holotype, BMNH *F. adult (seen 1929 by L.F. Graham), from Hermannsburg, N.T. and Killalpanima, S.A. (as "Central Australia Hermannsburg, Killalpanima").

Distribution: Lake Eyre basin, S.A., N.T.; only published localities Killalpanina, Hermannsburg. Ecology: larva - sedentary, soil, predator : adult - volant, burrowing; prey Orthoptera.

Tachytes approximatus Turner, 1908

Tachytes approximatus Turner, R.E. (1908). Notes on the Australian fossorial wasps of the family Sphegidae, with descriptions of new species. *Proc. Zool. Soc. Lond.* **1908**: 457–535 pl xxvi [483]. Type data: holotype, BMNH *F. adult (seen 1929 by L.F. Graham), from Mackay, Qld.

Distribution: NE coastal, Qld.; only published localities Mackay and Cairns. Ecology: larva - sedentary, soil, predator : adult - volant, burrowing; prey Orthoptera.

Tachytes codonocarpi Pulawski, 1975

Tachytes codonocarpi Pulawski, W. (1975). Two new species of Larrinae (Hym., Sphecidae) from Australia and Tunisia. *Pol. Pismo Entomol.* **45**: 165–169 [165]. Type data: holotype, SAMA *M. adult, from Everard Park Station, S.A.

Distribution: Lake Eyre basin, W plateau, S.A.; only published localities Everard Park Station and Lake Hart. Ecology: larva - sedentary, soil, predator : adult - volant, burrowing; prey Orthoptera.

Tachytes dispersus Turner, 1916

Tachytes dispersus Turner, R.E. (1916). Notes on fossorial Hymenoptera. XXI. On the Australian Larrinae of the genus *Tachytes. Ann. Mag. Nat. Hist. (8)* **17**: 299–306 [304]. Type data: holotype, BMNH *F. adult (seen 1929 by L.F. Graham), from Baudin Is., W.A.

Distribution: NE coastal, SW coastal, NW coastal, N coastal, Qld., W.A., N.T. Ecology: larva - sedentary, soil, predator : adult - volant, burrowing; prey Orthoptera.

Tachytes fatalis Turner, 1916

Tachytes fatalis Turner, R.E. (1916). Notes on fossorial Hymenoptera. XXI. On the Australian Larrinae of the genus *Tachytes. Ann. Mag. Nat. Hist. (8)* **17**: 299–306 [303]. Type data: holotype, BMNH *F. adult (seen 1929 by L.F. Graham), from Toowoomba, Qld.

Distribution: NE coastal, Qld.; type locality only. Ecology: larva - sedentary, soil, predator : adult - volant, burrowing; prey Orthoptera.

Tachytes formosissimus Turner, 1908

Tachytes formosissimus Turner, R.E. (1908). Notes on the Australian fossorial wasps of the family Sphegidae, with descriptions of new species. *Proc. Zool. Soc. Lond.* **1908**: 457–535 [482 pl xxvi fig 6]. Type data: holotype, BMNH *F. adult (seen 1929 by L.F. Graham), from Mackay, Qld.

Distribution: NE coastal, Qld.; type locality only. Ecology: larva - sedentary, soil, predator : adult - volant, burrowing; prey Orthoptera.

Tachytes mitis Turner, 1916

Tachytes mitis Turner, R.E. (1916). Notes on fossorial Hymenoptera. XXI. On the Australian Larrinae of the genus *Tachytes*. *Ann. Mag. Nat. Hist. (8)* **17**: 299–306 [301]. Type data: holotype, BMNH *F. adult (seen 1929 by L.F. Graham), from Kalamunda, W.A.

Distribution: NE coastal, SW coastal, Qld., W.A.; only published localities Townsville and Kalamunda. Ecology: larva - sedentary, soil, predator : adult - volant, burrowing; prey Orthoptera.

Tachytes plutocraticus Turner, 1910

Tachytes plutocraticus Turner, R.E. (1910). Additions to our knowledge of the fossorial wasps of Australia. *Proc. Zool. Soc. Lond.* **1910**: 253–356 [348]. Type data: holotype, BMNH *F. adult (seen 1929 by L.F. Graham), from Townsville, Qld.

Distribution: NE coastal, Qld.; type locality only. Ecology: larva - sedentary, soil, predator : adult - volant, burrowing; prey Orthoptera.

Tachytes relucens Turner, 1916

Tachytes relucens Turner, R.E. (1916). Notes on fossorial Hymenoptera. XXI. On the Australian Larrinae of the genus *Tachytes*. *Ann. Mag. Nat. Hist. (8)* **17**: 299–306 [300]. Type data: holotype, BMNH *F. adult (seen 1929 by L.F. Graham), from Mackay, Qld.

Distribution: NE coastal, Qld.; only published localities Mackay and Cape York. Ecology: larva - sedentary, soil, predator : adult - volant, burrowing; prey Orthoptera.

Tachytes rubellus Turner, 1908

Tachytes rubellus Turner, R.E. (1908). Notes on the Australian fossorial wasps of the family Sphegidae, with descriptions of new species. *Proc. Zool. Soc. Lond.* **1908**: 457–535 pl xxvi [482]. Type data: holotype, BMNH *M. adult (seen 1929 by L.F. Graham), from Port Darwin, N.T.

Distribution: S Gulfs, N coastal, S.A., N.T.; only published localities Adelaide and Darwin. Ecology: larva - sedentary, soil, predator : adult - volant, burrowing; prey Orthoptera.

Tachytes sulcatus Turner, 1916

Tachytes sulcatus Turner, R.E. (1916). Notes on fossorial Hymenoptera. XXI. On the Australian Larrinae of the genus *Tachytes*. *Ann. Mag. Nat. Hist. (8)* **17**: 299–306 [304]. Type data: holotype, BMNH *F. adult (seen 1929 by L.F. Graham), from Busselton, Cottesloe, W.A.

Distribution: SW coastal, W.A.; type locality only. Ecology: larva - sedentary, soil, predator : adult - volant, burrowing; prey Orthoptera.

Tachytes tachyrrhostus Saussure, 1855

Tachytes tachyrrhostus Saussure, H. de (1855). Mélanges hyménoptérologiques. *Mém. Soc. Phys. Hist. Nat. Genève* **14**: 1–67 [18]. Type data: holotype, GMNH *M. adult, from Australia (as New Holland).

Distribution: N.S.W., Vic., Tas.; published localities as states only. Ecology: larva - sedentary, soil, predator : adult - volant, burrowing; prey Orthoptera. Biological references: Schulz, W.A. (1911). Zweihundert alte Hymenopteren. *Zool. Ann. Würzburg* **4**: 1–220 (redescription); Turner, R.E. (1916). Notes on fossorial Hymenoptera. XXI. On the Australian Larrinae of the genus *Tachytes*. *Ann. Mag. Nat. Hist. (8)* **17**: 299–306 (identity doubtful).

Tachysphex Kohl, 1883

Tachysphex Kohl, F.F. (1883). Ueber neue Grabwespen des Mediterrangebietes. *Dtsch. Entomol. Z.* **27**:161–186 [166]. Type species *Tachysphex filicornis* Kohl, 1883 by subsequent designation, see Bingham, C.T. (1897). *The Fauna of British India, Including Ceylon and Burma*. Hymenoptera. Vol. I Wasps and Bees. London : Taylor & Francis xxx 579 pp. [192].

This group is found worldwide, see Bohart, R.M. & Menke, A.S. (1976). *Sphecid Wasps of the World : a Generic Revision*. Berkeley : Univ. California Press ix 695 pp. [267].

Tachysphex aborigenus Pulawski, 1977

Tachysphex aborigenus Pulawski, W. (1977). A synopsis of *Tachysphex* Kohl (Hym., Sphecidae) of Australia and Oceania. *Pol. Pismo Entomol.* **47**: 203–332 [312]. Type data: holotype, BMNH *F. adult, from Perth, W.A.

Distribution: NE coastal, Murray-Darling basin, Lake Eyre basin, W plateau, SW coastal, N coastal, Qld., N.S.W., S.A., W.A., N.T. Ecology: larva - sedentary, soil, predator : adult - volant, burrowing.

Tachysphex brevicornis Pulawski, 1977

Tachysphex brevicornis Pulawski, W. (1977). A synopsis of *Tachysphex* Kohl (Hym., Sphecidae) of Australia and Oceania. *Pol. Pismo Entomol.* **47**: 203–332 [263]. Type data: holotype, ANIC M. adult, from Bessie Spring, 8 km ESE of Cape Crawford, N.T.

Distribution: N coastal, N Gulf, N.T.; only published localities Bessie Spring, Borroloola, Caranbirini and Pine Creek. Ecology: larva - sedentary, soil, predator : adult - volant, burrowing.

Tachysphex buccalis Pulawski, 1977

Tachysphex buccalis Pulawski, W. (1977). A synopsis of *Tachysphex* Kohl (Hym., Sphecidae) of Australia and Oceania. *Pol. Pismo Entomol.* **47**: 203-332 [280]. Type data: holotype, ANIC M. adult, from Tomkinson Ranges, Mt. Davies and vicinity, S.A.

Distribution: W plateau, S.A.; type locality only. Ecology: larva - sedentary, soil, predator : adult - volant, burrowing.

Tachysphex circulans Pulawski, 1977

Tachysphex circulans Pulawski, W. (1977). A synopsis of *Tachysphex* Kohl (Hym., Sphecidae) of Australia and Oceania. *Pol. Pismo Entomol.* **47**: 203-332 [224]. Type data: holotype, BMNH *F. adult, from Yallingup, W.A.

Distribution: SW coastal, NW coastal, W.A.; only published localities Yallingup, Yanchep and Wiluna. Ecology: larva - sedentary, soil, predator : adult - volant, burrowing.

Tachysphex contrarius Pulawski, 1977

Tachysphex contrarius Pulawski, W. (1977). A synopsis of *Tachysphex* Kohl (Hym., Sphecidae) of Australia and Oceania. *Pol. Pismo Entomol.* **47**: 203-332 [241]. Type data: holotype, BMNH *F. adult, from Kalamunda, W.A.

Distribution: SE coastal, Murray-Darling basin, S Gulfs, SW coastal, NW coastal, N Gulf, Qld., N.S.W., A.C.T., Vic., Tas., S.A., W.A., N.T. Ecology: larva - sedentary, soil, predator : adult - volant, burrowing.

Tachysphex depressiventris Turner, 1916

Tachysphex depressiventris Turner, R.E. (1916). Notes on fossorial Hymenoptera. XX. On some Larrinae in the British Museum. *Ann. Mag. Nat. Hist. (8)* **17**: 248-259 [256]. Type data: lectotype, BMNH *F. adult, from Yallingup, W.A., designation by Pulawski, W. (1977). A synopsis of *Tachysphex* Kohl (Hym., Sphecidae) of Australia and Oceania. *Pol. Pismo Entomol.* **47**: 203-332 [270].

Distribution: NE coastal, SE coastal, Murray-Darling basin, SW coastal, N coastal, N Gulf, Qld., N.S.W., A.C.T., Vic., Tas., W.A., N.T.; also on New Guinea. Ecology: larva - sedentary, soil, predator : adult - volant, burrowing; prey Blattodea. Biological references: Evans, H.E., Matthews, R.W. & Pulawski, W. (1976). Notes on the nests and prey of four Australian species of *Tachysphex* Kohl, with description of a new species (Hymenoptera : Sphecidae). *J. Aust. Entomol. Soc.* **15**: 441-445 (nest, prey); Pulawski, W. (1977). A synopsis of *Tachysphex* Kohl (Hym., Sphecidae) of Australia and Oceania. *Pol. Pismo Entomol.* **47**: 203-332 (redescription); Alcock, J. (1980). Notes on the reproductive behaviour of some Australian solitary wasps (Hymenoptera : Sphecidae, *Tachysphex* and *Exeirus*). *J. Aust. Entomol. Soc.* **19**: 259-262 (nest, prey).

Tachysphex discrepans Turner, 1915

Tachysphex discrepans Turner, R.E. (1915). Notes on fossorial Hymenoptera. XVI. On the Thynnidae, Scoliidae and Crabronidae of Tasmania. *Ann. Mag. Nat. Hist. (8)* **15**: 537-559 [555]. Type data: lectotype, BMNH *F. adult (seen 1929 by L.F. Graham), from Eaglehawk Neck, Tas., designation by Pulawski, W. (1977). A synopsis of *Tachysphex* Kohl (Hym., Sphecidae) of Australia and Oceania. *Pol. Pismo Entomol.* **47**: 203-332 [264].

Distribution: SE coastal, Murray-Darling basin, N.S.W., A.C.T., Vic., Tas. Ecology: larva - sedentary, soil, predator : adult - volant, burrowing. Biological references: Pulawski, W. (1977). A synopsis of *Tachysphex* Kohl (Hym., Sphecidae) of Australia and Oceania. *Pol. Pismo Entomol.* **47**: 203-332 (redescription).

Tachysphex eucalypticus Pulawski, 1977

Tachysphex eucalypticus Pulawski, W. (1977). A synopsis of *Tachysphex* Kohl (Hym., Sphecidae) of Australia and Oceania. *Pol. Pismo Entomol.* **47**: 203-332 [251]. Type data: holotype, ANIC F. adult, from 5-15 mi S of Rainbow, Vic.

Distribution: Murray-Darling basin, W plateau, NW coastal, Vic., S.A., W.A.; only published localities near Rainbow, Mt. Davies and Carnarvon. Ecology: larva - sedentary, soil, predator : adult - volant, burrowing.

Tachysphex fanuiensis Cheesman, 1928

Taxonomic decision of Pulawski, W. (1977). A synopsis of *Tachysphex* Kohl (Hym., Sphecidae) of Australia and Oceania. *Pol. Pismo Entomol.* **47**: 203-332 [292].

Tachysphex fanuiensis corallinus Pulawski, 1977

Tachysphex fanuiensis corallinus Pulawski, W. (1977). A synopsis of *Tachysphex* Kohl (Hym., Sphecidae) of Australia and Oceania. *Pol. Pismo Entomol.* **47**: 203-332 [295]. Type data: holotype, ANIC F. adult, from Heron Is., Qld.

Distribution: NE coastal, SE coastal, N coastal, N Gulf, Qld., N.S.W., W.A., N.T.; also on New Guinea and the Solomon Ils. Ecology: larva - sedentary, soil, predator : adult - volant, burrowing.

Tachysphex fanuiensis howeanus Pulawski, 1977

Tachysphex fanuiensis howeanus Pulawski, W. (1977). A synopsis of *Tachysphex* Kohl (Hym., Sphecidae) of Australia and Oceania. *Pol. Pismo Entomol.* **47**: 203-332 [294]. Type data: holotype, ANIC F. adult, from Lord Howe Is.

Distribution: Lord Howe Is.; type locality only. Ecology: larva - sedentary, soil, predator : adult - volant, burrowing.

Tachysphex foliaceus Pulawski, 1977

Tachysphex foliaceus Pulawski, W. (1977). A synopsis of *Tachysphex* Kohl (Hym., Sphecidae) of Australia and Oceania. *Pol. Pismo Entomol.* **47**: 203-332 [267]. Type data: holotype, ANIC F. adult, from Cooper Creek, 11 km S by W of Nimbuwah Rock, N.T.

Distribution: NE coastal, N coastal, Qld., N.T.; only published localities near Cairns, Duaringa, Cooper Creek (12°17'S, 133°20'E) and Tumbling Waters. Ecology: larva - sedentary, soil, predator : adult - volant, burrowing.

Tachysphex fortior Turner, 1908

Tachysphex fortior Turner, R.E. (1908). Notes on the Australian fossorial wasps of the family Sphegidae, with descriptions of new species. *Proc. Zool. Soc. Lond.* **1908**: 457-535 pl xxvi [486]. Type data: holotype, BMNH *F. adult (seen 1929 by L.F. Graham), from SW Australia.

Distribution: SW coastal, NW coastal, W.A.; only published localities Dongara, near Geraldton, Mingenew, Yanchep and Yallingup. Ecology: larva - sedentary, soil, predator : adult - volant, burrowing. Biological references: Pulawski, W. (1977). A synopsis of *Tachysphex* Kohl (Hym., Sphecidae) of Australia and Oceania. *Pol. Pismo Entomol.* **47**: 203-332 (redescription).

Tachysphex galeatus Pulawski, 1977

Tachysphex galeatus Pulawski, W. (1977). A synopsis of *Tachysphex* Kohl (Hym., Sphecidae) of Australia and Oceania. *Pol. Pismo Entomol.* **47**: 203-332 [223]. Type data: holotype, ANIC F. adult, from 15 mi W of Bowen, Qld.

Distribution: NE coastal, Murray-Darling basin, Lake Eyre basin, Qld., N.S.W., N.T. Ecology: larva - sedentary, soil, predator : adult - volant, burrowing.

Tachysphex hypoleius (Smith, 1856)

Tachytes hypoleius Smith, F. (1856). *Catalogue of Hymenopterous Insects in the Collection of the British Museum.* Part IV, Sphegidae, Larridae and Crabronidae. pp. 207-497 London : British Museum [302]. Type data: holotype, BMNH *F. adult (seen 1929 by L.F. Graham), from Swan River, W.A.

Distribution: SW coastal, NW coastal, W.A.; only published localities Swan River and Wururga. Ecology: larva - sedentary, soil, predator : adult - volant, burrowing. Biological references: Turner, R.E. (1908). Notes on the Australian fossorial wasps of the family Sphegidae, with descriptions of new species. *Proc. Zool. Soc. Lond.* **1908**: 457-535 pl xxvi (generic placement); Pulawski, W. (1977). A synopsis of *Tachysphex* Kohl (Hym., Sphecidae) of Australia and Oceania. *Pol. Pismo Entomol.* **47**: 203-332 (redescription).

Tachysphex imbellis Turner, 1908

Tachysphex imbellis Turner, R.E. (1908). Notes on the Australian fossorial wasps of the family Sphegidae, with descriptions of new species. *Proc. Zool. Soc. Lond.* **1908**: 457-535 pl xxvi [485]. Type data: holotype, BMNH *F. adult (seen 1929 by L.F. Graham), from Mackay, Qld.

Tachysphex adelaidae Turner, R.E. (1910). New fossorial Hymenoptera from Australia. *Trans. Entomol. Soc. Lond.* **1910**: 407-429 pl 50 [425]. Type data: holotype, BMNH *F. adult (seen 1929 by L.F. Graham), from Adelaide, S.A.

Taxonomic decision of Pulawski, W. (1977). A synopsis of *Tachysphex* Kohl (Hym., Sphecidae) of Australia and Oceania. *Pol. Pismo Entomol.* **47**: 203-332 [249].

Distribution: NE coastal, S Gulfs, Qld., S.A.; only published localities Mackay and Adelaide. Ecology: larva - sedentary, soil, predator : adult - volant, burrowing.

Tachysphex mackayensis Turner, 1908

Tachytes australis Saussure, H. de (1867). Hymenoptera. in, Reise der österreichischen Fregatte Novara um die Erde in den Jahren 1857, 1858, 1859 unter den Befehlen des Commodore B. von Wüllerstorf-Urbair. Zoologischer theil. Vol. 2(2) 138 pp. Wien : K-K Hof- und Staatsdrückerei [68] [*non Tachytes australis* Saussure, 1855]. Type data: lectotype, NHMW *M. adult, from Sydney, N.S.W., designation by Pulawski, W. (1977). A synopsis of *Tachysphex* Kohl (Hym., Sphecidae) of Australia and Oceania. *Pol. Pismo Entomol.* **47**: 203-332 [297].

Tachysphex mackayensis Turner, R.E. (1908). Notes on the Australian fossorial wasps of the family Sphegidae, with descriptions of new species. *Proc. Zool. Soc. Lond.* **1908**: 457-535 pl xxvi [487]. Type data: holotype, BMNH *F. adult (seen 1929 by L.F. Graham), from Mackay, Qld.

Taxonomic decision of Pulawski, W. (1977). A synopsis of *Tachysphex* Kohl (Hym., Sphecidae) of Australia and Oceania. *Pol. Pismo Entomol.* **47**: 203-332 [297].

Distribution: NE coastal, SE coastal, Murray-Darling basin, S Gulfs, SW coastal, NW coastal, N coastal, N Gulf, Qld., N.S.W., A.C.T., Vic., Tas., S.A., W.A., N.T. Ecology: larva - sedentary, soil, predator : adult - volant, burrowing; prey Blattodea. Biological references: Evans, H.E., Matthews, R.W. & Pulawski, W. (1976). Notes on the nests and prey of four Australian species of *Tachysphex* Kohl, with description of a new species (Hymenoptera : Sphecidae). *J. Aust. Entomol. Soc.* **15**: 441-445 (prey).

Tachysphex maculipennis Pulawski, 1977

Tachysphex maculipennis Pulawski, W. (1977). A synopsis of *Tachysphex* Kohl (Hym., Sphecidae) of Australia and Oceania. *Pol. Pismo Entomol.* **47**: 203-332 [324]. Type data: holotype, ANIC F. adult, from 13 mi NE Geraldton, W.A.

Distribution: NE coastal, Murray-Darling basin, Lake Eyre basin, W plateau, SW coastal, NW coastal, N coastal, Qld., N.S.W., S.A., W.A., N.T. Ecology: larva - sedentary, soil, predator : adult - volant, burrowing.

Tachysphex maximus Pulawski, 1977

Tachysphex maximus Pulawski, W. (1977). A synopsis of *Tachysphex* Kohl (Hym., Sphecidae) of Australia and Oceania. *Pol. Pismo Entomol.* **47**: 203–332 [287]. Type data: holotype, ANIC F. adult, from 40 mi N of Broken Hill, N.S.W.

Distribution: Lake Eyre basin, Bulloo River basin, Qld., N.S.W., S.A.; only published localities Nocatunga, near Broken Hill and Wilpena Pound. Ecology: larva - sedentary, soil, predator : adult - volant, burrowing.

Tachysphex multifasciatus Pulawski, 1976

Tachysphex multifasciatus Pulawski, W. *in* Evans, H.E., Matthews, R.W. & Pulawski, W. (1976). Notes on the nests and prey of four Australian species of *Tachysphex* Kohl, with description of a new species (Hymenoptera : Sphecidae). *J. Aust. Entomol. Soc.* **15**: 441–445 [443]. Type data: holotype, ANIC M. adult, from Mt. Isa and vicinity, Qld.

Distribution: N coastal, N Gulf, Qld., W.A., N.T.; only published localities Mt. Isa, Drysdale River, Humpty Doo and Manbulloo Station - Katherine. Ecology: larva - sedentary, soil, predator : adult - volant, burrowing; prey Mantodea. Biological references: Evans, H.E., Matthews, R.W. & Pulawski, W. (1976). Notes on the nests and prey of four Australian species of *Tachysphex* Kohl, with description of a new species (Hymenoptera : Sphecidae). *J. Aust. Entomol. Soc.* **15**: 441–445 (nest, prey); Pulawski, W. (1977). A synopsis of *Tachysphex* Kohl (Hym., Sphecidae) of Australia and Oceania. *Pol. Pismo Entomol.* **47**: 203–332 (redescription).

Tachysphex nefarius Pulawski, 1977

Tachysphex nefarius Pulawski, W. (1977). A synopsis of *Tachysphex* Kohl (Hym., Sphecidae) of Australia and Oceania. *Pol. Pismo Entomol.* **47**: 203–332 [327]. Type data: holotype, ANIC F. adult, from Bluff Range, Biggenden, Qld.

Distribution: NE coastal, Qld.; type locality only. Ecology: larva - sedentary, soil, predator : adult - volant, burrowing.

Tachysphex novarae (Saussure, 1867)

Tachytes novarae Saussure, H. de (1867). Hymenoptera. *in, Reise der österreichischen Fregatte Novara um die Erde in den Jahren 1857, 1858, 1859 unter den Befehlen des Commodore B. von Wüllerstorf-Urbair*. Zoologischer Theil. Vol. 2(2) 138 pp. Wien : K-K Hof- und Staatsdrückerei [69]. Type data: syntypes (probable), NHMW *F. adult, from Nicobar Ils.

Distribution: NE coastal, Qld.; also in New Guinea and Oriental region. Ecology: larva - sedentary, soil, predator : adult - volant, burrowing. Biological references: Kohl, F.F. (1884). Die Gattungen und Arten der Larriden Autorum. *Verh. Zool.-Bot. Ges. Wien* **34**: 171–268, 327–454 (generic placement); Pulawski, W. (1977). A synopsis of *Tachysphex* Kohl (Hym., Sphecidae) of Australia and Oceania. *Pol. Pismo Entomol.* **47**: 203–332 (redescription).

Tachysphex pacificus Turner, 1908

Tachysphex pacificus Turner, R.E. (1908). Notes on the Australian fossorial wasps of the family Sphegidae, with descriptions of new species. *Proc. Zool. Soc. Lond.* **1908**: 457–535 pl xxvi [491]. Type data: holotype, BMNH *F. adult (seen 1929 by L.F. Graham), from Melbourne, Vic.

Distribution: NE coastal, SE coastal, Murray-Darling basin, SW coastal, Qld., N.S.W., A.C.T., Vic., Tas., W.A. Ecology: larva - sedentary, soil, predator : adult - volant, burrowing. Biological references: Pulawski, W. (1977). A synopsis of *Tachysphex* Kohl (Hym., Sphecidae) of Australia and Oceania. *Pol. Pismo Entomol.* **47**: 203–332 (redescription).

Tachysphex paucispina Pulawski, 1977

Tachysphex paucispina Pulawski, W. (1977). A synopsis of *Tachysphex* Kohl (Hym., Sphecidae) of Australia and Oceania. *Pol. Pismo Entomol.* **47**: 203–332 [230]. Type data: holotype, BMNH *F. adult, from Dongara, W.A.

Distribution: SW coastal, NW coastal, W.A.; only published localities Dongara and Nannup. Ecology: larva - sedentary, soil, predator : adult - volant, burrowing.

Tachysphex persistans Turner, 1916

Tachysphex persistans Turner, R.E. (1916). Notes on fossorial Hymenoptera. XX. On some Larrinae in the British Museum. *Ann. Mag. Nat. Hist. (8)* **17**: 248–259 [256]. Type data: holotype, BMNH *F. adult (seen 1929 by L.F. Graham), from Yallingup, W.A.

Distribution: NE coastal, SE coastal, Murray-Darling basin, S Gulfs, Lake Eyre basin, SW coastal, W plateau, N coastal, Qld., N.S.W., A.C.T., Vic., Tas., S.A., W.A., N.T. Ecology: larva - sedentary, soil, predator : adult - volant, burrowing. Biological references: Pulawski, W. (1977). A synopsis of *Tachysphex* Kohl (Hym., Sphecidae) of Australia and Oceania. *Pol. Pismo Entomol.* **47**: 203–332 (redescription).

Tachysphex pilosulus Turner, 1908

Tachysphex pilosulus Turner, R.E. (1908). Notes on the Australian fossorial wasps of the family Sphegidae, with descriptions of new species. *Proc. Zool. Soc. Lond.* **1908**: 457–535 pl xxvi [488]. Type data: lectotype, BMNH *F. adult, from Cape York, Qld., designation by Pulawski, W.

(1977). A synopsis of *Tachysphex* Kohl (Hym., Sphecidae) of Australia and Oceania. *Pol. Pismo Entomol.* **47**: 203–332 [320].

Distribution: NE coastal, SE coastal, Murray-Darling basin, Lake Eyre basin, W plateau, SW coastal, NW coastal, N coastal, N Gulf, Qld., N.S.W., Vic., S.A., W.A., N.T. Ecology: larva - sedentary, soil, predator : adult - volant, burrowing; prey Mantodea. Biological references: Evans, H.E., Matthews, R.W. & Pulawski, W. (1976). Notes on the nests and prey of four Australian species of *Tachysphex* Kohl, with description of a new species (Hymenoptera : Sphecidae). *J. Aust. Entomol. Soc.* **15**: 441–445 (nest, prey); Pulawski, W. (1977). A synopsis of *Tachysphex* Kohl (Hym., Sphecidae) of Australia and Oceania. *Pol. Pismo Entomol.* **47**: 203–332 (redescription); Alcock, J. (1980). Notes on the reproductive behaviour of some Australian solitary wasps (Hymenoptera : Sphecidae, *Tachysphex* and *Exeirus*). *J. Aust. Entomol. Soc.* **19**: 259–262 (nest, prey).

Tachysphex platypus Pulawski, 1977

Tachysphex platypus Pulawski, W. (1977). A synopsis of *Tachysphex* Kohl (Hym., Sphecidae) of Australia and Oceania. *Pol. Pismo Entomol.* **47**: 203–332 [253]. Type data: holotype, ANIC F. adult, from Marchagee, W.A.

Distribution: SW coastal, W.A.; type locality only. Ecology: larva - sedentary, soil, predator : adult - volant, burrowing.

Tachysphex pleuralis Pulawski, 1977

Tachysphex pleuralis Pulawski, W. (1977). A synopsis of *Tachysphex* Kohl (Hym., Sphecidae) of Australia and Oceania. *Pol. Pismo Entomol.* **47**: 203–332 [279]. Type data: holotype, ANIC M. adult, from McArthur River, 48 km SW by S of Borroloola, N.T.

Distribution: Lake Eyre basin, N Gulf, S.A., N.T.; only published localities Everard Park Station and McArthur River. Ecology: larva - sedentary, soil, predator : adult - volant, burrowing.

Tachysphex politus Pulawski, 1977

Tachysphex politus Pulawski, W. (1977). A synopsis of *Tachysphex* Kohl (Hym., Sphecidae) of Australia and Oceania. *Pol. Pismo Entomol.* **47**: 203–332 [277]. Type data: holotype, ANIC F. adult, from 40 mi N of Broken Hill, N.S.W.

Distribution: Bulloo River basin, Lake Eyre basin, Qld., N.S.W.; only published localities Nockatunga, near Broken Hill and Tibooburra. Ecology: larva - sedentary, soil, predator : adult - volant, burrowing.

Tachysphex proteus Pulawski, 1977

Tachysphex proteus Pulawski, W. (1977). A synopsis of *Tachysphex* Kohl (Hym., Sphecidae) of Australia and Oceania. *Pol. Pismo Entomol.* **47**: 203–332 [255]. Type data: holotype, NMV *F. adult, from Wilkur, Vic.

Distribution: Murray-Darling basin, Vic.; type locality only. Ecology: larva - sedentary, soil, predator : adult - volant, burrowing.

Tachysphex pubescens Pulawski, 1977

Tachysphex pubescens Pulawski, W. (1977). A synopsis of *Tachysphex* Kohl (Hym., Sphecidae) of Australia and Oceania. *Pol. Pismo Entomol.* **47**: 203–332 [248]. Type data: holotype, ANIC F. adult, from Wyperfeld Natl. Park, 25 mi N of Rainbow, Vic.

Distribution: Murray-Darling basin, Vic.; type locality only. Ecology: larva - sedentary, soil, predator : adult - volant, burrowing.

Tachysphex pugnator Turner, 1908

Tachysphex pugnator Turner, R.E. (1908). Notes on the Australian fossorial wasps of the family Sphegidae, with descriptions of new species. *Proc. Zool. Soc. Lond.* **1908**: 457–535 pl xxvi [491]. Type data: holotype, BMNH *F. adult (seen 1929 by L.F. Graham), from Adelaide, S.A.

Distribution: NE coastal, SE coastal, Murray-Darling basin, S Gulfs, Lake Eyre basin, W plateau, SW coastal, NW coastal, N coastal, Qld., N.S.W., Vic., S.A., W.A., N.T. Ecology: larva - sedentary, soil, predator : adult - volant, burrowing. Biological references: Pulawski, W. (1977). A synopsis of *Tachysphex* Kohl (Hym., Sphecidae) of Australia and Oceania. *Pol. Pismo Entomol.* **47**: 203–332 (redescription).

Tachysphex puncticeps Cameron, 1903

Tachysphex puncticeps Cameron, P. (1903). Descriptions of nineteen new species of Larridae, *Odynerus* and Apidae from Barrackpore. *Trans. Entomol. Soc. Lond.* **1903**: 117–132 [127]. Type data: holotype, OUM *F. adult, from Bengal, India.

Tachysphex rugidorsatus Turner, R.E. (1915). Notes on fossorial Hymenoptera. XVI. On the Thynnidae, Scoliidae and Crabronidae of Tasmania. *Ann. Mag. Nat. Hist. (8)* **15**: 537–559 [556]. Type data: holotype, BMNH 21.216 *F. adult (seen 1929 by L.F. Graham), from Eaglehawk Neck, Tas.

Taxonomic decision of Pulawski, W. (1975). Synonymical notes on Larrinae and Astatinae (Hymenoptera : Sphecidae). *J. Wash. Acad. Sci.* **64**: 308–323 [311].

Distribution: NE coastal, SE coastal, Murray-Darling basin, S Gulfs, Lake Eyre basin, W plateau, SW coastal, NW coastal, N coastal, N Gulf, Qld., N.S.W., A.C.T., Vic., Tas., S.A., W.A., N.T.; also in India and Oriental Region. Ecology: larva - sedentary, soil, predator : adult - volant, burrowing. Biological references: Pulawski, W.

(1977). A synopsis of *Tachysphex* Kohl (Hym., Sphecidae) of Australia and Oceania. *Pol. Pismo Entomol.* **47**: 203–332 (redescription).

Tachysphex rhynchocephalus Pulawski, 1977

Tachysphex rhynchocephalus Pulawski, W. (1977). A synopsis of *Tachysphex* Kohl (Hym., Sphecidae) of Australia and Oceania. *Pol. Pismo Entomol.* **47**: 203–332 [281]. Type data: holotype, ANIC M. adult, from 30 mi S Mt. Davies, S.A.

Distribution: W plateau, S.A.; only published localities Mt. Davies and Musgrave Park. Ecology: larva - sedentary, soil, predator : adult - volant, burrowing.

Tachysphex stimulator Turner, 1916

Tachysphex stimulator Turner, R.E. (1916). Notes on fossorial Hymenoptera. XX. On some Larrinae in the British Museum. *Ann. Mag. Nat. Hist.* (8) **17**: 248–259 [257]. Type data: lectotype, BMNH *F. adult, from Yallingup, W.A., designation by Pulawski, W. (1977). A synopsis of *Tachysphex* Kohl (Hym., Sphecidae) of Australia and Oceania. *Pol. Pismo Entomol.* **47**: 203–332 [227].

Distribution: Murray-Darling basin, S Gulfs, SW coastal, NW coastal, N coastal, N Gulf, Vic., S.A., W.A., N.T. Ecology: larva - sedentary, soil, predator : adult - volant, burrowing. Biological references: Pulawski, W. (1977). A synopsis of *Tachysphex* Kohl (Hym., Sphecidae) of Australia and Oceania. *Pol. Pismo Entomol.* **47**: 203–332 (redescription).

Tachysphex subopacus Turner, 1910

Tachysphex debilis Turner, R.E. (1908). Notes on the Australian fossorial wasps of the family Sphegidae, with descriptions of new species. *Proc. Zool. Soc. Lond.* **1908**: 457–535 pl xxvi [490] [*non Tachysphex debilis* Perez, 1907]. Type data: holotype, BMNH *F. adult (seen 1929 by L.F. Graham), from Cairns, Qld.

Tachysphex subopacus Turner, R.E. (1910). Additions to our knowledge of the fossorial wasps of Australia. *Proc. Zool. Soc. Lond.* **1910**: 253–356 [348] [*nom. nov.* for *Tachysphex debilis* Turner, 1908].

Distribution: NE coastal, Qld.; only published localities Cairns and Kuranda. Ecology: larva - sedentary, soil, predator : adult - volant, burrowing. Biological references: Pulawski, W. (1977). A synopsis of *Tachysphex* Kohl (Hym., Sphecidae) of Australia and Oceania. *Pol. Pismo Entomol.* **47**: 203–332 (redescription).

Tachysphex tenuis Turner, 1908

Tachysphex tenuis Turner, R.E. (1908). Notes on the Australian fossorial wasps of the family Sphegidae, with descriptions of new species. *Proc. Zool. Soc. Lond.* **1908**: 457–535 pl xxvi [489]. Type data: holotype, BMNH *F. adult (seen 1929 by L.F. Graham), from Port Darwin, N.T.

Distribution: NE coastal, Murray-Darling basin, SW coastal, N coastal, N Gulf, Qld., A.C.T., S.A., W.A., N.T. Ecology: larva - sedentary, soil, predator : adult - volant, burrowing. Biological references: Pulawski, W. (1977). A synopsis of *Tachysphex* Kohl (Hym., Sphecidae) of Australia and Oceania. *Pol. Pismo Entomol.* **47**: 203–332 (redescription).

Tachysphex tenuisculptus Pulawski, 1977

Tachysphex tenuisculptus Pulawski, W. (1977). A synopsis of *Tachysphex* Kohl (Hym., Sphecidae) of Australia and Oceania. *Pol. Pismo Entomol.* **47**: 203–332 [315]. Type data: holotype, ANIC F. adult, from Musgrave Park and vicinity (Amata), S.A.

Distribution: NE coastal, Murray-Darling basin, Lake Eyre basin, W plateau, SW coastal, NW coastal, N coastal, N Gulf, Qld., N.S.W., S.A., W.A., N.T. Ecology: larva - sedentary, soil, predator : adult - volant, burrowing.

Tachysphex truncatifrons Turner, 1908

Tachysphex truncatifrons Turner, R.E. (1908). Notes on the Australian fossorial wasps of the family Sphegidae, with descriptions of new species. *Proc. Zool. Soc. Lond.* **1908**: 457–535 pl xxvi [484]. Type data: syntypes (probable), OUM (not found by Pulawski in 1974) *F. adult, from Qld.

Distribution: doubtfully Australia. Ecology: larva - sedentary, soil, predator : adult - volant, burrowing. Biological references: Pulawski, W. (1977). A synopsis of *Tachysphex* Kohl (Hym., Sphecidae) of Australia and Oceania. *Pol. Pismo Entomol.* **47**: 203–332 (holotype not found, doubtful if Australian).

Tachysphex vardyi Pulawski, 1977

Tachysphex vardyi Pulawski, W. (1977). A synopsis of *Tachysphex* Kohl (Hym., Sphecidae) of Australia and Oceania. *Pol. Pismo Entomol.* **47**: 203–332 [233]. Type data: holotype, BMNH *adult, from South Perth, W.A.

Distribution: Murray-Darling basin, SW coastal, N coastal, Vic., W.A., N.T. Ecology: larva - sedentary, soil, predator : adult - volant, burrowing.

Tachysphex vividus Pulawski, 1977

Tachysphex vividus Pulawski, W. (1977). A synopsis of *Tachysphex* Kohl (Hym., Sphecidae) of Australia and Oceania. *Pol. Pismo Entomol.* **47**: 203–332 [274]. Type data: holotype, ANIC M. adult, from 25 mi W of Nyngan, N.S.W.

Distribution: Murray-Darling basin, Lake Eyre basin, N.S.W., S.A.; only published localities near Nyngan, Packsaddle and near Wilpena Pound. Ecology: larva - sedentary, soil, predator : adult - volant, burrowing.

Tachysphex walkeri Turner, 1908

Tachysphex walkeri Turner, R.E. (1908). Notes on the Australian fossorial wasps of the family Sphegidae, with descriptions of new species. *Proc. Zool. Soc. Lond.* **1908**: 457-535 pl xxvi [487]. Type data: holotype, BMNH *F. adult, from Sandy Islet, Long Reef, Qld.

Distribution: NE coastal, SE coastal, Murray-Darling basin, Lake Eyre basin, W plateau, NW coastal, N coastal, N Gulf, Qld., N.S.W., A.C.T., Vic., S.A., W.A., N.T. Ecology: larva - sedentary, soil, predator : adult - volant, burrowing; prey Blattodea. Biological references: Pulawski, W. (1977). A synopsis of *Tachysphex* Kohl (Hym., Sphecidae) of Australia and Oceania. *Pol. Pismo Entomol.* **47**: 203-332 (redescription); Alcock, J. (1980). Notes on the reproductive behaviour of some Australian solitary wasps (Hymenoptera : Sphecidae, *Tachysphex* and *Exeirus*). *J. Aust. Entomol. Soc.* **19**: 259-262 (nest, prey).

Aha Menke, 1977

Aha Menke, A.S. (1977). *Aha*, a new genus of Australian Sphecidae, and a revised key to the world genera of the tribe Miscophini (Hymenoptera, Larrinae). *Pol. Pismo Entomol.* **47**: 671-681 [672]. Type species *Aha ha* Menke, 1977 by original designation.

Aha evansi Menke, 1977

Aha evansi Menke, A.S. (1977). *Aha*, a new genus of Australian Sphecidae, and a revised key to the world genera of the tribe Miscophini (Hymenoptera, Larrinae). *Pol. Pismo Entomol.* **47**: 671-681 [677]. Type data: holotype, ANIC M. adult, from 12-21 mi N of Ouyen, Vic.

Distribution: Murray-Darling basin, Vic., S.A.; only published localities vicinity of Ouyen, Sherlock. Ecology: larva - sedentary, predator: adult - volant. Biological references: Lomholdt, O.C. (1980). The female *Aha evansi* Menke, 1977 (Hymenoptera : Sphecidae, Larrinae). *Entomol. Scand.* **11**: 241-244 (description of female).

Aha ha Menke, 1977

Aha ha Menke, A.S. (1977). *Aha*, a new genus of Australian Sphecidae, and a revised key to the world genera of the tribe Miscophini (Hymenoptera, Larrinae). *Pol. Pismo Entomol.* **47**: 671-681 [674]. Type data: holotype, ANIC M. adult, from Kununurra and vicinity, W.A.

Distribution: N coastal, W.A.; only published locality Kununurra and vicinity. Ecology: larva - sedentary, predator: adult - volant.

Lyroda Say, 1837

Lyroda Say, T. (1837). Descriptions of new species of North American Hymenoptera and observations on some already described. *Bost. J. Nat. Hist.* **1**: 361-416 [372] [described with subgeneric rank in *Lyrops* Illiger, 1807]. Type species *Lyrops (Lyroda) subita* Say, 1837 by subsequent designation, see Patton, W.H. (1881). Some characters useful in the study of the Sphecidae. *Proc. Bost. Soc. Nat. Hist.* **20**: 378-385. Compiled from secondary source: Pate, V.S.L. (1937). The generic names of the sphecoid wasps and their type species (Hymenoptera : Aculeata). *Mem. Am. Entomol. Soc.* **9**: 1-103.

This group is found worldwide, see Bohart, R.M. & Menke, A.S. (1976). *Sphecid Wasps of the World : a Generic Revision*. Berkeley : Univ. California Press ix 695 pp. [295].

Lyroda errans (Turner, 1936)

Gastrosericus errans Turner, R.E. (1936). Notes on fossorial Hymenoptera. XLV. On new sphegid wasps from Australia. *Ann. Mag. Nat. Hist. (10)* **18**: 533-545 [544]. Type data: syntypes (probable), ?BMNH *M. adult, from Yanchep, W.A.

Distribution: SW coastal, W.A.; type locality only. Ecology: larva - sedentary, soil, predator : adult - volant, burrowing; prey Orthoptera, nest may be in pre-existing cavity, prey of *Bembix moma* Evans and Matthews. Biological references: Menke, A.S. (1977). *Aha*, a new genus of Australian Sphecidae, and a revised key to the world genera of the tribe Miscophini (Hymenoptera, Larrinae). *Pol. Pismo Entomol.* **47**: 671-681 (generic placement); Evans, H.E. & Matthews, R.W. (1973). Systematics and nesting behavior of Australian *Bembix* sand wasps (Hymenoptera, Sphecidae). *Mem. Am. Entomol. Inst.* **20**: iv 387 pp. (as prey).

Lyroda michaelseni Schulz, 1908

Taxonomic decision of Turner, R.E. (1914). New fossorial Hymenoptera from Australia and Tasmania. *Proc. Linn. Soc. N.S.W.* **38**: 608-623

Lyroda michaelseni michaelseni Schulz, 1908

Lyroda michaelseni Schulz, W.A. (1908). Fossores. pp. 447-488 *in* Michaelsen, W. & Hartmeyer, R. (eds.) *Die Fauna Südwest-Australiens*. Vol. I Lfg. 13 Jena : G. Fischer [479]. Type data: syntypes (probable), ZMB *F.,M. adult, from Denham, W.A.

Distribution: NW coastal, W.A.; type locality only. Ecology: larva - sedentary, soil, predator : adult - volant, burrowing; prey Orthoptera, nest may be in pre-existing cavity.

Lyroda michaelseni tasmanica Turner, 1914

Lyroda michaelseni tasmanica Turner, R.E. (1914). New fossorial Hymenoptera from Australia and Tasmania. *Proc. Linn. Soc. N.S.W.* **38**: 608-623 [621]. Type data: syntypes (probable), whereabouts unknown, from Eaglehawk Neck, Tas.

Distribution: Tas.; type locality only. Ecology: larva - sedentary, soil, predator : adult - volant, burrowing; prey Orthoptera, nest may be in pre-existing cavity.

Lyroda minima Turner, 1936

Lyroda minima Turner, R.E. (1936). Notes on fossorial Hymenoptera. XLV. On new sphegid wasps from Australia. *Ann. Mag. Nat. Hist. (10)* **18**: 533-545 [545]. Type data: syntypes (probable), ?BMNH *adult, from Yanchep, W.A.

Distribution: SW coastal, W.A.; type locality only. Ecology: larva - sedentary, soil, predator : adult - volant, burrowing; prey Orthoptera, nest may be in pre-existing cavity.

Lyroda queenslandensis Turner, 1916

Lyroda queenslandensis Turner, R.E. (1916). Notes on fossorial Hymenoptera. XXIII. *Ann. Mag. Nat. Hist. (8)* **18**: 277-288 [285]. Type data: holotype, BMNH *M. adult (seen 1929 by L.F. Graham), from Bundaberg, Qld.

Distribution: NE coastal, Qld.; type locality only. Ecology: larva - sedentary, soil, predator : adult - volant, burrowing; prey Orthoptera, nest may be in pre-existing cavity.

Sericophorus Smith, 1851

Taxonomic decision of Bohart, R.M. & Menke, A.S. (1976). *Sphecid Wasps of the World : a Generic Revision*. Berkeley : Univ. California Press ix 695 pp. [44].

Sericophorus (Sericophorus) Smith, 1851

Sericophorus Smith, F. (1851). Descriptions of some new species of exotic Hymenoptera in the British Museum and other collections. *Ann. Mag. Nat. Hist. (2)* **7**: 28-33 [32]. Type species *Sericophorus chalybaeus* Smith, 1851 by monotypy.

Tachyrrhostus Saussure, H. de (1855). Mélanges hyménoptérologiques. *Mém. Soc. Phys. Hist. Nat. Genève* **14**: 1-67 [24]. Type species *Tachyrrhostus cyaneus* Saussure, 1855 by subsequent designation, see Pate, V.S.L. (1937). The generic names of the sphecoid wasps and their type species. *Mem. Am. Entomol. Soc.* **9**: 1-103 [63].

This group is also found in New Guinea, see Bohart, R.M. & Menke, A.S. (1976). *Sphecid Wasps of the World : a Generic Revision*. Berkeley : Univ. California Press ix 695 pp. [299].

Sericophorus (Sericophorus) aliceae Turner, 1936

Sericophorus aliceae Turner, R.E. (1936). Notes on fossorial Hymenoptera. XLV. On new sphegid wasps from Australia. *Ann. Mag. Nat. Hist. (10)* **18**: 533-545 [542]. Type data: holotype, BMNH *F. adult, from Mingenew, W.A.

Distribution: NW coastal, W.A.; type locality only. Ecology: larva - sedentary, soil, predator : adult - volant, burrowing; prey adult Diptera.

Sericophorus (Sericophorus) bicolor Smith, 1873

Sericophorus bicolor Smith, F. (1873). Descriptions of new species of fossorial Hymenoptera in the collection of the British Museum. *Ann. Mag. Nat. Hist. (4)* **12**: 291-300, 402-415 [405]. Type data: holotype, BMNH *F. adult (seen 1929 by L.F. Graham), from Swan River, W.A.

Distribution: Lake Eyre basin, SW coastal, Qld., W.A.; only published localities Coopers Creek and Swan River. Ecology: larva - sedentary, soil, predator : adult - volant, burrowing; prey adult Diptera.

Sericophorus (Sericophorus) brisbanensis Rayment, 1955

Sericophorus brisbanensis Rayment, T. (1955). Taxonomy, morphology and biology of sericophorine wasps. With diagnoses of two new genera and descriptions of forty new species and six subspecies. *Mem. Natl. Mus. Vict.* **19**: 11-105 pls 1-11 [26]. Type data: holotype, QM T7507 *M. adult, from Brisbane, Qld.

Distribution: NE coastal, Qld.; type locality only. Ecology: larva - sedentary, soil, predator : adult - volant, burrowing; prey adult Diptera.

Sericophorus (Sericophorus) carinatus Rayment, 1955

Sericophorus carinatus Rayment, T. (1955). Taxonomy, morphology and biology of sericophorine wasps. With diagnoses of two new genera and descriptions of forty new species and six subspecies. *Mem. Natl. Mus. Vict.* **19**: 11-105 pls 1-11 [26]. Type data: holotype, NMV 6776 *F. adult, from Sandringham, Vic.

Distribution: SE coastal, Vic.; type locality only. Ecology: larva - sedentary, soil, predator : adult - volant, burrowing; prey adult Diptera.

Sericophorus (Sericophorus) castaneus Rayment, 1955

Sericophorus castaneus Rayment, T. (1955). Taxonomy, morphology and biology of sericophorine wasps. With diagnoses of two new genera and descriptions of forty new species and six subspecies. *Mem. Natl. Mus. Vict.* **19**: 11-105 pls 1-11 [27]. Type data: holotype, QM F. adult, from Brisbane, Qld.

Distribution: NE coastal, Qld.; type locality only. Ecology: larva - sedentary, soil, predator : adult - volant, burrowing; prey adult Diptera.

Sericophorus (Sericophorus) chalybaeus Smith, 1851

Taxonomic decision of Bohart, R.M. & Menke, A.S. (1976). *Sphecid Wasps of the World : a Generic Revision*. Berkeley : Univ. California Press ix 695 pp. [302].

Sericophorus (Sericophorus) chalybaeus chalybaeus Smith, 1851

Sericophorus chalybeus Smith, F. (1851). Descriptions of some new species of exotic Hymenoptera in the British Museum and other collections. *Ann. Mag. Nat. Hist.* (2) **7**: 28–33 [32]. Type data: holotype, BMNH *F. adult (seen 1929 by L.F. Graham), from Australia (as New Holland).

Tachyrrhostus cyaneus Saussure, H. de (1855). Mélanges hyménoptérologiques. *Mém. Soc. Phys. Hist. Nat. Genève* **14**: 1–67 [26]. Type data: holotype, GMNH *F. adult, from Australia (as New Holland).

Distribution: SE coastal, SW coastal, N.S.W., Vic., Tas., W.A.; only published localities Leura, Lane Cove, Gorae West, Eaglehawk Neck and Kalamunda. Ecology: larva - sedentary, soil, predator : adult - volant, burrowing; prey adult Diptera. Biological references: Rayment, T. (1955). Biology of two hunting wasps. The specific descriptions of a new species and one allotype of *Sericophorus* and a new blowfly *Pollenia*. *Aust. Zool.* **12**: 132–141 pl 19 (nest, prey).

Sericophorus (Sericophorus) chalybaeus fulleri Rayment, 1955

Sericophorus chalybaeus fulleri Rayment, T. (1955). Taxonomy, morphology and biology of sericophorine wasps. With diagnoses of two new genera and descriptions of forty new species and six subspecies. *Mem. Natl. Mus. Vict.* **19**: 11–105 pls 1–11 [29]. Type data: holotype, ANIC F. adult, from Blundells, A.C.T.

Distribution: Murray-Darling basin, A.C.T.; type locality only. Ecology: larva - sedentary, soil, predator : adult - volant, burrowing; prey adult Diptera.

Sericophorus (Sericophorus) claviger (Kohl, 1892)

Taxonomic decision of Bohart, R.M. & Menke, A.S. (1976). *Sphecid Wasps of the World : a Generic Revision*. Berkeley : Univ. California Press ix 695 pp. [302].

Sericophorus (Sericophorus) claviger claviger (Kohl, 1892)

Tachyrhostus claviger Kohl, F.F. (1892). Neue Hymenopterenformen. *Ann. Naturh. Hofmus. Wien* **7**:197–234 [229 pl 13 figs 10,18]. Type data: syntypes (probable), NHMW *F. adult, from Australia.

Distribution: SE coastal, N.S.W., Vic.; only published localities Lane Cove, Woollahra, Mt. Victoria, Gippsland and Mordialloc. Ecology: larva - sedentary, soil, predator : adult - volant, burrowing; prey adult Diptera. Biological references: Dalla Torre, C.G. de (1897). *Catalogus Hymenopterorum Hucusque Descriptorum Systematicus et Synonymicus*. Fossores (Sphegidae). Leipzig : G. Engelmann Vol. 8 viii 749 pp. (generic placement).

Sericophorus (Sericophorus) claviger burnsiellus Rayment, 1955

Sericophorus claviger burnsiellus Rayment, T. (1955). Taxonomy, morphology and biology of sericophorine wasps. With diagnoses of two new genera and descriptions of forty new species and six subspecies. *Mem. Natl. Mus. Vict.* **19**: 11–105 pls 1–11 [30]. Type data: holotype, NMV 6778 M. adult, from Chelsea, Vic.

Distribution: SE coastal, Vic.; only published localities Chelsea, Cavendish and Lilydale. Ecology: larva - sedentary, soil, predator : adult - volant, burrowing; prey adult Diptera.

Sericophorus (Sericophorus) cliffordi Rayment, 1955

Sericophorus cliffordi Rayment, T. (1955). Taxonomy, morphology and biology of sericophorine wasps. With diagnoses of two new genera and descriptions of forty new species and six subspecies. *Mem. Natl. Mus. Vict.* **19**: 11–105 pls 1–11 [30]. Type data: holotype, NMV 6780 *F. adult, from Gorae West, Vic.

Distribution: SE coastal, Vic.; only published localities Gorae West and Portland. Ecology: larva - sedentary, soil, predator : adult - volant, burrowing; prey adult Diptera. Biological references: Rayment, T. (1955). Biology of two hunting wasps. The specific descriptions of a new species and one allotype of *Sericophorus* and a new blowfly *Pollenia*. *Aust. Zool.* **12**: 132–141 pl 19 (prey).

Sericophorus (Sericophorus) cockerelli Menke, 1976

Sericophorus hackeri Rayment, T. (1955). Taxonomy, morphology and biology of sericophorine wasps. With diagnoses of two new genera and descriptions of forty new species and six subspecies. *Mem. Natl. Mus. Vict.* **19**: 11–105 pls 1–11 [34] [*non Zoyphium hackeri* Cockerell, 1932]. Type data: holotype, QM *M. adult, from Brisbane, Qld.

Sericophorus (Sericophorus) cockerelli Bohart, R.M. & Menke, A.S. (1976). *Sphecid Wasps of the World : a Generic Revision*. Berkeley : Univ. California Press ix 695 pp. [302] [*nom. nov.* for *Sericophorus hackeri* Rayment, 1955].

Sericophorus (Sericophorus) cyanophilus Rayment, 1955

Sericophorus cyanophilus Rayment, T. (1955). Taxonomy, morphology and biology of sericophorine wasps. With diagnoses of two new genera and descriptions of forty new species and six subspecies. *Mem. Natl. Mus. Vict.* **19**: 11–105 pls 1–11 [31]. Type data: holotype, QM F. adult, from Stanthorpe, Qld.

Distribution: Murray-Darling basin, Qld.; type locality only. Ecology: larva - sedentary, soil, predator : adult - volant, burrowing; prey adult Diptera.

Sericophorus (Sericophorus) elegantior Rayment, 1955

Sericophorus elegantior Rayment, T. (1955). Taxonomy, morphology and biology of sericophorine wasps. With diagnoses of two new genera and descriptions of forty new species and six subspecies. *Mem. Natl. Mus. Vict.* **19**: 11–105 pls 1–11 [32]. Type data: holotype, NMV 6786 *F. adult, from Bolgart, W.A.

Distribution: SW coastal, W.A.; type locality only. Ecology: larva - sedentary, soil, predator : adult - volant, burrowing; prey adult Diptera.

Sericophorus (Sericophorus) froggatti Rayment, 1955

Sericophorus froggatti Rayment, T. (1955). Taxonomy, morphology and biology of sericophorine wasps. With diagnoses of two new genera and descriptions of forty new species and six subspecies. *Mem. Natl. Mus. Vict.* **19**: 11–105 pls 1–11 [33]. Type data: holotype, DARI *F. adult, from Mittagong, N.S.W.

Distribution: SE coastal, N.S.W.; type locality only. Ecology: larva - sedentary, soil, predator : adult - volant, burrowing; prey adult Diptera.

Sericophorus (Sericophorus) funebris Turner, 1907

Sericophorus funebris Turner, R.E. (1907). New species of Sphegidae from Australia. *Ann. Mag. Nat. Hist.* (7) **19**: 268–276 [276]. Type data: holotype, BMNH *F. adult (seen 1929 by L.F. Graham), from Mackay, Qld.

Distribution: NE coastal, Qld.; type locality only. Ecology: larva - sedentary, soil, predator : adult - volant, burrowing; prey adult Diptera.

Sericophorus (Sericophorus) gracilis Rayment, 1955

Sericophorus gracilis Rayment, T. (1955). Taxonomy, morphology and biology of sericophorine wasps. With diagnoses of two new genera and descriptions of forty new species and six subspecies. *Mem. Natl. Mus. Vict.* **19**: 11–105 pls 1–11 [33]. Type data: holotype, NMV T6788 *F. adult, from Glen Aplin, Qld.

Distribution: Murray-Darling basin, Qld.; type locality only. Ecology: larva - sedentary, soil, predator : adult - volant, burrowing; prey adult Diptera.

Sericophorus (Sericophorus) inornatus Rayment, 1955

Sericophorus inornatus Rayment, T. (1955). Taxonomy, morphology and biology of sericophorine wasps. With diagnoses of two new genera and descriptions of forty new species and six subspecies. *Mem. Natl. Mus. Vict.* **19**: 11–105 pls 1–11 [34]. Type data: holotype, QM T7506 *F. adult, from Wynyard, Qld."; Wynyard, Tas. was meant.

Distribution: Tas.; type locality only. Ecology: larva - sedentary, soil, predator : adult - volant, burrowing; prey adult Diptera.

Sericophorus (Sericophorus) lilacinus Rayment, 1955

Sericophorus lilacinus Rayment, T. (1955). Taxonomy, morphology and biology of sericophorine wasps. With diagnoses of two new genera and descriptions of forty new species and six subspecies. *Mem. Natl. Mus. Vict.* **19**: 11–105 pls 1–11 [35]. Type data: holotype, QM *F. adult, from Wynyard, Qld."; possibly Wynyard, Tas. was meant.

Distribution: Tas.; type locality only. Ecology: larva - sedentary, soil, predator : adult - volant, burrowing; prey adult Diptera.

Sericophorus (Sericophorus) littoralis Rayment, 1955

Sericophorus littoralis Rayment, T. (1955). Taxonomy, morphology and biology of sericophorine wasps. With diagnoses of two new genera and descriptions of forty new species and six subspecies. *Mem. Natl. Mus. Vict.* **19**: 11–105 pls 1–11 [35]. Type data: holotype, SAMA *F. adult, from Ardrossan, S.A.

Distribution: S Gulfs, S.A.; type locality only. Ecology: larva - sedentary, soil, predator : adult - volant, burrowing; prey adult Diptera.

Sericophorus (Sericophorus) metallescens Rayment, 1955

Sericophorus metallescens Rayment, T. (1955). Taxonomy, morphology and biology of sericophorine wasps. With diagnoses of two new genera and descriptions of forty new species and six subspecies. *Mem. Natl. Mus. Vict.* **19**: 11–105 pls 1–11 [36]. Type data: holotype, NMV 6790 *M. adult, from Fraser Park, N.S.W.

Distribution: SE coastal, N.S.W.; type locality only. Ecology: larva - sedentary, soil, predator : adult - volant, burrowing; prey adult Diptera.

Sericophorus (Sericophorus) minutus Rayment, 1955

Sericophorus minutus Rayment, T. (1955). Taxonomy, morphology and biology of sericophorine wasps. With diagnoses of two new genera and descriptions of forty new species and six subspecies. *Mem. Natl. Mus. Vict.* **19**: 11–105 pls 1–11 [37]. Type data: holotype, ANIC M. adult, from 20 mi SE of Bourke, N.S.W.

Distribution: Murray-Darling basin, N.S.W.; type locality only. Ecology: larva - sedentary, soil, predator : adult - volant, burrowing; prey adult Diptera.

Sericophorus (Sericophorus) nigror Rayment, 1955

Sericophorus nigror Rayment, T. (1955). Taxonomy, morphology and biology of sericophorine wasps. With diagnoses of two new genera and descriptions of forty new species and six subspecies. *Mem. Natl. Mus. Vict.* **19**: 11–105 pls 1–11 [37]. Type data: holotype, SAMA *F. adult, from Lucindale, S.A.

Distribution: Murray-Darling basin, S.A.; type locality only. Ecology: larva - sedentary, soil, predator : adult - volant, burrowing; prey adult Diptera.

Sericophorus (Sericophorus) niveifrons Rayment, 1955

Sericophorus niveifrons Rayment, T. (1955). Taxonomy, morphology and biology of sericophorine wasps. With diagnoses of two new genera and descriptions of forty new species and six subspecies. *Mem. Natl. Mus. Vict.* **19**: 11–105 pls 1–11 [38]. Type data: holotype, ANIC F. adult, from near Nocatunga, Qld.

Distribution: Bulloo River basin, Qld.; type locality only. Ecology: larva - sedentary, soil, predator : adult - volant, burrowing; prey adult Diptera.

Sericophorus (Sericophorus) occidentalis Rayment, 1955

Sericophorus occidentalis Rayment, T. (1955). Taxonomy, morphology and biology of sericophorine wasps. With diagnoses of two new genera and descriptions of forty new species and six subspecies. *Mem. Natl. Mus. Vict.* **19**: 11–105 pls 1–11 [38]. Type data: holotype, WAM 37-3929 *F. adult, from Narrogin, W.A.

Distribution: SW coastal, W.A.; type locality only. Ecology: larva - sedentary, soil, predator : adult - volant, burrowing; prey adult Diptera.

Sericophorus (Sericophorus) patongensis Rayment, 1955

Sericophorus patongensis Rayment, T. (1955). Taxonomy, morphology and biology of sericophorine wasps. With diagnoses of two new genera and descriptions of forty new species and six subspecies. *Mem. Natl. Mus. Vict.* **19**: 11–105 pls 1–11 [39]. Type data: holotype, AM F. adult, from Patonga, N.S.W.

Distribution: SE coastal, N.S.W.; type locality only. Ecology: larva - sedentary, soil, predator : adult - volant, burrowing; prey adult Diptera.

Sericophorus (Sericophorus) pescotti Rayment, 1955

Sericophorus pescotti Rayment, T. (1955). Taxonomy, morphology and biology of sericophorine wasps. With diagnoses of two new genera and descriptions of forty new species and six subspecies. *Mem. Natl. Mus. Vict.* **19**: 11–105 pls 1–11 [39]. Type data: holotype, NMV 6791 *F. adult, from Sandringham, Vic.

Distribution: NE coastal, SE coastal, Qld., Vic.; only published localities Brisbane and Sandringham. Ecology: larva - sedentary, soil, predator : adult - volant, burrowing; prey adult Diptera.

Sericophorus (Sericophorus) raymenti Menke, 1976

Sericophorus rufipes Rayment, T. (1955). Taxonomy, morphology and biology of sericophorine wasps. With diagnoses of two new genera and descriptions of forty new species and six subspecies. *Mem. Natl. Mus. Vict.* **19**: 11–105 pls 1–11 [47] [*non Zoyphium rufipes* Rohwer, 1911]. Type data: holotype, SAMA *F. adult, from Tas.

Sericophorus (Sericophorus) raymenti Menke A.S. (1976). in Bohart, R.M. & Menke, A.S. (1976). *Sphecid Wasps of the World : a Generic Revision*. Berkeley : Univ. California Press ix 695 pp. [302] [*nom. nov.* for *Sericophorus rufipes* Rayment, 1955].

Distribution: SE coastal, Murray-Darling basin, N.S.W., Vic., Tas.; only published localities Fraser Park, Woollahra, Kerang and Mt. Victoria. Ecology: larva - sedentary, soil, predator : adult - volant, burrowing; prey adult Diptera.

Sericophorus (Sericophorus) relucens Smith, 1856

Taxonomic decision of Turner, R.E. (1914). Notes on fossorial Hymenoptera. XIII. A revision of the Paranyssoninae. *Ann. Mag. Nat. Hist.* (8) **14**: 337–359 [359] and subspecies arrangement according to Rayment, T. (1955). Taxonomy, morphology and biology of sericophorine wasps. With diagnoses of two new genera and descriptions of forty new species and six subspecies. *Mem. Natl. Mus. Vict.* **19**: 11–105 pls 1–11.

Sericophorus (Sericophorus) relucens relucens Smith, 1856

Sericophorus relucens Smith, F. (1856). *Catalogue of Hymenopterous Insects in the Collection of the British Museum.* Part IV, Sphegidae, Larridae and Crabronidae. pp. 207–497 London : British Museum [357]. Type data: holotype, BMNH *F. adult (seen 1929 by L.F. Graham), from Adelaide, S.A.

Zoyphium rufipes Rohwer, S.A. (1911). Descriptions of new species of wasps with notes on described species. *Proc. U.S. Natl. Mus.* **40**(1837): 551–587 [585]. Type data: holotype, USNM 13767 *F. adult, from Duaringa, Dawson district, Qld.

Distribution: NE coastal, SE coastal, Murray-Darling basin, Lake Eyre basin, S Gulfs, SW coastal, N coastal, Qld., N.S.W., A.C.T., Vic., S.A., W.A., N.T. Ecology: larva - sedentary, soil, predator : adult - volant, burrowing; prey adult Diptera. Biological references: Matthews, R.W. & Evans, H.E. (1971). Biological notes on two species of *Sericophorus* from Australia (Hymenoptera : Sphecidae). *Psyche Camb.* **77**: 413–429 (nest, prey).

Sericophorus (Sericophorus) relucens nigricornis Rayment, 1955

Sericophorus relucens nigricornis Rayment, T. (1955). Taxonomy, morphology and biology of sericophorine wasps. With diagnoses of two new genera and descriptions of forty new species and six subspecies. *Mem. Natl. Mus. Vict.* **19**: 11–105 pls 1–11 [43]. Type data: syntypes, ANIC F. adult, from 25 mi E of Durham Downs, Qld.

Distribution: Murray-Darling basin, N coastal, Qld., N.S.W., Vic., W.A., N.T. Ecology: larva - sedentary, soil, predator : adult - volant, burrowing; prey adult Diptera.

Sericophorus (Sericophorus) relucens ruficornis Rayment, 1955

Sericophorus relucens ruficornis Rayment, T. (1955). Taxonomy, morphology and biology of sericophorine wasps. With diagnoses of two new genera and descriptions of forty new species and six subspecies. *Mem. Natl. Mus. Vict.* **19**: 11–105 pls 1–11 [44]. Type data: holotype, NMV 6805 *F. adult, from Tennant Creek, N.T.

Distribution: Lake Eyre basin, N coastal, Qld., N.S.W., Vic., S.A., W.A., N.T. Ecology: larva - sedentary, soil, predator : adult - volant, burrowing; prey adult Diptera.

Sericophorus (Sericophorus) rufobasalis Rayment, 1955

Sericophorus rufobasalis Rayment, T. (1955). Taxonomy, morphology and biology of sericophorine wasps. With diagnoses of two new genera and descriptions of forty new species and six subspecies. *Mem. Natl. Mus. Vict.* **19**: 11–105 pls 1–11 [46]. Type data: holotype, QM *M. adult whereabouts unknown, QM specimen labeled "cotype", QM specimen labeled "paratype", from Brisbane, Qld.

Distribution: NE coastal, SE coastal, Murray-Darling basin, Qld., N.S.W. Ecology: larva - sedentary, soil, predator : adult - volant, burrowing; prey adult Diptera.

Sericophorus (Sericophorus) rufotibialis Rayment, 1955

Sericophorus rufotibialis Rayment, T. (1955). Taxonomy, morphology and biology of sericophorine wasps. With diagnoses of two new genera and descriptions of forty new species and six subspecies. *Mem. Natl. Mus. Vict.* **19**: 11–105 pls 1–11 [47]. Type data: holotype, ANIC M. adult, from Blundells, A.C.T.

Distribution: Murray-Darling basin, A.C.T.; type locality only. Ecology: larva - sedentary, soil, predator : adult - volant, burrowing; prey adult Diptera.

Sericophorus (Sericophorus) rugosus Rayment, 1955

Sericophorus rugosus Rayment, T. (1955). Taxonomy, morphology and biology of sericophorine wasps. With diagnoses of two new genera and descriptions of forty new species and six subspecies. *Mem. Natl. Mus. Vict.* **19**: 11–105 pls 1–11 [48]. Type data: holotype, QM F. adult, from Brisbane, Qld.

Distribution: NE coastal, Qld.; type locality only. Ecology: larva - sedentary, soil, predator : adult - volant, burrowing; prey adult Diptera.

Sericophorus (Sericophorus) sculpturatus Rayment, 1955

Sericophorus sculpturatus Rayment, T. (1955). Biology of two hunting wasps. The specific descriptions of a new species and one allotype of *Sericophorus* and a new blowfly *Pollenia*. *Aust. Zool.* **12**: 132–141 pl 19 [133 pl 19 figs 1–12]. Type data: holotype, ANIC M. adult, from Busselton, W.A.

Distribution: SW coastal, W.A.; type locality only. Ecology: larva - sedentary, soil, predator : adult - volant, burrowing; prey adult Diptera.

Sericophorus (Sericophorus) spryi Rayment, 1955

Sericophorus spryi Rayment, T. (1955). Taxonomy, morphology and biology of sericophorine wasps. With diagnoses of two new genera and descriptions of forty new species and six subspecies. *Mem. Natl. Mus. Vict.* **19**: 11–105 pls 1–11 [48]. Type data: holotype, NMV 6793 *M. adult, from Chelsea, Vic.

Distribution: SE coastal, Vic.; only published localities Chelsea, Cavendish and Mordialloc. Ecology: larva - sedentary, soil, predator : adult - volant, burrowing; prey adult Diptera.

Sericophorus (Sericophorus) subviridis Rayment, 1955

Sericophorus subviridis Rayment, T. (1955). Taxonomy, morphology and biology of sericophorine wasps. With diagnoses of two new genera and descriptions of forty new species and six subspecies. *Mem. Natl. Mus. Vict.* **19**: 11–105 pls 1–11 [49]. Type data: holotype, NMV T6795 *F. adult, from Victoria Valley, Vic.

Distribution: SE coastal, Vic.; type locality only. Ecology: larva - sedentary, soil, predator : adult - volant, burrowing; prey adult Diptera.

Sericophorus (Sericophorus) sydneyi Rayment, 1955

Sericophorus sydneyi Rayment, T. (1955). Taxonomy, morphology and biology of sericophorine wasps. With diagnoses of two new genera and descriptions of forty new species and six subspecies. *Mem. Natl. Mus. Vict.* **19**: 11–105 pls 1–11 [50]. Type data: holotype, NMV 6796 *F. adult, from Bolgart, W.A.

Distribution: SW coastal, W.A.; type locality only. Ecology: larva - sedentary, soil, predator : adult - volant, burrowing; prey adult Diptera. Biological references: Matthews, R.W. & Evans, H.E. (1971). Biological notes on two species of *Sericophorus* from Australia (Hymenoptera : Sphecidae). *Psyche Camb.* **77**: 413–429 (prey).

Sericophorus (Sericophorus) tallongensis Rayment, 1955

Sericophorus tallongensis Rayment, T. (1955). Taxonomy, morphology and biology of sericophorine wasps. With diagnoses of two new genera and descriptions of forty new species and six subspecies. *Mem. Natl. Mus. Vict.* **19**: 11–105 pls 1–11 [50]. Type data: holotype, NMV 6798 *M. adult, from Tallong, N.S.W.

Distribution: SE coastal, N.S.W.; type locality only. Ecology: larva - sedentary, soil, predator : adult - volant, burrowing; prey adult Diptera.

Sericophorus (Sericophorus) teliferopodus Rayment, 1955

Taxonomic decision of Rayment, T. (1955). Taxonomy, morphology and biology of sericophorine wasps. With diagnoses of two new genera and descriptions of forty new species and six subspecies. *Mem. Natl. Mus. Vict.* **19**: 11–105 pls 1–11 [52].

Sericophorus (Sericophorus) teliferopodus teliferopodus Rayment, 1955

Sericophorus teliferopodus Rayment, T. (1955). Taxonomy, morphology and biology of sericophorine wasps. With diagnoses of two new genera and descriptions of forty new species and six subspecies. *Mem. Natl. Mus. Vict.* **19**: 11–105 pls 1–11 [51]. Type data: holotype, NMV 6800 *adult, from Sandringham, Vic.

Distribution: SE coastal, Vic.; only published localities Sandringham, Warburton and Cheltenham. Ecology: larva - sedentary, soil, predator : adult - volant, burrowing; prey adult Diptera.

Sericophorus (Sericophorus) teliferopodus okiellus Rayment, 1955

Sericophorus teliferopodus okiellus Rayment, T. (1955). Taxonomy, morphology and biology of sericophorine wasps. With diagnoses of two new genera and descriptions of forty new species and six subspecies. *Mem. Natl. Mus. Vict.* **19**: 11–105 pls 1–11 [52]. Type data: holotype, NMV 7602 F. adult, from Melton, Vic.

Distribution: SE coastal, Vic.; type locality only. Ecology: larva - sedentary, soil, predator : adult - volant, burrowing; prey adult Diptera.

Sericophorus (Sericophorus) victoriensis Rayment, 1953

Sericophorus victoriensis Rayment, T. (1953). New bees and wasps. Part XXI. Parasites on sericophorine wasps. *Vict. Nat.* **70**: 123–127 [124]. Type data: holotype, NMV 6803 *F. adult, ANIC M. adult, from Portland, Vic.

Distribution: SE coastal, Vic., S.A.; only published localities Portland, Mt. Richmond and Mt. Gambier. Ecology: larva - sedentary, soil, predator : adult - volant, burrowing; prey adult Diptera. Biological references: Rayment, T. (1955). Biology of two hunting wasps. The specific descriptions of a new species and one allotype of *Sericophorus* and a new blowfly *Pollenia*. *Aust. Zool.* **12**: 132–141 pl 19 (description of male, nest, prey); Rayment, T. (1959). Hyperparasitism by a minute fly and the specific description of a new species. *Aust. Zool.* **12**: 330–333 pl 39 (parasite, nest).

Sericophorus (Sericophorus) violaceus Rayment, 1955

Sericophorus violaceus Rayment, T. (1955). Taxonomy, morphology and biology of sericophorine wasps. With diagnoses of two new genera and descriptions of forty new species and six subspecies. *Mem. Natl. Mus. Vict.* **19**: 11–105 pls 1–11 [53]. Type data: holotype, SAMA *adult, from Torrens Gorge, S.A.

Distribution: S Gulfs, S.A.; type locality only. Ecology: larva - sedentary, soil, predator : adult - volant, burrowing; prey adult Diptera.

Sericophorus (Sericophorus) viridis (Saussure, 1855)

Taxonomic decision of Rayment, T. (1955). Taxonomy, morphology and biology of sericophorine wasps. With diagnoses of two new genera and descriptions of forty new species and six subspecies. *Mem. Natl. Mus. Vict.* **19**: 11–105 pls 1–111 [54].

Sericophorus (Sericophorus) viridis viridis (Saussure, 1855)

Tachyrrhostus viridis Saussure, H. de (1855). Mélanges hyménoptérologiques. *Mém. Soc. Phys. Hist. Nat. Genève* **14**: 1–67 [25]. Type data: holotype, GMNH *F. adult, from Australia (as New Holland).

Distribution: SE coastal, Murray-Darling basin, SW coastal, N.S.W., A.C.T., Vic., W.A. Ecology: larva - sedentary, soil, predator : adult - volant, burrowing; prey adult Diptera (Calliphoridae). Biological references: Matthews, R.W. & Evans, H.E. (1971). Biological notes on two species of *Sericophorus* from Australia (Hymenoptera : Sphecidae). *Psyche Camb.* **77**: 413-429 (nest, prey).

Sericophorus (Sericophorus) viridis roddi Rayment, 1955

Sericophorus viridis roddi Rayment, T. (1955). Taxonomy, morphology and biology of sericophorine wasps. With diagnoses of two new genera and descriptions of forty new species and six subspecies. *Mem. Natl. Mus. Vict.* **19**: 11-105 pls 1-11 [55]. Type data: holotype, NMV 6804 F. adult, from Fraser Park, N.S.W.

Distribution: SE coastal, N.S.W., Vic.; only published localities Fraser Park, Shoalhaven and Cheltenham. Ecology: larva - sedentary, soil, predator : adult - volant, burrowing; prey adult Diptera.

Sericophorus (Zoyphidium) Pate, 1937

Zoyphium Kohl, F.F. (1893). *Zoyphium*, eine neue Hymenopterengattung. *Verh. Zool.-Bot. Ges. Wien* **43**: 569-572 [569] [*non Zoyphium* Agassiz, 1847]. Type species *Zoyphium sericeum* Kohl, 1893 by monotypy.

Zoyphidium Pate, V.S.L. (1937). The generic names of the sphecoid wasps and their type species. *Mem. Am. Entomol. Soc.* **9**: 1-103 [68] [*nom. nov.* for *Zoyphium* Kohl, 1893].

Anacrucis Rayment, T. (1955). Taxonomy, morphology and biology of sericophorine wasps. With diagnoses of two new genera and descriptions of forty new species and six subspecies. *Mem. Natl. Mus. Vict.* **19**: 11-105 pls 1-11 [55]. Type species *Anacrucis laevigata* Rayment, 1955 by original designation.

Sericophorus (Zoyphidium) affinis (Hacker and Cockerell, 1922)

Zoyphium affine Hacker, H. & Cockerell, T.D.A. (1922). Some Australian wasps of the genera *Zoyphium* and *Arpactus*. *Mem. Qd. Mus.* **7**: 283-290 [289]. Type data: syntypes, QM *18F.+M. adult, from Brisbane, Qld.

Distribution: NE coastal, Qld.; type locality only. Ecology: larva - sedentary, soil, predator : adult - volant, burrowing; prey adult Diptera.

Sericophorus (Zoyphidium) argyreus (Hacker and Cockerell, 1922)

Zoyphium argyreum Hacker, H. & Cockerell, T.D.A. (1922). Some Australian wasps of the genera *Zoyphium* and *Arpactus*. *Mem. Qd. Mus.* **7**: 283-290 [286]. Type data: syntypes, QM T2693 *F. adult (labeled "type"), whereabouts of other syntype unknown, from Birkdale near Brisbane, Qld.

Distribution: NE coastal, Qld.; type locality only. Ecology: larva - sedentary, soil, predator : adult - volant, burrowing; prey adult Diptera.

Sericophorus (Zoyphidium) asperithorax (Rayment, 1955)

Anacrucis asperithorax Rayment, T. (1955). Taxonomy, morphology and biology of sericophorine wasps. With diagnoses of two new genera and descriptions of forty new species and six subspecies. *Mem. Natl. Mus. Vict.* **19**: 11-105 pls 1-11 [57]. Type data: holotype, NMV 6772 F. adult, from Bolgart, W.A.

Distribution: SW coastal, W.A.; type locality only. Ecology: larva - sedentary, soil, predator : adult - volant, burrowing; prey adult Diptera.

Sericophorus (Zoyphidium) cingulatus (Rayment, 1955)

Anacrucis cingulata Rayment, T. (1955). Taxonomy, morphology and biology of sericophorine wasps. With diagnoses of two new genera and descriptions of forty new species and six subspecies. *Mem. Natl. Mus. Vict.* **19**: 11-105 pls 1-11 [57]. Type data: holotype, DARI *F. adult, from Milthorpe, N.S.W.

Distribution: Murray-Darling basin, N.S.W.; type locality only. Ecology: larva - sedentary, soil, predator : adult - volant, burrowing; prey adult Diptera.

Sericophorus (Zoyphidium) clypeatus (Rayment, 1955)

Anacrucis clypeata Rayment, T. (1955). Taxonomy, morphology and biology of sericophorine wasps. With diagnoses of two new genera and descriptions of forty new species and six subspecies. *Mem. Natl. Mus. Vict.* **19**: 11-105 pls 1-11 [58]. Type data: holotype, NMV 6774 *F. adult, from S. Yarra, Vic.

Distribution: SE coastal, Vic.; type locality only. Ecology: larva - sedentary, soil, predator : adult - volant, burrowing; prey adult Diptera.

Sericophorus (Zoyphidium) collaris (Hacker and Cockerell, 1922)

Zoyphium collare Hacker, H. & Cockerell, T.D.A. (1922). Some Australian wasps of the genera *Zoyphium* and *Arpactus*. *Mem. Qd. Mus.* **7**: 283-290 [287]. Type data: holotype, QM T2687 *F. adult, from Birkdale near Brisbane, Qld.

Distribution: NE coastal, Qld.; type locality only. Ecology: larva - sedentary, soil, predator : adult - volant, burrowing; prey adult Diptera.

Sericophorus (Zoyphidium) crassicornis (Cockerell, 1914)

Zoyphium crassicorne Cockerell, T.D.A. (1914). A new fossorial wasp from Queensland. *Can. Entomol.* **46**: 271-272 [271]. Type data: syntypes, QM 2696 2 F. adult, from Brisbane, Qld.

Distribution: NE coastal, Qld.; type locality only. Ecology: larva - sedentary, soil, predator : adult - volant, burrowing; prey adult Diptera.

Sericophorus (Zoyphidium) dipteroides Turner, 1907

Sericophorus dipteroides Turner, R.E. (1907). New species of Sphegidae from Australia. *Ann. Mag. Nat. Hist. (7)* **19**: 268–276 [275]. Type data: holotype, BMNH *F. adult (seen 1929 by L.F. Graham), from Cairns, Qld.

Distribution: NE coastal, Qld.; type locality only. Ecology: larva - sedentary, soil, predator : adult - volant, burrowing; prey adult Diptera. Biological references: Turner, R.E. (1914). Notes on fossorial Hymenoptera. XIII. A revision of the Paranyssoninae. *Ann. Mag. Nat. Hist. (8)* **14**: 337–359 (taxonomy).

Sericophorus (Zoyphidium) doddi (Turner, 1912)

Zoyphium doddi Turner, R.E. (1912). Notes on fossorial Hymenoptera. IX. On some new species from the Australian and Austro-Malayan regions. *Ann. Mag. Nat. Hist. (8)* **10**: 48–63 [59]. Type data: holotype, BMNH *M. adult (seen 1929 by L.F. Graham), from Cairns, Qld.

Distribution: NE coastal, Qld.; type locality only. Ecology: larva - sedentary, soil, predator : adult - volant, burrowing; prey adult Diptera.

Sericophorus (Zoyphidium) emarginatus (Hacker and Cockerell, 1922)

Zoyphium emarginatum Hacker, H., & Cockerell, T.D.A. (1922). Some Australian wasps of the genera *Zoyphium* and *Arpactus. Mem. Qd. Mus.* **7**: 283–290 [285]. Type data: syntypes, QM 2690 *3F., 1 M. adult, from Brisbane, Qld.

Distribution: NE coastal, Qld.; type locality only. Ecology: larva - sedentary, soil, predator : adult - volant, burrowing; prey adult Diptera.

Sericophorus (Zoyphidium) erythrosoma (Turner, 1908)

Zoyphium erythrosoma Turner, R.E. (1908). Notes on the Australian fossorial wasps of the family Sphegidae, with descriptions of new species. *Proc. Zool. Soc. Lond.* **1908**: 457–535 pl xxvi [493]. Type data: holotype, BMNH *F. adult (seen 1929 by L.F. Graham), from Mackay, Qld.

Distribution: NE coastal, Qld.; only published localities Mackay, Townsville and Brisbane. Ecology: larva - sedentary, soil, predator : adult - volant, burrowing; prey adult Diptera.

Sericophorus (Zoyphidium) ferrugineus (Rayment, 1955)

Anacrucis ferruginea Rayment, T. (1955). Taxonomy, morphology and biology of sericophorine wasps. With diagnoses of two new genera and descriptions of forty new species and six subspecies. *Mem. Natl. Mus. Vict.* **19**: 11–105 pls 1–11 [59]. Type data: holotype, DARI *F. adult, from Wauchope, N.S.W.

Distribution: SE coastal, N.S.W.; type locality only. Ecology: larva - sedentary, soil, predator : adult - volant, burrowing; prey adult Diptera.

Sericophorus (Zoyphidium) flavofasciatus (Turner, 1916)

Zoyphium flavofasciatum Turner, R.E. (1916). Notes on fossorial Hymenoptera. XIX. On new species from Australia. *Ann. Mag. Nat. Hist. (8)* **17**: 116–136 [126]. Type data: holotype, BMNH *M. adult (seen 1929 by L.F. Graham), from Brisbane, Qld.

Distribution: NE coastal, Qld.; type locality only. Ecology: larva - sedentary, soil, predator : adult - volant, burrowing; prey adult Diptera. Biological references: Hacker, H. & Cockerell, T.D.A. (1922). Some Australian wasps of the genera *Zoyphium* and *Arpactus. Mem. Qd. Mus.* **7**: 283–290 (description of female).

Sericophorus (Zoyphidium) frontalis (Turner, 1908)

Zoyphium frontale Turner, R.E. (1908). Notes on the Australian fossorial wasps of the family Sphegidae, with descriptions of new species. *Proc. Zool. Soc. Lond.* **1908**: 457–535 pl xxvi [496]. Type data: holotype, BMNH *F. adult (seen 1929 by L.F. Graham), from Mackay, Qld.

Distribution: NE coastal, Qld.; type locality only. Ecology: larva - sedentary, soil, predator : adult - volant, burrowing; prey adult Diptera.

Sericophorus (Zoyphidium) fuscipennis (Hacker and Cockerell, 1922)

Zoyphium fuscipenne Hacker, H. & Cockerell, T.D.A. (1922). Some Australian wasps of the genera *Zoyphium* and *Arpactus. Mem. Qd. Mus.* **7**: 283–290 [289]. Type data: holotype, QM T2692 *F. adult, from Wedge Is., Tas.

Distribution: Tas.; type locality only. Ecology: larva - sedentary, soil, predator : adult - volant, burrowing; prey adult Diptera.

Sericophorus (Zoyphidium) hackeri (Cockerell, 1932)

Zoyphium hackeri Cockerell, T.D.A. (1932). Some wasps of the genus *Zoyphium. Mem. Qd. Mus.* **10**: 117–118 [117]. Type data: holotype, QM T4020 *F. adult, from Kuranda, Qld.

Distribution: NE coastal, Qld.; type locality only. Ecology: larva - sedentary, soil, predator : adult - volant, burrowing; prey adult Diptera.

Sericophorus (Zoyphidium) humilis (Cockerell, 1932)

Zoyphium humile Cockerell, T.D.A. (1932). Some wasps of the genus *Zoyphium*. *Mem. Qd. Mus.* **10**: 117–118 [117]. Type data: holotype, QM T4019 *F. adult, from Bribie Is., Qld.

Distribution: NE coastal, Qld.; type locality only. Ecology: larva - sedentary, soil, predator : adult - volant, burrowing; prey adult Diptera.

Sericophorus (Zoyphidium) iridipennis (Turner, 1914)

Zoyphium iridipenne Turner, R.E. (1914). Notes on fossorial Hymenoptera. XIII. A revision of the Paranyssoninae. *Ann. Mag. Nat. Hist.* (8) **14**: 337–359 [356]. Type data: holotype, BMNH *F. adult (seen 1929 by L.F. Graham), from Eaglehawk Neck, Tas.

Distribution: Tas.; type locality only. Ecology: larva - sedentary, soil, predator : adult - volant, burrowing; prey adult Diptera.

Sericophorus (Zoyphidium) kohlii (Turner, 1908)

Zoyphium kohlii Turner, R.E. (1908). Notes on the Australian fossorial wasps of the family Sphegidae, with descriptions of new species. *Proc. Zool. Soc. Lond.* **1908**: 457–535 pl xxvi [495]. Type data: holotype, BMNH *F. adult (seen 1929 by L.F. Graham), from Mackay, Qld.

Distribution: NE coastal, Qld.; type locality only. Ecology: larva - sedentary, soil, predator : adult - volant, burrowing; prey adult Diptera.

Sericophorus (Zoyphidium) laevigatus (Rayment, 1955)

Anacrucis laevigata Rayment, T. (1955). Taxonomy, morphology and biology of sericophorine wasps. With diagnoses of two new genera and descriptions of forty new species and six subspecies. *Mem. Natl. Mus. Vict.* **19**: 11–105 pls 1–11 [56]. Type data: holotype, NMV T6769 *F. adult, from Frankston, Vic.

Distribution: SE coastal, Vic.; type locality only. Ecology: larva - sedentary, soil, predator : adult - volant, burrowing; prey adult Diptera.

Sericophorus (Zoyphidium) niger (Hacker and Cockerell, 1922)

Zoyphium nigrum Hacker, H. & Cockerell, T.D.A. (1922). Some Australian wasps of the genera *Zoyphium* and *Arpactus*. *Mem. Qd. Mus.* **7**: 283–290 [286]. Type data: syntypes, QM T2689 *3M.+F. adult, from Brisbane and Caloundra, Qld.

Distribution: NE coastal, Qld.; type locality only. Ecology: larva - sedentary, soil, predator : adult - volant, burrowing; prey adult Diptera.

Sericophorus (Zoyphidium) ornatus (Hacker and Cockerell, 1922)

Zoyphium ornatum Hacker, H. & Cockerell, T.D.A. (1922). Some Australian wasps of the genera *Zoyphium* and *Arpactus*. *Mem. Qd. Mus.* **7**: 283–290 [287]. Type data: syntypes, QM T2697 *M.,F. adult, from Birkdale near Brisbane, Qld.

Distribution: NE coastal, Qld.; type locality only. Ecology: larva - sedentary, soil, predator : adult - volant, burrowing; prey adult Diptera.

Sericophorus (Zoyphidium) punctuosus (Rayment, 1955)

Anacrucis punctuosa Rayment, T. (1955). Taxonomy, morphology and biology of sericophorine wasps. With diagnoses of two new genera and descriptions of forty new species and six subspecies. *Mem. Natl. Mus. Vict.* **19**: 11–105 pls 1–11 [59]. Type data: holotype, ANIC F. adult, from Nedlands, W.A.

Distribution: SW coastal, W.A.; type locality only. Ecology: larva - sedentary, soil, predator : adult - volant, burrowing; prey adult Diptera (Tachinidae).

Sericophorus (Zoyphidium) pusillus (Hacker and Cockerell, 1922)

Zoyphium pusillum Hacker, H. & Cockerell, T.D.A. (1922). Some Australian wasps of the genera *Zoyphium* and *Arpactus*. *Mem. Qd. Mus.* **7**: 283–290 [288]. Type data: holotype, QM T2691 *M. adult, from Brisbane, Qld.

Distribution: NE coastal, Qld.; type locality only. Ecology: larva - sedentary, soil, predator : adult - volant, burrowing; prey adult Diptera.

Sericophorus (Zoyphidium) rufoniger (Turner, 1908)

Zoyphium rufonigrum Turner, R.E. (1908). Notes on the Australian fossorial wasps of the family Sphegidae, with descriptions of new species. *Proc. Zool. Soc. Lond.* **1908**: 457–535 [494 pl xxvi fig 8]. Type data: holotype, BMNH *M. adult (seen 1929 by L.F. Graham), from Darwin, N.T.

Distribution: N coastal, N.T.; type locality only. Ecology: larva - sedentary, soil, predator : adult - volant, burrowing; prey adult Diptera.

Sericophorus (Zoyphidium) sericeus (Kohl, 1893)

Zoyphium sericeum Kohl, F.F. (1893). *Zoyphium*, eine neue Hymenopterengattung. *Verh. Zool.-Bot. Ges. Wien* **43**: 569–572 [571]. Type data: holotype, ZMB *F. adult, from Adelaide, S.A.

Distribution: S Gulfs, S.A.; type locality only. Ecology: larva - sedentary, soil, predator : adult - volant, burrowing; prey adult Diptera.

Sericophorus (Zoyphidium) splendidus (Hacker and Cockerell, 1922)

Zoyphium splendidum Hacker, H. & Cockerell, T.D.A. (1922). Some Australian wasps of the genera *Zoyphium* and *Arpactus*. *Mem. Qd. Mus.* **7**: 283–290 [288]. Type data: syntypes, QM T2694 *8M.+F. adult, from Brisbane, Qld.

Distribution: NE coastal, Qld.; type locality only. Ecology: larva - sedentary, soil, predator : adult - volant, burrowing; prey adult Diptera.

Sericophorus (Zoyphidium) striatulus (Rayment, 1955)

Anacrucis striatula Rayment, T. (1955). Taxonomy, morphology and biology of sericophorine wasps. With diagnoses of two new genera and descriptions of forty new species and six subspecies. *Mem. Natl. Mus. Vict.* **19**: 11–105 pls 1–11 [56]. Type data: holotype, NMV T6770 *F. adult, from Jamberoo, N.S.W.

Distribution: SE coastal, N.S.W.; only published localities Jamberoo and Mt. Kiera. Ecology: larva - sedentary, soil, predator : adult - volant, burrowing; prey adult Diptera.

Sericophorus (Zoyphidium) tuberculatus (Turner, 1936)

Zoyphium tuberculatum Turner, R.E. (1936). Notes on fossorial Hymenoptera. XLV. On new sphegid wasps from Australia. *Ann. Mag. Nat. Hist. (10)* **18**: 533–545 [543]. Type data: holotype, BMNH *F. adult, from Dedari, W.A.

Distribution: W plateau, W.A.; type locality only. Ecology: larva - sedentary, soil, predator : adult - volant, burrowing; prey adult Diptera.

Sphodrotes Kohl, 1889

Sphodrotes Kohl, F.F. (1889). Neue Gattungen aus der Hymenopteren-Familie der Sphegiden. *Ann. Naturh. Hofmus. Wien* **4**: 188–196 [188]. Type species *Sphodrotes punctuosa* Kohl, 1889 by monotypy.

Sphodrotes acuticollis Lomholdt, 1983

Sphodrotes acuticollis Lomholdt, O. (1983). A revision of *Sphodrotes* Kohl, 1889 (Hymenoptera, Sphecoidea, Larridae). *Steenstrupia* **9**: 85–116 [96]. Type data: holotype, BMNH *M. adult, from north Queensland.

Distribution: Qld.; type locality only. Ecology: larva - sedentary, predator : adult - volant, burrowing; prey Hemiptera.

Sphodrotes cygnorum Turner, 1910

Sphodrotes cygnorum Turner, R.E. (1910). Additions to our knowledge of the fossorial wasps of Australia. *Proc. Zool. Soc. Lond.* **1910**: 253–356 [349]. Type data: holotype, BMNH *F. adult (seen 1929 by L.F. Graham), from Claremont, W.A.

Distribution: SW coastal, NW coastal, W.A. Ecology: larva - sedentary, soil, predator : adult - volant, burrowing; prey Hemiptera. Biological references: Turner, R.E. (1914). Notes on fossorial Hymenoptera. XIII. A revision of the Paranyssoninae. *Ann. Mag. Nat. Hist. (8)* **14**: 337–359 [345] (redescription); Lomholdt, O. (1983). A revision of *Sphodrotes* Kohl, 1889 (Hymenoptera, Sphecoidea, Larridae). *Steenstrupia* **9**: 85–116 (redescription, distribution).

Sphodrotes dearmata Lomholdt, 1983

Sphodrotes dearmata Lomholdt, O. (1983). A revision of *Sphodrotes* Kohl, 1889 (Hymenoptera, Sphecoidea, Larridae). *Steenstrupia* **9**: 85–116 [96]. Type data: holotype, ANIC F. adult, from 10 (mi) W of Mullewa, W.A.

Distribution: SW coastal, NW coastal, W.A.; only published localities Darlington and Denham. Ecology: larva - sedentary, predator : adult - volant, burrowing; prey Hemiptera.

Sphodrotes marginalis Turner, 1914

Sphodrotes marginalis Turner, R.E. (1914). Notes on fossorial Hymenoptera. XIII. A revision of the Paranyssoninae. *Ann. Mag. Nat. Hist. (8)* **14**: 337–359 [346]. Type data: lectotype, BMNH *F. adult, from Maryborough, Qld., designation by Lomholdt, O. (1983). A revision of *Sphodrotes* Kohl, 1889 (Hymenoptera, Sphecoidea, Larridae). *Steenstrupia* **9**: 85–116.

Distribution: NE coastal, Qld. Ecology: larva - sedentary, soil, predator : adult - volant, burrowing; prey Hemiptera. Biological references: Lomholdt, O. (1983). A revision of *Sphodrotes* Kohl, 1889 (Hymenoptera, Sphecoidea, Larridae). *Steenstrupia* **9**: 85–116 (redescription, distribution).

Sphodrotes nemoralis Evans, 1973

Sphodrotes nemoralis Evans, H.E. (1973). Observations on the nests and prey of *Sphodrotes nemoralis* sp.n. (Hymenoptera : Sphecidae). *J. Aust. Entomol. Soc.* **12**: 311–314 [311]. Type data: holotype, ANIC F. adult, from Kuranda and vicinity, Qld.

Distribution: NE coastal, N Gulf, Qld. Ecology: larva - sedentary, soil, predator : adult - volant, burrowing; prey Hemiptera (immature Pentatomoidea). Biological references: Evans, H.E. (1973). Observations on the nests and prey of *Sphodrotes nemoralis* sp.n. (Hymenoptera : Sphecidae). *J. Aust. Entomol. Soc.* **12**: 311–314; Lomholdt, O. (1983). A revision of *Sphodrotes* Kohl, 1889 (Hymenoptera, Sphecoidea, Larridae). *Steenstrupia* **9**: 85–116 (redescription, distribution).

Sphodrotes occidentalis Lomholdt, 1983

Sphodrotes occidentalis Lomholdt, O. (1983). A revision of *Sphodrotes* Kohl, 1889 (Hymenoptera, Sphecoidea, Larridae). *Steenstrupia* **9**: 85–116 [102]. Type data: holotype, WAM 83/566 *F. adult, from Busselton, W.A.

Distribution: SW coastal, W.A.; only published localities Busselton, Perth and Esperance. Ecology: larva - sedentary, predator : adult - volant, burrowing; prey Hemiptera.

Sphodrotes ordinaria Lomholdt, 1983

Sphodrotes ordinaria Lomholdt, O. (1983). A revision of *Sphodrotes* Kohl, 1889 (Hymenoptera, Sphecoidea, Larridae). *Steenstrupia* **9**: 85–116 [89]. Type data: holotype, H.E. Evans collection *F. adult, from 12–17 mi E Alice Springs, N.T.

Distribution: NE coastal, Murray-Darling basin, SE coastal, Lake Eyre basin, Qld., N.S.W., Vic., N.T. Ecology: larva - sedentary, predator : adult - volant, burrowing; prey Hemiptera.

Sphodrotes prima Lomholdt, 1983

Sphodrotes prima Lomholdt, O. (1983). A revision of *Sphodrotes* Kohl, 1889 (Hymenoptera, Sphecoidea, Larridae). *Steenstrupia* **9**: 85–116 [103]. Type data: holotype, ANIC F. adult, from Carnamah, W.A.

Distribution: Lake Eyre basin, SW coastal, NW coastal, N.T., W.A. Ecology: larva - sedentary, predator : adult - volant, burrowing; prey Hemiptera.

Sphodrotes punctuosa Kohl, 1889

Sphodrotes punctuosa Kohl, F.F. (1889). Neue Gattungen aus der Hymenopteren-Familie der Sphegiden. *Ann. Naturh. Hofmus. Wien* **4**: 188–196 [189 pl 8 figs 1,10,13,24]. Type data: holotype, NHMW *M. adult, from N.S.W.

Distribution: SE coastal, Murray-Darling basin, N.S.W., Vic., Tas. Ecology: larva - sedentary, predator : adult - volant; prey Hemiptera. Biological references: Turner, R.E. (1914). Notes on fossorial Hymenoptera. XIII. A revision of the Paranyssoninae. *Ann. Mag. Nat. Hist. (8)* **14**: 337–359 (redescription); Lomholdt, O. (1983). A revision of *Sphodrotes* Kohl, 1889 (Hymenoptera, Sphecoidea, Larridae). *Steenstrupia* **9**: 85–116 (redescription, distribution).

Sphodrotes rubra Lomholdt, 1983

Sphodrotes rubra Lomholdt, O. (1983). A revision of *Sphodrotes* Kohl, 1889 (Hymenoptera, Sphecoidea, Larridae). *Steenstrupia* **9**: 85–116 [94]. Type data: holotype, ANIC F. adult, from 44–45 km NE of Andado HS, Simpson Desert, N.T.

Distribution: Murray-Darling basin, Lake Eyre basin, N.S.W, S.A., N.T. Ecology: larva - sedentary, predator : adult - volant, burrowing; prey Hemiptera.

Sphodrotes rubricata Turner, 1910

Sphodrotes rubricatus Turner, R.E. (1910). New fossorial Hymenoptera from Australia. *Trans. Entomol. Soc. Lond.* **1910**: 407–429 [426 pl 50 fig 13]. Type data: lectotype, BMNH *F. adult, from Adelaide, S.A., designation by Lomholdt, O. (1983). A revision of *Sphodrotes* Kohl, 1889 (Hymenoptera, Sphecoidea, Larridae). *Steenstrupia* **9**: 85–116.

Sphodrotes pilosellus Turner, R.E. (1910). New fossorial Hymenoptera from Australia. *Trans. Entomol. Soc. Lond.* **1910**: 407–429 [427 pl 50]. Type data: holotype, BMNH *M. adult (seen 1929 by L.F. Graham), from Cairns, Qld.

Taxonomic decision of Lomholdt, O. (1983). A revision of *Sphodrotes* Kohl, 1889 (Hymenoptera, Sphecoidea, Larridae). *Steenstrupia* **9**: 85–116 [91].

Distribution: Murray-Darling basin, Lake Eyre basin, SW coastal, N coastal, S Gulfs, Qld, N.S.W, S.A., W.A., N.T. Ecology: larva - sedentary, predator : adult - volant; prey Hemiptera.

Sphodrotes splendens Lomholdt, 1983

Sphodrotes splendens Lomholdt, O. (1983). A revision of *Sphodrotes* Kohl, 1889 (Hymenoptera, Sphecoidea, Larridae). *Steenstrupia* **9**: 85–116 [111]. Type data: holotype, ANIC F. adult, from 8 km S of Coulomb Point, West Kimberley, W.A.

Distribution: Lake Eyre basin, N coastal, N Gulf, W.A., N.T., Qld.; only published localities near Coulomb Point, near Cape Bertholet, Avon Downs and near Nappamerry. Ecology: larva - sedentary, predator : adult - volant, burrowing; prey Hemiptera.

Larrisson Menke, 1967

Larrisson Menke, A.S. (1967). New genera of Old World Sphecidae (Hymenoptera). *Entomol. News* **78**: 29–35 [29]. Type species *Sericophorus abnormis* Turner, 1914 by monotypy and original designation.

Larrisson abnormis (Turner, 1914)

Sericophorus abnormis Turner, R.E. (1914). Notes on fossorial Hymenoptera. XIII. A revision of the Paranyssoninae. *Ann. Mag. Nat. Hist. (8)* **14**: 337–359 [352]. Type data: holotype, BMNH *M. (seen 1929 by L.F. Graham), from Yallingup, W.A.

Distribution: Lake Eyre basin, SW coastal, S.A., W.A.; only published localities near Musgrave Park, Edeowie Homestead and Yallingup. Ecology: larva - sedentary, predator : adult - volant. Biological references: Menke, A.S. (1979). A review of the genus *Larrisson* Menke (Hymenoptera : Sphecidae). *Aust. J. Zool.* **27**: 453–463 (redescription, generic placement).

Larrisson azyx Menke, 1979

Larrisson azyx Menke, A.S. (1979). A review of the genus *Larrisson* Menke (Hymenoptera : Sphecidae). *Aust. J. Zool.* **27**: 453-463 [460]. Type data: holotype, ANIC M. adult, from Kununurra and vicinity (Lily Creek), W.A.

Distribution: N coastal, W.A.; type locality only. Ecology: larva - sedentary, predator : adult - volant; prey of *Bembix moma* Evans and Matthews. Biological references: Evans, H.E. & Matthews, R.W. (1973). Systematics and nesting behavior of Australian *Bembix* sand wasps (Hymenoptera, Sphecidae). *Mem. Am. Entomol. Inst.* **20**: iv 387 pp. (as *Larrison* sp., as prey).

Larrisson nedymus Menke, 1979

Larrisson nedymus Menke, A.S. (1979). A review of the genus *Larrisson* Menke (Hymenoptera : Sphecidae). *Aust. J. Zool.* **27**: 453-463 [455]. Type data: holotype, ANIC F. adult, from Nilemah Station 50 mi S of Denham, W.A.

Distribution: NW coastal, W.A.; type locality only. Ecology: larva - sedentary, predator : adult - volant.

Larrisson rieki Menke, 1979

Larrisson rieki Menke, A.S. (1979). A review of the genus *Larrisson* Menke (Hymenoptera : Sphecidae). *Aust. J. Zool.* **27**: 453-463 [457]. Type data: holotype, UCDC *M. adult, from 10 mi W Mullewa, W.A.

Distribution: NW coastal, W.A.; type locality only. Ecology: larva - sedentary, predator : adult - volant.

Nitela Latreille, 1809

Nitela Latreille, P.A. (1809). *Genera Crustaceorum et Insectorum secundem ordinem naturalem in familias disposita, iconibus exemplisque plurimis explicata.* Paris : A. Koenig Vol. 4 397 pp. [77]. Type species *Nitela spinolae* Latreille, 1809 by monotypy.

This group is found worldwide, see Bohart, R.M. & Menke, A.S. (1976). *Sphecid Wasps of the World : a Generic Revision.* Berkeley : Univ. California Press ix 695 pp. [322].

Nitela australiensis Schulz, 1908

Nitela australiensis Schulz, W.A. (1908). Fossores. pp. 447-488 *in* Michaelsen, W. & Hartmeyer, R. (eds.) *Die Fauna Südwest-Australiens.* Vol. I Lfg. 13 Jena : G. Fischer [483]. Type data: syntypes (probable), ZMB *F. adult, from Denham, W.A.

Nitela nigricans Turner, R.E. (1910). New fossorial Hymenoptera from Australia. *Trans. Entomol. Soc. Lond.* **1910**: 407-429 pl 50 [428]. Type data: holotype, BMNH *F. adult (seen 1929 by L.F. Graham), from Bundaberg, Qld.

Taxonomic decision of Turner, R.E. (1916). Notes on fossorial Hymenoptera. XXIII. *Ann. Mag. Nat. Hist. (8)* **18**: 277-288 [286].

Distribution: NE coastal, SW coastal, NW coastal, Qld., Tas., W.A.; only published localities Bundaberg, Eaglehawk Neck, Yallingup and Denham. Ecology: larva - sedentary, arboreal, predator : adult - volant; nest in existing holes in wood.

Nitela kurandae Turner, 1908

Nitela kurandae Turner, R.E. (1908). Notes on the Australian fossorial wasps of the family Sphegidae, with descriptions of new species. *Proc. Zool. Soc. Lond.* **1908**: 457-535 pl xxvi [508]. Type data: holotype, BMNH *F. adult (seen 1929 by L.F. Graham), from Cairns, Qld.

Distribution: NE coastal, Qld.; only published localities Cairns, Kuranda, Bundaberg and Caloundra. Ecology: larva - sedentary, arboreal, predator : adult - volant; nest in existing holes in wood.

Nitela sculpturata Turner, 1916

Nitela reticulata Turner, R.E. (1908). Notes on the Australian fossorial wasps of the family Sphegidae, with descriptions of new species. *Proc. Zool. Soc. Lond.* **1908**: 457-535 pl xxvi [508] [*non Nitela reticulata* Ducke, 1908]. Type data: holotype, BMNH *F. adult (seen 1929 by L.F. Graham), from Mackay, Qld.

Nitela sculpturata Turner, R.E. (1916). Notes on fossorial Hymenoptera, xxiv. On the genus *Nitela* Latr. *Ann. Mag. Nat. Hist. (8)* **18**: 343-345 [343] [*nom. nov.* for *Nitela reticulata* Turner, 1908].

Distribution: NE coastal, Qld.; type locality only. Ecology: larva - sedentary, arboreal, predator : adult - volant; nest in existing holes in wood.

Auchenophorus Turner, 1907

Auchenophorus Turner, R.E. (1907). New species of Sphegidae from Australia. *Ann. Mag. Nat. Hist. (7)* **19**: 268-276 [270]. Type species *Auchenophorus coruscans* Turner, 1907 by original designation.

Auchenophorus aeneus Turner, 1907

Auchenophorus aeneus Turner, R.E. (1907). New species of Sphegidae from Australia. *Ann. Mag. Nat. Hist. (7)* **19**: 268-276 [271]. Type data: holotype, BMNH *F. adult (seen 1929 by L.F. Graham), from Mackay, Qld.

Distribution: NE coastal, Qld.; only published localities Mackay, Cairns and Kuranda. Ecology: larva - sedentary, predator: adult - volant.

Auchenophorus coruscans Turner, 1907

Auchenophorus coruscans Turner, R.E. (1907). New species of Sphegidae from Australia. *Ann. Mag. Nat. Hist. (7)* **19**: 268-276 [271]. Type data: holotype, BMNH *F. adult (seen 1929 by L.F. Graham), from Mackay, Qld.

Distribution: NE coastal, Qld.; type locality only. Ecology: larva - sedentary, predator: adult - volant. Biological references: Turner, R.E. (1910) Additions to our knowledge of the fossorial wasps of Australia. *Proc. Zool. Soc. Lond.* **1910**: 253–356 [pl 32 fig 15].

Auchenophorus fulvicornis Turner, 1907

Auchenophorus fulvicornis Turner, R.E. (1907). New species of Sphegidae from Australia. *Ann. Mag. Nat. Hist. (7)* **19**: 268–276 [272]. Type data: holotype, BMNH *M. adult (seen 1929 by L.F. Graham), from Kuranda, Qld.

Distribution: NE coastal, Qld.; only published localities Kuranda, Cairns. Ecology: larva - sedentary, predator: adult - volant. Biological references: Turner, R.E. (1916). Notes on fossorial Hymenoptera. XIII. On some Australian genera. *Ann. Mag. Nat. Hist. (8)* **18**: 277–288 (description of female).

Pison Jurine, 1808

Taxonomic decision of Bohart, R.M. & Menke, A.S. (1976). *Sphecid Wasps of the World : a Generic Revision*. Berkeley : Univ. California Press ix 695 pp. [45].

Pison (Pison) Jurine, 1808

Pison Jurine, L. (1808). *in* Spinola, M. *Insectorum Liguriae Species Novae aut Rariores*, quas in agro ligustico nuper detexit, descripsit, et iconibus illustravit Maximilianus Spinola, adjecto catalogo specierum auctoribus iam enumeratarum, quae in eadem regione passim occurrunt. Genuae : Gravier Vol. 2 ii 262 pp. [255]. Type species *Pison jurinei* Spinola, 1808 by monotypy. Compiled from secondary source: Bohart, R.M. & Menke, A.S. (1976). *Sphecid Wasps of the World : a Generic Revision*. Berkeley : Univ. California Press ix 695 pp.

This group is found worldwide except America north of Mexico, see Bohart, R.M. & Menke, A.S. (1976). *Sphecid Wasps of the World : a Generic Revision*. Berkeley : Univ. California Press ix 695 pp. [332].

Pison (Pison) aberrans Turner, 1908

Pison (Parapison) aberrans Turner, R.E. (1908). Notes on the Australian fossorial wasps of the family Sphegidae, with descriptions of new species. *Proc. Zool. Soc. Lond.* **1908**: 457–535 pl xxvi [519]. Type data: holotype, BMNH *M. adult (seen 1929 by L.F. Graham), from Mackay, Qld.

Distribution: NE coastal, Qld.; type locality only. Ecology: larva - sedentary, predator : adult - volant; prey spiders, nest using mud or in ground. Biological references: Bohart, R.M. & Menke, A.S. (1976). *Sphecid Wasps of the World : a Generic Revision*. Berkeley : Univ. California Press ix 695 pp. (generic placement).

Pison (Pison) areniferum Evans, 1981

Pison (Pison) areniferum Evans, H.E. (1981). Biosystematics of ground-nesting species of *Pison* in Australia (Hymenoptera : Sphecidae : Trypoxylini). *Proc. Entomol. Soc. Wash.* **83**: 421–427 [422]. Type data: holotype, QM T8498 *F. adult, from Amby, Qld.

Distribution: Murray-Darling basin, Qld.; type locality only. Ecology: larva - sedentary, soil, predator : adult - volant, burrowing; prey spiders, nest in ground.

Pison (Pison) auratum Shuckard, 1838

Pison auratus Shuckard, W.E. (1838). Descriptions of new exotic aculeate Hymenoptera. *Trans. Entomol. Soc. Lond.* **2**: 68–82 [78]. Type data: holotype, BMNH *F. adult (seen 1929 by L.F. Graham), from Cape of Good Hope"; label is in error, species is Australian, *teste* Smith, F. (1869). Descriptions of new species of the genus *Pison*, and a synonymic list of those already described. *Trans. Entomol. Soc. Lond.* **1869**: 289–300.

Distribution: N coastal, W.A., N.T. Ecology: larva - sedentary, predator : adult - volant; prey spiders, uses abandoned nests of *Sceliphron laetum*. Biological references: Naumann, I.D. (1983). The biology of mud nesting Hymenoptera (and their associates) and Isoptera in rock shelters of the Kakadu Region, Northern Territory. *Aust. Nat. Parks & Wildlife Service Spec. Publ. 10* pp. 127–189 (biology).

Pison (Pison) aureosericeum Rohwer, 1915

Pison aureosericeum Rohwer, S.A. (1915). Descriptions of new species of Hymenoptera. *Proc. U.S. Natl. Mus.* **49**(2105): 205–249 [246]. Type data: holotype, USNM 14254 *adult, from Duaringa, Qld.

Distribution: NE coastal, Qld., Vic.; only published localities Duaringa, Mackay, Kuranda and Vic. Ecology: larva - sedentary, predator : adult - volant; prey spiders, nest in old eumenid nests. Biological references: Hacker, H. (1918). Entomological contributions. *Mem. Qd. Mus.* **6**: 106–111 pls 31–32 (nest).

Pison (Pison) aurifex Smith, 1869

Pison aurifex Smith, F. (1869). Descriptions of new species of the genus *Pison*; and a synonymic list of those previously described. *Trans. Entomol. Soc. Lond.* **1869**: 289–300 [293]. Type data: holotype, BMNH *F. adult (seen 1929 by L.F. Graham), from Australia.

Distribution: N coastal, N.T.; only published localities Burnside and Marrakai. Ecology: larva - sedentary, predator : adult - volant; prey spiders, nest using mud or in ground. Biological references: Turner, R.E. (1916). Notes on the wasps of the genus *Pison*, and some allied genera. *Proc. Zool. Soc. Lond.* **1916**: 591–629 (redescription).

Pison (Pison) auriventre Turner, 1908

Pison auriventre Turner, R.E. (1908). Notes on the Australian fossorial wasps of the family Sphegidae, with descriptions of new species. *Proc. Zool. Soc. Lond.* **1908**: 457–535 pl xxvi [512]. Type data: holotype, BMNH *F. adult (seen 1929 by L.F. Graham), from Vic.

Distribution: NE coastal, Murray-Darling basin, Qld., N.S.W., A.C.T., Vic. Ecology: larva - sedentary, soil, predator : adult - volant, burrowing; prey spiders, nest in ground. Biological references: Evans, H.E. (1981). Biosystematics of ground-nesting species of *Pison* in Australia (Hymenoptera : Sphecidae : Trypoxylini). *Proc. Entomol. Soc. Wash.* **83**: 421–427 (redescription, biology).

Pison (Pison) barbatum Evans, 1981

Pison (Pison) barbatum Evans, H.E. (1981). Biosystematics of ground-nesting species of *Pison* in Australia (Hymenoptera : Sphecidae : Trypoxylini). *Proc. Entomol. Soc. Wash.* **83**: 421–427 [424]. Type data: holotype, QM T8500 *F. adult, from Port Douglas, Qld.

Distribution: NE coastal, Qld.; type locality only. Ecology: larva - sedentary, soil, predator : adult - volant, burrowing; prey spiders, nest in ground.

Pison (Pison) basale Smith, 1869

Pison basalis Smith, F. (1869). Descriptions of new species of the genus *Pison*; and a synonymic list of those previously described. *Trans. Entomol. Soc. Lond.* **1869**: 289–300 [292]. Type data: holotype, BMNH *F. adult (seen 1929 by L.F. Graham), from Australia.

Distribution: NE coastal, Qld.; only published localities Australia and Mackay. Ecology: larva - sedentary, predator : adult - volant; prey spiders, nest using mud or in ground.

Pison (Pison) caliginosum Turner, 1908

Pison (Parapison) caliginosum Turner, R.E. (1908). Notes on the Australian fossorial wasps of the family Sphegidae, with descriptions of new species. *Proc. Zool. Soc. Lond.* **1908**: 457–535 pl xxvi [518]. Type data: holotype, BMNH *F. adult (seen 1929 by L.F. Graham), from Kuranda, Qld.

Distribution: NE coastal, Qld.; only published localities Kuranda and near Cairns. Ecology: larva - sedentary, predator : adult - volant; prey spiders, nest using mud or in ground. Biological references: Bohart, R.M. & Menke, A.S. (1976). *Sphecid Wasps of the World : a Generic Revision*. Berkeley : Univ. California Press ix 695 pp. (generic placement).

Pison (Pison) ciliatum Evans, 1981

Pison (Pison) ciliatum Evans, H.E. (1981). Biosystematics of ground-nesting species of *Pison* in Australia (Hymenoptera : Sphecidae : Trypoxylini). *Proc. Entomol. Soc. Wash.* **83**: 421–427 [423]. Type data: holotype, QM T8499 *F. adult, from Amby, Qld.

Distribution: Murray-Darling basin, Qld.; type locality only. Ecology: larva - sedentary, soil, predator : adult - volant, burrowing; prey spiders, nest in ground.

Pison (Pison) congenerum Turner, 1916

Pison congener Turner, R.E. (1916). Notes on the wasps of the genus *Pison*, and some allied genera. *Proc. Zool. Soc. Lond.* **1916**: 591–629 [607]. Type data: holotype, BMNH *F. adult (seen 1929 by L.F. Graham), from Yallingup, W.A.

Distribution: SW coastal, W.A.; type locality only. Ecology: larva - sedentary, predator : adult - volant; prey spiders, nest using mud or in ground.

Pison (Pison) decipiens Smith, 1869

Pison decipiens Smith, F. (1869). Descriptions of new species of the genus *Pison*; and a synonymic list of those previously described. *Trans. Entomol. Soc. Lond.* **1869**: 289–300 [295]. Type data: holotype, BMNH *M. adult (seen 1929 by L.F. Graham), from Champion Bay, W.A.

Distribution: SE coastal, NW coastal, N.S.W., W.A.; only published localities Homebush and Champion Bay. Ecology: larva - sedentary, arboreal, predator : adult - volant; prey spiders, mudnest. Biological references: Froggatt, W.W. (1894). On the nests and habits of Australian Vespidae and Larridae. *Proc. Linn. Soc. N.S.W.* (2) **9**: 27–34 (biology).

Pison (Pison) deperditum Turner, 1917

Pison (Pisonitus) deperditum Turner, R.E. (1917). Notes on fossorial Hymenoptera. XXV. On new Sphecoidea in the British Museum. *Ann. Mag. Nat. Hist.* (8) **19**: 104–113 [109]. Type data: holotype, BMNH *F. adult (seen 1929 by L.F. Graham), from Darwin, N.T.

Distribution: N coastal, N.T.; type locality only. Ecology: larva - sedentary, predator : adult - volant; prey spiders, nest using mud or in ground. Biological references: Bohart, R.M. & Menke, A.S. (1976). *Sphecid Wasps of the World : a Generic Revision*. Berkeley : Univ. California Press ix 695 pp. (generic placement).

Pison (Pison) dimidiatum Smith, 1869

Pison dimidiatus Smith, F. (1869). Descriptions of new species of the genus *Pison*; and a synonymic list of those previously described. *Trans. Entomol. Soc. Lond.* **1869**: 289–300 [295]. Type data: holotype, BMNH *M. adult (seen 1929 by L.F. Graham), from Champion Bay, W.A.

Distribution: NW coastal, W.A.; type locality only. Ecology: larva - sedentary, predator : adult - volant; prey spiders, nest using mud or in ground.

Pison (Pison) dives Turner, 1916

Pison dives Turner, R.E. (1916). Notes on the wasps of the genus *Pison*, and some allied genera. *Proc. Zool. Soc. Lond.* **1916**: 591–629 [608]. Type data: holotype, BMNH *F. adult (seen 1929 by L.F. Graham), from Kuranda, Qld.

Distribution: NE coastal, Qld.; type locality only. Ecology: larva - sedentary, predator : adult - volant; prey spiders, nest using mud or in ground.

Pison (Pison) erythrocerum Kohl, 1884

Parapison ruficornis Smith, F. (1869). Descriptions of new species of the genus *Pison*; and a synonymic list of those previously described. *Trans. Entomol. Soc. Lond.* **1869**: 289–300 [300] [*non Parapison ruficornis* Smith, 1856]. Type data: holotype, BMNH *F. adult (seen 1929 by L.F. Graham), from Australia.
Pison erythrocerum Kohl, F.F. (1884). Die Gattungen und Arten der Larriden Autorum. *Verh. Zool.-Bot. Ges. Wien* **34**: 171–268, 327–454 [186] [*nom. nov.* for *Parapison ruficornis* Smith, 1869].

Distribution: NE coastal, Qld.; only published localities Mackay, Kuranda and Australia. Ecology: larva - sedentary, predator : adult - volant; prey spiders, nest using mud or in ground.

Pison (Pison) erythrogastrum Rohwer, 1915

Pison (Parapison) erythrogastrum Rohwer, S.A. (1915). Descriptions of new species of Hymenoptera. *Proc. U.S. Natl. Mus.* **49**(2105) : 205–249 [247]. Type data: holotype, USNM 14255 *F. adult, from Duaringa, Qld.

Distribution: NE coastal, SW coastal, Qld., W.A.; only published localities Duaringa and Kalamunda. Ecology: larva - sedentary, predator : adult - volant; prey spiders, nest using mud or in ground. Biological references: Bohart, R.M. & Menke, A.S. (1976). *Sphecid Wasps of the World : a Generic Revision*. Berkeley : Univ. California Press ix 695 pp. [335] (generic placement).

Pison (Pison) exclusum Turner, 1916

Pison (Parapison) exclusum Turner, R.E. (1916). Notes on fossorial Hymenoptera. XIX. On new species from Australia. *Ann. Mag. Nat. Hist. (8)* **17**: 116–136 [127]. Type data: holotype, BMNH *M. adult (seen 1929 by L.F. Graham), from Brisbane, Qld.

Distribution: NE coastal, Murray-Darling basin, Qld., Vic.; only published localities Brisbane, Horsham. Ecology: larva - sedentary, predator : adult - volant; prey spiders, nest using mud or in ground. Biological references: Bohart, R.M. & Menke, A.S. (1976). *Sphecid Wasps of the World : a Generic Revision*. Berkeley : Univ. California Press ix 695 pp. (generic placement).

Pison (Pison) exornatum Turner, 1916

Pison exornatum Turner, R.E. (1916). Notes on the wasps of the genus *Pison*, and some allied genera. *Proc. Zool. Soc. Lond.* **1916**: 591–629 [614]. Type data: holotype, BMNH *F. adult (seen 1929 by L.F. Graham), from Mackay, Qld.

Distribution: NE coastal, Qld.; type locality only. Ecology: larva - sedentary, predator : adult - volant; prey spiders, nest using mud or in ground.

Pison (Pison) exultans Turner, 1916

Pison exultans Turner, R.E. (1916). Notes on the wasps of the genus *Pison*, and some allied genera. *Proc. Zool. Soc. Lond.* **1916**: 591–629 [615]. Type data: holotype, BMNH *M. adult (seen 1929 by L.F. Graham), from Vic.

Distribution: Vic.; type locality only as Vic. Ecology: larva - sedentary, predator : adult - volant; prey spiders, nest using mud or in ground.

Pison (Pison) fenestratum Smith, 1869

Pison nitidus Smith, F. (1868). Descriptions of aculeate Hymenoptera from Australia. *Trans. Entomol. Soc. Lond.* **1868**: 231–258 [248] [*non Pison nitidus* Smith, 1858]. Type data: holotype, BMNH *F. adult (seen 1929 by L.F. Graham), from Champion Bay, W.A.
Pison fenestratus Smith, F. (1869). Descriptions of new species of the genus *Pison*; and a synonymic list of those previously described. *Trans. Entomol. Soc. Lond.* **1869**: 289–300 [291] [*nom. nov.* for *Pison nitidus* Smith, 1868].

Distribution: Lake Eyre basin, SW coastal, NW coastal, N.T., W.A.; only published localities Hermannsburg, Yallingup and Champion Bay. Ecology: larva - sedentary, predator : adult - volant; prey spiders, nest using mud or in ground.

Pison (Pison) festivum Smith, 1869

Pison festivus Smith, F. (1869). Descriptions of new species of the genus *Pison*; and a synonymic list of those previously described. *Trans. Entomol. Soc. Lond.* **1869**: 289–300 [296]. Type data: holotype, BMNH *F. adult (seen 1929 by L.F. Graham), from Champion Bay, W.A.

Distribution: SW coastal, NW coastal, W.A.; only published localities Swan River and Champion Bay. Ecology: larva - sedentary, predator : adult - volant; prey spiders, nest using mud or in ground.

Pison (Pison) fraterculus Turner, 1916

Pison fraterculus Turner, R.E. (1916). Notes on the wasps of the genus *Pison*, and some allied genera. *Proc. Zool. Soc. Lond.* **1916**: 591–629 [610]. Type data: holotype, BMNH *F. adult (seen 1929 by L.F. Graham), from Mackay, Qld.

Distribution: NE coastal, Qld.; type locality only. Ecology: larva - sedentary, predator : adult - volant; prey spiders, nest using mud or in ground.

Pison (Pison) fuscipenne Smith, 1869

Pison fuscipennis Smith, F. (1869). Descriptions of new species of the genus *Pison*; and a synonymic list of those previously described. *Trans. Entomol. Soc. Lond.* **1869**: 289–300 [294]. Type data: holotype, BMNH *F. adult (seen 1929 by L.F. Graham), from Champion Bay, W.A.

Distribution: NW coastal, W.A.; type locality only. Ecology: larva - sedentary, predator : adult - volant; prey spiders, nest using mud or in ground.

Pison (Pison) ignavum Turner, 1908

Pison ignavum Turner, R.E. (1908). Notes on the Australian fossorial wasps of the family Sphegidae, with descriptions of new species. *Proc. Zool. Soc. Lond.* **1908**: 457–535 pl xxvi [511]. Type data: holotype, BMNH *F. adult (seen 1929 by L.F. Graham), from Mackay, Cairns, Qld.

Distribution: NE coastal, SE coastal, Qld., Vic.; only published localities Mackay, Cairns, Kuranda, Brisbane area and Melbourne, also on many Pacific islands. Ecology: larva - sedentary, arboreal, predator : adult - volant; prey spiders, mudnest on leaf of bush. Biological references: Evans, H.E., Matthews, R.W. & Hook, A. (1981). Notes on the nests and prey of six species of *Pison* in Australia (Hymenoptera : Sphecidae). *Psyche Camb.* **87**: 221–230 (nest).

Pison (Pison) inconspicuum Turner, 1916

Pison inconspicuum Turner, R.E. (1916). Notes on the wasps of the genus *Pison*, and some allied genera. *Proc. Zool. Soc. Lond.* **1916**: 591–629 [612]. Type data: holotype, BMNH *M. adult (seen 1929 by L.F. Graham), from Mundaring Weir, W.A.

Distribution: SW coastal, W.A.; type locality only. Ecology: larva - sedentary, predator : adult - volant; prey spiders, nest using mud or in ground.

Pison (Pison) infumatum Turner, 1908

Pison infumatum Turner, R.E. (1908). Notes on the Australian fossorial wasps of the family Sphegidae, with descriptions of new species. *Proc. Zool. Soc. Lond.* **1908**: 457–535 pl xxvi [510]. Type data: holotype, BMNH *F. adult (seen 1929 by L.F. Graham), from Port Darwin, N.T.

Distribution: N coastal, N.T.; type locality only. Ecology: larva - sedentary, predator : adult - volant; prey spiders, nest using mud or in ground.

Pison (Pison) iridipenne Smith, 1879

Pison iridipenne Smith, F. (1879). Descriptions of new species of aculeate Hymenoptera collected by the Rev. Thos. Blackburn in the Sandwich Islands. *J. Linn. Soc. Lond. Zool.* **14**: 674–685 [676]. Type data: syntypes (probable), ?BMNH *adult, from Honolulu.

Distribution: NE coastal, Qld.; Australian localities are doubtful, species widely distributed on Pacific islands and Philippines. Ecology: larva - sedentary, predator : adult - volant; prey spiders, nest using mud in pre-existing cavity in wood, or in ground. Biological references: Krobein, K.V. (1949). The Aculeate Hymenoptera of Micronesia. I Scoliidae, Mutillidae, Pompilidae and Sphecidae. *Proc. Hawaii. Entomol. Soc.* **13**: 367–410 (queries Australian records).

Pison (Pison) lutescens Turner, 1916

Pison lutescens Turner, R.E. (1916). Notes on the wasps of the genus *Pison*, and some allied genera. *Proc. Zool. Soc. Lond.* **1916**: 591–629 [604]. Type data: holotype, BMNH *F. adult, from Mundaring Weir, W.A.

Distribution: SW coastal, W.A.; type locality only. Ecology: larva - sedentary, predator : adult - volant; prey spiders, nest using mud or in ground.

Pison (Pison) mandibulatum Turner, 1916

Pison mandibulatum Turner, R.E. (1916). Notes on the wasps of the genus *Pison*, and some allied genera. *Proc. Zool. Soc. Lond.* **1916**: 591–629 [605]. Type data: holotype, BMNH *F. adult (seen 1929 by L.F. Graham), from Yallingup, W.A.

Distribution: SW coastal, W.A.; type locality only. Ecology: larva - sedentary, predator : adult - volant; prey spiders, nest using mud or in ground.

Pison (Pison) marginatum Smith, 1856

Pison marginatus Smith, F. (1856). *Catalogue of Hymenopterous Insects in the Collection of the British Museum. Part IV, Sphegidae, Larridae and Crabronidae.* pp. 207–497 London : British Museum [314]. Type data: holotype, BMNH *F. adult (seen 1929 by L.F. Graham), from Hunter River, N.S.W.

Distribution: NE coastal, SE coastal, Murray-Darling, Qld., N.S.W., A.C.T., Vic., S.A.; only published localities S.A., Mackay, Hunter River, Canberra area and Melbourne. Ecology: larva - sedentary, predator : adult - volant; prey spiders, nest using mud in pre-existing cavity in wood (trap nest). Biological references: Evans, H.E., Matthews, R.W. & Hook, A. (1981). Notes on the nests and prey of six species of *Pison* in Australia (Hymenoptera : Sphecidae). *Psyche Camb.* **87**: 221–230 (nest, prey).

Pison (Pison) melanocephalum Turner, 1908

Pison melanocephalum Turner, R.E. (1908). Notes on the Australian fossorial wasps of the family Sphegidae, with descriptions of new species. *Proc. Zool. Soc. Lond.* **1908**: 457–535 [515 pl xxvi fig 12]. Type data: holotype, BMNH *F. adult (seen 1929 by L.F. Graham), from Cairns, Qld.

Distribution: NE coastal, Qld.; type locality only. Ecology: larva - sedentary, predator : adult - volant; prey spiders, nest using mud or in ground.

Pison (Pison) meridionale Turner, 1916

Pison meridionale Turner, R.E. (1916). Notes on the wasps of the genus *Pison*, and some allied genera. *Proc.*

Zool. Soc. Lond. **1916**: 591-629 [611]. Type data: holotype, BMNH *M. adult (seen 1929 by L.F. Graham), from Adelaide, S.A.

Distribution: S Gulfs, S.A.; type locality only. Ecology: larva - sedentary, predator : adult - volant; prey spiders, nest using mud or in ground.

Pison (Pison) noctulum Turner, 1908

Pison (Parapison) noctulum Turner, R.E. (1908). Notes on the Australian fossorial wasps of the family Sphegidae, with descriptions of new species. *Proc. Zool. Soc. Lond.* **1908**: 457-535 pl xxvi [516]. Type data: holotype, BMNH *F. adult (seen 1929 by L.F. Graham), from Mackay, Qld.

Distribution: NE coastal, Qld.; only published localities Mackay and Kuranda. Ecology: larva - sedentary, predator : adult - volant; prey spiders, nest using mud or in ground. Biological references: Bohart, R.M. & Menke, A.S. (1976). *Sphecid Wasps of the World : a Generic Revision*. Berkeley : Univ. California Press ix 695 pp. (generic placement).

Pison (Pison) peletieri Le Guillou, 1841

Pison peletieri Le Guillou, E.J.F. (1841). Description de 20 espèces nouvelles appartenant à diverses familles d'Hyménoptères. *Rev. Zool.* **4**: 322-325 [324]. Type data: syntypes (probable), MNHP *F. adult, from N. Australia (as Aust. sept.).

Distribution: Northern Australia type locality only. Ecology: larva - sedentary, predator : adult - volant; prey spiders, nest using mud or in ground.

Pison (Pison) perplexum Smith, 1856

Pison perplexus Smith, F. (1856). *Catalogue of Hymenopterous Insects in the Collection of the British Museum.* Part IV, Sphegidae, Larridae and Crabronidae. pp. 207-497 London : British Museum [314]. Type data: holotype, BMNH *M. adult (seen 1929 by L.F. Graham), from Australia.

Distribution: SW coastal, Vic., W.A.; only published localities Vic. and S. Perth. Ecology: larva - sedentary, predator : adult - volant; prey spiders, nest using mud in pre-existing cavity in wood. Biological references: Roth, H.L. (1885). Notes on the habits of some Australian Hymenoptera Aculeata. (With description of a new species by W.F. Kirby). *J. Linn. Soc. (Zool.)* **18**: 318-328 (biology); Turner, R.E. (1910). Additions to our knowledge of the fossorial wasps of Australia. *Proc. Zool. Soc. Lond.* **1910**: 253-356 (description of female).

Pison (Pison) pertinax Turner, 1908

Pison (Parapison) pertinax Turner, R.E. (1908). Notes on the Australian fossorial wasps of the family Sphegidae, with descriptions of new species. *Proc. Zool. Soc. Lond.* **1908**: 457-535 pl xxvi [517]. Type data: holotype, BMNH *F. adult (seen 1929 by L.F. Graham), from Mackay, Qld.

Distribution: NE coastal, Qld.; only published localities Mackay, Cairns and Brisbane. Ecology: larva - sedentary, predator : adult - volant; prey spiders, nest using mud or in ground. Biological references: Bohart, R.M. & Menke, A.S. (1976). *Sphecid Wasps of the World : a Generic Revision*. Berkeley : Univ. California Press ix 695 pp. (generic placement).

Pison (Pison) priscum Turner, 1908

Pison insulare priscum Turner, R.E. (1908). Notes on the Australian fossorial wasps of the family Sphegidae, with descriptions of new species. *Proc. Zool. Soc. Lond.* **1908**: 457-535 pl xxvi [510]. Type data: holotype, BMNH *F. adult (seen 1929 by L.F. Graham), from Mackay, Qld.

Distribution: NE coastal, Qld.; type locality only. Ecology: larva - sedentary, predator : adult - volant; prey spiders, nest using mud or in ground.

Pison (Pison) pulchrinum Turner, 1916

Pison pulchrinum Turner, R.E. (1916). Notes on the wasps of the genus *Pison*, and some allied genera. *Proc. Zool. Soc. Lond.* **1916**: 591-629 [613]. Type data: holotype, BMNH *F. adult (seen 1929 by L.F. Graham), from Mackay, Kuranda, Qld.

Distribution: NE coastal, Qld.; type locality only. Ecology: larva - sedentary, predator : adult - volant; prey spiders, nest using mud or in ground.

Pison (Pison) punctulatum Kohl, 1884

Pison punctulatum Kohl, F.F. (1884). Neue Hymenopteren in den Sammlungen des k.k. zoologischen Hof-Cabinetes zu Wien. ii. *Verh. Zool.-Bot. Ges. Wien* **33**: 331-386 [336]. Type data: syntypes (probable), NHMW *adult, from Peak Downs, Qld.

Distribution: NE coastal, Qld., N.S.W.; only published localities N.S.W., Peak Downs, Mackay. Ecology: larva - sedentary, predator : adult - volant; prey spiders, nest using mud or in ground.

Pison (Pison) ruficorne Smith, 1856

Pison (Pisonitus) ruficornis Smith, F. (1856). *Catalogue of Hymenopterous Insects in the Collection of the British Museum.* Part IV, Sphegidae, Larridae and Crabronidae. pp. 207-497 London : British Museum [315]. Type data: holotype, BMNH *F. adult (seen 1929 by L.F. Graham), from MacIntyre River, Qld.

Distribution: NE coastal, N coastal, Qld., Vic., W.A.; only published localities Vic., Macintyre River, Mackay, Kuranda and Ord River. Ecology: larva - sedentary, predator : adult - volant; prey spiders, nest using mud or in ground. Biological references: Bohart, R.M. & Menke, A.S. (1976). *Sphecid Wasps of the World : a Generic Revision*.

Berkeley : Univ. California Press ix 695 pp. (generic placement); Turner, R.E. (1908). Notes on the Australian fossorial wasps of the family Sphegidae, with descriptions of new species. *Proc. Zool. Soc. Lond.* **1908**: 457–535 pl xxvi (redescription).

Pison (Pison) rufipes Shuckard, 1838

Pison (Pisonitus) rufipes Shuckard, W.E. (1838). Descriptions of new exotic aculeate Hymenoptera. *Trans. Entomol. Soc. Lond.* **2**: 68–82 [79]. Type data: holotype, BMNH *F. adult (seen 1929 by L.F. Graham), from Tas. (as Van Diemen's Land).

Distribution: NE coastal, SE coastal, SW coastal, Qld., N.S.W., Vic., Tas., W.A.; also in New Caledonia. Ecology: larva - sedentary, arboreal, predator : adult - volant; prey spiders, mudnest. Biological references: Bohart, R.M. & Menke, A.S. 1976). *Sphecid Wasps of the World : a Generic Revision*. Berkeley : Univ. California Press ix 695 pp. (generic placement; Evans, H.E., Matthews, R.W. & Hook, A. (1981). Notes on the nests and prey of six species of *Pison* in Australia (Hymenoptera : Sphecidae). *Psyche Camb.* **87**: 221–230 (nest, prey).

Pison (Pison) scabrum Turner, 1908

Pison scabrum Turner, R.E. (1908). Notes on the Australian fossorial wasps of the family Sphegidae, with descriptions of new species. *Proc. Zool. Soc. Lond.* **1908**: 457–535 pl xxvi [509]. Type data: holotype, BMNH *F. adult (seen 1929 by L.F. Graham), from Mackay, Qld.

Distribution: NE coastal, Qld.; type locality only. Ecology: larva - sedentary, predator : adult - volant; prey spiders, nest using mud or in ground.

Pison (Pison) separatum Smith, 1869

Pison separatus Smith, F. (1869). Descriptions of new species of the genus *Pison*; and a synonymic list of those previously described. *Trans. Entomol. Soc. Lond.* **1869**: 289–300 [294]. Type data: holotype, BMNH *M. adult (seen 1929 by L.F. Graham), from Champion Bay, W.A.

Distribution: NW coastal, W.A.; type locality only. Ecology: larva - sedentary, predator : adult - volant; prey spiders, nest using mud or in ground.

Pison (Pison) simillimum Smith, 1869

Pison simillimus Smith, F. (1869). Descriptions of new species of the genus *Pison*; and a synonymic list of those previously described. *Trans. Entomol. Soc. Lond.* **1869**: 289–300 [292]. Type data: holotype, BMNH *M. adult (seen 1929 by L.F. Graham), from Australia.

Distribution: NE coastal, SE coastal, Qld., Vic.; only published localities E Australia, Brisbane and Vic. Ecology: larva - sedentary, predator : adult - volant; prey spiders, nest using mud or in ground.

Pison (Pison) simulans Turner, 1915

Pison (Parapison) simulans Turner, R.E. (1915). Notes on fossorial Hymenoptera. XVI. On the Thynnidae, Scoliidae and Crabronidae of Tasmania. *Ann. Mag. Nat. Hist.* (8) **15**: 537–559 [559]. Type data: holotype, BMNH *M. adult (seen 1929 by L.F. Graham), from Eaglehawk Neck, Tas.

Distribution: Tas.; type locality only. Ecology: larva - sedentary, predator : adult - volant; prey spiders, nest using mud or in ground. Biological references: Bohart, R.M. & Menke, A.S. (1976). *Sphecid Wasps of the World : a Generic Revision*. Berkeley : Univ. California Press ix 695 pp. (generic placement).

Pison (Pison) spinolae Shuckard, 1838

Pison spinolae Shuckard, W.E. (1838). Descriptions of new exotic aculeate Hymenoptera. *Trans. Entomol. Soc. Lond.* **2**: 68–82 [76]. Type data: holotype, BMNH *F. adult (seen 1929 by L.F. Graham), from Sydney, N.S.W.

Pison australis Saussure, H. de (1855). Mélanges hyménoptérologiques. *Mém. Soc. Phys. Hist. Nat. Genève* **14**: 1–67 [11]. Type data: syntypes (probable), whereabouts unknown, from Australia (as New Holland).

Pison tasmanicus Smith, F. (1856). *Catalogue of Hymenopterous Insects in the Collection of the British Museum.* Part IV, Sphegidae, Larridae and Crabronidae. pp. 207–497 London : British Museum [316]. Type data: holotype, BMNH *M. adult (seen 1929 by L.F. Graham), from Tas.

Taxonomic decision of Smith, F. (1869). Descriptions of new species of the genus *Pison*, and a synonymic list of those previously described. *Trans. Entomol. Soc. Lond.* **1869**: 289–300 [290].

Distribution: NE coastal, SE coastal, Murray-Darling basin, S Gulfs, N coastal, Qld., N.S.W., A.C.T., Tas., S.A., W.A., Norfolk Is.; also New Zealand. Ecology: larva - sedentary, predator : adult - volant; prey spiders, nest using mud in pre-existing cavity in wood. Biological references: Cowley, D.R. (1962). Aspects of the biology of the immature stages of *Pison spinolae* Shuckard (Hymenoptera : Sphecidae). *Trans. R. Soc. N.Z.* **1**: 355–363; Callan, E.M. (1977). *Macrosiagon diversiceps* (Coleoptera : Rhipiphoridae) reared from a sphecid wasp, with notes on other species. *Aust. Entomol. Mag.* **4**: 45–47 (parasite); Evans, H.E., Matthews, R.W. & Hook, A. (1981). Notes on the nests and prey of six species of *Pison* in Australia (Hymenoptera : Sphecidae). *Psyche Camb.* **87**: 221–230 (nest, prey).

Pison (Pison) strenuum Turner, 1916

Pison strenuum Turner, R.E. (1916). Notes on the wasps of the genus *Pison*, and some allied genera. *Proc. Zool. Soc. Lond.* **1916**: 591–629 [606]. Type data: holotype, BMNH *F. adult (seen 1929 by L.F. Graham), from Yallingup, S.Perth, W.A.

Distribution: SW coastal, W.A.; type locality only. Ecology: larva - sedentary, predator : adult - volant; prey spiders, nest using mud or in ground.

Pison (Pison) tenebrosum Turner, 1908

Pison (Parapison) tenebrosum Turner, R.E. (1908). Notes on the Australian fossorial wasps of the family Sphegidae, with descriptions of new species. *Proc. Zool. Soc. Lond.* **1908**: 457-535 pl xxvi [518]. Type data: holotype, BMNH *F. adult (seen 1929 by L.F. Graham), from Mackay, Qld.

Distribution: NE coastal, Qld.; type locality only. Ecology: larva - sedentary, predator : adult - volant; prey spiders, nest using mud or in ground. Biological references: Bohart, R.M. & Menke, A.S. (1976). *Sphecid Wasps of the World : a Generic Revision.* Berkeley : Univ. California Press ix 695 pp. (generic placement).

Pison (Pison) tibiale Smith, 1869

Pison tibialis Smith, F. (1869). Descriptions of new species of the genus *Pison*; and a synonymic list of those previously described. *Trans. Entomol. Soc. Lond.* **1869**: 289-300 [292]. Type data: holotype, BMNH *M. adult (seen 1929 by L.F. Graham), from W.A.

Distribution: NE coastal, Murray-Darling basin, SW coastal, Qld., N.S.W., W.A.; only published localities Brisbane, Springsure, Coolabah and Kalamunda. Ecology: larva - sedentary, predator : adult - volant; prey spiders, nest using mud or in ground.

Pison (Pison) vestitum Smith, 1856

Pison vestitus Smith, F. (1856). *Catalogue of Hymenopterous Insects in the Collection of the British Museum.* Part IV, Sphegidae, Larridae and Crabronidae. pp. 207-497 London : British Museum [315]. Type data: holotype, BMNH *F. adult (seen 1929 by L.F. Graham), from Australia.

Distribution: only published localities Australia and E Australia. Ecology: larva - sedentary, predator : adult - volant; prey spiders, nest using mud or in ground.

Pison (Pison) virosum Turner, 1908

Pison virosum Turner, R.E. (1908). Notes on the Australian fossorial wasps of the family Sphegidae, with descriptions of new species. *Proc. Zool. Soc. Lond.* **1908**: 457-535 pl xxvi [513]. Type data: holotype, BMNH *F. adult (seen 1929 by L.F. Graham), from Mackay, Qld.

Distribution: NE coastal, Qld.; type locality only. Ecology: larva - sedentary, predator : adult - volant; prey spiders, nest using mud or in ground.

Pison (Pison) westwoodii Shuckard, 1838

Pison westwoodii Shuckard, W.E. (1838). Descriptions of new exotic aculeate Hymenoptera. *Trans. Entomol. Soc. Lond.* **2**: 68-82 [77]. Type data: syntypes (probable), OUM or BMNH *F. adult, from Tas. (as Van Diemen's Land).

Pison obliquus Smith, F. (1856). *Catalogue of Hymenopterous Insects in the Collection of the British Museum.* Part IV, Sphegidae, Larridae and Crabronidae. pp. 207-497 London : British Museum [316]. Type data: syntypes (probable), OUM *F. adult, from Tas.

Taxonomic decision of Turner, R.E. (1916). Notes on the wasps of the genus *Pison*, and some allied genera. *Proc. Zool. Soc. Lond.* **1916**: 591-629 [608].

Distribution: NE coastal, Murray-Darling basin, SW coastal, N coastal, Qld., A.C.T., Tas., W.A., N.T. Ecology: larva - sedentary, predator : adult - volant; prey spiders, nest using mud in pre-existing cavity in wood. Biological references: Evans, H.E., Matthews, R.W. & Hook, A. (1981). Notes on the nests and prey of six species of *Pison* in Australia (Hymenoptera : Sphecidae). *Psyche Camb.* **87**: 221-230 (nest).

Pison (Pisonoides) Smith, 1858

Pisonoides Smith, F. (1858). Catalogue of the hymenopterous insects collected at Sarawak, Borneo; Mount Ophir, Malacca; and Singapore, by A.R. Wallace. *J. Linn. Soc. Lond. Zool* **2**: 42-130 [104] [described with subgeneric rank in *Pison* Jurine, 1808]. Type species *Pison obliteratum* Smith, 1858 by monotypy.

This group is also found in the Oriental Region, see Bohart, R.M. & Menke, A.S. (1976). *Sphecid Wasps of the World : a Generic Revision.* Berkeley : Univ. California Press ix 695 pp. [332].

Pison (Pisonoides) difficile Turner, 1908

Pison (Aulacophilus) difficile Turner, R.E. (1908). Notes on the Australian fossorial wasps of the family Sphegidae, with descriptions of new species. *Proc. Zool. Soc. Lond.* **1908**: 457-535 pl xxvi [520]. Type data: holotype, BMNH *F. adult (seen 1929 by L.F. Graham), from Mackay or Cairns, Qld.

Distribution: NE coastal, Qld.; only published localities Mackay and Cairns. Ecology: larva - sedentary, predator : adult - volant; prey spiders, nest using mud or in ground. Biological references: Turner, R.E. (1916). Notes on the wasps of the genus *Pison*, and some allied genera. *Proc. Zool.Soc. Lond.* **1916**: 591-629 (generic placement).

Pison (Pisonoides) icarioides Turner, 1908

Pison (Aulacophilus) icarioides Turner, R.E. (1908). Notes on the Australian fossorial wasps of the family Sphegidae, with descriptions of new species. *Proc. Zool. Soc. Lond.* **1908**: 457-535 [521 pl xxvi fig 13]. Type data: holotype, BMNH *F. adult (seen 1929 by L.F. Graham), from Mackay, Qld.

Distribution: NE coastal, Qld.; only published localities Mackay, Cairns and Brisbane. Ecology: larva - sedentary, predator : adult - volant; prey spiders, nest using mud or in ground. Biological references: Turner, R.E. (1916). Notes on the wasps of the genus *Pison*, and some allied genera. *Proc. Zool. Soc. Lond.* **1916**: 591–629 (generic placement).

Trypoxylon Latreille, 1796

Taxonomic decision of Bohart, R.M. & Menke, A.S. (1976). *Sphecid Wasps of the World : a Generic Revision*. Berkeley : Univ. California Press ix 695 pp. [45].

Trypoxylon (Trypoxylon) Latreille, 1796

Trypoxylon Latreille, P.A. (1796). *Précis des Caractères Génériques des Insectes,* disposés dans un ordre naturel. Bordeaux : Brive xiv 208 pp. [121]. Type species *Sphex figulus* Linnaeus, 1758 by subsequent designation, see Latreille, P.A. (1802–1803). *Histoire Naturelle, Générale et Particulière des Crustacés et des Insectes.* Paris : F. Dufart Vol. 3 xii 13+467 pp. [339]. Compiled from secondary source: Bohart, R.M. & Menke, A.S. (1976). *Sphecid Wasps of the World : a Generic Revision.* Berkeley : Univ. California Press ix 695 pp. [45].

This group is found worldwide, see Bohart, R.M. & Menke, A.S. (1976). *Sphecid Wasps of the World : a Generic Revision.* Berkeley : Univ. California Press ix 695 pp. [339].

Trypoxylon (Trypoxylon) albitarsatum Tsuneki, 1977

Taxonomic decision of Tsuneki, K. (1981). Studies on the genus *Trypoxylon* Latreille of the Oriental and Australian regions (Hymenoptera Sphecidae) VIII. Species from New Guinea and South Pacific Islands. IX. Species from Australia. *Spec. Publ. Japan Hymenopterists Ass. 14* 106 pp. [46].

Trypoxylon (Trypoxylon) albitarsatum huonense Tsuneki, 1977

Trypoxylon huonense Tsuneki, K. (1977). Some *Trypoxylon* species from the southwestern Pacific (Hymenoptera, Sphecidae, Larrinae). *Spec. Publ. Japan Hymenopterists Ass. 6* 20 pp. [14]. Type data: holotype, MNH *M. adult, from Huon Gulf, New Guinea.

Distribution: NE coastal, Qld.; only published localities Cairns, Nerada and Kuranda, also on New Guinea and Misool Is. Ecology: larva - sedentary, predator : adult - volant; prey spiders.

Trypoxylon (Trypoxylon) bituberculatum Tsuneki, 1977

Trypoxylon bituberculatum Tsuneki, K. (1977). Some *Trypoxylon* species from the southwestern Pacific (Hymenoptera, Sphecidae, Larrinae). *Spec. Publ. Japan Hymenopterists Ass. 6* 20 pp. [17]. Type data: holotype, MNH *F. adult, from Huon Gulf, New Guinea.

Distribution: NE coastal, Qld.; only published localities Cape York, Kuranda, Meringa and Cairns, also on New Guinea and Misool Is. Ecology: larva - sedentary, predator : adult - volant; prey spiders. Biological references: Tsuneki, K. (1981). Studies on the genus *Trypoxylon* Latreille of the Oriental and Australian regions (Hymenoptera Sphecidae) VIII. Species from New Guinea and South Pacific Islands. IX. Species from Australia. *Spec. Publ. Japan Hymenopterists Ass. 14* 106 pp. (redescription, Australian records).

Trypoxylon (Trypoxylon) eximium Smith, 1859

Trypoxylon eximium Smith, F. (1859). Catalogue of hymenopterous insects collected by Mr. A.R. Wallace at the islands of Aru and Key. *J. Linn. Soc. Lond. Zool.* **3**: 132–178 [161]. Type data: lectotype, OUM *F. adult, from Aru, designation by Tsuneki, K. (1978). Studies on the genus *Trypoxylon* Latreille of the Oriental and Australian regions (Hymenoptera, Sphecidae) II. Revision of the type series of the species described by F. Smith, P. Cameron, C.G. Nurse, W.H. Ashmead, R.E. Turner and O.W. Richards. *Spec. Publ. Japan Hymenopterists Ass. 8* 84 pp. [9].

Distribution: NE coastal, Qld.; only published localities Cape York, Gordonvale and Kuranda, also in New Guinea, Indonesia and Bismarck Archipelago. Ecology: larva - sedentary, predator : adult - volant; prey spiders. Biological references: Tsuneki, K. (1981). Studies on the genus *Trypoxylon* Latreille of the Oriental and Australian regions (Hymenoptera Sphecidae) VIII. Species from New Guinea and South Pacific Islands. IX. Species from Australia. *Spec. Publ. Japan Hymenopterists Ass. 14* 106 pp. (redescription, Australan records).

Trypoxylon (Trypoxylon) flavipes Tsuneki, 1979

Trypoxylon flavipes Tsuneki, K. (1979). Studies on the genus *Trypoxylon* Latreille of the Oriental and Australian regions (Hymenoptera, Sphecidae). III. Species from the Indian subcontinent including southeast Asia. *Spec. Publ. Japan Hymenopterists Ass. 9.* 178 pp. [24]. Type data: holotype, BPBM *F. adult, from Laos.

Distribution: NE coastal, Qld.; only published locality Mt. Tamborine, also in Ceylon, Laos, Borneo, Philippines, Sarawak and New Guinea. Ecology: larva - sedentary, predator : adult - volant; prey spiders. Biological references: Tsuneki, K. (1981). Studies on the genus *Trypoxylon* Latreille of the Oriental and Australian regions (Hymenoptera Sphecidae) VIII. Species from New Guinea and South Pacific Islands. IX. Species from Australia. *Spec. Publ. Japan Hymenopterists Ass. 14* 106 pp. (redescription, Australian record).

Trypoxylon (Trypoxylon) mindanaonis Tsuneki, 1976

Trypoxylon mindanaonis Tsuneki, K. (1976). Sphecoidea taken by the Noona Dan Expedition in the Philippine Islands (Insecta, Hymenoptera). *Steenstrupia* **4**: 33–120 [84]. Type data: holotype, ZMUC *F. adult, from Mindanao.

Distribution: NE coastal, Qld.; only published locality Brisbane, also in Philippines, Singapore, Borneo, Java and New Guinea. Ecology: larva - sedentary, predator : adult - volant; prey spiders. Biological references: Tsuneki, K. (1981). Studies on the genus *Trypoxylon* Latreille of the Oriental and Australian regions (Hymenoptera Sphecidae) VIII. Species from New Guinea and South Pacific Islands. IX. Species from Australia. *Spec. Publ. Japan Hymenopterists Ass.* **14** 106 pp. (redescription, Australian record).

Trypoxylon (Trypoxylon) papuanum Tsuneki, 1977

Trypoxylon papuanum Tsuneki, K. (1977). Some *Trypoxylon* species from the southwestern Pacific (Hymenoptera, Sphecidae, Larrinae). *Spec. Publ. Japan Hymenopterists Ass.* **6** 20 pp. [2]. Type data: holotype, MNH *F. adult, from New Guinea.

Distribution: NE coastal, N coastal, Qld., N.T.; also in Philippines, New Guinea, Bismarck Archipelago and Solomon Ils. Ecology: larva - sedentary, predator : adult - volant; prey spiders. Biological references: Tsuneki, K. (1981). Studies on the genus *Trypoxylon* Latreille of the Oriental and Australian regions (Hymenoptera Sphecidae) VIII. Species from New Guinea and South Pacific Islands. IX. Species from Australia. *Spec. Publ. Japan Hymenopterists Ass.* **14** 106 pp. (redescription, Australian records).

Trypoxylon (Trypoxylon) schmiedeknechti Kohl, 1906

Taxonomic decision of Tsuneki, K. (1978). Studies on the genus *Trypoxylon* Latreille of the Oriental and Australian regions (Hymenoptera, Sphecidae) I. Group of *Trypoxylon scutatum* Chevrier with some species from Madagascar and the adjacent islands. *Spec. Publ. Japan Hymenopterists Ass.* **7** 87 pp. [21].

Trypoxylon (Trypoxylon) schmiedeknechti connexum Turner, 1908

Trypoxylon connexum Turner, R.E. (1908). Notes on the Australian fossorial wasps of the family Sphegidae, with descriptions of new species. *Proc. Zool. Soc. Lond.* **1908**: 457–535 pl xxvi [522]. Type data: lectotype, BMNH *F. adult, from Mackay, Qld., designation by Tsuneki, K. (1978). Studies on the genus *Trypoxylon* Latreille of the Oriental and Australian regions (Hymenoptera, Sphecidae) II. Revision of the type series of the species described by F. Smith, P. Cameron, C.G. Nurse, W.H. Ashmead, R.E. Turner and O.W. Richards. *Spec. Publ. Japan Hymenopterists Ass.* **8** 84 pp. [70].

Distribution: NE coastal, SE coastal, N coastal, N Gulf, Qld., N.S.W., Vic., W.A., N.T.; also in New Guinea, Bismarck Archipelago and Sumba Is. Ecology: larva - sedentary, predator : adult - volant; prey spiders. Biological references: Tsuneki, K. (1981). Studies on the genus *Trypoxylon* Latreille of the Oriental and Australian regions (Hymenoptera Sphecidae) VIII. Species from New Guinea and South Pacific Islands. IX. Species from Australia. *Spec. Publ. Japan Hymenopterists Ass.* **14** 106 pp. (redescription).

Rhopalum Stephens, 1829

Taxonomic decision of Leclercq, J. (1978). Crabroniens du genre *Rhopalum* Stephens trouvés en Australie (Hymenoptera Sphecidae). *Bull. Soc. R. Sci. Liège* **47**: 352–362 [352].

Rhopalum (Rhopalum) Stephens, 1829

Rhopalum Stephens, J.F. (1829). *The Nomenclature of British Insects;* together with their synonyms : being a compendious list of such species as are contained in the systematic catalogue of British insects, and of those discovered subsequently to its publication; forming a guide to their classification, etc. London : Baldwin & Craddock 68 pp. [34]. Type species *Crabro rufiventris* Panzer, 1799 by subsequent designation, see Curtis, J. (1834-1837). *British Entomology;* being illustrations and descriptions of the genera of insects found in Great Britain and Ireland : containing coloured figures from nature of the most rare and beautiful species, and in many instances of the plants upon which they are found. London : John Curtis Vol. 11 pls 482–529. Compiled from secondary source: International Commission on Zoological Nomenclature (1978). Opinion 1106. Conservation of the generic name *Rhopalum* Stephens, 1829 (Insecta, Hymenoptera). *Bull. Zool. Nomen.* **34**: 237–239 [237].

This group is found worldwide except for the Ethiopian Region, see Bohart, R.M. & Menke, A.S. (1976). *Sphecid Wasps of the World : a Generic Revision.* Berkeley : Univ. California Press ix 695 pp. [387].

Rhopalum (Rhopalum) calixtum Leclercq, 1957

Rhopalum (Rhopalum) calixtum Leclercq, J. (1957). Le genre *Rhopalum* en Australie. *Bull. Ann. Soc. R. Entomol. Belg.* **93**: 177–232 [231]. Type data: holotype, ANIC F. adult, from Brisbane, Qld.

Distribution: NE coastal, Murray-Darling basin, Qld., N.S.W., A.C.T.; only published localities Brisbane, N.S.W., Canberra and Black Mt. Ecology: larva - sedentary, predator : adult - volant.

Rhopalum (Rhopalum) dineurum Leclercq, 1957

Rhopalum (Rhopalum) dineurum Leclercq, J. (1957). Le genre *Rhopalum* en Australie. *Bull. Ann. Soc. R. Entomol. Belg.* **93**: 177–232 [223]. Type data: holotype, BMNH *M. adult, from Eaglehawk Neck, Tas.

Distribution: SE coastal, N.S.W., Vic., Tas.; only published localities Illawarra, Nunawading, Baxter, Sand Hills and Eaglehawk Neck. Ecology: larva - sedentary, predator : adult - volant.

Rhopalum (Rhopalum) eucalypti Turner, 1915

Rhopalum eucalypti Turner, R.E. (1915). Notes on fossorial Hymenoptera. XV. New Australian Crabronidae. *Ann. Mag. Nat. Hist.* (8) **15**: 62–96 [90]. Type data: holotype, BMNH *F. adult (seen 1929 by L.F. Graham), from Eaglehawk Neck, Tas.

Distribution: Tas.; type locality only. Ecology: larva - sedentary, predator : adult - volant. Biological references: Leclercq, J. (1957). Le genre *Rhopalum* en Australie. *Bull. Ann. Soc. R. Entomol. Belg.* **93**: 177–232 (redescription).

Rhopalum (Rhopalum) grahami Leclercq, 1957

Rhopalum (Rhopalum) grahami Leclercq, J. (1957). Le genre *Rhopalum* en Australie. *Bull. Ann. Soc. R. Entomol. Belg.* **93**: 177–232 [227]. Type data: holotype, ANIC F. adult, from Black Mt., A.C.T.

Distribution: Murray-Darling basin, A.C.T., Vic.; only published localities Black Mt. and Sandhill Lake. Ecology: larva - sedentary, predator : adult - volant.

Rhopalum (Rhopalum) macrocephalum Turner, 1915

Rhopalum macrocephalus Turner, R.E. (1915). Notes on fossorial Hymenoptera. XV. New Australian Crabronidae. *Ann. Mag. Nat. Hist.* (8) **15**: 62–96 [86]. Type data: holotype, BMNH *F. adult (seen 1929 by L.F. Graham), from Caloundra, Qld.

Distribution: NE coastal, Murray-Darling basin, SE coastal, Qld., N.S.W., Vic.; only published localities Caloundra, Stanthorpe, near Batemans Bay and Murrabit. Ecology: larva - sedentary, predator : adult - volant. Biological references: Leclercq, J. (1957). Le genre *Rhopalum* en Australie. *Bull. Ann. Soc. R. Entomol. Belg.* **93**: 177–232 (redescription).

Rhopalum (Rhopalum) tenuiventre (Turner, 1908)

Crabro (Rhopalum) tenuiventris Turner, R.E. (1908). Notes on the Australian fossorial wasps of the family Sphegidae, with descriptions of new species. *Proc. Zool. Soc. Lond.* **1908**: 457–535 pl xxvi [524]. Type data: holotype, BMNH *F. adult (seen 1929 by L.F. Graham), from Mackay, Qld.

Distribution: NE coastal, Murray-Darling basin, Qld., A.C.T.; only published localities Brisbane, Mackay and A.C.T. Ecology: larva - sedentary, predator : adult - volant. Biological references: Leclercq, J. (1957). Le genre *Rhopalum* en Australie. *Bull. Ann. Soc. R. Entomol. Belg.* **93**: 177–232 (redescription).

Rhopalum (Rhopalum) tepicum Leclercq, 1957

Rhopalum (Rhopalum) tepicum Leclercq, J. (1957). Le genre *Rhopalum* en Australie. *Bull. Ann. Soc. R. Entomol. Belg.* **93**: 177–232 [226]. Type data: holotype, BMNH *M. adult, from Mt. Wellington, Tas.

Distribution: Tas.; type locality only. Ecology: larva - sedentary, predator : adult - volant.

Rhopalum (Rhopalum) testaceum Turner, 1917

Rhopalum testaceum Turner, R.E. (1917). Notes on fossorial Hymenoptera. XXV. On new Sphecoidea in the British Museum. *Ann. Mag. Nat. Hist.* (8) **19**: 104–113 [108]. Type data: holotype, BMNH *F. adult (seen 1929 by L.F. Graham), from Kuranda, Qld.

Distribution: NE coastal, Qld.; type locality only. Ecology: larva - sedentary, predator : adult - volant. Biological references: Leclercq, J. (1957). Le genre *Rhopalum* en Australie. *Bull. Ann. Soc. R. Entomol. Belg.* **93**: 177–232 (redescription).

Rhopalum (Rhopalum) transiens (Turner, 1908)

Crabro (Rhopalum) transiens Turner, R.E. (1908). Notes on the Australian fossorial wasps of the family Sphegidae, with descriptions of new species. *Proc. Zool. Soc. Lond.* **1908**: 457–535 pl xxvi [525]. Type data: holotype, BMNH *M. adult (seen 1929 by L.F. Graham), from Vic.

Distribution: Murray-Darling basin, N.S.W., Vic.; only published localities SE Bourke and Vic. Ecology: larva - sedentary, predator : adult - volant. Biological references: Leclercq, J. (1957). Le genre *Rhopalum* en Australie. *Bull. Ann. Soc. R. Entomol. Belg.* **93**: 177–232 (redescription).

Rhopalum (Rhopalum) tuberculicorne Turner, 1917

Rhopalum tuberculicorne Turner, R.E. (1917). Notes on fossorial Hymenoptera. XXV. On new Sphecoidea in the British Museum. *Ann. Mag. Nat. Hist.* (8) **19**: 104–113 [107]. Type data: holotype, BMNH *M. adult (seen 1929 by L.F. Graham), from Caloundra, Qld.

Distribution: NE coastal, Qld.; type locality only. Ecology: larva - sedentary, predator : adult - volant. Biological references: Leclercq, J. (1957). Le genre *Rhopalum* en Australie. *Bull. Ann. Soc. R. Entomol. Belg.* **93**: 177–232 (redescription).

Rhopalum (Rhopalum) verutum (Rayment, 1932)

Dasyproctus verutus Rayment, T. (1932). The flycatcher of the reeds. A new crabronid wasp. *Vict. Nat.* **48**:

171-174 [173] [subgeneric placement tentative]. Type data: holotype, ANIC 7576 M. adult, from Ferntree Gully, Vic.

Distribution: SE coastal, Vic.; type locality only. Ecology: larva - sedentary, arboreal, predator : adult - volant; prey adult Diptera, nest in plant stems. Biological references: Leclercq, J. (1972). Crabroniens du genre *Dasyproctus* trouvés en Asie et en Océanie. *Bull. Soc. R. Sci. Liège* **41**: 101-122 (taxonomy); Naumann, I.D. (1983). The systematic position of *Dasyproctus verutus* Rayment (Hymenoptera: Sphecidae: Crabroninae). *J. Aust.. Entomol. Soc.* **22**: 349-351 (generic placement).

Rhopalum (Corynopus) Lepeletier and Brullé, 1834

Corynopus Lepeletier de Saint-Fargeau, A. & Brullé, A. (1834). Monographie du genre *Crabro* F., de la famille des Hyménoptères fouisseurs. *Ann. Soc. Entomol. Fr.* **3**: 683-810 [802]. Type species *Crabro tibialis* Fabricius, 1798 by monotypy. Compiled from secondary source: Bohart, R.M. & Menke, A.S. (1976). *Sphecid Wasps of the World : a Generic Revision.* Berkeley : Univ. California Press ix 695 pp. [47].

This group is also found in the Holarctic and Neotropical Regions, see Bohart, R.M. & Menke, A.S. (1976). *Sphecid Wasps of the World : a Generic Revision.* Berkeley : Univ. California Press ix 695 pp. [387].

Rhopalum (Corynopus) anteum Leclercq, 1957

Rhopalum (Rhopalum) anteum Leclercq, J. (1957). Le genre *Rhopalum* en Australie. *Bull. Ann. Soc. R. Entomol. Belg.* **93**: 177-232 [215]. Type data: holotype, BMNH *M. adult, from Port Phillip, Vic.

Distribution: SE coastal, Vic.; only published localities Port Phillip, Belcombe Heights, Cheltenham and Barwon Heads. Ecology: larva - sedentary, predator : adult - volant. Biological references: Leclercq, J. (1978). Crabroniens du genre *Rhopalum* Stephens trouvés en Australie (Hymenoptera Sphecidae). *Bull. Soc. R. Sci. Liège* **47**: 352-362 (generic placement).

Rhopalum (Corynopus) australiae Leclercq, 1957

Rhopalum (Rhopalum) australiae Leclercq, J. (1957). Le genre *Rhopalum* en Australie. *Bull. Ann. Soc. R. Entomol. Belg.* **93**: 177-232 [210]. Type data: holotype, ANIC F. adult, from Blundells, A.C.T.

Distribution: Murray-Darling basin, A.C.T.; type locality only. Ecology: larva - sedentary, predator : adult - volant. Biological references: Leclercq, J. (1978). Crabroniens du genre *Rhopalum* Stephens trouvés en Australie (Hymenoptera Sphecidae). *Bull. Soc. R. Sci. Liège* **47**: 352-362 (generic placement).

Rhopalum (Corynopus) coriolum Leclercq, 1957

Rhopalum (Rhopalum) coriolum Leclercq, J. (1957). Le genre *Rhopalum* en Australie. *Bull. Ann. Soc. R. Entomol. Belg.* **93**: 177-232 [210]. Type data: holotype, NMV *F. adult, from Mt. Victoria, N.S.W.

Distribution: SE coastal, N.S.W.; type locality only. Ecology: larva - sedentary, predator : adult - volant. Biological references: Leclercq, J. (1978). Crabroniens du genre *Rhopalum* Stephens trouvés en Australie (Hymenoptera Sphecidae). *Bull. Soc. R. Sci. Liège* **47**: 352-362 (generic placement).

Rhopalum (Corynopus) dedarum Leclercq, 1957

Rhopalum (Rhopalum) dedarum Leclercq, J. (1957). Le genre *Rhopalum* en Australie. *Bull. Ann. Soc. R. Entomol. Belg.* **93**: 177-232 [189]. Type data: holotype, BMNH *M. adult, from Dedari, W.A.

Distribution: NE coastal, SE coastal, Murray-Darling basin, SW coastal, Qld., N.S.W., A.C.T., Vic., W.A. Ecology: larva - sedentary, predator : adult - volant. Biological references: Leclerg, J. (1978). Crabroniens du genre *Rhopalum* Stephens trouvés en Australie (Hymenoptera : Sphecidae). *Bull. Soc. R. Sci. Liège* **47**: 352-362 (taxonomy, distribution).

Rhopalum (Corynopus) evansianum Leclercq, 1978

Rhopalum (Corynopus) evansianum Leclercq, J. (1978). Crabroniens du genre *Rhopalum* Stephens trouvés en Australie (Hymenoptera Sphecidae). *Bull. Soc. R. Sci. Liège* **47**: 352-362 [357]. Type data: holotype, MCZ *F. adult, from Packsaddle, N.S.W.

Distribution: Lake Eyre basin, N.S.W.; only published localities Packsaddle and Wilcannia. Ecology: larva - sedentary, predator : adult - volant.

Rhopalum (Corynopus) evictum Leclercq, 1978

Rhopalum (Corynopus) evictum Leclercq, J. (1978). Crabroniens du genre *Rhopalum* Stephens trouvés en Australie (Hymenoptera Sphecidae). *Bull. Soc. R. Sci. Liège* **47**: 352-362 [358]. Type data: holotype, MCZ *M. adult, from Wyperfeld Natl. Park, Vic.

Distribution: Murray-Darling basin, N.S.W., Vic.; only published localities Wentworth and Wyperfeld Natl. Park. Ecology: larva - sedentary, predator : adult - volant.

Rhopalum (Corynopus) famicum Leclercq, 1978

Rhopalum (Corynopus) famicum Leclercq, J. (1978). Crabroniens du genre *Rhopalum* Stephens trouvés en Australie (Hymenoptera Sphecidae). *Bull. Soc. R. Sci. Liège* **47**: 352-362 [359]. Type data: holotype, MCZ *F. adult, from 10 mi N Mt. Magnet, W.A.

Distribution: W plateau, W.A.; only published localities Mt. Magnet and Wiluna. Ecology: larva - sedentary, predator : adult - volant.

Rhopalum (Corynopus) frenchii (Turner, 1908)

Crabro (Rhopalum) frenchii Turner, R.E. (1908). Notes on the Australian fossorial wasps of the family Sphegidae, with descriptions of new species. *Proc. Zool. Soc. Lond.* **1908**: 457–535 pl xxvi [526]. Type data: holotype, BMNH *F. adult (seen 1929 by L.F. Graham), from Vic.

Distribution: SE coastal, Murray-Darling basin, SW coastal, N.S.W., A.C.T., Vic., Tas., W.A. Ecology: larva - sedentary, predator : adult - volant. Biological references: Leclercq, J. (1957). Le genre *Rhopalum* en Australie. *Bull. Ann. Soc. R. Entomol. Belg.* **93**: 177–232 (redescription); Tsuneki, K. (1977). On the crabronine wasps of the southern Pacific and Australia (Hymenoptera, Sphecidae). *Spec. Publ. Japan Hymenopterists Ass. 3* 27 pp. (description of variety); Leclercq, J. (1978). Crabroniens du genre *Rhopalum* Stephens trouvés en Australie (Hymenoptera Sphecidae). *Bull. Soc. R. Sci. Liège* **47**: 352–362 (redescription, generic placement).

Rhopalum (Corynopus) harpax Leclercq, 1957

Rhopalum (Rhopalum) harpax Leclercq, J. (1957). Le genre *Rhopalum* en Australie. *Bull. Ann. Soc. R. Entomol. Belg.* **93**: 177–232 [202]. Type data: holotype, NMV 532 *M. adult, from Melbourne, Vic.

Distribution: SE coastal, Vic.; type locality only. Ecology: larva - sedentary, predator : adult - volant. Biological references: Leclercq, J. (1978). Crabroniens du genre *Rhopalum* Stephens trouvés en Australie (Hymenoptera Sphecidae). *Bull. Soc. R. Sci. Liège* **47**: 352–362 (generic placement).

Rhopalum (Corynopus) kerangi Leclercq, 1957

Rhopalum (Rhopalum) kerangi Leclercq, J. (1957). Le genre *Rhopalum* en Australie. *Bull. Ann. Soc. R. Entomol. Belg.* **93**: 177–232 [213]. Type data: holotype, NMV *F. adult, from Kerang, Vic.

Distribution: Murray-Darling basin, Vic.; type locality only. Ecology: larva - sedentary, predator : adult - volant. Biological references: Leclercq, J. (1978). Crabroniens du genre *Rhopalum* Stephens trouvés en Australie (Hymenoptera Sphecidae). *Bull. Soc. R. Sci. Liège* **47**: 352–362 (generic placement).

Rhopalum (Corynopus) kuehlhorni Leclercq, 1957

Rhopalum (Rhopalum) kuehlhorni Leclercq, J. (1957). Le genre *Rhopalum* en Australie. *Bull. Ann. Soc. R. Entomol. Belg.* **93**: 177–232 [203]. Type data: holotype, ZSM *F. adult, from S.A.

Distribution: Murray-Darling basin, A.C.T.; only published localities Canberra and Black Mt. Ecology: larva - sedentary, predator : adult - volant. Biological references: Leclercq, J. (1978). Crabroniens du genre *Rhopalum* Stephens trouvés en Australie (Hymenoptera Sphecidae). *Bull. Soc. R. Sci. Liège* **47**: 352–362 (generic placement).

Rhopalum (Corynopus) littorale Turner, 1915

Rhopalum littorale Turner, R.E. (1915). Notes on fossorial Hymenoptera. XV. New Australian Crabronidae. *Ann. Mag. Nat. Hist. (8)* **15**: 62–96 [91]. Type data: holotype, BMNH *F. adult (seen 1929 by L.F. Graham), from Yallingup, W.A.

Distribution: Murray-Darling basin, SE coastal, SW coastal, N.S.W., A.C.T., Vic., Tas., W.A. Ecology: larva - sedentary, predator : adult - volant. Biological references: Leclercq, J. (1957). Le genre *Rhopalum* en Australie. *Bull. Ann. Soc. R. Entomol. Belg.* **93**: 177–232 (redescription); Leclercq, J. (1978). Crabroniens du genre *Rhopalum* Stephens trouvés en Australie (Hymenoptera Sphecidae). *Bull. Soc. R. Sci. Liège* **47**: 352–362 (generic placement).

Rhopalum (Corynopus) neboissi Leclercq, 1957

Rhopalum (Rhopalum) neboissi Leclercq, J. (1957). Le genre *Rhopalum* en Australie. *Bull. Ann. Soc. R. Entomol. Belg.* **93**: 177–232 [214]. Type data: holotype, NMV *F. adult, from Nunawading, Vic.

Distribution: SE coastal, Murray-Darling basin, A.C.T., Vic.; only published localities Canberra, Corin Dam and Nunawading. Ecology: larva - sedentary, predator : adult - volant. Biological references: Leclercq, J. (1978). Crabroniens du genre *Rhopalum* Stephens trouvés en Australie (Hymenoptera Sphecidae). *Bull. Soc. R. Sci. Liège* **47**: 352–362 (generic placement).

Rhopalum (Corynopus) notogeum Leclercq, 1957

Rhopalum (Rhopalum) notogeum Leclercq, J. (1957). Le genre *Rhopalum* en Australie. *Bull. Ann. Soc. R. Entomol. Belg.* **93**: 177–232 [216]. Type data: holotype, BMNH *M. adult, from Dongarra, W.A.

Distribution: SW coastal, NW coastal, W.A.; only published localities Dongarra, Perth and Maya. Ecology: larva - sedentary, predator : adult - volant. Biological references: Leclercq, J. (1978). Crabroniens du genre *Rhopalum* Stephens trouvés en Australie (Hymenoptera Sphecidae). *Bull. Soc. R. Sci. Liège* **47**: 352–362 (generic placement).

Rhopalum (Corynopus) taeniatum Leclercq, 1957

Rhopalum (Rhopalum) taeniatum Leclercq, J. (1957). Le genre *Rhopalum* en Australie. *Bull. Ann. Soc. R. Entomol. Belg.* **93**: 177–232 [225]. Type data: holotype, BMNH *M. adult, from Kuranda, Qld.

Distribution: NE coastal, Qld.; type locality only. Ecology: larva - sedentary, predator : adult - volant. Biological references: Leclercq, J. (1978). Crabroniens du genre *Rhopalum* Stephens trouvés en Australie (Hymenoptera Sphecidae). *Bull. Soc. R. Sci. Liège* **47**: 352–362 (generic placement).

Rhopalum (Corynopus) tubarum Leclercq, 1957

Rhopalum (Rhopalum) tubarum Leclercq, J. (1957). Le genre *Rhopalum* en Australie. *Bull. Ann. Soc. R. Entomol. Belg.* **93**: 177–232 [218]. Type data: holotype, ANIC M. adult, from Brisbane, Qld.

Distribution: NE coastal, Qld.; only published localities Brisbane, Tamborine and "Nummbah" (not found on map). Ecology: larva - sedentary, predator : adult - volant. Biological references: Leclercq, J. (1978). Crabroniens du genre *Rhopalum* Stephens trouvés en Australie (Hymenoptera Sphecidae). *Bull. Soc. R. Sci. Liège* **47**: 352–362 (generic placement).

Rhopalum (Corynopus) variitarse Turner, 1915

Rhopalum variitarse Turner, R.E. (1915). Notes on fossorial Hymenoptera. XV. New Australian Crabronidae. *Ann. Mag. Nat. Hist.* (8) **15**: 62–96 [89]. Type data: holotype, BMNH *F. adult (seen 1929 by L.F. Graham), from Mt. Wellington and Eaglehawk Neck," Tas.

Distribution: Murray-Darling basin, SE coastal, N.S.W., A.C.T. Vic., Tas. Ecology: larva - sedentary, soil, predator : adult - volant, burrowing; prey adult Diptera. Biological references: Leclercq, J. (1957). Le genre *Rhopalum* en Australie. *Bull. Ann. Soc. R. Entomol. Belg.* **93**: 177–232 (redescription); Evans, H.E. & Matthews, R.W. (1971). Notes on the prey and nests of some Australian Crabronini (Hymenoptera : Sphecidae). *J. Aust. Entomol. Soc.* **10**: 1–4 (nest, prey); Leclercq, J. (1978). Crabroniens du genre *Rhopalum* Stephens trouvés en Australie (Hymenoptera Sphecidae). *Bull. Soc. R. Sci. Liège* **47**: 352–362 (generic placement).

Rhopalum (Corynopus) xenum Leclercq, 1957

Rhopalum (Rhopalum) xenum Leclercq, J. (1957). Le genre *Rhopalum* en Australie. *Bull. Ann. Soc. R. Entomol. Belg.* **93**: 177–232 [217]. Type data: holotype, BMNH *M. adult, from Dongarra, W.A.

Distribution: SE coastal, SW coastal, NW coastal, N.S.W., W.A.; only published localities Sydney, Dongarra, Yallingup, Carnac Is. and near Mandurah. Ecology: larva - sedentary, predator : adult - volant. Biological references: Leclercq, J. (1978). Crabroniens du genre *Rhopalum* Stephens trouvés en Australie (Hymenoptera Sphecidae). *Bull. Soc. R. Sci. Liège* **47**: 352–362 (generic placement).

Rhopalum (Notorhopalum) Leclercq, 1978

Notorhopalum Leclercq, J. (1978). Crabroniens du genre *Rhopalum* Stephens trouvés en Australie (Hymenoptera Sphecidae). *Bull. Soc. R. Sci. Liège* **47**: 352–362 [354] [described with subgeneric rank in *Rhopalum* Stephens, 1829]. Type species *Rhopalum (Notorhopalum) carnegiacum* Leclercq, 1978 by monotypy and original designation.

Rhopalum (Notorhopalum) carnegiacum Leclercq, 1978

Rhopalum (Notorhopalum) carnegiacum Leclercq, J. (1978). Crabroniens du genre *Rhopalum* Stephens trouvés en Australie (Hymenoptera Sphecidae). *Bull. Soc. R. Sci. Liège* **47**: 352–362 [356]. Type data: holotype, NMV *F. adult, from Carnegie, Vic.

Distribution: SE coastal, Vic.; type locality only. Ecology: larva - sedentary, predator : adult - volant.

Podagritus Spinola, 1851

Taxonomic decision of Bohart, R.M. & Menke, A.S. (1976). *Sphecid Wasps of the World : a Generic Revision*. Berkeley : Univ. California Press ix 695 pp. [47].

Podagritus (Echuca) Pate, 1944

Echuca Pate, V.S.L. (1944). Conspectus of the genera of pemphilidine wasps (Hymenoptera : Sphecidae). *Am. Midl. Nat.* **31**: 329–384 [353] [described with subgeneric rank in *Podagritus* Spinola, 1851]. Type species *Crabro tricolor* Smith, 1856 by original designation.

Podagritus (Echuca) alevinus Leclercq, 1957

Podagritus (Echuca) alevinus Leclercq, J. (1957). Recherches systématiques et taxonomiques sur le genre *Podagritus*. I. Sur onze espèces australiennes et une espèce des îles Fidji. *Bull. Inst. R. Sci. Nat. Belg.* **33**(15): 1–7 [3]. Type data: holotype, NMV T7970 *F. adult, from Frankston, Vic.

Distribution: SE coastal, Vic.; type locality only. Ecology: larva - sedentary, soil, predator : adult - volant, burrowing; prey adult Diptera.

Podagritus (Echuca) aliciae (Turner, 1915)

Rhopalum aliciae Turner, R.E. (1915). Notes on fossorial Hymenoptera. XV. New Australian Crabronidae. *Ann. Mag. Nat. Hist.* (8) **15**: 62–96 [90]. Type data: holotype, BMNH *F. adult (seen 1929 by L.F. Graham), from Yallingup, W.A.

Distribution: SW coastal, W.A.; type locality only. Ecology: larva - sedentary, soil, predator : adult - volant, burrowing; prey adult Diptera. Biological references: Leclercq, J. (1950). Notes systématiques sur les Crabroniens pédonculés (Hymenoptera Sphecidae). *Bull. Inst. R. Sci. Nat. Belg.* **26**(15): 1–19 (generic placement); Leclercq, J. (1955). Révision des *Podagritus* (Spinola, 1851) australiens (Hym. Sphecidae, Crabroninae). *Bull. Ann. Soc. R. Entomol. Belg.* **91**: 305-330 (redescription).

Podagritus (Echuca) anerus Leclercq, 1955

Podagritus (Echuca) anerus Leclercq, J. (1955). Révision des *Podagritus* (Spinola, 1851) australiens (Hym. Sphecidae, Crabroninae). *Bull. Ann. Soc. R. Entomol. Belg.* **91**: 305-330 [323]. Type data: holotype, NMV 271 *M. adult, from Seaford, Vic. (as Scaford).

Distribution: SE coastal, Vic.; type locality only. Ecology: larva - sedentary, soil, predator : adult - volant, burrowing; prey adult Diptera.

Podagritus (Echuca) australiensis Tsuneki, 1977

Podagritus (Echuca) australiensis Tsuneki, K. (1977). On the crabronine wasps of the southern Pacific and Australia (Hymenoptera, Sphecidae). *Spec. Publ. Japan Hymenopterists Ass.* **3** 27 pp. [5]. Type data: syntypes (probable), MNH * adult, from Mt. Victoria, N.S.W.

Distribution: SE coastal, N.S.W.; type locality only. Ecology: larva - sedentary, soil, predator : adult - volant, burrowing; prey adult Diptera.

Podagritus (Echuca) burnsi Leclercq, 1955

Podagritus (Echuca) burnsi Leclercq, J. (1955). Révision des *Podagritus* (Spinola, 1851) australiens (Hym. Sphecidae, Crabroninae). *Bull. Ann. Soc. R. Entomol. Belg.* **91**: 305-330 [317]. Type data: holotype, NMV 268 *F. adult, from Melbourne, Vic.

Distribution: SE coastal, Murray-Darling basin, Vic.; only published localities Melbourne and Kiata. Ecology: larva - sedentary, soil, predator : adult - volant, burrowing; prey adult Diptera.

Podagritus (Echuca) carolus Leclercq, 1955

Podagritus (Echuca) carolus Leclercq, J. (1955). Révision des *Podagritus* (Spinola, 1851) australiens (Hym. Sphecidae, Crabroninae). *Bull. Ann. Soc. R. Entomol. Belg.* **91**: 305-330 [319]. Type data: holotype, BMNH *F. adult, from Killalpanima, 100 mi E of Lake Eyre, S.A.

Distribution: Lake Eyre basin, S.A.; type locality only. Ecology: larva - sedentary, soil, predator : adult - volant, burrowing; prey adult Diptera.

Podagritus (Echuca) cornigerum (Tsuneki, 1977)

Rhopalum (Rhopalum) cornigerum Tsuneki, K. (1977). On the crabronine wasps of the southern Pacific and Australia (Hymenoptera, Sphecidae). *Spec. Publ. Japan Hymenopterists Ass.* **3** 27 pp. [3]. Type data: syntypes (probable), MNH *M. adult, from Sydney, N.S.W.

Distribution: SE coastal, Murray-Darling basin, N.S.W., A.C.T.; only published localities Sydney, Woolahra, Georges River and near Canberra. Ecology: larva - sedentary, soil, predator : adult - volant, burrowing; prey adult Diptera. Biological references: Leclercq, J. (1978). Crabroniens du genre *Rhopalum* Stephens trouvés en Australie (Hymenoptera Sphecidae). *Bull. Soc. R. Sci. Liège* **47**: 352-362 (generic placement).

Podagritus (Echuca) cygnorum (Turner, 1915)

Rhopalum cygnorum Turner, R.E. (1915). Notes on fossorial Hymenoptera. XV. New Australian Crabronidae. *Ann. Mag. Nat. Hist. (8)* **15**: 62-96 [88]. Type data: holotype, BMNH *F. adult (seen 1929 by L.F. Graham), from Kings Park, Perth, W.A.

Distribution: SW coastal, W.A.; type locality only. Ecology: larva - sedentary, soil, predator : adult - volant, burrowing; prey adult Diptera. Biological references: Leclercq, J. (1950). Notes systématiques sur les Crabroniens pédonculés (Hymenoptera Sphecidae). *Bull. Inst. R. Sci. Nat. Belg.* **26**(15): 1-19 (generic placement).

Podagritus (Echuca) doreeni Leclercq, 1955

Podagritus (Echuca) doreeni Leclercq, J. (1955). Révision des *Podagritus* (Spinola, 1851) australiens (Hym. Sphecidae, Crabroninae). *Bull. Ann. Soc. R. Entomol. Belg.* **91**: 305-330 [313]. Type data: holotype, NMV *F. adult, from Trentham, Vic.

Distribution: Murray-Darling basin, SE coastal, A.C.T., Vic.; only published localities Blundells and Trentham. Ecology: larva - sedentary, soil, predator : adult - volant, burrowing; prey adult Diptera.

Podagritus (Echuca) edgarus Leclercq, 1957

Podagritus (Echuca) edgarus Leclercq, J. (1957). Recherches systématiques et taxonomiques sur le genre *Podagritus*. I. Sur onze espèces australiennes et une espèce des îles Fidji. *Bull. Inst. R. Sci. Nat. Belg.* **33**(15): 1-7 [5]. Type data: holotype, ANIC M. adult, from 35 mi NW of Nyngan, N.S.W.

Distribution: Murray-Darling basin, N.S.W.; type locality only. Ecology: larva - sedentary, soil, predator : adult - volant, burrowing; prey adult Diptera.

Podagritus (Echuca) imbelle (Turner, 1915)

Rhopalum tricolor imbelle Turner, R.E. (1915). Notes on fossorial Hymenoptera. XV. New Australian Crabronidae. *Ann. Mag. Nat. Hist. (8)* **15**: 62-96 [92]. Type data: holotype, BMNH *F. adult, from SW Australia.

Distribution: SW coastal, W.A.; only published localities Yallingup, Bunbury and near Nannup. Ecology: larva - sedentary, soil, predator : adult - volant, burrowing; prey adult Diptera. Biological references: Leclercq, J. (1950). Notes systématiques sur les Crabroniens pédonculés (Hymenoptera Sphecidae). *Bull. Inst. R. Sci. Nat. Belg.* **26**(15): 1-19 (generic placement).

Podagritus (Echuca) kiatae Leclercq, 1955

Podagritus (Echuca) kiatae Leclercq, J. (1955). Révision des *Podagritus* (Spinola, 1851) australiens (Hym. Sphecidae, Crabroninae). *Bull. Ann. Soc. R. Entomol. Belg.* **91**: 305–330 [329]. Type data: holotype, NMV 270 *F. adult, from Kiata, Vic.

Distribution: Murray-Darling basin, S Gulfs, SW coastal, Vic., Tas., S.A., W.A.; only published localities Kiata, Westerway, Iron Knob and Nannup. Ecology: larva - sedentary, soil, predator : adult - volant, burrowing; prey adult Diptera.

Podagritus (Echuca) krombeini Leclercq, 1955

Podagritus (Echuca) krombeini Leclercq, J. (1955). Révision des *Podagritus* (Spinola, 1851) australiens (Hym. Sphecidae, Crabroninae). *Bull. Ann. Soc. R. Entomol. Belg.* **91**: 305–330 [320]. Type data: holotype, USNM *M. adult, from Sydney, N.S.W.

Distribution: SE coastal, N.S.W.; only published localities Sydney and Mt. Victoria. Ecology: larva - sedentary, soil, predator : adult - volant, burrowing; prey adult Diptera.

Podagritus (Echuca) leptospermi (Turner, 1915)

Rhopalum leptospermi Turner, R.E. (1915). Notes on fossorial Hymenoptera. XV. New Australian Crabronidae. *Ann. Mag. Nat. Hist.* (8) **15**: 62–96 [87]. Type data: holotype, BMNH *F. adult (seen 1929 by L.F. Graham), from Yallingup, Warren River, W.A.

Distribution: NE coastal, Murray-Darling basin, SE coastal, SW coastal, Qld., N.S.W., A.C.T., Vic., Tas., W.A. Ecology: larva - sedentary, soil, predator : adult - volant, burrowing; prey adult Diptera Therevidae. Biological references: Leclercq, J. (1950). Notes systématiques sur les Crabroniens pédonculés (Hymenoptera Sphecidae). *Bull. Inst. R. Sci. Nat. Belg.* **26**(15): 1–19 (generic placement); Evans, H.E. & Matthews, R.W. (1971). Notes on the prey and nests of some Australian Crabronini (Hymenoptera : Sphecidae). *J. Aust. Entomol. Soc.* **10**: 1–4 (nest, prey).

Podagritus (Echuca) marcellus Leclercq, 1955

Podagritus (Echuca) marcellus Leclercq, J. (1955). Révision des *Podagritus* (Spinola, 1851) australiens (Hym. Sphecidae, Crabroninae). *Bull. Ann. Soc. R. Entomol. Belg.* **91**: 305–330 [323]. Type data: syntypes (probable), BMNH *adult, from Dongarra, W.A.

Distribution: SW coastal, NW coastal, W.A.; only published localities Dongarra and Yallingup. Ecology: larva - sedentary, soil, predator : adult - volant, burrowing; prey adult Diptera.

Podagritus (Echuca) mullewanus Leclercq, 1970

Podagritus (Echuca) mullewanus Leclercq, J. (1970). Quelques *Podagritus* d'Australie et d'Amérique du Sud (Hymenoptera Sphecidae, Crabroninae). *Bull. Rech. Agron. Gembloux* (ns) **5**: 271–280 [273]. Type data: holotype, ANIC F. adult, from 10 mi W of Mullewa, W.A.

Distribution: S Gulfs, NW coastal, S.A., W.A.; only published localities near Kimba and near Mullewa. Ecology: larva - sedentary, soil, predator : adult - volant, burrowing; prey adult Diptera.

Podagritus (Echuca) peratus Leclercq, 1955

Podagritus (Echuca) peratus Leclercq, J. (1955). Révision des *Podagritus* (Spinola, 1851) australiens (Hym. Sphecidae, Crabroninae). *Bull. Ann. Soc. R. Entomol. Belg.* **91**: 305–330 [326]. Type data: syntypes (probable), BMNH *M. adult, from Yallingup, W.A.

Distribution: SW coastal, W plateau, W.A.; only published localities Yallingup, Southern Cross and Coolgardie. Ecology: larva - sedentary, soil, predator : adult - volant, burrowing; prey adult Diptera. Biological references: Leclercq, J. (1970). Quelques *Podagritus* d'Australie et d'Amérique du Sud (Hymenoptera Sphecidae, Crabroninae). *Bull. Rech. Agron. Gembloux* (ns) **5**: 271–280 (redescription).

Podagritus (Echuca) rieki Leclercq, 1957

Podagritus (Echuca) rieki Leclercq, J. (1957). Recherches systématiques et taxonomiques sur le genre *Podagritus*. I. Sur onze espèces australiennes et une espèce des îles Fidji. *Bull. Inst. R. Sci. Nat. Belg.* **33**(15): 1–7 [1]. Type data: holotype, ANIC F. adult, from Blundells, A.C.T.

Distribution: Murray-Darling basin, SE coastal, A.C.T., N.S.W., Vic.; only published localities Blundells, Kosciusko, Nicholson River and Kangaroo Ground. Ecology: larva - sedentary, soil, predator : adult - volant, burrowing; prey adult Diptera.

Podagritus (Echuca) tricolor (Smith, 1856)

Crabro tricolor Smith, F. (1856). *Catalogue of Hymenopterous Insects in the Collection of the British Museum. Part IV, Sphegidae, Larridae and Crabronidae.* pp. 207–497 London : British Museum [394]. Type data: neotype, BMNH *M. adult, subsequent designation by Leclercq, J. (1955). Révision des *Podagritus* (Spinola, 1851) australiens (Hym. Sphecidae, Crabroninae). *Bull. Ann. Soc. R. Entomol. Belg.* **91**: 305–330, from Eaglehawk Neck, Tas.

Crabro (Rhopalum) militaris Turner, R.E. (1908). Notes on the Australian fossorial wasps of the family Sphegidae, with descriptions of new species. *Proc. Zool. Soc. Lond.* **1908**: 457–535 pl xxvi [523]. Type data: holotype, BMNH *M. adult (seen 1929 by L.F. Graham), from Vic.

Taxonomic decision of Turner, R.E. (1915). Notes on fossorial Hymenoptera. XV. New Australian Crabronidae. *Ann. Mag. Nat. Hist.* (8) **15**: 62–96 [92].

Distribution: SE coastal, Murray-Darling basin, SW coastal, N.S.W., A.C.T., Vic., Tas., W.A. Ecology: larva - sedentary, soil, predator : adult - volant, burrowing; prey adult Diptera. Biological references: Pate, V.S.L. (1944). Conspectus of the genera of pemphilidine wasps (Hymenoptera : Sphecidae). *Am. Midl. Nat.* **31**: 329-384 (generic placement); Leclercq, J. (1955). Révision des *Podagritus* (Spinola, 1851) australiens (Hym. Sphecidae, Crabroninae). *Bull. Ann. Soc. R. Entomol. Belg.* **91**: 305-330 (redescription).

Podagritus (Echuca) yarrowi Leclercq, 1955

Podagritus (Echuca) yarrowi Leclercq, J. (1955). Révision des *Podagritus* (Spinola, 1851) australiens (Hym. Sphecidae, Crabroninae). *Bull. Ann. Soc. R. Entomol. Belg.* **91**: 305-330 [328]. Type data: syntypes (probable), BMNH *adult, from Dongarra, W.A.

Distribution: SW coastal, NW coastal, W.A.; only published localities Dongarra and Yallingup. Ecology: larva - sedentary, soil, predator : adult - volant, burrowing; prey adult Diptera.

Notocrabro Leclercq, 1951

Notocrabro Leclercq, J. (1951). Notes systématiques sur quelques Crabroniens (Hymenoptera Sphecidae) américains, orientaux et australiens. *Bull. Ann. Soc. R. Entomol. Belg.* **87**: 31-56 [52]. Type species *Crabro (Rhopalum) idoneus* Turner, 1908 by monotypy and original designation.

Spinocrabro Leclercq, J. (1954). *Monographie Systématique, Phylogénétique et Zoogéographique des Hyménoptères Crabroniens*. Liège : Lejeunia Press 371 pp. [209] [unnecessary nom. nov. for *Notocrabro* Leclercq, 1951].

Taxonomic decision of Leclerq, J. (1978). Crabroniens du genre *Rhopalum* Stephens trouvés en Australie (Hymenoptera Sphecidae). *Bull. Soc. R. Sci. Liège* **47**: 352-362 [47].

Notocrabro idoneus (Turner, 1908)

Crabro (Rhopalum) idoneus Turner, R.E. (1908). Notes on the Australian fossorial wasps of the family Sphegidae, with descriptions of new species. *Proc. Zool. Soc. Lond.* **1908**: 457-535 pl xxvi [527]. Type data: holotype, BMNH *F. adult (seen 1929 by L.F. Graham), from Mackay, Qld.

Rhopalum spinulifer Turner, R.E. (1918). Notes on fossorial Hymenoptera. XXXII. On new species in the British Museum. *Ann. Mag. Nat. Hist. (9)* **1**: 86-96 [93]. Type data: holotype, BMNH *M. adult (seen 1929 by L.F. Graham), from Kuranda, Qld.

Taxonomic decision of Leclercq, J. (1954). *Monographie Systématique, Phylogénétique et Zoogéographique des Hyménoptères Crabroniens*. Liège : Lejeunia Press 371 pp. [209].

Distribution: NE coastal, Qld.; only published localities Mackay, Kuranda, Herberton and Cape York. Ecology: larva - sedentary, predator : adult - volant. Biological references: Leclercq, J. (1974). Crabroniens d'Australie (Hymenoptera Sphecidae Crabroninae). *Bull. Ann. Soc. R. Entomol. Belg.* **110**: 37-57 (generic placement).

Notocrabro micheneri Leclercq, 1974

Notocrabro micheneri Leclercq, J. (1974). Crabroniens d'Australie (Hymenoptera Sphecidae Crabroninae). *Bull. Ann. Soc. R. Entomol. Belg.* **110**: 37-57 [41]. Type data: holotype, ANIC M. adult, from Binna Burra, Lamington Natl. Park, Qld.

Distribution: NE coastal, Qld.; type locality only. Ecology: larva - sedentary, predator : adult - volant.

Pseudoturneria Leclercq, 1954

Turneriola Leclercq, J. (1951). Notes systématiques sur quelques Crabroniens (Hymenoptera Sphecidae) américains, orientaux et australiens. *Bull. Ann. Soc. R. Entomol. Belg.* **87**: 31-56 [54] [*non Turneriola* China, 1933]. Type species *Crabro perlucidus* Turner, 1908 by monotypy and original designation.

Pseudoturneria Leclercq, J. (1954). *Monographie Systématique, Phylogénétique et Zoogéographique des Hyménoptères Crabroniens*. Liège : Lejeunia Press 371 pp. [208] [*nom. nov.* for *Turneriola* Leclercq, 1951].

Pseudoturneria couloni Leclercq, 1974

Pseudoturneria couloni Leclercq, J. (1974). Crabroniens d'Australie (Hymenoptera Sphecidae Crabroninae). *Bull. Ann. Soc. R. Entomol. Belg.* **110**: 37-57 [42]. Type data: holotype, ZMB *F. adult, from Port Philip, Vic.

Distribution: SE coastal, Vic.; type locality only. Ecology: larva - sedentary, predator : adult - volant.

Pseudoturneria perlucida (Turner, 1908)

Crabro perlucidus Turner, R.E. (1908). Notes on the Australian fossorial wasps of the family Sphegidae, with descriptions of new species. *Proc. Zool. Soc. Lond.* **1908**: 457-535 pl xxvi [529 pl xxvi fig 15]. Type data: holotype, BMNH *F. adult (seen 1929 by L.F. Graham), from Mackay, Qld.

Distribution: NE coastal, Qld.; type locality only. Ecology: larva - sedentary, predator : adult - volant. Biological references: Leclerq, J. (1951). Notes systématiques sur quelques Crabroniens (Hymenoptera Sphecidae) américains, orientaux et australiens. *Bull. Ann. Soc. R. Entomol. Belg.* **87**: 31-56 (taxonomy).

Pseudoturneria territorialis Leclercq, 1974

Pseudoturneria territorialis Leclercq, J. (1974). Crabroniens d'Australie (Hymenoptera Sphecidae

Crabroninae). *Bull. Ann. Soc. R. Entomol. Belg.* **110**: 37-57 [44]. Type data: holotype, ANIC M. adult, from Blundells, A.C.T.

Distribution: Murray-Darling basin, SE coastal, N.S.W., A.C.T., Vic.; only published localities Brown Mt., Blundells, Corin Dam and Blackburn.

Piyuma Pate, 1944

Piyuma Pate, V.S.L. (1944). Conspectus of the genera of pemphilidine wasps (Hymenoptera : Sphecidae). *Am. Midl. Nat.* **31**: 329-384 [356]. Type species *Piyuma koxinga* Pate, 1944 (= *Crabro prosopoides* Turner, 1908) by original designation.

This group is also found in the Oriental Region, see Bohart, R.M. & Menke, A.S. (1976). *Sphecid Wasps of the World : a Generic Revision*. Berkeley : Univ. California Press ix 695 pp. [409].

Piyuma prosopoides (Turner, 1908)

Crabro prosopoides Turner, R.E. (1908). Notes on the Australian fossorial wasps of the family Sphegidae, with descriptions of new species. *Proc. Zool. Soc. Lond.* **1908**: 457-535 pl xxvi [528]. Type data: holotype, BMNH *F. adult (seen 1929 by L.F. Graham), from Mackay, Qld.

Distribution: NE coastal, Qld.; only published localities Mackay, Townsville, Kuranda and Brisbane, also Borneo, Taiwan and Philippines. Ecology: larva - sedentary, arboreal, predator : adult - volant; prey mainly small adult Diptera, nest in pre-existing hole in wood. Biological references: Pate, V.S.L. (1944). Conspectus of the genera of pemphilidine wasps (Hymenoptera : Sphecidae). *Am. Midl. Nat.* **31**: 329-384; Bohart, R.M. & Menke, A.S. (1976). *Sphecid Wasps of the World : a Generic Revision*. Berkeley : Univ. California Press ix 695 pp. (biology, list of synonyms).

Chimiloides Leclercq, 1951

Chimiloides Leclercq, J. (1951). Notes systématiques sur quelques Crabroniens (Hymenoptera Sphecidae) américains, orientaux et australiens. *Bull. Ann. Soc. R. Entomol. Belg.* **87**: 31-56 [50]. Type species *Crabro nigromaculatus* Smith, 1868 by original designation.

Chimiloides doddii (Turner, 1908)

Crabro doddii Turner, R.E. (1908). Notes on the Australian fossorial wasps of the family Sphegidae, with descriptions of new species. *Proc. Zool. Soc. Lond.* **1908**: 457-535 pl xxvi [529]. Type data: holotype, BMNH *F. adult (seen 1929 by L.F. Graham), from Townsville, Qld.
Crabro erythrogaster Turner, R.E. (1910). New fossorial Hymenoptera from Australia. *Trans. Entomol. Soc. Lond.* **1910**: 407-429 pl 50 [429]. Type data: holotype, BMNH *M. adult (seen 1929 by L.F. Graham), from Bundaberg, Qld.

Taxonomic decision of Leclercq, J. (1954). *Monographie Systématique, Phylogénétique et Zoogéographique des Hyménoptères Crabroniens*. Liège : Lejeunia Press 371 pp. [212].

Distribution: NE coastal, Murray-Darling basin, SE coastal, Qld., Vic.; only published localities Townsville, near Collinsville, Stanthorpe and Hamilton. Ecology: larva - sedentary, predator : adult - volant. Biological references: Leclercq, J. (1951). Notes systématiques sur quelques Crabroniens (Hymenoptera Sphecidae) américains, orientaux et australiens. *Bull. Ann. Soc. R. Entomol. Belg.* **87**: 31-56 (generic placement).

Chimiloides nigromaculatus (Smith, 1868)

Crabro nigromaculatus Smith, F. (1868). Descriptions of aculeate Hymenoptera from Australia. *Trans. Entomol. Soc. Lond.* **1868**: 231-258 [249]. Type data: holotype, BMNH *M. adult (seen 1929 by L.F. Graham), from Moreton Bay, Qld.

Distribution: NE coastal, SW coastal, NW coastal, Qld., W.A.; only published localities Moreton Bay, Merredin, Carnarvon. Ecology: larva - sedentary, predator : adult - volant. Biological references: Leclercq, J. (1951). Notes systématiques sur quelques Crabroniens (Hymenoptera Sphecidae) américains, orientaux et australiens. *Bull. Ann. Soc. R. Entomol. Belg.* **87**: 31-56 (generic placement).

Chimiloides piliferus Leclercq, 1954

Chimiloides piliferus Leclercq, J. (1954). *Monographie Systématique, Phylogénétique et Zoogéographique des Hyménoptères Crabroniens*. Liège : Lejeunia Press 371 pp. [212]. Type data: holotype, BMNH *M. adult, from Townsville, Qld.

Distribution: NE coastal, Murray-Darling basin, S Gulfs, SW coastal, NW coastal, N coastal, Qld., Vic., S.A., W.A., N.T. Ecology: larva - sedentary, predator, adult - volant.

Neodasyproctus Arnold, 1926

Neodasyproctus Arnold, G. (1926). The Sphegidae of South Africa. Part VII. *Ann. Transvaal Mus.* **11**: 338-376 [373] [described with subgeneric rank in *Thyreopus* Lepeletier and Brullé, 1834]. Type species *Thyreopus (Neodasyproctus) kohli* Brauns, 1926 *in* Arnold, *loc. cit.* (1926) by monotypy.

This group is also found in the Ethiopian Region and Fiji, see Bohart, R.M. & Menke, A.S. (1976). *Sphecid Wasps of the World : a Generic Revision*. Berkeley : Univ. California Press ix 695 pp. [418].

Neodasyproctus veitchi (Turner, 1917)

Crabro veitchi Turner, R.E. (1917). New species of Hymenoptera in the British Museum. *Trans. Entomol. Soc. Lond.* **1917**: 53-84 [84]. Type data: holotype, BMNH *F. adult (seen 1929 by L.F. Graham), from Fiji.

Distribution: Fiji, no Australian locality specified. Ecology: larva - sedentary, predator : adult - volant. Biological references: Leclercq, J. (1950). Notes systématiques sur les Crabroniens pédonculés (Hymenoptera Sphecidae). *Bull. Inst. R. Sci. Nat. Belg.* **26**(15): 1–19 (generic placement); Leclercq, J. (1951). Sur quelques *Neodasyproctus* (Arnold, 1926) nouveaux ou peu connus (Hymenoptera, Sphecidae, Crabroninae). *Rev. Zool. Bot. Afr.* **44**: 333–337 (redescription).

Dasyproctus Lepeletier and Brullé, 1834

Dasyproctus Lepeletier de Saint-Fargeau, A. & Brullé, A. (1834). Monographie du genre *Crabro* F., de la famille des Hyménoptères fouisseurs. *Ann. Soc. Entomol. Fr.* **3**: 683–810 [801]. Type species *Dasyproctus bipunctatus* Lepeletier and Brullé, 1834 by monotypy. Compiled from secondary source: Bohart, R.M. & Menke, A.S. (1976). *Sphecid Wasps of the World : a Generic Revision.* Berkeley : Univ. California Press ix 695 pp.

This group is also found in the Ethiopian and Oriental Regions, see Bohart, R.M. & Menke, A.S. (1976). *Sphecid Wasps of the World : a Generic Revision.* Berkeley : Univ. California Press ix 695 pp. [419].

Dasyproctus australgilis Leclercq, 1972

Dasyproctus australgilis Leclercq, J. (1972). Crabroniens du genre *Dasyproctus* trouvés en Asie et en Océanie. *Bull. Soc. R. Sci. Liège* **41**: 101–122 [113]. Type data: holotype, BMNH *adult, from Mackay, Qld.

Distribution: NE coastal, Qld.; only published localities Mackay, Westwood and Meringa. Ecology: larva - sedentary, arboreal, predator : adult - volant; prey adult Diptera, nest in plant stems.

Dasyproctus burnettianus Turner, 1912

Dasyproctus burnettianus Turner, R.E. (1912). Notes on fossorial Hymenoptera. IX. On some new species from the Australian and Austro-Malayan regions. *Ann. Mag. Nat. Hist. (8)* **10**: 48–63 [62]. Type data: holotype, BMNH *F. adult (seen 1929 by L.F. Graham), from Bundaberg, Qld.

Distribution: NE coastal, Qld.; type locality only. Ecology: larva - sedentary, arboreal, predator : adult - volant; prey adult Diptera, nest in plant stems. Biological references: Leclercq, J. (1956). Les *Dasyproctus* (Lepeletier de St-Fargeau et Brullé 1834) du sud-est asiatique et de l'Océanie (Hym. Sphecidae Crabroninae). *Bull. Ann. Soc. R. Entomol. Belg.* **92**: 139–167 (redescription).

Dasyproctus conator (Turner, 1908)

Crabro (Rhopalum) conator Turner, R.E. (1908). Notes on the Australian fossorial wasps of the family Sphegidae, with descriptions of new species. *Proc. Zool. Soc. Lond.* **1908**: 457–535 pl xxvi [526]. Type data: holotype, BMNH *M. adult (seen 1929 by L.F. Graham), from Cooktown, Qld.

Distribution: NE coastal, Qld.; only published localities Cooktown and Kuranda. Ecology: larva - sedentary, arboreal, predator : adult - volant; prey adult Diptera, nest in plant stems. Biological references: Turner, R.E. (1912). Notes on fossorial Hymenoptera. IX. On some new species from the Australian and Austro-Malayan regions. *Ann. Mag. Nat. Hist. (8)* **10**: 48–63 (generic placement); Leclercq, J. (1972). Crabroniens du genre *Dasyproctus* trouvés en Asie et en Océanie. *Bull. Soc. R. Sci. Liège* **41**: 101–122 (taxonomy).

Dasyproctus expectatus Turner, 1912

Dasyproctus expectatus Turner, R.E. (1912). Notes on fossorial Hymenoptera. IX. On some new species from the Australian and Austro-Malayan regions. *Ann. Mag. Nat. Hist. (8)* **10**: 48–63 [60]. Type data: holotype, BMNH *F. adult (seen 1929 by L.F. Graham), from Sydney, N.S.W.

Distribution: NE coastal, SE coastal, Qld., N.S.W., Vic.; only published localities Fitzroy Is., Sydney and Mooroopna. Ecology: larva - sedentary, arboreal, predator : adult - volant; prey adult Diptera, nest in plant stems. Biological references: Leclercq, J. (1972). Crabroniens du genre *Dasyproctus* trouvés en Asie et en Océanie. *Bull. Soc. R. Sci. Liège* **41**: 101–122 (taxonomy).

Dasyproctus yorki Leclercq, 1956

Dasyproctus yorki Leclercq, J. (1956). Les *Dasyproctus* (Lepeletier de St-Fargeau et Brullé 1834) du sud-est asiatique et de l'Océanie (Hym. Sphecidae Crabroninae). *Bull. Ann. Soc. R. Entomol. Belg.* **92**: 139–167 [157]. Type data: holotype, NHMW *F. adult, from Cape York, Qld.

Distribution: NE coastal, N coastal, Qld., W.A.; only published localities Cape York, Fitzroy Is., Dunk Is. and Wyndham. Ecology: larva - sedentary, arboreal, predator : adult - volant; prey adult Diptera, nest in plant stems. Biological references: Leclercq, J. (1972). Crabroniens du genre *Dasyproctus* trouvés en Asie et en Océanie. *Bull. Soc. R. Sci. Liège* **41**: 101–122 (distribution).

Williamsita Pate, 1947

Taxonomic decision of Bohart, R.M. & Menke, A.S. (1976). *Sphecid Wasps of the World : a Generic Revision.* Berkeley : Univ. California Press ix 695 pp. [49].

Williamsita (Androcrabro) Leclercq, 1950

Androcrabro Leclercq, J. (1950). Sur les Crabroniens orientaux et australiens rangés par R.E. Turner (1912–1915) dans le genre *Crabro* (subgenus *Solenius*). *Bull. Ann. Soc. Entomol. Belg.* **86**: 191–198 [192] [described with subgeneric rank in *Williamsita* Pate, 1947]. Type species *Crabro neglectus* Smith, 1868 by original designation.

This group is also found in New Caledonia, see Bohart, R.M. & Menke, A.S. (1976). *Sphecid Wasps of the World : a Generic Revision*. Berkeley : Univ. California Press ix 695 pp. [421].

Williamsita (Androcrabro) bivittata (Turner, 1908)

Crabro bivittatus Turner, R.E. (1908). Notes on the Australian fossorial wasps of the family Sphegidae, with descriptions of new species. *Proc. Zool. Soc. Lond.* **1908**: 457-535 pl xxvi [534]. Type data: holotype, BMNH *F. adult (seen 1929 by L.F. Graham), from Vic.

Distribution: NE coastal, SE coastal, Murray-Darling basin, S Gulfs, Qld., N.S.W., A.C.T., Vic., Tas., S.A. Ecology: larva - sedentary, predator : adult - volant; nest in log, prey adult Diptera (Calliphoridae). Biological references: Leclercq, J. (1950). Sur les Crabroniens orientaux et australiens rangés par R.E. Turner (1912-1915) dans le genre *Crabro* (subgenus *Solenius*). *Bull. Ann. Soc. Entomol. Belg.* **86**: 191-198 (generic placement); Evans, H.E. & Matthews, R.W. (1971). Notes on the prey and nests of some Australian Crabronini (Hymenoptera : Sphecidae). *J. Aust. Entomol. Soc.* **10**: 1-4 (nest, prey).

Williamsita (Androcrabro) bushiella Leclercq, 1974

Williamsita bushiella Leclercq, J. (1974). Crabroniens d'Australie (Hymenoptera Sphecidae Crabroninae). *Bull. Ann. Soc. R. Entomol. Belg.* **110**: 37-57 [52]. Type data: holotype, MCZ *F. adult, from 10.2 mi W of Nundroo, S.A.

Distribution: W plateau, S.A.; type locality only. Ecology: larva - sedentary, predator : adult - volant; prey adult Diptera.

Williamsita (Androcrabro) manifestata (Turner, 1915)

Crabro (Solenius) manifestatus Turner, R.E. (1915). Notes on fossorial Hymenoptera. XV. New Australian Crabronidae. *Ann. Mag. Nat. Hist. (8)* **15**: 62-96 [95]. Type data: holotype, BMNH *F. adult (seen 1929 by L.F. Graham), from Kalamunda, W.A.

Distribution: SW coastal, W.A.; only published localities Kalamunda, Waroona, Perth and Bunbury. Ecology: larva - sedentary, predator : adult - volant; prey adult Diptera. Biological references: Leclercq, J. (1950). Sur les Crabroniens orientaux et australiens rangés par R.E. Turner (1912-1915) dans le genre *Crabro* (subgenus *Solenius*). *Bull. Ann. Soc. Entomol. Belg.* **86**: 191-198 (generic placement).

Williamsita (Androcrabro) neglecta (Smith, 1868)

Crabro neglectus Smith, F. (1868). Descriptions of aculeate Hymenoptera from Australia. *Trans. Entomol. Soc. Lond.* **1868**: 231-258 [249]. Type data: holotype, BMNH *M. adult (seen 1929 by L.F. Graham), from S.A.

Distribution: Tas., (S.A.); only published localities Scottsdale and S. Aust. Ecology: larva - sedentary, predator : adult - volant; prey adult Diptera. Biological references: Leclercq, J. (1950). Sur les Crabroniens orientaux et australiens rangés par R.E. Turner (1912-1915) dans le genre *Crabro* (subgenus *Solenius*). *Bull. Ann. Soc. Entomol. Belg.* **86**: 191-198 (generic placement).

Williamsita (Androcrabro) ordinaria (Turner, 1908)

Crabro ordinarius Turner, R.E. (1908). Notes on the Australian fossorial wasps of the family Sphegidae, with descriptions of new species. *Proc. Zool. Soc. Lond.* **1908**: 457-535 pl xxvi [532]. Type data: holotype, BMNH *F. adult (seen 1929 by L.F. Graham), from Mackay, Qld.

Distribution: NE coastal, N coastal, Qld., N.S.W., N.T. Ecology: larva - sedentary, predator : adult - volant; prey adult Diptera. Biological references: Leclercq, J. (1950). Sur les Crabroniens orientaux et australiens rangés par R.E. Turner (1912-1915) dans le genre *Crabro* (subgenus *Solenius*). *Bull. Ann. Soc. Entomol. Belg.* **86**: 191-198 (generic placement).

Williamsita (Androcrabro) riekiella Leclercq, 1974

Williamsita riekiella Leclercq, J. (1974). Crabroniens d'Australie (Hymenoptera Sphecidae Crabroninae). *Bull. Ann. Soc. R. Entomol. Belg.* **110**: 37-57 [53]. Type data: holotype, ANIC F. adult, from National Park, N.S.W. (probably Royal National Park S of Sydney).

Distribution: SE coastal, N.S.W.; type locality only. Ecology: larva - sedentary, predator : adult - volant; prey adult Diptera.

Williamsita (Androcrabro) smithiensis Leclercq, 1954

Crabro tridentatus Smith, F. (1868). Descriptions of aculeate Hymenoptera from Australia. *Trans. Entomol. Soc. Lond.* **1868**: 231-258 [250] [*non Crabo tridentatus* Fabricius, 1775]. Type data: holotype, BMNH *F. adult (seen 1929 by L.F. Graham), from Moreton Bay, Qld.

Williamsita (Androcrabro) smithiensis Leclercq, J. (1954). *Monographie Systématique, Phylogénétique et Zoogéographique des Hyménoptères Crabroniens*. Liège : Lejeunia Press 371pp. [263] [*nom. nov.* for *Crabro tridentatus* Smith, 1868].

Distribution: NE coastal, SE coastal, Murray-Darling basin, Qld., N.S.W., A.C.T., Vic. Ecology: larva - sedentary, predator : adult - volant; prey adult Diptera.

Williamsita (Androcrabro) tasmanica (Smith, 1856)

Crabro tasmanicus Smith, F. (1856). *Catalogue of Hymenopterous Insects in the Collection of the British Museum.* Part IV, Sphegidae, Larridae and Crabronidae. pp. 207–497 London : British Museum [425]. Type data: holotype, BMNH *M. adult (seen 1929 by L.F. Graham), from Tas.

Distribution: Tas. Ecology: larva - sedentary, predator : adult - volant; prey adult Diptera. Biological references: Leclercq, J. (1950). Sur les Crabroniens orientaux et australiens rangés par R.E. Turner (1912–1915) dans le genre *Crabro* (subgenus *Solenius*). *Bull. Ann. Soc. Entomol. Belg.* **86**: 191–198 (generic placement).

Williamsita (Androcrabro) vedetta Leclercq, 1974

Williamsita vedetta Leclercq, J. (1974). Crabroniens d'Australie (Hymenoptera Sphecidae Crabroninae). *Bull. Ann. Soc. R. Entomol. Belg.* **110**: 37–57 [55]. Type data: holotype, NMV 4476 *F. adult, from Bunbury, W.A.

Distribution: SW coastal, W.A.; type locality only. Ecology: larva - sedentary, predator : adult - volant; prey adult Diptera.

Ectemnius Dahlbom, 1845

Taxonomic decision of Leclercq, J. (1954). *Monographie Systématique, Phylogénétique et Zoogéographique des Hyménoptères Crabroniens.* Liège : Lejeunia Press 371 pp. [264].

Ectemnius (Hypocrabro) Ashmead, 1899

Hypocrabro Ashmead, W.H. (1899). Classification of the entomophilous wasps, or the superfamily Sphegoidea. *Can. Entomol.* **31**: 145–155, 161–174, 212–225, 238–251, 291–300, 322–330, 345–357 [168]. Type species *Crabro decemmaculatus* Say, 1823 by original designation.

This group is also found in the Holarctic, Neotropical and Oriental Regions, see Bohart, R.M. & Menke, A.S. (1976). *Sphecid Wasps of the World : a Generic Revision.* Berkeley : Univ. California Press ix 695 pp. [422].

Ectemnius (Hypocrabro) hebetescens (Turner, 1908)

Crabro hebetescens Turner, R.E. (1908). Notes on the Australian fossorial wasps of the family Sphegidae, with descriptions of new species. *Proc. Zool. Soc. Lond.* **1908**: 457–535 pl xxvi [530]. Type data: holotype, BMNH *F. adult (seen 1929 by L.F. Graham), from Mackay, Qld.

Distribution: NE coastal, Qld.; type locality only. Ecology: larva - sedentary, predator : adult - volant; prey adult Diptera. Biological references: Leclercq, J. (1950). Sur les Crabroniens orientaux et australiens rangés par R.E. Turner (1912–1915) dans le genre *Crabro* (subgenus *Solenius*). *Bull. Ann. Soc. Entomol. Belg.* **86**: 191–198 (generic placement).

Ectemnius (Hypocrabro) mackayensis (Turner, 1908)

Crabro mackayensis Turner, R.E. (1908). Notes on the Australian fossorial wasps of the family Sphegidae, with descriptions of new species. *Proc. Zool. Soc. Lond.* **1908**: 457–535 pl xxvi [532]. Type data: holotype, BMNH *F. adult (seen 1929 by L.F. Graham), from Mackay, Qld.

Distribution: NE coastal, Qld.; type locality only. Ecology: larva - sedentary, predator : adult - volant; prey adult Diptera. Biological references: Leclercq, J. (1950). Sur les Crabroniens orientaux et australiens rangés par R.E. Turner (1912–1915) dans le genre *Crabro* (subgenus *Solenius*). *Bull. Ann. Soc. Entomol. Belg.* **86**: 191–198 (generic placement).

Ectemnius (Hypocrabro) reginellus Leclercq, 1954

Crabro cinctus Turner, R.E. (1908). Notes on the Australian fossorial wasps of the family Sphegidae, with descriptions of new species. *Proc. Zool. Soc. Lond.* **1908**: 457–535 [531 pl xxvi fig 14] [*non Crabo cinctus* Rossi, 1791]. Type data: holotype, BMNH *F. adult (seen 1929 by L.F. Graham), from Mackay, Qld.

Ectemnius (Hypocrabro) reginellus Leclercq, J. (1954). *Monographie Systématique, Phylogénétique et Zoogéographique des Hyménoptères Crabroniens.* Liège : Lejeunia Press 371 pp. [268] [*nom. nov.* for *Crabro cinctus* Turner, 1908].

Distribution: NE coastal, Qld.; only published localities Mackay, Westwood, Brisbane, Ayr and Collinsville. Ecology: larva - sedentary, predator : adult - volant; prey adult Diptera.

Ectemnius (Cameronitus) Leclercq, 1950

Cameronitus Leclercq, J. (1950). Notes systématiques sur les Crabroniens pédonculés (Hymenoptera Sphecidae). *Bull. Inst. R. Sci. Nat. Belg.* **26**(15): 1–19 [14] [described with subgeneric rank in *Ectemnius* Dahlbom, 1845]. Type species *Crabro menyllus* Cameron, 1905 by original designation.

This group is also found in the Palearctic and Oriental Regions, see Bohart, R.M. & Menke, A.S. (1976). *Sphecid Wasps of the World : a Generic Revision.* Berkeley : Univ. California Press ix 695 pp. [422].

Ectemnius (Cameronitus) conglobatus (Turner, 1908)

Crabro conglobatus Turner, R.E. (1908). Notes on the Australian fossorial wasps of the family Sphegidae, with descriptions of new species. *Proc. Zool. Soc. Lond.* **1908**: 457–535 pl xxvi [533]. Type data: holotype, BMNH *F. adult (seen 1929 by L.F. Graham), from Mackay, Qld.

Distribution: NE coastal, Qld.; only published localities Mackay and Kuranda. Ecology: larva - sedentary, predator : adult - volant; prey adult Diptera. Biological references: Leclercq, J. (1954). *Monographie Systématique, Phylogénétique et Zoogéographique des Hyménoptères Crabroniens.* Liège : Lejeunia Press 371 pp. (generic placement).

Lestica Billberg, 1820

Taxonomic decision of Leclercq, J. (1954). *Monographie Systématique, Phylogénétique et Zoogéographique des Hyménoptères Crabroniens.* Liège : Lejeunia Press 371 pp. [291].

Lestica (Solenius) Lepeletier and Brullé, 1834

Solenius Lepeletier de Saint-Fargeau, A. & Brullé, A. (1834). Monographie du genre *Crabro* F., de la famille des Hyménoptères fouisseurs. *Ann. Soc. Entomol. Fr.* **3**: 683–810 [713]. Type species *Solenius interruptus* Lepeletier and Brullé, 1834 by subsequent designation, see International Commission on Zoological Nomenclature (1974). Opinion 1015. *Solenius* Lepeletier & Brullé, 1834 (Insecta, Hymenoptera) : designation of a type-species under the plenary powers. *Bull. Zool. Nomen.* **31**: 16–18. Compiled from secondary source: Bohart, R.M. & Menke, A.S. (1976). *Sphecid Wasps of the World : a Generic Revision.* Berkeley : Univ. California Press ix 695 pp.

This group is found worldwide, see Bohart, R.M. & Menke, A.S. (1976). *Sphecid Wasps of the World : a Generic Revision.* Berkeley : Univ. California Press ix 695 pp. [428].

Lestica (Solenius) relicta Leclercq, 1951

Lestica (Solenius) relicta Leclercq, J. (1951). Sur trois espèces de *Lestica (Solenius)* (Hym., Sphecidae, Crabroninae). *Bull. Ann. Soc. R. Entomol. Belg.* **87**: 169–173 [169]. Type data: holotype, GMNH *F. adult, from Australia.

Distribution: no Australian locality specified. Ecology: larva - sedentary, predator : adult - volant; prey adult Lepidoptera.

Acanthostethus Smith, 1869

Acanthostethus Smith, F. (1869). Descriptions of new genera and species of exotic Hymenoptera. *Trans. Entomol. Soc. Lond.* **1869**: 301–311 [306]. Type species *Acanthostethus basalis* Smith, 1869 by monotypy.

Acanthostethus brisbanensis (Turner, 1915)

Nysson (Acanthostethus) brisbanensis Turner, R.E. (1915). Notes on fossorial Hymenoptera. XV. New Australian Crabronidae. *Ann. Mag. Nat. Hist. (8)* **15**: 62–96 [81]. Type data: holotype, BMNH *F. adult (seen 1929 by L.F. Graham), from Brisbane, Qld.

Distribution: NE coastal, Qld.; type locality only. Ecology: larva - sedentary, soil, predator : adult - volant; cleptoparasite.

Acanthostethus confertus (Turner, 1915)

Nysson (Acanthostethus) confertus Turner, R.E. (1915). Notes on fossorial Hymenoptera. XV. New Australian Crabronidae. *Ann. Mag. Nat. Hist. (8)* **15**: 62–96 [82]. Type data: holotype, BMNH *M. adult (seen 1929 by L.F. Graham), from Cairns, Qld.

Distribution: NE coastal, Qld.; type locality only. Ecology: larva - sedentary, soil, predator : adult - volant; cleptoparasite.

Acanthostethus gilberti (Turner, 1915)

Nysson (Acanthostethus) gilberti Turner, R.E. (1915). Notes on fossorial Hymenoptera. XV. New Australian Crabronidae. *Ann. Mag. Nat. Hist. (8)* **15**: 62–96 [84]. Type data: holotype, BMNH *F. adult (seen 1929 by L.F. Graham), from Cairns, Qld.

Distribution: NE coastal, Qld.; type locality only. Ecology: larva - sedentary, soil, predator : adult - volant; cleptoparasite.

Acanthostethus hentyi (Rayment, 1953)

Nysson hentyi Rayment, T. (1953). New bees and wasps. Part XXI. Parasites on sericophorine wasps. *Vict. Nat.* **70**: 123–127 [124 figs on 126]. Type data: holotype, ANIC M. adult, from Cape Nelson Road (Portland), Vic.

Distribution: SE coastal, Vic.; type locality only. Ecology: larva - sedentary, predator : adult - volant ; cleptoparasite. Biological references: Bohart, R.M. & Menke, A.S. (1976). *Sphecid Wasps of the World : a Generic Revision.* Berkeley : Univ. California Press ix 695 pp. (generic placement).

Acanthostethus minimus (Turner, 1915)

Nysson (Acanthostethus) minimus Turner, R.E. (1915). Notes on fossorial Hymenoptera. XV. New Australian Crabronidae. *Ann. Mag. Nat. Hist. (8)* **15**: 62–96 [83]. Type data: holotype, BMNH *M. adult (seen 1929 by L.F. Graham), from Kuranda, Qld.

Distribution: NE coastal, Qld.; only published localities Kuranda and Cairns. Ecology: larva - sedentary, soil, predator : adult - volant; cleptoparasite.

Acanthostethus moerens (Turner, 1915)

Nysson (Acanthostethus) moerens Turner, R.E. (1915). Notes on fossorial Hymenoptera. XV. New Australian Crabronidae. *Ann. Mag. Nat. Hist. (8)* **15**: 62–96 [83]. Type data: holotype, BMNH *M. adult (seen 1929 by L.F. Graham), from Yallingup, W.A.

Distribution: SW coastal, W.A.; type locality only. Ecology: larva - sedentary, soil, predator : adult - volant; cleptoparasite.

Acanthostethus mysticus (Gerstäcker, 1867)

Nysson mysticus Gerstäcker, A. (1867). Die Arten der Gattung *Nysson* Latr. *Abh. Naturf. Ges. Halle* **10**: 71–122 [112]. Type data: syntypes (probable), ZMB *M. adult, from Swan River, W.A.

Acanthostethus basalis Smith, F. (1869). Descriptions of new genera and species of exotic Hymenoptera. *Trans. Entomol. Soc. Lond.* **1869**: 301–311 [307 pl 6 fig 3]. Type data: holotype, BMNH *F. adult (seen 1929 by L.F. Graham), from Australia.

Taxonomic decision of Handlirsch, A. (1887). Monographie der mit *Nysson* und *Bembex* verwandten Grabwespen. *Sber. Akad. Wiss. Wien, Math.-Nat. Kl.* **95**: 246–421 [328].

Distribution: SW coastal, N coastal, W.A., S.A.; only published localities Swan River, Ord River and S.A. Ecology: larva - sedentary, soil, predator : adult - volant; cleptoparasite. Biological references: Handlirsch, A. (1895). Nachträge und Schlusswort zur Monographie des mit *Nysson* und *Bembex* verwandten Grabwespen. *Sber. Akad. Wiss. Wien* **104**: 801–1079 pls 1–2 (generic placement).

Acanthostethus nudiventris (Turner, 1915)

Nysson (Acanthostethus) nudiventris Turner, R.E. (1915). Notes on fossorial Hymenoptera. XV. New Australian Crabronidae. *Ann. Mag. Nat. Hist. (8)* **15**: 62–96 [81]. Type data: holotype, BMNH *M. adult (seen 1929 by L.F. Graham), from Yallingup, W.A.

Distribution: SW coastal, W.A.; type locality only. Ecology: larva - sedentary, soil, predator : adult - volant; cleptoparasite.

Acanthostethus obliteratus (Turner, 1910)

Nysson (Acanthostethus) obliteratus Turner, R.E. (1910). Additions to our knowledge of the fossorial wasps of Australia. *Proc. Zool. Soc. Lond.* **1910**: 253–356 [350]. Type data: holotype, BMNH *M. adult (seen 1929 by L.F. Graham), from South Perth, W.A.

Distribution: SW coastal, W.A.; only published localities South Perth and Perth. Ecology: larva - sedentary, soil, predator : adult - volant; cleptoparasite.

Acanthostethus portlandensis (Rayment, 1953)

Nysson portlandensis Rayment, T. (1953). New bees and wasps. Part XXI. Parasites on sericophorine wasps. *Vict. Nat.* **70**: 123–127 [124 figs on 126]. Type data: holotype, ANIC F. adult, from Cape Nelson Road (Portland), Vic.

Distribution: SE coastal, Murray-Darling basin, Vic., A.C.T., S.A.; only published localities Portland, Black Mt. and Mt. Gambier. Ecology: larva - sedentary, soil, predator : adult - volant ; cleptoparasite. Biological references: Matthews, R.W. & Evans, H.E. (1971). Biological notes on two species of *Sericophorus* from Australia (Hymenoptera : Sphecidae). *Psyche Camb.* **77**: 413–429 (host, generic placement).

Acanthostethus punctatissimus (Turner, 1908)

Nysson (Acanthostethus) punctatissimus Turner, R.E. (1908). Notes on the Australian fossorial wasps of the family Sphegidae, with descriptions of new species. *Proc. Zool. Soc. Lond.* **1908**: 457–535 [505 pl xxvi fig 9]. Type data: holotype, BMNH *F. adult (seen 1929 by L.F. Graham), from Mackay, Qld.

Distribution: NE coastal, Qld.; type locality only. Ecology: larva - sedentary, soil, predator : adult - volant; cleptoparasite.

Acanthostethus saussurei (Handlirsch, 1887)

Nysson saussurei Handlirsch, A. (1887). Monographie der mit *Nysson* und *Bembex* verwandten Grabwespen. *Sber. Akad. Wiss. Wien, Math.-Nat. Kl.* **95**: 246–421 [332 pl 4 fig 14]. Type data: syntypes (probable), GMNH *M.,F. adult, from S.A.

Distribution: S.A.; type locality only. Ecology: larva - sedentary, soil, predator : adult - volant; cleptoparasite. Biological references: Handlirsch, A. (1895). Nachträge und Schlusswort zur Monographie des mit *Nysson* und *Bembex* verwandten Grabwespen. *Sber. Akad. Wiss. Wien* **104**: 801–1079 pls 1–2 (generic placement).

Acanthostethus spiniger (Turner, 1908)

Nysson (Acanthostethus) spiniger Turner, R.E. (1908). Notes on the Australian fossorial wasps of the family Sphegidae, with descriptions of new species. *Proc. Zool. Soc. Lond.* **1908**: 457–535 pl xxvi [507]. Type data: holotype, BMNH *F. adult (seen 1929 by L.F. Graham), from Mackay, Qld.

Distribution: NE coastal, Qld.; type locality only. Ecology: larva - sedentary, soil, predator : adult - volant; cleptoparasite.

Acanthostethus tasmanicus (Turner, 1915)

Nysson (Acanthostethus) tasmanicus Turner, R.E. (1915). Notes on fossorial Hymenoptera. XV. New Australian Crabronidae. *Ann. Mag. Nat. Hist. (8)* **15**: 62–96 [80]. Type data: holotype, BMNH *F. adult (seen 1929 by L.F. Graham), from Eaglehawk Neck and Mt Wellington, Tas.

Distribution: Murray-Darling basin, A.C.T., Tas.; only published localities Corin Dam, Eaglehawk Neck and Mt. Wellington. Ecology: larva - sedentary, soil, predator : adult - volant; cleptoparasite. Biological references: Evans, H.E. & Matthews, R.W. (1971). Nesting behaviour and larval stages of some Australian nyssonine sand wasps (Hymenoptera : Sphecidae). *Aust. J. Zool.* **19**: 293–310 (?host).

Acanthostethus triangularis (Turner, 1940)

Nysson (Acanthostethus) triangularis Turner, R.E. (1940). Notes on Fossorial Hymenoptera. XLIX. On new Australian species. *Ann. Mag. Nat. Hist. (11)* **5**: 96-105 [104]. Type data: holotype, BMNH *F. adult, from Mingenew, W.A.

Distribution: NW coastal, W.A.; type locality only. Ecology: larva - sedentary, soil, predator : adult - volant; cleptoparasite.

Clitemnestra Spinola, 1851

Clitemnestra Spinola, M. (1851). Himenópteros. pp. 153-569 *in* Gay, C. *Historia Física y Política de Chile*. Zoologia. Paris : Maulde & Revon Vol. 6 572 pp. [341]. Type species *Arpactus (Clitemnestra) gayi* Spinola, 1851 by monotypy and original designation.

Miscothyris Smith, F. (1869). Descriptions of new genera and species of exotic Hymenoptera. *Trans. Entomol. Soc. Lond.* **1869**: 301-311 [307]. Type species *Miscothyris thoracicus* Smith, 1869 by monotypy.

Astaurus Rayment, T. (1955). Taxonomy, morphology and biology of sericophorine wasps. With diagnoses of two new genera and descriptions of forty new species and six subspecies. *Mem. Natl. Mus. Vict.* **19**: 11-105 pls 1-11 [60]. Type species *Astaurus hylaeoides* Rayment, 1955 by original designation.

Taxonomic decision of Bohart, R.M. & Menke, A.S. (1976). *Sphecid Wasps of the World : a Generic Revision*. Berkeley : Univ. California Press ix 695 pp. [51].

This group is also found in Chile, see Bohart, R.M. & Menke, A.S. (1976). *Sphecid Wasps of the World : a Generic Revision*. Berkeley : Univ. California Press ix 695 pp. [485].

Clitemnestra duboulayi (Turner, 1908)

Gorytes duboulayi Turner, R.E. (1908). Notes on the Australian fossorial wasps of the family Sphegidae, with descriptions of new species. *Proc. Zool. Soc. Lond.* **1908**: 457-535 pl xxvi [496]. Type data: holotype, BMNH *F. adult (seen 1929 by L.F. Graham), from NW coast of Australia (probably Nicol Bay).

Distribution: Murray-Darling basin, NW coastal, W.A., Vic.; only published localities Champion Bay and Rutherglen. Ecology: larva - sedentary, soil, predator : adult - volant; prey Hemiptera. Biological references: Turner, R.E. (1912). Notes on fossorial Hymenoptera. IX. On some new species from the Australian and Austro-Malayan regions. *Ann. Mag. Nat. Hist. (8)* **10**: 48-63 (generic placement).

Clitemnestra guttatulus (Turner, 1936)

Arpactus (Miscothyris) guttatulus Turner, R.E. (1936). Notes on fossorial Hymenoptera. XLV. On new sphegid wasps from Australia. *Ann. Mag. Nat. Hist. (10)* **18**: 533-545 [541]. Type data: holotype, ?BMNH *F. adult, from 10 mi S Perth, W.A.

Distribution: SW coastal, W.A.; type locality only. Ecology: larva - sedentary, soil, predator : adult - volant; prey Hemiptera. Biological references: Bohart, R.M. & Menke, A.S. (1976). *Sphecid Wasps of the World : a Generic Revision*. Berkeley : Univ. California Press ix 695 pp. (generic placement).

Clitemnestra lucidulus (Turner, 1908)

Gorytes lucidulus Turner, R.E. (1908). Notes on the Australian fossorial wasps of the family Sphegidae, with descriptions of new species. *Proc. Zool. Soc. Lond.* **1908**: 457-535 pl xxvi [498 pl xxvi fig 11]. Type data: holotype, BMNH *F. adult (seen 1929 by L.F. Graham), from Mackay, Qld.

Distribution: NE coastal, Qld.; only published localities Mackay, Cairns. Ecology: larva - sedentary, soil, predator : adult - volant; prey Hemiptera. Biological references: Turner, R.E. (1912). Notes on fossorial Hymenoptera. IX. On some new species from the Australian and Austro-Malayan regions. *Ann. Mag. Nat. Hist. (8)* **10**: 48-63 (generic placement).

Clitemnestra megalophthalmus (Handlirsch, 1895)

Gorytes (Miscothyris) megalophthalmus Handlirsch, A. (1895). Nachträge und Schlusswort zur Monographie des mit *Nysson* und *Bembex* verwandten Grabwespen. *Sber. Akad. Wiss. Wien* **104**: 801-1079 pls 1-2 [862 pl 1 figs 1-11]. Type data: holotype, ZMB *M. adult, from Australia.

Distribution: no Australian locality specified. Ecology: larva - sedentary, soil, predator : adult - volant; prey Hemiptera. Biological references: Bohart, R.M. & Menke, A.S. (1976). *Sphecid Wasps of the World : a Generic Revision*. Berkeley : Univ. California Press ix 695 pp. (generic placement).

Clitemnestra mimetica (Cockerell, 1915)

Miscothyris lucidulus mimeticus Cockerell, T.D.A. (1915). A wasp resembling a bee (Hym.). *Entomol. News* **26**: 268 [268]. Type data: syntypes (probable), Cockerell collection *adult, from Gilgai, N.S.W.

Distribution: Murray-Darling basin, N.S.W.; type locality only. Ecology: larva - sedentary, soil, predator : adult - volant; prey Hemiptera. Biological references: Bohart, R.M. & Menke, A.S. (1976). *Sphecid Wasps of the World : a Generic Revision*. Berkeley : Univ. California Press ix 695 pp. (generic placement).

Clitemnestra perlucidus (Turner, 1916)

Miscothyris perlucidus Turner, R.E. (1916). Notes on fossorial Hymenoptera. XXIII. *Ann. Mag. Nat. Hist. (8)* **18**: 277-288 [278]. Type data: holotype, BMNH *adult (seen 1929 by L.F. Graham), from Kuranda, Qld.

Distribution: NE coastal, Qld.; type locality only. Ecology: larva - sedentary, soil, predator : adult - volant; prey Hemiptera. Biological references: Bohart, R.M. & Menke, A.S. (1976). *Sphecid Wasps of the World : a Generic Revision*. Berkeley : Univ. California Press ix 695 pp. (generic placement).

Clitemnestra plomleyi (Turner, 1940)

Arpactus (Miscothyris) plomleyi Turner, R.E. (1940). Notes on Fossorial Hymenoptera. XLIX. On new Australian species. *Ann. Mag. Nat. Hist. (11)* **5**: 96–105 [105]. Type data: syntypes (probable), BMNH *adult, from Barrington Tops, N.S.W.

Astaurus hylaeoides Rayment, T. (1955). Taxonomy, morphology and biology of sericophorine wasps. With diagnoses of two new genera and descriptions of forty new species and six subspecies. *Mem. Natl. Mus. Vict.* **19**: 11–105 pls 1–11 [61 pl 7]. Type data: holotype, NMV *M. adult, paratype ANIC M. adult, from Gorae West, Vic.

Taxonomic decision of Evans, H.E. & Matthews, R.W. (1971). Nesting behaviour and larval stages of some Australian nyssonine sand wasps (Hymenoptera : Sphecidae). *Aust. J. Zool.* **19**: 293–310 [299].

Distribution: Murray-Darling basin, SE coastal, N.S.W., A.C.T., Vic.; only published localities Barrington Tops, Corin Dam, Mt. Franklin and Gorae West. Ecology: larva - sedentary, soil, predator : adult - volant; prey Homoptera (Cicadellidae), nest in pre-existing hole.

Clitemnestra sanguinolentus (Turner, 1908)

Gorytes sanguinolentus Turner, R.E. (1908). Notes on the Australian fossorial wasps of the family Sphegidae, with descriptions of new species. *Proc. Zool. Soc. Lond.* **1908**: 457–535 [497 pl xxvi fig 10]. Type data: holotype, BMNH *F. adult (seen 1929 by L.F. Graham), from Mackay, Qld.

Distribution: NE coastal, Qld.; type locality only. Ecology: larva - sedentary, soil, predator : adult - volant; prey Hemiptera. Biological references: Turner, R.E. (1912). Notes on fossorial Hymenoptera. IX. On some new species from the Australian and Austro-Malayan regions. *Ann. Mag. Nat. Hist. (8)* **10**: 48–63 (generic placement).

Clitemnestra tenuicornis (Rayment, 1955)

Astaurus tenuicornis Rayment, T. (1955). Taxonomy, morphology and biology of sericophorine wasps. With diagnoses of two new genera and descriptions of forty new species and six subspecies. *Mem. Natl. Mus. Vict.* **19**: 11–105 pls 1–11 [62]. Type data: holotype, NMV M. adult, from Cheltenham, N.S.W.

Distribution: SE coastal, N.S.W.; type locality only. Ecology: larva - sedentary, soil, predator : adult - volant; prey Hemiptera.

Clitemnestra thoracicus (Smith, 1869)

Miscothyris thoracicus Smith, F. (1869). Descriptions of new genera and species of exotic Hymenoptera. *Trans. Entomol. Soc. Lond.* **1869**: 301–311 [308 pl 6 fig 5]. Type data: holotype, BMNH *F. adult (seen 1929 by L.F. Graham), from Champion Bay, W.A.

Distribution: NW coastal, W.A.; type locality only. Ecology: larva - sedentary, soil, predator : adult - volant; prey Hemiptera. Biological references: Bohart, R.M. & Menke, A.S. (1976). *Sphecid Wasps of the World : a Generic Revision*. Berkeley : Univ. California Press ix 695 pp. (generic placement).

Argogorytes Ashmead, 1899

Argogorytes Ashmead, W.H. (1899). Classification of the entomophilous wasps, or the superfamily Sphegoidea. *Can. Entomol.* **31**: 322–330 [324]. Type species *Gorytes carbonarius* Smith, 1856 by monotypy.

This group is found in all continental faunal regions except the Ethiopian, see Bohart, R.M. & Menke, A.S. (1976). *Sphecid Wasps of the World : a Generic Revision*. Berkeley : Univ. California Press ix 695 pp. [491]. Species now known not to occur in Australia: *Argogorytes stenopygus* (Handlirsch, 1895) erroneously recorded from Australia, see Bohart & Menke (*op. cit.*).

Argogorytes crucigera (Hacker and Cockerell, 1922)

Arpactus crucigera Hacker, H. & Cockerell, T.D.A. (1922). Some Australian wasps of the genera *Zoyphium* and *Arpactus*. *Mem. Qd. Mus.* **7**: 283–290 [289]. Type data: holotype, QM HY2698 *F. adult, from Brisbane, Qld.

Distribution: NE coastal, Qld.; type locality only. Ecology: larva - sedentary, soil, predator : adult - volant, burrowing; prey Hemiptera. Biological references: Bohart, R.M. & Menke, A.S. (1976). *Sphecid Wasps of the World : a Generic Revision*. Berkeley : Univ. California Press ix 695 pp. (generic placement).

Argogorytes rubrosignatus (Turner, 1915)

Arpactus rubrosignatus Turner, R.E. (1915). Notes on fossorial Hymenoptera. XV. New Australian Crabronidae. *Ann. Mag. Nat. Hist. (8)* **15**: 62–96 [77]. Type data: holotype, BMNH *F. adult (seen 1929 by L.F. Graham), from between Yallingup and Busselton, W.A.

Distribution: SW coastal, W.A.; type locality only. Ecology: larva - sedentary, soil, predator : adult - volant, burrowing; prey Hemiptera. Biological references: Bohart, R.M. & Menke, A.S. (1976). *Sphecid Wasps of the World : a Generic Revision*. Berkeley : Univ. California Press ix 695 pp. (generic placement).

Argogorytes rufomixtus (Turner, 1914)

Gorytes rufomixtus Turner, R.E. (1914). New fossorial Hymenoptera from Australia and Tasmania. *Proc. Linn. Soc. N.S.W.* **38**: 608–623 [620]. Type data: syntypes (probable), AM *F. adult, from Jindabyne, N.S.W.

Distribution: NE coastal, SE coastal, Qld., N.S.W.; only published localities Brisbane and Jindabyne. Ecology: larva - sedentary, soil, predator : adult - volant, burrowing; prey Hemiptera. Biological references: Hacker, H. & Cockerell, T.D.A. (1922). Some Australian wasps of the genera *Zoyphium* and *Arpactus*. *Mem. Qd. Mus.* **7**: 283–290 (description of male); Bohart, R.M. & Menke, A.S. (1976). *Sphecid Wasps of the World : a Generic Revision*. Berkeley : Univ. California Press ix 695 pp. (generic placement).

Argogorytes secernendus (Turner, 1915)

Arpactus secernendus Turner, R.E. (1915). Notes on fossorial Hymenoptera. XV. New Australian Crabronidae. *Ann. Mag. Nat. Hist. (8)* **15**: 62–96 [78]. Type data: holotype, BMNH *F. adult (seen 1929 by L.F. Graham), from SE Australia.

Distribution: ; type locality only as SE Australia. Ecology: larva - sedentary, soil, predator : adult - volant, burrowing; prey Hemiptera. Biological references: Bohart, R.M. & Menke, A.S. (1976). *Sphecid Wasps of the World : a Generic Revision*. Berkeley : Univ. California Press ix 695 pp. (generic placement).

Austrogorytes Bohart, 1967

Austrogorytes Bohart, R.M. (1967). New genera of Gorytini (Hymenoptera : Sphecidae : Nyssoninae). *Pan-Pac. Entomol.* **43**: 155–161 [155]. Type species *Gorytes bellicosus* Smith, 1862 by original designation.

Austrogorytes aurantiacus (Turner, 1915)

Arpactus aurantiacus Turner, R.E. (1915). Notes on fossorial Hymenoptera. XV. New Australian Crabronidae. *Ann. Mag. Nat. Hist. (8)* **15**: 62–96 [71]. Type data: holotype, SAMA *M. adult, from Ankertell, W.A.

Distribution: W plateau, W.A.; type locality only. Ecology: larva - sedentary, soil, predator : adult - volant, burrowing; prey Hemiptera. Biological references: Bohart, R.M. (1967). New genera of Gorytini (Hymenoptera : Sphecidae : Nyssoninae). *Pan-Pac. Entomol.* **43**: 155–161 (generic placement).

Austrogorytes bellicosus (Smith, 1862)

Gorytes bellicosus Smith, F. (1862). Descriptions of new species of Australian Hymenoptera, and of a species of *Formica* from New Zealand. *Trans. Entomol. Soc. Lond. (3)* **1**: 53–62 [55]. Type data: holotype, BMNH *F. adult (seen 1929 by L.F. Graham), from Adelaide, S.A.

Gorytes dizonus Handlirsch, A. (1895). Nachträge und Schlusswort zur Monographie des mit *Nysson* und *Bembex* verwandten Grabwespen. *Sber. Akad. Wiss. Wien* **104**: 801–1079 pls 1–2 [873]. Type data: syntypes (probable), SMNS *M. adult, from Vic.

Taxonomic decision of Turner, R.E. (1915). Notes on fossorial Hymenoptera. XV. New Australian Crabronidae. *Ann. Mag. Nat. Hist. (8)* **15**: 62–96 [70].

Distribution: Murray-Darling basin, S Gulfs, A.C.T., Vic., S.A.; only published localities Corin Dam, S of Canberra, Vic. and Adelaide. Ecology: larva - sedentary, soil, predator : adult - volant, burrowing; prey Hemiptera (Eurymelidae). Biological references: Bohart, R.M. (1967). New genera of Gorytini (Hymenoptera : Sphecidae : Nyssoninae). *Pan-Pac. Entomol.* **43**: 155–161 (generic placement); Evans, H.E. & Matthews, R.W. (1971). Nesting behaviour and larval stages of some Australian nyssonine sand wasps (Hymenoptera : Sphecidae). *Aust. J. Zool.* **19**: 293–310 (nest, prey, larva).

Austrogorytes browni (Turner, 1936)

Arpactus browni Turner, R.E. (1936). Notes on fossorial Hymenoptera. XLV. On new sphegid wasps from Australia. *Ann. Mag. Nat. Hist. (10)* **18**: 533–545 [540]. Type data: syntypes (probable), BMNH *M. adult, from Dedari, W.A.

Distribution: W plateau, W.A.; type locality only. Ecology: larva - sedentary, soil, predator : adult - volant, burrowing; prey Hemiptera. Biological references: Bohart, R.M. (1967). New genera of Gorytini (Hymenoptera : Sphecidae : Nyssoninae). *Pan-Pac. Entomol.* **43**: 155–161 (generic placement).

Austrogorytes chrysozonus (Turner, 1915)

Arpactus chrysozonus Turner, R.E. (1915). Notes on fossorial Hymenoptera. XV. New Australian Crabronidae. *Ann. Mag. Nat. Hist. (8)* **15**: 62–96 [72]. Type data: holotype, BMNH *F. adult (seen 1929 by L.F. Graham), from Brisbane, Qld.

Distribution: NE coastal, Qld.; type locality only. Ecology: larva - sedentary, soil, predator : adult - volant, burrowing; prey Hemiptera. Biological references: Bohart, R.M. (1967). New genera of Gorytini (Hymenoptera : Sphecidae : Nyssoninae). *Pan-Pac. Entomol.* **43**: 155–161 (generic placement).

Austrogorytes ciliatus (Handlirsch, 1895)

Gorytes ciliatus Handlirsch, A. (1895). Nachträge und Schlusswort zur Monographie des mit *Nysson* und *Bembex* verwandten Grabwespen. *Sber. Akad. Wiss. Wien* **104**: 801–1079 pls 1–2 [874]. Type data: holotype, ZMB *M. adult, from Adelaide, S.A.

Distribution: S Gulfs, NW coastal, S.A., W.A.; only published localities Adelaide and Champion Bay. Ecology: larva - sedentary, soil, predator : adult - volant, burrowing; prey Hemiptera.

Biological references: Bohart, R.M. (1967). New genera of Gorytini (Hymenoptera : Sphecidae : Nyssoninae). *Pan-Pac. Entomol.* **43**: 155–161 (generic placement).

Austrogorytes consuetipes (Turner, 1915)

Arpactus consuetipes Turner, R.E. (1915). Notes on fossorial Hymenoptera. XV. New Australian Crabronidae. *Ann. Mag. Nat. Hist. (8)* **15**: 62–96 [77]. Type data: holotype, BMNH *M. adult (seen 1929 by L.F. Graham), from N.S.W.

Distribution: N.S.W.; type locality only as N.S.W. Ecology: larva - sedentary, soil, predator : adult - volant, burrowing; prey Hemiptera. Biological references: Bohart, R.M. (1967). New genera of Gorytini (Hymenoptera : Sphecidae : Nyssoninae). *Pan-Pac. Entomol.* **43**: 155–161 (generic placement).

Austrogorytes cygnorum (Turner, 1908)

Gorytes cygnorum Turner, R.E. (1908). Notes on the Australian fossorial wasps of the family Sphegidae, with descriptions of new species. *Proc. Zool. Soc. Lond.* **1908**: 457–535 pl xxvi [500]. Type data: holotype, BMNH *M. adult (seen 1929 by L.F. Graham), from Swan River, W.A.

Distribution: SW coastal, W.A.; type locality only. Ecology: larva - sedentary, soil, predator : adult - volant, burrowing; prey Hemiptera. Biological references: Bohart, R.M. (1967). New genera of Gorytini (Hymenoptera : Sphecidae : Nyssoninae). *Pan-Pac. Entomol.* **43**: 155–161 (generic placement).

Austrogorytes frenchii (Turner, 1908)

Gorytes frenchii Turner, R.E. (1908). Notes on the Australian fossorial wasps of the family Sphegidae, with descriptions of new species. *Proc. Zool. Soc. Lond.* **1908**: 457–535 pl xxvi [501]. Type data: holotype, BMNH *M. adult (seen 1929 by L.F. Graham), from Vic.

Distribution: SE coastal, N.S.W., Vic.; only published localities Sydney and Vic. Ecology: larva - sedentary, soil, predator : adult - volant, burrowing; prey Hemiptera. Biological references: Bohart, R.M. (1967). New genera of Gorytini (Hymenoptera : Sphecidae : Nyssoninae). *Pan-Pac. Entomol.* **43**: 155–161 (generic placement).

Austrogorytes obesus (Turner, 1915)

Arpactus obesus Turner, R.E. (1915). Notes on fossorial Hymenoptera. XV. New Australian Crabronidae. *Ann. Mag. Nat. Hist. (8)* **15**: 62–96 [74]. Type data: holotype, BMNH *M. adult (seen 1929 by L.F. Graham), from Yallingup, W.A.

Distribution: SW coastal, W.A.; type locality only. Ecology: larva - sedentary, soil, predator : adult - volant, burrowing; prey Hemiptera. Biological references: Bohart, R.M. (1967). New genera of Gorytini (Hymenoptera : Sphecidae : Nyssoninae). *Pan-Pac. Entomol.* **43**: 155–161 (generic placement).

Austrogorytes perkinsi (Turner, 1912)

Gorytes perkinsi Turner, R.E. (1912). Notes on fossorial Hymenoptera. IX. On some new species from the Australian and Austro-Malayan regions. *Ann. Mag. Nat. Hist. (8)* **10**: 48–63 [57]. Type data: holotype, BMNH *F. adult (seen 1929 by L.F. Graham), from Cairns, Qld.

Distribution: NE coastal, Qld.; type locality only. Ecology: larva - sedentary, soil, predator : adult - volant, burrowing; prey Hemiptera. Biological references: Bohart, R.M. (1967). New genera of Gorytini (Hymenoptera : Sphecidae : Nyssoninae). *Pan-Pac. Entomol.* **43**: 155–161 (generic placement).

Austrogorytes pretiosus (Turner, 1915)

Arpactus pretiosus Turner, R.E. (1915). Notes on fossorial Hymenoptera. XV. New Australian Crabronidae. *Ann. Mag. Nat. Hist. (8)* **15**: 62–96 [75]. Type data: holotype, BMNH *M. adult (seen 1929 by L.F. Graham), from Yallingup, W.A.

Distribution: SW coastal, W.A.; type locality only. Ecology: larva - sedentary, soil, predator : adult - volant, burrowing; prey Hemiptera. Biological references: Bohart, R.M. (1967). New genera of Gorytini (Hymenoptera : Sphecidae : Nyssoninae). *Pan-Pac. Entomol.* **43**: 155–161 (generic placement).

Austrogorytes spinicornis (Turner, 1915)

Arpactus spinicornis Turner, R.E. (1915). Notes on fossorial Hymenoptera. XV. New Australian Crabronidae. *Ann. Mag. Nat. Hist. (8)* **15**: 62–96 [76]. Type data: holotype, SAMA *M. adult, from Beverley, W.A.

Distribution: SW coastal, W.A.; type locality only. Ecology: larva - sedentary, soil, predator : adult - volant, burrowing; prey Hemiptera. Biological references: Bohart, R.M. (1967). New genera of Gorytini (Hymenoptera : Sphecidae : Nyssoninae). *Pan-Pac. Entomol.* **43**: 155–161 (generic placement).

Austrogorytes spryi (Turner, 1915)

Arpactus spryi Turner, R.E. (1915). Notes on fossorial Hymenoptera. XV. New Australian Crabronidae. *Ann. Mag. Nat. Hist. (8)* **15**: 62–96 [73]. Type data: holotype, BMNH *M. adult (seen 1929 by L.F. Graham), from Mordialloc, Vic.

Distribution: SE coastal, Vic.; type locality only. Ecology: larva - sedentary, soil, predator : adult - volant, burrowing; prey Hemiptera. Biological references: Bohart, R.M. (1967). New genera of Gorytini (Hymenoptera : Sphecidae : Nyssoninae). *Pan-Pac. Entomol.* **43**: 155–161 (generic placement).

Austrogorytes tarsatus (Smith, 1856)

Gorytes tarsatus Smith, F. (1856). *Catalogue of Hymenopterous Insects in the Collection of the British Museum.* Part IV, Sphegidae, Larridae and Crabronidae. pp. 207-497 London : British Museum [366]. Type data: holotype, BMNH *M. adult (seen 1929 by L.F. Graham), from Adelaide, S.A.

Gorytes eximius Smith, F. (1862). Descriptions of new species of Australian Hymenoptera, and of a species of *Formica* from New Zealand. *Trans. Entomol. Soc. Lond. (3)* **1**: 53-62 [55]. Type data: holotype, BMNH *F. adult (seen 1929 by L.F. Graham), from Adelaide, S.A.

Taxonomic decision of Handlirsch, A. (1889). Monographie der mit *Nysson* und *Bembex* verwandten Grabwespen. iii. *Sber. Akad. Wiss. Wien, Math.-Nat. Kl.* **97**: 316-565 [543].

Distribution: S Gulfs, S.A.; type locality only. Ecology: larva - sedentary, soil, predator : adult - volant, burrowing; prey Hemiptera. Biological references: Bohart, R.M. (1967). New genera of Gorytini (Hymenoptera : Sphecidae : Nyssoninae). *Pan-Pac. Entomol.* **43**: 155-161 (generic placement).

Exeirus Shuckard, 1838

Exeirus Shuckard, W.E. (1838). Descriptions of new exotic aculeate Hymenoptera. *Trans. Entomol. Soc. Lond.* **2**: 68-82 [71]. Type species *Exeirus lateritius* Shuckard, 1838 by monotypy.

Exeirus lateritius Shuckard, 1838

Exeirus lateritius Shuckard, W.E. (1838). Descriptions of new exotic aculeate Hymenoptera. *Trans. Entomol. Soc. Lond.* **2**: 68-82 [72 pl 8 fig 2]. Type data: syntypes (probable), whereabouts unknown, from Sydney, N.S.W. and Tas.

Sphecius lanio Stal, C. (1857). Nya arter af Sphegidae. *Ofvers. K. Vetensk Akad. Forh.* **14**: 63-64 [64]. Type data: syntypes (probable), NHRM *adult, from Australia (as New Holland).

Taxonomic decision of Handlirsch, A. (1888). Monographie der mit *Nysson* und *Bembex* verwandten Grabwespen. ii. *Sber. Akad. Wiss. Wien, Math.-Nat. Kl.* **96**: 219-311 [306].

Distribution: Murray-Darling basin, SE coastal, N.S.W., Tas. Ecology: larva - sedentary, soil, predator : adult - volant, burrowing; prey Homoptera (Cicadidae). Biological references: McCulloch, A.R. (1923). War in the garden. *Aust. Mus. Mag.* **1**: 209-212 (behaviour); Alcock, J. (1980). Notes on the reproductive behaviour of some Australian solitary wasps (Hymenoptera : Sphecidae, *Tachysphex* and *Exeirus*). *J. Aust. Entomol. Soc.* **19**: 259-262 (mating behaviour).

Sphecius Dahlbom, 1843

Taxonomic decision of Bohart, R.M. & Menke, A.S. (1976). *Sphecid Wasps of the World : a Generic Revision.* Berkeley : Univ. California Press ix 695 pp. [52].

Sphecius (Nothosphecius) Pate, 1936

Nothosphecius Pate, V.S.L. (1936). Studies in the nyssonine wasps. II. The subgenera of *Sphecius* (Hymenoptera : Sphecidae : Gorytini). *Bull. Brooklyn Entomol. Soc.* **31**: 198-200 [199] [described with subgeneric rank in *Sphecius* Dahlbom, 1845]. Type species *Sphecius grandidieri* Saussure, 1887 by original designation.

This group is also found in the Ethiopian and Oriental Regions, see Bohart, R.M. & Menke, A.S. (1976). *Sphecid Wasps of the World : a Generic Revision.* Berkeley : Univ. California Press ix 695 pp. [509].

Sphecius (Nothosphecius) pectoralis (Smith, 1856)

Stizus pectoralis Smith, F. (1856). *Catalogue of Hymenopterous Insects in the Collection of the British Museum.* Part IV, Sphegidae, Larridae and Crabronidae. pp. 207-497 London : British Museum [337]. Type data: syntypes, BMNH *F,M adult (seen 1929 by L.F. Graham), from Australia.

Distribution: NE coastal, Qld.; only published localities Rockhampton and Yeppoon. Ecology: larva - sedentary, soil, predator : adult - volant, burrowing; prey Homoptera (Cicadidae). Biological references: Bohart, R.M. & Menke, A.S. (1976). *Sphecid Wasps of the World : a Generic Revision.* Berkeley : Univ. California Press ix 695 pp. (generic placement); Evans, H.E. & Matthews, R.W. (1971). Nesting behaviour and larval stages of some Australian nyssonine sand wasps (Hymenoptera : Sphecidae). *Aust. J. Zool.* **19**: 293-310 (nest, larva, prey).

Ammatomus Costa, 1859

Ammatomus Costa, A. (1859). Immenotteri aculeati, famiglia degli Sfecidei. Nyssonidea. pp. 1-56 *in* Costa, O.G. & A. (1829-1886). *Fauna dell Regno di Napoli.* Napoli 11 vols. [36]. Type species *Gorytes coarctatus* Spinola, 1808 by monotypy. Compiled from secondary source: Bohart, R.M. & Menke, A.S. (1976). *Sphecid Wasps of the World : a Generic Revision.* Berkeley : Univ. California Press ix 695 pp. [53].

This group is also found in the Palearctic, Ethiopian and Oriental Regions, see Bohart, R.M. & Menke, A.S. (1976). *Sphecid Wasps of the World : a Generic Revision.* Berkeley : Univ. California Press ix 695 pp. [512].

Ammatomus austrinus (Bingham, 1912)

Gorytes austrinus Bingham, C.T. (1912). South African and Australian Aculeate Hymenoptera in the Oxford Museum. *Trans. Entomol. Soc. Lond.* **1912**: 375-383 [376]. Type data: syntypes (probable), OUM *F. adult, from Hunters River, N.S.W. (as New Holland) or Tas. (as Van Dieman's Land).

Distribution: (SE coastal), (N.S.W.), (Tas.); type locality only. Ecology: larva - sedentary, soil, predator : adult - volant, burrowing. Biological references: Bohart, R.M. & Menke, A.S. (1976). *Sphecid Wasps of the World : a Generic Revision.* Berkeley : Univ. California Press ix 695 pp. (generic placement).

Ammatomus decoratus (Handlirsch, 1889)

Gorytes ornatus Smith, F. (1868). Descriptions of aculeate Hymenoptera from Australia. *Trans. Entomol. Soc. Lond.* **1868**: 231-258 [248] [*non Gorytes ornatus* Smith, 1856]. Type data: syntypes (probable), ?BMNH *F. adult.

Gorytes decoratus Handlirsch, A. (1889). Monographie der mit *Nysson* und *Bembex* verwandten Grabwespen. iii. *Sber. Akad. Wiss. Wien, Math.-Nat. Kl.* **97**: 316-565 [542] [*nom. nov.* for *Gorytes ornatus* Smith, 1868]. from Champion Bay, W.A.

Distribution: NW coastal, W.A.; type locality only. Ecology: larva - sedentary, soil, predator : adult - volant, burrowing. Biological references: Turner, R.E. (1912). Notes on fossorial Hymenoptera. x. On new species from the oriental and ethiopian regions. *Ann. Mag. Nat. Hist. (8)* **10**: 361-377 (generic placement).

Ammatomus icarioides (Turner, 1908)

Gorytes icarioides Turner, R.E. (1908). Notes on the Australian fossorial wasps of the family Sphegidae, with descriptions of new species. *Proc. Zool. Soc. Lond.* **1908**: 457-535 pl xxvi [499]. Type data: syntypes (probable), whereabouts unknown, from Mackay, Qld.

Distribution: NE coastal, Qld.; only published localities Mackay, Townsville and near Brisbane. Ecology: larva - sedentary, soil, predator : adult - volant, burrowing; prey Hemiptera (Flatidae and Eurybrachidae). Biological references: Turner, R.E. (1912). Notes on fossorial Hymenoptera. x. On new species from the oriental and ethiopian regions. *Ann. Mag. Nat. Hist. (8)* **10**: 361-377 (generic placement); Hook, A. (1981). Nesting biology of *Tanyoprymnus moneduloides* and *Ammatomus icarioides*. *Ann. Entomol. Soc. Am.* **74**: 409-411; Evans, H.E. (1983). The larva of *Ammatomus icarioides* (Turner) (Hymenoptera, Sphecidae, Nyssoninae). *Pan-Pac. Entomol.* **59**: 52-54.

Bembecinus Costa, 1859

Bembecinus Costa, A. (1859). Immenotteri aculeati, famiglia degli Sfecidei. Nyssonidea pp. 1-56 *in* Costa, O.G. & A. (1829-1886). *Fauna dell Regno di Napoli.* Napoli 11 vols. [4]. Type species *Bembecinus meridionalis* Costa, 1859 by monotypy.

This group is found worldwide, see Bohart, R.M. & Menke, A.S. (1976). *Sphecid Wasps of the World : a Generic Revision.* Berkeley : Univ. California Press ix 695 pp. [529]. Species now known not to occur in Australia: *Bembecinus monodon* (Handlirsch, 1895) described in *Stizus*, erroneously recorded from Australia (misreading of type locality "Afr. austral." as "Africa, Australia").

Bembecinus antipodum (Handlirsch, 1892)

Stizus antipodum Handlirsch, A. (1892). Monographie der mit *Nysson* und *Bembex* verwandten Grabwespen. vi. *Sber. Akad. Wiss. Wien, Math.-Nat. Kl.* **101**: 25-205 [52]. Type data: syntypes (probable), ZMH (evidently destroyed in war) F. adult, from Sydney, N.S.W.

Distribution: SE coastal, N.S.W.; only published localities Sydney and Kuringai Chase. Ecology: larva - sedentary, soil, predator : adult - volant, burrowing; prey Homoptera. Biological references: Evans, H.E. & Matthews, R.W. (1971). Nesting behaviour and larval stages of some Australian nyssonine sand wasps (Hymenoptera : Sphecidae). *Aust. J. Zool.* **19**: 293-310 (nest, prey, generic placement).

Bembecinus egens (Handlirsch, 1892)

Stizus egens Handlirsch, A. (1892). Monographie der mit *Nysson* und *Bembex* verwandten Grabwespen. vi. *Sber. Akad. Wiss. Wien, Math.-Nat. Kl.* **101**: 25-205 [51]. Type data: holotype, GMHN *F. adult, from Port Denison, Qld.

Distribution: NE coastal, Qld.; only published localities Cairns, Yeppoon, near Mackay and Port Denison. Ecology: larva - sedentary, soil, predator : adult - volant, burrowing; prey Homoptera. Biological references: Evans, H.E. & Matthews, R.W. (1971). Nesting behaviour and larval stages of some Australian nyssonine sand wasps (Hymenoptera : Sphecidae). *Aust. J. Zool.* **19**: 293-310 (nest, prey, generic placement).

Bembecinus gorytoides (Handlirsch, 1895)

Stizus gorytoides Handlirsch, A. (1895). Nachträge und Schlusswort zur Monographie des mit *Nysson* und *Bembex* verwandten Grabwespen. *Sber. Akad. Wiss. Wien* **104**: 801-1079 pls 1-2 [985 pl 2 figs 5-6]. Type data: syntypes (probable), GMNH *F. adult, from Pacific Ocean.

Distribution: this species is found in Fiji, the record from Australia may be an error. Ecology: larva - sedentary, soil, predator : adult - volant, burrowing; prey Homoptera. Biological references:

Lohrmann, E. (1943). Die Grabwespengruppe der Stizinen. *Mitt. Münch. Entomol. Ges.* **33**: 188–208 (generic placement).

Bembecinus hirtula (Smith, 1856)

Larra hirtula Smith, F. (1856). *Catalogue of Hymenopterous Insects in the Collection of the British Museum.* Part IV, Sphegidae, Larridae and Crabronidae. pp. 207–497 London : British Museum [344]. Type data: holotype, BMNH *M. adult (seen 1929 by L.F. Graham), from Adelaide, S.A.

Distribution: Murray-Darling basin, S Gulfs, N.S.W., A.C.T., S.A.; (N Australia) only published localities Peak Hill, Wilcannia, Murrumbidgee River, Adelaide. Ecology: larva - sedentary, soil, predator : adult - volant, burrowing; prey Homoptera. Biological references: Lohrmann, E. (1943). Die Grabwespengruppe der Stizinen. *Mitt. Münch. Entomol. Ges.* **33**: 188–208 (generic placement); Evans, H.E. & Matthews, R.W. (1971). Nesting behaviour and larval stages of some Australian nyssonine sand wasps (Hymenoptera : Sphecidae). *Aust. J. Zool.* **19**: 293–310 (nest, prey, behaviour).

Bembecinus signatus (Handlirsch, 1892)

Stizus signatus Handlirsch, A. (1892). Monographie der mit *Nysson* und *Bembex* verwandten Grabwespen. vi. *Sber. Akad. Wiss. Wien, Math.-Nat. Kl.* **101**: 25–205 [53]. Type data: syntypes (probable), ZMH (evidently destroyed in war) M. adult, from Sydney, N.S.W.

Distribution: SE coastal, N.S.W.; type locality only. Ecology: larva - sedentary, soil, predator : adult - volant, burrowing; prey Homoptera. Biological references: Bohart, R.M. & Menke, A.S. (1976). *Sphecid Wasps of the World : a Generic Revision.* Berkeley : Univ. California Press ix 695 pp. (generic placement).

Bembecinus turneri (Froggatt, 1917)

Stizus turneri Froggatt, W.W. (1917). "Policeman flies." Fossorial wasps that catch flies. *Agric. Gaz. N.S.W.* **28**: 667–669 [668]. Type data: syntypes (probable), DARI, from Hay, N.S.W.

Distribution: Murray-Darling basin, N.S.W.; type locality only. Ecology: larva - sedentary, soil, predator : adult - volant, burrowing; prey Diptera. Biological references: Lohrmann, E. (1943). Die Grabwespengruppe der Stizinen. *Mitt. Münch. Entomol. Ges.* **33**: 188–208 (generic placement).

Bembix Fabricius, 1775

Bembix Fabricius, J.C. (1775). *Systema Entomologiae, sistens insectorum classes, ordines, genera, species, adiectis synonymis, locis, descriptionibus, observationibus.* Flensburgi et Lipsiae : Kortii xxvii 832 pp. [xxiii (pages not numbered)]. Type species *Bembex rostrata* Fabricius, 1781 by subsequent designation, see Latreille, P.A. (1810). *Considérations Générales sur l'Ordre Naturel des Animaux Composant les Classes des Crustacès, des Arachnides, et des Insectes;* avec un Tableau Méthodique de leurs Genres, Disposés en Familles. Paris : F. Schoell 444 pp. Compiled from secondary source: Bohart, R.M. & Menke, A.S. (1976). *Sphecid Wasps of the World : a Generic Revision.* Berkeley : Univ. California Press ix 695 pp.

Bembex Fabricius, J.C. (1776). *Genera insectorum eorumque characteres naturales secundum numerum, figuram, situm et proportionem omnium partium oris, adiecta mantissa specierum nuper detectarum.* Chilonii : M.F. Bartschius 310 pp. [122] [emend. of *Bembix* Fabricius, 1775].

This group is found worldwide, see Bohart, R.M. & Menke, A.S. (1976). *Sphecid Wasps of the World : a Generic Revision.* Berkeley : Univ. California Press ix 695 pp. [543].

Bembix allunga Evans and Matthews, 1973

Bembix allunga Evans, H.E. & Matthews, R.W. (1973). Systematics and nesting behavior of Australian *Bembix* sand wasps (Hymenoptera, Sphecidae). *Mem. Am. Entomol. Inst.* **20**: iv 387 pp. [137]. Type data: holotype, ANIC M. adult, from Yeppoon, Qld.

Distribution: NE coastal, Murray-Darling basin, N coastal, N Gulf, Qld., N.S.W., W.A., N.T. Ecology: larva - sedentary, soil, predator : adult - volant, burrowing; prey adult Diptera, Neuroptera, Odonata, Homoptera. Biological references: Evans, H.E., Evans, M.A. & Hook, A. (1982). Observations on the nests and prey of Australian *Bembix* sand wasps (Hymenoptera : Sphecidae). *Aust. J. Zool.* **30**: 71–80 (nest, prey); Evans, H.E. (1982). Two new species of Australian *Bembix* sand wasps, with notes on other species of the genus (Hymenoptera, Sphecidae). *Aust. Entomol. Mag.* **9**: 7–12 (distribution).

Bembix atrifrons Smith, 1856

Bembex atrifrons Smith, F. (1856). *Catalogue of Hymenopterous Insects in the Collection of the British Museum.* Part IV, Sphegidae, Larridae and Crabronidae. pp. 207–497 London : British Museum [327]. Type data: holotype, BMNH 21.1,694 *F. adult (seen 1929 by L.F. Graham), from Swan River, W.A.

Bembex flavilabris Smith, F. (1873). Descriptions of new species of fossorial Hymenoptera in the collection of the British Museum. *Ann. Mag. Nat. Hist.* (4) **12**: 291–300, 402–415 [299]. Type data: holotype, BMNH 21.1,695 *F. adult (seen 1929 by L.F. Graham), from W.A.

Bembex funebris Turner, R.E. (1910). Additions to our knowledge of the fossorial wasps of Australia. *Proc. Zool. Soc. Lond.* **1910**: 253–356 [353]. Type data: holotype, BMNH 21.1,696 *M. adult (seen 1929 by L.F. Graham), from South Perth, W.A.

Taxonomic decision of Evans, H.E. & Matthews, R.W. (1973). Systematics and nesting behavior of Australian *Bembix* sand wasps (Hymenoptera, Sphecidae). *Mem. Am. Entomol. Inst.* **20**: iv 387 pp. [154].

Distribution: Murray-Darling basin, SE coastal, Lake Eyre basin, S Gulfs, W plateau, SW coastal, NW coastal, Qld., N.S.W., Vic., W.A., N.T.; S of Tropic of Capricorn, mainly in the dry interior. Ecology: larva - sedentary, soil, predator : adult - volant, burrowing; prey adult Diptera.

Bembix aureofasciata Turner, 1910

Bembex aureofasciata Turner, R.E. (1910). Additions to our knowledge of the fossorial wasps of Australia. *Proc. Zool. Soc. Lond.* **1910**: 253–356 [354 pl 32 fig 14]. Type data: holotype, BMNH 21.1,697 *M. adult (seen 1929 by L.F. Graham), from South Perth, W.A.

Distribution: SW coastal, W plateau, NW coastal, W.A. Ecology: larva - sedentary, soil, predator : adult - volant, burrowing. Biological references: Evans, H.E. & Matthews, R.W. (1973). Systematics and nesting behavior of Australian *Bembix* sand wasps (Hymenoptera, Sphecidae). *Mem. Am. Entomol. Inst.* **20**: iv 387 pp. (redescription).

Bembix baringa Evans and Matthews, 1973

Bembix baringa Evans, H.E. & Matthews, R.W. (1973). Systematics and nesting behavior of Australian *Bembix* sand wasps (Hymenoptera, Sphecidae). *Mem. Am. Entomol. Inst.* **20**: iv 387 pp. [47]. Type data: holotype, ANIC M. adult, from Wiluna, W.A.

Distribution: W plateau, W.A.; type locality only. Ecology: larva - sedentary, soil, predator : adult - volant, burrowing.

Bembix berontha Evans and Matthews, 1973

Bembix berontha Evans, H.E. & Matthews, R.W. (1973). Systematics and nesting behavior of Australian *Bembix* sand wasps (Hymenoptera, Sphecidae). *Mem. Am. Entomol. Inst.* **20**: iv 387 pp. [176]. Type data: holotype, ANIC M. adult, from Nilemah Station 50 mi S of Denham, W.A.

Distribution: NW coastal, W plateau, W.A.; only published localities Nilemah Station (50 mi S Denham), Carnarvon and Wiluna. Ecology: larva - sedentary, soil, predator : adult - volant, burrowing.

Bembix boaliri Evans and Matthews, 1973

Bembix boaliri Evans, H.E. & Matthews, R.W. (1973). Systematics and nesting behavior of Australian *Bembix* sand wasps (Hymenoptera, Sphecidae). *Mem. Am. Entomol. Inst.* **20**: iv 387 pp. [309]. Type data: holotype, SAMA *F. adult, from 9 mi E of Everard Park, S.A.

Distribution: Lake Eyre basin, S.A.; type locality only. Ecology: larva - sedentary, soil, predator : adult - volant, burrowing.

Bembix boonamin Evans and Matthews, 1973

Bembix boonamin Evans, H.E. & Matthews, R.W. (1973). Systematics and nesting behavior of Australian *Bembix* sand wasps (Hymenoptera, Sphecidae). *Mem. Am. Entomol. Inst.* **20**: iv 387 pp. [282]. Type data: holotype, SAMA *M. adult, from Meekatharra-Billiluna Pool, Canning Stock Route, W.A.

Distribution: NW coastal, W plateau, W.A.; only published localities Meekatharra - Billiluna Pool, 80 km W Roy Hill. Ecology: larva - sedentary, soil, predator : adult - volant, burrowing.

Bembix buntor Evans and Matthews, 1973

Bembix buntor Evans, H.E. & Matthews, R.W. (1973). Systematics and nesting behavior of Australian *Bembix* sand wasps (Hymenoptera, Sphecidae). *Mem. Am. Entomol. Inst.* **20**: iv 387 pp. [87]. Type data: holotype, BMNH *M. adult, from Bolgart, W.A.

Distribution: SW coastal, NW coastal, W plateau, W.A.; only published localities Bolgart, Yericoin, Anketell. Ecology: larva - sedentary, soil, predator : adult - volant, burrowing.

Bembix burando Evans and Matthews, 1973

Bembix burando Evans, H.E. & Matthews, R.W. (1973). Systematics and nesting behavior of Australian *Bembix* sand wasps (Hymenoptera, Sphecidae). *Mem. Am. Entomol. Inst.* **20**: iv 387 pp. [112]. Type data: holotype, BPBM *F. adult, from Cooktown, Qld.

Distribution: NE coastal, Qld.; only published localities Cooktown and "Yanyarandi" (not found on map). Ecology: larva - sedentary, soil, predator : adult - volant, burrowing.

Bembix burraburra Evans and Matthews, 1973

Bembix burraburra Evans, H.E. & Matthews, R.W. (1973). Systematics and nesting behavior of Australian *Bembix* sand wasps (Hymenoptera, Sphecidae). *Mem. Am. Entomol. Inst.* **20**: iv 387 pp. [168]. Type data: holotype, ANIC M. adult, from 15 mi E of Yalgoo, W.A.

Distribution: NW coastal, W.A.; type locality only. Ecology: larva - sedentary, soil, predator : adult - volant, burrowing.

Bembix carripan Evans and Matthews, 1973

Bembix carripan Evans, H.E. & Matthews, R.W. (1973). Systematics and nesting behavior of Australian *Bembix* sand wasps (Hymenoptera, Sphecidae). *Mem. Am. Entomol. Inst.* **20**: iv 387 pp. [129]. Type data: holotype, SAMA *F. adult, from Meekatharra - Billiluna Pool, W.A.

Distribution: SW coastal, W plateau, W.A.; only published localities Meekatharra - Billiluna Pool and Southern Cross. Ecology: larva - sedentary, soil, predator : adult - volant, burrowing.

Bembix cooba Evans and Matthews, 1973

Bembix cooba Evans, H.E. & Matthews, R.W. (1973). Systematics and nesting behavior of Australian *Bembix* sand wasps (Hymenoptera, Sphecidae). *Mem. Am. Entomol. Inst.* **20**: iv 387 pp. [171]. Type data: holotype, ANIC M. adult, from Packsaddle Mt., N.S.W.

Distribution: Murray-Darling basin, Lake Eyre basin, N.S.W., Vic.; only published localities Packsaddle Mt., Wilcannia, Clifton Downs and Rainbow. Ecology: larva - sedentary, soil, predator : adult - volant, burrowing; prey adult Diptera.

Bembix coonundura Evans and Matthews, 1973

Bembix coonundura Evans, H.E. & Matthews, R.W. (1973). Systematics and nesting behavior of Australian *Bembix* sand wasps (Hymenoptera, Sphecidae). *Mem. Am. Entomol. Inst.* **20**: iv 387 pp. [150]. Type data: holotype, BMNH *M. adult, from Yanchep, W.A.

Distribution: SW coastal, W plateau, W.A.; only published localities Yanchep, Yallingup, Busselton, Jurien Bay and Wiluna (Lake Violet). Ecology: larva - sedentary, soil, predator : adult - volant, burrowing; prey adult Odonata. Biological references: Wheeler, W.M. & Dow, R. (1933). Unusual prey of *Bembix*. *Psyche Camb.* **40**: 57–59 (prey).

Bembix cursitans Handlirsch, 1894

Bembex cursitans Handlirsch, A. (1894). Monographie der mit *Nysson* und *Bembex* verwandten Grabwespen. vii. (Schluss). *Sber. Akad. Wiss. Wien* **102**: 657–942 [762 pl 1 fig 30, pl 6 fig 27]. Type data: holotype, ZMB *F. adult, from Australia (as New Holland).

Distribution: SW coastal, NW coastal, W.A. Ecology: larva - sedentary, soil, predator : adult - volant, burrowing; prey adult Diptera. Biological references: Evans, H.E. & Matthews, R.W. (1973). Systematics and nesting behavior of Australian *Bembix* sand wasps (Hymenoptera, Sphecidae). *Mem. Am. Entomol. Inst.* **20**: iv 387 pp. (nest, prey).

Bembix egens Handlirsch, 1894

Bembex egens Handlirsch, A. (1894). Monographie der mit *Nysson* und *Bembex* verwandten Grabwespen. vii. (Schluss). *Sber. Akad. Wiss. Wien* **102**: 657–942 [753 pl 1 fig 25, pl 3 fig 24, pl 4 fig 12, pl 5 fig 38, pl 6 fig 22]. Type data: syntypes (probable), whereabouts unknown, from Australia.

Distribution: Lake Eyre basin, W plateau, NW coastal, W.A., N.T. Ecology: larva - sedentary, soil, predator : adult - volant, burrowing. Biological references: Evans, H.E. & Matthews, R.W. (1973). Systematics and nesting behavior of Australian *Bembix* sand wasps (Hymenoptera, Sphecidae). *Mem. Am. Entomol. Inst.* **20**: iv 387 pp. (redescription).

Bembix eleebana Evans and Matthews, 1973

Bembix eleebana Evans, H.E. & Matthews, R.W. (1973). Systematics and nesting behavior of Australian *Bembix* sand wasps (Hymenoptera, Sphecidae). *Mem. Am. Entomol. Inst.* **20**: iv 387 pp. [85]. Type data: holotype, ANIC F. adult, from Dorrigo, N.S.W.

Distribution: SE coastal, Lake Eyre basin, N.S.W., (Vic.), N.T.; only published localities Dorrigo, ?Melbourne and Alice Springs. Ecology: larva - sedentary, soil, predator : adult - volant, burrowing.

Bembix flavifrons Smith, 1856

Bembex flavifrons Smith, F. (1856). *Catalogue of Hymenopterous Insects in the Collection of the British Museum. Part IV, Sphegidae, Larridae and Crabronidae.* pp. 207–497 London : British Museum [324]. Type data: holotype, BMNH 21.1,689 *F. adult (seen 1929 by L.F. Graham), from Australia.

Bembex saussurei Handlirsch, A. (1894). Monographie der mit *Nysson* und *Bembex* verwandten Grabwespen. vii. (Schluss). *Sber. Akad. Wiss. Wien* **102**: 657–942 [873 pl 4 fig 17]. Type data: lectotype, ZMB *F. adult, from Adelaide, S.A., designation by Evans, H.E. & Matthews, R.W. (1973). Systematics and nesting behavior of Australian *Bembix* sand wasps (Hymenoptera, Sphecidae). *Mem. Am. Entomol. Inst.* **20**: iv 387 pp.

Taxonomic decision of Turner, R.E. (1908). Notes on the Australian fossorial wasps of the family Sphegidae, with descriptions of new species. *Proc. Zool. Soc. Lond.* **1908**: 457–535 pl xxvi [505].

Distribution: NE coastal, Murray-Darling basin, SE coastal, S Gulfs, Qld., N.S.W., Vic., S.A. Ecology: larva - sedentary, soil, predator : adult - volant, burrowing. Biological references: Evans, H.E. & Matthews, R.W. (1973). Systematics and nesting behavior of Australian *Bembix* sand wasps (Hymenoptera, Sphecidae). *Mem. Am. Entomol. Inst.* **20**: iv 387 pp. (redescription).

Bembix flavipes Smith, 1856

Bembex flavipes Smith, F. (1856). *Catalogue of Hymenopterous Insects in the Collection of the British Museum. Part IV, Sphegidae, Larridae and Crabronidae.* pp. 207–497 London : British Museum [325]. Type data: holotype, BMNH 21.1,699 *F. adult (seen 1929 by L.F. Graham), from Australia.

Bembix brevis Lohrmann, E. (1942). Neue *Bembix*-Arten des Wiener Naturhistorischen Museums. *Ann. Naturh. Mus. Wien* **52**: 155–160 [160]. Type data: holotype, NHMW *F. adult, from Cooktown, Qld.

Taxonomic decision of Evans, H.E. & Matthews, R.W. (1973). Systematics and nesting behavior of Australian *Bembix* sand wasps (Hymenoptera, Sphecidae). *Mem. Am. Entomol. Inst.* **20**: iv 387 pp. [246].

Distribution: NE coastal, Murray-Darling basin, NW coastal, N coastal, N Gulf, Qld., N.S.W., W.A., N.T. Ecology: larva - sedentary, soil, predator : adult - volant, burrowing; prey adult Apidae (*Trigona essingtoni* Cockerell) Evans, H.E., Evans, M.A. & Hook, A. (1982).

Observations on the nests and prey of Australian *Bembix* sand wasps (Hymenoptera : Sphecidae). *Aust. J. Zool.* **30**: 71–80 (nest, prey).

Bembix flaviventris Smith, 1873

Bembex flaviventris Smith, F. (1873). Descriptions of new species of fossorial Hymenoptera in the collection of the British Museum. *Ann. Mag. Nat. Hist. (4)* **12**: 291–300, 402–415 [299]. Type data: holotype, BMNH 21.1,702 *F. adult (seen 1929 by L.F. Graham), from Swan River, W.A.

Bembex calcarina Handlirsch, A. (1894). Monographie der mit *Nysson* und *Bembex* verwandten Grabwespen. vii. (Schluss). *Sber. Akad. Wiss. Wien* **102**: 657–942 [754 pl 1 fig 26, pl 4 figs 13–14, pl 6 fig 23]. Type data: holotype, ZMB *M. adult, from Adelaide, S.A.

Taxonomic decision of Turner, R.E. (1910). Additions to our knowledge of the fossorial wasps of Australia. *Proc. Zool. Soc. Lond.* **1910**: 253–356 [351].

Distribution: Murray-Darling basin, Lake Eyre basin, S Gulfs, W plateau, SW coastal, NW coastal, Qld., N.S.W., Vic., S.A., W.A. Ecology: larva - sedentary, soil, predator : adult - volant, burrowing; prey adult Diptera. Biological references: Evans, H.E. & Matthews, R.W. (1973). Systematics and nesting behavior of Australian *Bembix* sand wasps (Hymenoptera, Sphecidae). *Mem. Am. Entomol. Inst.* **20**: iv 387 pp. (redescription, nest, prey).

Bembix furcata Erichson, 1842

Bembex furcata Erichson, W.F. (1842). Beitrag zur Fauna von Vandiemensland mit besonderer Rücksicht auf die geographische Verbreitung der Insekten. *Arch. Naturg.* **8**: 83–287 [266]. Type data: lectotype, ZMB *M. adult, from Tas., designation by Evans, H.E. & Matthews, R.W. (1973). Systematics and nesting behavior of Australian *Bembix* sand wasps (Hymenoptera, Sphecidae). *Mem. Am. Entomol. Inst.* **20**: iv 387 pp.

Distribution: Murray-Darling basin, SE coastal,S Gulfs, SW coastal, NW coastal, Qld., N.S.W., Vic., Tas., S.A., W.A. Ecology: larva - sedentary, soil, predator : adult - volant, burrowing; prey adult Diptera. Biological references: Evans, H.E. & Matthews, R.W. (1973). Systematics and nesting behavior of Australian *Bembix* sand wasps (Hymenoptera, Sphecidae). *Mem. Am. Entomol. Inst.* **20**: iv 387 pp. (redescription, nest, prey).

Bembix gelane Evans and Matthews, 1973

Bembix gelane Evans, H.E. & Matthews, R.W. (1973). Systematics and nesting behavior of Australian *Bembix* sand wasps (Hymenoptera, Sphecidae). *Mem. Am. Entomol. Inst.* **20**: iv 387 pp. [232]. Type data: holotype, ANIC M. adult, from 17 mi W of Morven, Qld.

Distribution: Lake Eyre basin, Qld.; only published localities near Morven and Charleville. Ecology: larva - sedentary, soil, predator : adult - volant, burrowing.

Bembix ginjulla Evans and Matthews, 1973

Bembix ginjulla Evans, H.E. & Matthews, R.W. (1973). Systematics and nesting behavior of Australian *Bembix* sand wasps (Hymenoptera, Sphecidae). *Mem. Am. Entomol. Inst.* **20**: iv 387 pp. [188]. Type data: holotype, ANIC M. adult, from 4 mi E of Wilcannia, N.S.W.

Distribution: Murray-Darling basin, N.S.W., Vic.; only published localities Wilcannia, Caldwell and Wilkur. Ecology: larva - sedentary, soil, predator : adult - volant, burrowing.

Bembix goyarra Evans and Matthews, 1973

Bembix goyarra Evans, H.E. & Matthews, R.W. (1973). Systematics and nesting behavior of Australian *Bembix* sand wasps (Hymenoptera, Sphecidae). *Mem. Am. Entomol. Inst.* **20**: iv 387 pp. [84]. Type data: holotype, WAM 71-804 *M. adult, from Kathleen Valley, W.A.

Distribution: W plateau, W.A.; type locality only. Ecology: larva - sedentary, soil, predator : adult - volant, burrowing.

Bembix gunamarra Evans and Matthews, 1973

Bembix gunamarra Evans, H.E. & Matthews, R.W. (1973). Systematics and nesting behavior of Australian *Bembix* sand wasps (Hymenoptera, Sphecidae). *Mem. Am. Entomol. Inst.* **20**: iv 387 pp. [68]. Type data: holotype, ANIC M. adult, from Yeppoon, Qld.

Distribution: NE coastal, N coastal, Qld., N.T. Ecology: larva - sedentary, soil, predator : adult - volant, burrowing; prey adult Diptera.

Bembix gurindji Evans, 1982

Bembix gurindji Evans, H.E. (1982). Two new species of Australian *Bembix* sand wasps, with notes on other species of the genus (Hymenoptera, Sphecidae). *Aust. Entomol. Mag.* **9**: 7–12 [7]. Type data: holotype, ANIC M. adult, from Beebom Crossing, Daly River, on Tipperary Station, N.T.

Distribution: N coastal, W.A., N.T. Ecology: larva - sedentary, soil, predator : adult - volant, burrowing.

Bembix hokarra Evans and Matthews, 1973

Bembix hokarra Evans, H.E. & Matthews, R.W. (1973). Systematics and nesting behavior of Australian *Bembix* sand wasps (Hymenoptera, Sphecidae). *Mem. Am. Entomol. Inst.* **20**: iv 387 pp. [94]. Type data: holotype, CAS *M. adult, from Milly Milly, W.A.

Distribution: NW coastal, N coastal, W.A.; only published localities Milly Milly, Mileura and Louisa Downs. Ecology: larva - sedentary, soil, predator : adult - volant, burrowing.

Bembix kamulla Evans and Matthews, 1973

Bembix kamulla Evans, H.E. & Matthews, R.W. (1973). Systematics and nesting behavior of Australian *Bembix* sand wasps (Hymenoptera, Sphecidae). *Mem. Am. Entomol. Inst.* **20**: iv 387 pp. [77]. Type data: holotype, AM M. adult, from Caldwell, N.S.W.

Distribution: Murray-Darling basin, Lake Eyre basin, SW coastal, Qld., N.S.W., (Vic.), W.A. Ecology: larva - sedentary, soil, predator : adult - volant, burrowing; prey adult Neuroptera, Diptera. Biological references: Evans, H.E., Evans, M.A. & Hook, A. (1982). Observations on the nests and prey of Australian *Bembix* sand wasps (Hymenoptera : Sphecidae). *Aust. J. Zool.* **30**: 71–80 (nest, prey).

Bembix kora Evans and Matthews, 1973

Bembix kora Evans, H.E. & Matthews, R.W. (1973). Systematics and nesting behavior of Australian *Bembix* sand wasps (Hymenoptera, Sphecidae). *Mem. Am. Entomol. Inst.* **20**: iv 387 pp. [143]. Type data: holotype, ANIC M. adult, from 50 mi S of Nappamerry, Qld.

Distribution: NE coastal, Murray-Darling basin, Lake Eyre basin, Qld.; only published localities near Napamerry, Cunnamulla and Rockhampton. Ecology: larva - sedentary, soil, predator : adult - volant, burrowing.

Bembix kununurra Evans and Matthews, 1973

Bembix kununurra Evans, H.E. & Matthews, R.W. (1973). Systematics and nesting behavior of Australian *Bembix* sand wasps (Hymenoptera, Sphecidae). *Mem. Am. Entomol. Inst.* **20**: iv 387 pp. [284]. Type data: holotype, ANIC M. adult (unnumbered), from Kununurra and vicinity, W.A.

Distribution: N coastal, W.A.; type locality only. Ecology: larva - sedentary, soil, predator : adult - volant, burrowing; prey adult Diptera. Biological references: Evans, H.E., Evans, M.A. & Hook, A. (1982). Observations on the nests and prey of Australian *Bembix* sand wasps (Hymenoptera : Sphecidae). *Aust. J. Zool.* **30**: 71–80 (nest, prey); Evans, H.E. (1982). Two new species of Australian *Bembix* sand wasps, with notes on other species of the genus (Hymenoptera, Sphecidae). *Aust. Entomol. Mag.* **9**: 7–12 (description of female).

Bembix lamellata Handlirsch, 1894

Bembex lamellata Handlirsch, A. (1894). Monographie der mit *Nysson* und *Bembex* verwandten Grabwespen. vii. (Schluss). *Sber. Akad. Wiss. Wien* **102**: 657–942 [842 pl 2 fig 36, pl 7 fig 26]. Type data: holotype, ZMB *M. adult, from Adelaide, S.A.

Distribution: NE coastal, Murray-Darling basin, Lake Eyre basin, S Gulfs, W plateau, NW coastal, Qld., N.S.W., Vic., S.A., W.A., N.T. Ecology: larva - sedentary, soil, predator : adult - volant, burrowing; prey adult Diptera. Biological references: Evans, H.E. & Matthews, R.W. (1973). Systematics and nesting behavior of Australian *Bembix* sand wasps (Hymenoptera, Sphecidae). *Mem. Am. Entomol. Inst.* **20**: iv 387 pp. (redescription, nest, prey).

Bembix latifasciata Turner, 1912

Bembex latifasciata Turner, R.E. (1912). Notes on fossorial Hymenoptera. IX. On some new species from the Australian and Austro-Malayan regions. *Ann. Mag. Nat. Hist. (8)* **10**: 48–63 [57]. Type data: holotype, BMNH 21.1,692 *M. adult (seen 1929 by L.F. Graham), from Strelley River, W.A.

Distribution: Lake Eyre basin, W plateau, NW coastal, N.S.W., W.A., N.T.; localities other than type locality are uncertain. Ecology: larva - sedentary, soil, predator : adult - volant, burrowing. Biological references: Evans, H.E. & Matthews, R.W. (1973). Systematics and nesting behavior of Australian *Bembix* sand wasps (Hymenoptera, Sphecidae). *Mem. Am. Entomol. Inst.* **20**: iv 387 pp. (redescription).

Bembix leeuwinensis Turner, 1915

Bembex leeuwinensis Turner, R.E. (1915). Notes on fossorial Hymenoptera. XVIII. On the Australian species of *Bembex*. *Ann. Mag. Nat. Hist. (8)* **16**: 434–447 [446]. Type data: holotype, BMNH 21.1,698 *M. adult (seen 1929 by L.F. Graham), from Yallingup, W.A.

Distribution: SW coastal, W.A.; only published localities Yallingup and Busselton. Ecology: larva - sedentary, soil, predator : adult - volant, burrowing. Biological references: Evans, H.E. & Matthews, R.W. (1973). Systematics and nesting behavior of Australian *Bembix* sand wasps (Hymenoptera, Sphecidae). *Mem. Am. Entomol. Inst.* **20**: iv 387 pp. (redescription).

Bembix littoralis Turner, 1908

Bembex littoralis Turner, R.E. (1908). Notes on the Australian fossorial wasps of the family Sphegidae, with descriptions of new species. *Proc. Zool. Soc. Lond.* **1908**: 457–535 pl xxvi [502]. Type data: holotype, BMNH 21.1,700 *M. adult (seen 1929 by L.F. Graham), from Darwin, N.T.

Bembix flavolatera Parker, J.B. (1929). A generic revision of the fossorial wasps of the tribes Stizini and Bembicini with notes and descriptions of new species. *Proc. U.S. Natl. Mus.* **75**(2776): 1–203 [134]. Type data: holotype, ZMB *M. adult, from Hermannsburg, N.T.

Taxonomic decision of Evans, H.E. & Matthews, R.W. (1973). Systematics and nesting behavior of Australian *Bembix* sand wasps (Hymenoptera, Sphecidae). *Mem. Am. Entomol. Inst.* **20**: iv 387 pp. [268].

Distribution: NE coastal, SE coastal, Murray-Darling basin, Lake Eyre basin, S Gulfs, W plateau, SW coastal, NW coastal, N coastal, N Gulf, Qld., N.S.W., Vic., S.A., W.A., N.T. Ecology: larva - sedentary, soil, predator : adult - volant, burrowing; prey adult Diptera. Biological references: Evans, H.E. & Matthews, R.W. (1973). Systematics and nesting behavior of Australian *Bembix* sand wasps (Hymenoptera, Sphecidae). *Mem. Am. Entomol. Inst.* **20**: iv 387 pp. (redescription, nest, prey).

Bembix lobimana Handlirsch, 1894

Bembex lobimana Handlirsch, A. (1894). Monographie der mit *Nysson* und *Bembex* verwandten Grabwespen. vii. (Schluss). *Sber. Akad. Wiss. Wien* **102**: 657-942 [755 pl 1 fig 27, pl 4 fig 15, pl 6 fig 24]. Type data: lectotype, GMHN *M. adult, from N.S.W., designation by Evans, H.E. & Matthews, R.W. (1973). Systematics and nesting behavior of Australian *Bembix* sand wasps (Hymenoptera, Sphecidae). *Mem. Am. Entomol. Inst.* **20**: iv 387 pp.

Distribution: Murray-Darling basin, S Gulfs, NW coastal, N.S.W., S.A., W.A.; only published localities N.S.Wales, Nov. Holl., Austr. merid., False Bay (Eyre Peninsula) and Port Vincent. Ecology: larva - sedentary, soil, predator : adult - volant, burrowing. Biological references: Evans, H.E. & Matthews, R.W. (1973). Systematics and nesting behavior of Australian *Bembix* sand wasps (Hymenoptera, Sphecidae). *Mem. Am. Entomol. Inst.* **20**: iv 387 pp. (redescription).

Bembix loorea Evans and Matthews, 1973

Bembix loorea Evans, H.E. & Matthews, R.W. (1973). Systematics and nesting behavior of Australian *Bembix* sand wasps (Hymenoptera, Sphecidae). *Mem. Am. Entomol. Inst.* **20**: iv 387 pp. [308]. Type data: holotype, ANIC F. adult, from near Reedy Rockhole, Amadeus Basin, N.T.

Distribution: W plateau, N.T.; type locality only. Ecology: larva - sedentary, soil, predator : adult - volant, burrowing.

Bembix magarra Evans and Matthews, 1973

Bembix magarra Evans, H.E. & Matthews, R.W. (1973). Systematics and nesting behavior of Australian *Bembix* sand wasps (Hymenoptera, Sphecidae). *Mem. Am. Entomol. Inst.* **20**: iv 387 pp. [233]. Type data: holotype, QM T7139 *F. adult, from South Hedland, W.A.

Distribution: Murray-Darling basin, NW coastal, Vic., W.A.; only published localities Wyperfeld Natl. Park and South Hedland. Ecology: larva - sedentary, soil, predator : adult - volant, burrowing.

Bembix maliki Evans and Matthews, 1973

Bembix maliki Evans, H.E. & Matthews, R.W. (1973). Systematics and nesting behavior of Australian *Bembix* sand wasps (Hymenoptera, Sphecidae). *Mem. Am. Entomol. Inst.* **20**: iv 387 pp. [96]. Type data: holotype, SAMA *M. adult, from Coward Springs, 82 mi NW of Maree, S.A.

Distribution: Lake Eyre basin, S.A.; type locality only. Ecology: larva - sedentary, soil, predator : adult - volant, burrowing.

Bembix mareeba Evans and Matthews, 1973

Bembix mareeba Evans, H.E. & Matthews, R.W. (1973). Systematics and nesting behavior of Australian *Bembix* sand wasps (Hymenoptera, Sphecidae). *Mem. Am. Entomol. Inst.* **20**: iv 387 pp. [42]. Type data: holotype, ANIC M. adult, from Granite Creek, 21 mi S of Miriam Vale, Qld.

Distribution: NE coastal, Qld.; Brisbane to Cape York. Ecology: larva - sedentary, soil, predator : adult - volant, burrowing; prey adult Diptera. Biological references: Evans, H.E., Evans, M.A. & Hook, A. (1982). Observations on the nests and prey of Australian *Bembix* sand wasps (Hymenoptera : Sphecidae). *Aust. J. Zool.* **30**: 71-80 (nest, prey).

Bembix marhra Evans and Matthews, 1973

Bembix marhra Evans, H.E. & Matthews, R.W. (1973). Systematics and nesting behavior of Australian *Bembix* sand wasps (Hymenoptera, Sphecidae). *Mem. Am. Entomol. Inst.* **20**: iv 387 pp. [212]. Type data: holotype, SAMA *M. adult, from Wolf Creek, W.A.

Distribution: N coastal, W.A.; type locality only. Ecology: larva - sedentary, soil, predator : adult - volant, burrowing.

Bembix marsupiata Handlirsch, 1894

Bembex marsupiata Handlirsch, A. (1894). Monographie der mit *Nysson* und *Bembex* verwandten Grabwespen. vii. (Schluss). *Sber. Akad. Wiss. Wien* **102**: 657-942 [757 pl 6 fig 25]. Type data: lectotype, NHMW *F. adult, from Swan River, W.A., designation by Evans, H.E. & Matthews, R.W. (1973). Systematics and nesting behavior of Australian *Bembix* sand wasps (Hymenoptera, Sphecidae). *Mem. Am. Entomol. Inst.* **20**: iv 387 pp.

Distribution: SW coastal, W.A.; two specimens labelled "N. S. Wales", this locality is queried. Ecology: larva - sedentary, soil, predator : adult - volant, burrowing. Biological references: Evans, H.E. & Matthews, R.W. (1973). Systematics and nesting behavior of Australian *Bembix* sand wasps (Hymenoptera, Sphecidae). *Mem. Am. Entomol. Inst.* **20**: iv 387 pp. (redescription).

Bembix mianga Evans and Matthews, 1973

Bembix mianga Evans, H.E. & Matthews, R.W. (1973). Systematics and nesting behavior of Australian *Bembix* sand wasps (Hymenoptera, Sphecidae). *Mem. Am. Entomol. Inst.* **20**: iv 387 pp. [213]. Type data: holotype, ANIC M. adult, from 3 mi W of Wentworth, N.S.W.

Distribution: Murray-Darling basin, Lake Eyre basin, W plateau, NW coastal, Qld., N.S.W., S.A., W.A., N.T. Ecology: larva - sedentary, soil, predator : adult - volant, burrowing; prey adult Diptera.

Bembix minya Evans and Matthews, 1973

Bembix minya Evans, H.E. & Matthews, R.W. (1973). Systematics and nesting behavior of Australian *Bembix* sand wasps (Hymenoptera, Sphecidae). *Mem. Am. Entomol. Inst.* **20**: iv 387 pp. [147]. Type data: holotype, ANIC M. adult, from 5-15 mi S of Rainbow, Vic.

Distribution: Murray-Darling basin, S Gulfs, Qld., Vic., S.A. Ecology: larva - sedentary, soil, predator : adult - volant, burrowing.

Bembix mokari Evans and Matthews, 1973

Bembix mokari Evans, H.E. & Matthews, R.W. (1973). Systematics and nesting behavior of Australian *Bembix* sand wasps (Hymenoptera, Sphecidae). *Mem. Am. Entomol. Inst.* **20**: iv 387 pp. [310]. Type data: holotype, SAMA *F. adult, from Coward Springs, 82 mi NW of Maree, S.A.

Distribution: Lake Eyre basin, S.A., (Vic.); only published localities Coward Springs, ?Vic. Ecology: larva - sedentary, soil, predator : adult - volant, burrowing.

Bembix moma Evans and Matthews, 1973

Bembix moma Evans, H.E. & Matthews, R.W. (1973). Systematics and nesting behavior of Australian *Bembix* sand wasps (Hymenoptera, Sphecidae). *Mem. Am. Entomol. Inst.* **20**: iv 387 pp. [195]. Type data: holotype, ANIC M. adult, from Kununurra and vicinity, W.A.

Distribution: Murray-Darling basin, Lake Eyre basin, W plateau, SW coastal, NW coastal, N coastal, N Gulf, Qld., N.S.W., Vic., S.A., W.A., N.T. Ecology: larva - sedentary, soil, predator : adult - volant, burrowing; prey adult Hymenoptera.

Bembix moonga Evans and Matthews, 1973

Bembix moonga Evans, H.E. & Matthews, R.W. (1973). Systematics and nesting behavior of Australian *Bembix* sand wasps (Hymenoptera, Sphecidae). *Mem. Am. Entomol. Inst.* **20**: iv 387 pp. [167]. Type data: holotype, ANIC M. adult, from Wiluna, W.A.

Distribution: W plateau, W.A.; type locality only. Ecology: larva - sedentary, soil, predator : adult - volant, burrowing.

Bembix mundurra Evans and Matthews, 1973

Bembix mundurra Evans, H.E. & Matthews, R.W. (1973). Systematics and nesting behavior of Australian *Bembix* sand wasps (Hymenoptera, Sphecidae). *Mem. Am. Entomol. Inst.* **20**: iv 387 pp. [179]. Type data: holotype, WAM 73-644 *M. adult, from 6 mi W of Mullewa, W.A.

Distribution: Lake Eyre basin, SW coastal, NW coastal, W plateau, N.S.W., W.A. Ecology: larva - sedentary, soil, predator : adult - volant, burrowing; prey adult Diptera.

Bembix munta Evans and Matthews, 1973

Bembix munta Evans, H.E. & Matthews, R.W. (1973). Systematics and nesting behavior of Australian *Bembix* sand wasps (Hymenoptera, Sphecidae). *Mem. Am. Entomol. Inst.* **20**: iv 387 pp. [210]. Type data: holotype, ANIC M. adult, from 60 mi S of Carnarvon, W.A.

Distribution: NW coastal, W.A.; type locality only. Ecology: larva - sedentary, soil, predator : adult - volant, burrowing.

Bembix musca Handlirsch, 1894

Bembex musca Handlirsch, A. (1894). Monographie der mit *Nysson* und *Bembex* verwandten Grabwespen. vii. (Schluss). *Sber. Akad. Wiss. Wien* **102**: 657-942 [844 pl 2 fig 35, pl 7 fig 27]. Type data: holotype, GMNH *M. adult, from Rockhampton, Qld.

Bembex mackayensis Turner, R.E. (1910). Additions to our knowledge of the fossorial wasps of Australia. *Proc. Zool. Soc. Lond.* **1910**: 253-356 [351]. Type data: holotype, BMNH 21.1,701 *M. adult (seen 1929 by L.F. Graham), from Mackay and Cairns, Qld.

Taxonomic decision of Evans, H.E. & Matthews, R.W. (1973). Systematics and nesting behavior of Australian *Bembix* sand wasps (Hymenoptera, Sphecidae). *Mem. Am. Entomol. Inst.* **20**: iv 387 pp. [251].

Distribution: NE coastal, SE coastal, Murray-Darling basin, N coastal, Qld., N.S.W., N.T. Ecology: larva - sedentary, soil, predator : adult - volant, burrowing; prey adult Hymenoptera (*Trigona carbonaria* Smith) Apidae Evans, H.E., Evans, M.A. & Hook, A. (1982). Observations on the nests and prey of Australian *Bembix* sand wasps (Hymenoptera : Sphecidae). *Aust. J. Zool.* **30**: 71-80 (nest, prey).

Bembix nigropectinata Turner, 1936

Bembex nigropectinatus Turner, R.E. (1936). Notes on fossorial Hymenoptera. XLV. On new sphegid wasps from Australia. *Ann. Mag. Nat. Hist. (10)* **18**: 533-545 [538]. Type data: holotype, BMNH 21.1,691 *M. adult, from Mingenew, W.A.

Distribution: NW coastal, W.A.; only published localities Mingenew, Maya and Nilemah Station. Ecology: larva - sedentary, soil, predator : adult - volant, burrowing. Biological references: Evans, H.E. & Matthews, R.W. (1973). Systematics and nesting behavior of Australian *Bembix* sand wasps (Hymenoptera, Sphecidae). *Mem. Am. Entomol. Inst.* **20**: iv 387 pp. (redescription).

Bembix obiri Evans, 1982

Bembix obiri Evans, H.E. (1982). Two new species of Australian *Bembix* sand wasps, with notes on other species of the genus (Hymenoptera, Sphecidae). *Aust. Entomol. Mag.* **9**: 7–12 [10]. Type data: holotype, ANIC M. adult, from 16 km E by N of Mt. Cahill, N.T.

Distribution: N coastal, N.T.; type locality only. Ecology: larva - sedentary, soil, predator : adult - volant, burrowing.

Bembix octosetosa Lohrmann, 1942

Bembix octosetosa Lohrmann, E. (1942). Neue *Bembix*-Arten des Wiener Naturhistorischen Museums. *Ann. Naturh. Mus. Wien* **52**: 155–160 [158]. Type data: holotype, NHMW *F. adult, from Rockhampton, Qld.

Distribution: NE coastal, Murray-Darling basin, Qld., N.S.W., Vic. Ecology: larva - sedentary, soil, predator : adult - volant, burrowing; prey adult Diptera. Biological references: Evans, H.E. & Matthews, R.W. (1973). Systematics and nesting behavior of Australian *Bembix* sand wasps (Hymenoptera, Sphecidae). *Mem. Am. Entomol. Inst.* **20**: iv 387 pp. (redescription); Evans, H.E., Evans, M.A. & Hook, A. (1982). Observations on the nests and prey of Australian *Bembix* sand wasps (Hymenoptera : Sphecidae). *Aust. J. Zool.* **30**: 71–80 (nest, prey).

Bembix olba Evans and Matthews, 1973

Bembix olba Evans, H.E. & Matthews, R.W. (1973). Systematics and nesting behavior of Australian *Bembix* sand wasps (Hymenoptera, Sphecidae). *Mem. Am. Entomol. Inst.* **20**: iv 387 pp. [133]. Type data: holotype, SAMA *M. adult, from near Alice Springs, N.T.

Distribution: Lake Eyre basin, NW coastal, W plateau, Qld., S.A., W.A., N.T. Ecology: larva - sedentary, soil, predator : adult - volant, burrowing.

Bembix oomborra Evans and Matthews, 1973

Bembix oomborra Evans, H.E. & Matthews, R.W. (1973). Systematics and nesting behavior of Australian *Bembix* sand wasps (Hymenoptera, Sphecidae). *Mem. Am. Entomol. Inst.* **20**: iv 387 pp. [83]. Type data: holotype, BMNH *M. adult, from Darwin, N.T.

Distribution: N coastal, N.T.; type locality only. Ecology: larva - sedentary, soil, predator : adult - volant, burrowing.

Bembix ourapilla Evans and Matthews, 1973

Bembix ourapilla Evans, H.E. & Matthews, R.W. (1973). Systematics and nesting behavior of Australian *Bembix* sand wasps (Hymenoptera, Sphecidae). *Mem. Am. Entomol. Inst.* **20**: iv 387 pp. [235]. Type data: holotype, ANIC M. adult, from Nilemah Station 50 mi S of Denham, W.A.

Distribution: NW coastal, W.A.; only published localities Nilemah Station and 68 mi N of Northampton. Ecology: larva - sedentary, soil, predator : adult - volant, burrowing.

Bembix palmata Smith, 1856

Bembex palmata Smith, F. (1856). *Catalogue of Hymenopterous Insects in the Collection of the British Museum.* Part IV, Sphegidae, Larridae and Crabronidae. pp. 207–497 London : British Museum [325]. Type data: holotype, BMNH 21.1,686 *M. adult (seen 1929 by L.F. Graham), from ?Australia.

Bembex tridentifera Smith, F. (1873). Descriptions of new species of fossorial Hymenoptera in the collection of the British Museum. *Ann. Mag. Nat. Hist.* (4) **12**: 291–300, 402–415 [298]. Type data: holotype, BMNH 21.1,687 *F. adult (seen 1929 by L.F. Graham), from Moreton Bay, Qld.

Taxonomic decision of Handlirsch, A. (1894). Monographie der mit *Nysson* und *Bembex* verwandten Grabwespen. vii. (Schluss). *Sber. Akad. Wiss. Wien* **102**: 657–942 [751].

Distribution: NE coastal, SE coastal, Murray-Darling basin, S Gulfs, N coastal, Qld., N.S.W., A.C.T., Vic., S.A., N.T. Ecology: larva - sedentary, soil, predator : adult - volant, burrowing; prey adult Diptera. Biological references: Evans, H.E. & Matthews, R.W. (1973). Systematics and nesting behavior of Australian *Bembix* sand wasps (Hymenoptera, Sphecidae). *Mem. Am. Entomol. Inst.* **20**: iv 387 pp. (redescription, nest, prey).

Bembix palona Evans and Matthews, 1973

Bembix palona Evans, H.E. & Matthews, R.W. (1973). Systematics and nesting behavior of Australian *Bembix* sand wasps (Hymenoptera, Sphecidae). *Mem. Am. Entomol. Inst.* **20**: iv 387 pp. [190]. Type data: holotype, ANIC M. adult, from 5 mi N of Nocatunga, Qld.

Distribution: Murray-Darling basin, Bulloo River basin, Lake Eyre basin, S Gulfs, NW coastal, Qld., N.S.W., S.A., W.A., N.T. Ecology: larva - sedentary, soil, predator : adult - volant, burrowing.

Bembix pectinipes Handlirsch, 1894

Bembex palmata Smith, F. (1873). Descriptions of new species of fossorial Hymenoptera in the collection of the British Museum. *Ann. Mag. Nat. Hist.* (4) **12**: 291–300, 402–415 [298] [*non Bembex palmata* Smith, 1856]. Type data: holotype, BMNH 21.1,688 *M. adult (seen 1929 by L.F. Graham), from N. Australia.

Bembex pectinipes Handlirsch, A. (1894). Monographie der mit *Nysson* und *Bembex* verwandten Grabwespen. vii. (Schluss). *Sber. Akad. Wiss. Wien* **102**: 657–942 [875 pl 4 fig 18]. Type data: holotype, ZMH (destroyed in war), from Qld.

Taxonomic decision of Turner, R.E. (1908). Notes on the Australian fossorial wasps of the family Sphegidae, with descriptions of new species. *Proc. Zool. Soc. Lond.* **1908**: 457–535 pl xxvi [505].

Distribution: NE coastal, Lake Eyre basin, N coastal, Qld., S.A., N.T. Ecology: larva - sedentary, soil, predator : adult - volant, burrowing; prey adult Diptera. Biological references: Evans, H.E. & Matthews, R.W. (1973). Systematics and nesting behavior of Australian *Bembix* sand wasps (Hymenoptera, Sphecidae). *Mem. Am. Entomol. Inst.* **20**: iv 387 pp. (redescription, prey).

Bembix pikati Evans and Matthews, 1973

Bembix pikati Evans, H.E. & Matthews, R.W. (1973). Systematics and nesting behavior of Australian *Bembix* sand wasps (Hymenoptera, Sphecidae). *Mem. Am. Entomol. Inst.* **20**: iv 387 pp. [230]. Type data: holotype, SAMA *M. adult, from 76 mi W of Musgrave Park, S.A.

Distribution: Lake Eyre basin, W plateau, S.A.; only published localities Musgrave Park and vicinity and Everard Park. Ecology: larva - sedentary, soil, predator : adult - volant, burrowing.

Bembix pillara Evans and Matthews, 1973

Bembix pillara Evans, H.E. & Matthews, R.W. (1973). Systematics and nesting behavior of Australian *Bembix* sand wasps (Hymenoptera, Sphecidae). *Mem. Am. Entomol. Inst.* **20**: iv 387 pp. [226]. Type data: holotype, ANIC M. adult, from Lake Way, 10 mi S of Wiluna, W.A.

Distribution: W plateau, NW coastal, W.A.; only published localities Lake Way, Nilemah Station and Wiluna. Ecology: larva - sedentary, soil, predator : adult - volant, burrowing.

Bembix promontorii Lohrmann, 1942

Bembix promontorii Lohrmann, E. (1942). Neue *Bembix*-Arten des Wiener Naturhistorischen Museums. *Ann. Naturh. Mus. Wien* **52**: 155–160 [157]. Type data: holotype, NHMW *M. adult, from Cape York, Qld.

Distribution: NE coastal, SE coastal, Qld., N.S.W. Ecology: larva - sedentary, soil, predator : adult - volant, burrowing. Biological references: Evans, H.E. & Matthews, R.W. 1973. Systematics and nesting behavior of Australian *Bembix* sand wasps (Hymenoptera, Sphecidae). *Mem. Am. Entomol. Inst.* **20**: iv 387 pp. (redescription); Evans, H.E., Evans, M.A. & Hook, A. (1982). Observations on the nests and prey of Australian *Bembix* sand wasps (Hymenoptera : Sphecidae). *Aust. J. Zool.* **30**: 71–80 (nest).

Bembix pulka Evans and Matthews, 1973

Bembix pulka Evans, H.E. & Matthews, R.W. (1973). Systematics and nesting behavior of Australian *Bembix* sand wasps (Hymenoptera, Sphecidae). *Mem. Am. Entomol. Inst.* **20**: iv 387 pp. [166]. Type data: holotype, SAMA *F. adult, from 30 mi S Mt. Davies, Tompkinson Ranges, S.A.

Distribution: W plateau, S.A.; type locality only. Ecology: larva - sedentary, soil, predator : adult - volant, burrowing.

Bembix severa Smith, 1873

Bembex severa Smith, F. (1873). Descriptions of new species of fossorial Hymenoptera in the collection of the British Museum. *Ann. Mag. Nat. Hist.* (4) **12**: 291–300, 402–415 [297]. Type data: holotype, BMNH 21.1,693 *F. adult (seen 1929 by L.F. Graham), from Swan River, W.A.

Distribution: Murray-Darling basin, Lake Eyre basin, S Gulfs, W plateau, SW coastal, NW coastal, N coastal, N.S.W., Vic., S.A., W.A. Ecology: larva - sedentary, soil, predator : adult - volant, burrowing. Biological references: Evans, H.E. & Matthews, R.W. (1973). Systematics and nesting behavior of Australian *Bembix* sand wasps (Hymenoptera, Sphecidae). *Mem. Am. Entomol. Inst.* **20**: iv 387 pp. (redescription).

Bembix thooma Evans and Matthews, 1973

Bembix thooma Evans, H.E. & Matthews, R.W. (1973). Systematics and nesting behavior of Australian *Bembix* sand wasps (Hymenoptera, Sphecidae). *Mem. Am. Entomol. Inst.* **20**: iv 387 pp. [183]. Type data: holotype, ANIC F. adult, from Lake Violet, 6 mi S of Wiluna, W.A.

Distribution: Murray-Darling basin, Lake Eyre basin, W plateau, Qld., N.S.W., Vic., W.A., N.T. Ecology: larva - sedentary, soil, predator : adult - volant, burrowing; prey adult Hymenoptera (Thynninae). Biological references: Evans, H.E. (1982). Two new species of Australian *Bembix* sand wasps, with notes on other species of the genus (Hymenoptera, Sphecidae). *Aust. Entomol. Mag.* **9**: 7–12 (distribution).

Bembix tibooburra Evans and Matthews, 1973

Bembix tibooburra Evans, H.E. & Matthews, R.W. (1973). Systematics and nesting behavior of Australian *Bembix* sand wasps (Hymenoptera, Sphecidae). *Mem. Am. Entomol. Inst.* **20**: iv 387 pp. [193]. Type data: holotype, SAMA *M. adult, from Everard Park, S.A.

Distribution: Lake Eyre basin, W plateau, NW coastal, Qld., N.S.W., S.A., W.A. Ecology: larva - sedentary, soil, predator : adult - volant, burrowing.

Bembix trepida Handlirsch, 1894

Bembex trepida Handlirsch, A. (1894). Monographie der mit *Nysson* und *Bembex* verwandten Grabwespen. vii. (Schluss). *Sber. Akad. Wiss. Wien* **102**: 657–942 [759 pl 1 fig 28, pl 4 fig 16, pl 6 fig 26]. Type data: lectotype, NHMW *M. adult, from Australia (as New Holland), designation by Evans, H.E. & Matthews, R.W. (1973). Systematics and nesting behavior of Australian *Bembix* sand wasps (Hymenoptera, Sphecidae). *Mem. Am. Entomol. Inst.* **20**: iv 387 pp.

Bembix victoriensis Lohrmann, E. (1942). Neue *Bembix*-Arten des Wiener Naturhistorischen Museums. *Ann. Naturh. Mus. Wien* **52**: 155–160 [159]. Type data: holotype, NHMW *F. adult, from Vic.

Taxonomic decision of Evans, H.E. & Matthews, R.W. (1973). Systematics and nesting behavior of Australian *Bembix* sand wasps (Hymenoptera, Sphecidae). *Mem. Am. Entomol. Inst.* **20**: iv 387 pp.

Distribution: NE coastal, SE coastal, Murray-Darling basin, S Gulfs, Qld., N.S.W., A.C.T., Vic., S.A. Ecology: larva - sedentary, soil, predator : adult - volant, burrowing; prey adult Diptera.

Bembix tuberculiventris Turner, 1908

Bembex tuberculiventris Turner, R.E. (1908). Notes on the Australian fossorial wasps of the family Sphegidae, with descriptions of new species. *Proc. Zool. Soc. Lond.* **1908**: 457–535 pl xxvi [503]. Type data: holotype, BMNH 21.1,703 *M. adult (seen 1929 by L.F. Graham), from Cooktown, Qld.

Distribution: NE coastal, Murray-Darling basin, Lake Eyre basin, S Gulfs, NW coastal, N coastal, N Gulf, Qld., N.S.W., Vic., S.A., W.A., N.T. Ecology: larva - sedentary, soil, predator : adult - volant, burrowing; prey adult Hymenoptera (Colletidae, Halictidae). Biological references: Evans, H.E. & Matthews, R.W. (1973). Systematics and nesting behavior of Australian *Bembix* sand wasps (Hymenoptera, Sphecidae). *Mem. Am. Entomol. Inst.* **20**: iv 387 pp. (redescription, nest, prey); Evans, H.E., Evans, M.A. & Hook, A. (1982). Observations on the nests and prey of Australian *Bembix* sand wasps (Hymenoptera : Sphecidae). *Aust. J. Zool.* **30**: 71–80 (nest, prey).

Bembix uloola Evans and Matthews, 1973

Bembix uloola Evans, H.E. & Matthews, R.W. (1973). Systematics and nesting behavior of Australian *Bembix* sand wasps (Hymenoptera, Sphecidae). *Mem. Am. Entomol. Inst.* **20**: iv 387 pp. [208]. Type data: holotype, ANIC M. adult, from 34 mi W of Wiluna, W.A.

Distribution: W plateau, W.A.; type locality only. Ecology: larva - sedentary, soil, predator : adult - volant, burrowing.

Bembix undeneya Evans and Matthews, 1973

Bembix undeneya Evans, H.E. & Matthews, R.W. (1973). Systematics and nesting behavior of Australian *Bembix* sand wasps (Hymenoptera, Sphecidae). *Mem. Am. Entomol. Inst.* **20**: iv 387 pp. [142]. Type data: holotype, CAS *M. adult, from MacDonald Downs, N.T.

Distribution: Murray-Darling basin, Lake Eyre basin, Qld., N.S.W., Vic., N.T. Ecology: larva - sedentary, soil, predator : adult - volant, burrowing.

Bembix variabilis Smith, 1856

Bembex variabilis Smith, F. (1856). *Catalogue of Hymenopterous Insects in the Collection of the British Museum.* Part IV, Sphegidae, Larridae and Crabronidae. pp. 207–497 London : British Museum [325]. Type data: holotype, BMNH 21.1,705 *F. adult (seen 1929 by L.F. Graham), from Hunter River, W.A. and Port Essington, N.T.".

Bembex raptor Smith, F. (1856). *Catalogue of Hymenopterous Insects in the Collection of the British Museum.* Part IV, Sphegidae, Larridae and Crabronidae. pp. 207–497 London : British Museum [326]. Type data: holotype, BMNH 21.1,704 *M. adult (seen 1929 by L.F. Graham), from Adelaide, S.A.

Bembex crabroniformis Smith, F. (1873). Descriptions of new species of fossorial Hymenoptera in the collection of the British Museum. *Ann. Mag. Nat. Hist.* (4) **12**: 291–300, 402–415 [296]. Type data: holotype, BMNH 21.1,706 *M. adult (seen 1929 by L.F. Graham), from Port Essington, N.T.

Taxonomic decision of Handlirsch, A. (1894). Monographie der mit *Nysson* und *Bembex* verwandten Grabwespen. vii. (Schluss). *Sber. Akad. Wiss. Wien* **102**: 657–942 [846,901].

Distribution: NE coastal, SE coastal, Murray-Darling basin, Lake Eyre basin, S Gulfs, W plateau, SW coastal, NW coastal, N coastal, N Gulf, Qld., N.S.W., A.C.T., Vic., S.A., W.A., N.T.; also found on Timor. Ecology: larva - sedentary, soil, predator : adult - volant, burrowing; prey adult Diptera. Biological references: Evans, H.E. & Matthews, R.W. (1973). Systematics and nesting behavior of Australian *Bembix* sand wasps (Hymenoptera, Sphecidae). *Mem. Am. Entomol. Inst.* **20**: iv 387 pp. (redescription, nest, prey).

Bembix vespiformis Smith, 1856

Bembex vespiformis Smith, F. (1856). *Catalogue of Hymenopterous Insects in the Collection of the British Museum.* Part IV, Sphegidae, Larridae and Crabronidae. pp. 207–497 London : British Museum [327]. Type data: holotype, BMNH 21.1,690 *M. adult (seen 1929 by L.F. Graham), from Adelaide, S.A.

Distribution: NE coastal, Murray-Darling basin, S Gulfs, Lake Eyre basin, W plateau, SW coastal, Qld., N.S.W., Vic., S.A., W.A. Ecology: larva - sedentary, soil, predator : adult - volant, burrowing; prey adult Diptera. Biological references: Evans, H.E. & Matthews, R.W. (1973). Systematics and nesting behavior of Australian *Bembix* sand wasps (Hymenoptera, Sphecidae). *Mem. Am. Entomol. Inst.* **20**: iv 387 pp. (redescription, nest, prey).

Bembix wadamiri Evans and Matthews, 1973

Bembix wadamiri Evans, H.E. & Matthews, R.W. (1973). Systematics and nesting behavior of Australian *Bembix* sand wasps (Hymenoptera, Sphecidae). *Mem.*

Am. Entomol. Inst. **20**: iv 387 pp. [145]. Type data: holotype, BPBM *M. adult, from Maningrida, Arnhem Land, N.T.

Distribution: N coastal, W.A., N.T. Ecology: larva - sedentary, soil, predator : adult - volant, burrowing.

Bembix wangoola Evans and Matthews, 1973

Bembix wangoola Evans, H.E. & Matthews, R.W. (1973). Systematics and nesting behavior of Australian *Bembix* sand wasps (Hymenoptera, Sphecidae). *Mem. Am. Entomol. Inst.* **20**: iv 387 pp. [285]. Type data: holotype, ANIC M. adult, from Packsaddle Mt., N.S.W.

Distribution: Murray-Darling basin, Lake Eyre basin, W plateau, N.S.W., S.A. Ecology: larva - sedentary, soil, predator : adult - volant, burrowing.

Bembix wanna Evans and Matthews, 1973

Bembix wanna Evans, H.E. & Matthews, R.W. (1973). Systematics and nesting behavior of Australian *Bembix* sand wasps (Hymenoptera, Sphecidae). *Mem. Am. Entomol. Inst.* **20**: iv 387 pp. [149]. Type data: holotype, NMV *M. adult, from Vic.

Distribution: Murray-Darling basin, SE coastal, Vic.; only published localities Williamstown, Macedon and Gunbower. Ecology: larva - sedentary, soil, predator : adult - volant, burrowing.

Bembix warawara Evans and Matthews, 1973

Bembix warawara Evans, H.E. & Matthews, R.W. (1973). Systematics and nesting behavior of Australian *Bembix* sand wasps (Hymenoptera, Sphecidae). *Mem. Am. Entomol. Inst.* **20**: iv 387 pp. [169]. Type data: holotype, AM M. adult, from Calumet, 26 mi NE Binnaway, N.S.W.

Distribution: Murray-Darling basin, Qld., N.S.W., Vic.; only published localities Cunnamulla, Calumet and 80 mi. N of Melbourne. Ecology: larva - sedentary, soil, predator : adult - volant, burrowing.

Bembix weema Evans and Matthews, 1973

Bembix weema Evans, H.E. & Matthews, R.W. (1973). Systematics and nesting behavior of Australian *Bembix* sand wasps (Hymenoptera, Sphecidae). *Mem. Am. Entomol. Inst.* **20**: iv 387 pp. [80]. Type data: holotype, AM M. adult (unnumbered), from Wydgee Station near Mt. Magnet, W.A.

Distribution: W plateau, W.A.; type locality only. Ecology: larva - sedentary, soil, predator : adult - volant, burrowing.

Bembix wilcannia Evans and Matthews, 1973

Bembix wilcannia Evans, H.E. & Matthews, R.W. (1973). Systematics and nesting behavior of Australian *Bembix* sand wasps (Hymenoptera, Sphecidae). *Mem.*

Am. Entomol. Inst. **20**: iv 387 pp. [135]. Type data: holotype, ANIC M. adult, from 10 mi W of Wilcannia, N.S.W.

Distribution: Murray-Darling basin, Lake Eyre basin, Qld., N.S.W.; only published localities Wilcannia, Packsaddle Mt. and 57 mi NW Quilpie. Ecology: larva - sedentary, soil, predator : adult - volant, burrowing; prey adult Neuroptera.

Bembix wiluna Evans and Matthews, 1973

Bembix wiluna Evans, H.E. & Matthews, R.W. (1973). Systematics and nesting behavior of Australian *Bembix* sand wasps (Hymenoptera, Sphecidae). *Mem. Am. Entomol. Inst.* **20**: iv 387 pp. [89]. Type data: holotype, ANIC M. adult, from Wiluna, W.A.

Distribution: Murray-Darling basin, Lake Eyre basin, W plateau, Qld., N.S.W., S.A., W.A., N.T. Ecology: larva - sedentary, soil, predator : adult - volant, burrowing. Biological references: Evans, H.E. (1982). Two new species of Australian *Bembix* sand wasps, with notes on other species of the genus (Hymenoptera, Sphecidae). *Aust. Entomol. Mag.* **9**: 7-12 (distribution).

Bembix wollowra Evans and Matthews, 1973

Bembix wollowra Evans, H.E. & Matthews, R.W. (1973). Systematics and nesting behavior of Australian *Bembix* sand wasps (Hymenoptera, Sphecidae). *Mem. Am. Entomol. Inst.* **20**: iv 387 pp. [175]. Type data: holotype, CAS *M. adult, from Milly Milly, W.A.

Distribution: NW coastal, W.A.; type locality only. Ecology: larva - sedentary, soil, predator : adult - volant, burrowing.

Bembix wolpa Evans and Matthews, 1973

Bembix wolpa Evans, H.E. & Matthews, R.W. (1973). Systematics and nesting behavior of Australian *Bembix* sand wasps (Hymenoptera, Sphecidae). *Mem. Am. Entomol. Inst.* **20**: iv 387 pp. [311]. Type data: holotype, ANIC F. adult, from 16 mi W of Alice Springs, N.T.

Distribution: Lake Eyre basin, SW coastal, W.A., N.T.; only published localities Beverley and 16 mi W of Alice Springs. Ecology: larva - sedentary, soil, predator : adult - volant, burrowing.

Bembix wowine Evans and Matthews, 1973

Bembix wowine Evans, H.E. & Matthews, R.W. (1973). Systematics and nesting behavior of Australian *Bembix* sand wasps (Hymenoptera, Sphecidae). *Mem. Am. Entomol. Inst.* **20**: iv 387 pp. [81]. Type data: holotype, ANIC M. adult, from Nilemah Station 50 mi S of Denham, W.A.

Distribution: NW coastal, W.A.; type locality only. Ecology: larva - sedentary, soil, predator : adult - volant, burrowing.

Bembix yalta Evans and Matthews, 1973

Bembix yalta Evans, H.E. & Matthews, R.W. (1973). Systematics and nesting behavior of Australian *Bembix* sand wasps (Hymenoptera, Sphecidae). *Mem. Am. Entomol. Inst.* **20**: iv 387 pp. [206]. Type data: holotype, BMNH *M. adult, from Bolgart, W.A.

Distribution: SW coastal, W.A.; type locality only. Ecology: larva - sedentary, soil, predator : adult - volant, burrowing.

Bembix yunkara Evans and Matthews, 1973

Bembix yunkara Evans, H.E. & Matthews, R.W. (1973). Systematics and nesting behavior of Australian *Bembix* sand wasps (Hymenoptera, Sphecidae). *Mem. Am. Entomol. Inst.* **20**: iv 387 pp. [92]. Type data: holotype, ANIC M. adult, from Nilemah Station 50 mi S of Denham, W.A.

Distribution: NW coastal, W.A.; type locality only. Ecology: larva - sedentary, soil, predator : adult - volant, burrowing.

Cerceris Latreille, 1802

Cerceris Latreille, P.A. (1802). *Histoire Naturelle, Générale et Particulière des Crustacés et des Insectes.* Paris : F. Dufart Vol. 3 xii 13+467 pp. [367]. Type species *Philanthus ornatus* Fabricius, 1790 by subsequent designation, see Latreille, P.A. (1810). *Considérations Générales sur l'Ordre Naturel des Animaux Composant les Classes des Crustacès, des Arachnides, et des Insectes;* avec un Tableau Méthodique de leurs Genres, Disposés en Familles. Paris : F. Schoell 444 pp. Compiled from secondary source: Bohart, R.M. & Menke, A.S. (1976). *Sphecid Wasps of the World : a Generic Revision.* Berkeley : Univ. California Press ix 695 pp.

This group is found worldwide, see Bohart, R.M. & Menke, A.S. (1976). *Sphecid Wasps of the World : a Generic Revision.* Berkeley : Univ. California Press ix 695 pp. [575].

Cerceris adae Turner, 1936

Cerceris adae Turner, R.E. (1936). Notes on fossorial Hymenoptera. XLV. On new sphegid wasps from Australia. *Ann. Mag. Nat. Hist. (10)* **18**: 533–545 [535]. Type data: syntypes (probable), ?BMNH *adult, from 10 mi SW Perth, W.A.

Distribution: SW coastal, W.A.; type locality only. Ecology: larva - sedentary, soil, predator : adult - volant, burrowing.

Cerceris alastoroides Turner, 1914

Cerceris alastoroides Turner, R.E. (1914). New fossorial Hymenoptera from Australia and Tasmania. *Proc. Linn. Soc. N.S.W.* **38**: 608–623 [619]. Type data: holotype, NMV *F. adult, from Borroloola, N.T.

Distribution: N Gulf, N.T.; type locality only. Ecology: larva - sedentary, soil, predator : adult - volant, burrowing.

Cerceris anthicivora Evans, 1982

Cerceris anthicivora Evans, H.E. (1982). The genus *Cerceris* in eastern Australia (Hymenoptera : Sphecidae). *Trans. Am. Entomol. Soc.* **107**: 299–380 [369]. Type data: holotype, ANIC F. adult, from Elizabeth River 40 km S Darwin, N.T.

Distribution: N Gulf, N coastal, Qld., N.T. Ecology: larva - sedentary, soil, predator : adult - volant, burrowing; prey adult Coleoptera (Anthicidae). Biological references: Evans, H.E. & Hook, A.W. (1982). Communal nesting in Australian *Cerceris* digger wasps. pp. 159–163 *in* Breed, M.D., Michener, C.D. & Evans, H.E. (eds.) *The Biology Social Insects.* Proceedings of the Ninth Congress of the International Union for the Study of Social Insects, Boulder, Colorado, August 1982. Boulder : Westview Press (biology).

Cerceris antipodes Smith, 1856

Cerceris antipodes Smith, F. (1856). *Catalogue of Hymenopterous Insects in the Collection of the British Museum.* Part IV, Sphegidae, Larridae and Crabronidae. pp. 207–497 London : British Museum [451]. Type data: holotype, BMNH *F. adult (seen 1929 by L.F. Graham), from Australia.

Cerceris saeva Smith, F. (1873). Descriptions of new species of fossorial Hymenoptera in the collection of the British Museum. *Ann. Mag. Nat. Hist. (4)* **12**: 291–300, 402–415 [414]. Type data: holotype, BMNH *F. adult (seen 1929 by L.F. Graham), from Lower Plenty, Vic.

Cerceris brisbanensis Cockerell, T.D.A. (1930). Wasps of the genus *Cerceris* in the Queensland Museum. *Mem. Qd. Mus.* **10**: 32–36 [34]. Type data: syntypes, QM *F. adult, whereabouts of others unknown, from Darra, Qld.

Cerceris goodwini Cockerell, T.D.A. (1930). Wasps of the genus *Cerceris* in the Queensland Museum. *Mem. Qd. Mus.* **10**: 32–36 [35]. Type data: holotype, QM T4017 *F. adult, from Stanthorpe, Qld.

Cerceris ziegleri Rayment, T. (1947). New bees and wasps - Part IV. A new *Cerceris* wasp, and some small chrysomelid beetles. *Vict. Nat.* **63**: 256–260 [256]. Type data: holotype, ANIC M. adult, from Emerald, Vic.

Cerceris sculleni Tsuneki, K. (1968). On some *Cerceris* from Australia, with a tentative key to the Australian species (Hymenoptera, Sphecidae). *Etizenia* **28**: 1–32 [15]. Type data: holotype, AMNH *F. adult, from N.S.W.

Taxonomic decision of Evans, H.E. (1982). The genus *Cerceris* in eastern Australia (Hymenoptera : Sphecidae). *Trans. Am. Entomol. Soc.* **107**: 299–380 [356].

Distribution: NE coastal, Murray-Darling basin, SE coastal, S Gulfs, Lake Eyre basin, Qld., N.S.W., A.C.T., Vic., S.A. Ecology: larva - sedentary, soil, predator : adult - volant, burrowing; prey adult Coleoptera (Chrysomelidae). Biological references: Turner, R.E. (1912). A revision of the Australian species of the genus *Cerceris*. (Hymenoptera.). *Proc. Linn. Soc. N.S.W.* **36**: 644–678 (redescription); Evans, H.E. &

Matthews, R.W. (1970). Notes on the nests and prey of Australian wasps of the genus *Cerceris* (Hymenoptera : Sphecidae). *J. Aust. Entomol. Soc.* **9**: 153–156 (nest, prey); Alcock, J. (1980). Communal nesting in an Australian solitary wasp, *Cerceris antipodes* Smith (Hymenoptera, Sphecidae). *J. Aust. Entomol. Soc.* **19**: 223–228 (nest).

Cerceris armigera Turner, 1917

Cerceris armigera Turner, R.E. (1917). Notes on fossorial Hymenoptera. XXV. On new Sphecoidea in the British Museum. *Ann. Mag. Nat. Hist. (8)* **19**: 104–113 [104]. Type data: holotype, BMNH *F. adult (seen 1929 by L.F. Graham), from Darra, Qld.

Cerceris armigera rufofusca Turner, R.E. (1936). Notes on fossorial Hymenoptera. XLV. On new sphegid wasps from Australia. *Ann. Mag. Nat. Hist. (10)* **18**: 533–545 [538]. Type data: holotype, BMNH *F. adult, from Mundaring Weir, W.A.

Taxonomic decision of Evans, H.E. (1982). The genus *Cerceris* in eastern Australia (Hymenoptera : Sphecidae). *Trans. Am. Entomol. Soc.* **107**: 299–380 [360].

Distribution: NE coastal, Murray-Darling basin, Lake Eyre basin, S Gulfs, SW coastal, Qld., N.S.W., S.A., W.A. Ecology: larva - sedentary, soil, predator : adult - volant, burrowing. Biological references: Evans, H.E. & Hook, A.W. (1982). Communal nesting in Australian *Cerceris* Digger Wasps. pp. 159–163 *in* Breed, M.D., Michener, C.D. & Evans, H.E. (eds.) (1982). *The Biology of Social Insects.* Proceedings of the Ninth Congress of the International Union for the Study of Social Insects, Boulder, Colorado, August 1982. Boulder : Westview Press (biology).

Cerceris aurantiaca Smith, 1873

Cerceris aurantiaca Smith, F. (1873). Descriptions of new species of fossorial Hymenoptera in the collection of the British Museum. *Ann. Mag. Nat. Hist. (4)* **12**: 291–300, 402–415 [414]. Type data: holotype, BMNH *F. adult (seen 1929 by L.F. Graham), from S.A.

Distribution: Murray-Darling basin, Lake Eyre basin, S Gulfs, (N.S.W.), Qld., Vic., S.A. Ecology: larva - sedentary, soil, predator : adult - volant, burrowing. Biological references: Turner, R.E. (1912). A revision of the Australian species of the genus *Cerceris*. (Hymenoptera.). *Proc. Linn. Soc. N.S.W.* **36**: 644–678 (redescription); Tsuneki, K. (1968). On some *Cerceris* from Australia, with a tentative key to the Australian species (Hymenoptera, Sphecidae). *Etizenia* **28**: 1–32 (redescription); Evans, H.E. (1982). The genus *Cerceris* in eastern Australia (Hymenoptera : Sphecidae). *Trans. Am. Entomol. Soc.* **107**: 299–380 (redescription, distribution).

Cerceris australis Saussure, 1855

Cerceris australis Saussure, H. de (1855). Mélanges hyménoptérologiques. *Mém. Soc. Phys. Hist. Nat. Genève* **14**: 1–67 [6]. Type data: syntypes (probable), whereabouts unknown, from Tas.

Cerceris nigrocincta Smith, F. (1856). *Catalogue of Hymenopterous Insects in the Collection of the British Museum. Part IV, Sphegidae, Larridae and Crabronidae.* pp. 207–497 London : British Museum [450]. Type data: syntypes (probable), BMNH *F. adult (seen 1929 by L.F. Graham), from Adelaide, S.A.

[*Cerceris sculleni*] Tsuneki, K. (1968). On some *Cerceris* from Australia, with a tentative key to the Australian species (Hymenoptera, Sphecidae). *Etizenia* **28**: 1–32 [15] [male paratype not *Cerceris antipodes* Smith, 1856 *vide* Evans, H.E. (1982). *op. cit.*]. from N.S.W.

Taxonomic decision of Evans, H.E. (1982). The genus *Cerceris* in eastern Australia (Hymenoptera : Sphecidae). *Trans. Am. Entomol. Soc.* **107**: 299–380 [348].

Distribution: NE coastal, SE coastal, Murray-Darling basin, S Gulfs, (Tas.), Qld., N.S.W., A.C.T., Vic., S.A. Ecology: larva - sedentary, soil, predator : adult - volant, burrowing; prey adult Coleoptera (Scarabaeidae). Biological references: Tsuneki, K. (1968). On some *Cerceris* from Australia, with a tentative key to the Australian species (Hymenoptera, Sphecidae). *Etizenia* **28**: 1–32 (redescription); Evans, H.E. & Matthews, R.W. (1970). Notes on the nests and prey of Australian wasps of the genus *Cerceris* (Hymenoptera : Sphecidae). *J. Aust. Entomol. Soc.* **9**: 153–156 (nest, prey); Evans, H.E. & Hook, A.W. (1982). Communal nesting in the digger wasp *Cerceris australis* (Hymenoptera : Sphecidae). *Aust. J. Zool.* **30**: 557–568 (biology).

Cerceris balteata Evans, 1982

Cerceris balteata Evans, H.E. (1982). The genus *Cerceris* in eastern Australia (Hymenoptera : Sphecidae). *Trans. Am. Entomol. Soc.* **107**: 299–380 [314]. Type data: holotype, ANIC F. adult, from Bendora, A.C.T.

Distribution: SE coastal, Murray-Darling basin, N.S.W., A.C.T., Vic. Ecology: larva - sedentary, soil, predator : adult - volant, burrowing.

Cerceris calida Turner, 1915

Cerceris calida Turner, R.E. (1915). Notes on fossorial Hymenoptera. XV. New Australian Crabronidae. *Ann. Mag. Nat. Hist. (8)* **15**: 62–96 [66]. Type data: holotype, BMNH *F. adult (seen 1929 by L.F. Graham), from Kuranda, Qld.

Distribution: NE coastal, N coastal, Qld., N.T. Ecology: larva - sedentary, soil, predator : adult - volant, burrowing. Biological references: Evans, H.E. (1982). The genus *Cerceris* in eastern Australia (Hymenoptera : Sphecidae). *Trans. Am. Entomol. Soc.* **107**: 299–380 (redescription, distribution).

Cerceris cucullata Bingham, 1912

Cerceris cucullata Bingham, C.T. (1912). South African and Australian Aculeate Hymenoptera in the Oxford Museum. *Trans. Entomol. Soc. Lond.* **1912**: 375–383 [375]. Type data: holotype, OUM *M. adult, from Port Essington, N.T.

Distribution: N coastal, N.T.; type locality only. Ecology: larva - sedentary, soil, predator : adult - volant, burrowing.

Cerceris cunnamulla Evans, 1982

Cerceris cunnamulla Evans, H.E. (1982). The genus *Cerceris* in eastern Australia (Hymenoptera : Sphecidae). *Trans. Am. Entomol. Soc.* **107**: 299–380 [366]. Type data: holotype, ANIC F. adult, from 30 km S Cunnamulla, Qld.

Distribution: Lake Eyre basin, Murray-Darling basin, N coastal, S Gulfs, Qld., N.S.W., S.A., W.A., N.T. Ecology: larva - sedentary, soil, predator : adult - volant, burrowing.

Cerceris darrensis Cockerell, 1930

Cerceris darrensis Cockerell, T.D.A. (1930). Wasps of the genus *Cerceris* in the Queensland Museum. *Mem. Qd. Mus.* **10**: 32–36 [34]. Type data: holotype, QM T4016 *F. adult, from Darra, Qld.

Distribution: NE coastal, (Murray-Darling basin), (N.S.W.), Qld.; only known localities Brisbane and (near Cobar and Nyngan). Ecology: larva - sedentary, soil, predator : adult - volant, burrowing. Biological references: Evans, H.E. (1982). The genus *Cerceris* in eastern Australia (Hymenoptera : Sphecidae). *Trans. Am. Entomol. Soc.* **107**: 299–380 (redescription, distribution).

Cerceris dedariensis Turner, 1936

Cerceris dedariensis Turner, R.E. (1936). Notes on fossorial Hymenoptera. XLV. On new sphegid wasps from Australia. *Ann. Mag. Nat. Hist. (10)* **18**: 533–545 [534]. Type data: holotype, BMNH *F. adult, from Dedari, W.A.

Distribution: W plateau; type locality only. Ecology: larva - sedentary, soil, predator : adult - volant, burrowing.

Cerceris euchroma Turner, 1910

Cerceris euchroma Turner, R.E. (1910). New fossorial Hymenoptera from Australia. *Trans. Entomol. Soc. Lond.* **1910**: 407–429 pl 50 [424 pl 50 fig 11]. Type data: holotype, BMNH *F. adult (seen 1929 by L.F. Graham), from Cairns, Qld.

Distribution: NE coastal, (Murray-Darling basin), (Lake Eyre basin), Qld., N.S.W., Vic., S.A. Ecology: larva - sedentary, soil, predator : adult - volant, burrowing. Biological references: Turner, R.E. (1912). A revision of the Australian species of the genus *Cerceris*. (Hymenoptera.). *Proc. Linn. Soc. N.S.W.* **36**: 644–678 (redescription); Evans, H.E. (1982). The genus *Cerceris* in eastern Australia (Hymenoptera : Sphecidae). *Trans. Am. Entomol. Soc.* **107**: 299–380 (redescription, distribution).

Cerceris eungella Evans, 1982

Cerceris eungella Evans, H.E. (1982). The genus *Cerceris* in eastern Australia (Hymenoptera : Sphecidae). *Trans. Am. Entomol. Soc.* **107**: 299–380 [353]. Type data: holotype, QM T8489 *F. adult, from Eungella Natl. Park, Qld.

Distribution: NE coastal, Murray-Darling basin, Lake Eyre basin, Qld., N.S.W., N.T. Ecology: larva - sedentary, soil, predator : adult - volant, burrowing.

Cerceris exleyae Evans, 1982

Cerceris exleyae Evans, H.E. (1982). The genus *Cerceris* in eastern Australia (Hymenoptera : Sphecidae). *Trans. Am. Entomol. Soc.* **107**: 299–380 [376]. Type data: holotype, QM T8494 *F. adult, from Longreach, Qld.

Distribution: Lake Eyre basin, Murray-Darling basin, Qld., N.S.W.; only published localities Longreach and Yarrawin. Ecology: larva - sedentary, soil, predator : adult - volant, burrowing.

Cerceris fluvialis Smith, 1873

Cerceris fluvialis Smith, F. (1873). Descriptions of new species of fossorial Hymenoptera in the collection of the British Museum. *Ann. Mag. Nat. Hist. (4)* **12**: 291–300, 402–415 [412]. Type data: holotype, BMNH *F. adult (seen 1929 by L.F. Graham), from Swan River, W.A.

Distribution: Murray-Darling basin, Lake Eyre basin, SW coastal, Qld., N.S.W., S.A., W.A. Ecology: larva - sedentary, soil, predator : adult - volant, burrowing. Biological references: Turner, R.E. (1912). A revision of the Australian species of the genus *Cerceris*. (Hymenoptera.). *Proc. Linn. Soc. N.S.W.* **36**: 644–678 (redescription); Evans, H.E. (1982). The genus *Cerceris* in eastern Australia (Hymenoptera : Sphecidae). *Trans. Am. Entomol. Soc.* **107**: 299–380.

Cerceris forficata Evans, 1982

Cerceris forficata Evans, H.E. (1982). The genus *Cerceris* in eastern Australia (Hymenoptera : Sphecidae). *Trans. Am. Entomol. Soc.* **107**: 299–380 [328]. Type data: holotype, SAMA *F. adult, from New Kalamurina Station, Warburton River, S.A.

Distribution: Lake Eyre basin, S.A.; type locality only. Ecology: larva - sedentary, soil, predator : adult - volant, burrowing.

Cerceris froggatti Turner, 1912

Cerceris froggatti Turner, R.E. (1912). A revision of the Australian species of the genus *Cerceris*. (Hymenoptera).

Proc. Linn. Soc. N.S.W. **36**: 644-678 [668]. Type data: holotype, BMNH *F. adult (seen 1929 by L.F. Graham), from Rockhampton, Qld.

Distribution: NE coastal, Murray-Darling basin, Qld., (N.S.W.). Ecology: larva - sedentary, soil, predator : adult - volant, burrowing. Biological references: Tsuneki, K. (1968). On some *Cerceris* from Australia, with a tentative key to the Australian species (Hymenoptera, Sphecidae). *Etizenia* **28**: 1-32 (description of male possibly belonging to this species); Evans, H.E. (1982). The genus *Cerceris* in eastern Australia (Hymenoptera : Sphecidae). *Trans. Am. Entomol. Soc.* **107**: 299-380 (redescription, distribution); Evans, H.E. & Hook, A.W. (1982). Communal nesting in Australian *Cerceris* Digger Wasps. pp. 159-163 *in* Breed, M.D., Michener, C.D. & Evans, H.E. (eds.) (1982). *The Biology of Social Insects*. Proceedings of the Ninth Congress of the International Union for the Study of Social Insects, Boulder, Colorado, August 1982. Boulder, : Westview Press (biology).

Cerceris gilberti Turner, 1916

Cerceris gilberti Turner, R.E. (1916). Notes on fossorial Hymenoptera. XXIII. *Ann. Mag. Nat. Hist. (8)* **18**: 277-288 [277]. Type data: holotype, BMNH *F. adult (seen 1929 by L.F. Graham), from Mackay, Qld.

Distribution: NE coastal, SE coastal, Murray Darling basin, S Gulfs, N coastal, Qld., N.S.W, Vic., S.A., W.A.; type locality only. Ecology: larva - sedentary, soil, predator : adult - volant, burrowing. Biological references: Evans, H.E. (1982). The genus *Cerceris* in eastern Australia (Hymenoptera : Sphecidae). *Trans. Am. Entomol. Soc.* **107**: 299-380.

Cerceris gilesi Turner, 1910

Cerceris gilesi Turner, R.E. (1910). Additions to our knowledge of the fossorial wasps of Australia. *Proc. Zool. Soc. Lond.* **1910**: 253-356 [346 pl 32 fig 13]. Type data: holotype, BMNH *F. adult (seen 1929 by L.F. Graham), from Claremont, W.A.

Distribution: SW coastal, W.A.; only published localities Claremont and Yallingup. Ecology: larva - sedentary, soil, predator : adult - volant, burrowing. Biological references: Turner, R.E. (1912). A revision of the Australian species of the genus *Cerceris*. (Hymenoptera.). *Proc. Linn. Soc. N.S.W.* **36**: 644-678 (redescription).

Cerceris goddardi Cockerell, 1930

Cerceris goddardi Cockerell, T.D.A. (1930). Wasps of the genus *Cerceris* in the Queensland Museum. *Mem. Qd. Mus.* **10**: 32-36 [33]. Type data: holotype, QM T4015 *F. adult, from Dunk Is., Qld.

Cerceris venusta Auctorum [*non Cerceris venusta* Smith, 1873]; for explanation see Evans, H.E. (1982). The genus *Cerceris* in eastern Australia (Hymenoptera : Sphecidae). *Trans. Am. Entomol. Soc.* **107**: 299-380 (351)].

Cerceris insulicola Tsuneki, K. (1968). On some *Cerceris* from Australia, with a tentative key to the Australian species (Hymenoptera, Sphecidae). *Etizenia* **28**: 1-32 [27]. Type data: syntypes (probable), USNM *M. adult, from Prince of Wales Is., Qld.

Taxonomic decision of Evans, H.E. (1982). The genus *Cerceris* in eastern Australia (Hymenoptera : Sphecidae). *Trans. Am. Entomol. Soc.* **107**: 299-380 [351].

Distribution: NE coastal, SE coastal, N coastal, Qld., N.S.W., N.T. Ecology: larva - sedentary, soil, predator : adult - volant, burrowing. Biological references: Evans, H.E. & Hook, A.W. (1982). Communal nesting in Australian *Cerceris* Digger Wasps. pp. 159-163 *in* Breed, M.D., Michener, C.D. & Evans, H.E. (eds.) (1982). *The Biology of Social Insects*. Proceedings of the Ninth Congress of the International Union for the Study of Social Insects, Boulder, Colorado, August 1982. Boulder : Westview Press (biology).

Cerceris hackeriana Cockerell, 1930

Cerceris hackeriana Cockerell, T.D.A. (1930). Wasps of the genus *Cerceris* in the Queensland Museum. *Mem. Qd. Mus.* **10**: 32-36 [32]. Type data: holotype, QM T4014 *F. adult, from Tooloom, N.S.W.

Distribution: SE coastal, Qld., N.S.W.; only published localities Mt. Tamborine, Kangaroo Valley, Upper Allyn River, Tooloom and Ebor. Ecology: larva - sedentary, soil, predator : adult - volant, burrowing. Biological references: Evans, H.E. (1982). The genus *Cerceris* in eastern Australia (Hymenoptera : Sphecidae). *Trans. Am. Entomol. Soc.* **107**: 299-380 (redescription, distribution).

Cerceris inexpectata Turner, 1908

Cerceris inexpectata Turner, R.E. (1908). Notes on the Australian fossorial wasps of the family Sphegidae, with descriptions of new species. *Proc. Zool. Soc. Lond.* **1908**: 457-535 [469 pl xxvi fig 5]. Type data: holotype, BMNH *F. adult (seen 1929 by L.F. Graham), from Mackay, Qld.

Distribution: NE coastal, Qld.; only published localities Mackay, Townsville, Mareeba and Cairns. Ecology: larva - sedentary, soil, predator : adult - volant, burrowing. Biological references: Turner, R.E. (1912). A revision of the Australian species of the genus *Cerceris*. (Hymenoptera.). *Proc. Linn. Soc. N.S.W.* **36**: 644-678 (description of male); Evans, H.E. (1982). The genus *Cerceris* in eastern Australia (Hymenoptera : Sphecidae). *Trans. Am. Entomol. Soc.* **107**: 299-380.

Cerceris iridis Evans, 1982

Cerceris iridis Evans, H.E. (1982). The genus *Cerceris* in eastern Australia (Hymenoptera : Sphecidae). *Trans. Am. Entomol. Soc.* **107**: 299-380 [339]. Type data: holotype, ANIC *F. adult, from Dividing Range, 15 km W of Captain Billy Creek, Cape York, Qld.

Distribution: NE coastal, Qld.; only published localities on Cape York. Ecology: larva - sedentary, soil, predator : adult - volant, burrowing.

Cerceris labeculata Turner, 1908

Cerceris labeculata Turner, R.E. (1908). Notes on the Australian fossorial wasps of the family Sphegidae, with descriptions of new species. *Proc. Zool. Soc. Lond.* **1908**: 457-535 pl xxvi [470]. Type data: holotype, BMNH *F. adult (seen 1929 by L.F. Graham), from Cairns, Qld.

Distribution: NE coastal, SE coastal, Qld., N.S.W. Ecology: larva - sedentary, soil, predator : adult - volant, burrowing. Biological references: Evans, H.E. (1982). The genus *Cerceris* in eastern Australia (Hymenoptera : Sphecidae). *Trans. Am. Entomol. Soc.* **107**: 299-380 (redescription, distribution).

Cerceris latiberbis Tsuneki, 1968

Cerceris latiberbis Tsuneki, K. (1968). On some *Cerceris* from Australia, with a tentative key to the Australian species (Hymenoptera, Sphecidae). *Etizenia* **28**: 1-32 [19]. Type data: holotype, USNM *M. adult, from Prince of Wales Is., Qld.

Distribution: NE coastal, Qld.; type locality only. Ecology: larva - sedentary, soil, predator : adult - volant, burrowing. Biological references: Evans, H.E. (1982). The genus *Cerceris* in eastern Australia (Hymenoptera : Sphecidae). *Trans. Am. Entomol. Soc.* **107**: 299-380 (redescription).

Cerceris listrognatha Evans, 1982

Cerceris listrognatha Evans, H.E. (1982). The genus *Cerceris* in eastern Australia (Hymenoptera : Sphecidae). *Trans. Am. Entomol. Soc.* **107**: 299-380 [333]. Type data: holotype, NMV *F. adult, from Claudie River (Cape York), Qld.

Distribution: NE coastal, Murray-Darling basin, Qld., N.S.W. Ecology: larva - sedentary, soil, predator : adult - volant, burrowing.

Cerceris luculenta Evans, 1982

Cerceris luculenta Evans, H.E. (1982). The genus *Cerceris* in eastern Australia (Hymenoptera : Sphecidae). *Trans. Am. Entomol. Soc.* **107**: 299-380 [363]. Type data: holotype, ANIC F. adult, from 27 N Bourke, N.S.W.

Distribution: Murray-Darling basin, N.S.W.; type locality only. Ecology: larva - sedentary, soil, predator : adult - volant, burrowing.

Cerceris megacantha Evans, 1982

Cerceris megacantha Evans, H.E. (1982). The genus *Cerceris* in eastern Australia (Hymenoptera : Sphecidae). *Trans. Am. Entomol. Soc.* **107**: 299-380 [317]. Type data: holotype, QM T8493 *F. adult, from Blunder Creek, Brisbane, Qld.

Distribution: NE coastal, Qld. Ecology: larva - sedentary, soil, predator : adult - volant, burrowing. Biological references: Evans, H.E. & Hook, A.W. (1982). Communal nesting in Australian *Cerceris* digger wasps. pp 159-163 *in* Breed, M.D., Michener, C.D. & Evans, H.E. (eds.) (1982). *The Biology of Social Insects*. Proceedings of the Ninth Congress of the International Union for the Study of Social Insects, Boulder, Colorado, August 1982. Boulder : Westview Press (biology).

Cerceris merredinensis Turner, 1936

Cerceris merredinensis Turner, R.E. (1936). Notes on fossorial Hymenoptera. XLV. On new sphegid wasps from Australia. *Ann. Mag. Nat. Hist. (10)* **18**: 533-545 [536]. Type data: holotype, BMNH *F. adult, from Merredin, W.A.

Distribution: SW coastal, W.A.; type locality only. Ecology: larva - sedentary, soil, predator : adult - volant, burrowing.

Cerceris minuscula Turner, 1910

Cerceris minuscula Turner, R.E. (1910). Additions to our knowledge of the fossorial wasps of Australia. *Proc. Zool. Soc. Lond.* **1910**: 253-356 [347]. Type data: syntypes (probable), BMNH *F,M adult, from Mackay, Qld.,..

Distribution: NE coastal, Murray-Darling basin, SE coastal, Lake Eyre basin, SW coastal, N coastal, Qld., N.S.W., A.C.T., Vic., S.A., W.A., N.T.; (subspecies described from Melanesia). Ecology: larva - sedentary, soil, predator : adult - volant, burrowing. Biological references: Tsuneki, K. (1968). On some *Cerceris* from Australia, with a tentative key to the Australian species (Hymenoptera, Sphecidae). *Etizenia* **28**: 1-32 (redescription); Krombein, K.V. (1969). A revision of the Melanesian wasps of the genus *Cerceris* Latr. (Hymenoptera : Sphecidae). *Smithson. Contrib. Zool.* 22: 1-36 (Melanesian subspecies); Evans, H.E. (1982). The genus *Cerceris* in eastern Australia (Hymenoptera : Sphecidae). *Trans. Am. Entomol. Soc.* **107**: 299-380 (redescription, distribution); Evans, H.E. & Hook, A.W. (1982). Communal nesting in Australian *Cerceris* Digger Wasps. pp. 159-163 *in* Breed, M.D., Michener, C.D. & Evans, H.E. (eds.) (1982). *The Biology of Social Insects*. Proceedings of the Ninth Congress of the International Union for the Study of Social Insects, Boulder, Colorado, August 1982. Boulder : Westview Press (biology).

Cerceris multiguttata Turner, 1908

Cerceris multiguttata Turner, R.E. (1908). Notes on the Australian fossorial wasps of the family Sphegidae, with descriptions of new species. *Proc. Zool. Soc. Lond.* **1908**: 457–535 pl xxvi [471]. Type data: holotype, BMNH *F. adult (seen 1929 by L.F. Graham), from Mackay to Cairns, Qld.

Distribution: NE coastal, Qld. Ecology: larva - sedentary, soil, predator : adult - volant, burrowing. Biological references: Turner, R.E. (1912). A revision of the Australian species of the genus *Cerceris*. (Hymenoptera.). *Proc. Linn. Soc. N.S.W.* **36**: 644–678 (description of male); Evans, H.E. (1982). The genus *Cerceris* in eastern Australia (Hymenoptera : Sphecidae). *Trans. Am. Entomol. Soc.* **107**: 299–380 (redescription, distribution).

Cerceris opposita Smith, 1873

Cerceris opposita Smith, F. (1873). Descriptions of new species of fossorial Hymenoptera in the collection of the British Museum. *Ann. Mag. Nat. Hist.* (4) **12**: 291–300, 402–415 [413]. Type data: holotype, BMNH *F. adult (seen 1929 by L.F. Graham), from Lower Plenty, Vic.

Distribution: NE coastal, Murray-Darling basin, SE coastal, Qld., N.S.W., A.C.T., Vic. Ecology: larva - sedentary, soil, predator : adult - volant, burrowing; prey adult Coleoptera (Chrysomelidae). Biological references: Turner, R.E. (1912). A revision of the Australian species of the genus *Cerceris*. (Hymenoptera.). *Proc. Linn. Soc. N.S.W.* **36**: 644–678 (description of male); Evans, H.E. & Matthews, R.W. (1970). Notes on the nests and prey of Australian wasps of the genus *Cerceris* (Hymenoptera : Sphecidae). *J. Aust. Entomol. Soc.* **9**: 153–156 (nest, prey); Evans, H.E. (1982). The genus *Cerceris* in eastern Australia (Hymenoptera : Sphecidae). *Trans. Am. Entomol. Soc.* **107**: 299–380 (redescription, distribution).

Cerceris osculata Evans, 1982

Cerceris osculata Evans, H.E. (1982). The genus *Cerceris* in eastern Australia (Hymenoptera : Sphecidae). *Trans. Am. Entomol. Soc.* **107**: 299–380 [361]. Type data: holotype, ANIC F. adult, from 35 mi S Nappamerry, Qld.

Distribution: Lake Eyre basin, Murray-Darling basin, Qld.; only published localities near Nappamerry and Cunnamulla. Ecology: larva - sedentary, soil, predator : adult - volant, burrowing.

Cerceris perkinsi Turner, 1910

Cerceris perkinsi Turner, R.E. (1910). New fossorial Hymenoptera from Australia. *Trans. Entomol. Soc. Lond.* **1910**: 407–429 [423 pl 50 fig 12]. Type data: holotype, BMNH *F. adult (seen 1929 by L.F. Graham), from Cairns, Qld.

Distribution: NE coastal, Murray-Darling basin, Lake Eyre basin, N coastal, Qld., N.S.W., S.A., N.T. Ecology: larva - sedentary, soil, predator : adult - volant, burrowing. Biological references: Turner, R.E. (1912). A revision of the Australian species of the genus *Cerceris*. (Hymenoptera.). *Proc. Linn. Soc. N.S.W.* **36**: 644–678 (redescription); Evans, H.E. (1982). The genus *Cerceris* in eastern Australia (Hymenoptera : Sphecidae). *Trans. Am. Entomol. Soc.* **107**: 299–380 (redescription, distribution).

Cerceris praedura Turner, 1908

Cerceris praedura Turner, R.E. (1908). Notes on the Australian fossorial wasps of the family Sphegidae, with descriptions of new species. *Proc. Zool. Soc. Lond.* **1908**: 457–535 pl xxvi [472]. Type data: holotype, BMNH *F. adult (seen 1929 by L.F. Graham), from Mackay , Qld.

Cerceris koala Tsuneki, K. (1968). On some *Cerceris* from Australia, with a tentative key to the Australian species (Hymenoptera, Sphecidae). *Etizenia* **28**: 1–32 [17]. Type data: holotype, USNM *F. adult, from Prince of Wales Is., Qld.

Taxonomic decision of Evans, H.E. (1982). The genus *Cerceris* in eastern Australia (Hymenoptera : Sphecidae). *Trans. Am. Entomol. Soc.* **107**: 299–380 [372].

Distribution: NE coastal, Qld. Ecology: larva - sedentary, soil, predator : adult - volant, burrowing. Biological references: Turner, R.E. (1912). A revision of the Australian species of the genus *Cerceris*. (Hymenoptera.). *Proc. Linn. Soc. N.S.W.* **36**: 644–678 (redescription).

Cerceris raymenti Turner, 1936

Cerceris raymenti Turner, R.E. (1936). Notes on fossorial Hymenoptera. XLV. On new sphegid wasps from Australia. *Ann. Mag. Nat. Hist.* (10) **18**: 533–545 [537]. Type data: holotype, BMNH *F. adult, from Gunbower Is., Murray River, Vic.

Distribution: Murray-Darling basin, S Gulfs, Qld., N.S.W., Vic., S.A. Ecology: larva - sedentary, soil, predator : adult - volant, burrowing. Biological references: Evans, H.E. (1982). The genus *Cerceris* in eastern Australia (Hymenoptera : Sphecidae). *Trans. Am. Entomol. Soc.* **107**: 299–380 (redescription, distribution).

Cerceris ropalidioides Evans, 1982

Cerceris ropalidioides Evans, H.E. (1982). The genus *Cerceris* in eastern Australia (Hymenoptera : Sphecidae). *Trans. Am. Entomol. Soc.* **107**: 299–380 [336]. Type data: holotype, ANIC F. adult, from Mt. Molloy, Qld.

Distribution: NE coastal, Qld.; only published localities Mt. Molloy and Mareeba. Ecology: larva - sedentary, soil, predator : adult - volant, burrowing.

Cerceris sedula Evans, 1982

Cerceris sedula Evans, H.E. (1982). The genus *Cerceris* in eastern Australia (Hymenoptera : Sphecidae). *Trans. Am. Entomol. Soc.* **107**: 299–380 [320]. Type data: holotype, ANIC F. adult, from Wyperfeld Natl. Park, Vic.

Distribution: Murray-Darling basin, Vic., S.A.; only published localities Wyperfeld National Park, Yaapeet, near Ouyen and near Swan Reach. Ecology: larva - sedentary, soil, predator : adult - volant, burrowing.

Cerceris siccata Evans, 1982

Cerceris siccata Evans, H.E. (1982). The genus *Cerceris* in eastern Australia (Hymenoptera : Sphecidae). *Trans. Am. Entomol. Soc.* **107**: 299–380 [355]. Type data: holotype, ANIC F. adult, from Packsaddle, 179 km N of Broken Hill, N.S.W.

Distribution: Murray-Darling basin, Lake Eyre basin, N.S.W., Vic., N.T. Ecology: larva - sedentary, soil, predator : adult - volant, burrowing.

Cerceris spilota Evans, 1982

Cerceris spilota Evans, H.E. (1982). The genus *Cerceris* in eastern Australia (Hymenoptera : Sphecidae). *Trans. Am. Entomol. Soc.* **107**: 299–380 [374]. Type data: holotype, ANIC F. adult, from Clifton Downs, N.S.W.

Distribution: Murray-Darling basin, Lake Eyre basin, N.S.W., S.A. Ecology: larva - sedentary, soil, predator : adult - volant, burrowing.

Cerceris spinipleuris Turner, 1918

Cerceris varipes Smith, F. (1873). Descriptions of new species of fossorial Hymenoptera in the collection of the British Museum. *Ann. Mag. Nat. Hist. (4)* **12**: 291–300, 402–415 [413] [*non Cerceris varipes* Smith, 1858]. Type data: holotype, BMNH *F. adult (seen 1929 by L.F. Graham), from Adelaide, S.A.

Cerceris spinipleuris Turner, R.E. (1918). Notes on fossorial Hymenoptera. XXXII. On new species in the British Museum. *Ann. Mag. Nat. Hist. (9)* **1**: 86–96 [91] [*nom. nov.* for *Cerceris varipes* Smith 1873].

Distribution: S Gulfs, S.A.; type locality only. Ecology: larva - sedentary, soil, predator : adult - volant, burrowing. Biological references: Turner, R.E. (1912). A revision of the Australian species of the genus *Cerceris*. (Hymenoptera.). *Proc. Linn. Soc. N.S.W.* **36**: 644–678 (redescription); Evans, H.E. (1982). The genus *Cerceris* in eastern Australia (Hymenoptera : Sphecidae). *Trans. Am. Entomol. Soc.* **107**: 299–380 (redescription).

Cerceris trifida Evans, 1982

Cerceris trifida Evans, H.E. (1982). The genus *Cerceris* in eastern Australia (Hymenoptera : Sphecidae). *Trans. Am. Entomol. Soc.* **107**: 299–380 [375]. Type data: holotype, ANIC F. adult, from 30 (mi) S Cunnamulla, Qld.

Distribution: Murray-Darling basin, Lake Eyre basin, Qld., N.S.W., Vic., S.A., N.T. Ecology: larva - sedentary, soil, predator : adult - volant, burrowing.

Cerceris unispinosa Turner, 1917

Cerceris unispinosa Turner, R.E. (1917). Notes on fossorial Hymenoptera. XXV. On new Sphecoidea in the British Museum. *Ann. Mag. Nat. Hist. (8)* **19**: 104–113 [105]. Type data: holotype, BMNH *F. adult (seen 1929 by L.F. Graham), from Darra, Qld.

Distribution: NE coastal, Qld.; only published localities Darra, Brisbane and Tamborine. Ecology: larva - sedentary, soil, predator : adult - volant, burrowing. Biological references: Evans, H.E. (1982). The genus *Cerceris* in eastern Australia (Hymenoptera : Sphecidae). *Trans. Am. Entomol. Soc.* **107**: 299–380 (redescription, distribution); Evans, H.E. & Hook, A.W. (1982). Communal nesting in Australian *Cerceris* Digger Wasps. pp. 159–163 *in* Breed, M.D., Michener, C.D. & Evans, H.E. (eds.) (1982). *The Biology of Social Insects*. Proceedings of the Ninth Congress of the International Union for the Study of Social Insects, Boulder, Colorado, August 1982. Boulder : Westview Press (biology).

Cerceris venusta Smith, 1873

Cerceris venusta Smith, F. (1873). Descriptions of new species of fossorial Hymenoptera in the collection of the British Museum. *Ann. Mag. Nat. Hist. (4)* **12**: 291–300, 402–415 [413]. Type data: holotype, BMNH *F. adult (seen 1929 by L.F. Graham), from Qld.

Distribution: NE coastal, N Gulf, Qld., N.S.W., N.T. Ecology: larva - sedentary, soil, predator : adult - volant, burrowing; prey adult Coleoptera (Scarabaeidae). Biological references: Turner, R.E. (1912). A revision of the Australian species of the genus *Cerceris*. (Hymenoptera.). *Proc. Linn. Soc. N.S.W.* **36**: 644–678 (description of male); Krombein, K.V. (1969). A revision of the Melanesian wasps of the genus *Cerceris* Latr. (Hymenoptera : Sphecidae). *Smithson. Contrib. Zool.* **22**: 1–36 (Melanesian subspecies); Krombein, K.V. (1973). Systematic and distributional notes on Melanesian *Cerceris* (Hymenoptera : Sphecidae). *Proc. Entomol. Soc. Wash.* **75**: 464–467 (prey); Evans, H.E. (1982). The genus *Cerceris* in eastern Australia (Hymenoptera : Sphecidae). *Trans. Am. Entomol. Soc.* **107**: 299–380.

Cerceris victrix Turner, 1910

Cerceris victrix Turner, R.E. (1910). New fossorial Hymenoptera from Australia. *Trans. Entomol. Soc. Lond.*

1910: 407–429 pl 50 [422 pl 50 fig 10]. Type data: holotype, BMNH *F. adult (seen 1929 by L.F. Graham), from Moreton Bay, Qld.

Distribution: NE coastal, Qld.; only published localities Moreton Bay and Cairns. Ecology: larva - sedentary, soil, predator : adult - volant, burrowing. Biological references: Evans, H.E. (1982). The genus *Cerceris* in eastern Australia (Hymenoptera : Sphecidae). *Trans. Am. Entomol. Soc.* **107**: 229–380 (redescription).

Cerceris windorum Tsuneki, 1968

Cerceris windorum Tsuneki, K. (1968). On some *Cerceris* from Australia, with a tentative key to the Australian species (Hymenoptera, Sphecidae). *Etizenia* **28**: 1–32 [20]. Type data: holotype, USNM *F. adult, from Prince of Wales Is., Qld.

Distribution: NE coastal, Murray-Darling basin, Lake Eyre basin, Qld., N.T. Ecology: larva - sedentary, soil, predator : adult - volant, burrowing. Biological references: Evans, H.E. & Mathews, R.W. (1970). Notes on the nests and prey of Australian wasps of the genus *Cerceris* (Hymenoptera : Sphecidae). *J. Aust. Entomol. Soc.* **9**: 153–156 (biology, as *Cerceris minuscula*); Evans, H.E. (1982). The genus *Cerceris* in eastern Australia (Hymenoptera : Sphecidae). *Trans. Am. Entomol. Soc.* **107**: 229–380 (redescription, distribution).

Cerceris xanthura Evans, 1982

Cerceris xanthura Evans, H.E. (1982). The genus *Cerceris* in eastern Australia (Hymenoptera : Sphecidae). *Trans. Am. Entomol. Soc.* **107**: 299–380 [345]. Type data: holotype, QM T8490 *F. adult, from Blunder Creek, Brisbane, Qld.

Distribution: NE coastal, Bulloo River basin, Lake Eyre basin, Murray-Darling basin, N Gulf, Qld., N.S.W., Vic., S.A. Ecology: larva - sedentary, soil, predator : adult - volant, burrowing. Biological references: Evans, H.E. & Hook, A.W. (1982). Communal nesting in Australian *Cerceris* digger wasps. pp. 159–163 *in* Breed, M.D., Michener, C.D. & Evans, H.E. (eds.) (1982). *The Biology of Social Insects*. Proceedings of the Ninth Congress of the International Union for the Study of Social Insects, Boulder, Colorado, August 1982. Boulder : Westview Press (biology).

Incertae sedis

Alyson tomentosum Macleay, W.S. (1826). Annulosa. Catalogue of insects, collected by Captain King, R.N. pp. 438–469 *in* King, P.P. *Narrative of a Survey of the Inter-tropical and Western Coasts of Australia Performed between the Years 1818 and 1822*. London : John Murray Vol. 2 [457] [the identity of this species is unknown]. Type data: syntypes (probable), whereabouts unknown, from Australia (round coast).

APPENDIX I

ABBREVIATIONS AND SYMBOLS

For definitions of nomenclatural terms which appear throughout the *Catalogue*, the reader should refer to the text and glossary of the 1961 edn. of the International Code of Zoological Nomenclature.

A.C.T	Australian Capital Territory
alt.	altitude
Art.	Article
E	east, eastern
ed./eds.	editor/editors
edn.	edition
emend.	emendation
F.	female
fasc.	fascicule
fig/figs	figure/figures
ft	feet
ICZN	International Code of Zoological Nomenclature
Is./Ils.	Island/Islands
km	kilometre
livr.	livraison
m	metre
M.	male
mi	mile
ms	manuscript
Mt./Mts.	Mount, Mountain/Mountains
N	north, northern
Nat.	Natural
Natl.	National
no.	number
nom. nov.	*nomen novum*
nom. nud.	*nomen nudum*
ns	new series
N.S.W.	New South Wales
N.T.	Northern Territory
pl/pls	plate/plates
pro	for
pt/pts	part/parts
Qd./Qld.	Queensland
S	south, southern
S.A.	South Australia
ser.	series
sp.	species
Tas./Tasm.	Tasmania
var./Var.	variety
Vic./Vict.	Victoria
vol./Vol.	volume
W	west, western
W	worker in the Formicidae, with reference to type specimen(s)
W.A.	Western Australia
[*name*]	square brackets enclosing a valid or available name indicate a qualification of the use of that name in the context in which it appears.
*	appears only with reference to type specimen information and indicates that the author has not seen the specimen(s).

APPENDIX II
MUSEUM ACRONYMS

AM	Australian Museum, Sydney, N.S.W., Australia
ANIC	Australian National Insect Collection, CSIRO Div. of Entomology, Canberra, A.C.T., Australia
BIE	Instituto di Entomologia, Bologna, Italy
BMNH	British Museum (Natural History) London, U.K
BPBM	Bernice P. Bishop Museum, Honolulu, Hawaii, U.S.A
CAS	California Academy of Sciences, San Francisco, Calif., U.S.A
DARI	Insect Collection, Dept. of Agriculture, Rydalmere, N.S.W., Australia
DEIB	Deutsch Entomologie Institute di Berlin, Federal Republic of Germany
ETHZ	Eidgenössische Technische Hochschule, Zürich, Switzerland
GMNH	Museum d'Histoire Naturelle, Genève, Switzerland
LS	Linnaean Society, London, U.K.
MCG	Museo Civico di Storia Natural "Giacomo Doria", Genoa, Italy
MCZ	Museum of Comparative Zoology, Harvard Univ., Cambridge, Mass., U.S.A.
MGH	Museum Godeffroy, Hamburg, Federal Republic of Germany
MNH	Musei Nationalis Hungarici, Budapest, Hungary
MNHP	Museum National d'Histoire Naturelle, Paris, France
MZUT	Museo Zoologia, Universita, Torino, Italy
NHMW	Naturhistorisches Museum, Wien, Austria
NHRM	Naturhistoriske Riksmuseum, Stockholm, Sweden
NMV	Museum of Victoria, Melbourne, Vic., Australia
OUM	Oxford University Museum, Oxford, U.K.
QM	Queensland Museum, Fortitude Valley, Qld., Australia
RIB	Institut Royal de Sciences Naturelle de Belgigue, Bruxelles, Belgium
RMNH	Rijksmuseum van Natuurlijke Historie, Leiden, Netherlands
SAMA	South Australia Museum, Adelaide, S.A., Australia
SMNS	Stadtliches Museum für Naturkunde, Stuttgart, Federal Republic of Germany
UCDC	University of California, Davis, Calif., U.S.A.
USNM	United States National Museum, Washington, D.C., U.S.A.
UZM	Universitetets Kobenhaven, Denmark
WAM	Western Australia Museum, Perth, W.A., Australia
ZMA	Universiteit van Amsterdam, Netherlands
ZMB	Museum für Naturkunde an der Universitaet Humbolt zu Berlin, German Democratic Republic
ZMH	Zoologische Museum für Hamburg, Federal Republic of Germany
ZSM	Zoologisches Sammlung des Bayerischen Staates, München, Federal Republic of Germany

TAXONOMIC INDEX
FORMICOIDEA

abdominalis, Podomyrma 80
abdominalis pulchra, Podomyrma 80
aberrans, Amblyopone 18
aberrans, Cerapachys 23
aberrans formosa, Myrmecia (Promyrmecia) 12
aberrans haematosticta, Myrmecia (Promyrmecia) 12
aberrans maura, Myrmecia (Promyrmecia) 12
aberrans, Myrmecia 6
aberrans, Phyracaces 23
aberrans sericata, Myrmecia (Promyrmecia) 10
aberrans taylori, Myrmecia (Promyrmecia) 10
abruptus, Prolasius 142
abstinens abstinens, Mayriella 65
abstinens hackeri, Mayriella 65
abstinens, Mayriella 65
abstinens, Mayriella abstinens 65
abstinens venustula, Mayriella 66
Acanthoclinea 92
Acantholepis (Acrostigma) australis 145
Acantholepis (Acrostigma) bosii 145
Acantholepis (Acrostigma) clivispina 146
Acantholepis (Acrostigma) froggatti 146
Acantholepis kirbii 95
Acantholepis mamillatus 102
Acantholepis (Stigmacros) aemula 145
Acantholepis (Stigmacros) aemula intacta 147
Acantholepis (Stigmacros) foreli 146
Acantholepis (Stigmacros) fossulata 146
Acantholepis (Stigmacros) medioreticulata 147
Acantholepis (Stigmacros) occidentalis 147
Acantholepis (Stigmacros) pilosella 147
Acantholepis tuberculatus 101
Acanthoponera (Anacanthoponera) imbellis scabra 30
Acanthoponera (Anacanthoponera) leae 30
Acanthoponera imbellis 30
(Acanthoponera) imbellis hilare, Ectatomma 30
Acanthoponera nigra 30
Acanthoponera occidentalis 30
aciculata, Ectatomma 40
aciculata, Rhytidoponera 40
aciculata, Stigmacros 145
aciculata, Stigmacros (Cyrtostigmacros) 145
Acropyga 2 , 107
Acropyga indistincta 107
Acropyga moluccana 107
Acropyga moluccana australis 107
Acropyga myops 107
Acrostigma 144
(Acrostigma) australis, Acantholepis 145
(Acrostigma) bosii, Acantholepis 145
(Acrostigma) clivispina, Acantholepis 146
(Acrostigma) froggatti, Acantholepis 146
acuminata, Notoncus foreli 125
acuta, Stigmacros 145

acuta, Stigmacros (Stigmacros) 145
acuticostata, Bothroponera excavata 21
acuticostata, Ponera (Bothroponera) excavata 21
acutidens, Odontomachus ruficeps 36
adami, Camponotus 109
adamus, Cerapachys 24
adamus, Cerapachys (Phyracaces) 24
adelaidae adelaidae, Podomyrma 80
adelaidae brevidentata, Podomyrma 80
adelaidae, Myrmica 80
adelaidae obscurior, Podomyrma 80
adelaidae, Podomyrma 80
adelaidae, Podomyrma adelaidae 80
adlerzi, Leptogenys conigera 33
adlerzi, Leptogenys (Lobopelta) conigera 33
Adlerzia 2 , 53
Adlerzia froggatti 53
(Adlerzia) froggatti, Monomorium 53
adlerzii, Prenolepis 121
aegra, Polyrhachis (Chariomyrma) gab 134
aegra, Polyrhachis gab 134
aemula, Acantholepis (Stigmacros) 145
aemula intacta, Acantholepis (Stigmacros) 147
aemula, Stigmacros 145
aeneogaster, Iridomyrmex discors 98
aeneopilosus aeneopilosus, Camponotus 109
aeneopilosus, Camponotus 109
aeneopilosus, Camponotus aeneopilosus 109
aeneopilosus flavidopubescens, Camponotus 109
aeneovirens, Formica 122
aeneovirens, Melophorus 122
aeneum, Liometopum 101
Aenictus 2 , 52
Aenictus aratus 52
Aenictus ceylonicus 52
Aenictus deuqueti 53
Aenictus exiguus 53
Aenictus hilli 53
Aenictus philiporum 53
Aenictus turneri 52
aequalis, Bothriomyrmex pusillus 92
aerea, Polyrhachis hookeri 135
aeschyle, Polyrhachis 131
aeschyle, Polyrhachis (Hedomyrma) 131
aesopus, Camponotus arcuatus 109
aesopus, Melophorus turneri 124
affinis australis, Pheidologeton 79
affinis, Myrmecia 16
afflatus, Camponotus 109
afflatus, Camponotus (Myrmosaga) 109
agilis, Formica 115
agilis, Iridomyrmex 96
ajax, Meranoplus mars 68
albertisi, Calomyrmex 108

albertisi, Camponotus 108
albipes cedarensis, Technomyrmex 106
albipes, Formica (Tapinoma) 106
albipes, Technomyrmex 3, 106
albitarsus, Iridomyrmex 96
albopilosus albopilosus, Calomyrmex 108
albopilosus, Calomyrmex 108
albopilosus, Calomyrmex albopilosus 108
albopilosus, Camponotus 108
albopilosus wienandsi, Calomyrmex 108
albopilosus wienandsi, Camponotus (Calomyrmex) 108
alinodis, Colobostruma 59
alinodis, Epopostruma 59
alligator, Orectognathus 72
Amblyopone 2, 18
Amblyopone aberrans 18
Amblyopone australis 18
Amblyopone australis fortis 19
Amblyopone australis foveolata 19
Amblyopone australis minor 19
Amblyopone castaneus 35
Amblyopone clarki 19
Amblyopone exigua 19
Amblyopone ferruginea 19
Amblyopone ferruginea longidens 20
Amblyopone (Fulakora) gracilis 19
Amblyopone (Fulakora) lucida 20
Amblyopone (Fulakora) punctulata 20
Amblyopone gingivalis 19
Amblyopone glauerti 19
Amblyopone gracilis 19
Amblyopone hackeri 19
Amblyopone leae 19
Amblyopone longidens 20
Amblyopone lucida 20
Amblyopone mandibularis 19
Amblyopone mercovichi 20
Amblyopone michaelseni 20
Amblyopone obscura 18
Amblyopone punctulata 20
Amblyopone smithi 20
Amblyopone wilsoni 20
ammon ammon, Polyrhachis 131
ammon angusta, Polyrhachis 131
ammon angustata, Polyrhachis 131
ammon, Formica 131
ammon, Polyrhachis 131
ammon, Polyrhachis ammon 131
ammonoeides ammonoeides, Polyrhachis 131
ammonoeides crawleyi, Polyrhachis 131
ammonoeides crawleyi, Polyrhachis (Hagiomyrma) 131
ammonoeides, Polyrhachis 131
ammonoeides, Polyrhachis ammonoeides 131
amperei, Camponotus cinereus 110
amperei, Camponotus (Myrmocamelus) cinereus 110
ampla ampla, Pheidole 74
ampla mackayensis, Pheidole 74
ampla norfolkensis, Pheidole 79
ampla parallela, Pheidole 79
ampla parviceps, Pheidole 74
ampla perthensis, Pheidole 75
ampla, Pheidole 74
ampla, Pheidole ampla 74
ampla, Pheidole variabilis 74
ampla yarrensis, Pheidole 79

amyoti, Formica 149
(Anacanthoponera) imbellis scabra, Acanthoponera 30
(Anacanthoponera) leae, Acanthoponera 30
analis, Myrmecia 6
anceps, Formica 96
anceps, Iridomyrmex 96
anceps, Rhytidoponera 40
andrei, Chalcoponera victoriae 42
andromache, Polyrhachis 136
andrynicum, Tetramorium 89
anguliceps, Polyrhachis 131
anguliceps, Polyrhachis (Hedomyrma) 131
angusta, Polyrhachis ammon 131
angustata, Polyrhachis ammon 131
angustatus, Cerapachys 24
angustatus, Phyracaces 24
angusticeps, Leptomyrmex varians 105
angusticornis, Dolichoderus 93
angusticornis, Dolichoderus (Hypoclinea) 93
angustinoda, Leptogenys 32
angustinoda, Leptogenys (Lobopelta) 32
Anisopheidole 54
Anisopheidole antipodum 54
anitae, Leptogenys 32
anitae, Leptogenys (Lobopelta) 32
annectens, Notoncus gilberti 126
Anochetus 20
Anochetus armstrongi 20
Anochetus graeffei 20
Anochetus paripungens 20
Anochetus rectangularis 21
Anochetus rectangularis diabolus 21
Anochetus turneri 21
Anochetus turneri latunei 21
Anoplolepis 107
Anoplolepis longipes 3, 107
antennata, Prolasius 142
antennatus howensis, Orectognathus 73
antennatus, Orectognathus 72
antennatus, Prolasius 142
antennatus septentrionalis, Orectognathus 72
anthracina anthracina, Pheidole 75
anthracina grandii, Pheidole 75
anthracina orba, Pheidole 75
anthracina, Pheidole 75
anthracina, Pheidole anthracina 75
anthracina, Stigmacros 145
anthracina, Stigmacros (Campostigmacros) 145
antipodum, Anisopheidole 54
antipodum, Atta 54
antipodum, Tetramorium 90
antonii, Technomyrmex bicolor 106
Aphaenogaster 54
Aphaenogaster barbigula 54
Aphaenogaster longiceps 54
Aphaenogaster (Nystalomyrma) barbigula 54
Aphaenogaster poultoni 54
Aphaenogaster pythia 55
Aphantolepis 106
Aphantolepis quadricolor 106
appendiculata appendiculata, Polyrhachis 131
appendiculata, Polyrhachis 131
appendiculata, Polyrhachis appendiculata 131
appendiculata schoopae, Polyrhachis 131
araneoides arcuata, Rhytidoponera 40

araneoides, Ponera 40
araneoides, Rhytidoponera 40
aratus, Aenictus 52
arcadius, Iridomyrmex 96
arcuata, Formica 131
arcuata, Polyrhachis 131
arcuata, Rhytidoponera araneoides 40
arcuatus aesopus, Camponotus 109
arcuatus arcuatus, Camponotus 109
arcuatus, Camponotus 109
arcuatus, Camponotus arcuatus 109
argentosa, Polyrhachis daemeli 133
armstrongi, Anochetus 20
armstrongi, Camponotus 110
armstrongi, Camponotus (Myrmogonia) 110
armstrongi, Chelaner 56
armstrongi, Dolichoderus 93
armstrongi, Dolichoderus (Hypoclinea) 93
armstrongi, Monomorium (Holcomyrmex) 56
armstrongi, Stigmacros 145
armstrongi, Stigmacros (Cyrtostigmacros) 145
arnoldi, Myrmecia 6
arnoldi, Rhytidoponera (Chalcoponera) 40
asper, Sphinctomyrmex 50
aspera, Camponotus cruentatus 112
aspera, Camponotus (Myrmosericus) cruentatus 112
aspera, Ponera metallica 40
aspera, Rhytidoponera 40
aspera scabrior, Rhytidoponera (Chalcoponera) 48
astuta, Bothroponera 21
astuta, Pachycondyla 21
atalanta, Cardiocondyla nuda 55
athertonensis athertonensis, Pheidole 75
athertonensis cedarensis, Pheidole 75
athertonensis, Myrmecia auriventris 6
athertonensis, Pheidole 75
athertonensis, Pheidole athertonensis 75
athertonensis tambourinensis, Pheidole 75
(Atopogyne) eurydice, Cremastogaster 61
atrata, Myrmecia 6
atriscapa, Myrmecia 6
Atta antipodum 54
Atta spinoda 77
Atta vigilans 79
augustulus, Camponotus (Colobopsis) fictor 113
augustulus, Camponotus fictor 113
aurata, Ponera (Ectatomma) 41
aurata, Rhytidoponera 41
aurea aurea, Polyrhachis 131
aurea depilis, Polyrhachis 132
aurea, Polyrhachis 131
aurea, Polyrhachis aurea 131
aurea, Polyrhachis guerini 131
aureolus aureolus, Meranoplus 66
aureolus doddi, Meranoplus 66
aureolus linae, Meranoplus 66
aureolus, Meranoplus 66
aureolus, Meranoplus aureolus 66
aureorufa, Myrmecia mandibularis 12
auriventris athertonensis, Myrmecia 6
auriventris, Myrmecia 6
aurocincta, Formica 110
aurocinctus, Camponotus 110
aurofasciatus, Camponotus (Myrmamblys) 115
australe colosseensis, Diacamma 29

australe, Diacamma 29
australe levis, Diacamma 29
australe, Tetramorium 89
australiae, Polyrhachis connectens 139
australiae, Polyrhachis relucens 139
australicum, Monomorium 70
australicum, Monomorium subcoecum 70
australis, Acantholepis (Acrostigma) 145
australis, Acropyga moluccana 107
australis, Amblyopone 18
australis australis, Crematogaster 60
australis australis, Pheidologeton 79
australis chillagoensis, Cremastogaster 60
australis chillagoensis, Crematogaster 60
australis, Colobostruma 59
australis, Cremastogaster 60
australis, Crematogaster 60
australis, Crematogaster australis 60
australis, Dolichoderus 93
australis, Echinopla 121
australis edentata, Syscia 24
australis, Eurhopalothrix 64
australis, Formica 29
australis fortis, Amblyopone 19
australis foveolata, Amblyopone 19
australis, Leptothorax 65
australis, Leptothorax (Goniothorax) 65
australis, Lioponera longitarsus 26
australis, Mesoponera 35
australis minor, Amblyopone 19
australis mjobergi, Pheidologeton 79
australis, Pheidologeton 79
australis, Pheidologeton affinis 79
australis, Pheidologeton australis 79
australis, Platythyrea pusilla 37
australis, Polyrhachis 137
australis, Ponera melanaria 35
australis, Pseudolasius 144
australis, Stigmacros 145
australis, Strumigenys szalayi 88
australis sycites, Cremastogaster 60
australis sycites, Crematogaster 60
australis, Syscia 24
australis, Triglyphothrix (Xiphomyrmex) striatidens 91
azureus, Iridomyrmex bicknelli 96

bagoti, Melophorus 122
ballaratensis, Iridomyrmex itinerans 100
barbata, Bothroponera 21
barbata, Myrmecia (Promyrmecia) fulvipes 10
barbata, Pachycondyla (Bothroponera) 21
barbigula, Aphaenogaster 54
barbigula, Aphaenogaster (Nystalomyrma) 54
bardus, Camponotus walkeri 121
barnardi, Polyrhachis 132
barnardi, Polyrhachis (Myrmhopla) 132
barnardi, Rhytidoponera 41
barretti, Meranoplus 66
barretti, Monomorium fraterculus 70
barretti, Monomorium (Lampromyrmex) fraterculus 70
barretti, Polyrhachis 132
barretti, Polyrhachis (Hedomyrma) 132
barretti, Rhytidoponera 41
barretti, Stigmacros 145

basalis, Podomyrma 80
basirufa, Myrmecia vindex 17
basirufus, Leptomyrmex erythrocephalus 103
baucis, Pheidole bos 75
bedoti, Polyrhachis 132
belisaria, Solenopsis 86
belisarius, Solenopsis 86
bellicosa, Polyrhachis 132
bellicosus, Camponotus intrepidus 115
bellicosus, Polyrhachis 132
Belonopelta darwinii 52
bendigensis, Camponotus suffusus 120
bicarinata, Myrmica 89
bicarinatum, Tetramorium 89
bicknelli azureus, Iridomyrmex 96
bicknelli bicknelli, Iridomyrmex 96
bicknelli brunneus, Iridomyrmex 96
bicknelli, Iridomyrmex 96
bicknelli, Iridomyrmex bicknelli 96
bicknelli lutea, Iridomyrmex 96
bicknelli splendidus, Iridomyrmex 96
bicolor antonii, Technomyrmex 106
bicolor, Cerapachys 24
bicolor, Eubothroponera 38
bicolor, Phyracaces 24
bicolor, Technomyrmex 106
biconvexus, Iridomyrmex 96
bicorne, Monomorium 56
bicornis, Chelaner 56
bidens, Discothyrea 29
bidentata, Leptogenys 32
bidentata, Leptogenys (Lobopelta) 32
bidentata, Turneria 107
bidentata, Turnesia 107
bigenus, Camponotus 110
bigenus, Camponotus (Myrmocamelus) 110
bimaculata brevidentata, Podomyrma 80
bimaculata obscurior, Podomyrma 80
bimaculata, Podomyrma 80
bimaculatus, Opisthopsis pictus 128
binodis, Cerapachys 24
binodis, Cerapachys (Phyracaces) 24
biroi, Gnamptogenys 30
biroi, Melophorus 122
biroi, Stictoponera 30
bispinosa, Froggattella kirbii 95
bispinosa, Froggattella kirbyi 95
bispinosa, Podomyrma 80
bogischi, Chelaner subapterus 59
bogischi, Monomorium subapterum 59
bombalensis, Pheidole proxima 77
borealis, Rhytidoponera 41
borealis, Rhytidoponera (Chalcoponera) numeensis 41
bos baucis, Pheidole 75
bos bos, Pheidole 75
bos eubos, Pheidole 75
bos, Pheidole 75
bos, Pheidole bos 75
bosii, Acantholepis (Acrostigma) 145
bosii, Stigmacros 145
Bothriomyrmex 2 , 92
Bothriomyrmex flavus 92
Bothriomyrmex pusillus 92
Bothriomyrmex pusillus aequalis 92
Bothriomyrmex pusillus pusillus 92

Bothriomyrmex scissor 92
Bothriomyrmex wilsoni 92
Bothroponera 21
Bothroponera astuta 21
Bothroponera barbata 21
(Bothroponera) barbata, Pachycondyla 21
Bothroponera denticulata 21
Bothroponera dubitata 21
(Bothroponera) dubitata, Ponera 21
Bothroponera excavata 21
Bothroponera excavata acuticostata 21
(Bothroponera) excavata acuticostata, Ponera 21
Bothroponera excavata excavata 21
Bothroponera mayri 22
Bothroponera piliventris 22
Bothroponera piliventris intermedia 22
(Bothroponera) piliventris intermedia, Ponera 22
Bothroponera piliventris piliventris 22
Bothroponera piliventris regularis 22
(Bothroponera) piliventris regularis, Pachycondyla 22
Bothroponera porcata 22
(Bothroponera) porcata, Ponera 22
Bothroponera sublaevis 22
Bothroponera sublaevis kurandensis 22
(Bothroponera) sublaevis kurandensis, Pachycondyla 22
Bothroponera sublaevis murina 22
(Bothroponera) sublaevis murina, Pachycondyla 22
Bothroponera sublaevis reticulata 22
(Bothroponera) sublaevis reticulata, Ponera 22
Bothroponera sublaevis rubicunda 22
Bothroponera sublaevis sublaevis 22
(Bothroponera) tasmaniensis, Pachycondyla 38
Brachyponera 23
Brachyponera croceicornis 23
Brachyponera lutea 23
Brachyponera lutea clara 23
(Brachyponera) lutea clara, Euponera 23
Brachyponera lutea lutea 23
(Brachyponera) luteipes croceicornis, Euponera 23
(Brachyponera) luteipes inops, Euponera 23
(Brachyponera) rufonigra, Euponera 52
brachytera, Stigmacros 145
brachytera, Stigmacros (Campostigmacros) 145
braueri braueri, Paratrechina 129
braueri glabrior, Paratrechina 129
braueri glabrior, Prenolepis 129
braueri, Paratrechina 129
braueri, Paratrechina braueri 129
braueri, Prenolepis 129
brevicollis, Cerapachys 24
brevicollis, Phyracaces 24
brevicornis, Pheidole 75
brevidentata, Podomyrma adelaidae 80
brevidentata, Podomyrma bimaculata 80
brevinoda, Myrmecia 6
brevinoda, Myrmecia forficata 6
brevior, Rhytidoponera cerastes 48
brevis, Cerapachys 24
brevis, Phyracaces 24
brevispina, Stigmacros 145
brevispina, Stigmacros (Stigmacros) 145
brooksi, Stigmacros 145
brooksi, Stigmacros (Cyrtostigmacros) 145
broomense, Monomorium 70
broomense, Monomorium (Mitara) laeve 70

broomense, Tapinoma minutum 106
broomensis, Cremastogaster laeviceps 61
broomensis, Crematogaster laeviceps 61
broomensis, Tapinoma minutum 106
browni, Mesostruma 69
brunea, Melophorus (Melophorus) 122
brunea, Prolasius 142
bruneus, Melophorus 122
bruneus, Prolasius 142
brunnea, Chalcoponera 41
brunneiceps, Leptomyrmex erythrocephalus 103
brunneus, Iridomyrmex bicknelli 96
brunnicornis, Eusphinctus (Nothosphinctus) 50
brunnipes, Eubothroponera 37
brunnipes, Platythyrea 37

caeciliae, Rhytidoponera (Chalcoponera) 45
caelatinoda, Myrmecia (Promyrmecia) fulvipes 12
cairnsiana, Pheidole 76
cairnsiana, Pheidole javana 76
callima, Myrmecia 7
callima, Promyrmecia 7
Calomyrmex 108
Calomyrmex albertisi 108
Calomyrmex albopilosus 108
Calomyrmex albopilosus albopilosus 108
Calomyrmex albopilosus wienandsi 108
(Calomyrmex) albopilosus wienandsi, Camponotus 108
Calomyrmex glauerti 108
Calomyrmex impavidus 108
Calomyrmex purpureus 108
Calomyrmex purpureus purpureus 108
Calomyrmex purpureus smaragdina 108
Calomyrmex similis 108
Calomyrmex splendidus 108
Calomyrmex splendidus mutans 108
(Calomyrmex) splendidus mutans, Camponotus 108
Calomyrmex splendidus splendidus 108
Calomyrmex splendidus viridiventris 109
caloratus, Camponotus (Colobopsis) gasseri 114
caloratus, Camponotus gasseri 114
Calyptomyrmex 55
Calyptomyrmex schraderi 55
cameratus, Camponotus 110
cameratus, Camponotus (Myrmogonia) 110
camillae, Mystrium 35
(Campomyrma) flavibasis, Polyrhachis 133
(Campomyrma) gravis, Polyrhachis 134
(Campomyrma) hirsuta quinquedentata, Polyrhachis 135
(Campomyrma) leae cedarensis, Polyrhachis 137
(Campomyrma) longipes, Polyrhachis 137
(Campomyrma) patiens, Polyrhachis 138
(Campomyrma) polymnia maculata, Polyrhachis 138
(Campomyrma) prometheus, Polyrhachis 138
(Campomyrma) sidnica perthensis, Polyrhachis 140
(Campomyrma) sidnica tambourinensis, Polyrhachis 140
(Campomyrma) thalia io, Polyrhachis 141
(Campomyrma) zimmerae, Polyrhachis 141
Camponotus 1 , 109
Camponotus adami 109
Camponotus aeneopilosus 109
Camponotus aeneopilosus aeneopilosus 109
Camponotus aeneopilosus flavidopubescens 109
Camponotus afflatus 109

Camponotus albertisi 108
Camponotus albopilosus 108
Camponotus arcuatus 109
Camponotus arcuatus aesopus 109
Camponotus arcuatus arcuatus 109
Camponotus armstrongi 110
Camponotus aurocinctus 110
Camponotus bigenus 110
Camponotus (Calomyrmex) albopilosus wienandsi 108
Camponotus (Calomyrmex) splendidus mutans 108
Camponotus cameratus 110
Camponotus capito 110
Camponotus capito capito 110
Camponotus capito ebeninithorax 110
Camponotus carazzii 126
Camponotus ceriseipes 110
Camponotus chalceoides 110
Camponotus chalceus 110
Camponotus cinereus 110
Camponotus cinereus amperei 110
Camponotus cinereus cinereus 110
Camponotus cinereus notterae 111
Camponotus claripes 111
Camponotus claripes claripes 111
Camponotus claripes elegans 111
Camponotus claripes inverallensis 111
Camponotus claripes marcens 111
Camponotus claripes minimus 111
Camponotus claripes nudimalis 111
Camponotus claripes orbiculatopunctatus 111
Camponotus claripes piperatus 111
Camponotus (Colobopsis) fictor 113
Camponotus (Colobopsis) fictor augustulus 113
Camponotus (Colobopsis) gasseri caloratus 114
Camponotus (Colobopsis) gasseri lysias 114
Camponotus (Colobopsis) gasseri obtusitruncatus 114
Camponotus (Colobopsis) howensis 115
Camponotus (Colobopsis) sanguinifrons 118
Camponotus consectator 111
Camponotus consobrinus 111
Camponotus cowlei 112
Camponotus crenatus 112
Camponotus cruentatus 112
Camponotus cruentatus aspera 112
Camponotus denticulatus 112
Camponotus dimidiatus 112
Camponotus (Dinomyrmex) rufus 118
Camponotus (Dinomyrmex) simulator 119
Camponotus (Dinomyrmex) spinitarsus 119
Camponotus (Dinomyrmex) subnitidus longinodis 119
Camponotus discors 112
Camponotus discors discors 112
Camponotus discors yarrabahensis 112
Camponotus dorycus 112
Camponotus dorycus confusus 112
Camponotus dromas 112
Camponotus ectatommoides 125
Camponotus emeryi 133
Camponotus ephippium 112
Camponotus ephippium ephippium 113
Camponotus ephippium narses 113
Camponotus eremicus 113
Camponotus erythropus 113
Camponotus esau 113
Camponotus evae 113

Camponotus evae evae 113
Camponotus evae zeuxis 113
Camponotus extensus 113
Camponotus fictor 113
Camponotus fictor augustulus 113
Camponotus fictor fictor 113
Camponotus fieldeae 113
Camponotus fieldellus 114
Camponotus froggatti 114
Camponotus gasseri 114
Camponotus gasseri caloratus 114
Camponotus gasseri gasseri 114
Camponotus gasseri lysias 114
Camponotus gasseri obtusitruncatus 114
Camponotus gibbinotus 114
Camponotus gouldianus 114
Camponotus hartogi 114
Camponotus horni 114
Camponotus howensis 115
Camponotus impavidus 108
Camponotus inflatus 115
Camponotus innexus 115
Camponotus insipidus 115
Camponotus intrepidus 115
Camponotus intrepidus bellicosus 115
Camponotus intrepidus intrepidus 115
Camponotus janeti 115
Camponotus latrunculus 115
Camponotus latrunculus latrunculus 115
Camponotus latrunculus victoriensis 115
Camponotus leae 116
Camponotus lividicoxis 116
Camponotus lownei 116
Camponotus maculatus 116
Camponotus maculatus discors 112
Camponotus maculatus humilior 116
Camponotus magnus 115
Camponotus michaelseni 116
Camponotus midas 116
Camponotus molossus 116
Camponotus myoporus 116
Camponotus (Myrmamblys) aurofasciatus 115
Camponotus (Myrmocamelus) bigenus 110
Camponotus (Myrmocamelus) cinereus amperei 110
Camponotus (Myrmocamelus) dromas 112
Camponotus (Myrmocamelus) ephippium narses 113
Camponotus (Myrmocamelus) esau 113
Camponotus (Myrmocamelus) pellax 118
Camponotus (Myrmogonia) armstrongi 110
Camponotus (Myrmogonia) cameratus 110
Camponotus (Myrmogonia) eremicus 113
Camponotus (Myrmogonia) evae zeuxis 113
Camponotus (Myrmogonia) oetkeri voltai 117
Camponotus (Myrmogonia) punctiventris 118
Camponotus (Myrmogonia) sanguinea 118
Camponotus (Myrmogonia) tumidus 120
Camponotus (Myrmophyma) capito ebeninithorax 110
Camponotus (Myrmophyma) ceriseipes 110
Camponotus (Myrmophyma) chalceoides 110
Camponotus (Myrmophyma) claripes minima 111
Camponotus (Myrmophyma) claripes orbiculatopunctatus 111
Camponotus (Myrmophyma) claripes piperatus 111
Camponotus (Myrmophyma) darlingtoni 120
Camponotus (Myrmophyma) lividicoxis 116
Camponotus (Myrmophyma) nitidiceps 117
Camponotus (Myrmophyma) rottnesti 120
Camponotus (Myrmophyma) tristis 120
Camponotus (Myrmosaga) afflatus 109
Camponotus (Myrmosaga) chalceus 110
Camponotus (Myrmosaga) erythropus 113
Camponotus (Myrmosaga) ferruginipes 114
Camponotus (Myrmosaulus) scutellus 121
Camponotus (Myrmosaulus) versicolor 120
Camponotus (Myrmosericus) cruentatus aspera 112
Camponotus (Myrmosphincta) leae 116
Camponotus (Myrmosphincta) whitei 121
Camponotus (Myrmoturba) latrunculus 115
Camponotus (Myrmoturba) latrunculus victoriensis 115
Camponotus (Myrmoturba) maculatus yarrabahensis 112
Camponotus (Myrmoturba) villosa 120
Camponotus nigriceps 116
Camponotus nigriceps clarior 117
Camponotus nigriceps lividipes 117
Camponotus nigriceps nigriceps 116
Camponotus nigriceps obniger 117
Camponotus nigriceps pallidiceps 117
Camponotus nigroaeneus 117
Camponotus nigroaeneus divus 117
Camponotus nigroaeneus nigroaeneus 117
Camponotus nitidiceps 117
Camponotus novaehollandiae 117
Camponotus oetkeri 117
Camponotus oetkeri oetkeri 117
Camponotus oetkeri voltai 117
Camponotus oxleyi 117
Camponotus pellax 118
Camponotus podenzanai 126
Camponotus postcornutus 118
Camponotus punctiventris 118
Camponotus purpureus 108
Camponotus reticulatus 118, 119
Camponotus reticulatus mackayensis 118
Camponotus rubiginosus 118
Camponotus rufus 118
Camponotus sanguinea 118
Camponotus sanguinifrons 118
Camponotus schencki 119
Camponotus scratius 118
Camponotus scratius nuntius 118
Camponotus scratius scratius 118
Camponotus semicarinatus 118
Camponotus similis 108
Camponotus simulator 119
Camponotus spenceri 119
Camponotus spinitarsus 119
Camponotus splendidus 108
Camponotus sponsorum 119
Camponotus subnitidus 119
Camponotus subnitidus famelicus 119
Camponotus subnitidus longinodis 119
Camponotus subnitidus subnitidus 119
Camponotus suffusus 119
Camponotus suffusus bendigensis 120
Camponotus suffusus suffusus 119
Camponotus (Tanaemyrmex) myoporus 116
Camponotus (Tanaemyrmex) postcornutus 118
Camponotus (Tanaemyrmex) spenceri 119
Camponotus (Tanaemyrmex) tricoloratus 120
Camponotus tasmani 120

Camponotus testaceipes 120
Camponotus tricoloratus 120
Camponotus tristis 120
Camponotus tumidus 120
Camponotus versicolor 120
Camponotus villosus 120
Camponotus vitreus 120
Camponotus walkeri 121
Camponotus walkeri bardus 121
Camponotus walkeri walkeri 121
Camponotus whitei 121
Camponotus wiederkehri 121
Camponotus wiederkehri lucidior 121
Camponotus wiederkehri wiederkehri 121
Campostigmacros 144
(Campostigmacros) anthracina, Stigmacros 145
(Campostigmacros) brachytera, Stigmacros 145
(Campostigmacros) epinotalis, Stigmacros 146
(Campostigmacros) marginata, Stigmacros 147
(Campostigmacros) nitida, Stigmacros 147
(Campostigmacros) stanleyi, Stigmacros 148
candida, Melophorus turneri 124
candidus, Melophorus turneri 124
capitale, Tetramorium 89
capitalis, Xiphomyrmex 89
capitatus minor, Notoncus 125
capitatus, Notoncus 125
capito, Camponotus 110
capito, Camponotus capito 110
capito capito, Camponotus 110
capito ebeninithorax, Camponotus 110
capito ebeninithorax, Camponotus (Myrmophyma) 110
carazzii, Camponotus 126
carazzii, Notostigma 126
carbonaria, Chalcoponera metallica 44
cardigaster, Myrmecia 7
Cardiocondyla 55
Cardiocondyla nuda 55
Cardiocondyla nuda atalanta 55
Cardiocondyla nuda nereis 55
carinata, Rhytidoponera 41
caro, Ectatomma (Rhytidoponera) cristatum 40
carteri, Myrmecorhynchus 124
castanea, Myopopone 35
castanea, Podomyrma 84
castanea, Rhytidoponera 47
castanea, Stigmacros 146
castanea, Stigmacros (Cyrtostigmacros) 146
castaneus, Amblyopone 35
castaneus, Phyracaces 24
castrae, Iridomyrmex detectus 102
castrae, Iridomyrmex purpureus 102
cataulacoidea, Polyrhachis 132
cedarensis, Pheidole athertonensis 75
cedarensis, Polyrhachis (Campomyrma) leae 137
cedarensis, Polyrhachis leae 137
cedarensis, Rhytidoponera (Chalcoponera) victoriae 49
cedarensis, Technomyrmex albipes 106
cedaris, Sphinctomyrmex 50
cedaris, Sphinctomyrmex (Eusphinctus) fallax 50
celaena, Myrmecia 7
celaena, Promyrmecia 7
centrale, Monomorium 56
centralis, Chelaner 56
centralis, Leptogenys conigera 33

centralis, Leptogenys (Lobopelta) conigera 33
cephalicum, Tapinoma (Micromyrma) minutum 106
cephalicum, Tapinoma minutum 106
cephalotes, Myrmecia 7
cephalotes, Odontomachus 36
cephalotes, Platythyrea 37
cephalotes, Promyrmecia 7
Cerapachys 23
Cerapachys aberrans 23
Cerapachys adamus 24
Cerapachys angustatus 24
Cerapachys bicolor 24
Cerapachys binodis 24
Cerapachys brevicollis 24
Cerapachys brevis 24
Cerapachys clarki 24
Cerapachys constrictus 24
Cerapachys crassus 24
Cerapachys edentatus 24
Cerapachys elegans 25
Cerapachys emeryi 25 , 51
Cerapachys emeryi clarus 50
Cerapachys fervidus 25
Cerapachys ficosus 25
Cerapachys flammeus 25
Cerapachys gilesi 25
Cerapachys grandis 25
Cerapachys greavesi 25
Cerapachys gwynethae 26
Cerapachys heros 26
Cerapachys incontentus 26
Cerapachys jovis 26
Cerapachys larvatus 26
Cerapachys latus 26
Cerapachys longitarsus 26
Cerapachys macrops 26
Cerapachys mjobergi 26
Cerapachys mullewanus 26
Cerapachys nigriventris 27
Cerapachys (Phyracaces) adamus 24
Cerapachys (Phyracaces) binodis 24
Cerapachys (Phyracaces) jovis 26
Cerapachys (Phyracaces) mjobergi 26
Cerapachys (Phyracaces) singularis rotula 28
Cerapachys (Phyracaces) sjostedti 28
Cerapachys (Phyracaces) turneri 28
Cerapachys picipes 27
Cerapachys pictus 27
Cerapachys piliventris 27
Cerapachys potteri 27
Cerapachys princeps 27
Cerapachys punctatissimus 27
Cerapachys ruficornis 27
Cerapachys rugulinodis 27
Cerapachys senescens 27
Cerapachys simmonsae 28
Cerapachys singularis 28
Cerapachys sjostedti 28
Cerapachys turneri 28
Cerapachys varians 28
cerastes brevior, Rhytidoponera 48
cerastes, Rhytidoponera 41
ceriseipes, Camponotus 110
ceriseipes, Camponotus (Myrmophyma) 110
cerornata, Colobostruma 59

ceylonica, Typhlatta 52
ceylonicus, Aenictus 52
chalceoides, Camponotus 110
chalceoides, Camponotus (Myrmophyma) 110
chalceus, Camponotus 110
chalceus, Camponotus (Myrmosaga) 110
chalchas, Polyrhachis 132
Chalcoponera 40
(Chalcoponera) arnoldi, Rhytidoponera 40
(Chalcoponera) aspera scabrior, Rhytidoponera 48
Chalcoponera brunnea 41
(Chalcoponera) caeciliae, Rhytidoponera 45
(Chalcoponera) clarki, Rhytidoponera 42
(Chalcoponera) dubia, Rhytidoponera 43
(Chalcoponera) fastuosa, Rhytidoponera 42
Chalcoponera flavipes 43
Chalcoponera hilli 42
(Chalcoponera) lamellinodis, Rhytidoponera 44
Chalcoponera metallica carbonaria 44
(Chalcoponera) metallica inornata, Rhytidoponera 44
(Chalcoponera) metallica purpurascens, Rhytidoponera 45
(Chalcoponera) metallica varians, Rhytidoponera 45
(Chalcoponera) numeensis borealis, Rhytidoponera 41
Chalcoponera pulchra 45
Chalcoponera victoriae andrei 42
(Chalcoponera) victoriae cedarensis, Rhytidoponera 49
(Chalcoponera) victoriae maledicta, Rhytidoponera 44
Chalcoponera viridis 49
chalybaea, Rhytidoponera 41
chalybaea, Rhytidoponera impressa 41
Chariomyrma 130
(Chariomyrma) gab aegra, Polyrhachis 134
(Chariomyrma) gab tripellis, Polyrhachis 134
Chariostigmacros 144
(Chariostigmacros) hirsuta, Stigmacros 147
chasei chasei, Iridomyrmex 97
chasei chasei, Myrmecia 7
chasei concolor, Iridomyrmex 97
chasei, Cremastogaster laeviceps 61
chasei, Crematogaster laeviceps 61
chasei, Iridomyrmex 96 , 97
chasei, Iridomyrmex chasei 97
chasei ludlowi, Myrmecia 7
chasei, Myrmecia 7
chasei, Myrmecia chasei 7
chasei, Podomyrma 80
chasei yalgooensis, Iridomyrmex 97
Chelaner 55
Chelaner armstrongi 56
Chelaner bicornis 56
Chelaner centralis 56
Chelaner falcatus 56
Chelaner flavigaster 56
Chelaner flavipes 56
Chelaner foreli 56
Chelaner gilberti 56
Chelaner gilberti gilberti 56
Chelaner gilberti mediorubrus 56
Chelaner howensis 56
Chelaner insolescens 57
Chelaner insularis 57
Chelaner kiliani 57
Chelaner kiliani kiliani 57
Chelaner kiliani obscurellus 57

Chelaner kiliani tambourinensis 57
Chelaner leae 57
Chelaner longiceps 57
Chelaner macareaveyi 57
Chelaner occidaneus 57
Chelaner rothsteini 57
Chelaner rothsteini doddi 58
Chelaner rothsteini humilior 58
Chelaner rothsteini leda 58
Chelaner rothsteini rothsteini 58
Chelaner rothsteini squamigena 58
Chelaner rothsteini tostum 58
Chelaner rubriceps 58
Chelaner rubriceps cinctus 58
Chelaner rubriceps extreminigrus 58
Chelaner rubriceps rubriceps 58
Chelaner rubriceps rubrus 58
Chelaner sanguinolentus 58
Chelaner sculpturatus 58
Chelaner sordidus 59
Chelaner sordidus nigriventris 59
Chelaner sordidus sordidus 59
Chelaner subapterus 59
Chelaner subapterus bogischi 59
Chelaner subapterus subapterus 59
Chelaner turneri 59
Chelaner whitei 59
chelifer, Leptogenys 32
chelifer, Pseudoponera 32
chillagoensis, Cremastogaster australis 60
chillagoensis, Crematogaster australis 60
chnoopyx, Rhytidoponera 41
christae, Dacryon 81
christae, Podomyrma 81
chrysogaster, Myrmecia 7
chrysogaster, Promyrmecia 7
chrysothorax, Polyrhachis 132
chrysothorax, Polyrhachis (Hedomyrma) 132
cinctum, Monomorium rubriceps 58
cinctus, Chelaner rubriceps 58
cinereus amperei, Camponotus 110
cinereus amperei, Camponotus (Myrmocamelus) 110
cinereus, Camponotus 110
cinereus, Camponotus cinereus 110
cinereus cinereus, Camponotus 110
cinereus notterae, Camponotus 111
clara, Brachyponera lutea 23
clara, Euponera (Brachyponera) lutea 23
clarior, Camponotus nigriceps 117
clarior, Cremastogaster laeviceps 61
clarior, Crematogaster laeviceps 61
claripes, Camponotus 111
claripes, Camponotus claripes 111
claripes claripes, Camponotus 111
claripes elegans, Camponotus 111
claripes inverallensis, Camponotus 111
claripes marcens, Camponotus 111
claripes minima, Camponotus (Myrmophyma) 111
claripes minimus, Camponotus 111
claripes nudimalis, Camponotus 111
claripes orbiculatopunctatus, Camponotus 111
claripes orbiculatopunctatus, Camponotus (Myrmophyma) 111
claripes piperatus, Camponotus 111
claripes piperatus, Camponotus (Myrmophyma) 111

clarithorax, Iridomyrmex glaber 99
clarki, Amblyopone 19
clarki, Cerapachys 24
clarki clarki, Plagiolepis 130
clarki, Dolichoderus 93
clarki, Euponera (Trachymesopus) 52
clarki impasta, Plagiolepis 130
clarki, Leptogenys 32
clarki, Leptomyrmex erythrocephalus 103
clarki, Myrmecia 7
clarki, Orectognathus 72
clarki, Phyracaces 24
clarki, Plagiolepis 129 , 130
clarki, Plagiolepis clarki 130
clarki, Podomyrma 81
clarki, Prolasius 142
clarki, Pseudopodomyrma 81
clarki, Rhytidoponera 41
clarki, Rhytidoponera (Chalcoponera) 42
clarki, Solenopsis 87
clarki, Stigmacros 146
clarki, Stigmacros (Cyrtostigmacros) 146
clarki, Trachymesopus 52
Clarkistruma 59
clarus, Cerapachys emeryi 50
clarus mjobergi, Sphinctomyrmex 51
clarus, Phyracaces 27
clarus, Sphinctomyrmex 50
clavicornis, Ponera 38
cleopatra, Polyrhachis 132
clio, Polyrhachis 132
clitellarius, Iridomyrmex nitidus 101
clivispina, Acantholepis (Acrostigma) 146
clivispina, Stigmacros 146
clotho, Polyrhachis 132
clusor, Dolichoderus 93
cnemidatus, Leptomyrmex erythrocephalus 103
coarctata mackayensis, Ponera 31
coccinatus, Orectognathus 72
Codiomyrmex flagellatus 64
Codiomyrmex semicomptus 64
(Colobopsis) fictor augustulus, Camponotus 113
(Colobopsis) fictor, Camponotus 113
Colobopsis gasseri 114
(Colobopsis) gasseri caloratus, Camponotus 114
(Colobopsis) gasseri lysias, Camponotus 114
(Colobopsis) gasseri obtusitruncatus, Camponotus 114
(Colobopsis) howensis, Camponotus 115
Colobopsis rufifrons semicarinata 118
(Colobopsis) sanguinifrons, Camponotus 118
Colobostruma 59
Colobostruma alinodis 59
Colobostruma australis 59
Colobostruma cerornata 59
Colobostruma elliotti 60
Colobostruma froggatti 60
Colobostruma leae 60
(Colobostruma) leae, Epopostruma 60
Colobostruma nancyae 60
Colobostruma papulata 60
colosseensis, Diacamma australe 29
comata, Myrmecia 7
comata, Polyrhachis 134
concentrica concentrica, Pheidole 76
concentrica, Pheidole 76

concentrica, Pheidole concentrica 76
concentrica recurva, Pheidole 76
concolor, Iridomyrmex chasei 97
conficta, Pheidole 76
confusa, Rhytidoponera 42
confusum, Tetramorium 89
confusus, Camponotus dorycus 112
congrua, Hypoponera 31
congrua, Ponera 31
conifer, Iridomyrmex 97
conigera adlerzi, Leptogenys 33
conigera adlerzi, Leptogenys (Lobopelta) 33
conigera centralis, Leptogenys 33
conigera centralis, Leptogenys (Lobopelta) 33
conigera conigera, Leptogenys 32
conigera exigua, Leptogenys 33
conigera exigua, Leptogenys (Lobopelta) 33
conigera, Leptogenys 32
conigera, Leptogenys conigera 32
conigera, Lobopelta 32
conigera mutans, Leptogenys 33
conigera mutans, Leptogenys (Lobopelta) 33
connectens australiae, Polyrhachis 139
consectator, Camponotus 111
consectator, Formica 111
consobrina, Formica 111
consobrinus, Camponotus 111
constans, Melophorus 122
constricta, Phyracaces 24
constricta, Polyrhachis 132
constrictus, Cerapachys 24
contemta, Polyrhachis 133
continentis, Iridomyrmex mattiroloi 100
continentis, Pheidole tasmaniensis 78
convergens, Podomyrma 81
convexa opacior, Rhytidoponera 49
convexa, Rhytidoponera 42
convexa rufiventris, Rhytidoponera 47
convexiuscula, Hypoponera 31
convexiuscula, Ponera trigona 31
convexum, Ectatomma 42
convexum gemma, Ectatomma (Rhytidoponera) 49
convexum nodiferum, Ectatomma (Rhytidoponera) 46
convexum rufescens, Ectatomma (Rhytidoponera) 47
convexum spatiatum, Ectatomma (Rhytidoponera) 45
convexum violaceum, Ectatomma (Rhytidoponera) 49
convexus, Prolasius 142
cordata, Formica 97
cordata, Myrmecia 7
cordatus, Iridomyrmex 97
cordatus stewartii, Iridomyrmex 97
coriarius magnus, Odontomachus 36
coriarius obscura, Odontomachus 36
coriarius, Odontomachus 36
coriarius semicircularis, Odontomachus 36
corniger corniger, Oligomyrmex 72
corniger, Oligomyrmex 72
corniger, Oligomyrmex corniger 72
corniger parvicornis, Oligomyrmex 72
cornigera, Cremastogaster 60
cornigera, Crematogaster 60
cornuta fusciventris, Rhytidoponera 48
cornuta, Rhytidoponera 42
cornutum, Ectatomma (Rhytidoponera) 42
cornutum taurus, Ectatomma (Rhytidoponera) 48

cowlei, Camponotus 112
crassicornis, Discothyrea 29
crassinoda, Myrmecia 14
crassinode, Ectatomma (Rhytidoponera) 42
crassinodis, Rhytidoponera 42
crassus, Cerapachys 24
crassus, Phyracaces 24
crawleyella, Polyrhachis 134
crawleyi, Meranoplus 68
crawleyi, Polyrhachis ammonoeides 131
crawleyi, Polyrhachis (Hagiomyrma) ammonoeides 131
Cremastogaster 60
Cremastogaster (Atopogyne) eurydice 61
Cremastogaster australis 60
Cremastogaster australis chillagoensis 60
Cremastogaster australis sycites 60
Cremastogaster cornigera 60
Cremastogaster frivola sculpticeps 61
Cremastogaster frivolus 61
Cremastogaster fusca 61
Cremastogaster kutteri 61
Cremastogaster laeviceps broomensis 61
Cremastogaster laeviceps chasei 61
Cremastogaster laeviceps clarior 61
Cremastogaster longiceps 62
Cremastogaster mjobergi 62
Cremastogaster pallidipes 62
Cremastogaster pallipes 62
Cremastogaster pythia 62
Cremastogaster rufotestacea 63
Cremastogaster scita 63
Cremastogaster scita mixta 63
Cremastogaster sordidula dispar 61
Cremastogaster sordidula froggatti 62
Cremastogaster sordidula gilberti 62
Cremastogaster sordidula queenslandica 62
Cremastogaster sordidula rogans 63
Crematogaster 60
Crematogaster australis 60
Crematogaster australis australis 60
Crematogaster australis chillagoensis 60
Crematogaster australis sycites 60
Crematogaster cornigera 60
Crematogaster dispar 61
Crematogaster eurydice 61
Crematogaster frivola 61
Crematogaster frivola frivola 61
Crematogaster frivola sculpticeps 61
Crematogaster froggatti scabrula 63
Crematogaster fusca 61
Crematogaster kutteri 61
Crematogaster laeviceps 61
Crematogaster laeviceps broomensis 61
Crematogaster laeviceps chasei 61
Crematogaster laeviceps clarior 61
Crematogaster laeviceps laeviceps 61
Crematogaster longiceps 61
Crematogaster longiceps curticeps 62
Crematogaster longiceps longiceps 62
Crematogaster mjobergi 62
Crematogaster (Orthocrema) rufotestacea dentinasis 63
Crematogaster pallida 62
Crematogaster pallidus 62
Crematogaster pallipes 62
Crematogaster perthensis 62
Crematogaster piceus 62
Crematogaster pythia 62
Crematogaster queenslandica 62
Crematogaster queenslandica froggatti 62
Crematogaster queenslandica gilberti 62
Crematogaster queenslandica queenslandica 62
Crematogaster queenslandica rogans 63
Crematogaster queenslandica scabrula 63
Crematogaster rufotestacea 63
Crematogaster rufotestacea dentinasis 63
Crematogaster rufotestacea rufotestacea 63
Crematogaster scita 63
Crematogaster scita mixta 63
Crematogaster scita scita 63
Crematogaster whitei 63
Crematogaster xerophila 63
Crematogaster xerophila exigua 63
Crematogaster xerophila xerophila 63
crenatus, Camponotus 112
cristata, Rhytidoponera 42
cristata yorkensis, Rhytidoponera 49
cristatum caro, Ectatomma (Rhytidoponera) 40
cristatum, Ectatomma 42
cristatum punctiventris, Ectatomma (Rhytidoponera) 46
cristulatum, Ectatomma (Rhytidoponera) metallicum 48
croceicornis, Brachyponera 23
croceicornis, Euponera (Brachyponera) luteipes 23
croesus, Rhytidoponera 42
crudelis, Myrmecia 16
cruentata, Formica 112
cruentatus aspera, Camponotus 112
cruentatus aspera, Camponotus (Myrmosericus) 112
cruentatus, Camponotus 112
Cryptocerus pubescens 68
Cryptopone 28
Cryptopone rotundiceps 28
cupreata, Polyrhachis hermione 135
curticeps, Crematogaster longiceps 62
curtus, Melophorus 122
curvispina, Meranoplus puryi 68
cyaneus, Iridomyrmex 97
cydista, Myrmecia 8
cydista, Promyrmecia 8
(Cyrtomyrma) doddi, Polyrhachis 133
(Cyrtomyrma) mackayi, Polyrhachis 137
(Cyrtomyrma) nox, Polyrhachis 138
(Cyrtomyrma) rastellata yorkana, Polyrhachis 139
(Cyrtomyrma) townsvillei, Polyrhachis 141
Cyrtostigmacros 144
(Cyrtostigmacros) aciculata, Stigmacros 145
(Cyrtostigmacros) armstrongi, Stigmacros 145
(Cyrtostigmacros) brooksi, Stigmacros 145
(Cyrtostigmacros) castanea, Stigmacros 146
(Cyrtostigmacros) clarki, Stigmacros 146
(Cyrtostigmacros) extreminigra, Stigmacros 146
(Cyrtostigmacros) ferruginea, Stigmacros 146
(Cyrtostigmacros) flava, Stigmacros 146
(Cyrtostigmacros) glauerti, Stigmacros 146
(Cyrtostigmacros) lanaris, Stigmacros 147
(Cyrtostigmacros) major, Stigmacros 147
(Cyrtostigmacros) proxima, Stigmacros 148
(Cyrtostigmacros) sordida, Stigmacros 148
(Cyrtostigmacros) striata, Stigmacros 148
cyrus, Ectatomma (Rhytidoponera) 41
Dacryon 79

Dacryon christae 81
Dacryon ferruginea 81
Dacryon kitschneri 82
Dacryon liber 83
Dacryon marginatus 83
Dacryon mjobergi 83
Dacryon nitida 83
Dacryon omniparens 84
Dacryon turneri 84
daemeli argentosa, Polyrhachis 133
daemeli daemeli, Polyrhachis 133
daemeli exlex, Polyrhachis 133
daemeli exlex, Polyrhachis (Hedomyrma) 133
daemeli, Polyrhachis 133
daemeli, Polyrhachis daemeli 133
darlingtoni, Camponotus (Myrmophyma) 120
darlingtoni darlingtoni, Leptomyrmex 103
darlingtoni fascigaster, Leptomyrmex 103
darlingtoni jucundus, Leptomyrmex 103
darlingtoni, Leptogenys 33
darlingtoni, Leptogenys (Lobopelta) 33
darlingtoni, Leptomyrmex 103
darlingtoni, Leptomyrmex darlingtoni 103
darlingtoni, Orectognathus 73
darwinianum fida, Tapinoma (Doleromyrma) 97
darwinianum, Tapinoma (Doleromyrma) 97
darwinianus darwinianus, Iridomyrmex 97
darwinianus fida, Iridomyrmex 97
darwinianus, Iridomyrmex 97
darwinianus, Iridomyrmex darwinianus 97
darwinianus leae, Iridomyrmex 97
darwinii, Belonopelta 52
darwinii, Trachymesopus 52
deceptum, Tetramorium 89
decipians, Myrmecia 8
decipiens, Leptomyrmex erythrocephalus 104
decora, Hypoponera 31
decora, Ponera 31
degener, Polyrhachis sokolova 140
delbruckii, Podomyrma 81
delicata, Polyrhachis 139
densestrigosa densestrigosa, **Podomyrma** 81
densestrigosa, Podomyrma 81
densestrigosa, Podomyrma densestrigosa 81
densestrigosa teres, Podomyrma 81
dentata, Notoncus foreli 125
dentatus, Dolichoderus 93
dentatus, Dolichoderus doriae 93
denticulata, Bothroponera 21
denticulatus, Camponotus 112
dentinasis, Crematogaster (Orthocrema) rufotestacea 63
dentinasis, Crematogaster rufotestacea 63
dentinodis, Eubothroponera 37
dentinodis, Platythyrea 37
depilis, Iridomyrmex itinerans 100
depilis, Polyrhachis aurea 132
depressiceps depressiceps, Prolasius 142
depressiceps, Melophorus 142
depressiceps, Prolasius 142
depressiceps, Prolasius depressiceps 142
depressiceps similis, Prolasius 142
deserticola deserticola, Pheidole 76
deserticola foveifrons, Pheidole 76
deserticola, Pheidole 76
deserticola, Pheidole deserticola 76

desertorum, Myrmecia 8
desertorum, Myrmecia vindex 8
detecta, Formica 101
detectus castrae, Iridomyrmex 102
detectus sanguinea, Iridomyrmex 102
detectus viridiaeneus, Iridomyrmex 102
deuqueti, Aenictus 53
diabolus, Anochetus rectangularis 21
Diacamma 28
Diacamma australe 29
Diacamma australe colosseensis 29
Diacamma australe levis 29
diadela, Trapeziopelta 35
diadematus diadematus, Opisthopsis 127
diadematus dubius, Opisthopsis 127
diadematus, Opisthopsis 127
diadematus, Opisthopsis diadematus 127
Diceratoclinea 92
dichospila, Myrmecia 8
dichospila, Myrmecia (Promyrmecia) 8
dichrous, Meranoplus 66
dimidiata, Myrmecia 8
dimidiatus, Camponotus 112
dimidiatus, Meranoplus 66
diminuta, Leptogenys 33
diminuta, Ponera 33
diminuta yarrabahna, Leptogenys 33
diminuta yarrabahna, Leptogenys (Lobopelta) 33
(Dinomyrmex) rufus, Camponotus 118
(Dinomyrmex) simulator, Camponotus 119
(Dinomyrmex) spinitarsus, Camponotus 119
(Dinomyrmex) subnitidus longinodis, Camponotus 119
Diodontolepis 29
discors aeneogaster, Iridomyrmex 98
discors, Camponotus 112
discors, Camponotus discors 112
discors, Camponotus maculatus 112
discors discors, Camponotus 112
discors discors, Iridomyrmex 97
discors, Iridomyrmex 97
discors, Iridomyrmex discors 97
discors obscurior, Iridomyrmex 98
discors occipitalis, Iridomyrmex 98
discors yarrabahensis, Camponotus 112
Discothyrea 29
Discothyrea bidens 29
Discothyrea crassicornis 29
Discothyrea leae 29
Discothyrea turtoni 29
Discothyrea velutina 29
dispar, Cremastogaster sordidula 61
dispar, Crematogaster 61
dispar, Liomyrmex (Machomyrma) 65
dispar, Machomyrma 65
dispar, Myrmecia 8
dispar, Promyrmecia 8
diversus diversus, Meranoplus 66
diversus duyfkeni, Meranoplus 67
diversus, Meranoplus 66
diversus, Meranoplus diversus 66
diversus oxleyi, Meranoplus 67
diversus unicolor, Meranoplus 67
divus, Camponotus nigroaeneus 117
dixoni, Myrmecia 8
dixoni, Promyrmecia 8

dixoni, Rhytidoponera 45
doddi, Chelaner rothsteini 58
doddi, Meranoplus aureolus 66
doddi, Monomorium (Paraholcomyrmex) rothsteini 58
doddi, Onychomyrmex 37
doddi, Pheidole longiceps 77
doddi, Polyrhachis 133
doddi, Polyrhachis (Cyrtomyrma) 133
Doleromyrma 95
(Doleromyrma) darwinianum fida, Tapinoma 97
(Doleromyrma) darwinianum, Tapinoma 97
dolichocephala, Pheidole 79
DOLICHODERINAE 2 , 92
Dolichoderus 92
Dolichoderus angusticornis 93
Dolichoderus armstrongi 93
Dolichoderus australis 93
Dolichoderus clarki 93
Dolichoderus clusor 93
Dolichoderus dentatus 93
Dolichoderus doriae 93
Dolichoderus doriae dentatus 93
Dolichoderus doriae extensispina 93
Dolichoderus extensispinus 93
Dolichoderus formosus 93
Dolichoderus glauerti 93
Dolichoderus goudiei 94
Dolichoderus (Hypoclinea) angusticornis 93
Dolichoderus (Hypoclinea) armstrongi 93
Dolichoderus (Hypoclinea) formosus 93
Dolichoderus (Hypoclinea) glauerti 93
Dolichoderus (Hypoclinea) goudiei 94
Dolichoderus (Hypoclinea) nigricornis 94
Dolichoderus (Hypoclinea) occidentalis 94
Dolichoderus (Hypoclinea) parvus 94
Dolichoderus (Hypoclinea) reflexus 94
Dolichoderus (Hypoclinea) scabridus ruficornis 94
Dolichoderus (Hypoclinea) tristis 93
Dolichoderus (Hypoclinea) ypsilon nigra 95
Dolichoderus (Hypoclinea) ypsilon rufotibialis 95
Dolichoderus kirbyi 95
Dolichoderus nigricornis 94
Dolichoderus occidentalis 94
Dolichoderus parvus 94
Dolichoderus reflexus 94
Dolichoderus scabridus 94
Dolichoderus scabridus ruficornis 94
Dolichoderus scabridus scabridus 94
Dolichoderus scabridus ypsilon 94
Dolichoderus scrobiculatus 94
Dolichoderus turneri 94
Dolichoderus ypsilon 94
Dolichoderus ypsilon nigra 95
Dolichoderus ypsilon rufotibialis 95
Dolichoderus ypsilon ypsilon 94
domestica, Iridomyrmex rufoniger 102
donisthorpei, Monomorium 70
donisthorpei, Monomorium (Mitara) 70
doriae dentatus, Dolichoderus 93
doriae, Dolichoderus 93
doriae extensispina, Dolichoderus 93
dorycus, Camponotus 112
dorycus confusus, Camponotus 112
dorycus, Formica 112
DORYLINAE 2 , 52

Dorylozelus 32
Dorylozelus mjobergi 34
douglasi, Rhytidoponera 42 , 43
dromas, Camponotus 112
dromas, Camponotus (Myrmocamelus) 112
dromus, Iridomyrmex 98
dromus, Phyracaces 25
dubia, Rhytidoponera 43
dubia, Rhytidoponera (Chalcoponera) 43
dubitata, Bothroponera 21
dubitata, Ponera (Bothroponera) 21
dubius, Opisthopsis diadematus 127
duchaussoyi, Eusphinctus 50
duchaussoyi, Sphinctomyrmex 50
duyfkeni, Meranoplus diversus 67

ebenina, Leptogenys 33
ebenina, Leptogenys (Lobopelta) 33
ebeninithorax, Camponotus capito 110
ebeninithorax, Camponotus (Myrmophyma) capito 110
eccentrica, Mesostruma 69
Echinopla 121
Echinopla australis 121
Echinopla turneri 121
Echinopla turneri pictipes 122
Echinopla turneri turneri 121
Ectatomma (Acanthoponera) imbellis hilare 30
Ectatomma aciculata 40
(Ectatomma) aurata, Ponera 41
Ectatomma convexum 42
Ectatomma cristatum 42
Ectatomma impressum 44
Ectatomma impressum purpureum 46
Ectatomma mayri 45
Ectatomma nudatum 46
Ectatomma punctata 46
Ectatomma (Rhytidoponera) convexum gemma 49
Ectatomma (Rhytidoponera) convexum nodiferum 46
Ectatomma (Rhytidoponera) convexum rufescens 47
Ectatomma (Rhytidoponera) convexum spatiatum 45
Ectatomma (Rhytidoponera) convexum violaceum 49
Ectatomma (Rhytidoponera) cornutum 42
Ectatomma (Rhytidoponera) cornutum taurus 48
Ectatomma (Rhytidoponera) crassinode 42
Ectatomma (Rhytidoponera) cristatum caro 40
Ectatomma (Rhytidoponera) cristatum punctiventris 46
Ectatomma (Rhytidoponera) cyrus 41
Ectatomma (Rhytidoponera) haeckeli 44
Ectatomma (Rhytidoponera) impressum splendidum 47
Ectatomma (Rhytidoponera) maniae 45
Ectatomma (Rhytidoponera) mayri glabrius 45
Ectatomma (Rhytidoponera) metallicum cristulatum 48
Ectatomma (Rhytidoponera) metallicum modestum 49
Ectatomma (Rhytidoponera) metallicum obscurum 41
Ectatomma (Rhytidoponera) metallicum scrobiculatum 49
Ectatomma (Rhytidoponera) reticulatum 47
Ectatomma (Rhytidoponera) rothneyi 46
Ectatomma (Rhytidoponera) scaberrimum 47
Ectatomma (Rhytidoponera) socrus 48
Ectatomma (Rhytidoponera) spoliatum 48
Ectatomma (Rhytidoponera) tenue 48
Ectatomma (Rhytidoponera) turneri 49
Ectatomma (Rhytidoponera) victoriae 49

Ectatomma scabrum 48
Ectatomma socialis 23
ectatommoides, Camponotus 125
ectatommoides, Notoncus 125
Ectomomyrmex 29
Ectomomyrmex ruficornis 30
edentata, Syscia australis 24
edentatus, Cerapachys 24
elegans, Camponotus claripes 111
elegans, Cerapachys 25
elegans, Myrmecia 8
elegans, Phyracaces 25
elegans, Polyrhachis terpischore 141
elegans, Polyrhachis terpischore 141
elegans, Promyrmecia 8
elegans, Stigmacros 146
elegantulus, Orectognathus 73
elliotti, Colobostruma 60
elliotti, Epitritus 60
elliptica, Hypoponera 31
elliptica, Ponera truncata 31
elongata grossestriata, Podomyrma 82
elongata, Podomyrma 81
emdeni, Strumigenys 87
emeryi, Camponotus 133
emeryi, Cerapachys 25 , 51
emeryi clarus, Cerapachys 50
emeryi, Iridomyrmex 98
emeryi myops, Sphinctomyrmex 51
emeryi, Myrmecorhynchus 125
emeryi, Phyracaces 25
emmae, Epitritus 85
cmmae, Quadristruma 3 , 85
enigmatica, Rhytidoponera 43
enormis, Notoncus 125
ephippium, Camponotus 112
ephippium, Camponotus ephippium 113
ephippium ephippium, Camponotus 113
ephippium, Formica 113
ephippium narses, Camponotus 113
ephippium narses, Camponotus (Myrmocamelus) 113
epinotalis, Stigmacros 146
epinotalis, Stigmacros (Campostigmacros) 146
Epitritus elliotti 60
Epitritus emmae 85
Epopostruma 63
Epopostruma alinodis 59
Epopostruma (Colobostruma) leae 60
Epopostruma froggatti 60
Epopostruma frosti 63
Epopostruma monstrosa 64
Epopostruma quadrispinosa 64
Epopostruma quadrispinosa quadrispinosa 64
(Epopostruma) quadrispinosa, Strumigenys 64
(Epopostruma) turneri, Strumigenys 69
erato, Polyrhachis 133
eremicus, Camponotus 113
eremicus, Camponotus (Myrmogonia) 113
eremita, Rhytidoponera 43
Erimelophorus 122
erythrocephala, Formica 103
erythrocephalus basirufus, Leptomyrmex 103
erythrocephalus brunneiceps, Leptomyrmex 103
erythrocephalus clarki, Leptomyrmex 103
erythrocephalus cnemidatus, Leptomyrmex 103

erythrocephalus decipiens, Leptomyrmex 104
erythrocephalus erythrocephalus, Leptomyrmex 103
erythrocephalus, Leptomyrmex 103
erythrocephalus, Leptomyrmex erythrocephalus 103
erythrocephalus mandibularis, Leptomyrmex 104
erythrocephalus rufithorax, Leptomyrmex 104
erythrocephalus unctus, Leptomyrmex 104
erythrocephalus venustus, Leptomyrmex 104
erythropus, Camponotus 113
erythropus, Camponotus (Myrmosaga) 113
erythropygus, Pristomyrmex 85
esau, Camponotus 113
esau, Camponotus (Myrmocamelus) 113
esuriens, Myrmecia 8
eubos, Pheidole bos 75
Eubothroponera 37
Eubothroponera bicolor 38
Eubothroponera brunnipes 37
Eubothroponera dentinodis 37
Eubothroponera micans 37
Eubothroponera reticulata 38
Eubothroponera septentrionalis 38
eudoxia, Myrmecia forficata 7
eupoecila, Myrmecia 9
eupoecila, Promyrmecia 9
Euponera (Brachyponera) lutea clara 23
Euponera (Brachyponera) luteipes croceicornis 23
Euponera (Brachyponera) luteipes inops 23
Euponera (Brachyponera) rufonigra 52
Euponera (Trachymesopsus) rotundiceps 28
Euponera (Trachymesopus) clarki 52
Euponera (Trachymesopus) pachynoda 52
Eurhopalothrix 64
Eurhopalothrix australis 64
Eurhopalothrix procera 64
eurydice, Cremastogaster (Atopogyne) 61
eurydice, Crematogaster 61
Eusphinctus duchaussoyi 50
Eusphinctus (Eusphinctus) hackeri 50
Eusphinctus (Eusphinctus) occidentalis 51
(Eusphinctus) fallax cedaris, Sphinctomyrmex 50
(Eusphinctus) fallax hedwigae, Sphinctomyrmex 51
(Eusphinctus) fallax, Sphinctomyrmex 51
Eusphinctus fulvipes 51
(Eusphinctus) hackeri, Eusphinctus 50
Eusphinctus hirsutus 51
Eusphinctus (Nothosphinctus) brunnicornis 50
Eusphinctus (Nothosphinctus) fulvidus 50
Eusphinctus (Nothosphinctus) manni 50
Eusphinctus (Nothosphinctus) nigricans 51
Eusphinctus (Nothosphinctus) septentrionalis 51
Eusphinctus (Nothosphinctus) silaceus 50
(Eusphinctus) occidentalis, Eusphinctus 51
(Eusphinctus) steinheili, Sphinctomyrmex 51
euterpe, Polyrhachis 133
evae, Camponotus 113
evae, Camponotus evae 113
evae evae, Camponotus 113
evae zeuxis, Camponotus 113
evae zeuxis, Camponotus (Myrmogonia) 113
excavata acuticostata, Bothroponera 21
excavata acuticostata, Ponera (Bothroponera) 21
excavata, Bothroponera 21
excavata, Bothroponera excavata 21
excavata excavata, Bothroponera 21

excavata, Myrmecia 9
excavata, Promyrmecia 9
excavatus, Meranoplus 67
excisa excisa, Leptogenys 33
excisa, Leptogenys 33
excisa, Leptogenys excisa 33
excisa, Lobopelta 33
excisa major, Leptogenys 33
excisa major, Leptogenys (Lobopelta) 33
exedra, Ponera 38
exigua, Amblyopone 19
exigua, Crematogaster xerophila 63
exigua, Leptogenys conigera 33
exigua, Leptogenys (Lobopelta) conigera 33
exigua, Myrmecia 9
exigua, Plagiolepis 130
exigua, Promyrmecia 9
exigua quadrimaculata, Plagiolepis 130
exiguus, Aenictus 53
exlex, Polyrhachis daemeli 133
exlex, Polyrhachis (Hedomyrma) daemeli 133
exolympica, Mesostruma 69
exsanguis, Iridomyrmex 98
extensispina, Dolichoderus doriae 93
extensispinus, Dolichoderus 93
extensus, Camponotus 113
extreminigra, Stigmacros 146
extreminigra, Stigmacros (Cyrtostigmacros) 146
extreminigrum, Monomorium rubriceps 58
extreminigrus, Chelaner rubriceps 58
exulans, Polyrhachis 133
exulans, Polyrhachis (Myrmhopla) 133

falcata, Schizopelta 56
falcatus, Chelaner 56
fallax cedaris, Sphinctomyrmex (Eusphinctus) 50
fallax fallax, Leptogenys 33
fallax fortior, Leptogenys 34
fallax fortior, Leptogenys (Lobopelta) 34
fallax hedwigae, Sphinctomyrmex (Eusphinctus) 51
fallax, Leptogenys 33
fallax, Leptogenys fallax 33
fallax, Lobopelta 33
fallax, Myrmecia 14
fallax, Sphinctomyrmex (Eusphinctus) 51
famelicus, Camponotus subnitidus 119
fasciata, Myrmecia 9
fascigaster, Leptomyrmex darlingtoni 103
fastuosa, Rhytidoponera (Chalcoponera) 42
femorata, Myrmecia (Pristomyrmecia) piliventris 10
femorata, Podomyrma 81
femorata, Polyrhachis 133
femoratus, Polyrhachis 133
fenestratus, Meranoplus 67
ferocior, Strumigenys 87
ferruginea, Amblyopone 19
ferruginea, Dacryon 81
ferruginea longidens, Amblyopone 20
ferruginea, Myrmecia 9
ferruginea, Myrmecia nigriceps 9
ferruginea, Podomyrma 81
ferruginea, Rhytidoponera 43
ferruginea, Stigmacros 146
ferruginea, Stigmacros (Cyrtostigmacros) 146

ferrugineus, Meranoplus 67
ferruginipes, Camponotus (Myrmosaga) 114
fervidus, Cerapachys 25
fervidus, Phyracaces 25
fici, Phyracaces 25
ficosus, Cerapachys 25
ficosus, Phyracaces 25
fictor augustulus, Camponotus 113
fictor augustulus, Camponotus (Colobopsis) 113
fictor, Camponotus 113
fictor, Camponotus (Colobopsis) 113
fictor, Camponotus fictor 113
fictor fictor, Camponotus 113
fida, Iridomyrmex darwinianus 97
fida, Tapinoma (Doleromyrma) darwinianum 97
fieldeae, Camponotus 113
fieldellus, Camponotus 114
fieldi fieldi, Melophorus 122
fieldi major, Melophorus 122
fieldi, Melophorus 122
fieldi, Melophorus fieldi 122
fieldi, Monomorium 70
fieldi, Monomorium (Martia) 70
fieldi propinqua, Melophorus 123
flagellatus, Codiomyrmex 64
flagellatus, Glamyromyrmex 64
flammeus, Cerapachys 25
flammeus, Phyracaces 25
flammicollis, Myrmecia 9
flava, Rhytidoponera 41
flava, Stigmacros 146
flava, Stigmacros (Cyrtostigmacros) 146
flavescens, Phyracaces 25
flavibasis, Polyrhachis 133
flavibasis, Polyrhachis (Campomyrma) 133
flavicoma flavicoma, Myrmecia 9
flavicoma minuscula, Myrmecia 9
flavicoma, Myrmecia 9
flavicoma, Myrmecia flavicoma 9
flavicornis flavicornis, Prolasius 142
flavicornis minor, Prolasius 142
flavicornis, Prolasius 142
flavicornis, Prolasius flavicornis 142
flavicornis, Rhytidoponera 43
flavidiscus, Prolasius 142
flavidopubescens, Camponotus aeneopilosus 109
flavigaster, Chelaner 56
flavigaster, Xiphomyrmex 56
flavinodis, Stigmacros 146
flavipes, Chalcoponera 43
flavipes, Chelaner 56
flavipes, Hypoclinea 98
flavipes, Iridomyrmex 98
flavipes, Monomorium (Notomyrmex) 56
flavipes, Rhytidoponera 43
flavus, Bothriomyrmex 92
flavus, Iridomyrmex 98
flindersi, Rhytidoponera 43
foetans, Iridomyrmex 96
forceps, Myrmecia 9
forceps obscuriceps, Myrmecia 9
foreli, Acantholepis (Stigmacros) 146
foreli acuminata, Notoncus 125
foreli, Chelaner 56
foreli dentata, Notoncus 125

foreli, Monomorium (Holcomyrmex) 56
foreli, Notoncus 125
foreli, Notostigma 126
foreli, Rhytidoponera 43
foreli subdentata, Notoncus 125
forficata brevinoda, Myrmecia 6
forficata eudoxia, Myrmecia 7
forficata, Formica 10
forficata, Myrmecia 10
forficata rubra, Myrmecia 10
Formica aeneovirens 122
Formica agilis 115
Formica ammon 131
Formica amyoti 149
Formica anceps 96
Formica arcuata 131
Formica aurocincta 110
Formica australis 29
Formica consectator 111
Formica consobrina 111
Formica cordata 97
Formica cruentata 112
Formica detecta 101
Formica dorycus 112
Formica ephippium 113
Formica erythrocephala 103
Formica forficata 10
Formica gracilis 99
Formica gulosa 11
Formica hexacantha 135
Formica inequalis 149
Formica intrepida 115
Formica itincrans 100
Formica latreillii 136
Formica longipes 107
Formica macrocephala 100
Formica maculata 116
Formica minuta 149
Formica (Myrmecopsis) respiciens 128
Formica nigriceps 116
Formica nigriventris 104
Formica nigroaenea 117
Formica nitida 116
Formica nitidissima 143
Formica piliventris 119
Formica procidua 101
Formica purpurea 101
Formica purpurescens 149
Formica rastellata 139
Formica relucens 139
Formica rufonigra 102
Formica sexspinosa 140
Formica smaragdina 127
Formica smithii 101
Formica suffusa 119
Formica (Tapinoma) albipes 106
Formica terebrans 120
Formica testaceipes 120
Formica virescens 127
Formica viridis 127
Formica vitrea 120
FORMICIDAE 5
FORMICINAE 2, 107
FORMICOIDEA 1
formicoides, Melophorus 143

formicoides, Prolasius nitidissimus 143
formosa, Myrmecia maura 12
formosa, Myrmecia (Promyrmecia) aberrans 12
formosa, Myrmica 81
formosa, Podomyrma 81
formosus, Dolichoderus 93
formosus, Dolichoderus (Hypoclinea) 93
fornicatus, Iridomyrmex 98
fortior, Leptogenys fallax 34
fortior, Leptogenys (Lobopelta) fallax 34
fortirugis, Podomyrma 82
fortis, Amblyopone australis 19
fossulata, Acantholepis (Stigmacros) 146
foveifrons, Pheidole deserticola 76
foveolata, Amblyopone australis 19
foveolata, Rhytidoponera 43
foveolatus, Polyrhachis 94
foveolatus, Pristomyrmex 85
fraterculus barretti, Monomorium 70
fraterculus barretti, Monomorium (Lampromyrmex) 70
fraterculus fraterculus, Monomorium 70
fraterculus, Monomorium 70
fraterculus, Monomorium fraterculus 70
fraterculus, Monomorium (Mitara) laeve 70
fraudatrix, Melophorus iridescens 123
frenchi, Turneria 107
friedae, Strumigenys 87
frivola, Crematogaster 61
frivola, Crematogaster frivola 61
frivola frivola, Crematogaster 61
frivola sculpticeps, Cremastogaster 61
frivola sculpticeps, Crematogaster 61
frivolus, Cremastogaster 61
Froggattella 95
Froggattella kirbii 95
Froggattella kirbii bispinosa 95
Froggattella kirbii ianthina 95
Froggattella kirbii kirbii 95
Froggattella kirbii laticeps 95
Froggattella kirbii lutescens 95
Froggattella kirbii nigripes 95
Froggattella kirbyi bispinosa 95
Froggattella kirbyi ianthina 95
Froggattella kirbyi laticeps 95
Froggattella kirbyi lutescens 95
Froggattella kirbyi nigripes 95
Froggattella latispina 95
froggatti, Acantholepis (Acrostigma) 146
froggatti, Adlerzia 53
froggatti, Camponotus 114
froggatti, Colobostruma 60
froggatti, Cremastogaster sordidula 62
froggatti, Crematogaster queenslandica 62
froggatti, Epopostruma 60
froggatti imbecilis, Sphinctomyrmex 50
froggatti, Iridomyrmex 98
froggatti, Leptomyrmex 104
froggatti, Melophorus iridescens 123
froggatti, Meranoplus 67
froggatti, Monomorium (Adlerzia) 53
froggatti, Myrmecia 10
froggatti, Pheidole 54
froggatti, Polyrhachis 133
froggatti scabrula, Crematogaster 63
froggatti, Solenopsis 87

froggatti, Sphinctomyrmex 50
froggatti, Stigmacros 146
frontalis, Pheidole longiceps 77
frosti, Epopostruma 63
frosti, Hexadaceton 63
fucosa, Myrmecia 10
fucosa, Myrmecia (Promyrmecia) 10
(Fulakora) gracilis, Amblyopone 19
(Fulakora) lucida, Amblyopone 20
(Fulakora) punctulata, Amblyopone 20
fulgida, Myrmecia 10
fuliginosa, Rhytidoponera 43
fulviculis, Myrmecia 10
fulviculis, Myrmecia (Pristomyrmecia) fulvipes 10
fulvidus, Eusphinctus (Nothosphinctus) 50
fulvihirtus, Melophorus 123
fulvipes barbata, Myrmecia (Promyrmecia) 10
fulvipes caelatinoda, Myrmecia (Promyrmecia) 12
fulvipes, Eusphinctus 51
fulvipes fulviculis, Myrmecia (Pristomyrmecia) 10
fulvipes gilberti, Myrmecia 10
fulvipes, Myrmecia 10
fusca, Cremastogaster 61
fusca, Crematogaster 61
fuscipes, Myrmecia 8
fuscipes, Polyrhachis 133
fuscipes, Tetramorium 89
fuscipes, Xiphomyrmex turneri 89
fusciventris, Iridomyrmex gracilis 99
fusciventris, Rhytidoponera cornuta 48
fusciventris, Solenopsis 87

gab aegra, Polyrhachis 134
gab aegra, Polyrhachis (Chariomyrma) 134
gab gab, Polyrhachis 134
gab, Polyrhachis 134
gab, Polyrhachis gab 134
gab, Polyrhachis guerini 134
gab senilis, Polyrhachis 134
gab tripellis, Polyrhachis (Chariomyrma) 134
gasseri caloratus, Camponotus 114
gasseri caloratus, Camponotus (Colobopsis) 114
gasseri, Camponotus 114
gasseri, Camponotus gasseri 114
gasseri, Colobopsis 114
gasseri gasseri, Camponotus 114
gasseri lysias, Camponotus 114
gasseri lysias, Camponotus (Colobopsis) 114
gasseri obtusitruncatus, Camponotus 114
gasseri obtusitruncatus, Camponotus (Colobopsis) 114
gellibrandi, Pheidole 76
gemma, Ectatomma (Rhytidoponera) convexum 49
gibbinotus, Camponotus 114
gigas, Myrmecia pyriformis 7
gilberti annectens, Notoncus 126
gilberti, Chelaner 56
gilberti, Chelaner gilberti 56
gilberti, Cremastogaster sordidula 62
gilberti, Crematogaster queenslandica 62
gilberti gilberti, Chelaner 56
gilberti gracilior, Notoncus 126
gilberti, Iridomyrmex 98
gilberti mediorubra, Monomorium 56
gilberti mediorubrus, Chelaner 56

gilberti, Monomorium 56
gilberti, Myrmecia 10
gilberti, Myrmecia fulvipes 10
gilberti, Notoncus 126
gilesi, Cerapachys 25
gilesi, Phyracaces 25
gingivalis, Amblyopone 19
glaber clarithorax, Iridomyrmex 99
glaber glaber, Iridomyrmex 99
glaber, Iridomyrmex 3 , 99
glaber, Iridomyrmex glaber 99
glabra, Hypoclinea 99
glabrinota, Polyrhachis 134
glabrinotum, Polyrhachis (Myrmhopla) 134
glabrior, Paratrechina braueri 129
glabrior, Prenolepis braueri 129
glabrius, Ectatomma (Rhytidoponera) mayri 45
Glamyromyrmex 64
Glamyromyrmex flagellatus 64
Glamyromyrmex semicomptus 64
glauerti, Amblyopone 19
glauerti, Calomyrmex 108
glauerti, Dolichoderus 93
glauerti, Dolichoderus (Hypoclinea) 93
glauerti, Lithomyrmex 19
glauerti, Stigmacros 146
glauerti, Stigmacros (Cyrtostigmacros) 146
Gnamptogenys 30
Gnamptogenys biroi 30
godeffroyi, Strumigenys 87
(Goniothorax) australis, Leptothorax 65
goudiei, Dolichoderus 94
goudiei, Dolichoderus (Hypoclinea) 94
goudiei, Promyrmecia 17
gouldianus, Camponotus 114
gracilior, Notoncus gilberti 126
gracilis, Amblyopone 19
gracilis, Amblyopone (Fulakora) 19
gracilis, Formica 99
gracilis fusciventris, Iridomyrmex 99
gracilis gracilis, Iridomyrmex 99
gracilis, Iridomyrmex 99
gracilis, Iridomyrmex gracilis 99
gracilis mayri, Iridomyrmex 99
gracilis minor, Iridomyrmex 99
gracilis, Myrmecia 15
gracilis nugenti, Podomyrma 82
gracilis, Podomyrma 82
gracilis rubriceps, Iridomyrmex 99
gracilis spurcus, Iridomyrmex 99
graeffei, Anochetus 20
grandii, Pheidole anthracina 75
grandis, Cerapachys 25
grandis, Phyracaces 25
gratiosa, Myrmecia 11
gratiosa, Myrmecina 82
gratiosa, Podomyrma 82
gravis, Polyrhachis 134
gravis, Polyrhachis (Campomyrma) 134
greavesi, Cerapachys 25
greavesi, Myrmecia 11
greavesi, Phyracaces 25
greavesi, Probolomyrmex 39
greavesi, Promyrmecia 11
greavesi, Rhytidoponera 43

greavesi, Teratomyrmex 149
gregoryi, Rhytidoponera 44
grossestriata, Podomyrma 82
grossestriata, Podomyrma elongata 82
guerini aurea, Polyrhachis 131
guerini gab, Polyrhachis 134
guerini guerini, Polyrhachis 134
guerini lata, Polyrhachis 134
guerini pallescens, Polyrhachis 134
guerini, Polyrhachis 134
guerini, Polyrhachis guerini 134
guerini vermiculosa, Polyrhachis 134
gulosa, Formica 11
gulosa, Myrmecia 11
gulosa obscurior, Myrmecia 11
guttulata, Strumigenys 87
gwynethae, Cerapachys 26
gwynethae, Neophyracaces 26

hackeri, Amblyopone 19
hackeri, Eusphinctus (Eusphinctus) 50
hackeri, Leptogenys 34
hackeri, Leptogenys (Lobopelta) 34
hackeri, Leptomyrmex nigriventris 104
hackeri, Mayriella abstinens 65
haddoni haddoni, Opisthopsis 127
haddoni, Opisthopsis 127
haddoni, Opisthopsis haddoni 127
haddoni rufoniger, Opisthopsis 127
haeckeli, Ectatomma (Rhytidoponera) 44
haeckeli, Rhytidoponera 44
hacmatosticta, Myrmecia (Promyrmecia) aberrans 12
Hagiomyrma 130
(Hagiomyrma) ammonoeides crawleyi, Polyrhachis 131
Hagiostigmacros 144
(Hagiostigmacros) punctatissima, Stigmacros 148
(Hagiostigmacros) spinosa, Stigmacros 148
Halmamyrmecia 6
harderi, Myrmecia 11
harderi swalei, Myrmecia 16
hartmeyeri, Iridomyrmex 99
hartmeyeri, Pheidole 76
hartogi, Camponotus 114
hector, Polyrhachis 136
hecuba, Polyrhachis 134
hedleyi, Onychomyrmex 37
Hedomyrma 130
(Hedomyrma) aeschyle, Polyrhachis 131
(Hedomyrma) anguliceps, Polyrhachis 131
(Hedomyrma) barretti, Polyrhachis 132
(Hedomyrma) chrysothorax, Polyrhachis 132
(Hedomyrma) daemeli exlex, Polyrhachis 133
(Hedomyrma) humerosa, Polyrhachis 136
(Hedomyrma) kershawi, Polyrhachis 136
(Hedomyrma) machaon, Polyrhachis 137
(Hedomyrma) mjobergi, Polyrhachis 138
hedwigae, Sphinctomyrmex (Eusphinctus) fallax 51
heinlethii heinlethii, Polyrhachis 135
heinlethii, Polyrhachis 135
heinlethii, Polyrhachis heinlethii 135
heinlethii sophiae, Polyrhachis 135
hellenae, Prolasius 143
hemiflavus hemiflavus, Prolasius 143
hemiflavus, Prolasius 143

hemiflavus, Prolasius hemiflavus 143
hemiflavus wilsoni, Prolasius 143
hemiphaeum, Monomorium (Notomyrmex) 57
herbertonensis, Hypoponera 31
herbertonensis, Ponera pruinosa 31
hermione cupreata, Polyrhachis 135
hermione hermione, Polyrhachis 135
hermione, Polyrhachis 135
hermione, Polyrhachis hermione 135
heros, Cerapachys 26
heros, Phyracaces 26
hestia, Polyrhachis semipolita 140
Heteroponera 30
Heteroponera imbellis 30
Heteroponera leae 30
Heteroponera relicta 30
hexacantha, Formica 135
hexacantha, Polyrhachis 135
Hexadaceton 63
Hexadaceton frosti 63
hickmani, Notoncus 126
hilare, Ectatomma (Acanthoponera) imbellis 30
hilli, Aenictus 53
hilli, Chalcoponera 42
hilli, Meranoplus 67
hilli, Myrmecia 11
hilli, Promyrmecia 11
hilli, Rhytidoponera 44
hirsuta hirsuta, Polyrhachis 135
hirsuta, Myrmecia 11
hirsuta, Polyrhachis 135
hirsuta, Polyrhachis hirsuta 135
hirsuta quinquedentata, Polyrhachis 135
hirsuta quinquedentata, Polyrhachis (Campomyrma) 135
hirsuta, Stigmacros 147
hirsuta, Stigmacros (Chariostigmacros) 147
hirsutus, Eusphinctus 51
hirsutus hirsutus, Meranoplus 67
hirsutus, Melophorus 123
hirsutus, Meranoplus 67
hirsutus, Meranoplus hirsutus 67
hirsutus minor, Meranoplus 67
hirsutus, Pseudonotoncus 144
hirsutus rugosa, Meranoplus 67
(Holcomyrmex) armstrongi, Monomorium 56
(Holcomyrmex) foreli, Monomorium 56
(Holcomyrmex) niger, Monomorium 57
(Holcomyrmex) whitei, Monomorium 59
hookeri aerea, Polyrhachis 135
hookeri hookeri, Polyrhachis 135
hookeri lownei, Polyrhachis 135
hookeri obscura, Polyrhachis 135
hookeri, Polyrhachis 135
hookeri, Polyrhachis hookeri 135
horni, Camponotus 114
hospes, Meranoplus 67
howense, Monomorium (Notomyrmex) 56
howensis, Camponotus 115
howensis, Camponotus (Colobopsis) 115
howensis, Chelaner 56
howensis, Orectognathus 73
howensis, Orectognathus antennatus 73
humerosa, Polyrhachis 136
humerosa, Polyrhachis (Hedomyrma) 136
humilior, Camponotus maculatus 116

humilior, Chelaner rothsteini 58
humilior, Monomorium rothsteini 58
(Hypoclinea) angusticornis, Dolichoderus 93
(Hypoclinea) armstrongi, Dolichoderus 93
Hypoclinea flavipes 98
(Hypoclinea) formosus, Dolichoderus 93
Hypoclinea glabra 99
(Hypoclinea) glauerti, Dolichoderus 93
(Hypoclinea) goudiei, Dolichoderus 94
(Hypoclinea) nigricornis, Dolichoderus 94
(Hypoclinea) occidentalis, Dolichoderus 94
(Hypoclinea) parvus, Dolichoderus 94
(Hypoclinea) reflexus, Dolichoderus 94
(Hypoclinea) scabridus ruficornis, Dolichoderus 94
Hypoclinea scrobiculata 94
(Hypoclinea) tristis, Dolichoderus 93
(Hypoclinea) ypsilon nigra, Dolichoderus 95
(Hypoclinea) ypsilon rufotibialis, Dolichoderus 95
Hypoponera 31
Hypoponera congrua 31
Hypoponera convexiuscula 31
Hypoponera decora 31
Hypoponera elliptica 31
Hypoponera herbertonensis 31
Hypoponera mackayensis 31
Hypoponera mina 31
Hypoponera queenslandensis 31
Hypoponera rectidens 31
Hypoponera scitula 32
Hypoponera sulciceps 32

ianthina, Froggattella kirbii 95
ianthina, Froggattella kirbyi 95
ilia ilia, Monomorium 70
ilia lamingtonense, Monomorium 70
ilia lamingtonensis, Monomorium (Mitara) 70
ilia, Monomorium 70
ilia, Monomorium ilia 70
ilia, Monomorium (Martia) 70
imbecilis, Sphinctomyrmex 50
imbecilis, Sphinctomyrmex froggatti 50
imbellis, Acanthoponera 30
imbellis, Heteroponera 30
imbellis hilare, Ectatomma (Acanthoponera) 30
imbellis scabra, Acanthoponera (Anacanthoponera) 30
impasta, Plagiolepis clarki 130
impavidus, Calomyrmex 108
impavidus, Camponotus 108
impressa chalybaea, Rhytidoponera 41
impressa, Rhytidoponera 44
impressa, Stigmacros 147
impressa, Stigmacros (Stigmacros) 147
impressiceps, Pheidole 76
impressum, Ectatomma 44
impressum purpureum, Ectatomma 46
impressum splendidum, Ectatomma (Rhytidoponera) 47
impressum, Tetramorium 89
impressus, Xiphomyrmex 89
incerta, Iridomyrmex rufoniger 102
Incertae sedis 149
incertus, Iridomyrmex rufoniger 102
incisa, Rhytidoponera 44
inconspicua inconspicua, Polyrhachis 136
inconspicua, Polyrhachis 136

inconspicua, Polyrhachis inconspicua 136
inconspicua subnitens, Polyrhachis 136
inconspicuus, Phyracaces 26
incontentus, Cerapachys 26
incurvata, Pheidole 76
indistincta, Acropyga 107
inequalis, Formica 149
inermis, Podomyrma 82
inermis, Stigmacros 147
inermis, Stigmacros (Pseudostigmacros) 147
infima, Myrmecia 11
infima, Myrmecia picta 11
inflatus, Camponotus 115
innexus, Camponotus 115
innocens innocens, Iridomyrmex 99
innocens, Iridomyrmex 99
innocens, Iridomyrmex innocens 99
innocens malandanus, Iridomyrmex 99
inops, Euponera (Brachyponera) luteipes 23
inornata, Rhytidoponera 44
inornata, Rhytidoponera (Chalcoponera) metallica 44
inquilina, Myrmecia 11
insculpta, Solenopsis 87
insculptus, Solenopsis 87
insipidus, Camponotus 115
insolescens, Chelaner 57
insolescens, Monomorium (Notomyrmex) 57
insularis, Chelaner 57
insularis, Melophorus 123
insularis, Monomorium (Notomyrmex) 57
intacta, Acantholepis (Stigmacros) aemula 147
intacta, Stigmacros 147
integrum, Tapinoma minutum 106
intermedia, Bothroponera piliventris 22
intermedia, Ponera (Bothroponera) piliventris 22
intrepida, Formica 115
intrepidus bellicosus, Camponotus 115
intrepidus, Camponotus 115
intrepidus, Camponotus intrepidus 115
intrepidus intrepidus, Camponotus 115
intricata, Leptogenys 34
intricata, Leptogenys (Lobopelta) 34
inverallensis, Camponotus claripes 111
io, Polyrhachis (Campomyrma) thalia 141
io, Polyrhachis thalia 141
iridescens fraudatrix, Melophorus 123
iridescens froggatti, Melophorus 123
iridescens iridescens, Melophorus 123
iridescens, Melophorus 123
iridescens, Melophorus iridescens 123
iridescens, Myrmecocystus 123
Iridomyrmex 1 , 3 , 95
Iridomyrmex agilis 96
Iridomyrmex albitarsus 96
Iridomyrmex anceps 96
Iridomyrmex arcadius 96
Iridomyrmex bicknelli 96
Iridomyrmex bicknelli azureus 96
Iridomyrmex bicknelli bicknelli 96
Iridomyrmex bicknelli brunneus 96
Iridomyrmex bicknelli lutea 96
Iridomyrmex bicknelli splendidus 96
Iridomyrmex biconvexus 96
Iridomyrmex chasei 96 , 97
Iridomyrmex chasei chasei 97

Iridomyrmex chasei concolor 97
Iridomyrmex chasei yalgooensis 97
Iridomyrmex conifer 97
Iridomyrmex cordatus 97
Iridomyrmex cordatus stewartii 97
Iridomyrmex cyaneus 97
Iridomyrmex darwinianus 97
Iridomyrmex darwinianus darwinianus 97
Iridomyrmex darwinianus fida 97
Iridomyrmex darwinianus leae 97
Iridomyrmex detectus castrae 102
Iridomyrmex detectus sanguinea 102
Iridomyrmex detectus viridiaeneus 102
Iridomyrmex discors 97
Iridomyrmex discors aeneogaster 98
Iridomyrmex discors discors 97
Iridomyrmex discors obscurior 98
Iridomyrmex discors occipitalis 98
Iridomyrmex dromus 98
Iridomyrmex emeryi 98
Iridomyrmex exsanguis 98
Iridomyrmex flavipes 98
Iridomyrmex flavus 98
Iridomyrmex foetans 96
Iridomyrmex fornicatus 98
Iridomyrmex froggatti 98
Iridomyrmex gilberti 98
Iridomyrmex glaber 3 , 99
Iridomyrmex glaber clarithorax 99
Iridomyrmex glaber glaber 99
Iridomyrmex gracilis 99
Iridomyrmex gracilis fusciventris 99
Iridomyrmex gracilis gracilis 99
Iridomyrmex gracilis mayri 99
Iridomyrmex gracilis minor 99
Iridomyrmex gracilis rubriceps 99
Iridomyrmex gracilis spurcus 99
Iridomyrmex hartmeyeri 99
Iridomyrmex innocens 99
Iridomyrmex innocens innocens 99
Iridomyrmex innocens malandanus 99
Iridomyrmex itinerans 100
Iridomyrmex itinerans ballaratensis 100
Iridomyrmex itinerans depilis 100
Iridomyrmex itinerans itinerans 100
Iridomyrmex itinerans perthensis 100
Iridomyrmex longiceps 100
Iridomyrmex macrocephalus 100
Iridomyrmex mattiroloi 100
Iridomyrmex mattiroloi continentis 100
Iridomyrmex mattiroloi mattiroloi 100
Iridomyrmex mattiroloi parcens 100
Iridomyrmex mattiroloi splendens 100
Iridomyrmex mjobergi 100
Iridomyrmex nitida 101
Iridomyrmex nitidiceps 100
Iridomyrmex nitidus 101
Iridomyrmex nitidus clitellarius 101
Iridomyrmex nitidus nitidus 101
Iridomyrmex nitidus queenslandensis 101
Iridomyrmex obscurus 101
Iridomyrmex prociduus 101
Iridomyrmex punctatissimus 101
Iridomyrmex purpureus 101
Iridomyrmex purpureus castrae 102

Iridomyrmex purpureus purpureus 101
Iridomyrmex purpureus sanguinea 102
Iridomyrmex purpureus viridiaeneus 102
Iridomyrmex rostrinotus 98
Iridomyrmex rufoniger 102
Iridomyrmex rufoniger domestica 102
Iridomyrmex rufoniger incerta 102
Iridomyrmex rufoniger incertus 102
Iridomyrmex rufoniger pallidus 102
Iridomyrmex rufoniger rufoniger 102
Iridomyrmex rufoniger septentrionalis 102
Iridomyrmex rufoniger suchieri 102
Iridomyrmex rufoniger victorianus 102
Iridomyrmex vicina 103
Iridomyrmex viridigaster 103
(Ischnomyrmex) longiceps ruginota, Stenamma 54
ithona, Polyrhachis 136
ithonus, Polyrhachis 136
itinerans ballaratensis, Iridomyrmex 100
itinerans depilis, Iridomyrmex 100
itinerans, Formica 100
itinerans, Iridomyrmex 100
itinerans, Iridomyrmex itinerans 100
itinerans itinerans, Iridomyrmex 100
itinerans perthensis, Iridomyrmex 100

jacksoniana, Polyrhachis 136
janeti, Camponotus 115
javana cairnsiana, Pheidole 76
jocosus, Opisthopsis 127
jocosus, Technomyrmex 106
jovis, Cerapachys 26
jovis, Cerapachys (Phyracaces) 26
jucundus, Leptomyrmex darlingtoni 103

kanangra, Orectognathus 73
katerinae, Stenothorax 53
kershawi, Polyrhachis 136
kershawi, Polyrhachis (Hedomyrma) 136
kiliani, Chelaner 57
kiliani, Chelaner kiliani 57
kiliani kiliani, Chelaner 57
kiliani, Monomorium 57
kiliani obscurella, Monomorium 57
kiliani obscurellus, Chelaner 57
kiliani tambourinensis, Chelaner 57
kiliani tambourinensis, Monomorium 57
kimberleyensis, Sima punctulata 18
kimberleyensis, Tetraponera punctulata 18
kirbii, Acantholepis 95
kirbii bispinosa, Froggattella 95
kirbii, Froggattella 95
kirbii, Froggattella kirbii 95
kirbii ianthina, Froggattella 95
kirbii kirbii, Froggattella 95
kirbii laticeps, Froggattella 95
kirbii lutescens, Froggattella 95
kirbii nigripes, Froggattella 95
kirbyi bispinosa, Froggattella 95
kirbyi, Dolichoderus 95
kirbyi ianthina, Froggattella 95
kirbyi laticeps, Froggattella 95
kirbyi lutescens, Froggattella 95

kirbyi nigripes, Froggattella 95
kitschneri, Dacryon 82
kitschneri, Podomyrma 82
kraepelini, Podomyrma 82
kurandensis, Bothroponera sublaevis 22
kurandensis, Pachycondyla (Bothroponera) sublaevis 22
kurandensis, Rhytidoponera 44
kutteri, Cremastogaster 61
kutteri, Crematogaster 61

lachesis, Polyrhachis 136
laciniosa malandensis, Rhytidoponera 47
laeve broomense, Monomorium (Mitara) 70
laeve fraterculus, Monomorium (Mitara) 70
laeve laeve, Monomorium 71
laeve, Monomorium 71
laeve, Monomorium laeve 71
laeve nigrius, Monomorium 71
laeve nigrius, Monomorium (Mitara) 71
laeviceps broomensis, Cremastogaster 61
laeviceps broomensis, Crematogaster 61
laeviceps chasei, Cremastogaster 61
laeviceps chasei, Crematogaster 61
laeviceps clarior, Cremastogaster 61
laeviceps clarior, Crematogaster 61
laeviceps, Crematogaster 61
laeviceps, Crematogaster laeviceps 61
laeviceps laeviceps, Crematogaster 61
laeviceps, Pseudomyrma 18
laeviceps, Tetraponera 18
laevigata, Mesostruma 69
laevinodis, Promyrmecia 12
laevissima, Podomyrma 82
laevissima, Polyrhachis 137
lamellinodis, Rhytidoponera 44
lamellinodis, Rhytidoponera (Chalcoponera) 44
lamingtonense, Monomorium ilia 70
lamingtonensis, Monomorium (Mitara) ilia 70
(Lampromyrmex) fraterculus barretti, Monomorium 70
lampros, Podomyrma 82
lanaris, Stigmacros 147
lanaris, Stigmacros (Cyrtostigmacros) 147
lanuginosa, Triglyphothrix 91
lanuginosum, Tetramorium 91
larvatus, Cerapachys 26
larvatus, Phyracaces 26
lata, Polyrhachis guerini 134
laticephalum, Tetramorium 89
laticeps, Froggattella kirbii 95
laticeps, Froggattella kirbyi 95
laticeps, Melophorus 123
laticeps, Rhytidoponera 44
latigena, Pheidole variabilis 78
latispina, Froggattella 95
latreillii, Formica 136
latreillii, Polyrhachis 136
latrunculus, Camponotus 115
latrunculus, Camponotus latrunculus 115
latrunculus, Camponotus (Myrmoturba) 115
latrunculus latrunculus, Camponotus 115
latrunculus victoriensis, Camponotus 115
latrunculus victoriensis, Camponotus (Myrmoturba) 115
latunei, Anochetus turneri 21
latus, Cerapachys 26

leae, Acanthoponera (Anacanthoponera) 30
leae, Amblyopone 19
leae, Camponotus 116
leae, Camponotus (Myrmosphincta) 116
leae cedarensis, Polyrhachis 137
leae cedarensis, Polyrhachis (Campomyrma) 137
leae, Chelaner 57
leae, Colobostruma 60
leae, Discothyrea 29
leae, Epopostruma (Colobostruma) 60
leae, Heteroponera 30
leae, Iridomyrmex darwinianus 97
leae leae, Polyrhachis 136
leae, Lordomyrma 65
leae, Metapone 69
leae, Monomorium 57
leae norfolkensis, Ponera 38
leae oculata, Ponera 38
leae, Phyracaces 25
leae, Polyrhachis 136
leae, Polyrhachis leae 136
leae, Ponera 38
leae, Strumigenys 88
leda, Chelaner rothsteini 58
leda, Monomorium rothsteini 58
lepidus, Opisthopsis pictus 128
Leptanilla 2, 53
Leptanilla swani 53
LEPTANILLINAE 2
Leptogenys 32
Leptogenys angustinoda 32
Leptogenys anitae 32
Leptogenys bidentata 32
Leptogenys chelifer 32
Leptogenys clarki 32
Leptogenys conigera 32
Leptogenys conigera adlerzi 33
Leptogenys conigera centralis 33
Leptogenys conigera conigera 32
Leptogenys conigera exigua 33
Leptogenys conigera mutans 33
Leptogenys darlingtoni 33
Leptogenys diminuta 33
Leptogenys diminuta yarrabahna 33
Leptogenys ebenina 33
Leptogenys excisa 33
Leptogenys excisa excisa 33
Leptogenys excisa major 33
Leptogenys fallax 33
Leptogenys fallax fallax 33
Leptogenys fallax fortior 34
Leptogenys hackeri 34
Leptogenys intricata 34
Leptogenys (Lobopelta) angustinoda 32
Leptogenys (Lobopelta) anitae 32
Leptogenys (Lobopelta) bidentata 32
Leptogenys (Lobopelta) conigera adlerzi 33
Leptogenys (Lobopelta) conigera centralis 33
Leptogenys (Lobopelta) conigera exigua 33
Leptogenys (Lobopelta) conigera mutans 33
Leptogenys (Lobopelta) darlingtoni 33
Leptogenys (Lobopelta) diminuta yarrabahna 33
Leptogenys (Lobopelta) ebenina 33
Leptogenys (Lobopelta) excisa major 33
Leptogenys (Lobopelta) fallax fortior 34

Leptogenys (Lobopelta) hackeri 34
Leptogenys (Lobopelta) intricata 34
Leptogenys (Lobopelta) magna 34
Leptogenys (Lobopelta) mjobergi 34
Leptogenys (Lobopelta) neutralis 34
Leptogenys magna 34
Leptogenys mjobergi 34
Leptogenys neutralis 34
Leptogenys (Odontopelta) turneri longensis 34
Leptogenys sjostedti 34
Leptogenys tricosa 34
Leptogenys turneri 34
Leptogenys turneri longensis 34
Leptogenys turneri turneri 34
Leptomyrmex 103
Leptomyrmex darlingtoni 103
Leptomyrmex darlingtoni darlingtoni 103
Leptomyrmex darlingtoni fascigaster 103
Leptomyrmex darlingtoni jucundus 103
Leptomyrmex erythrocephalus 103
Leptomyrmex erythrocephalus basirufus 103
Leptomyrmex erythrocephalus brunneiceps 103
Leptomyrmex erythrocephalus clarki 103
Leptomyrmex erythrocephalus cnemidatus 103
Leptomyrmex erythrocephalus decipiens 104
Leptomyrmex erythrocephalus erythrocephalus 103
Leptomyrmex erythrocephalus mandibularis 104
Leptomyrmex erythrocephalus rufithorax 104
Leptomyrmex erythrocephalus unctus 104
Leptomyrmex erythrocephalus venustus 104
Leptomyrmex froggatti 104
Leptomyrmex mjobergi 104
Leptomyrmex nigriventris 104
Leptomyrmex nigriventris hackeri 104
Leptomyrmex nigriventris nigriventris 104
Leptomyrmex nigriventris tibialis 104
Leptomyrmex unicolor 105
Leptomyrmex varians 105
Leptomyrmex varians angusticeps 105
Leptomyrmex varians quadricolor 105
Leptomyrmex varians rothneyi 105
Leptomyrmex varians ruficeps 105
Leptomyrmex varians rufipes 105
Leptomyrmex varians varians 105
Leptomyrmex wiburdi 105
Leptomyrmex wiburdi pictus 105
Leptomyrmex wiburdi wiburdi 105
Leptothorax 64
Leptothorax australis 65
Leptothorax (Goniothorax) australis 65
Leptothorax nudus 55
levior, Polyrhachis 137
levior, Rhytidoponera punctata 42
levis, Diacamma australe 29
liber, Dacryon 83
libra, Podomyrma 83
lienosus, Opisthopsis 127
linae, Meranoplus aureolus 66
Liometopum aeneum 101
Liomyrmex (Machomyrma) dispar 65
Lioponera longitarsus 26
Lioponera longitarsus australis 26
lippulum, Monomorium 54
liteae, Pheidole 76
Lithomyrmex 18
Lithomyrmex glauerti 19
lividicoxis, Camponotus 116
lividicoxis, Camponotus (Myrmophyma) 116
lividipes, Camponotus nigriceps 117
(Lobopelta) angustinoda, Leptogenys 32
(Lobopelta) anitae, Leptogenys 32
(Lobopelta) bidentata, Leptogenys 32
Lobopelta conigera 32
(Lobopelta) conigera adlerzi, Leptogenys 33
(Lobopelta) conigera centralis, Leptogenys 33
(Lobopelta) conigera exigua, Leptogenys 33
(Lobopelta) conigera mutans, Leptogenys 33
(Lobopelta) darlingtoni, Leptogenys 33
(Lobopelta) diminuta yarrabahna, Leptogenys 33
(Lobopelta) ebenina, Leptogenys 33
Lobopelta excisa 33
(Lobopelta) excisa major, Leptogenys 33
Lobopelta fallax 33
(Lobopelta) fallax fortior, Leptogenys 34
(Lobopelta) hackeri, Leptogenys 34
(Lobopelta) intricata, Leptogenys 34
(Lobopelta) magna, Leptogenys 34
(Lobopelta) mjobergi, Leptogenys 34
(Lobopelta) neutralis, Leptogenys 34
lombokensis, Polyrhachis 137
lombokensis yarrabahensis, Polyrhachis 137
lombokensis yarrabahensis, Polyrhachis (Myrmatopa) 137
longensis, Leptogenys (Odontopelta) turneri 34
longensis, Leptogenys turneri 34
longiceps, Aphaenogaster 54
longiceps, Chelaner 57
longiceps, Cremastogaster 62
longiceps, Crematogaster 61
longiceps, Crematogaster longiceps 62
longiceps curticeps, Crematogaster 62
longiceps doddi, Pheidole 77
longiceps frontalis, Pheidole 77
longiceps, Iridomyrmex 100
longiceps longiceps, Crematogaster 62
longiceps longiceps, Pheidole 76
longiceps, Monomorium (Notomyrmex) 57
longiceps, Myrmica 54
longiceps, Pheidole 76
longiceps, Pheidole longiceps 76
longiceps ruginota, Stenamma (Ischnomyrmex) 54
longidens, Amblyopone 20
longidens, Amblyopone ferruginea 20
longinodis, Camponotus (Dinomyrmex) subnitidus 119
longinodis, Camponotus subnitidus 119
longinodis, Myrmecia 12
longipes, Anoplolepis 3, 107
longipes, Formica 107
longipes, Polyrhachis (Campomyrma) 137
longitarsus australis, Lioponera 26
longitarsus, Cerapachys 26
longitarsus, Lioponera 26
Lordomyrma 65
Lordomyrma leae 65
Lordomyrma punctiventris 65
Lordomyrma rugosa 84
loweryi, Mesostruma 69
lownei, Camponotus 116
lownei, Polyrhachis hookeri 135
lucida, Amblyopone 20

lucida, Amblyopone (Fulakora) 20
lucida, Myrmecia 10
lucidior, Camponotus wiederkehri 121
lucidula, Plagiolepis 130
ludius ludius, Melophorus 123
ludius, Melophorus 123
ludius, Melophorus ludius 123
ludius sulla, Melophorus 123
ludlowi, Myrmecia chasei 7
lutea, Brachyponera 23
lutea, Brachyponera lutea 23
lutea clara, Brachyponera 23
lutea clara, Euponera (Brachyponera) 23
lutea, Iridomyrmex bicknelli 96
lutea lutea, Brachyponera 23
lutea, Myrmecia 8
lutea, Ponera 23
luteiforceps, Myrmecia 12
luteiforceps, Promyrmecia 12
luteipes croceicornis, Euponera (Brachyponera) 23
luteipes inops, Euponera (Brachyponera) 23
lutescens, Froggattella kirbii 95
lutescens, Froggattella kirbyi 95
lydiae, Polyrhachis schenkii 139
lysias, Camponotus (Colobopsis) gasseri 114
lysias, Camponotus gasseri 114
lysistrata, Polyrhachis 137
lysistrata, Polyrhachis (Myrmothrinax) 137

macareaveyi, Chelaner 57
machaon, Polyrhachis 137
machaon, Polyrhachis (Hedomyrma) 137
Machomyrma 65
Machomyrma dispar 65
(Machomyrma) dispar, Liomyrmex 65
Machomyrma silvestrii 53
mackayensis, Camponotus reticulatus 118
mackayensis, Hypoponera 31
mackayensis, Pheidole ampla 74
mackayensis, Ponera coarctata 31
mackayi, Polyrhachis 137
mackayi, Polyrhachis (Cyrtomyrma) 137
macrocephala, Formica 100
macrocephalus, Iridomyrmex 100
macrophthalma, Podomyrma 83
macrops, Cerapachys 26
macrops, Neophyracaces 26
macrops, Nothomyrmecia 5
macropus, Polyrhachis 137
maculata, Formica 116
maculata, Polyrhachis (Campomyrma) polymnia 138
maculata, Polyrhachis polymnia 138
maculatus, Camponotus 116
maculatus discors, Camponotus 112
maculatus humilior, Camponotus 116
maculatus yarrabahensis, Camponotus (Myrmoturba) 112
maculiventris, Podomyrma micans 83
magna, Leptogenys 34
magna, Leptogenys (Lobopelta) 34
magnus, Camponotus 115
magnus, Odontomachus coriarius 36
major, Leptogenys excisa 33
major, Leptogenys (Lobopelta) excisa 33
major, Melophorus fieldi 122
major, Opisthopsis 128
major, Stigmacros 147
major, Stigmacros (Cyrtostigmacros) 147
malandanus, Iridomyrmex innocens 99
malandensis, Myrmecia tarsata 15
malandensis, Prionogenys podenzanai 39
malandensis, Rhytidoponera laciniosa 47
maledicta, Rhytidoponera 44
maledicta, Rhytidoponera (Chalcoponera) victoriae 44
maloni, Promyrmecia 11
mamillatus, Acantholepis 102
mandibularis, Amblyopone 19
mandibularis aureorufa, Myrmecia 12
mandibularis, Leptomyrmex erythrocephalus 104
mandibularis, Myrmecia 12
mandibularis postpetiolaris, Myrmecia (Promyrmecia) 12
maniae, Ectatomma (Rhytidoponera) 45
maniae, Rhytidoponera 45
manni, Eusphinctus (Nothosphinctus) 50
manni norfolkensis, Oligomyrmex 72
marcens, Camponotus claripes 111
marginata, Podomyrma 83
marginata, Stigmacros 147
marginata, Stigmacros (Campostigmacros) 147
marginatus, Dacryon 83
marius, Melophorus 123
marmorata, Promyrmecia 17
mars ajax, Meranoplus 68
mars mars, Meranoplus 68
mars, Meranoplus 68
mars, Meranoplus mars 68
(Martia) fieldi, Monomorium 70
(Martia) ilia, Monomorium 70
mattiroloi continentis, Iridomyrmex 100
mattiroloi, Iridomyrmex 100
mattiroloi, Iridomyrmex mattiroloi 100
mattiroloi mattiroloi, Iridomyrmex 100
mattiroloi parcens, Iridomyrmex 100
mattiroloi splendens, Iridomyrmex 100
maura formosa, Myrmecia 12
maura maura, Myrmecia 12
maura, Myrmecia 12
maura, Myrmecia maura 12
maura, Myrmecia (Promyrmecia) aberrans 12
maurus, Opisthopsis 128
mayri, Bothroponera 22
mayri, Ectatomma 45
mayri glabrius, Ectatomma (Rhytidoponera) 45
mayri, Iridomyrmex gracilis 99
mayri, Rhytidoponera 45
mayri, Strumigenys 87
Mayriella 65
Mayriella abstinens 65
Mayriella abstinens abstinens 65
Mayriella abstinens hackeri 65
Mayriella abstinens venustula 66
Mayriella overbecki 65
Mayriella spinosior 66
mediana, Rhytidoponera rothneyi 46
mediofusca, Pheidole variabilis 78
medioreticulata, Acantholepis (Stigmacros) 147
medioreticulata, Stigmacros 147
mediorubra, Monomorium gilberti 56
mediorubra, Myrmecia pilosula 7
mediorubrus, Chelaner gilberti 56

megalops, Tetramorium 89
melanaria australis, Ponera 35
melleum, Tetramorium 86
melleus, Rhoptromyrmex 86
Melophorus 122
Melophorus aeneovirens 122
Melophorus bagoti 122
Melophorus biroi 122
(Melophorus) brunea, Melophorus 122
Melophorus bruneus 122
Melophorus constans 122
Melophorus curtus 122
Melophorus depressiceps 142
Melophorus fieldi 122
Melophorus fieldi fieldi 122
Melophorus fieldi major 122
Melophorus fieldi propinqua 123
Melophorus formicoides 143
Melophorus fulvihirtus 123
Melophorus hirsutus 123
Melophorus insularis 123
Melophorus iridescens 123
Melophorus iridescens fraudatrix 123
Melophorus iridescens froggatti 123
Melophorus iridescens iridescens 123
Melophorus laticeps 123
Melophorus ludius 123
Melophorus ludius ludius 123
Melophorus ludius sulla 123
Melophorus marius 123
Melophorus (Melophorus) brunea 122
Melophorus mjobergi 123
Melophorus omniparens 124
Melophorus pillipes 124
Melophorus potteri 124
Melophorus scipio 124
Melophorus spinisquamis 126
Melophorus turneri 124
Melophorus turneri aesopus 124
Melophorus turneri candida 124
Melophorus turneri candidus 124
Melophorus turneri perthensis 124
Melophorus turneri turneri 124
Melophorus wheeleri 124
Meranoplus 66
Meranoplus aureolus 66
Meranoplus aureolus aureolus 66
Meranoplus aureolus doddi 66
Meranoplus aureolus linae 66
Meranoplus barretti 66
Meranoplus crawleyi 68
Meranoplus dichrous 66
Meranoplus dimidiatus 66
Meranoplus diversus 66
Meranoplus diversus diversus 66
Meranoplus diversus duyfkeni 67
Meranoplus diversus oxleyi 67
Meranoplus diversus unicolor 67
Meranoplus excavatus 67
Meranoplus fenestratus 67
Meranoplus ferrugineus 67
Meranoplus froggatti 67
Meranoplus hilli 67
Meranoplus hirsutus 67
Meranoplus hirsutus hirsutus 67
Meranoplus hirsutus minor 67
Meranoplus hirsutus rugosa 67
Meranoplus hospes 67
Meranoplus mars 68
Meranoplus mars ajax 68
Meranoplus mars mars 68
Meranoplus minimus 68
Meranoplus minor 68
Meranoplus mjobergi 68
Meranoplus oceanicus 68
Meranoplus pubescens 68
Meranoplus puryi 68
Meranoplus puryi curvispina 68
Meranoplus puryi puryi 68
Meranoplus similis 68
Meranoplus testudineus 68
mercovichi, Amblyopone 20
Mesoponera 34
Mesoponera australis 35
Mesostruma 68
Mesostruma browni 69
Mesostruma eccentrica 69
Mesostruma exolympica 69
Mesostruma laevigata 69
Mesostruma loweryi 69
Mesostruma turneri 69
metallica aspera, Ponera 40
metallica carbonaria, Chalcoponera 44
metallica inornata, Rhytidoponera (Chalcoponera) 44
metallica, Ponera 45
metallica purpurascens, Rhytidoponera (Chalcoponera) 45
metallica, Rhytidoponera 45
metallica tasmaniensis, Rhytidoponera 48
metallica varians, Rhytidoponera (Chalcoponera) 45
metallicum cristulatum, Ectatomma (Rhytidoponera) 48
metallicum modestum, Ectatomma (Rhytidoponera) 49
metallicum obscurum, Ectatomma (Rhytidoponera) 41
metallicum scrobiculatum, Ectatomma (Rhytidoponera) 49
Metapone 69
Metapone leae 69
Metapone mjobergi 69
Metapone tillyardi 69
Metapone tricolor 69
micans, Eubothroponera 37
micans maculiventris, Podomyrma 83
micans micans, Podomyrma 83
micans micans, Polyrhachis 137
micans ops, Polyrhachis 137
micans, Platythyrea 37
micans, Podomyrma 83
micans, Podomyrma micans 83
micans, Polyrhachis 137
micans, Polyrhachis micans 137
micans, Rhytidoponera 45
micans sericeiventris, Podomyrma 80
michaelseni, Amblyopone 20
michaelseni, Camponotus 116
michaelseni michaelseni, Myrmecia 12
michaelseni, Myrmecia 12
michaelseni, Myrmecia michaelseni 12
michaelseni overbecki, Myrmecia 13
michaelseni perthensis, Myrmecia 12
michaelseni queenslandica, Myrmecia 12

michaelseni rugosa, Myrmecia (Promyrmecia) 15
(Micromyrma) minutum cephalicum, Tapinoma 106
(Micromyrma) rottnestense, Tapinoma 106
micron, Monomorium 71
midas, Camponotus 116
midas, Myrmecia 13
mina, Hypoponera 31
mina, Ponera 31
minima, Camponotus (Myrmophyma) claripes 111
minimus, Camponotus claripes 111
minimus, Meranoplus 68
minor, Amblyopone australis 19
minor, Iridomyrmex gracilis 99
minor, Meranoplus 68
minor, Meranoplus hirsutus 67
minor, Notoncus capitatus. 125
minor, Prolasius flavicornis 142
minor, Stigmacros 147
minor, Stigmacros (Stigmacros) 147
minuscula, Myrmecia flavicoma 9
minuta, Formica 149
minutula, Paratrechina 129
minutula, Prenolepis 129
minutum broomense, Tapinoma 106
minutum broomensis, Tapinoma 106
minutum cephalicum, Tapinoma 106
minutum cephalicum, Tapinoma (Micromyrma) 106
minutum integrum, Tapinoma 106
minutum minutum, Tapinoma 105
minutum, Tapinoma 105
minutum, Tapinoma minutum 105
mirabilis, Rhytidoponera 46
Mitara 70
(Mitara) donisthorpei, Monomorium 70
(Mitara) ilia lamingtonensis, Monomorium 70
(Mitara) laeve broomense, Monomorium 70
(Mitara) laeve fraterculus, Monomorium 70
(Mitara) laeve nigrius, Monomorium 71
(Mitara) sydneyense nigella, Monomorium 71
mixta, Cremastogaster scita 63
mixta, Crematogaster scita 63
mjobergi, Cerapachys 26
mjobergi, Cerapachys (Phyracaces) 26
mjobergi, Cremastogaster 62
mjobergi, Crematogaster 62
mjobergi, Dacryon 83
mjobergi, Dorylozelus 34
mjobergi, Iridomyrmex 100
mjobergi, Leptogenys 34
mjobergi, Leptogenys (Lobopelta) 34
mjobergi, Leptomyrmex 104
mjobergi, Melophorus 123
mjobergi, Meranoplus 68
mjobergi, Metapone 69
mjobergi, Myrmecia 13
mjobergi, Notoncus 125
mjobergi, Oligomyrmex 72
mjobergi, Onychomyrmex 37
mjobergi, Orectognathus 73
mjobergi, Pheidole 77
mjobergi, Pheidole (Pheidolacanthinus) 77
mjobergi, Pheidologeton australis 79
mjobergi, Podomyrma 83
mjobergi, Polyrhachis 138
mjobergi, Polyrhachis (Hedomyrma) 138

mjobergi, Ponera 28
mjobergi, Prenolepis 143
mjobergi, Sphinctomyrmex 51
mjobergi, Sphinctomyrmex clarus 51
mjobergi unicolor, Orectognathus 73
mjoebergella, Prenolepis 143
mjoebergella, Prolasius 143
modestum, Ectatomma (Rhytidoponera) metallicum 49
moestus, Opisthopsis respiciens 128
molossus, Camponotus 116
moluccana, Acropyga 107
moluccana australis, Acropyga 107
Monomorium 70
Monomorium (Adlerzia) froggatti 53
Monomorium australicum 70
Monomorium bicorne 56
Monomorium broomense 70
Monomorium centrale 56
Monomorium donisthorpei 70
Monomorium fieldi 70
Monomorium fraterculus 70
Monomorium fraterculus barretti 70
Monomorium fraterculus fraterculus 70
Monomorium gilberti 56
Monomorium gilberti mediorubra 56
Monomorium (Holcomyrmex) armstrongi 56
Monomorium (Holcomyrmex) foreli 56
Monomorium (Holcomyrmex) niger 57
Monomorium (Holcomyrmex) whitei 59
Monomorium ilia 70
Monomorium ilia ilia 70
Monomorium ilia lamingtonense 70
Monomorium kiliani 57
Monomorium kiliani obscurella 57
Monomorium kiliani tambourinensis 57
Monomorium laeve 71
Monomorium laeve laeve 71
Monomorium laeve nigrius 71
Monomorium (Lampromyrmex) fraterculus barretti 70
Monomorium leae 57
Monomorium lippulum 54
Monomorium (Martia) fieldi 70
Monomorium (Martia) ilia 70
Monomorium micron 71
Monomorium (Mitara) donisthorpei 70
Monomorium (Mitara) ilia lamingtonensis 70
Monomorium (Mitara) laeve broomense 70
Monomorium (Mitara) laeve fraterculus 70
Monomorium (Mitara) laeve nigrius 71
Monomorium (Mitara) sydneyense nigella 71
Monomorium (Notomyrmex) flavipes 56
Monomorium (Notomyrmex) hemiphaeum 57
Monomorium (Notomyrmex) howense 56
Monomorium (Notomyrmex) insolescens 57
Monomorium (Notomyrmex) insularis 57
Monomorium (Notomyrmex) longiceps 57
Monomorium (Notomyrmex) sanguinolentum 58
Monomorium (Notomyrmex) sculpturatum 58
Monomorium occidaneus 57
Monomorium (Parаholcomyrmex) rothsteini doddi 58
Monomorium rothsteini 58
Monomorium rothsteini humilior 58
Monomorium rothsteini leda 58
Monomorium rothsteini squamigena 58
Monomorium rothsteini tostum 58

Monomorium rubriceps 58
Monomorium rubriceps cinctum 58
Monomorium rubriceps extreminigrum 58
Monomorium rubriceps rubra 58
Monomorium sordidum 59
Monomorium sordidum nigriventris 59
Monomorium subapterum 59
Monomorium subapterum bogischi 59
Monomorium subcoecum australicum 70
Monomorium sydneyense 71
Monomorium sydneyense nigellum 71
Monomorium sydneyense sydneyense 71
monstrosa, Epopostruma 64
muckeli, Podomyrma 83
mullewanus, Cerapachys 26
mullewanus, Phyracaces 26
murina, Bothroponera sublaevis 22
murina, Myrmecia 15
murina, Pachycondyla (Bothroponera) sublaevis 22
musgravei, Myrmecorhynchus 125
mutans, Calomyrmex splendidus 108
mutans, Camponotus (Calomyrmex) splendidus 108
mutans, Leptogenys conigera 33
mutans, Leptogenys (Lobopelta) conigera 33
Myopias 35
Myopias tasmaniensis 35
Myopopone 35
Myopopone castanea 35
myoporus, Camponotus 116
myoporus, Camponotus (Tanaemyrmex) 116
myops, Acropyga 107
myops, Pheidole 54
myops, Sphinctomyrmex 51
myops, Sphinctomyrmex emeryi 51
(Myrmamblys) aurofasciatus, Camponotus 115
(Myrmatopa) lombokensis yarrabahensis, Polyrhachis 137
Myrmecia 2 , 3 , 6
Myrmecia aberrans 6
Myrmecia affinis 16
Myrmecia analis 6
Myrmecia arnoldi 6
Myrmecia atrata 6
Myrmecia atriscapa 6
Myrmecia auriventris 6
Myrmecia auriventris athertonensis 6
Myrmecia brevinoda 6
Myrmecia callima 7
Myrmecia cardigaster 7
Myrmecia celaena 7
Myrmecia cephalotes 7
Myrmecia chasei 7
Myrmecia chasei chasei 7
Myrmecia chasei ludlowi 7
Myrmecia chrysogaster 7
Myrmecia clarki 7
Myrmecia comata 7
Myrmecia cordata 7
Myrmecia crassinoda 14
Myrmecia crudelis 16
Myrmecia cydista 8
Myrmecia decipians 8
Myrmecia desertorum 8
Myrmecia dichospila 8
Myrmecia dimidiata 8
Myrmecia dispar 8
Myrmecia dixoni 8
Myrmecia elegans 8
Myrmecia esuriens 8
Myrmecia eupoecila 9
Myrmecia excavata 9
Myrmecia exigua 9
Myrmecia fallax 14
Myrmecia fasciata 9
Myrmecia ferruginea 9
Myrmecia flammicollis 9
Myrmecia flavicoma 9
Myrmecia flavicoma flavicoma 9
Myrmecia flavicoma minuscula 9
Myrmecia forceps 9
Myrmecia forceps obscuriceps 9
Myrmecia forficata 10
Myrmecia forficata brevinoda 6
Myrmecia forficata eudoxia 7
Myrmecia forficata rubra 10
Myrmecia froggatti 10
Myrmecia fucosa 10
Myrmecia fulgida 10
Myrmecia fulviculis 10
Myrmecia fulvipes 10
Myrmecia fulvipes gilberti 10
Myrmecia fuscipes 8
Myrmecia gilberti 10
Myrmecia gracilis 15
Myrmecia gratiosa 11
Myrmecia greavesi 11
Myrmecia gulosa 11
Myrmecia gulosa obscurior 11
Myrmecia harderi 11
Myrmecia harderi swalei 16
Myrmecia hilli 11
Myrmecia hirsuta 11
Myrmecia infima 11
Myrmecia inquilina 11
Myrmecia longinodis 12
Myrmecia lucida 10
Myrmecia lutea 8
Myrmecia luteiforceps 12
Myrmecia mandibularis 12
Myrmecia mandibularis aureorufa 12
Myrmecia maura 12
Myrmecia maura formosa 12
Myrmecia maura maura 12
Myrmecia michaelseni 12
Myrmecia michaelseni michaelseni 12
Myrmecia michaelseni overbecki 13
Myrmecia michaelseni perthensis 12
Myrmecia michaelseni queenslandica 12
Myrmecia midas 13
Myrmecia mjobergi 13
Myrmecia murina 15
Myrmecia nigra 13
Myrmecia nigriceps 13
Myrmecia nigriceps ferruginea 9
Myrmecia nigriscapa 13
Myrmecia nigriventris 16
Myrmecia nigrocincta 13
Myrmecia nobilis 13
Myrmecia occidentalis 13
Myrmecia opaca 13

Myrmecia paucidens 16
Myrmecia pavida 14
Myrmecia petiolata 14
Myrmecia picta 14
Myrmecia picta infima 11
Myrmecia picta nigra 13
Myrmecia picticeps 14
Myrmecia piliventris 14
Myrmecia piliventris rectidens 14
Myrmecia pilosula 14
Myrmecia pilosula mediorubra 7
Myrmecia potteri 14
Myrmecia princeps 8
Myrmecia (Pristomyrmecia) fulvipes fulviculis 10
Myrmecia (Pristomyrmecia) piliventris femorata 10
Myrmecia (Pristomyrmecia) regina 11
Myrmecia (Promyrmecia) aberrans formosa 12
Myrmecia (Promyrmecia) aberrans haematosticta 12
Myrmecia (Promyrmecia) aberrans maura 12
Myrmecia (Promyrmecia) aberrans sericata 10
Myrmecia (Promyrmecia) aberrans taylori 10
Myrmecia (Promyrmecia) dichospila 8
Myrmecia (Promyrmecia) fucosa 10
Myrmecia (Promyrmecia) fulvipes barbata 10
Myrmecia (Promyrmecia) fulvipes caelatinoda 12
Myrmecia (Promyrmecia) mandibularis postpetiolaris 12
Myrmecia (Promyrmecia) michaelseni rugosa 15
Myrmecia pulchra 14
Myrmecia pumilio 17
Myrmecia pyriformis 15
Myrmecia pyriformis gigas 7
Myrmecia regularis 15
Myrmecia rowlandi 15
Myrmecia rubicunda 15
Myrmecia rubripes 15
Myrmecia rufinodis 15
Myrmecia rufonigra 17
Myrmecia rugosa 15
Myrmecia sanguinea 15
Myrmecia simillima 16
Myrmecia singularis 9
Myrmecia spadicea 16
Myrmecia subfasciata 16
Myrmecia suttoni 16
Myrmecia swalei 16
Myrmecia tarsata 16
Myrmecia tarsata malandensis 15
Myrmecia tarsata rowlandi 15
Myrmecia tasmaniensis 8
Myrmecia tepperi 17
Myrmecia testaceipes 17
Myrmecia tricolor 16
Myrmecia tricolor rogeri 16
Myrmecia urens 17
Myrmecia varians 17
Myrmecia vindex 17
Myrmecia vindex basirufa 17
Myrmecia vindex desertorum 8
Myrmecia vindex vindex 17
Myrmecia walkeri 9
MYRMECIINAE 2 , 6
Myrmecina 71
Myrmecina gratiosa 82
Myrmecina rugosa 71

Myrmecocystus iridescens 123
Myrmecopsis 127
(Myrmecopsis) respiciens, Formica 128
Myrmecorhynchus 124
Myrmecorhynchus carteri 124
Myrmecorhynchus emeryi 125
Myrmecorhynchus musgravei 125
Myrmecorhynchus nitidus 125
Myrmecorhynchus rufithorax 125
(Myrmhopla) barnardi, Polyrhachis 132
(Myrmhopla) exulans, Polyrhachis 133
(Myrmhopla) glabrinotum, Polyrhachis 134
Myrmica adelaidae 80
Myrmica bicarinata 89
Myrmica formosa 81
Myrmica longiceps 54
Myrmica oblonga 92
Myrmica simillima 90
MYRMICINAE 2 , 53
Myrmocamelus 109
(Myrmocamelus) bigenus, Camponotus 110
(Myrmocamelus) cinereus amperei, Camponotus 110
(Myrmocamelus) dromas, Camponotus 112
(Myrmocamelus) ephippium narses, Camponotus 113
(Myrmocamelus) esau, Camponotus 113
(Myrmocamelus) pellax, Camponotus 118
(Myrmogonia) armstrongi, Camponotus 110
(Myrmogonia) cameratus, Camponotus 110
(Myrmogonia) eremicus, Camponotus 113
(Myrmogonia) evae zeuxis, Camponotus 113
(Myrmogonia) oetkeri voltai, Camponotus 117
(Myrmogonia) punctiventris, Camponotus 118
(Myrmogonia) sanguinea, Camponotus 118
(Myrmogonia) tumidus, Camponotus 120
Myrmophyma 109
(Myrmophyma) capito ebeninithorax, Camponotus 110
(Myrmophyma) ceriseipes, Camponotus 110
(Myrmophyma) chalceoides, Camponotus 110
(Myrmophyma) claripes minima, Camponotus 111
(Myrmophyma) claripes orbiculatopunctatus, Camponotus 111
(Myrmophyma) claripes piperatus, Camponotus 111
(Myrmophyma) darlingtoni, Camponotus 120
(Myrmophyma) lividicoxis, Camponotus 116
(Myrmophyma) nitidiceps, Camponotus 117
(Myrmophyma) rottnesti, Camponotus 120
(Myrmophyma) tristis, Camponotus 120
(Myrmosaga) afflatus, Camponotus 109
(Myrmosaga) chalceus, Camponotus 110
(Myrmosaga) erythropus, Camponotus 113
(Myrmosaga) ferruginipes, Camponotus 114
(Myrmosaulus) scutellus, Camponotus 121
(Myrmosaulus) versicolor, Camponotus 120
(Myrmosericus) cruentatus aspera, Camponotus 112
(Myrmosphincta) leae, Camponotus 116
(Myrmosphincta) whitei, Camponotus 121
(Myrmothrinax) lysistrata, Polyrhachis 137
(Myrmoturba) latrunculus, Camponotus 115
(Myrmoturba) latrunculus victoriensis, Camponotus 115
(Myrmoturba) maculatus yarrabahensis, Camponotus 112
(Myrmoturba) villosa, Camponotus 120
Mystrium 35
Mystrium camillae 35
nana, Paratrechina 129

nana, Paratrechina (Nylanderia) 129
nancyae, Colobostruma 60
nanus, Orectognathus 73
narses, Camponotus ephippium 113
narses, Camponotus (Myrmocamelus) ephippium 113
Neoamblyopone 18
Neophyracaces 23
Neophyracaces gwynethae 26
Neophyracaces macrops 26
Neophyracaces piliventris 27
Neophyracaces potteri 27
nereis, Cardiocondyla nuda 55
neutralis, Leptogenys 34
neutralis, Leptogenys (Lobopelta) 34
newmani, Phyracaces 25
nigella, Monomorium (Mitara) sydneyense 71
nigellum, Monomorium sydneyense 71
niger, Monomorium (Holcomyrmex) 57
niger, Prolasius 143
nigra, Acanthoponera 30
nigra, Dolichoderus (Hypoclinea) ypsilon 95
nigra, Dolichoderus ypsilon 95
nigra, Myrmecia 13
nigra, Myrmecia picta 13
nigra, Rhytidoponera 42
nigricans, Eusphinctus (**Nothosphinctus**) 51
nigricans, Sphinctomyrmex 51
nigriceps, Camponotus 116
nigriceps, Camponotus **nigriceps** 116
nigriceps clarior, Camponotus 117
nigriceps ferruginea, Myrmecia 9
nigriceps, Formica 116
nigriceps lividipes, Camponotus 117
nigriceps, Myrmecia 13
nigriceps nigriceps, Camponotus 116
nigriceps obniger, Camponotus 117
nigriceps pallidiceps, Camponotus 117
nigricornis, Dolichoderus 94
nigricornis, Dolichoderus (Hypoclinea) 94
nigripes, Froggattella kirbii 95
nigripes, Froggattella kirbyi 95
nigriscapa, Myrmecia 13
nigrius, Monomorium laeve 71
nigrius, Monomorium (Mitara) laeve 71
nigriventris, Cerapachys 27
nigriventris, Chelaner sordidus 59
nigriventris, Formica 104
nigriventris hackeri, Leptomyrmex 104
nigriventris, Leptomyrmex 104
nigriventris, Leptomyrmex nigriventris 104
nigriventris, Monomorium sordidum 59
nigriventris, Myrmecia 16
nigriventris nigriventris, Leptomyrmex 104
nigriventris, Orectognathus 73
nigriventris, Phyracaces 27
nigriventris, Prolasius 143
nigriventris tibialis, Leptomyrmex 104
nigroaenea, Formica 117
nigroaeneus, Camponotus 117
nigroaeneus, Camponotus nigroaeneus 117
nigroaeneus divus, Camponotus 117
nigroaeneus nigroaeneus, Camponotus 117
nigrocincta, Myrmecia 13
nitida, Dacryon 83
nitida, Formica 116

nitida, Iridomyrmex 101
nitida, Podomyrma 83
nitida, Rhytidoponera 46
nitida, Stigmacros 147
nitida, Stigmacros (Campostigmacros) 147
nitidiceps, Camponotus 117
nitidiceps, Camponotus (Myrmophyma) 117
nitidiceps, Iridomyrmex 100
nitidissima, Formica 143
nitidissimus formicoides, Prolasius 143
nitidissimus nitidissimus, Prolasius 143
nitidissimus, Prolasius 143
nitidissimus, Prolasius nitidissimus 143
nitidus clitellarius, Iridomyrmex 101
nitidus, Iridomyrmex 101
nitidus, Iridomyrmex nitidus 101
nitidus, Myrmecorhynchus 125
nitidus nitidus, Iridomyrmex 101
nitidus queenslandensis, Iridomyrmex 101
nobilis, Myrmecia 13
nobilis, Promyrmecia 13
nodifera, Rhytidoponera 46
nodiferum, Ectatomma (Rhytidoponera) convexum 46
norfolkensis, Oligomyrmex 72
norfolkensis, Oligomyrmex manni 72
norfolkensis, Pheidole ampla 79
norfolkensis, Ponera leae 38
Nothomyrmecia 2 , 3 , 5
Nothomyrmecia macrops 5
NOTHOMYRMECIINAE 2 , 5
Nothosphinctus 49
(Nothosphinctus) brunnicornis, Eusphinctus 50
(Nothosphinctus) fulvidus, Eusphinctus 50
(Nothosphinctus) manni, Eusphinctus 50
(Nothosphinctus) nigricans, Eusphinctus 51
(Nothosphinctus) septentrionalis, Eusphinctus 51
(Nothosphinctus) silaceus, Eusphinctus 50
(Notomyrmex) flavipes, Monomorium 56
(Notomyrmex) hemiphaeum, Monomorium 57
(Notomyrmex) howense, Monomorium 56
(Notomyrmex) insolescens, Monomorium 57
(Notomyrmex) insularis, Monomorium 57
(Notomyrmex) longiceps, Monomorium 57
(Notomyrmex) sanguinolentum, Monomorium 58
(Notomyrmex) sculpturatum, Monomorium 58
Notoncus 125
Notoncus capitatus 125
Notoncus capitatus minor 125
Notoncus ectatommoides 125
Notoncus enormis 125
Notoncus foreli 125
Notoncus foreli acuminata 125
Notoncus foreli dentata 125
Notoncus foreli subdentata 125
Notoncus gilberti 126
Notoncus gilberti annectens 126
Notoncus gilberti gracilior 126
Notoncus hickmani 126
Notoncus mjobergi 125
Notoncus politus 126
Notoncus rodwayi 125
Notoncus rotundiceps 126
Notoncus spinisquamis 126
Notostigma 126
Notostigma carazzii 126

Notostigma foreli 126
Notostigma podenzanai 126
Notostigma sanguinea 126
notterae, Camponotus cinereus 111
novaehollandiae, Camponotus 117
novemdentata, Podomyrma 83
nox, Polyrhachis 138
nox, Polyrhachis (Cyrtomyrma) 138
nuda atalanta, Cardiocondyla 55
nuda, Cardiocondyla 55
nuda nereis, Cardiocondyla 55
nuda, Podomyrma 83
nudata, Rhytidoponera 46
nudatum, Ectatomma 46
nudimalis, Camponotus claripes 111
nudus, Leptothorax 55
nugenti, Podomyrma gracilis 82
numeensis borealis, Rhytidoponera (Chalcoponera) 41
nuntius, Camponotus scratius 118
(Nylanderia) nana, Paratrechina 129
(Nylanderia) tasmaniensis, Prenolepis 129
nynganensis, Plagiolepis 130
Nystalomyrma 54
(Nystalomyrma) barbigula, Aphaenogaster 54

oblonga, Myrmica 92
oblonga, Vollenhovia 92
obniger, Camponotus nigriceps 117
obscura, Amblyopone 18
obscura, Odontomachus coriarius 36
obscura, Paratrechina 129
obscura, Podomyrma 83
obscura, Polyrhachis hookeri 135
obscura, Prenolepis 129
obscura vaga, Prenolepis 129
obscurella, Monomorium kiliani 57
obscurellus, Chelaner kiliani 57
obscuriceps, Myrmecia forceps 9
obscurior, Iridomyrmex discors 98
obscurior, Myrmecia gulosa 11
obscurior, Podomyrma adelaidae 80
obscurior, Podomyrma bimaculata 80
obscurum, Ectatomma (Rhytidoponera) metallicum 41
obscurus, Iridomyrmex 101
obtusitruncatus, Camponotus (Colobopsis) gasseri 114
obtusitruncatus, Camponotus gasseri 114
occidaneus, Chelaner 57
occidaneus, Monomorium 57
occidentalis, Acantholepis (Stigmacros) 147
occidentalis, Acanthoponera 30
occidentalis, Dolichoderus 94
occidentalis, Dolichoderus (Hypoclinea) 94
occidentalis, Eusphinctus (Eusphinctus) 51
occidentalis, Myrmecia 13
occidentalis, Promyrmecia 13
occidentalis, Rhytidoponera 45
occidentalis, Sphinctomyrmex 51
occidentalis, Stigmacros 147
occipitalis, Iridomyrmex discors 98
oceanicus, Meranoplus 68
ocior, Pheidole variabilis 78
Octella 71
(Octella) pachycerus, Oligomyrmex 72
octodentata, Podomyrma 84

oculata, Ponera 149
oculata, Ponera leae 38
ocyma, Pheidole variabilis 78
odae, Podomyrma 84
Odontomachus 36
Odontomachus cephalotes 36
Odontomachus coriarius 36
Odontomachus coriarius magnus 36
Odontomachus coriarius obscura 36
Odontomachus coriarius semicircularis 36
Odontomachus ruficeps 36
Odontomachus ruficeps acutidens 36
Odontomachus ruficeps rubriceps 36
Odontomachus ruficeps rufescens 36
Odontomachus ruficeps turneri 36
Odontomachus septentrionalis 36
Odontomachus sharpei 36
Odontomachus turneri 36
Odontomyrmex 84
Odontomyrmex quadridentatus 85
(Odontomyrmex) quadridentatus queenslandensis, Pristomyrmex 85
Odontopelta 32
(Odontopelta) turneri longensis, Leptogenys 34
Oecophylla 126
Oecophylla smaragdina 127
oetkeri, Camponotus 117
oetkeri, Camponotus oetkeri 117
oetkeri oetkeri, Camponotus 117
oetkeri voltai, Camponotus 117
oetkeri voltai, Camponotus (Myrmogonia) 117
Oligomyrmex 71
Oligomyrmex corniger 72
Oligomyrmex corniger corniger 72
Oligomyrmex corniger parvicornis 72
Oligomyrmex manni norfolkensis 72
Oligomyrmex mjobergi 72
Oligomyrmex norfolkensis 72
Oligomyrmex (Octella) pachycerus 72
Oligomyrmex pachycerus 72
omniparens, Dacryon 84
omniparens, Melophorus 124
omniparens, Podomyrma 84
Onychomyrmex 36
Onychomyrmex doddi 37
Onychomyrmex hedleyi 37
Onychomyrmex mjobergi 37
opaca, Myrmecia 13
opaca, Prionopelta 39
opaca, Promyrmecia 13
opaca, Strumigenys 88
opacior, Rhytidoponera convexa 49
opaciventris, Pheidole 77
Opisthopsis 127
Opisthopsis diadematus 127
Opisthopsis diadematus diadematus 127
Opisthopsis diadematus dubius 127
Opisthopsis haddoni 127
Opisthopsis haddoni haddoni 127
Opisthopsis haddoni rufoniger 127
Opisthopsis jocosus 127
Opisthopsis lienosus 127
Opisthopsis major 128
Opisthopsis maurus 128
Opisthopsis pictus 128

Opisthopsis pictus bimaculatus 128
Opisthopsis pictus lepidus 128
Opisthopsis pictus palliatus 128
Opisthopsis pictus pictus 128
Opisthopsis respiciens 128
Opisthopsis respiciens moestus 128
Opisthopsis respiciens respiciens 128
Opisthopsis rufithorax 128
ops, Polyrhachis micans 137
orba, Pheidole anthracina 75
orbiculatopunctatus, Camponotus claripes 111
orbiculatopunctatus, Camponotus (Myrmophyma) claripes 111
orbis, Rhopalothrix 86
Orectognathus 72
Orectognathus alligator 72
Orectognathus antennatus 72
Orectognathus antennatus howensis 73
Orectognathus antennatus septentrionalis 72
Orectognathus clarki 72
Orectognathus coccinatus 72
Orectognathus darlingtoni 73
Orectognathus elegantulus 73
Orectognathus howensis 73
Orectognathus kanangra 73
Orectognathus mjobergi 73
Orectognathus mjobergi unicolor 73
Orectognathus nanus 73
Orectognathus nigriventris 73
Orectognathus parvispinus 73
Orectognathus perplexus 88
Orectognathus phyllobates 73
Orectognathus robustus 74
Orectognathus rostratus 74
Orectognathus satan 74
Orectognathus sexspinosus 74
Orectognathus versicolor 74
ornata, Polyrhachis 138
ornatum, Tetramorium 90
(Orthocrema) rufotestacea dentinasis, Crematogaster 63
overbecki, Mayriella 65
overbecki, Myrmecia michaelseni 13
overbecki overbecki, Podomyrma 84
overbecki, Peronomyrmex 74
overbecki, Podomyrma 84
overbecki, Podomyrma overbecki 84
overbecki varicolor, Podomyrma 84
oxleyi, Camponotus 117
oxleyi, Meranoplus diversus 67

pachycerus, Oligomyrmex 72
pachycerus, Oligomyrmex (Octella) 72
Pachycondyla astuta 21
Pachycondyla (Bothroponera) barbata 21
Pachycondyla (Bothroponera) piliventris regularis 22
Pachycondyla (Bothroponera) sublaevis kurandensis 22
Pachycondyla (Bothroponera) sublaevis murina 22
Pachycondyla (Bothroponera) tasmaniensis 38
Pachycondyla piliventris 22
pachynoda, Euponera (Trachymesopus) 52
pachynoda, Trachymesopus 52
pacificum, Tetramorium 90
pacificum validiusculum, Tetramorium 91
pallescens, Polyrhachis guerini 134

palliatus, Opisthopsis pictus 128
pallida, Crematogaster 62
pallidiceps, Camponotus nigriceps 117
pallidipes, Cremastogaster 62
pallidus, Crematogaster 62
pallidus, Iridomyrmex rufoniger 102
pallidus, Prolasius 143
pallipes, Cremastogaster 62
pallipes, Crematogaster 62
papuanum, Proceratium 40
papulata, Colobostruma 60
(Paraholcomyrmex) rothsteini doddi, Monomorium 58
parallela, Pheidole ampla 79
parallela, Platythyrea 37
parallela, Ponera 37
Paranomopone 30
Paranomopone relicta 30
Paratrechina 128
Paratrechina braueri 129
Paratrechina braueri braueri 129
Paratrechina braueri glabrior 129
Paratrechina minutula 129
Paratrechina nana 129
Paratrechina (Nylanderia) nana 129
Paratrechina obscura 129
Paratrechina rosae 129
Paratrechina tasmaniensis 129
Paratrechina vaga 129
parcens, Iridomyrmex mattiroloi 100
paripungens, Anochetus 20
parva, Platythyrea 37
parva, Podomyrma 81
parviceps, Pheidole ampla 74
parvicornis, Oligomyrmex corniger 72
parvispina, Pheidole variabilis 78
parvispinus, Orectognathus 73
parvus, Dolichoderus 94
parvus, Dolichoderus (Hypoclinea) 94
patiens, Polyrhachis 138
patiens, Polyrhachis (Campomyrma) 138
paucidens, Myrmecia 16
pavida, Myrmecia 14
pellax, Camponotus 118
pellax, Camponotus (Myrmocamelus) 118
penelope, Polyrhachis 138
peninsularis, Rhytidoponera 46
Peronomyrmex 74
Peronomyrmex overbecki 74
perplexa, Strumigenys 88
perplexus, Orectognathus 88
perstictus, Sphinctomyrmex 51
perthensis, Crematogaster 62
perthensis, Iridomyrmex itinerans 100
perthensis, Melophorus turneri 124
perthensis, Myrmecia michaelseni 12
perthensis, Pheidole ampla 75
perthensis, Polyrhachis (Campomyrma) sidnica 140
perthensis, Polyrhachis sidnica 140
petiolata, Myrmecia 14
petiolata, Rhytidoponera 45
(Pheidolacanthinus) mjobergi, Pheidole 77
Pheidole 3, 74
Pheidole ampla 74
Pheidole ampla ampla 74
Pheidole ampla mackayensis 74

Pheidole ampla norfolkensis 79
Pheidole ampla parallela 79
Pheidole ampla parviceps 74
Pheidole ampla perthensis 75
Pheidole ampla yarrensis 79
Pheidole anthracina 75
Pheidole anthracina anthracina 75
Pheidole anthracina grandii 75
Pheidole anthracina orba 75
Pheidole athertonensis 75
Pheidole athertonensis athertonensis 75
Pheidole athertonensis cedarensis 75
Pheidole athertonensis tambourinensis 75
Pheidole bos 75
Pheidole bos baucis 75
Pheidole bos bos 75
Pheidole bos eubos 75
Pheidole brevicornis 75
Pheidole cairnsiana 76
Pheidole concentrica 76
Pheidole concentrica concentrica 76
Pheidole concentrica recurva 76
Pheidole conficta 76
Pheidole deserticola 76
Pheidole deserticola deserticola 76
Pheidole deserticola foveifrons 76
Pheidole dolichocephala 79
Pheidole froggatti 54
Pheidole gellibrandi 76
Pheidole hartmeyeri 76
Pheidole impressiceps 76
Pheidole incurvata 76
Pheidole javana cairnsiana 76
Pheidole liteae 76
Pheidole longiceps 76
Pheidole longiceps doddi 77
Pheidole longiceps frontalis 77
Pheidole longiceps longiceps 76
Pheidole mjobergi 77
Pheidole myops 54
Pheidole opaciventris 77
Pheidole (Pheidolacanthinus) mjobergi 77
Pheidole platypus 77
Pheidole proxima 77
Pheidole proxima bombalensis 77
Pheidole proxima proxima 77
Pheidole proxima transversa 77
Pheidole pyriformis 77
Pheidole spinoda 77
Pheidole tasmaniensis 77
Pheidole tasmaniensis continentis 78
Pheidole tasmaniensis tasmaniensis 77
Pheidole trapezoidea 78
Pheidole turneri 78
Pheidole variabilis 78
Pheidole variabilis ampla 74
Pheidole variabilis latigena 78
Pheidole variabilis mediofusca 78
Pheidole variabilis ocior 78
Pheidole variabilis ocyma 78
Pheidole variabilis parvispina 78
Pheidole variabilis praedo 78
Pheidole variabilis redunca 78
Pheidole variabilis rugocciput 78
Pheidole variabilis rugosula 79

Pheidole variabilis variabilis 78
Pheidole vigilans 79
Pheidole wiesei 79
Pheidologeton 79
Pheidologeton affinis australis 79
Pheidologeton australis 79
Pheidologeton australis australis 79
Pheidologeton australis mjobergi 79
philiporum, Aenictus 53
phryne, Polyrhachis 138
phyllobates, Orectognathus 73
Phyracaces aberrans 23
(Phyracaces) adamus, Cerapachys 24
Phyracaces angustatus 24
Phyracaces bicolor 24
(Phyracaces) binodis, Cerapachys 24
Phyracaces brevicollis 24
Phyracaces brevis 24
Phyracaces castaneus 24
Phyracaces clarki 24
Phyracaces clarus 27
Phyracaces constricta 24
Phyracaces crassus 24
Phyracaces dromus 25
Phyracaces elegans 25
Phyracaces emeryi 25
Phyracaces fervidus 25
Phyracaces fici 25
Phyracaces ficosus 25
Phyracaces flammeus 25
Phyracaces flavescens 25
Phyracaces gilesi 25
Phyracaces grandis 25
Phyracaces greavesi 25
Phyracaces heros 26
Phyracaces inconspicuus 26
(Phyracaces) jovis, Cerapachys 26
Phyracaces larvatus 26
Phyracaces leae 25
(Phyracaces) mjobergi, Cerapachys 26
Phyracaces mullewanus 26
Phyracaces newmani 25
Phyracaces nigriventris 27
Phyracaces picipes 27
Phyracaces pictus 27
Phyracaces princeps 27
Phyracaces punctatissima 27
Phyracaces pygmaeus 26
Phyracaces reticulatus 26
Phyracaces ruficornis 27
Phyracaces rugulinodis 27
Phyracaces scrutator 25
Phyracaces senescens 27
Phyracaces simmonsae 28
(Phyracaces) singularis rotula, Cerapachys 28
(Phyracaces) sjostedti, Cerapachys 28
(Phyracaces) turneri, Cerapachys 28
Phyracaces varians 28
piceus, Crematogaster 62
picipes, Cerapachys 27
picipes, Phyracaces 27
picta infima, Myrmecia 11
picta, Myrmecia 14
picta nigra, Myrmecia 13
picticeps, Myrmecia 14

pictipes, Echinopla turneri 122
pictus bimaculatus, Opisthopsis 128
pictus, Cerapachys 27
pictus lepidus, Opisthopsis 128
pictus, Leptomyrmex wiburdi 105
pictus, Opisthopsis 128
pictus, Opisthopsis pictus 128
pictus palliatus, Opisthopsis 128
pictus, Phyracaces 27
pictus pictus, Opisthopsis 128
piliventris, Bothroponera 22
piliventris, Bothroponera piliventris 22
piliventris, Cerapachys 27
piliventris femorata, Myrmecia (Pristomyrmecia) 10
piliventris, Formica 119
piliventris intermedia, Bothroponera 22
piliventris intermedia, Ponera (Bothroponera) 22
piliventris, Myrmecia 14
piliventris, Neophyracaces 27
piliventris, Pachycondyla 22
piliventris piliventris, Bothroponera 22
piliventris rectidens, Myrmecia 14
piliventris regularis, Bothroponera 22
piliventris regularis, Pachycondyla (Bothroponera) 22
pillipes, Melophorus 124
pilosella, Acantholepis (Stigmacros) 147
pilosella, Stigmacros 147
pilosula mediorubra, Myrmecia 7
pilosula, Myrmecia 14
pilosula, Rhytidoponera 46
piperatus, Camponotus claripes 111
piperatus, Camponotus (Myrmophyma) claripes 111
Plagiolepis 129
Plagiolepis clarki 129 , 130
Plagiolepis clarki clarki 130
Plagiolepis clarki impasta 130
Plagiolepis exigua 130
Plagiolepis exigua quadrimaculata 130
Plagiolepis lucidula 130
Plagiolepis nynganensis 130
Plagiolepis squamulosa 130
platypus, Pheidole 77
Platythyrea 37
Platythyrea brunnipes 37
Platythyrea cephalotes 37
Platythyrea dentinodis 37
Platythyrea micans 37
Platythyrea parallela 37
Platythyrea parva 37
Platythyrea pusilla australis 37
Platythyrea turneri 38
podenzanai, Camponotus 126
podenzanai malandensis, Prionogenys 39
podenzanai, Notostigma 126
podenzanai podonzanai, Prionogenys 39
podenzanai, Prionogenys 39
Podomyrma 79
Podomyrma abdominalis 80
Podomyrma abdominalis pulchra 80
Podomyrma adelaidae 80
Podomyrma adelaidae adelaidae 80
Podomyrma adelaidae brevidentata 80
Podomyrma adelaidae obscurior 80
Podomyrma basalis 80
Podomyrma bimaculata 80

Podomyrma bimaculata brevidentata 80
Podomyrma bimaculata obscurior 80
Podomyrma bispinosa 80
Podomyrma castanea 84
Podomyrma chasei 80
Podomyrma christae 81
Podomyrma clarki 81
Podomyrma convergens 81
Podomyrma delbruckii 81
Podomyrma densestrigosa 81
Podomyrma densestrigosa densestrigosa 81
Podomyrma densestrigosa teres 81
Podomyrma elongata 81
Podomyrma elongata grossestriata 82
Podomyrma femorata 81
Podomyrma ferruginea 81
Podomyrma formosa 81
Podomyrma fortirugis 82
Podomyrma gracilis 82
Podomyrma gracilis nugenti 82
Podomyrma gratiosa 82
Podomyrma grossestriata 82
Podomyrma inermis 82
Podomyrma kitschneri 82
Podomyrma kraepelini 82
Podomyrma laevissima 82
Podomyrma lampros 82
Podomyrma libra 83
Podomyrma macrophthalma 83
Podomyrma marginata 83
Podomyrma micans 83
Podomyrma micans maculiventris 83
Podomyrma micans micans 83
Podomyrma micans sericeiventris 80
Podomyrma mjobergi 83
Podomyrma muckeli 83
Podomyrma nitida 83
Podomyrma novemdentata 83
Podomyrma nuda 83
Podomyrma obscura 83
Podomyrma octodentata 84
Podomyrma odae 84
Podomyrma omniparens 84
Podomyrma overbecki 84
Podomyrma overbecki overbecki 84
Podomyrma overbecki varicolor 84
Podomyrma parva 81
Podomyrma rugosa 84
Podomyrma striata 84
Podomyrma tricolor 84
Podomyrma turneri 84
podonzanai, Prionogenys podonzanai 39
politus, Notoncus 126
polymnia maculata, Polyrhachis 138
polymnia maculata, Polyrhachis (Campomyrma) 138
polymnia polymnia, Polyrhachis 138
polymnia, Polyrhachis 138
polymnia, Polyrhachis polymnia 138
Polyrhachis 130
Polyrhachis aeschyle 131
Polyrhachis ammon 131
Polyrhachis ammon ammon 131
Polyrhachis ammon angusta 131
Polyrhachis ammon angustata 131
Polyrhachis ammonoeides 131

Polyrhachis ammonoeides ammonoeides 131
Polyrhachis ammonoeides crawleyi 131
Polyrhachis andromache 136
Polyrhachis anguliceps 131
Polyrhachis appendiculata 131
Polyrhachis appendiculata appendiculata 131
Polyrhachis appendiculata schoopae 131
Polyrhachis arcuata 131
Polyrhachis aurea 131
Polyrhachis aurea aurea 131
Polyrhachis aurea depilis 132
Polyrhachis australis 137
Polyrhachis barnardi 132
Polyrhachis barretti 132
Polyrhachis bedoti 132
Polyrhachis bellicosa 132
Polyrhachis bellicosus 132
Polyrhachis (Campomyrma) flavibasis 133
Polyrhachis (Campomyrma) gravis 134
Polyrhachis (Campomyrma) hirsuta quinquedentata 135
Polyrhachis (Campomyrma) leae cedarensis 137
Polyrhachis (Campomyrma) longipes 137
Polyrhachis (Campomyrma) patiens 138
Polyrhachis (Campomyrma) polymnia maculata 138
Polyrhachis (Campomyrma) prometheus 138
Polyrhachis (Campomyrma) sidnica perthensis 140
Polyrhachis (Campomyrma) sidnica tambourinensis 140
Polyrhachis (Campomyrma) thalia io 141
Polyrhachis (Campomyrma) zimmerae 141
Polyrhachis cataulacoidea 132
Polyrhachis chalchas 132
Polyrhachis (Chariomyrma) gab aegra 134
Polyrhachis (Chariomyrma) gab tripellis 134
Polyrhachis chrysothorax 132
Polyrhachis cleopatra 132
Polyrhachis clio 132
Polyrhachis clotho 132
Polyrhachis comata 134
Polyrhachis connectens australiae 139
Polyrhachis constricta 132
Polyrhachis contemta 133
Polyrhachis crawleyella 134
Polyrhachis (Cyrtomyrma) doddi 133
Polyrhachis (Cyrtomyrma) mackayi 137
Polyrhachis (Cyrtomyrma) nox 138
Polyrhachis (Cyrtomyrma) rastellata yorkana 139
Polyrhachis (Cyrtomyrma) townsvillei 141
Polyrhachis daemeli 133
Polyrhachis daemeli argentosa 133
Polyrhachis daemeli daemeli 133
Polyrhachis daemeli exlex 133
Polyrhachis delicata 139
Polyrhachis doddi 133
Polyrhachis erato 133
Polyrhachis euterpe 133
Polyrhachis exulans 133
Polyrhachis femorata 133
Polyrhachis femoratus 133
Polyrhachis flavibasis 133
Polyrhachis foveolatus 94
Polyrhachis froggatti 133
Polyrhachis fuscipes 133
Polyrhachis gab 134
Polyrhachis gab aegra 134
Polyrhachis gab gab 134

Polyrhachis gab senilis 134
Polyrhachis glabrinota 134
Polyrhachis gravis 134
Polyrhachis guerini 134
Polyrhachis guerini aurea 131
Polyrhachis guerini gab 134
Polyrhachis guerini guerini 134
Polyrhachis guerini lata 134
Polyrhachis guerini pallescens 134
Polyrhachis guerini vermiculosa 134
Polyrhachis (Hagiomyrma) ammonoeides crawleyi 131
Polyrhachis hector 136
Polyrhachis hecuba 134
Polyrhachis (Hedomyrma) aeschyle 131
Polyrhachis (Hedomyrma) anguliceps 131
Polyrhachis (Hedomyrma) barretti 132
Polyrhachis (Hedomyrma) chrysothorax 132
Polyrhachis (Hedomyrma) daemeli exlex 133
Polyrhachis (Hedomyrma) humerosa 136
Polyrhachis (Hedomyrma) kershawi 136
Polyrhachis (Hedomyrma) machaon 137
Polyrhachis (Hedomyrma) mjobergi 138
Polyrhachis heinlethii 135
Polyrhachis heinlethii heinlethii 135
Polyrhachis heinlethii sophiae 135
Polyrhachis hermione 135
Polyrhachis hermione cupreata 135
Polyrhachis hermione hermione 135
Polyrhachis hexacantha 135
Polyrhachis hirsuta 135
Polyrhachis hirsuta hirsuta 135
Polyrhachis hirsuta quinquedentata 135
Polyrhachis hookeri 135
Polyrhachis hookeri aerea 135
Polyrhachis hookeri hookeri 135
Polyrhachis hookeri lownei 135
Polyrhachis hookeri obscura 135
Polyrhachis humerosa 136
Polyrhachis inconspicua 136
Polyrhachis inconspicua inconspicua 136
Polyrhachis inconspicua subnitens 136
Polyrhachis ithona 136
Polyrhachis ithonus 136
Polyrhachis jacksoniana 136
Polyrhachis kershawi 136
Polyrhachis lachesis 136
Polyrhachis laevissima 137
Polyrhachis latreillii 136
Polyrhachis leae 136
Polyrhachis leae cedarensis 137
Polyrhachis leae leae 136
Polyrhachis levior 137
Polyrhachis lombokensis 137
Polyrhachis lombokensis yarrabahensis 137
Polyrhachis lysistrata 137
Polyrhachis machaon 137
Polyrhachis mackayi 137
Polyrhachis macropus 137
Polyrhachis micans 137
Polyrhachis micans micans 137
Polyrhachis micans ops 137
Polyrhachis mjobergi 138
Polyrhachis (Myrmatopa) lombokensis yarrabahensis 137
Polyrhachis (Myrmhopla) barnardi 132
Polyrhachis (Myrmhopla) exulans 133

Polyrhachis (Myrmhopla) glabrinotum 134
Polyrhachis (Myrmothrinax) lysistrata 137
Polyrhachis nox 138
Polyrhachis ornata 138
Polyrhachis patiens 138
Polyrhachis penelope 138
Polyrhachis phryne 138
Polyrhachis polymnia 138
Polyrhachis polymnia maculata 138
Polyrhachis polymnia polymnia 138
Polyrhachis prometheus 138
Polyrhachis pseudothrinax 138
Polyrhachis punctiventris 138
Polyrhachis pyrrhus 139
Polyrhachis quadricuspis 139
Polyrhachis queenslandica 139
Polyrhachis rastellata 139
Polyrhachis rastellata yorkana 139
Polyrhachis relucens 139
Polyrhachis relucens australiae 139
Polyrhachis rowlandi 139
Polyrhachis schenkii 139
Polyrhachis schenkii lydiae 139
Polyrhachis schenkii schenkii 139
Polyrhachis schwiedlandi 139
Polyrhachis semiaurata 139
Polyrhachis semipolita 140
Polyrhachis semipolita hestia 140
Polyrhachis semipolita semipolita 140
Polyrhachis sempronia 140
Polyrhachis sexspinosa 140
Polyrhachis sidnica 140
Polyrhachis sidnica perthensis 140
Polyrhachis sidnica sidnica 140
Polyrhachis sidnica tambourinensis 140
Polyrhachis sokolova 140
Polyrhachis sokolova degener 140
Polyrhachis sokolova sokolova 140
Polyrhachis templi 140
Polyrhachis terpischore elegans 141
Polyrhachis terpsichore 140
Polyrhachis terpsichore elegans 141
Polyrhachis terpsichore rufifemur 141
Polyrhachis terpsichore terpsichore 140
Polyrhachis thais 141
Polyrhachis thalia 141
Polyrhachis thalia io 141
Polyrhachis thalia thalia 141
Polyrhachis thusnelda 141
Polyrhachis townsvillei 141
Polyrhachis trapezoidea 141
Polyrhachis tubifera 141
Polyrhachis turneri 141
Polyrhachis urania 141
Polyrhachis zimmerae 141
Ponera 38
Ponera araneoides 40
Ponera (Bothroponera) dubitata 21
Ponera (Bothroponera) excavata acuticostata 21
Ponera (Bothroponera) piliventris intermedia 22
Ponera (Bothroponera) porcata 22
Ponera (Bothroponera) sublaevis reticulata 22
Ponera clavicornis 38
Ponera coarctata mackayensis 31
Ponera congrua 31
Ponera decora 31
Ponera diminuta 33
Ponera (Ectatomma) aurata 41
Ponera exedra 38
Ponera leae 38
Ponera leae norfolkensis 38
Ponera leae oculata 38
Ponera lutea 23
Ponera melanaria australis 35
Ponera metallica 45
Ponera metallica aspera 40
Ponera mina 31
Ponera mjobergi 28
Ponera oculata 149
Ponera parallela 37
Ponera pruinosa herbertonensis 31
Ponera queenslandensis 31
Ponera rectidens 31
Ponera ruginoda 14
Ponera scitula 32
Ponera selenophora 38
Ponera sulciceps 32
Ponera trigona convexiuscula 31
Ponera truncata elliptica 31
PONERINAE 2, 18
porcata, Bothroponera 22
porcata, Ponera (Bothroponera) 22
postcornutus, Camponotus 118
postcornutus, Camponotus (Tanaemyrmex) 118
postpetiolaris, Myrmecia (Promyrmecia) mandibularis 12
potteri, Cerapachys 27
potteri, Melophorus 124
potteri, Myrmecia 14
potteri, Neophyracaces 27
potteri, Promyrmecia 14
poultoni, Aphaenogaster 54
praedo, Pheidole variabilis 78
Prenolepis adlerzii 121
Prenolepis braueri 129
Prenolepis braueri glabrior 129
Prenolepis minutula 129
Prenolepis mjobergi 143
Prenolepis mjoebergella 143
Prenolepis (Nylanderia) tasmaniensis 129
Prenolepis obscura 129
Prenolepis obscura vaga 129
Prenolepis rosae 129
princeps, Cerapachys 27
princeps, Myrmecia 8
princeps, Phyracaces 27
Prionogenys 39
Prionogenys podenzanai 39
Prionogenys podenzanai malandensis 39
Prionogenys podenzanai podonzanai 39
Prionopelta 39
Prionopelta opaca 39
Pristomyrmecia 6
(Pristomyrmecia) fulvipes fulviculis, Myrmecia 10
(Pristomyrmecia) piliventris femorata, Myrmecia 10
(Pristomyrmecia) regina, Myrmecia 11
Pristomyrmex 84
Pristomyrmex erythropygus 85
Pristomyrmex foveolatus 85
Pristomyrmex (Odontomyrmex) quadridentatus queenslandensis 85

Pristomyrmex quadridentatus 85
Pristomyrmex thoracicus 85
Pristomyrmex wheeleri 85
Pristomyrmex wilsoni 85
Probolomyrmex 39
Probolomyrmex greavesi 39
procera, Eurhopalothrix 64
procera, Rhopalothrix 64
Proceratium 39
Proceratium papuanum 40
Proceratium stictum 40
procidua, Formica 101
prociduus, Iridomyrmex 101
Prodiscothyrea 29
Prodiscothyrea velutina 29
Prolasius 142
Prolasius abruptus 142
Prolasius antennata 142
Prolasius antennatus 142
Prolasius brunea 142
Prolasius bruneus 142
Prolasius clarki 142
Prolasius convexus 142
Prolasius depressiceps 142
Prolasius depressiceps depressiceps 142
Prolasius depressiceps similis 142
Prolasius flavicornis 142
Prolasius flavicornis flavicornis 142
Prolasius flavicornis minor 142
Prolasius flavidiscus 142
Prolasius hellenae 143
Prolasius hemiflavus 143
Prolasius hemiflavus hemiflavus 143
Prolasius hemiflavus wilsoni 143
Prolasius mjoebergella 143
Prolasius niger 143
Prolasius nigriventris 143
Prolasius nitidissimus 143
Prolasius nitidissimus formicoides 143
Prolasius nitidissimus nitidissimus 143
Prolasius pallidus 143
Prolasius quadrata 143
Prolasius quadratus 143
Prolasius reticulata 143
Prolasius reticulatus 143
Prolasius robustus 144
Prolasius wheeleri 144
prometheus, Polyrhachis 138
prometheus, Polyrhachis (Campomyrma) 138
Promyrmecia 6
(Promyrmecia) aberrans formosa, Myrmecia 12
(Promyrmecia) aberrans haematosticta, Myrmecia 12
(Promyrmecia) aberrans maura, Myrmecia 12
(Promyrmecia) aberrans sericata, Myrmecia 10
(Promyrmecia) aberrans taylori, Myrmecia 10
Promyrmecia callima 7
Promyrmecia celaena 7
Promyrmecia cephalotes 7
Promyrmecia chrysogaster 7
Promyrmecia cydista 8
(Promyrmecia) dichospila, Myrmecia 8
Promyrmecia dispar 8
Promyrmecia dixoni 8
Promyrmecia elegans 8
Promyrmecia eupoecila 9

Promyrmecia excavata 9
Promyrmecia exigua 9
(Promyrmecia) fucosa, Myrmecia 10
(Promyrmecia) fulvipes barbata, Myrmecia 10
(Promyrmecia) fulvipes caelatinoda, Myrmecia 12
Promyrmecia goudiei 17
Promyrmecia greavesi 11
Promyrmecia hilli 11
Promyrmecia laevinodis 12
Promyrmecia luteiforceps 12
Promyrmecia maloni 11
(Promyrmecia) mandibularis postpetiolaris, Myrmecia 12
Promyrmecia marmorata 17
(Promyrmecia) michaelseni rugosa, Myrmecia 15
Promyrmecia nobilis 13
Promyrmecia occidentalis 13
Promyrmecia opaca 13
Promyrmecia potteri 14
Promyrmecia rubicunda 15
Promyrmecia ruginodis 16
Promyrmecia scabra 11
Promyrmecia shepherdi 17
Promyrmecia testaceipes 17
Promyrmecia wilsoni 17
pronotalis, Rhytidoponera 46
propinqua, Melophorus fieldi 123
Protamblyopone 18
Protholcomyrmex 55
proxima bombalensis, Pheidole 77
proxima, Pheidole 77
proxima, Pheidole proxima 77
proxima proxima, Pheidole 77
proxima, Stigmacros 148
proxima, Stigmacros (Cyrtostigmacros) 148
proxima transversa, Pheidole 77
pruinosa herbertonensis, Ponera 31
Pseudolasius 144
Pseudolasius australis 144
Pseudomyrma laeviceps 18
PSEUDOMYRMECINAE 2 , 17
Pseudonotoncus 144
Pseudonotoncus hirsutus 144
Pseudonotoncus turneri 144
Pseudopodomyrma 80
Pseudopodomyrma clarki 81
Pseudoponera chelifer 32
Pseudostigmacros 144
(Pseudostigmacros) inermis, Stigmacros 147
pseudothrinax, Polyrhachis 138
pubescens, Cryptocerus 68
pubescens, Meranoplus 68
pulchra, Chalcoponera 45
pulchra, Myrmecia 14
pulchra, Podomyrma abdominalis 80
pumilio, Myrmecia 17
punctata, Ectatomma 46
punctata levior, Rhytidoponera 42
punctata, Rhytidoponera 46
punctatissima, Phyracaces 27
punctatissima, Stigmacros 148
punctatissima, Stigmacros (Hagiostigmacros) 148
punctatissimus, Cerapachys 27
punctatissimus, Iridomyrmex 101
punctigera, Rhytidoponera 46
punctiventris, Camponotus 118

punctiventris, Camponotus (Myrmogonia) 118
punctiventris, Ectatomma (Rhytidoponera) cristatum 46
punctiventris, Lordomyrma 65
punctiventris, Polyrhachis 138
punctiventris, Rhytidoponera 46
punctulata, Amblyopone 20
punctulata, Amblyopone (Fulakora) 20
punctulata kimberleyensis, Sima 18
punctulata kimberleyensis, Tetraponera 18
punctulata punctulata, Tetraponera 18
punctulata, Tetraponera 18
punctulata, Tetraponera punctulata 18
purpurascens, Rhytidoponera (Chalcoponera) metallica 45
purpurea, Formica 101
purpurea, Rhytidoponera 46
purpurescens, Formica 149
purpureum, Ectatomma impressum 46
purpureus, Calomyrmex 108
purpureus, Calomyrmex purpureus 108
purpureus, Camponotus 108
purpureus castrae, Iridomyrmex 102
purpureus, Iridomyrmex 101
purpureus, Iridomyrmex purpureus 101
purpureus purpureus, Calomyrmex 108
purpureus purpureus, Iridomyrmex 101
purpureus sanguinea, Iridomyrmex 102
purpureus smaragdina, Calomyrmex 108
purpureus viridiaeneus, Iridomyrmex 102
puryi curvispina, Meranoplus 68
puryi, Meranoplus 68
puryi, Meranoplus puryi 68
puryi puryi, Meranoplus 68
pusilla australis, Platythyrea 37
pusilla, Stigmacros 148
pusilla, Stigmacros (Stigmacros) 148
pusillum, Tapinoma 92
pusillus aequalis, Bothriomyrmex 92
pusillus, Bothriomyrmex 92
pusillus, Bothriomyrmex pusillus 92
pusillus pusillus, Bothriomyrmex 92
pygmaeus, Phyracaces 26
pyriformis gigas, Myrmecia 7
pyriformis, Myrmecia 15
pyriformis, Pheidole 77
pyrrhus, Polyrhachis 139
pythia, Aphaenogaster 55
pythia, Cremastogaster 62
pythia, Crematogaster 62

quadrata, Prolasius 143
quadratus, Prolasius 143
quadriceps, Rhytidoponera 45
quadricolor, Aphantolepis 106
quadricolor, Leptomyrmex varians 105
quadricolor, Technomyrmex 106
quadricuspis, Polyrhachis 139
quadridentatus, Odontomyrmex 85
quadridentatus, Pristomyrmex 85
quadridentatus queenslandensis, Pristomyrmex (Odontomyrmex) 85
quadrimaculata, Plagiolepis exigua 130
quadrispinosa, Epopostruma 64
quadrispinosa, Epopostruma quadrispinosa 64
quadrispinosa quadrispinosa, Epopostruma 64
quadrispinosa, Strumigenys (Epopostruma) 64
Quadristruma 85
Quadristruma emmae 3 , 85
queenslandensis, Hypoponera 31
queenslandensis, Iridomyrmex nitidus 101
queenslandensis, Ponera 31
queenslandensis, Pristomyrmex (Odontomyrmex) quadridentatus 85
queenslandica, Cremastogaster sordidula 62
queenslandica, Crematogaster 62
queenslandica, Crematogaster queenslandica 62
queenslandica froggatti, Crematogaster 62
queenslandica gilberti, Crematogaster 62
queenslandica, Myrmecia michaelseni 12
queenslandica, Polyrhachis 139
queenslandica queenslandica, Crematogaster 62
queenslandica rogans, Crematogaster 63
queenslandica scabrula, Crematogaster 63
quinquedentata, Polyrhachis (Campomyrma) hirsuta 135
quinquedentata, Polyrhachis hirsuta 135
quinquedentata, Strumigenys 88

rastellata, Formica 139
rastellata, Polyrhachis 139
rastellata yorkana, Polyrhachis 139
rastellata yorkana, Polyrhachis (Cyrtomyrma) 139
rectangularis, Anochetus 21
rectangularis diabolus, Anochetus 21
rectangularis, Stigmacros 148
rectangularis, Stigmacros (Stigmacros) 148
rectidens, Hypoponera 31
rectidens, Myrmecia piliventris 14
rectidens, Ponera 31
recurva, Pheidole concentrica 76
redunca, Pheidole variabilis 78
reflexa, Rhytidoponera 47
reflexus, Dolichoderus 94
reflexus, Dolichoderus (Hypoclinea) 94
regina, Myrmecia (Pristomyrmecia) 11
regularis, Bothroponera piliventris 22
regularis, Myrmecia 15
regularis, Pachycondyla (Bothroponera) piliventris 22
relicta, Heteroponera 30
relicta, Paranomopone 30
relucens australiae, Polyrhachis 139
relucens, Formica 139
relucens, Polyrhachis 139
respiciens, Formica (Myrmecopsis) 128
respiciens moestus, Opisthopsis 128
respiciens, Opisthopsis 128
respiciens, Opisthopsis respiciens 128
respiciens respiciens, Opisthopsis 128
reticulata, Bothroponera sublaevis 22
reticulata, Eubothroponera 38
reticulata, Ponera (Bothroponera) sublaevis 22
reticulata, Prolasius 143
reticulata, Rhytidoponera 47
reticulata, Stigmacros 148
reticulatum, Ectatomma (Rhytidoponera) 47
reticulatus, Camponotus 118 , 119
reticulatus mackayensis, Camponotus 118
reticulatus, Phyracaces 26
reticulatus, Prolasius 143

Rhopalomastix 85
Rhopalomastix rothneyi 86
Rhopalothrix 86
Rhopalothrix orbis 86
Rhopalothrix procera 64
Rhoptromyrmex 86
Rhoptromyrmex melleus 86
Rhoptromyrmex wroughtonii 86
Rhytidoponera 40
Rhytidoponera aciculata 40
Rhytidoponera anceps 40
Rhytidoponera araneoides 40
Rhytidoponera araneoides arcuata 40
Rhytidoponera aspera 40
Rhytidoponera aurata 41
Rhytidoponera barnardi 41
Rhytidoponera barretti 41
Rhytidoponera borealis 41
Rhytidoponera carinata 41
Rhytidoponera castanea 47
Rhytidoponera cerastes 41
Rhytidoponera cerastes brevior 48
Rhytidoponera (Chalcoponera) arnoldi 40
Rhytidoponera (Chalcoponera) aspera scabrior 48
Rhytidoponera (Chalcoponera) caeciliae 45
Rhytidoponera (Chalcoponera) clarki 42
Rhytidoponera (Chalcoponera) dubia 43
Rhytidoponera (Chalcoponera) fastuosa 42
Rhytidoponera (Chalcoponera) lamellinodis 44
Rhytidoponera (Chalcoponera) metallica inornata 44
Rhytidoponera (Chalcoponera) metallica purpurascens 45
Rhytidoponera (Chalcoponera) metallica varians 45
Rhytidoponera (Chalcoponera) numeensis borealis 41
Rhytidoponera (Chalcoponera) victoriae cedarensis 49
Rhytidoponera (Chalcoponera) victoriae maledicta 44
Rhytidoponera chalybaea 41
Rhytidoponera chnoopyx 41
Rhytidoponera clarki 41
Rhytidoponera confusa 42
Rhytidoponera convexa 42
Rhytidoponera convexa opacior 49
Rhytidoponera convexa rufiventris 47
(Rhytidoponera) convexum gemma, Ectatomma 49
(Rhytidoponera) convexum nodiferum, Ectatomma 46
(Rhytidoponera) convexum rufescens, Ectatomma 47
(Rhytidoponera) convexum spatiatum, Ectatomma 45
(Rhytidoponera) convexum violaceum, Ectatomma 49
Rhytidoponera cornuta 42
Rhytidoponera cornuta fusciventris 48
(Rhytidoponera) cornutum, Ectatomma 42
(Rhytidoponera) cornutum taurus, Ectatomma 48
(Rhytidoponera) crassinode, Ectatomma 42
Rhytidoponera crassinodis 42
Rhytidoponera cristata 42
Rhytidoponera cristata yorkensis 49
(Rhytidoponera) cristatum caro, Ectatomma 40
(Rhytidoponera) cristatum punctiventris, Ectatomma 46
Rhytidoponera croesus 42
(Rhytidoponera) cyrus, Ectatomma 41
Rhytidoponera dixoni 45
Rhytidoponera douglasi 42 , 43
Rhytidoponera dubia 43
Rhytidoponera enigmatica 43
Rhytidoponera eremita 43

Rhytidoponera ferruginea 43
Rhytidoponera flava 41
Rhytidoponera flavicornis 43
Rhytidoponera flavipes 43
Rhytidoponera flindersi 43
Rhytidoponera foreli 43
Rhytidoponera foveolata 43
Rhytidoponera fuliginosa 43
Rhytidoponera greavesi 43
Rhytidoponera gregoryi 44
Rhytidoponera haeckeli 44
(Rhytidoponera) haeckeli, Ectatomma 44
Rhytidoponera hilli 44
Rhytidoponera impressa 44
Rhytidoponera impressa chalybaea 41
(Rhytidoponera) impressum splendidum, Ectatomma 47
Rhytidoponera incisa 44
Rhytidoponera inornata 44
Rhytidoponera kurandensis 44
Rhytidoponera laciniosa malandensis 47
Rhytidoponera lamellinodis 44
Rhytidoponera laticeps 44
Rhytidoponera maledicta 44
Rhytidoponera maniae 45
(Rhytidoponera) maniae, Ectatomma 45
Rhytidoponera mayri 45
(Rhytidoponera) mayri glabrius, Ectatomma 45
Rhytidoponera metallica 45
Rhytidoponera metallica tasmaniensis 48
(Rhytidoponera) metallicum cristulatum, Ectatomma 48
(Rhytidoponera) metallicum modestum, Ectatomma 49
(Rhytidoponera) metallicum obscurum, Ectatomma 41
(Rhytidoponera) metallicum scrobiculatum, Ectatomma 49
Rhytidoponera micans 45
Rhytidoponera mirabilis 46
Rhytidoponera nigra 42
Rhytidoponera nitida 46
Rhytidoponera nodifera 46
Rhytidoponera nudata 46
Rhytidoponera occidentalis 45
Rhytidoponera peninsularis 46
Rhytidoponera petiolata 45
Rhytidoponera pilosula 46
Rhytidoponera pronotalis 46
Rhytidoponera punctata 46
Rhytidoponera punctata levior 42
Rhytidoponera punctigera 46
Rhytidoponera punctiventris 46
Rhytidoponera purpurea 46
Rhytidoponera quadriceps 45
Rhytidoponera reflexa 47
Rhytidoponera reticulata 47
(Rhytidoponera) reticulatum, Ectatomma 47
(Rhytidoponera) rothneyi, Ectatomma 46
Rhytidoponera rothneyi mediana 46
Rhytidoponera rufescens 47
Rhytidoponera rufithorax 47
Rhytidoponera rufiventris 47
Rhytidoponera rufonigra 47
Rhytidoponera scaberrima 47
(Rhytidoponera) scaberrimum, Ectatomma 47
Rhytidoponera scabra 48
Rhytidoponera scabrior 48
Rhytidoponera socrus 48

(Rhytidoponera) socrus, Ectatomma 48
Rhytidoponera spoliata 48
(Rhytidoponera) spoliatum, Ectatomma 48
Rhytidoponera stridulator 45
Rhytidoponera tasmaniensis 48
Rhytidoponera taurus 48
(Rhytidoponera) tenue, Ectatomma 48
Rhytidoponera tenuis 48
Rhytidoponera trachypyx 48
Rhytidoponera turneri 49
(Rhytidoponera) turneri, Ectatomma 49
Rhytidoponera tyloxys 49
Rhytidoponera victoriae 49
(Rhytidoponera) victoriae, Ectatomma 49
Rhytidoponera violacea 49
Rhytidoponera viridis 49
Rhytidoponera yorkensis 49
robustus, Orectognathus 74
robustus, Prolasius 144
rodwayi, Notoncus 125
rogans, Cremastogaster sordidula 63
rogans, Crematogaster queenslandica 63
rogeri, Myrmecia tricolor 16
rosae, Paratrechina 129
rosae, Prenolepis 129
rostratus, Orectognathus 74
rostrinotus, Iridomyrmex 98
rothneyi, Ectatomma (Rhytidoponera) 46
rothneyi, Leptomyrmex varians 105
rothneyi mediana, Rhytidoponera 46
rothneyi, Rhopalomastix 86
rothsteini, Chelaner 57
rothsteini, Chelaner rothsteini 58
rothsteini doddi, Chelaner 58
rothsteini doddi, Monomorium (Paraholcomyrmex) 58
rothsteini humilior, Chelaner 58
rothsteini humilior, Monomorium 58
rothsteini leda, Chelaner 58
rothsteini leda, Monomorium 58
rothsteini, Monomorium 58
rothsteini rothsteini, Chelaner 58
rothsteini squamigena, Chelaner 58
rothsteini squamigena, Monomorium 58
rothsteini tostum, Chelaner 58
rothsteini tostum, Monomorium 58
rottnestense, Tapinoma 106
rottnestense, Tapinoma (Micromyrma) 106
rottnesti, Camponotus (Myrmophyma) 120
rotula, Cerapachys (Phyracaces) singularis 28
rotundiceps, Cryptopone 28
rotundiceps, Euponera (Trachymesopsus) 28
rotundiceps, Notoncus 126
rowlandi, Myrmecia 15
rowlandi, Myrmecia tarsata 15
rowlandi, Polyrhachis 139
rubicunda, Bothroponera sublaevis 22
rubicunda, Myrmecia 15
rubicunda, Promyrmecia 15
rubiginosus, Camponotus 118
rubra, Monomorium rubriceps 58
rubra, Myrmecia forficata 10
rubriceps, Chelaner 58
rubriceps, Chelaner rubriceps 58
rubriceps cinctum, Monomorium 58
rubriceps cinctus, Chelaner 58

rubriceps extreminigrum, Monomorium 58
rubriceps extreminigrus, Chelaner 58
rubriceps, Iridomyrmex gracilis 99
rubriceps, Monomorium 58
rubriceps, Odontomachus ruficeps 36
rubriceps rubra, Monomorium 58
rubriceps rubriceps, Chelaner 58
rubriceps rubrus, Chelaner 58
rubripes, Myrmecia 15
rubrus, Chelaner rubriceps 58
rufa, Stigmacros 148
rufa, Stigmacros (Stigmacros) 148
rufescens, Ectatomma (Rhytidoponera) convexum 47
rufescens, Odontomachus ruficeps 36
rufescens, Rhytidoponera 47
ruficeps acutidens, Odontomachus 36
ruficeps, Leptomyrmex varians 105
ruficeps, Odontomachus 36
ruficeps rubriceps, Odontomachus 36
ruficeps rufescens, Odontomachus 36
ruficeps turneri, Odontomachus 36
ruficornis, Cerapachys 27
ruficornis, Dolichoderus (Hypoclinea) scabridus 94
ruficornis, Dolichoderus scabridus 94
ruficornis, Ectomomyrmex 30
ruficornis, Phyracaces 27
rufifemur, Polyrhachis terpsichore 141
rufifrons semicarinata, Colobopsis 118
rufinodis, Myrmecia 15
rufipes, Leptomyrmex varians 105
rufithorax, Leptomyrmex erythrocephalus 104
rufithorax, Myrmecorhynchus 125
rufithorax, Opisthopsis 128
rufithorax, Rhytidoponera 47
rufiventris, Rhytidoponera 47
rufiventris, Rhytidoponera convexa 47
rufoniger domestica, Iridomyrmex 102
rufoniger incerta, Iridomyrmex 102
rufoniger incertus, Iridomyrmex 102
rufoniger, Iridomyrmex 102
rufoniger, Iridomyrmex rufoniger 102
rufoniger, Opisthopsis haddoni 127
rufoniger pallidus, Iridomyrmex 102
rufoniger rufoniger, Iridomyrmex 102
rufoniger septentrionalis, Iridomyrmex 102
rufoniger suchieri, Iridomyrmex 102
rufoniger victorianus, Iridomyrmex 102
rufonigra, Euponera (Brachyponera) 52
rufonigra, Formica 102
rufonigra, Myrmecia 17
rufonigra, Rhytidoponera 47
rufonigra, Trachymesopus 52
rufotestacea, Cremastogaster 63
rufotestacea, Crematogaster 63
rufotestacea, Crematogaster rufotestacea 63
rufotestacea dentinasis, Crematogaster 63
rufotestacea dentinasis, Crematogaster (Orthocrema) 63
rufotestacea rufotestacea, Crematogaster 63
rufotibialis, Dolichoderus (Hypoclinea) ypsilon 95
rufotibialis, Dolichoderus ypsilon 95
rufus, Camponotus 118
rufus, Camponotus (Dinomyrmex) 118
ruginoda, Ponera 14
ruginodis, Promyrmecia 16
ruginota, Stenamma (Ischnomyrmex) longiceps 54

rugocciput, Pheidole variabilis 78
rugosa, Lordomyrma 84
rugosa, Meranoplus hirsutus 67
rugosa, Myrmecia 15
rugosa, Myrmecia (Promyrmecia) michaelseni 15
rugosa, Myrmecina 71
rugosa, Podomyrma 84
rugosula, Pheidole variabilis 79
rugulinodis, Cerapachys 27
rugulinodis, Phyracaces 27

sanguinea, Camponotus 118
sanguinea, Camponotus (Myrmogonia) 118
sanguinea, Iridomyrmex detectus 102
sanguinea, Iridomyrmex purpureus 102
sanguinea, Myrmecia 15
sanguinea, Notostigma 126
sanguinifrons, Camponotus 118
sanguinifrons, Camponotus (Colobopsis) 118
sanguinolentum, Monomorium (Notomyrmex) 58
sanguinolentus, Chelaner 58
satan, Orectognathus 74
scaberrima, Rhytidoponera 47
scaberrimum, Ectatomma (Rhytidoponera) 47
scabra, Acanthoponera (Anacanthoponera) imbellis 30
scabra, Promyrmecia 11
scabra, Rhytidoponera 48
scabridus, Dolichoderus 94
scabridus, Dolichoderus scabridus 94
scabridus ruficornis, Dolichoderus 94
scabridus ruficornis, Dolichoderus (Hypoclinea) 94
scabridus scabridus, Dolichoderus 94
scabridus ypsilon, Dolichoderus 94
scabrior, Rhytidoponera 48
scabrior, Rhytidoponera (Chalcoponera) aspera 48
scabrula, Crematogaster froggatti 63
scabrula, Crematogaster queenslandica 63
scabrum, Ectatomma 48
schencki, Camponotus 119
schenkii lydiae, Polyrhachis 139
schenkii, Polyrhachis 139
schenkii, Polyrhachis schenkii 139
schenkii schenkii, Polyrhachis 139
Schizopelta 55
Schizopelta falcata 56
schoopae, Polyrhachis appendiculata 131
schraderi, Calyptomyrmex 55
schwiedlandi, Polyrhachis 139
scipio, Melophorus 124
scissor, Bothriomyrmex 92
scita, Cremastogaster 63
scita, Crematogaster 63
scita, Crematogaster scita 63
scita mixta, Cremastogaster 63
scita mixta, Crematogaster 63
scita scita, Crematogaster 63
scitula, Hypoponera 32
scitula, Ponera 32
scratius, Camponotus 118
scratius, Camponotus scratius 118
scratius nuntius, Camponotus 118
scratius scratius, Camponotus 118
scrobiculata, Hypoclinea 94
scrobiculatum, Ectatomma (Rhytidoponera) metallicum 49
scrobiculatus, Dolichoderus 94
scrutator, Phyracaces 25
sculpticeps, Cremastogaster frivola 61
sculpticeps, Crematogaster frivola 61
sculpturatum, Monomorium (Notomyrmex) 58
sculpturatus, Chelaner 58
scutellus, Camponotus (Myrmosaulus) 121
sedis, Incertae 149
selenophora, Ponera 38
semiaurata, Polyrhachis 139
semicarinata, Colobopsis rufifrons 118
semicarinatus, Camponotus 118
semicircularis, Odontomachus coriarius 36
semicomptus, Codiomyrmex 64
semicomptus, Glamyromyrmex 64
semipolita hestia, Polyrhachis 140
semipolita, Polyrhachis 140
semipolita, Polyrhachis semipolita 140
semipolita semipolita, Polyrhachis 140
sempronia, Polyrhachis 140
senescens, Cerapachys 27
senescens, Phyracaces 27
senilis, Polyrhachis gab 134
septentrionalis, Eubothroponera 38
septentrionalis, Eusphinctus (Nothosphinctus) 51
septentrionalis, Iridomyrmex rufoniger 102
septentrionalis, Odontomachus 36
septentrionalis, Orectognathus antennatus 72
septentrionalis, Sphinctomyrmex 51
sericata, Myrmecia (Promyrmecia) aberrans 10
sericeiventris, Podomyrma micans 80
sexspinosa, Formica 140
sexspinosa, Polyrhachis 140
sexspinosus, Orectognathus 74
sharpei, Odontomachus 36
shepherdi, Promyrmecia 17
sidnica perthensis, Polyrhachis 140
sidnica perthensis, Polyrhachis (Campomyrma) 140
sidnica, Polyrhachis 140
sidnica, Polyrhachis sidnica 140
sidnica sidnica, Polyrhachis 140
sidnica tambourinensis, Polyrhachis 140
sidnica tambourinensis, Polyrhachis (Campomyrma) 140
silaceus, Eusphinctus (Nothosphinctus) 50
silvestrii, Machomyrma 53
Sima punctulata kimberleyensis 18
similis, Calomyrmex 108
similis, Camponotus 108
similis, Meranoplus 68
similis, Prolasius depressiceps 142
simillima, Myrmecia 16
simillima, Myrmica 90
simillimum, Tetramorium 90
simmonsae, Cerapachys 28
simmonsae, Phyracaces 28
simulator, Camponotus 119
simulator, Camponotus (Dinomyrmex) 119
singularis, Cerapachys 28
singularis, Myrmecia 9
singularis rotula, Cerapachys (Phyracaces) 28
sjostedti, Cerapachys 28
sjostedti, Cerapachys (Phyracaces) 28
sjostedti, Leptogenys 34
sjostedti, Tetramorium 90

sjostedti, Tetramorium (Xiphomyrmex) 90
smaragdina, Calomyrmex purpureus 108
smaragdina, Formica 127
smaragdina, Oecophylla 127
smithi, Amblyopone 20
smithii, Formica 101
socialis, Ectatomma 23
socrus, Ectatomma (Rhytidoponera) 48
socrus, Rhytidoponera 48
sokolova degener, Polyrhachis 140
sokolova, Polyrhachis 140
sokolova, Polyrhachis sokolova 140
sokolova sokolova, Polyrhachis 140
Solenopsis 86
Solenopsis belisaria 86
Solenopsis belisarius 86
Solenopsis clarki 87
Solenopsis froggatti 87
Solenopsis fusciventris 87
Solenopsis insculpta 87
Solenopsis insculptus 87
sophiae, Polyrhachis heinlethii 135
sophiae, Technomyrmex 107
sordida, Stigmacros 148
sordida, Stigmacros (Cyrtostigmacros) 148
sordidula dispar, Cremastogaster 61
sordidula froggatti, Cremastogaster 62
sordidula gilberti, Cremastogaster 62
sordidula queenslandica, Cremastogaster 62
sordidula rogans, Cremastogaster 63
sordidum, Monomorium 59
sordidum nigriventris, Monomorium 59
sordidus, Chelaner 59
sordidus, Chelaner sordidus 59
sordidus nigriventris, Chelaner 59
sordidus sordidus, Chelaner 59
spadicea, Myrmecia 16
spatiatum, Ectatomma (Rhytidoponera) convexum 45
spenceri, Camponotus 119
spenceri, Camponotus (Tanaemyrmex) 119
Sphinctomyrmex 49
Sphinctomyrmex asper 50
Sphinctomyrmex cedaris 50
Sphinctomyrmex clarus 50
Sphinctomyrmex clarus mjobergi 51
Sphinctomyrmex duchaussoyi 50
Sphinctomyrmex emeryi myops 51
Sphinctomyrmex (Eusphinctus) fallax 51
Sphinctomyrmex (Eusphinctus) fallax cedaris 50
Sphinctomyrmex (Eusphinctus) fallax hedwigae 51
Sphinctomyrmex (Eusphinctus) steinheili 51
Sphinctomyrmex froggatti 50
Sphinctomyrmex froggatti imbecilis 50
Sphinctomyrmex imbecilis 50
Sphinctomyrmex mjobergi 51
Sphinctomyrmex myops 51
Sphinctomyrmex nigricans 51
Sphinctomyrmex occidentalis 51
Sphinctomyrmex perstictus 51
Sphinctomyrmex septentrionalis 51
Sphinctomyrmex steinheili 51
Sphinctomyrmex trux 52
Sphinctomyrmex turneri 52
spininode, Tetramorium 90
spinisquamis, Melophorus 126
spinisquamis, Notoncus 126
spinitarsus, Camponotus 119
spinitarsus, Camponotus (Dinomyrmex) 119
spinoda, Atta 77
spinoda, Pheidole 77
spinosa, Stigmacros 148
spinosa, Stigmacros (Hagiostigmacros) 148
spinosior, Mayriella 66
splendens, Iridomyrmex mattiroloi 100
splendidior, Tetramorium 90
splendidior, Xiphomyrmex striolatus 90
splendidum, Ectatomma (Rhytidoponera) impressum 47
splendidus, Calomyrmex 108
splendidus, Calomyrmex splendidus 108
splendidus, Camponotus 108
splendidus, Iridomyrmex bicknelli 96
splendidus mutans, Calomyrmex 108
splendidus mutans, Camponotus (Calomyrmex) 108
splendidus splendidus, Calomyrmex 108
splendidus viridiventris, Calomyrmex 109
spoliata, Rhytidoponera 48
spoliatum, Ectatomma (Rhytidoponera) 48
sponsorum, Camponotus 119
spurcus, Iridomyrmex gracilis 99
squamigena, Chelaner rothsteini 58
squamigena, Monomorium rothsteini 58
squamulosa, Plagiolepis 130
stanleyi, Stigmacros 148
stanleyi, Stigmacros (Campostigmacros) 148
steinheili, Sphinctomyrmex 51
steinheili, Sphinctomyrmex (Eusphinctus) 51
Stenamma (Ischnomyrmex) longiceps ruginota 54
Stenothorax 53
Stenothorax katerinae 53
stewartii, Iridomyrmex cordatus 97
Stictoponera biroi 30
stictum, Proceratium 40
Stigmacros 144
Stigmacros aciculata 145
Stigmacros acuta 145
(Stigmacros) acuta, Stigmacros 145
Stigmacros aemula 145
(Stigmacros) aemula, Acantholepis 145
(Stigmacros) aemula intacta, Acantholepis 147
Stigmacros anthracina 145
Stigmacros armstrongi 145
Stigmacros australis 145
Stigmacros barretti 145
Stigmacros bosii 145
Stigmacros brachytera 145
Stigmacros brevispina 145
(Stigmacros) brevispina, Stigmacros 145
Stigmacros brooksi 145
Stigmacros (Campostigmacros) anthracina 145
Stigmacros (Campostigmacros) brachytera 145
Stigmacros (Campostigmacros) epinotalis 146
Stigmacros (Campostigmacros) marginata 147
Stigmacros (Campostigmacros) nitida 147
Stigmacros (Campostigmacros) stanleyi 148
Stigmacros castanea 146
Stigmacros (Chariostigmacros) hirsuta 147
Stigmacros clarki 146
Stigmacros clivispina 146
Stigmacros (Cyrtostigmacros) aciculata 145
Stigmacros (Cyrtostigmacros) armstrongi 145

Stigmacros (Cyrtostigmacros) brooksi 145
Stigmacros (Cyrtostigmacros) castanea 146
Stigmacros (Cyrtostigmacros) clarki 146
Stigmacros (Cyrtostigmacros) extreminigra 146
Stigmacros (Cyrtostigmacros) ferruginea 146
Stigmacros (Cyrtostigmacros) flava 146
Stigmacros (Cyrtostigmacros) glauerti 146
Stigmacros (Cyrtostigmacros) lanaris 147
Stigmacros (Cyrtostigmacros) major 147
Stigmacros (Cyrtostigmacros) proxima 148
Stigmacros (Cyrtostigmacros) sordida 148
Stigmacros (Cyrtostigmacros) striata 148
Stigmacros elegans 146
Stigmacros epinotalis 146
Stigmacros extreminigra 146
Stigmacros ferruginea 146
Stigmacros flava 146
Stigmacros flavinodis 146
(Stigmacros) foreli, Acantholepis 146
(Stigmacros) fossulata, Acantholepis 146
Stigmacros froggatti 146
Stigmacros glauerti 146
Stigmacros (Hagiostigmacros) punctatissima 148
Stigmacros (Hagiostigmacros) spinosa 148
Stigmacros hirsuta 147
Stigmacros impressa 147
(Stigmacros) impressa, Stigmacros 147
Stigmacros inermis 147
Stigmacros intacta 147
Stigmacros lanaris 147
Stigmacros major 147
Stigmacros marginata 147
Stigmacros medioreticulata 147
(Stigmacros) medioreticulata, Acantholepis 147
Stigmacros minor 147
(Stigmacros) minor, Stigmacros 147
Stigmacros nitida 147
Stigmacros occidentalis 147
(Stigmacros) occidentalis, Acantholepis 147
Stigmacros pilosella 147
(Stigmacros) pilosella, Acantholepis 147
Stigmacros proxima 148
Stigmacros (Pseudostigmacros) inermis 147
Stigmacros punctatissima 148
Stigmacros pusilla 148
(Stigmacros) pusilla, Stigmacros 148
Stigmacros rectangularis 148
(Stigmacros) rectangularis, Stigmacros 148
Stigmacros reticulata 148
Stigmacros rufa 148
(Stigmacros) rufa, Stigmacros 148
Stigmacros sordida 148
Stigmacros spinosa 148
Stigmacros stanleyi 148
Stigmacros (Stigmacros) acuta 145
Stigmacros (Stigmacros) brevispina 145
Stigmacros (Stigmacros) impressa 147
Stigmacros (Stigmacros) minor 147
Stigmacros (Stigmacros) pusilla 148
Stigmacros (Stigmacros) rectangularis 148
Stigmacros (Stigmacros) rufa 148
Stigmacros (Stigmacros) wilsoni 148
Stigmacros striata 148
Stigmacros termitoxenus 148
Stigmacros wilsoni 148

(Stigmacros) wilsoni, Stigmacros 148
striata, Podomyrma 84
striata, Stigmacros 148
striata, Stigmacros (Cyrtostigmacros) 148
striatidens australis, Triglyphothrix (Xiphomyrmex) 91
strictum, Tetramorium 90
stridulator, Rhytidoponera 45
striolatum, Tetramorium 90
striolatus splendidior, Xiphomyrmex 90
striolatus, Tetramorium (Xiphomyrmex) viehmeyeri 90
Strumigenys 87
Strumigenys emdeni 87
Strumigenys (Epopostruma) quadrispinosa 64
Strumigenys (Epopostruma) turneri 69
Strumigenys ferocior 87
Strumigenys friedae 87
Strumigenys godeffroyi 87
Strumigenys guttulata 87
Strumigenys leae 88
Strumigenys mayri 87
Strumigenys opaca 88
Strumigenys perplexa 88
Strumigenys quinquedentata 88
Strumigenys szalayi 88
Strumigenys szalayi australis 88
Strumigenys xenos 88
subapterum bogischi, Monomorium 59
subapterum, Monomorium 59
subapterus bogischi, Chelaner 59
subapterus, Chelaner 59
subapterus, Chelaner subapterus 59
subapterus subapterus, Chelaner 59
subcoecum australicum, Monomorium 70
subdentata, Notoncus foreli 125
subfasciata, Myrmecia 16
sublaevis, Bothroponera 22
sublaevis, Bothroponera sublaevis 22
sublaevis kurandensis, Bothroponera 22
sublaevis kurandensis, Pachycondyla (Bothroponera) 22
sublaevis murina, Bothroponera 22
sublaevis murina, Pachycondyla (Bothroponera) 22
sublaevis reticulata, Bothroponera 22
sublaevis reticulata, Ponera (Bothroponera) 22
sublaevis rubicunda, Bothroponera 22
sublaevis sublaevis, Bothroponera 22
subnitens, Polyrhachis inconspicua 136
subnitidus, Camponotus 119
subnitidus, Camponotus subnitidus 119
subnitidus famelicus, Camponotus 119
subnitidus longinodis, Camponotus 119
subnitidus longinodis, Camponotus (Dinomyrmex) 119
subnitidus subnitidus, Camponotus 119
suchieri, Iridomyrmex rufoniger 102
suffusa, Formica 119
suffusus bendigensis, Camponotus 120
suffusus, Camponotus 119
suffusus, Camponotus suffusus 119
suffusus suffusus, Camponotus 119
sulciceps, Hypoponera 32
sulciceps, Ponera 32
sulla, Melophorus ludius 123
suttoni, Myrmecia 16
swalei, Myrmecia 16
swalei, Myrmecia harderi 16
swani, Leptanilla 53

sycites, Cremastogaster australis 60
sycites, Crematogaster australis 60
sydneyense, Monomorium 71
sydneyense, Monomorium sydneyense 71
sydneyense nigella, Monomorium (Mitara) 71
sydneyense nigellum, Monomorium 71
sydneyense sydneyense, Monomorium 71
Syscia australis 24
Syscia australis edentata 24
szalayi australis, Strumigenys 88
szalayi, Strumigenys 88

tambourinensis, Chelaner kiliani 57
tambourinensis, Monomorium kiliani 57
tambourinensis, Pheidole athertonensis 75
tambourinensis, Polyrhachis (Campomyrma) sidnica 140
tambourinensis, Polyrhachis sidnica 140
(Tanaemyrmex) myoporus, Camponotus 116
(Tanaemyrmex) postcornutus, Camponotus 118
(Tanaemyrmex) spenceri, Camponotus 119
(Tanaemyrmex) tricoloratus, Camponotus 120
Tapinoma 105
(Tapinoma) albipes, Formica 106
Tapinoma (Doleromyrma) darwinianum 97
Tapinoma (Doleromyrma) darwinianum fida 97
Tapinoma (Micromyrma) minutum cephalicum 106
Tapinoma (Micromyrma) rottnestense 106
Tapinoma minutum 105
Tapinoma minutum broomense 106
Tapinoma minutum broomensis 106
Tapinoma minutum cephalicum 106
Tapinoma minutum integrum 106
Tapinoma minutum minutum 105
Tapinoma pusillum 92
Tapinoma rottnestense 106
tarsata malandensis, Myrmecia 15
tarsata, Myrmecia 16
tarsata rowlandi, Myrmecia 15
tasmani, Camponotus 120
tasmaniensis continentis, Pheidole 78
tasmaniensis, Myopias 35
tasmaniensis, Myrmecia 8
tasmaniensis, Pachycondyla (Bothroponera) 38
tasmaniensis, Paratrechina 129
tasmaniensis, Pheidole 77
tasmaniensis, Pheidole tasmaniensis 77
tasmaniensis, Prenolepis (Nylanderia) 129
tasmaniensis, Rhytidoponera 48
tasmaniensis, Rhytidoponera metallica 48
tasmaniensis tasmaniensis, Pheidole 77
taurus, Ectatomma (Rhytidoponera) cornutum 48
taurus, Rhytidoponera 48
taylori, Myrmecia (Promyrmecia) aberrans 10
Technomyrmex 106
Technomyrmex albipes 3 , 106
Technomyrmex albipes cedarensis 106
Technomyrmex bicolor 106
Technomyrmex bicolor antonii 106
Technomyrmex jocosus 106
Technomyrmex quadricolor 106
Technomyrmex sophiae 107
templi, Polyrhachis 140
tenue, Ectatomma (Rhytidoponera) 48
tenuis, Rhytidoponera 48

tepperi, Myrmecia 17
Teratomyrmex 149
Teratomyrmex greavesi 149
terebrans, Formica 120
teres, Podomyrma densestrigosa 81
termitoxenus, Stigmacros 148
terpischore elegans, Polyrhachis 141
terpsichore elegans, Polyrhachis 141
terpsichore, Polyrhachis 140
terpsichore, Polyrhachis terpsichore 140
terpsichore rufifemur, Polyrhachis 141
terpsichore terpsichore, Polyrhachis 140
testaceipes, Camponotus 120
testaceipes, Formica 120
testaceipes, Myrmecia 17
testaceipes, Promyrmecia 17
testudineus, Meranoplus 68
Tetramorium 3 , 88
Tetramorium andrynicum 89
Tetramorium antipodum 90
Tetramorium australe 89
Tetramorium bicarinatum 89
Tetramorium capitale 89
Tetramorium confusum 89
Tetramorium deceptum 89
Tetramorium fuscipes 89
Tetramorium impressum 89
Tetramorium lanuginosum 91
Tetramorium laticephalum 89
Tetramorium megalops 89
Tetramorium melleum 86
Tetramorium ornatum 90
Tetramorium pacificum 90
Tetramorium pacificum validiusculum 91
Tetramorium simillimum 90
Tetramorium sjostedti 90
Tetramorium spininode 90
Tetramorium splendidior 90
Tetramorium strictum 90
Tetramorium striolatum 90
Tetramorium thalidum 91
Tetramorium turneri 91
Tetramorium validiusculum 91
Tetramorium viehmeyeri 91
Tetramorium (Xiphomyrmex) sjostedti 90
Tetramorium (Xiphomyrmex) turneri 91
Tetramorium (Xiphomyrmex) viehmeyeri 91
Tetramorium (Xiphomyrmex) viehmeyeri striolatus 90
Tetraponera 2 , 17
Tetraponera laeviceps 18
Tetraponera punctulata 18
Tetraponera punctulata kimberleyensis 18
Tetraponera punctulata punctulata 18
thais, Polyrhachis 141
thalia io, Polyrhachis 141
thalia io, Polyrhachis (Campomyrma) 141
thalia, Polyrhachis 141
thalia, Polyrhachis thalia 141
thalia thalia, Polyrhachis 141
thalidum, Tetramorium 91
Thlipsepinotus 109
thoracicus, Pristomyrmex 85
thusnelda, Polyrhachis 141
tibialis, Leptomyrmex nigriventris 104
tillyardi, Metapone 69

tostum, Chelaner rothsteini 58
tostum, Monomorium rothsteini 58
townsvillei, Polyrhachis 141
townsvillei, Polyrhachis (Cyrtomyrma) 141
(Trachymesopsus) rotundiceps, Euponera 28
Trachymesopus 52
Trachymesopus clarki 52
(Trachymesopus) clarki, Euponera 52
Trachymesopus darwinii 52
Trachymesopus pachynoda 52
(Trachymesopus) pachynoda, Euponera 52
Trachymesopus rufonigra 52
trachypyx, Rhytidoponera 48
transversa, Pheidole proxima 77
Trapeziopelta diadela 35
trapezoidea, Pheidole 78
trapezoidea, Polyrhachis 141
Trichomelophorus 122
tricolor, Metapone 69
tricolor, Myrmecia 16
tricolor, Podomyrma 84
tricolor rogeri, Myrmecia 16
tricoloratus, Camponotus 120
tricoloratus, Camponotus (Tanaemyrmex) 120
tricosa, Leptogenys 34
Triglyphothrix 91
Triglyphothrix lanuginosa 91
Triglyphothrix (Xiphomyrmex) striatidens australis 91
trigona convexiuscula, Ponera 31
tripellis, Polyrhachis (Chariomyrma) gab 134
tristis, Camponotus 120
tristis, Camponotus (Myrmophyma) 120
tristis, Dolichoderus (Hypoclinea) 93
truncata elliptica, Ponera 31
trux, Sphinctomyrmex 52
tuberculatus, Acantholepis 101
tubifera, Polyrhachis 141
tumidus, Camponotus 120
tumidus, Camponotus (Myrmogonia) 120
turneri, Aenictus 52
turneri aesopus, Melophorus 124
turneri, Anochetus 21
turneri candida, Melophorus 124
turneri candidus, Melophorus 124
turneri, Cerapachys 28
turneri, Cerapachys (Phyracaces) 28
turneri, Chelaner 59
turneri, Dacryon 84
turneri, Dolichoderus 94
turneri, Echinopla 121
turneri, Echinopla turneri 121
turneri, Ectatomma (Rhytidoponera) 49
turneri fuscipes, Xiphomyrmex 89
turneri latunei, Anochetus 21
turneri, Leptogenys 34
turneri, Leptogenys turneri 34
turneri longensis, Leptogenys 34
turneri longensis, Leptogenys (Odontopelta) 34
turneri, Melophorus 124
turneri, Melophorus turneri 124
turneri, Mesostruma 69
turneri, Odontomachus 36
turneri, Odontomachus ruficeps 36
turneri perthensis, Melophorus 124
turneri, Pheidole 78
turneri pictipes, Echinopla 122
turneri, Platythyrea 38
turneri, Podomyrma 84
turneri, Polyrhachis 141
turneri, Pseudonotoncus 144
turneri, Rhytidoponera 49
turneri, Sphinctomyrmex 52
turneri, Strumigenys (Epopostruma) 69
turneri, Tetramorium 91
turneri, Tetramorium (Xiphomyrmex) 91
turneri turneri, Echinopla 121
turneri turneri, Leptogenys 34
turneri turneri, Melophorus 124
turneri, Vollenhovia 59
Turneria 107
Turneria bidentata 107
Turneria frenchi 107
Turnesia bidentata 107
turtoni, Discothyrea 29
tyloxys, Rhytidoponera 49
Typhlatta ceylonica 52

unctus, Leptomyrmex erythrocephalus 104
unicolor, Leptomyrmex 105
unicolor, Meranoplus diversus 67
unicolor, Orectognathus mjobergi 73
urania, Polyrhachis 141
urens, Myrmecia 17

vaga, Paratrechina 129
vaga, Prenolepis obscura 129
validiusculum, Tetramorium 91
validiusculum, Tetramorium pacificum 91
variabilis ampla, Pheidole 74
variabilis latigena, Pheidole 78
variabilis mediofusca, Pheidole 78
variabilis ocior, Pheidole 78
variabilis ocyma, Pheidole 78
variabilis parvispina, Pheidole 78
variabilis, Pheidole 78
variabilis, Pheidole variabilis 78
variabilis praedo, Pheidole 78
variabilis redunca, Pheidole 78
variabilis rugocciput, Pheidole 78
variabilis rugosula, Pheidole 79
variabilis variabilis, Pheidole 78
varians angusticeps, Leptomyrmex 105
varians, Cerapachys 28
varians, Leptomyrmex 105
varians, Leptomyrmex varians 105
varians, Myrmecia 17
varians, Phyracaces 28
varians quadricolor, Leptomyrmex 105
varians, Rhytidoponera (Chalcoponera) metallica 45
varians rothneyi, Leptomyrmex 105
varians ruficeps, Leptomyrmex 105
varians rufipes, Leptomyrmex 105
varians varians, Leptomyrmex 105
varicolor, Podomyrma overbecki 84
velutina, Discothyrea 29
velutina, Prodiscothyrea 29
venustula, Mayriella abstinens 66
venustus, Leptomyrmex erythrocephalus 104

venustus, Xiphomyrmex viehmeyeri 91
vermiculosa, Polyrhachis guerini 134
versicolor, Camponotus 120
versicolor, Camponotus (Myrmosaulus) 120
versicolor, Orectognathus 74
vicina, Iridomyrmex 103
victoriae andrei, Chalcoponera 42
victoriae cedarensis, Rhytidoponera (Chalcoponera) 49
victoriae, Ectatomma (Rhytidoponera) 49
victoriae maledicta, Rhytidoponera (Chalcoponera) 44
victoriae, Rhytidoponera 49
victorianus, Iridomyrmex rufoniger 102
victoriensis, Camponotus latrunculus 115
victoriensis, Camponotus (Myrmoturba) latrunculus 115
viehmeyeri striolatus, Tetramorium (Xiphomyrmex) 90
viehmeyeri, Tetramorium 91
viehmeyeri, Tetramorium (Xiphomyrmex) 91
viehmeyeri venustus, Xiphomyrmex 91
vigilans, Atta 79
vigilans, Pheidole 79
villosa, Camponotus (Myrmoturba) 120
villosus, Camponotus 120
vindex basirufa, Myrmecia 17
vindex desertorum, Myrmecia 8
vindex, Myrmecia 17
vindex, Myrmecia vindex 17
vindex vindex, Myrmecia 17
violacea, Rhytidoponera 49
violaceum, Ectatomma (Rhytidoponera) convexum 49
virescens, Formica 127
viridiaeneus, Iridomyrmex detectus 102
viridiaeneus, Iridomyrmex purpureus 102
viridigaster, Iridomyrmex 103
viridis, Chalcoponera 49
viridis, Formica 127
viridis, Rhytidoponera 49
viridiventris, Calomyrmex splendidus 109
vitrea, Formica 120
vitreus, Camponotus 120
Vollenhovia 91
Vollenhovia oblonga 92
Vollenhovia turneri 59
voltai, Camponotus (Myrmogonia) oetkeri 117
voltai, Camponotus oetkeri 117

walkeri bardus, Camponotus 121
walkeri, Camponotus 121
walkeri, Camponotus walkeri 121
walkeri, Myrmecia 9
walkeri walkeri, Camponotus 121
wheeleri, Melophorus 124
wheeleri, Pristomyrmex 85
wheeleri, Prolasius 144
whitei, Camponotus 121
whitei, Camponotus (Myrmosphincta) 121
whitei, Chelaner 59
whitei, Crematogaster 63
whitei, Monomorium (Holcomyrmex) 59
wiburdi, Leptomyrmex 105
wiburdi, Leptomyrmex wiburdi 105
wiburdi pictus, Leptomyrmex 105
wiburdi wiburdi, Leptomyrmex 105

wiederkehri, Camponotus 121
wiederkehri, Camponotus wiederkehri 121
wiederkehri lucidior, Camponotus 121
wiederkehri wiederkehri, Camponotus 121
wienandsi, Calomyrmex albopilosus 108
wienandsi, Camponotus (Calomyrmex) albopilosus 108
wiesei, Pheidole 79
wilsoni, Amblyopone 20
wilsoni, Bothriomyrmex 92
wilsoni, Pristomyrmex 85
wilsoni, Prolasius hemiflavus 143
wilsoni, Promyrmecia 17
wilsoni, Stigmacros 148
wilsoni, Stigmacros (Stigmacros) 148
wroughtonii, Rhoptromyrmex 86

xenos, Strumigenys 88
xerophila, Crematogaster 63
xerophila, Crematogaster xerophila 63
xerophila exigua, Crematogaster 63
xerophila xerophila, Crematogaster 63
Xiphomyrmex capitalis 89
Xiphomyrmex flavigaster 56
Xiphomyrmex impressus 89
(Xiphomyrmex) sjostedti, Tetramorium 90
(Xiphomyrmex) striatidens australis, Triglyphothrix 91
Xiphomyrmex striolatus splendidior 90
Xiphomyrmex turneri fuscipes 89
(Xiphomyrmex) turneri, Tetramorium 91
(Xiphomyrmex) viehmeyeri striolatus, Tetramorium 90
(Xiphomyrmex) viehmeyeri, Tetramorium 91
Xiphomyrmex viehmeyeri venustus 91

yalgooensis, Iridomyrmex chasei 97
yarrabahensis, Camponotus discors 112
yarrabahensis, Camponotus (Myrmoturba) maculatus 112
yarrabahensis, Polyrhachis lombokensis 137
yarrabahensis, Polyrhachis (Myrmatopa) lombokensis 137
yarrabahna, Leptogenys diminuta 33
yarrabahna, Leptogenys (Lobopelta) diminuta 33
yarrensis, Pheidole ampla 79
yorkana, Polyrhachis (Cyrtomyrma) rastellata 139
yorkana, Polyrhachis rastellata 139
yorkensis, Rhytidoponera 49
yorkensis, Rhytidoponera cristata 49
ypsilon, Dolichoderus 94
ypsilon, Dolichoderus scabridus 94
ypsilon, Dolichoderus ypsilon 94
ypsilon nigra, Dolichoderus 95
ypsilon nigra, Dolichoderus (Hypoclinea) 95
ypsilon rufotibialis, Dolichoderus 95
ypsilon rufotibialis, Dolichoderus (Hypoclinea) 95
ypsilon ypsilon, Dolichoderus 94

Zasphinctus 49
zeuxis, Camponotus evae 113
zeuxis, Camponotus (Myrmogonia) evae 113
zimmerae, Polyrhachis 141
zimmerae, Polyrhachis (Campomyrma) 141

TAXONOMIC INDEX

SPHECOIDEA AND VESPOIDEA

abbreviata, Liris (Leptolarra) 234
abbreviata, Notogonia 234
abdita, Isodontia 226
abdita nugenti, Isodontia 226
abditus nugenti, Sphex (Isodontia) 226
aberrans, Pison (Parapison) 257
aberrans, Pison (Pison) 257
Abispa 160
(Abispa), Abispa 160
Abispa (Abispa) 160
Abispa (Abispa) australiana 161
Abispa (Abispa) ephippium 161
Abispa (Abispa) laticincta 161
Abispa (Abispa) splendida 161
Abispa (Abispa) splendida australis 161
Abispa (Abispa) splendida splendida 161
(Abispa) australiana, Abispa 161
(Abispa) australiana, Vespa 161
Abispa australis 161
(Abispa) ephippium, Abispa 161
(Abispa) laticincta, Abispa 161
Abispa meadewaldoensis 161
Abispa paragioides 199
(Abispa) splendida, Abispa 161
(Abispa) splendida australis, Abispa 161
(Abispa) splendida splendida, Abispa 161
abispoides, Rhynchium 175
abnormis, Alastor 182
abnormis, Larrisson 255
abnormis, Paralastor 182
abnormis, Sericophorus 255
aborigenus, Tachysphex 238
Acanthostethus 218 , 277
Acanthostethus basalis 278
Acanthostethus brisbanensis 277
(Acanthostethus) brisbanensis, Nysson 277
Acanthostethus confertus 277
(Acanthostethus) confertus, Nysson 277
Acanthostethus gilberti 277
(Acanthostethus) gilberti, Nysson 277
Acanthostethus hentyi 277
Acanthostethus minimus 277
(Acanthostethus) minimus, Nysson 277
Acanthostethus moerens 277
(Acanthostethus) moerens, Nysson 277
Acanthostethus mysticus 278
Acanthostethus nudiventris 278
(Acanthostethus) nudiventris, Nysson 278
Acanthostethus obliteratus 278
(Acanthostethus) obliteratus, Nysson 278
Acanthostethus portlandensis 278
Acanthostethus punctatissimus 278
(Acanthostethus) punctatissimus, Nysson 278
Acanthostethus saussurei 278

Acanthostethus spiniger 278
(Acanthostethus) spiniger, Nysson 278
Acanthostethus tasmanicus 278
(Acanthostethus) tasmanicus, Nysson 278
Acanthostethus triangularis 279
(Acanthostethus) triangularis, Nysson 279
Acarodynerus 162
Acarodynerus acarophilus 162
Acarodynerus clypeatus 162
Acarodynerus denticulatus 162
Acarodynerus dietrichianus 162
Acarodynerus dietrichianus dietrichianus 162
Acarodynerus dietrichianus rufocaudatus 162
Acarodynerus drewsenianus 162
Acarodynerus exarmatus 163
Acarodynerus legatus 163
Acarodynerus lunaris 163
Acarodynerus paleovariatus 163
Acarodynerus posttegulatus 163
Acarodynerus propodalaris 163
Acarodynerus quadrangolum 163
Acarodynerus queenslandicus 163
Acarodynerus spargovillensis 164
Acarodynerus spectrum 164
Acarodynerus triangulum 164
acarophilus, Acarodynerus 162
Acarozumia 164
Acarozumia amaliae 164
acuticollis, Sphodrotes 254
adae, Cerceris 296
adelaidae, Tachysphex 240
aeneus, Auchenophorus 256
aequifasciatus, Paralastor 182
aestuans, Tachytes 237
affine, Zoyphium 251
affinis, Aphelotoma 219
affinis, Sericophorus (Zoyphidium) 251
agitata, Liris (Leptolarra) 234
agitata, Notogonia 234
Aha 244
Aha evansi 244
Aha ha 244
ahasverus, Sphex (Sphex) 222
alariformis, Odynerus 198
alariformis, Parifodynerus 198
alaris, Odynerus (Leionotus) 201
alaris, Pseudepipona 201
Alastor abnormis 182
Alastor albocinctus 193
Alastor argentifrons 183
Alastor australis 184
Alastor carinatus 184
Alastor clotho 165
Alastor emarginatus 187

Alastor eriurgus 187
Alastor lachesis 189
Alastor (Paralastor) brunneus 184
Alastor (Paralastor) cruentatus 185
Alastor (Paralastor) flaviceps 188
Alastor (Paralastor) fraternus 188
Alastor (Paralastor) infernalis 189
Alastor (Paralastor) insularis 189
Alastor (Paralastor) lateritius 190
Alastor (Paralastor) maculiventris 190
Alastor (Paralastor) nautarum 191
Alastor (Paralastor) pusillus 193
Alastor (Paralastor) sanguineus 194
Alastor (Paralastor) smithii 195
Alastor (Paralastor) tridentatus 200
Alastor (Paralastor) vulneratus 197
Alastor (Paralastor) vulpinus 197
Alastor parca 192
Alastor picteti 192
Alastor punctulatus 193
Alastor similis 193
Alastor tasmaniensis 196
Alastor tuberculatus 196
alastoripennis, Odynerus (Ancistrocerus) 182
alastoripennis, Paralastor 182
Alastoroides 164
alastoroides, Cerceris 296
Alastoroides clotho 165
albifrons, Paralastor 182
albifrons, Vespa 182
albitarsatum huonense, Trypoxylon (Trypoxylon) 264
albitarsatum, Trypoxylon (Trypoxylon) 264
albocinctus, Alastor 193
albohirta, Isodontia 226
albohirtus, Sphex (Isodontia) 226
alecto, Larra 234
alecto, Larrada 234
alecto, Odynerus 181
alevinus, Podagritus (Echuca) 269
alexandriae, Odynerus (Rhynchium) angulatus 201
alexandriae, Paralastor 182
alexandriae, Pseudepipona angulata 201
aliceae, Sericophorus 245
aliceae, Sericophorus (Sericophorus) 245
aliciae, Macrocalymma 180
aliciae, Paragia (Paragia) decipiens 155
aliciae, Podagritus (Echuca) 269
aliciae, Rhopalum 269
Allorhynchium 165
Allorhynchium iridipenne 165
allunga, Bembix 285
Alyson tomentosum 303
amaliae, Acarozumia 164
amaliae, Nortonia 164
amator, Sphex 223
amicus, Antamenes (Australochilus) 167
amicus, Pachymenes 167
Ammatomus 283
Ammatomus austrinus 284
Ammatomus decoratus 284
Ammatomus icarioides 284
Ammophila 228
Ammophila ardens 229
Ammophila atripes 229
Ammophila aurifera 229

Ammophila clavus 229
Ammophila eyrensis 229
Ammophila impatiens 229
Ammophila instabilis 229
Ammophila (Parapsammophila) eremophila 228
Ammophila suspiciosa 228
Ampulex 220
Ampulex compressa 220
Ampulex micans 220
Ampulicinae 151
Anacrucis 251
Anacrucis asperithorax 251
Anacrucis cingulata 251
Anacrucis clypeata 251
Anacrucis ferruginea 252
Anacrucis laevigata 253
Anacrucis punctuosa 253
Anacrucis striatula 254
analis, Ischnocoelia robusta 179
Ancistroceroides 182
(Ancistroceroides) cruentus, Odynerus 185
(Ancistroceroides) sanguinolentus, Odynerus 200
Ancistrocerus 165
(Ancistrocerus) alastoripennis, Odynerus 182
Ancistrocerus fluvialis 165
(Ancistrocerus) fluvialis, Odynerus 165
Ancistrocerus (Subancistrocerus) monstricornis 207
(Ancistrocerus) vernalis, Odynerus 166
Androcrabro 274
(Androcrabro) bivittata, Williamsita 275
(Androcrabro) bushiella, Williamsita 275
(Androcrabro) manifestata, Williamsita 275
(Androcrabro) neglecta, Williamsita 275
(Androcrabro) ordinaria, Williamsita 275
(Androcrabro) riekiella, Williamsita 275
(Androcrabro) smithiensis, Williamsita 275
(Androcrabro) tasmanica, Williamsita 276
(Androcrabro) vedetta, Williamsita 276
(Androcrabro), Williamsita 274
anerus, Podagritus (Echuca) 270
angulata alexandriae, Pseudepipona 201
angulata angulata, Pseudepipona 201
angulata, Pseudepipona 201
angulata, Pseudepipona angulata 201
angulata, Riekia 154
angulatus alexandriae, Odynerus (Rhynchium) 201
angulatus, Odynerus 201
anguloides, Pseudalastor 199
anostreptus, Paralastor 183
Antamenes 166
Antamenes (Antamenes) 166
(Antamenes), Antamenes 166
Antamenes (Antamenes) pseudoneotropicus 166
Antamenes (Antamenes) tasmaniae 166
Antamenes (Antamenes) vernalis 166
Antamenes (Antamenes) vorticosus 166
Antamenes (Australochilus) 167
Antamenes (Australochilus) amicus 167
Antamenes (Australochilus) citreocinctus 167
Antamenes (Australochilus) ferrugineus 167
Antamenes (Australochilus) flavoniger 167
Antamenes (Australochilus) hackeri 167
Antamenes (Australochilus) hostilis 167
Antamenes (Australochilus) jugulatus 167
(Antamenes) pseudoneotropicus, Antamenes 166

(Antamenes) tasmaniae, Antamenes 166
(Antamenes) vernalis, Antamenes 166
(Antamenes) vorticosus, Antamenes 166
anteum, Rhopalum (Corynopus) 267
anteum, Rhopalum (Rhopalum) 267
anthicivora, Cerceris 296
antipodes, Cerceris 296
antipodum, Bembecinus 284
antipodum, Stizus 284
Aphelotoma 219
Aphelotoma affinis 219
Aphelotoma aterrima 220
Aphelotoma auricula 219
Aphelotoma auriventris 219
Aphelotoma fuscata 219
Aphelotoma melanogaster 219
Aphelotoma nigricula 219
Aphelotoma rufiventris 219
Aphelotoma striaticollis 219
Aphelotoma tasmanica 219
Aphelotoma tasmanica auriventris 219
Aphelotoma tasmanica tasmanica 219
apicalis, Eumenes 176
Apoidea 151
approximatus, Arpactophilus 230
approximatus, Austrostigmus 230
approximatus, Tachytes 237
arator, Arpactophilus 230
arator, Harpactophilus 230
arcuata arcuata, Delta 170
arcuata, Delta 170
arcuata, Delta arcuata 170
arcuata, Vespa 170
ardens, Ammophila 229
arenicola, Paralastor 183
areniferum, Pison (Pison) 257
argentatus, Sphex 222
argentatus, Sphex (Sphex) 222
argentifrons, Alastor 183
argentifrons, Paralastor 183
argentifrons, Sphex 225
Argogorytes 280
Argogorytes crucigera 280
Argogorytes rubrosignatus 280
Argogorytes rufomixtus 281
Argogorytes secernendus 281
Argogorytes stenopygus 280
argyreum, Zoyphium 251
argyreus, Sericophorus (Zoyphidium) 251
argyrias, Paralastor 183
armigera, Cerceris 297
armigera rufofusca, Cerceris 297
Arpactophilus 218 , 230
Arpactophilus approximatus 230
Arpactophilus arator 230
Arpactophilus dubius 231
Arpactophilus glabrellus 231
Arpactophilus kohlii 231
Arpactophilus queenslandensis 231
Arpactophilus reticulatus 231
Arpactophilus ruficollis 231
Arpactophilus steindachneri 231
Arpactophilus steindachneri deserticolus 231
Arpactophilus steindachneri steindachneri 231
Arpactophilus sulcatus 231

Arpactophilus tricolor 232
Arpactus aurantiacus 281
Arpactus browni 281
Arpactus chrysozonus 281
Arpactus consuetipes 282
Arpactus crucigera 280
Arpactus (Miscothyris) guttatulus 279
Arpactus (Miscothyris) plomleyi 280
Arpactus obesus 282
Arpactus pretiosus 282
Arpactus rubrosignatus 280
Arpactus secernendus 281
Arpactus spinicornis 282
Arpactus spryi 282
aruensis, Delta incola 171
aruensis, Eumenes (Delta) incola 171
asperithorax, Anacrucis 251
asperithorax, Sericophorus (Zoyphidium) 251
aspra, Pseudepipona 201
aspra, Pseudepipona (Pseudepipona) 201
assatus, Delta campaniformis 171
assatus, Eumenes campaniformis 171
Astata 233
Astata australasiae 233
Astaurus 279
Astaurus hylaeoides 280
Astaurus tenuicornis 280
aterrima, Aphelotoma 220
aterrima, Austrotoma 220
aterrimus, Paralastor 183
atrifrons, Bembex 285
atrifrons, Bembix 285
atripennis, Paralastor 183
atripes, Ammophila 229
atrum, Rhynchium 203
Auchenophorus 256
Auchenophorus aeneus 256
Auchenophorus coruscans 256
Auchenophorus fulvicornis 257
(Aulacophilus) difficile, Pison 263
(Aulacophilus) icarioides, Pison 263
aurantiaca, Cerceris 297
aurantiaca, Ischnocoelia robusta 179
aurantiacus, Arpactus 281
aurantiacus, Austrogorytes 281
aurantiopicta, Bidentodynerus bicolor 170
aurantiopictus, Odynerus (Rhynchium) bicolor 170
aurantiopilosella, Pseudepipona (Syneuodynerus) 207
aurantiopilosellus, Syneuodynerus 207
auratum, Pison (Pison) 257
auratus, Pison 257
aureofasciata, Bembex 286
aureofasciata, Bembix 286
aureosericeum, Pison 257
aureosericeum, Pison (Pison) 257
auricula, Aphelotoma 219
aurifera, Ammophila 229
aurifex, Pison 257
aurifex, Pison (Pison) 257
aurifex, Sphex 226
aurifrons, Isodontia 226
aurifrons, Sphex 226
auriventre, Pison 258
auriventre, Pison (Pison) 258
auriventris, Aphelotoma 219

auriventris, Aphelotoma tasmanica 219
aurocinctus, Odynerus 183
aurocinctus, Paralastor 183
auster, Paralastor 184
austragilis, Dasyproctus 274
australasiae, Astata 233
australense, Rhynchium 204
australensis, Ectopioglossa 174
australensis, Ectopioglossa polita 174
australensis, Eumenes (Pareumenes) 174
australensis, Montezumia 164
australensis, Pseudozethus 202
australiae, Rhopalum (Corynopus) 267
australiae, Rhopalum (Rhopalum) 267
australiana, Abispa (Abispa) 161
australiana, Vespa (Abispa) 161
australiensis, Nitela 256
australiensis, Podagritus (Echuca) 270
australis, Abispa 161
australis, Abispa (Abispa) splendida 161
australis, Alastor 184
australis australis, Paragia (Paragiella) 156
australis borealis, Paragia (Paragiella) 156
australis, Cerceris 297
australis, Elimus 174
australis, Liris 236
australis, Palmodes 227
australis, Paragia 156
australis, Paragia (Paragiella) 156
australis, Paragia (Paragiella) australis 156
australis, Paralastor 184
australis, Pison 262
australis, Spilomena 232
australis, Tachytes 236 , 240
Australochilus 167
(Australochilus) amicus, Antamenes 167
(Australochilus), Antamenes 167
(Australochilus) citreocinctus, Antamenes 167
(Australochilus) ferrugineus, Antamenes 167
(Australochilus) flavoniger, Antamenes 167
(Australochilus) hackeri, Antamenes 167
(Australochilus) hostilis, Antamenes 167
(Australochilus) jugulatus, Antamenes 167
Australodynerus 168
Australodynerus convexus 168
Australodynerus merredinensis 168
Australodynerus merredinensis everardensis 168
Australodynerus merredinensis merredinensis 168
Australodynerus merredinensis victoriensis 168
Australodynerus punctiventris 168
Australodynerus pusilloides 168
Australodynerus pusilloides impudicus 168
Australodynerus pusilloides pusilloides 168
Australodynerus pusillus 168
Australodynerus yanchepensis 206
Australodynerus yanchepensis nigrithorax 206
Australozethus 169
Australozethus continentalis 169
Australozethus occidentalis 169
Australozethus tasmaniensis 169
Australozethus tasmaniensis montanus 169
Australozethus tasmaniensis tasmaniensis 169
austrinus, Ammatomus 284
austrinus, Gorytes 284
austrinus, Polistes (Polistella) townsvillensis 213

Austrogorytes 281
Austrogorytes aurantiacus 281
Austrogorytes bellicosus 281
Austrogorytes browni 281
Austrogorytes chrysozonus 281
Austrogorytes ciliatus 281
Austrogorytes consuetipes 282
Austrogorytes cygnorum 282
Austrogorytes frenchii 282
Austrogorytes obesus 282
Austrogorytes perkinsi 282
Austrogorytes pretiosus 282
Austrogorytes spinicornis 282
Austrogorytes spryi 282
Austrogorytes tarsatus 283
Austrostigmus 230
Austrostigmus approximatus 230
Austrostigmus dubius 231
Austrostigmus glabrellus 231
Austrostigmus reticulatus 231
Austrostigmus ruficollis 231
Austrotoma 220
Austrotoma aterrima 220
azyx, Larrisson 256

balder, Polistes 209
balder, Polistes (Megapolistes) 209
balteata, Cerceris 297
balyi, Odynerus 202
bannitus, Sphex 225
barbatum, Pison (Pison) 258
baringa, Bembix 286
basale, Pison (Pison) 258
basalis, Acanthostethus 278
basalis, Pison 258
basilicus, Chlorion (Proterosphex) 223
basilicus, Sphex (Sphex) 223
basilissa, Liris (Leptolarra) 235
basilissa, Notogonia 235
bellicosus, Austrogorytes 281
bellicosus, Gorytes 281
Bembecinus 284
Bembecinus antipodum 284
Bembecinus egens 284
Bembecinus gorytoides 284
Bembecinus hirtula 285
Bembecinus monodon 284
Bembecinus signatus 285
Bembecinus turneri 285
Bembex 285
Bembex atrifrons 285
Bembex aureofasciata 286
Bembex calcarina 288
Bembex crabroniformis 294
Bembex cursitans 287
Bembex egens 287
Bembex flavifrons 287
Bembex flavilabris 285
Bembex flavipes 287
Bembex flaviventris 288
Bembex funebris 285
Bembex furcata 288
Bembex lamellata 289
Bembex latifasciata 289

Bembex leeuwinensis 289
Bembex littoralis 289
Bembex lobimana 290
Bembex mackayensis 291
Bembex marsupiata 290
Bembex musca 291
Bembex nigropectinatus 291
Bembex palmata 292
Bembex pectinipes 292
Bembex raptor 294
Bembex saussurei 287
Bembex severa 293
Bembex trepida 293
Bembex tridentifera 292
Bembex tuberculiventris 294
Bembex variabilis 294
Bembex vespiformis 294
Bembix 150, 218, 285
Bembix allunga 285
Bembix atrifrons 285
Bembix aureofasciata 286
Bembix baringa 286
Bembix berontha 286
Bembix boaliri 286
Bembix boonamin 286
Bembix brevis 287
Bembix buntor 286
Bembix burando 286
Bembix burraburra 286
Bembix carripan 286
Bembix cooba 286
Bembix coonundura 287
Bembix cursitans 287
Bembix egens 287
Bembix eleebana 287
Bembix flavifrons 287
Bembix flavipes 287
Bembix flaviventris 288
Bembix flavolatera 289
Bembix furcata 288
Bembix gelane 288
Bembix ginjulla 288
Bembix goyarra 288
Bembix gunamarra 288
Bembix gurindji 288
Bembix hokarra 288
Bembix kamulla 288
Bembix kora 289
Bembix kununurra 289
Bembix lamellata 289
Bembix latifasciata 289
Bembix leeuwinensis 289
Bembix littoralis 289
Bembix lobimana 290
Bembix loorea 290
Bembix magarra 290
Bembix maliki 290
Bembix mareeba 290
Bembix marhra 290
Bembix marsupiata 290
Bembix mianga 290
Bembix minya 291
Bembix mokari 291
Bembix moma 291
Bembix moonga 291
Bembix mundurra 291
Bembix munta 291
Bembix musca 291
Bembix nigropectinata 291
Bembix obiri 291
Bembix octosetosa 292
Bembix olba 292
Bembix oomborra 292
Bembix ourapilla 292
Bembix palmata 292
Bembix palona 292
Bembix pectinipes 292
Bembix pikati 293
Bembix pillara 293
Bembix promontorii 293
Bembix pulka 293
Bembix severa 293
Bembix thooma 293
Bembix tibooburra 293
Bembix trepida 293
Bembix tuberculiventris 294
Bembix uloola 294
Bembix undeneya 294
Bembix variabilis 294
Bembix vespiformis 294
Bembix victoriensis 294
Bembix wadamiri 294
Bembix wangoola 295
Bembix wanna 295
Bembix warawara 295
Bembix weema 295
Bembix wilcannia 295
Bembix wiluna 295
Bembix wollowra 295
Bembix wolpa 295
Bembix wowine 295
Bembix yalta 296
Bembix yunkara 296
bengalense, Chalybion 220
bengalensis, Pelopoeus 220
bernardii bernardii, Polistes (Polistella) 211
bernardii duplicinctus, Polistes 213
bernardii duplicinctus, Polistes (Polistella) 211
bernardii insulae, Polistes (Polistella) 211
bernardii, Polistes 211
bernardii, Polistes (Polistella) 211
bernardii, Polistes (Polistella) bernardii 211
bernardii richardsi, Polistes (Polistella) 211
berontha, Bembix 286
Bethyloidea 151
bicarinatus, Paralastor 184
bicincta, Eumenes 171
bicincta, Parodynerus 198
bicincta, Vespa 198
bicinctus, Delta 171
bicolor aurantiopicta, Bidentodynerus 170
bicolor aurantiopictus, Odynerus (Rhynchium) 170
bicolor bicolor, Bidentodynerus 169
bicolor, Bidentodynerus 169
bicolor, Bidentodynerus bicolor 169
bicolor flavescentulus, Bidentodynerus 170
bicolor flavescentulus, Odynerus (Rhynchium) 170
bicolor nigrocinctoides, Bidentodynerus 170
bicolor nigrocinctoides, Odynerus (Rhynchium) 170
bicolor, Odynerus 169

bicolor, Paragia 157
bicolor, Paragia (Paragiella) 157
bicolor, Sericophorus 245
bicolor, Sericophorus (Sericophorus) 245
bicoloratus, Odynerus 205
bicoloratus, Stenodyneriellus 205
bidens, Paragia 158
Bidentodynerus 169
Bidentodynerus bicolor 169
Bidentodynerus bicolor aurantiopicta 170
Bidentodynerus bicolor bicolor 169
Bidentodynerus bicolor flavescentulus 170
Bidentodynerus bicolor nigrocinctoides 170
bilobatus, Sphex 223
bilobatus, Sphex (Sphex) 223
bimaculata, Microglossa 232
bimaculata, Spilomena 232
bioculata, Polistes 214
bischoffi, Paralastor 184
bisulcatus, Odynerus (Leionotus) 166
bituberculatum, Trypoxylon 264
bituberculatum, Trypoxylon (Trypoxylon) 264
bivittata, Williamsita (Androcrabro) 275
bivittatus, Crabro 275
bizonatus, Odynerus 198
blumburyensis, Ischnocoelia occidentalis 179
boaliri, Bembix 286
boonamin, Bembix 286
borealis, Paragia (Paragiella) australis 156
brevicornis, Pachycoelius 181
brevicornis, Tachysphex 238
brevis, Bembix 287
brisbanensis, Acanthostethus 277
brisbanensis, Cerceris 296
brisbanensis, Nysson (Acanthostethus) 277
brisbanensis, Paralastor 184
brisbanensis, Sericophorus 245
brisbanensis, Sericophorus (Sericophorus) 245
brisbanensis, Stenodyneriellus 205
browni, Arpactus 281
browni, Austrogorytes 281
brunneus, Alastor (Paralastor) 184
brunneus, Paralastor 184
buccalis, Tachysphex 239
buntor, Bembix 286
burando, Bembix 286
burnettianus, Dasyproctus 274
burnsi, Podagritus (Echuca) 270
burnsiellus, Sericophorus claviger 246
burnsiellus, Sericophorus (Sericophorus) claviger 246
burraburra, Bembix 286
bushiella, Williamsita 275
bushiella, Williamsita (Androcrabro) 275

cabeti, Icaria 217
cabeti, Ropalidia (Icarielia) romandi 217
caementarium, Sceliphron (Sceliphron) 221
caementarium, Sphex 221
cairnensis, Leptomenoides 180
calcarina, Bembex 288
calida, Cerceris 297
calida, Paragia 157
calida, Paragia (Paragiella) 157
caliginosum, Pison (Parapison) 258

caliginosum, Pison (Pison) 258
calixtum, Rhopalum (Rhopalum) 265
Cameronitus 276
(Cameronitus) conglobatus, Ectemnius 277
(Cameronitus), Ectemnius 276
campaniformis assatus, Delta 171
campaniformis assatus, Eumenes 171
campaniformis campaniformis, Delta 171
campaniformis, Delta 171
campaniformis, Delta campaniformis 171
campaniformis, Vespa 171
canescens, Sphex 223
caprai, Paralastor 184
carbonaria, Sphex 223
carbonarius, Dolichurus 218
carbonicolor, Sphex 223
carbonicolor, Sphex (Sphex) 223
carinatus, Alastor 184
carinatus, Discoelius 181
carinatus, Pachycoelius 181
carinatus, Paralastor 184
carinatus, Sericophorus 245
carinatus, Sericophorus (Sericophorus) 245
carinifrons, Psenulus 230
carinifrons scutellatus, Psenulus 230
carnarvonensis, Stenodyneriellus 205
carnegiacum, Rhopalum (Notorhopalum) 269
carnowi, Ischnocoelia integra 178
carolus, Podagritus (Echuca) 270
carripan, Bembix 286
castaneus, Sericophorus 245
castaneus, Sericophorus (Sericophorus) 245
cavifemur, Pseudalastor 199
Celia 232
centrocontinentalis, Polistes humilis 212
centrocontinentalis, Polistes (Polistella) humilis 212
Cerceris 296
Cerceris adae 296
Cerceris alastoroides 296
Cerceris anthicivora 296
Cerceris antipodes 296
Cerceris armigera 297
Cerceris armigera rufofusca 297
Cerceris aurantiaca 297
Cerceris australis 297
Cerceris balteata 297
Cerceris brisbanensis 296
Cerceris calida 297
Cerceris cucullata 298
Cerceris cunnamulla 298
Cerceris darrensis 298
Cerceris dedariensis 298
Cerceris euchroma 298
Cerceris eungella 298
Cerceris exleyae 298
Cerceris fluvialis 298
Cerceris forficata 298
Cerceris froggatti 298
Cerceris gilberti 299
Cerceris gilesi 299
Cerceris goddardi 299
Cerceris goodwini 296
Cerceris hackeriana 299
Cerceris inexpectata 299
Cerceris insulicola 299

Cerceris iridis 300
Cerceris koala 301
Cerceris labeculata 300
Cerceris latiberbis 300
Cerceris listrognatha 300
Cerceris luculenta 300
Cerceris megacantha 300
Cerceris merredinensis 300
Cerceris minuscula 300
Cerceris multiguttata 301
Cerceris nigrocincta 297
Cerceris opposita 301
Cerceris osculata 301
Cerceris perkinsi 301
Cerceris praedura 301
Cerceris raymenti 301
Cerceris ropalidioides 301
Cerceris saeva 296
Cerceris sculleni 296, 297
Cerceris sedula 302
Cerceris siccata 302
Cerceris spilota 302
Cerceris spinipleuris 302
Cerceris trifida 302
Cerceris unispinosa 302
Cerceris varipes 302
Cerceris venusta 299, 302
Cerceris victrix 302
Cerceris windorum 303
Cerceris xanthura 303
Cerceris ziegleri 296
chalybaeus chalybaeus, Sericophorus (Sericophorus) 246
chalybaeus fulleri, Sericophorus 246
chalybaeus fulleri, Sericophorus (Sericophorus) 246
chalybaeus, Sericophorus (Sericophorus) 246
chalybaeus, Sericophorus (Sericophorus) chalybaeus 246
chalybeus, Sericophorus 246
Chalybion 220
Chalybion bengalense 220
chartergiformis, Pseudepipona 201
Chimiloides 273
Chimiloides doddii 273
Chimiloides nigromaculatus 273
Chimiloides piliferus 273
Chlorion (Proterosphex) basilicus 223
Chlorion (Proterosphex) rhodosoma 225
chrysonota, Larrada 235
chrysonota, Liris (Leptolarra) 235
chrysozonus, Arpactus 281
chrysozonus, Austrogorytes 281
ciliatum, Pison (Pison) 258
ciliatus, Austrogorytes 281
ciliatus, Gorytes 281
cinctus, Crabro 276
cingulata, Anacrucis 251
cingulatus, Sericophorus (Zoyphidium) 251
circulans, Tachysphex 239
citreocinctus, Antamenes (Australochilus) 167
citreocinctus, Odynerus (Leionotus) 167
clarior, Polistes humilis 212
claviger burnsiellus, Sericophorus 246
claviger burnsiellus, Sericophorus (Sericophorus) 246
claviger claviger, Sericophorus (Sericophorus) 246
claviger, Sericophorus (Sericophorus) 246
claviger, Sericophorus (Sericophorus) claviger 246

claviger, Tachyrhostus 246
clavigera, Sphex 222
clavus, Ammophila 229
clavus, Sphex 229
cliffordi, Sericophorus 246
cliffordi, Sericophorus (Sericophorus) 246
Clitemnestra 279
Clitemnestra duboulayi 279
Clitemnestra guttatulus 279
Clitemnestra lucidulus 279
Clitemnestra megalophthalmus 279
Clitemnestra mimetica 279
Clitemnestra perlucidus 279
Clitemnestra plomleyi 280
Clitemnestra sanguinolentus 280
Clitemnestra tenuicornis 280
Clitemnestra thoracicus 280
clotho, Alastor 165
clotho, Alastoroides 165
clypalaris, Pseudepipona 202
clypeata, Anacrucis 251
clypeatus, Acarodynerus 162
clypeatus, Odynerus (Leionotus) 162
clypeatus, Sericophorus (Zoyphidium) 251
clypeopunctatus, Paralastor 184
cockerelli, Sericophorus (Sericophorus) 246
codonocarpi, Tachytes 237
cognata, Sphex 223
cognatus, Sphex (Sphex) 223
Coleoptera 160
collare, Zoyphium 251
collaris, Sericophorus (Zoyphidium) 251
Collembola 218
Colletidae 151
commixta, Liris (Leptolarra) 235
commixta, Notogonia 235
commutatus, Paralastor 185
compressa, Ampulex 220
compressa, Sphex 220
comptus comptus, Paralastor 185
comptus, Paralastor 185
comptus, Paralastor comptus 185
comptus rubescens, Paralastor 185
conator, Crabro (Rhopalum) 274
conator, Dasyproctus 274
concinna, Paragia 159
concolor concolor, Pseudalastor 199
concolor, Odynerus (Leionotus) 199
concolor, Pseudalastor 199
concolor, Pseudalastor concolor 199
concolor rapax, Pseudalastor 200
confertus, Acanthostethus 277
confertus, Nysson (Acanthostethus) 277
confusa, Pseudabispa 199
confusus, Pseudozethus 202
congener, Pison 258
congenerum, Pison (Pison) 258
conglobatus, Crabro 277
conglobatus, Ectemnius (Cameronitus) 277
connexum, Trypoxylon 265
connexum, Trypoxylon (Trypoxylon) schmiedeknechti 265
conspiciendus, Paralastor 185
conspicuus, Paralastor 185
constrictus, Paralastor 185

consuetipes, Arpactus 282
consuetipes, Austrogorytes 282
continentalis, Australozethus 169
contrarius, Tachysphex 239
convexus, Australodynerus 168
cooba, Bembix 286
coonundura, Bembix 287
corallinus, Tachysphex fanuiensis 239
coriolum, Rhopalum (Corynopus) 267
coriolum, Rhopalum (Rhopalum) 267
cornigerum, Podagritus (Echuca) 270
cornigerum, Rhopalum (Rhopalum) 270
coruscans, Auchenophorus 256
Corynopus 267
(Corynopus) anteum, Rhopalum 267
(Corynopus) australiae, Rhopalum 267
(Corynopus) coriolum, Rhopalum 267
(Corynopus) dedarum, Rhopalum 267
(Corynopus) evansianum, Rhopalum 267
(Corynopus) evictum, Rhopalum 267
(Corynopus) famicum, Rhopalum 267
(Corynopus) frenchii, Rhopalum 268
(Corynopus) harpax, Rhopalum 268
(Corynopus) kerangi, Rhopalum 268
(Corynopus) kuehlhorni, Rhopalum 268
(Corynopus) littorale, Rhopalum 268
(Corynopus) neboissi, Rhopalum 268
(Corynopus) notogeum, Rhopalum 268
(Corynopus), Rhopalum 267
(Corynopus) taeniatum, Rhopalum 268
(Corynopus) tubarum, Rhopalum 269
(Corynopus) variitarse, Rhopalum 269
(Corynopus) xenum, Rhopalum 269
couloni, Pseudoturneria 272
Crabro bivittatus 275
Crabro cinctus 276
Crabro conglobatus 277
Crabro doddii 273
Crabro erythrogaster 273
Crabro hebetescens 276
Crabro mackayensis 276
Crabro neglectus 275
Crabro nigromaculatus 273
Crabro ordinarius 275
Crabro perlucidus 272
Crabro prosopoides 273
Crabro (Rhopalum) conator 274
Crabro (Rhopalum) frenchii 268
Crabro (Rhopalum) idoneus 272
Crabro (Rhopalum) militaris 271
Crabro (Rhopalum) tenuiventris 266
Crabro (Rhopalum) transiens 266
Crabro (Solenius) manifestatus 275
Crabro tasmanicus 276
Crabro tricolor 271
Crabro tridentatus 275
Crabro veitchi 273
crabroniformis, Bembex 294
crassicorne, Zoyphium 251
crassicornis, Sericophorus (Zoyphidium) 251
crassipes, Larrada 235
Cratolarra 233
(Cratolarra) femorata, Larra 233
(Cratolarra), Larra 233
crucigera, Argogorytes 280

crucigera, Arpactus 280
cruentatus, Alastor (Paralastor) 185
cruentatus, Paralastor 185
cruentus, Odynerus (Ancistroceroides) 185
cruentus, Paralastor 185
cucullata, Cerceris 298
cunnamulla, Cerceris 298
cursitans, Bembex 287
cursitans, Bembix 287
cyaneus, Tachyrrhostus 246
cyanophilus, Sericophorus 247
cyanophilus, Sericophorus (Sericophorus) 247
Cygnaea 155
(Cygnaea), Paragia 155
(Cygnaea) vespiformis, Paragia 155
cygnorum, Austrogorytes 282
cygnorum, Gorytes 282
cygnorum, Podagritus (Echuca) 270
cygnorum, Rhopalum 270
cygnorum, Sceliphron laetum 221
cygnorum, Sphodrotes 254

darnleyensis, Stenodyneriellus 205
darrensis, Cerceris 298
darwini, Ropalidia (Icariola) 214
darwinianus, Paralastor 186
darwiniensis, Sphex 223
darwiniensis, Sphex (Sphex) 223
Dasyproctus 274
Dasyproctus austragilis 274
Dasyproctus burnettianus 274
Dasyproctus conator 274
Dasyproctus expectatus 274
Dasyproctus verutus 266
Dasyproctus yorki 274
dearmata, Sphodrotes 254
debilis, Paralastor 186
debilis, Tachysphex 243
debilitatus, Paralastor 186
deceptor, Icaria 214
deceptor, Paragia 157
deceptor, Ropalidia (Icariola) 214
deceptrix, Paragia (Paragiella) 157
decipiens aliciae, Paragia (Paragia) 155
decipiens decipiens, Diemodynerus 173
decipiens decipiens, Paragia (Paragia) 155
decipiens, Diemodynerus 173
decipiens, Diemodynerus decipiens 173
decipiens, Odynerus (Leionotus) 173
decipiens, Paragia 155
decipiens, Paragia (Paragia) 155
decipiens, Paragia (Paragia) decipiens 155
decipiens, Pison 258
decipiens, Pison (Pison) 258
decipiens positus, Diemodynerus 173
decorata, Sphex 224
decoratum, Rhynchium 174
decoratus, Ammatomus 284
decoratus, Epiodynerus 174
decoratus, Gorytes 284
decoratus, Odynerus (Leionotus) 181
decoratus, Sphex (Sphex) 224
dedariensis, Cerceris 298
dedarum, Rhopalum (Corynopus) 267

dedarum, Rhopalum (Rhopalum) 267
Delta 170
Delta arcuata 170
Delta arcuata arcuata 170
Delta bicinctus 171
Delta campaniformis 171
Delta campaniformis assatus 171
Delta campaniformis campaniformis 171
Delta incola 171
Delta incola aruensis 171
(Delta) incola aruensis, Eumenes 171
Delta incola teleporus 171
Delta latreillei 172
Delta latreillei latreillei 172
Delta nigritarsis 172
Delta philantes 172
Delta transmarinum 172
Delta xanthurum 172
Delta xanthurum xanthurum 172
denticulatus, Acarodynerus 162
dentiger, Paralastor 186
deperditum, Pison (Pison) 258
deperditum, Pison (Pisonitus) 258
depressiventris, Tachysphex 239
deserticolus, Arpactophilus steindachneri 231
deserticolus, Harpactophilus steindachneri 231
despectus, Paralastor 186
diabolicus, Paralastor 186
diadema, Paralastor 186
Dicranorhina 237
Dicranorhina intaminata 237
diemensis, Diemodynerus 173
diemensis, Odynerus (Leionotus) 173
Diemodynerus 173
Diemodynerus decipiens 173
Diemodynerus decipiens decipiens 173
Diemodynerus decipiens positus 173
Diemodynerus diemensis 173
Diemodynerus pseudacarodynerus 173
Diemodynerus saucius 173
dietrichianus, Acarodynerus 162
dietrichianus, Acarodynerus dietrichianus 162
dietrichianus dietrichianus, Acarodynerus 162
dietrichianus, Odynerus (Odynerus) 162
dietrichianus rufocaudatus, Acarodynerus 162
difficile, Pison (Aulacophilus) 263
difficile, Pison (Pisonoides) 263
dimidiatum, Pison (Pison) 258
dimidiatus, Pison 258
dineurum, Rhopalum (Rhopalum) 266
Diptera 218
dipteroides, Sericophorus 252
dipteroides, Sericophorus (Zoyphidium) 252
Discoelius carinatus 181
Discoelius ecclesiasticus 177
Discoelius elongatus 178
Discoelius ephippium 202
Discoelius insignis 203
Discoelius spinosus 203
Discoelius verreauxii 203
discrepans, Tachysphex 239
dispersus, Tachytes 237
dives, Pison 259
dives, Pison (Pison) 259
dizonus, Gorytes 281

doddi, Metaparagia 155
doddi, Sericophorus (Zoyphidium) 252
doddi, Zoyphium 252
doddii, Chimiloides 273
doddii, Crabro 273
dolichocerus, Sphex (Sphex) 225
Dolichurus 218
Dolichurus carbonarius 218
dominula, Vespa 209
dominulus, Polistes (Polistes) 209
donatus, Paralastor 186
doreeni, Podagritus (Echuca) 270
drewseni, Odynerus 181
drewsenianus, Acarodynerus 162
dubiosus, Paralastor 186
dubius, Arpactophilus 231
dubius, Austrostigmus 231
duboulayi, Clitemnestra 279
duboulayi, Gorytes 279
duplicinctus, Polistes bernardii 213
duplicinctus, Polistes (Polistella) bernardii 211
dyscritias, Paralastor 187

eboraca, Ropalidia (Ropalidia) 213
ecclesiastica, Ischnocoelia 177
ecclesiasticus, Discoelius 177
Echuca 269
(Echuca) alevinus, Podagritus 269
(Echuca) aliciae, Podagritus 269
(Echuca) anerus, Podagritus 270
(Echuca) australiensis, Podagritus 270
(Echuca) burnsi, Podagritus 270
(Echuca) carolus, Podagritus 270
(Echuca) cornigerum, Podagritus 270
(Echuca) cygnorum, Podagritus 270
(Echuca) doreeni, Podagritus 270
(Echuca) edgarus, Podagritus 270
(Echuca) imbelle, Podagritus 270
(Echuca) kiatae, Podagritus 271
(Echuca) krombeini, Podagritus 271
(Echuca) leptospermi, Podagritus 271
(Echuca) marcellus, Podagritus 271
(Echuca) mullewanus, Podagritus 271
(Echuca) peratus, Podagritus 271
(Echuca), Podagritus 269
(Echuca) rieki, Podagritus 271
(Echuca) tricolor, Podagritus 271
(Echuca) yarrowi, Podagritus 272
Ectemnius 276
Ectemnius (Cameronitus) 276
Ectemnius (Cameronitus) conglobatus 277
Ectemnius (Hypocrabro) 276
Ectemnius (Hypocrabro) hebetescens 276
Ectemnius (Hypocrabro) mackayensis 276
Ectemnius (Hypocrabro) reginellus 276
Ectopioglossa 174
Ectopioglossa australensis 174
Ectopioglossa polita 174
Ectopioglossa polita australensis 174
edgarus, Podagritus (Echuca) 270
egens, Bembecinus 284
egens, Bembex 287
egens, Bembix 287
egens, Stizus 284

eleebana, Bembix 287
elegans, Paralastor 187
elegantior, Sericophorus 247
elegantior, Sericophorus (Sericophorus) 247
elegantula, Ropalidia (Icariola) 214
elegantula, Spilomena 232
Elimus 174
Elimus australis 174
Elimus ferrugineus 178
Elimus mackayensis 174
Elimus robustus 179
elongata, Ischnocoelia 178
elongatus, Discoelius 178
emarginatum, Zoyphium 252
emarginatus, Alastor 187
emarginatus, Paralastor 187
emarginatus, Sericophorus (Zoyphidium) 252
ephippioides, Pseudabispa 199
ephippium, Abispa (Abispa) 161
ephippium, Discoelius 202
ephippium, Pseudozethus 202
ephippium, Sphex 224
ephippium, Sphex (Sphex) 224
ephippium, Vespa 161
Epiodynerus 174 , 181
Epiodynerus decoratus 174
Epiodynerus nigrocinctus 175
Epiodynerus nigrocinctus nigrocinctus 175
Epiodynerus tamarinus 175
Epiodynerus tamarinus inviolatus 175
Epiodynerus tamarinus tamarinus 175
Epiodynerus tasmaniensis 175
Epsilon 160
eremophila, Ammophila (Parapsammophila) 228
eremophila, Parapsammophila 228
eriurgus, Alastor 187
eriurgus, Paralastor 187
ermineus, Sphex (Sphex) 224
errans, Gastrosericus 244
errans, Lyroda 244
erythrinus, Polistes 210
erythrinus, Polistes (Megapolistes) 210
erythrocerum, Pison 259
erythrocerum, Pison (Pison) 259
erythrogaster, Crabro 273
erythrogastrum, Pison (Parapison) 259
erythrogastrum, Pison (Pison) 259
erythrosoma, Sericophorus (Zoyphidium) 252
erythrosoma, Zoyphium 252
eucalypti, Rhopalum 266
eucalypti, Rhopalum (Rhopalum) 266
eucalypticus, Tachysphex 239
Euchalcomenes 176
Euchalcomenes gilberti 176
euchroma, Cerceris 298
euclidias, Paralastor 187
Eudiscoelius 176
Eudiscoelius gilberti 176
eugonias, Paralastor 187
Eumenes 176
Eumenes apicalis 176
Eumenes bicincta 171
Eumenes campaniformis assatus 171
Eumenes (Delta) incola aruensis 171
Eumenes (Eumenes) simplicilamellatus 176

Eumenes fluctuans 176
Eumenes incola teleporus 171
Eumenes latreillei 172
Eumenes nigritarsis 172
Eumenes (Pareumenes) australensis 174
Eumenes philantes 172
Eumenes simplicilamellatus 176
(Eumenes) simplicilamellatus, Eumenes 176
Eumenes transmarinus 172
Eumenes xanthura 172
EUMENIDAE 150 , 151 , 152 , 152
Eumeninae 151
eungella, Cerceris 298
Euodynerus 176
Euodynerus polyphemus 177
eurostoma, Ropalidia (Icariola) 214
eustomus, Paralastor 187
eutretus, Paralastor 188
evansi, Aha 244
evansianum, Rhopalum (Corynopus) 267
everardensis, Australodynerus merredinensis 168
evictum, Rhopalum (Corynopus) 267
exarmatus, Acarodynerus 163
exarmatus, Odynerus (Stenodynerus) 163
excellens, Paragia 158
excisus, Paralastor vulpinus 197
exclusum, Pison (Parapison) 259
exclusum, Pison (Pison) 259
Exeirus 283
Exeirus lateritius 283
exilis, Odynerus (Leionotus) 180
eximium, Trypoxylon 264
eximium, Trypoxylon (Trypoxylon) 264
eximius, Gorytes 283
exleyae, Cerceris 298
exornatum, Pison 259
exornatum, Pison (Pison) 259
expectatus, Dasyproctus 274
extraneus, Leptomenoides 180
extraneus, Odynerus (Leionotus) 180
exul, Polemistus 230
exultans, Pison 259
exultans, Pison (Pison) 259
eyrensis, Ammophila 229

facilis, Polistes 210
facilis, Polistes (Megapolistes) 210
fallax, Paralastor 188
famicum, Rhopalum (Corynopus) 267
fanuiensis corallinus, Tachysphex 239
fanuiensis howeanus, Tachysphex 239
fanuiensis, Tachysphex 239
fatalis, Tachytes 237
femorata, Larra (Cratolarra) 233
femoratus, Tachytes 233
fenestratum, Pison (Pison) 259
fenestratus, Pison 259
ferruginea, Anacrucis 252
ferruginea, Ischnocoelia 178
ferrugineus, Antamenes (Australochilus) 167
ferrugineus, Elimus 178
ferrugineus, Sericophorus (Zoyphidium) 252
festinans, Larrada 235
festinans, Liris (Leptolarra) 235

festivum, Pison (Pison) 259
festivus, Pison 259
finschii, Sphex 224
finschii, Sphex (Sphex) 224
flammiger flammiger, Flammodynerus 177
flammiger, Flammodynerus 177
flammiger, Flammodynerus flammiger 177
flammiger nigroflammeus, Flammodynerus 177
flammiger, Odynerus 177
Flammodynerus 177
Flammodynerus flammiger 177
Flammodynerus flammiger flammiger 177
Flammodynerus flammiger nigroflammeus 177
Flammodynerus pseudoloris 177
Flammodynerus subalaris 177
flavescentulus, Bidentodynerus bicolor 170
flavescentulus, Odynerus (Rhynchium) bicolor 170
flaviceps, Alastor (Paralastor) 188
flaviceps, Paralastor 188
flavifrons, Bembex 287
flavifrons, Bembix 287
flavilabris, Bembex 285
flavinoda, Ropalidia (Icariola) interrupta 215
flavinoda, Ropalidia variegata 215
flavipes, Bembex 287
flavipes, Bembix 287
flavipes, Trypoxylon 264
flavipes, Trypoxylon (Trypoxylon) 264
flaviventris, Bembex 288
flaviventris, Bembix 288
flavocinctus, Odynerus 166
flavofasciatum, Zoyphium 252
flavofasciatus, Sericophorus (Zoyphidium) 252
flavolatera, Bembix 289
flavoniger, Antamenes (Australochilus) 167
fluctuans, Eumenes 176
fluvialis, Ancistrocerus 165
fluvialis, Cerceris 298
fluvialis, Odynerus (Ancistrocerus) 165
foliaceus, Tachysphex 240
forficata, Cerceris 298
Formicoidea 151
formosa, Sphex 223
formosellus, Sphex 224
formosellus, Sphex (Sphex) 224
formosissimus, Tachytes 237
formosum, Sceliphron (Prosceliphron) 222
formosus, Pelopoeus 222
fortior, Tachysphex 240
frater, Paralastor 188
fraterculus, Pison 259
fraterculus, Pison (Pison) 259
fraternus, Alastor (Paralastor) 188
fraternus, Paralastor 188
frenchii, Austrogorytes 282
frenchii, Crabro (Rhopalum) 268
frenchii, Gorytes 282
frenchii, Rhopalum (Corynopus) 268
froggatti, Cerceris 298
froggatti, Paracrabro 232
froggatti, Sericophorus 247
froggatti, Sericophorus (Sericophorus) 247
frontale, Zoyphium 252
frontalis, Sericophorus (Zoyphidium) 252
fulleri, Sericophorus chalybaeus 246

fulleri, Sericophorus (Sericophorus) chalybaeus 246
fulva fulva, Ischnocoelia 178
fulva, Ischnocoelia 178
fulva, Ischnocoelia fulva 178
fulva major, Ischnocoelia 178
fulvicornis, Auchenophorus 257
fulvopruinosa, Ropalidia (Ropalidia) 213
fulvopruinosus, Odynerus (Leionotus) 213
fulvus, Stenolabus 178
fumipennis, Sphex 224
fumipennis, Sphex (Sphex) 224
funebris, Bembex 285
funebris, Sericophorus 247
funebris, Sericophorus (Sericophorus) 247
furcata, Bembex 288
furcata, Bembix 288
fuscata, Aphelotoma 219
fuscipenne, Pison (Pison) 259
fuscipenne, Zoyphium 252
fuscipennis, Pison 259
fuscipennis, Sericophorus (Zoyphidium) 252

galeatus, Tachysphex 240
gallicus, Polistes 209
Gastrosericus errans 244
gelane, Bembix 288
generosa, Paragia (Paragiella) 157
germanica, Vespa 208
germanica, Vespula (Paravespula) 208
gilberti, Acanthostethus 277
gilberti, Cerceris 299
gilberti, Euchalcomenes 176
gilberti, Eudiscoelius 176
gilberti, Nysson (Acanthostethus) 277
gilberti, Sphex 224
gilberti, Sphex (Sphex) 224
gilesi, Cerceris 299
ginjulla, Bembix 288
glabrellus, Arpactophilus 231
glabrellus, Austrostigmus 231
globosa, Sphex 227
globosus, Prionyx 227
goddardi, Cerceris 299
godeffroyi, Sphex 226
godeffroyi, Sphex (Sphex) sericeus 226
goodwini, Cerceris 296
Gorytes austrinus 284
Gorytes bellicosus 281
Gorytes ciliatus 281
Gorytes cygnorum 282
Gorytes decoratus 284
Gorytes dizonus 281
Gorytes duboulayi 279
Gorytes eximius 283
Gorytes frenchii 282
Gorytes icarioides 284
Gorytes lucidulus 279
Gorytes (Miscothyris) megalophthalmus 279
Gorytes ornatus 284
Gorytes perkinsi 282
Gorytes rufomixtus 281
Gorytes sanguinolentus 280
Gorytes tarsatus 283
gorytoides, Bembecinus 284

gorytoides, Stizus 284
goyarra, Bembix 288
gracilenta, Ropalidia (Icariola) 214
gracilis, Sericophorus 247
gracilis, Sericophorus (Sericophorus) 247
grahami, Rhopalum (Rhopalum) 266
gratiosa, Sphex 225
gratiosissimus, Sphex 225
gregaria gregaria, Ropalidia (Icariola) 214
gregaria, Icaria 215
gregaria, Ropalidia (Icariola) 214
gregaria, Ropalidia (Icariola) gregaria 214
gregaria spilocephala, Ropalidia (Icariola) 215
gunamarra, Bembix 288
gurindji, Bembix 288
guttatulus, Arpactus (Miscothyris) 279
guttatulus, Clitemnestra 279

ha, Aha 244
habilis, Paralastor 188
hackeri, Antamenes (Australochilus) 167
hackeri, Sericophorus 246
hackeri, Sericophorus (Zoyphidium) 252
hackeri, Zoyphium 252
hackeriana, Cerceris 299
haemorrhoidalis, Liris (Leptolarra) 235
haemorrhoidalis magnifica, Liris (Leptolarra) 235
Harpactophilus 230
Harpactophilus arator 230
Harpactophilus kohlii 231
Harpactophilus steindachneri deserticolus 231
Harpactophilus sulcatus 231
Harpactophilus tricolor 232
Harpactopus saevus 228
harpax, Rhopalum (Corynopus) 268
harpax, Rhopalum (Rhopalum) 268
hebetescens, Crabro 276
hebetescens, Ectemnius (Hypocrabro) 276
hentyi, Acanthostethus 277
hentyi, Nysson 277
hilaris, Paralastor 188
hirsuta, Paragia 157
hirsuta, Paragia (Paragiella) 157
hirsuta, Ropalidia (Icariola) 215
hirtula, Bembecinus 285
hirtula, Larra 285
histrio, Leptomenoides 180
hobartia, Spilomena 232
hokarra, Bembix 288
hostilis, Antamenes (Australochilus) 167
howeanus, Tachysphex fanuiensis 239
humile, Zoyphium 253
humilis centrocontinentalis, Polistes 212
humilis centrocontinentalis, Polistes (Polistella) 212
humilis clarior, Polistes 212
humilis humilis, Polistes (Polistella) 212
humilis, Polistes (Polistella) 211
humilis, Polistes (Polistella) humilis 212
humilis pseudoscach, Polistes 212
humilis, Sericophorus (Zoyphidium) 253
humilis synoecus, Polistes (Polistella) 212
humilis, Vespa 212
humilis xanthorrhoicus, Polistes 212
huonense, Trypoxylon 264

huonense, Trypoxylon (Trypoxylon) albitarsatum 264
hylaeoides, Astaurus 280
Hypocrabro 276
(Hypocrabro), Ectemnius 276
(Hypocrabro) hebetescens, Ectemnius 276
(Hypocrabro) mackayensis, Ectemnius 276
(Hypocrabro) reginellus, Ectemnius 276
hypoleius, Tachysphex 240
hypoleius, Tachytes 240

Icaria 213
Icaria cabeti 217
Icaria deceptor 214
Icaria gregaria 215
Icaria jucunda 215
Icaria plebeja 216
Icaria revolutionalis 216
Icaria socialistica 216
Icaria spilocephala 215
Icarielia 217
(Icarielia) nigrior, Ropalidia 217
(Icarielia) romandi cabeti, Ropalidia 217
(Icarielia) romandi romandi, Ropalidia 217
(Icarielia) romandi, Ropalidia 217
(Icarielia), Ropalidia 217
icarioides, Ammatomus 284
icarioides, Gorytes 284
icarioides, Paralastor 188
icarioides, Pison (Aulacophilus) 263
icarioides, Pison (Pisonoides) 263
Icariola 214
(Icariola) darwini, Ropalidia 214
(Icariola) deceptor, Ropalidia 214
(Icariola) elegantula, Ropalidia 214
(Icariola) eurostoma, Ropalidia 214
(Icariola) gracilenta, Ropalidia 214
(Icariola) gregaria gregaria, Ropalidia 214
(Icariola) gregaria, Ropalidia 214
(Icariola) gregaria spilocephala, Ropalidia 215
(Icariola) hirsuta, Ropalidia 215
(Icariola) interrupta flavinoda, Ropalidia 215
(Icariola) interrupta interrupta, Ropalidia 215
(Icariola) interrupta, Ropalidia 215
(Icariola) kurandae, Ropalidia 215
(Icariola) mackayensis, Ropalidia 215
(Icariola) marginata jucunda, Ropalidia 215
(Icariola) marginata, Ropalidia 215
(Icariola) mutabilis mutabilis, Ropalidia 216
(Icariola) mutabilis, Ropalidia 215
(Icariola) mutabilis torresiana, Ropalidia 216
(Icariola) plebiana, Ropalidia 216
(Icariola) proletaria, Ropalidia 216
(Icariola) revolutionalis, Ropalidia 216
(Icariola), Ropalidia 214
(Icariola) socialistica, Ropalidia 216
(Icariola) trichophthalma, Ropalidia 217
(Icariola) turneri, Ropalidia 217
idoneus, Crabro (Rhopalum) 272
idoneus, Notocrabro 272
ignavum, Pison 260
ignavum, Pison (Pison) 260
ignotus, Paralastor 188
imbelle, Podagritus (Echuca) 270
imbelle, Rhopalum tricolor 270

imbellis, Tachysphex 240
imitator, Paralastor 189
impatiens, Ammophila 229
imperialis, Sphex (Sphex) 226
impudicus, Australodynerus pusilloides 168
Incertae sedis 303
incola aruensis, Delta 171
incola aruensis, Eumenes (Delta) 171
incola, Delta 171
incola teleporus, Delta 171
incola teleporus, Eumenes 171
inconspicuum, Pison 260
inconspicuum, Pison (Pison) 260
indecoratus, Odynerus 181
inexpectata, Cerceris 299
infernalis, Alastor (Paralastor) 189
infernalis, Paralastor 189
infimus, Paralastor 189
infumatum, Pison 260
infumatum, Pison (Pison) 260
inornatus, Sericophorus 247
inornatus, Sericophorus (Sericophorus) 247
insignis, Discoelius 203
insignis, Pseudozethus 203
instabilis, Ammophila 229
insulae, Polistes (Polistella) bernardii 211
insulare priscum, Pison 261
insularis, Alastor (Paralastor) 189
insularis, Paralastor 189
insulicola, Cerceris 299
intaminata, Dicranorhina 237
intaminata, Piagetia 237
integer, Stenolabus 178
integra carnowi, Ischnocoelia 178
integra integra, Ischnocoelia 178
integra, Ischnocoelia 178
integra, Ischnocoelia integra 178
integra major, Ischnocoelia 178
interrupta flavinoda, Ropalidia (Icariola) 215
interrupta interrupta, Ropalidia (Icariola) 215
interrupta, Ropalidia (Icariola) 215
interrupta, Ropalidia (Icariola) interrupta 215
interrupta, Ropalidia variegata 215
interstitialis, Psenulus 230
inviolatus, Epiodynerus tamarinus 175
iridescens, Spilomena 232
iridipenne, Allorhynchium 165
iridipenne, Pison 260
iridipenne, Pison (Pison) 260
iridipenne, Rhynchium 165
iridipenne, Zoyphium 253
iridipennis, Sericophorus (Zoyphidium) 253
iridis, Cerceris 300
Ischnocoelia 177
Ischnocoelia ecclesiastica 177
Ischnocoelia elongata 178
Ischnocoelia ferruginea 178
Ischnocoelia fulva 178
Ischnocoelia fulva fulva 178
Ischnocoelia fulva major 178
Ischnocoelia integra 178
Ischnocoelia integra carnowi 178
Ischnocoelia integra integra 178
Ischnocoelia integra major 178
Ischnocoelia occidentalis 178 , 179

Ischnocoelia occidentalis blumburyensis 179
Ischnocoelia occidentalis occidentalis 179
Ischnocoelia polychroma 179
Ischnocoelia robusta 179
Ischnocoelia robusta analis 179
Ischnocoelia robusta aurantiaca 179
Ischnocoelia robusta robusta 179
Ischnocoelia robusta unicolor 179
Ischnocoelia xanthochroma 179
Isodontia 226
Isodontia abdita 226
Isodontia abdita nugenti 226
(Isodontia) abditus nugenti, Sphex 226
Isodontia albohirta 226
(Isodontia) albohirtus, Sphex 226
Isodontia aurifrons 226
Isodontia nigella 227
Isodontia obscurella 227
Isodontia vidua 227

jucunda, Icaria 215
jucunda, Ropalidia (Icariola) marginata 215
jugulatus, Antamenes (Australochilus) 167

kamulla, Bembix 288
Katamenes 176
kerangi, Rhopalum (Corynopus) 268
kerangi, Rhopalum (Rhopalum) 268
kiatae, Podagritus (Echuca) 271
koala, Cerceris 301
kohlii, Arpactophilus 231
kohlii, Harpactophilus 231
kohlii, Sericophorus (Zoyphidium) 253
kohlii, Zoyphium 253
kora, Bembix 289
krombeini, Podagritus (Echuca) 271
kuehlhorni, Rhopalum (Corynopus) 268
kuehlhorni, Rhopalum (Rhopalum) 268
kununurra, Bembix 289
kurandae, Nitela 256
kurandae, Ropalidia (Icariola) 215

labeculata, Cerceris 300
lachesis, Alastor 189
lachesis, Paralastor 189
laetum cygnorum, Sceliphron 221
laetum laetum, Sceliphron (Sceliphron) 221
laetum, Sceliphron (Sceliphron) 221
laetum, Sceliphron (Sceliphron) laetum 221
laetus, Paralastor 189
laetus, Pelopoeus 221
laevigata, Anacrucis 253
laevigatissimus, Polistes 211
laevigatissimus, Polistes (Stenopolistes) 211
laevigatus, Sericophorus (Zoyphidium) 253
lamellata, Bembex 289
lamellata, Bembix 289
lanio, Sphecius 283
Larra 233 , 234
Larra alecto 234
Larra (Cratolarra) 233
Larra (Cratolarra) femorata 233

Larra hirtula 285
Larra (Larra) 233
(Larra), Larra 233
Larra (Larra) melanocnemis 233
Larra melanocnemis 233
(Larra) melanocnemis, Larra 233
Larra psilocera 233
Larra scelesta 234
Larrada alecto 234
Larrada chrysonota 235
Larrada crassipes 235
Larrada festinans 235
Larrada nigripes 236
Larrisson 255
Larrisson abnormis 255
Larrisson azyx 256
Larrisson nedymus 256
Larrisson rieki 256
lateritius, Alastor (Paralastor) 190
lateritius, Exeirus 283
lateritius, Paralastor 190
latetergum, Ropalidia (Polistratus) 214
latiberbis, Cerceris 300
laticincta, Abispa (Abispa) 161
latifasciata, Bembex 289
latifasciata, Bembix 289
latreillei, Delta 172
latreillei, Delta latreillei 172
latreillei, Eumenes 172
latreillei latreillei, Delta 172
leeuwinensis, Bembex 289
leeuwinensis, Bembix 289
legatus, Acarodynerus 163
(Leionotus) alaris, Odynerus 201
(Leionotus) bisulcatus, Odynerus 166
(Leionotus) citreocinctus, Odynerus 167
(Leionotus) clypeatus, Odynerus 162
(Leionotus) concolor, Odynerus 199
(Leionotus) decipiens, Odynerus 173
(Leionotus) decoratus, Odynerus 181
(Leionotus) diemensis, Odynerus 173
(Leionotus) exilis, Odynerus 180
(Leionotus) extraneus, Odynerus 180
(Leionotus) fulvopruinosus, Odynerus 213
(Leionotus) macilentus, Odynerus 169
(Leionotus) nigrocinctus, Odynerus 175
(Leionotus) succinctus, Odynerus 202
(Leionotus) tamarinus, Odynerus 175
(Leionotus) tasmaniensis, Odynerus 175
Lepidoptera 150 , 160 , 208
leptias, Paralastor 190
Leptolarra 234
(Leptolarra) abbreviata, Liris 234
(Leptolarra) agitata, Liris 234
(Leptolarra) basilissa, Liris 235
(Leptolarra) chrysonota, Liris 235
(Leptolarra) commixta, Liris 235
(Leptolarra) festinans, Liris 235
(Leptolarra) haemorrhoidalis, Liris 235
(Leptolarra) haemorrhoidalis magnifica, Liris 235
(Leptolarra), Liris 234
(Leptolarra) obliquetruncata, Liris 236
(Leptolarra) recondita, Liris 236
(Leptolarra) regina, Liris 236
(Leptolarra) serena, Liris 236

(Leptolarra) spathulifera, Liris 236
Leptomenoides 180
Leptomenoides cairnensis 180
Leptomenoides extraneus 180
Leptomenoides histrio 180
Leptomenoides mackayensis 180
Leptomenoides pachymeniformis 180
Leptomenoides placidior 180
Leptomenoides pronotalis 180
leptospermi, Podagritus (Echuca) 271
leptospermi, Rhopalum 271
Lestica 277
Lestica (Solenius) 277
Lestica (Solenius) relicta 277
lilacinus, Sericophorus 247
lilacinus, Sericophorus (Sericophorus) 247
Liris 234 , 236
Liris australis 236
Liris (Leptolarra) 234
Liris (Leptolarra) abbreviata 234
Liris (Leptolarra) agitata 234
Liris (Leptolarra) basilissa 235
Liris (Leptolarra) chrysonota 235
Liris (Leptolarra) commixta 235
Liris (Leptolarra) festinans 235
Liris (Leptolarra) haemorrhoidalis 235
Liris (Leptolarra) haemorrhoidalis magnifica 235
Liris (Leptolarra) obliquetruncata 236
Liris (Leptolarra) recondita 236
Liris (Leptolarra) regina 236
Liris (Leptolarra) serena 236
Liris (Leptolarra) spathulifera 236
Liris (Liris) 234
(Liris), Liris 234
Liris (Liris) melania 234
Liris magnifica 235
Liris melania 234
(Liris) melania, Liris 234
Liris nigripes 236
listrognatha, Cerceris 300
littorale, Rhopalum 268
littorale, Rhopalum (Corynopus) 268
littoralis, Bembex 289
littoralis, Bembix 289
littoralis, Sericophorus 247
littoralis, Sericophorus (Sericophorus) 247
lobimana, Bembex 290
lobimana, Bembix 290
longebispinosa, Pseudepipona (Syneuodynerus) 207
longebispinosus, Syneuodynerus 207
longiceps, Spilomena 232
longifrons, Microglossa 233
longifrons, Spilomena 233
loorea, Bembix 290
lucidulus, Clitemnestra 279
lucidulus, Gorytes 279
lucidulus mimeticus, Miscothyris 279
luctuosa, Sphex 224
luctuosus, Sphex (Sphex) 224
luculenta, Cerceris 300
lunaris, Acarodynerus 163
luteiventris, Spilomena 233
lutescens, Pison 260
lutescens, Pison (Pison) 260
lutescens, Psen 230

Lyroda 218, 244
Lyroda errans 244
Lyroda michaelseni 244
Lyroda michaelseni michaelseni 244
Lyroda michaelseni tasmanica 244
Lyroda minima 245
Lyroda queenslandensis 245

macilentus, Odynerus (Leionotus) 169
mackayensis, Bembex 291
mackayensis, Crabro 276
mackayensis, Ectemnius (Hypocrabro) 276
mackayensis, Elimus 174
mackayensis, Leptomenoides 180
mackayensis, Paralastor 190
mackayensis, Ropalidia (Icariòla) 215
mackayensis, Tachysphex 240
Macrocalymma 180
Macrocalymma aliciae 180
Macrocalymma smithianum 181
macrocephalum, Rhopalum (Rhopalum) 266
macrocephalus, Rhopalum 266
maculata, Paragia 154
maculata, Rolandia 154
maculipennis, Tachysphex 240
maculiventris, Alastor (Paralastor) 190
maculiventris, Paralastor 190
magarra, Bembix 290
magdalena, Paragia 157
magdalena, Paragia (Paragiella) 157
magnifica, Liris 235
magnifica, Liris (Leptolarra) haemorrhoidalis 235
magnificum, Rhynchium 204
major, Ischnocoelia fulva 178
major, Ischnocoelia integra 178
malayanus, Polistes 210
maliki, Bembix 290
mandibulatum, Pison 260
mandibulatum, Pison (Pison) 260
manifestata, Williamsita (Androcrabro) 275
manifestatus, Crabro (Solenius) 275
manilae, Notogonia 235
marcellus, Podagritus (Echuca) 271
mareeba, Bembix 290
marginalis, Sphodrotes 254
marginata jucunda, Ropalidia (Icariola) 215
marginata, Ropalidia (Icariola) 215
marginatum, Pison (Pison) 260
marginatus, Pison 260
marhra, Bembix 290
marsupiata, Bembex 290
marsupiata, Bembix 290
MASARIDAE 150, 151, 154
Masarinae 151
maximus, Tachysphex 241
meadewaldoensis, Abispa 161
mediocris, Pachycoelius 181
medius, Paralastor 190
meesi, Paralastor xerophilus 197
megacantha, Cerceris 300
megalophthalmus, Clitemnestra 279
megalophthalmus, Gorytes (Miscothyris) 279
Megapolistes 209
(Megapolistes) balder, Polistes 209

(Megapolistes) erythrinus, Polistes 210
(Megapolistes) facilis, Polistes 210
(Megapolistes) olivaceus, Polistes 210
(Megapolistes), Polistes 209
(Megapolistes) schach, Polistes 210
(Megapolistes) tepidus, Polistes 210
(Megapolistes) tepidus tepidus, Polistes 210
melania, Liris 234
melania, Liris (Liris) 234
melanocephalum, Pison 260
melanocephalum, Pison (Pison) 260
melanocnemis, Larra 233
melanocnemis, Larra (Larra) 233
melanogaster, Aphelotoma 219
meridionale, Pison 260
meridionale, Pison (Pison) 260
merredinensis, Australodynerus 168
merredinensis, Australodynerus merredinensis 168
merredinensis, Cerceris 300
merredinensis everardensis, Australodynerus 168
merredinensis merredinensis, Australodynerus 168
merredinensis victoriensis, Australodynerus 168
mesochloroides, Paralastor mesochlorus 190
mesochlorus mesochloroides, Paralastor 190
mesochlorus mesochlorus, Paralastor 190
mesochlorus, Paralastor 190
mesochlorus, Paralastor mesochlorus 190
metallescens, Sericophorus 247
metallescens, Sericophorus (Sericophorus) 247
Metaparagia 155
Metaparagia doddi 155
Metaparagia pictifrons 155
metathoracicus, Odynerus 200
metathoracicus, Pseudalastor 200
mianga, Bembix 290
micans, Ampulex 220
michaelseni, Lyroda 244
michaelseni, Lyroda michaelseni 244
michaelseni michaelseni, Lyroda 244
michaelseni tasmanica, Lyroda 244
micheneri, Notocrabro 272
Microglossa 232
Microglossa bimaculata 232
Microglossa longifrons 233
Microglossa rufitarsus 233
Microglossella 232
microgonias, Paralastor 190
militaris, Crabro (Rhopalum) 271
mimetica, Clitemnestra 279
mimetica, Paragia (Paragiella) 157
mimeticus, Miscothyris lucidulus 279
mimulus, Sphex 225
mimulus, Sphex (Sphex) 225
mimus, Paralastor 190
mindanaonis, Trypoxylon 265
mindanaonis, Trypoxylon (Trypoxylon) 265
minima, Lyroda 245
minimus, Acanthostethus 277
minimus, Nysson (Acanthostethus) 277
minuscula, Cerceris 300
minutus, Sericophorus 248
minutus, Sericophorus (Sericophorus) 248
minya, Bembix 291
mirabile, Rhynchium 204
mirabile, Rygchium 204

Miscothyris 279
(Miscothyris) guttatulus, Arpactus 279
Miscothyris lucidulus mimeticus 279
(Miscothyris) megalophthalmus, Gorytes 279
Miscothyris perlucidus 279
(Miscothyris) plomleyi, Arpactus 280
Miscothyris thoracicus 280
mitis, Tachytes 238
modesta, Sphex 225
modestus, Sphex (Sphex) 225
moerens, Acanthostethus 277
moerens, Nysson (Acanthostethus) 277
mokari, Bembix 291
moma, Bembix 291
Monerebia 160
monodon, Bembecinus 284
monstricornis, Ancistrocerus (Subancistrocerus) 207
monstricornis, Subancistrocerus 207
montanus, Australozethus tasmaniensis 169
montanus, Pseudozethus tasmaniensis 169
Montezumia australensis 164
moonga, Bembix 291
morosa, Paragia 158
morosa, Paragia (Paragiella) 158
mullewanus, Podagritus (Echuca) 271
multicolor, Paralastor 191
multifasciatus, Tachysphex 241
multiguttata, Cerceris 301
mundurra, Bembix 291
munta, Bembix 291
musca, Bembex 291
musca, Bembix 291
mutabilis mutabilis, Ropalidia (Icariola) 216
mutabilis, Paralastor 191
mutabilis, Ropalidia (Icariola) 215
mutabilis, Ropalidia (Icariola) mutabilis 216
mutabilis torresiana, Ropalidia (Icariola) 216
mysticus, Acanthostethus 278
mysticus, Nysson 278

nasuta, Paragia 158
nasuta, Paragia (Paragiella) 158
nautarum, Alastor (Paralastor) 191
nautarum, Paralastor 191
neboissi, Rhopalum (Corynopus) 268
neboissi, Rhopalum (Rhopalum) 268
nedymus, Larrisson 256
nefarius, Tachysphex 241
neglecta, Williamsita (Androcrabro) 275
neglectus, Crabro 275
neglectus, Odynerus (Parodynerus) 191
neglectus, Paralastor 191
nemoralis, Sphodrotes 254
neochromus, Paralastor 191
Neodasyproctus 273
Neodasyproctus veitchi 273
Neuroptera 218
nigella, Isodontia 227
nigella, Sphex 227
niger, Sericophorus (Zoyphidium) 253
nigricans, Nitela 256
nigricornis, Sericophorus relucens 249
nigricornis, Sericophorus (Sericophorus) relucens 249
nigricula, Aphelotoma 219

nigrior, Ropalidia (Icarielia) 217
nigripes, Larrada 236
nigripes, Liris 236
nigritarsis, Delta 172
nigritarsis, Eumenes 172
nigrithorax, Australodynerus yanchepensis 206
nigrithorax, Stenodyneriellus yanchepensis 206
nigrocincta, Cerceris 297
nigrocinctoides, Bidentodynerus bicolor 170
nigrocinctoides, Odynerus (Rhynchium) bicolor 170
nigrocinctus, Epiodynerus 175
nigrocinctus, Epiodynerus nigrocinctus 175
nigrocinctus nigrocinctus, Epiodynerus 175
nigrocinctus, Odynerus (Leionotus) 175
nigroflammeus, Flammodynerus flammiger 177
nigrolimbatum, Rhynchium 204
nigromaculatus, Chimiloides 273
nigromaculatus, Crabro 273
nigropectinata, Bembix 291
nigropectinatus, Bembex 291
nigror, Sericophorus 248
nigror, Sericophorus (Sericophorus) 248
nigrum, Zoyphium 253
Nitela 256
Nitela australiensis 256
Nitela kurandae 256
Nitela nigricans 256
Nitela reticulata 256
Nitela sculpturata 256
nitidiventris, Sphex 225
nitidus, Pison 259
niveifrons, Sericophorus 248
niveifrons, Sericophorus (Sericophorus) 248
nocatunga, Riekia 154
noctulum, Pison (Parapison) 261
noctulum, Pison (Pison) 261
Nortonia amaliae 164
Nothosphecius 283
(Nothosphecius) pectoralis, Sphecius 283
(Nothosphecius), Sphecius 283
Notocrabro 272
Notocrabro idoneus 272
Notocrabro micheneri 272
notogeum, Rhopalum (Corynopus) 268
notogeum, Rhopalum (Rhopalum) 268
Notogonia abbreviata 234
Notogonia agitata 234
Notogonia basilissa 235
Notogonia commixta 235
Notogonia manilae 235
Notogonia obliquetruncata 236
Notogonia recondita 236
Notogonia regina 236
Notogonia retiaria 235
Notogonia serena 236
Notogonia spathulifera 236
Notorhopalum 269
(Notorhopalum) carnegiacum, Rhopalum 269
(Notorhopalum), Rhopalum 269
novarae, Tachysphex 241
novarae, Tachytes 241
novempunctatus, Stenodyneriellus 205
nudiventris, Acanthostethus 278
nudiventris, Nysson (Acanthostethus) 278
nugenti, Isodontia abdita 226

nugenti, Sphex (Isodontia) abditus 226
Nysson (Acanthostethus) brisbanensis 277
Nysson (Acanthostethus) confertus 277
Nysson (Acanthostethus) gilberti 277
Nysson (Acanthostethus) minimus 277
Nysson (Acanthostethus) moerens 277
Nysson (Acanthostethus) nudiventris 278
Nysson (Acanthostethus) obliteratus 278
Nysson (Acanthostethus) punctatissimus 278
Nysson (Acanthostethus) spiniger 278
Nysson (Acanthostethus) tasmanicus 278
Nysson (Acanthostethus) triangularis 279
Nysson hentyi 277
Nysson mysticus 278
Nysson portlandensis 278
Nysson saussurei 278

obesus, Arpactus 282
obesus, Austrogorytes 282
obiri, Bembix 291
obliquetruncata, Liris (Leptolarra) 236
obliquetruncata, Notogonia 236
obliquus, Pison 263
obliteratus, Acanthostethus 278
obliteratus, Nysson (Acanthostethus) 278
obscurella, Isodontia 227
obscurella, Sphex 227
occidentalis, Australozethus 169
occidentalis blumburyensis, Ischnocoelia 179
occidentalis, Ischnocoelia 178, 179
occidentalis, Ischnocoelia occidentalis 179
occidentalis occidentalis, Ischnocoelia 179
occidentalis, Paralastor 191
occidentalis, Sericophorus 248
occidentalis, Sericophorus (Sericophorus) 248
occidentalis, Sphodrotes 255
occidentata, Pseudepipona (Syneuodynerus) 207
occidentatus, Syneuodynerus 207
octosetosa, Bembix 292
Odonata 218
odynericornis, Paralastor 191
odyneripennis, Paralastor 191
odyneroides, Paragia 158
odyneroides, Paragia (Paragiella) 158
odyneroides, Paralastor 191
Odynerus 160, 181
Odynerus alariformis 198
Odynerus alecto 181
Odynerus (Ancistroceroides) cruentus 185
Odynerus (Ancistroceroides) sanguinolentus 200
Odynerus (Ancistrocerus) alastoripennis 182
Odynerus (Ancistrocerus) fluvialis 165
Odynerus (Ancistrocerus) vernalis 166
Odynerus angulatus 201
Odynerus aurocinctus 183
Odynerus balyi 202
Odynerus bicolor 169
Odynerus bicoloratus 205
Odynerus bizonatus 198
(Odynerus) dietrichianus, Odynerus 162
Odynerus drewseni 181
Odynerus flammiger 177
Odynerus flavocinctus 166
Odynerus indecoratus 181

Odynerus (Leionotus) alaris 201
Odynerus (Leionotus) bisulcatus 166
Odynerus (Leionotus) citreocinctus 167
Odynerus (Leionotus) clypeatus 162
Odynerus (Leionotus) concolor 199
Odynerus (Leionotus) decipiens 173
Odynerus (Leionotus) decoratus 181
Odynerus (Leionotus) diemensis 173
Odynerus (Leionotus) exilis 180
Odynerus (Leionotus) extraneus 180
Odynerus (Leionotus) fulvopruinosus 213
Odynerus (Leionotus) macilentus 169
Odynerus (Leionotus) nigrocinctus 175
Odynerus (Leionotus) succinctus 202
Odynerus (Leionotus) tamarinus 175
Odynerus (Leionotus) tasmaniensis 175
Odynerus metathoracicus 200
Odynerus (Odynerus) dietrichianus 162
Odynerus (Parodynerus) neglectus 191
Odynerus polyphemus 177
Odynerus pusillus 168
Odynerus (Rhynchium) angulatus alexandriae 201
Odynerus (Rhynchium) bicolor aurantiopictus 170
Odynerus (Rhynchium) bicolor flavescentulus 170
Odynerus (Rhynchium) bicolor nigrocinctoides 170
Odynerus (Rhynchium) rubroviolaceus 194
Odynerus saucius 173
Odynerus splendidus 161
Odynerus (Stenodyneroides) subalaris 198
Odynerus (Stenodynerus) exarmatus 163
Odynerus (Stenodynerus) posttegulatus 163
Odynerus (Stenodynerus) spargovillensis 164
Odynerus subalaris 177
Odynerus triangulum 164
okiellus, Sericophorus (Sericophorus) teliferopodus 250
okiellus, Sericophorus teliferopodus 250
olba, Bembix 292
olivacea, Vespa 210
olivaceus, Polistes (Megapolistes) 210
oloris, Paralastor 192
oomborra, Bembix 292
opposita, Cerceris 301
optabilis, Paralastor 192
opulenta, Sphex 223
Orancistrocerus 181
ordinaria, Sphodrotes 255
ordinaria, Williamsita (Androcrabro) 275
ordinarius, Crabro 275
ordinarius, Paralastor 192
orientalis, Paralastor 192
ornatum, Zoyphium 253
ornatus, Gorytes 284
ornatus, Sericophorus (Zoyphidium) 253
osculata, Cerceris 301
ourapilla, Bembix 292

Pachycoelius 181
Pachycoelius brevicornis 181
Pachycoelius carinatus 181
Pachycoelius mediocris 181
Pachymenes amicus 167
Pachymenes pseudoneotropicus 166
Pachymenes rectispina 166
Pachymenes tasmaniae 166

Pachymenes vorticosus 166
pachymeniformis, Leptomenoides 180
pacificus, Tachysphex 241
paganus, Pseudalastor tridentatus 200
paleovariatus, Acarodynerus 163
pallida, Pseudepipona 202
pallidus, Paralastor 192
palmata, Bembex 292
palmata, Bembix 292
Palmodes 227
Palmodes australis 227
Palmodes sagax 227
palona, Bembix 292
papuanum, Trypoxylon 265
papuanum, Trypoxylon (Trypoxylon) 265
Paracrabro 232
Paracrabro froggatti 232
Paragia 155
Paragia australis 156
Paragia bicolor 157
Paragia bidens 158
Paragia calida 157
Paragia concinna 159
Paragia (Cygnaea) 155
Paragia (Cygnaea) vespiformis 155
Paragia deceptor 157
Paragia decipiens 155
(Paragia) decipiens aliciae, Paragia 155
(Paragia) decipiens decipiens, Paragia 155
(Paragia) decipiens, Paragia 155
Paragia excellens 158
Paragia hirsuta 157
Paragia maculata 154
Paragia magdalena 157
Paragia morosa 158
Paragia nasuta 158
Paragia odyneroides 158
Paragia (Paragia) 155
(Paragia), Paragia 155
Paragia (Paragia) decipiens 155
Paragia (Paragia) decipiens aliciae 155
Paragia (Paragia) decipiens decipiens 155
Paragia (Paragia) schulthessi 156
Paragia (Paragia) smithii 156
Paragia (Paragia) tricolor 156
Paragia (Paragiella) 156
Paragia (Paragiella) australis 156
Paragia (Paragiella) australis australis 156
Paragia (Paragiella) australis borealis 156
Paragia (Paragiella) bicolor 157
Paragia (Paragiella) calida 157
Paragia (Paragiella) deceptrix 157
Paragia (Paragiella) generosa 157
Paragia (Paragiella) hirsuta 157
Paragia (Paragiella) magdalena 157
Paragia (Paragiella) mimetica 157
Paragia (Paragiella) morosa 158
Paragia (Paragiella) nasuta 158
Paragia (Paragiella) odyneroides 158
Paragia (Paragiella) perkinsi 158
Paragia (Paragiella) propodealis 158
Paragia (Paragiella) sobrina 158
Paragia (Paragiella) venusta 159
Paragia (Paragiella) walkeri 159
Paragia perkinsi 158

Paragia pictifrons 155
Paragia praedator 158
Paragia saussurii 156
Paragia schulthessi 156
(Paragia) schulthessi, Paragia 156
Paragia smithii 156
(Paragia) smithii, Paragia 156
Paragia sobrina 158
Paragia tricolor 156
(Paragia) tricolor, Paragia 156
Paragia venusta 159
Paragia vespiformis 155
Paragia walkeri 159
Paragiella 156
(Paragiella) australis australis, Paragia 156
(Paragiella) australis borealis, Paragia 156
(Paragiella) australis, Paragia 156
(Paragiella) bicolor, Paragia 157
(Paragiella) calida, Paragia 157
(Paragiella) deceptrix, Paragia 157
(Paragiella) generosa, Paragia 157
(Paragiella) hirsuta, Paragia 157
(Paragiella) magdalena, Paragia 157
(Paragiella) mimetica, Paragia 157
(Paragiella) morosa, Paragia 158
(Paragiella) nasuta, Paragia 158
(Paragiella) odyneroides, Paragia 158
(Paragiella), Paragia 156
(Paragiella) perkinsi, Paragia 158
(Paragiella) propodealis, Paragia 158
(Paragiella) sobrina, Paragia 158
(Paragiella) venusta, Paragia 159
(Paragiella) walkeri, Paragia 159
paragioides, Abispa 199
paragioides, Pseudabispa 199
Paralastor 160 , 182
Paralastor abnormis 182
Paralastor aequifasciatus 182
Paralastor alastoripennis 182
Paralastor albifrons 182
Paralastor alexandriae 182
Paralastor anostreptus 183
Paralastor arenicola 183
Paralastor argentifrons 183
Paralastor argyrias 183
Paralastor aterrimus 183
Paralastor atripennis 183
Paralastor aurocinctus 183
Paralastor auster 184
Paralastor australis 184
Paralastor bicarinatus 184
Paralastor bischoffi 184
Paralastor brisbanensis 184
Paralastor brunneus 184
(Paralastor) brunneus, Alastor 184
Paralastor caprai 184
Paralastor carinatus 184
Paralastor clypeopunctatus 184
Paralastor commutatus 185
Paralastor comptus 185
Paralastor comptus comptus 185
Paralastor comptus rubescens 185
Paralastor conspiciendus 185
Paralastor conspicuus 185
Paralastor constrictus 185

Paralastor cruentatus 185
(Paralastor) cruentatus, Alastor 185
Paralastor cruentus 185
Paralastor darwinianus 186
Paralastor debilis 186
Paralastor debilitatus 186
Paralastor dentiger 186
Paralastor despectus 186
Paralastor diabolicus 186
Paralastor diadema 186
Paralastor donatus 186
Paralastor dubiosus 186
Paralastor dyscritias 187
Paralastor elegans 187
Paralastor emarginatus 187
Paralastor eriurgus 187
Paralastor euclidias 187
Paralastor eugonias 187
Paralastor eustomus 187
Paralastor eutretus 188
Paralastor fallax 188
Paralastor flaviceps 188
(Paralastor) flaviceps, Alastor 188
Paralastor frater 188
Paralastor fraternus 188
(Paralastor) fraternus, Alastor 188
Paralastor habilis 188
Paralastor hilaris 188
Paralastor icarioides 188
Paralastor ignotus 188
Paralastor imitator 189
Paralastor infernalis 189
(Paralastor) infernalis, Alastor 189
Paralastor infimus 189
Paralastor insularis 189
(Paralastor) insularis, Alastor 189
Paralastor lachesis 189
Paralastor laetus 189
Paralastor lateritius 190
(Paralastor) lateritius, Alastor 190
Paralastor leptias 190
Paralastor mackayensis 190
Paralastor maculiventris 190
(Paralastor) maculiventris, Alastor 190
Paralastor medius 190
Paralastor mesochlorus 190
Paralastor mesochlorus mesochloroides 190
Paralastor mesochlorus mesochlorus 190
Paralastor microgonias 190
Paralastor mimus 190
Paralastor multicolor 191
Paralastor mutabilis 191
Paralastor nautarum 191
(Paralastor) nautarum, Alastor 191
Paralastor neglectus 191
Paralastor neochromus 191
Paralastor occidentalis 191
Paralastor odynericornis 191
Paralastor odyneripennis 191
Paralastor odyneroides 191
Paralastor oloris 192
Paralastor optabilis 192
Paralastor ordinarius 192
Paralastor orientalis 192
Paralastor pallidus 192

Paralastor parca 192
Paralastor petiolatus 192
Paralastor picteti 192
Paralastor placens 193
Paralastor plebeius 193
Paralastor princeps 193
Paralastor pseudochromus 193
Paralastor punctulatus 193
Paralastor pusillus 193
(Paralastor) pusillus, Alastor 193
Paralastor roseotinctus 194
Paralastor rubroviolaceus 194
Paralastor rufipes 194
Paralastor sanguineus 194
(Paralastor) sanguineus, Alastor 194
Paralastor saussurei 194
Paralastor semirufus 194
Paralastor simillimus 194
Paralastor simplex 194
Paralastor simulator 194
Paralastor smithii 195
(Paralastor) smithii, Alastor 195
Paralastor solitarius 195
Paralastor subhabilis 195
Paralastor submersus 195
Paralastor subobscurus 195
Paralastor suboloris 195
Paralastor subplebeius 195
Paralastor subpunctulatus 195
Paralastor summus 195
Paralastor synchromus 195
Paralastor tasmaniensis 196
Paralastor tricarinulatus 196
Paralastor tricolor 196
(Paralastor) tridentatus, Alastor 200
Paralastor tuberculatus 196
Paralastor victor 196
Paralastor viduus 196
Paralastor vulneratus 197
(Paralastor) vulneratus, Alastor 197
Paralastor vulpinus 197
(Paralastor) vulpinus, Alastor 197
Paralastor vulpinus excisus 197
Paralastor vulpinus vulpinus 197
Paralastor xanthochromus 197
Paralastor xanthus 197
Paralastor xerophilus 197
Paralastor xerophilus meesi 197
Paralastor xerophilus xerophilus 197
Paralastoroides 165
(Parapison) aberrans, Pison 257
(Parapison) caliginosum, Pison 258
(Parapison) erythrogastrum, Pison 259
(Parapison) exclusum, Pison 259
(Parapison) noctulum, Pison 261
(Parapison) pertinax, Pison 261
Parapison ruficornis 259
(Parapison) simulans, Pison 262
(Parapison) tenebrosum, Pison 263
Parapsammophila 228
Parapsammophila eremophila 228
(Parapsammophila) eremophila, Ammophila 228
Paravespula 208
(Paravespula) germanica, Vespula 208
(Paravespula), Vespula 208

(Paravespula) vulgaris, Vespula 209
parca, Alastor 192
parca, Paralastor 192
(Pareumenes) australensis, Eumenes 174
parificus, Parifodynerus 198
Parifodynerus 197
Parifodynerus alariformis 198
Parifodynerus parificus 198
Parodynerus 198
Parodynerus bicincta 198
(Parodynerus) neglectus, Odynerus 191
patongensis, Sericophorus 248
patongensis, Sericophorus (Sericophorus) 248
paucispina, Tachysphex 241
pectinipes, Bembex 292
pectinipes, Bembix 292
pectoralis, Sphecius (Nothosphecius) 283
pectoralis, Stizus 283
peletieri, Pison 261
peletieri, Pison (Pison) 261
Pelopoeus bengalensis 220
Pelopoeus formosus 222
Pelopoeus laetus 221
peratus, Podagritus (Echuca) 271
perkinsi, Austrogorytes 282
perkinsi, Cerceris 301
perkinsi, Gorytes 282
perkinsi, Paragia 158
perkinsi, Paragia (Paragiella) 158
perlucida, Pseudoturneria 272
perlucidus, Clitemnestra 279
perlucidus, Crabro 272
perlucidus, Miscothyris 279
perplexum, Pison (Pison) 261
perplexus, Pison 261
persistans, Tachysphex 241
pertinax, Pison (Parapison) 261
pertinax, Pison (Pison) 261
pescotti, Sericophorus 248
pescotti, Sericophorus (Sericophorus) 248
petiolatus, Paralastor 192
philantes, Delta 172
philantes, Eumenes 172
Piagetia intaminata 237
picteti, Alastor 192
picteti, Paralastor 192
picteti, Polistes tepidus 210
pictifrons, Metaparagia 155
pictifrons, Paragia 155
pikati, Bembix 293
piliferus, Chimiloides 273
pillara, Bembix 293
pilosellus, Sphodrotes 255
pilosulus, Tachysphex 241
Pison 218 , 257
(Pison) aberrans, Pison 257
(Pison) areniferum, Pison 257
Pison (Aulacophilus) difficile 263
Pison (Aulacophilus) icarioides 263
(Pison) auratum, Pison 257
Pison auratus 257
Pison aureosericeum 257
(Pison) aureosericeum, Pison 257
Pison aurifex 257
(Pison) aurifex, Pison 257

Pison auriventre 258
(Pison) auriventre, Pison 258
Pison australis 262
(Pison) barbatum, Pison 258
(Pison) basale, Pison 258
Pison basalis 258
(Pison) caliginosum, Pison 258
(Pison) ciliatum, Pison 258
Pison congener 258
(Pison) congenerum, Pison 258
Pison decipiens 258
(Pison) decipiens, Pison 258
(Pison) deperditum, Pison 258
(Pison) dimidiatum, Pison 258
Pison dimidiatus 258
Pison dives 259
(Pison) dives, Pison 259
Pison erythrocerum 259
(Pison) erythrocerum, Pison 259
(Pison) erythrogastrum, Pison 259
(Pison) exclusum, Pison 259
Pison exornatum 259
(Pison) exornatum, Pison 259
Pison exultans 259
(Pison) exultans, Pison 259
(Pison) fenestratum, Pison 259
Pison fenestratus 259
(Pison) festivum, Pison 259
Pison festivus 259
Pison fraterculus 259
(Pison) fraterculus, Pison 259
(Pison) fuscipenne, Pison 259
Pison fuscipennis 259
Pison ignavum 260
(Pison) ignavum, Pison 260
Pison inconspicuum 260
(Pison) inconspicuum, Pison 260
Pison infumatum 260
(Pison) infumatum, Pison 260
Pison insulare priscum 261
Pison iridipenne 260
(Pison) iridipenne, Pison 260
Pison lutescens 260
(Pison) lutescens, Pison 260
Pison mandibulatum 260
(Pison) mandibulatum, Pison 260
(Pison) marginatum, Pison 260
Pison marginatus 260
Pison melanocephalum 260
(Pison) melanocephalum, Pison 260
Pison meridionale 260
(Pison) meridionale, Pison 260
Pison nitidus 259
(Pison) noctulum, Pison 261
Pison obliquus 263
Pison (Parapison) aberrans 257
Pison (Parapison) caliginosum 258
Pison (Parapison) erythrogastrum 259
Pison (Parapison) exclusum 259
Pison (Parapison) noctulum 261
Pison (Parapison) pertinax 261
Pison (Parapison) simulans 262
Pison (Parapison) tenebrosum 263
Pison peletieri 261
(Pison) peletieri, Pison 261

(Pison) perplexum, Pison 261
Pison perplexus 261
(Pison) pertinax, Pison 261
Pison (Pison) 257
(Pison), Pison 257
Pison (Pison) aberrans 257
Pison (Pison) areniferum 257
Pison (Pison) auratum 257
Pison (Pison) aureosericeum 257
Pison (Pison) aurifex 257
Pison (Pison) auriventre 258
Pison (Pison) barbatum 258
Pison (Pison) basale 258
Pison (Pison) caliginosum 258
Pison (Pison) ciliatum 258
Pison (Pison) congenerum 258
Pison (Pison) decipiens 258
Pison (Pison) deperditum 258
Pison (Pison) dimidiatum 258
Pison (Pison) dives 259
Pison (Pison) erythrocerum 259
Pison (Pison) erythrogastrum 259
Pison (Pison) exclusum 259
Pison (Pison) exornatum 259
Pison (Pison) exultans 259
Pison (Pison) fenestratum 259
Pison (Pison) festivum 259
Pison (Pison) fraterculus 259
Pison (Pison) fuscipenne 259
Pison (Pison) ignavum 260
Pison (Pison) inconspicuum 260
Pison (Pison) infumatum 260
Pison (Pison) iridipenne 260
Pison (Pison) lutescens 260
Pison (Pison) mandibulatum 260
Pison (Pison) marginatum 260
Pison (Pison) melanocephalum 260
Pison (Pison) meridionale 260
Pison (Pison) noctulum 261
Pison (Pison) peletieri 261
Pison (Pison) perplexum 261
Pison (Pison) pertinax 261
Pison (Pison) priscum 261
Pison (Pison) pulchrinum 261
Pison (Pison) punctulatum 261
Pison (Pison) ruficorne 261
Pison (Pison) rufipes 262
Pison (Pison) scabrum 262
Pison (Pison) separatum 262
Pison (Pison) simillimum 262
Pison (Pison) simulans 262
Pison (Pison) spinolae 262
Pison (Pison) strenuum 262
Pison (Pison) tenebrosum 263
Pison (Pison) tibiale 263
Pison (Pison) vestitum 263
Pison (Pison) virosum 263
Pison (Pison) westwoodii 263
Pison (Pisonitus) deperditum 258
Pison (Pisonitus) ruficornis 261
Pison (Pisonitus) rufipes 262
Pison (Pisonoides) 263
Pison (Pisonoides) difficile 263
Pison (Pisonoides) icarioides 263
(Pison) priscum, Pison 261

Pison pulchrinum 261
(Pison) pulchrinum, Pison 261
Pison punctulatum 261
(Pison) punctulatum, Pison 261
(Pison) ruficorne, Pison 261
(Pison) rufipes, Pison 262
Pison scabrum 262
(Pison) scabrum, Pison 262
(Pison) separatum, Pison 262
Pison separatus 262
(Pison) simillimum, Pison 262
Pison simillimus 262
(Pison) simulans, Pison 262
Pison spinolae 262
(Pison) spinolae, Pison 262
Pison strenuum 262
(Pison) strenuum, Pison 262
Pison tasmanicus 262
(Pison) tenebrosum, Pison 263
(Pison) tibiale, Pison 263
Pison tibialis 263
(Pison) vestitum, Pison 263
Pison vestitus 263
Pison virosum 263
(Pison) virosum, Pison 263
Pison westwoodii 263
(Pison) westwoodii, Pison 263
(Pisonitus) deperditum, Pison 258
(Pisonitus) ruficornis, Pison 261
(Pisonitus) rufipes, Pison 262
Pisonoides 263
(Pisonoides) difficile, Pison 263
(Pisonoides) icarioides, Pison 263
(Pisonoides), Pison 263
Piyuma 273
Piyuma prosopoides 273
placens, Paralastor 193
placidior, Leptomenoides 180
platypus, Tachysphex 242
plebeius, Paralastor 193
plebeja, Icaria 216
plebiana, Ropalidia (Icariola) 216
pleuralis, Tachysphex 242
plomleyi, Arpactus (Miscothyris) 280
plomleyi, Clitemnestra 280
plutocraticus, Tachytes 238
Podagritus 269
Podagritus (Echuca) 269
Podagritus (Echuca) alevinus 269
Podagritus (Echuca) aliciae 269
Podagritus (Echuca) anerus 270
Podagritus (Echuca) australiensis 270
Podagritus (Echuca) burnsi 270
Podagritus (Echuca) carolus 270
Podagritus (Echuca) cornigerum 270
Podagritus (Echuca) cygnorum 270
Podagritus (Echuca) doreeni 270
Podagritus (Echuca) edgarus 270
Podagritus (Echuca) imbelle 270
Podagritus (Echuca) kiatae 271
Podagritus (Echuca) krombeini 271
Podagritus (Echuca) leptospermi 271
Podagritus (Echuca) marcellus 271
Podagritus (Echuca) mullewanus 271
Podagritus (Echuca) peratus 271

Podagritus (Echuca) rieki 271
Podagritus (Echuca) tricolor 271
Podagritus (Echuca) yarrowi 272
Podalonia 228
Podalonia tydei 228
Podalonia tydei suspiciosa 228
Polemistus 230
Polemistus exul 230
Polistella 211
(Polistella) bernardii bernardii, Polistes 211
(Polistella) bernardii duplicinctus, Polistes 211
(Polistella) bernardii insulae, Polistes 211
(Polistella) bernardii, Polistes 211
(Polistella) bernardii richardsi, Polistes 211
(Polistella) humilis centrocontinentalis, Polistes 212
(Polistella) humilis humilis, Polistes 212
(Polistella) humilis, Polistes 211
(Polistella) humilis synoecus, Polistes 212
(Polistella), Polistes 211
(Polistella) sgarambus, Polistes 212
(Polistella) townsvillensis austrinus, Polistes 213
(Polistella) townsvillensis, Polistes 212
(Polistella) townsvillensis townsvillensis, Polistes 212
(Polistella) variabilis, Polistes 213
Polistes 208 , 209 , 213
Polistes balder 209
Polistes bernardii 211
Polistes bernardii duplicinctus 213
Polistes bioculata 214
(Polistes) dominulus, Polistes 209
Polistes erythrinus 210
Polistes facilis 210
Polistes gallicus 209
Polistes humilis centrocontinentalis 212
Polistes humilis clarior 212
Polistes humilis pseudoscach 212
Polistes humilis xanthorrhoicus 212
Polistes laevigatissimus 211
Polistes malayanus 210
Polistes (Megapolistes) 209
Polistes (Megapolistes) balder 209
Polistes (Megapolistes) erythrinus 210
Polistes (Megapolistes) facilis 210
Polistes (Megapolistes) olivaceus 210
Polistes (Megapolistes) schach 210
Polistes (Megapolistes) tepidus 210
Polistes (Megapolistes) tepidus tepidus 210
Polistes (Polistella) 211
Polistes (Polistella) bernardii 211
Polistes (Polistella) bernardii bernardii 211
Polistes (Polistella) bernardii duplicinctus 211
Polistes (Polistella) bernardii insulae 211
Polistes (Polistella) bernardii richardsi 211
Polistes (Polistella) humilis 211
Polistes (Polistella) humilis centrocontinentalis 212
Polistes (Polistella) humilis humilis 212
Polistes (Polistella) humilis synoecus 212
Polistes (Polistella) sgarambus 212
Polistes (Polistella) townsvillensis 212
Polistes (Polistella) townsvillensis austrinus 213
Polistes (Polistella) townsvillensis townsvillensis 212
Polistes (Polistella) variabilis 213
Polistes (Polistes) 209
(Polistes), Polistes 209
Polistes (Polistes) dominulus 209
Polistes richardsi 211
Polistes romandi 217
Polistes sgarambus 212
Polistes (Stenopolistes) 210
Polistes (Stenopolistes) laevigatissimus 211
Polistes (Stenopolistes) riekii 211
Polistes synoecus 212 , 213
Polistes tasmaniensis 212
Polistes tepidus picteti 210
Polistes townsvillensis 211 , 212
Polistes tricolor 212
Polistes variabilis reginae 212
Polistinae 151
Polistratus 213
(Polistratus) latetergum, Ropalidia 214
(Polistratus), Ropalidia 213
polita australensis, Ectopioglossa 174
polita, Ectopioglossa 174
politus, Tachysphex 242
polychroma, Ischnocoelia 179
polyphemus, Euodynerus 177
polyphemus, Odynerus 177
Pompiloidea 151
portlandensis, Acanthostethus 278
portlandensis, Nysson 278
positus, Diemodynerus decipiens 173
posttegulatus, Acarodynerus 163
posttegulatus, Odynerus (Stenodynerus) 163
praedator, Paragia 158
praedura, Cerceris 301
praetexta, Sphex 226
pretiosus, Arpactus 282
pretiosus, Austrogorytes 282
prima, Sphodrotes 255
princeps, Paralastor 193
princeps, Sphex 222
Prionyx 227
Prionyx globosus 227
Prionyx saevus 228
priscum, Pison insulare 261
priscum, Pison (Pison) 261
proletaria, Ropalidia (Icariola) 216
promontorii, Bembix 293
pronotalis, Leptomenoides 180
propodalaris, Acarodynerus 163
propodealis, Paragia (Paragiella) 158
Prosceliphron 222
(Prosceliphron) formosum, Sceliphron 222
(Prosceliphron), Sceliphron 222
prosopoides, Crabro 273
prosopoides, Piyuma 273
(Proterosphex) basilicus, Chlorion 223
(Proterosphex) rhodosoma, Chlorion 225
proteus, Tachysphex 242
Psen lutescens 230
Psenulus 229
Psenulus carinifrons 230
Psenulus carinifrons scutellatus 230
Psenulus interstitialis 230
Psenulus scutellatus 230
Pseudabispa 199
Pseudabispa confusa 199
Pseudabispa ephippioides 199
Pseudabispa paragioides 199
pseudacarodynerus, Diemodynerus 173

Pseudalastor 199
Pseudalastor anguloides 199
Pseudalastor cavifemur 199
Pseudalastor concolor 199
Pseudalastor concolor concolor 199
Pseudalastor concolor rapax 200
Pseudalastor metathoracicus 200
Pseudalastor superbus 200
Pseudalastor tridentatus 200
Pseudalastor tridentatus paganus 200
Pseudalastor tridentatus septentrionalis 200
Pseudalastor tridentatus transgrediens 200
Pseudalastor tridentatus tridentatus 200
pseudancistrocerus, Pseudonortonia 206
pseudancistrocerus, Stenodyneriellus 206
Pseudepipona 201
Pseudepipona alaris 201
Pseudepipona angulata 201
Pseudepipona angulata alexandriae 201
Pseudepipona angulata angulata 201
Pseudepipona aspra 201
(Pseudepipona) aspra, Pseudepipona 201
Pseudepipona chartergiformis 201
Pseudepipona clypalaris 202
Pseudepipona pallida 202
Pseudepipona (Pseudepipona) aspra 201
Pseudepipona succincta 202
Pseudepipona succincta purgata 202
Pseudepipona succincta succincta 202
Pseudepipona (Syneuodynerus) aurantiopilosella 207
Pseudepipona (Syneuodynerus) longebispinosa 207
Pseudepipona (Syneuodynerus) occidentata 207
pseudochromus, Paralastor 193
pseudoloris, Flammodynerus 177
pseudoneotropicus, Antamenes (Antamenes) 166
pseudoneotropicus, Pachymenes 166
Pseudonortonia pseudancistrocerus 206
pseudoscach, Polistes humilis 212
pseudospinosus, Pseudozethus 203
Pseudoturneria 272
Pseudoturneria couloni 272
Pseudoturneria perlucida 272
Pseudoturneria territorialis 272
Pseudozethus 202
Pseudozethus australensis 202
Pseudozethus confusus 202
Pseudozethus ephippium 202
Pseudozethus insignis 203
Pseudozethus pseudospinosus 203
Pseudozethus spinosus 203
Pseudozethus tasmaniensis montanus 169
Pseudozethus verreauxii 203
psilocera, Larra 233
pubescens, Tachysphex 242
pugnator, Tachysphex 242
pulchrinum, Pison 261
pulchrinum, Pison (Pison) 261
pulka, Bembix 293
punctatissimus, Acanthostethus 278
punctatissimus, Nysson (Acanthostethus) 278
punctatissimus, Stenodyneriellus 206
puncticeps, Tachysphex 242
punctiventris, Australodynerus 168
punctulatum, Pison 261
punctulatum, Pison (Pison) 261

punctulatus, Alastor 193
punctulatus, Paralastor 193
punctum, Ropalidia 213
punctuosa, Anacrucis 253
punctuosa, Sphodrotes 255
punctuosus, Sericophorus (Zoyphidium) 253
purgata, Pseudepipona succincta 202
pusilloides, Australodynerus 168
pusilloides, Australodynerus pusilloides 168
pusilloides impudicus, Australodynerus 168
pusilloides pusilloides, Australodynerus 168
pusillum, Zoyphium 253
pusillus, Alastor (Paralastor) 193
pusillus, Australodynerus 168
pusillus, Odynerus 168
pusillus, Paralastor 193
pusillus, Sericophorus (Zoyphidium) 253

quadrangolum, Acarodynerus 163
queenslandensis, Arpactophilus 231
queenslandensis, Lyroda 245
queenslandensis, Stigmus 231
queenslandicus, Acarodynerus 163

rapax, Pseudalastor concolor 200
raptor, Bembex 294
raymenti, Cerceris 301
raymenti, Sericophorus (Sericophorus) 248
recondita, Liris (Leptolarra) 236
recondita, Notogonia 236
rectispina, Pachymenes 166
regina, Liris (Leptolarra) 236
regina, Notogonia 236
reginae, Polistes variabilis 212
reginellus, Ectemnius (Hypocrabro) 276
relicta, Lestica (Solenius) 277
relucens nigricornis, Sericophorus 249
relucens nigricornis, Sericophorus (Sericophorus) 249
relucens relucens, Sericophorus (Sericophorus) 248
relucens ruficornis, Sericophorus 249
relucens ruficornis, Sericophorus (Sericophorus) 249
relucens, Sericophorus 248
relucens, Sericophorus (Sericophorus) 248
relucens, Sericophorus (Sericophorus) relucens 248
relucens, Tachytes 238
resplendens, Sphex 225
resplendens, Sphex (Sphex) 225
retiaria, Notogonia 235
reticulata, Nitela 256
reticulatus, Arpactophilus 231
reticulatus, Austrostigmus 231
revolutionalis, Icaria 216
revolutionalis, Ropalidia (Icariola) 216
rhodosoma, Chlorion (Proterosphex) 225
rhodosoma, Sphex (Sphex) 225
Rhopalidia 213
Rhopalum 265
Rhopalum aliciae 269
(Rhopalum) anteum, Rhopalum 267
(Rhopalum) australiae, Rhopalum 267
(Rhopalum) calixtum, Rhopalum 265
(Rhopalum) conator, Crabro 274
(Rhopalum) coriolum, Rhopalum 267

(Rhopalum) cornigerum, Rhopalum 270
Rhopalum (Corynopus) 267
Rhopalum (Corynopus) anteum 267
Rhopalum (Corynopus) australiae 267
Rhopalum (Corynopus) coriolum 267
Rhopalum (Corynopus) dedarum 267
Rhopalum (Corynopus) evansianum 267
Rhopalum (Corynopus) evictum 267
Rhopalum (Corynopus) famicum 267
Rhopalum (Corynopus) frenchii 268
Rhopalum (Corynopus) harpax 268
Rhopalum (Corynopus) kerangi 268
Rhopalum (Corynopus) kuehlhorni 268
Rhopalum (Corynopus) littorale 268
Rhopalum (Corynopus) neboissi 268
Rhopalum (Corynopus) notogeum 268
Rhopalum (Corynopus) taeniatum 268
Rhopalum (Corynopus) tubarum 269
Rhopalum (Corynopus) variitarse 269
Rhopalum (Corynopus) xenum 269
Rhopalum cygnorum 270
(Rhopalum) dedarum, Rhopalum 267
(Rhopalum) dineurum, Rhopalum 266
Rhopalum eucalypti 266
(Rhopalum) eucalypti, Rhopalum 266
(Rhopalum) frenchii, Crabro 268
(Rhopalum) grahami, Rhopalum 266
(Rhopalum) harpax, Rhopalum 268
(Rhopalum) idoneus, Crabro 272
(Rhopalum) kerangi, Rhopalum 268
(Rhopalum) kuehlhorni, Rhopalum 268
Rhopalum leptospermi 271
Rhopalum littorale 268
(Rhopalum) macrocephalum, Rhopalum 266
Rhopalum macrocephalus 266
(Rhopalum) militaris, Crabro 271
(Rhopalum) neboissi, Rhopalum 268
(Rhopalum) notogeum, Rhopalum 268
Rhopalum (Notorhopalum) 269
Rhopalum (Notorhopalum) carnegiacum 269
Rhopalum (Rhopalum) 265
(Rhopalum), Rhopalum 265
Rhopalum (Rhopalum) anteum 267
Rhopalum (Rhopalum) australiae 267
Rhopalum (Rhopalum) calixtum 265
Rhopalum (Rhopalum) coriolum 267
Rhopalum (Rhopalum) cornigerum 270
Rhopalum (Rhopalum) dedarum 267
Rhopalum (Rhopalum) dineurum 266
Rhopalum (Rhopalum) eucalypti 266
Rhopalum (Rhopalum) grahami 266
Rhopalum (Rhopalum) harpax 268
Rhopalum (Rhopalum) kerangi 268
Rhopalum (Rhopalum) kuehlhorni 268
Rhopalum (Rhopalum) macrocephalum 266
Rhopalum (Rhopalum) neboissi 268
Rhopalum (Rhopalum) notogeum 268
Rhopalum (Rhopalum) taeniatum 268
Rhopalum (Rhopalum) tenuiventre 266
Rhopalum (Rhopalum) tepicum 266
Rhopalum (Rhopalum) testaceum 266
Rhopalum (Rhopalum) transiens 266
Rhopalum (Rhopalum) tubarum 269
Rhopalum (Rhopalum) tuberculicorne 266
Rhopalum (Rhopalum) verutum 266

Rhopalum (Rhopalum) xenum 269
Rhopalum spinulifer 272
(Rhopalum) taeniatum, Rhopalum 268
(Rhopalum) tenuiventre, Rhopalum 266
(Rhopalum) tenuiventris, Crabro 266
(Rhopalum) tepicum, Rhopalum 266
Rhopalum testaceum 266
(Rhopalum) testaceum, Rhopalum 266
(Rhopalum) transiens, Crabro 266
(Rhopalum) transiens, Rhopalum 266
Rhopalum tricolor imbelle 270
(Rhopalum) tubarum, Rhopalum 269
Rhopalum tuberculicorne 266
(Rhopalum) tuberculicorne, Rhopalum 266
Rhopalum variitarse 269
(Rhopalum) verutum, Rhopalum 266
(Rhopalum) xenum, Rhopalum 269
Rhynchium 203
Rhynchium abispoides 175
(Rhynchium) angulatus alexandriae, Odynerus 201
Rhynchium atrum 203
Rhynchium australense 204
(Rhynchium) bicolor aurantiopictus, Odynerus 170
(Rhynchium) bicolor flavescentulus, Odynerus 170
(Rhynchium) bicolor nigrocinctoides, Odynerus 170
Rhynchium decoratum 174
Rhynchium iridipenne 165
Rhynchium magnificum 204
Rhynchium mirabile 204
Rhynchium nigrolimbatum 204
Rhynchium rothi 204
(Rhynchium) rubroviolaceus, Odynerus 194
Rhynchium rufipes 204
Rhynchium superbum 205
rhynchocephalus, Tachysphex 243
richardsi, Polistes 211
richardsi, Polistes (Polistella) bernardii 211
rieki, Larrisson 256
rieki, Podagritus (Echuca) 271
Riekia 154
Riekia angulata 154
Riekia nocatunga 154
riekiella, Williamsita 275
riekiella, Williamsita (Androcrabro) 275
riekii, Polistes (Stenopolistes) 211
robusta analis, Ischnocoelia 179
robusta aurantiaca, Ischnocoelia 179
robusta, Ischnocoelia 179
robusta, Ischnocoelia robusta 179
robusta robusta, Ischnocoelia 179
robusta unicolor, Ischnocoelia 179
robustus, Elimus 179
roddi, Sericophorus (Sericophorus) viridis 251
roddi, Sericophorus viridis 251
Rolandia 154
Rolandia maculata 154
romandi cabeti, Ropalidia (Icarielia) 217
romandi, Polistes 217
romandi romandi, Ropalidia (Icarielia) 217
romandi, Ropalidia (Icarielia) 217
romandi, Ropalidia (Icarielia) romandi 217
Ropalidia 208 , 213
(Ropalidia) eboraca, Ropalidia 213
(Ropalidia) fulvopruinosa, Ropalidia 213
Ropalidia (Icarielia) 217

Ropalidia (Icarielia) nigrior 217
Ropalidia (Icarielia) romandi 217
Ropalidia (Icarielia) romandi cabeti 217
Ropalidia (Icarielia) romandi romandi 217
Ropalidia (Icariola) 214
Ropalidia (Icariola) darwini 214
Ropalidia (Icariola) deceptor 214
Ropalidia (Icariola) elegantula 214
Ropalidia (Icariola) eurostoma 214
Ropalidia (Icariola) gracilenta 214
Ropalidia (Icariola) gregaria 214
Ropalidia (Icariola) gregaria gregaria 214
Ropalidia (Icariola) gregaria spilocephala 215
Ropalidia (Icariola) hirsuta 215
Ropalidia (Icariola) interrupta 215
Ropalidia (Icariola) interrupta flavinoda 215
Ropalidia (Icariola) interrupta interrupta 215
Ropalidia (Icariola) kurandae 215
Ropalidia (Icariola) mackayensis 215
Ropalidia (Icariola) marginata 215
Ropalidia (Icariola) marginata jucunda 215
Ropalidia (Icariola) mutabilis 215
Ropalidia (Icariola) mutabilis mutabilis 216
Ropalidia (Icariola) mutabilis torresiana 216
Ropalidia (Icariola) plebiana 216
Ropalidia (Icariola) proletaria 216
Ropalidia (Icariola) revolutionalis 216
Ropalidia (Icariola) socialistica 216
Ropalidia (Icariola) trichophthalma 217
Ropalidia (Icariola) turneri 217
Ropalidia (Polistratus) 213
Ropalidia (Polistratus) latetergum 214
Ropalidia punctum 213
(Ropalidia), Ropalidia 213
Ropalidia (Ropalidia) 213
Ropalidia (Ropalidia) eboraca 213
Ropalidia (Ropalidia) fulvopruinosa 213
Ropalidia variegata flavinoda 215
Ropalidia variegata interrupta 215
ropalidioides, Cerceris 301
roseotinctus, Paralastor 194
rothi, Rhynchium 204
rubellus, Tachytes 238
rubescens, Paralastor comptus 185
rubra, Sphodrotes 255
rubricata, Sphodrotes 255
rubricatus, Sphodrotes 255
rubrosignatus, Argogorytes 280
rubrosignatus, Arpactus 280
rubroviolaceus, Odynerus (Rhynchium) 194
rubroviolaceus, Paralastor 194
ruficollis, Arpactophilus 231
ruficollis, Austrostigmus 231
ruficorne, Pison (Pison) 261
ruficornis, Parapison 259
ruficornis, Pison (Pisonitus) 261
ruficornis, Sericophorus relucens 249
ruficornis, Sericophorus (Sericophorus) relucens 249
rufipes, Paralastor 194
rufipes, Pison (Pison) 262
rufipes, Pison (Pisonitus) 262
rufipes, Rhynchium 204
rufipes, Sericophorus 248
rufipes, Vespa 204
rufipes, Zoyphium 249

rufitarsus, Microglossa 233
rufitarsus, Spilomena 233
rufiventris, Aphelotoma 219
rufobasalis, Sericophorus 249
rufobasalis, Sericophorus (Sericophorus) 249
rufocaudatus, Acarodynerus dietrichianus 162
rufofusca, Cerceris armigera 297
rufomixtus, Argogorytes 281
rufomixtus, Gorytes 281
rufoniger, Sericophorus (Zoyphidium) 253
rufonigrum, Zoyphium 253
rufotibialis, Sericophorus 249
rufotibialis, Sericophorus (Sericophorus) 249
rugidorsatus, Tachysphex 242
rugifer, Sphex (Sphex) 225
rugosus, Sericophorus 249
rugosus, Sericophorus (Sericophorus) 249
Rygchium mirabile 204
Rygchium superbum 205

saeva, Cerceris 296
saevus, Harpactopus 228
saevus, Prionyx 228
sagax, Palmodes 227
sanguineus, Alastor (Paralastor) 194
sanguineus, Paralastor 194
sanguinolentus, Clitemnestra 280
sanguinolentus, Gorytes 280
sanguinolentus, Odynerus (Ancistroceroides) 200
saucius, Diemodynerus 173
saucius, Odynerus 173
saussurei, Acanthostethus 278
saussurei, Bembex 287
saussurei, Nysson 278
saussurei, Paralastor 194
saussurii, Paragia 156
scabrum, Pison 262
scabrum, Pison (Pison) 262
scelesta, Larra 234
Sceliphron 221
(Sceliphron) caementarium, Sceliphron 221
Sceliphron laetum cygnorum 221
(Sceliphron) laetum laetum, Sceliphron 221
(Sceliphron) laetum, Sceliphron 221
Sceliphron (Prosceliphron) 222
Sceliphron (Prosceliphron) formosum 222
(Sceliphron), Sceliphron 221
Sceliphron (Sceliphron) 221
Sceliphron (Sceliphron) caementarium 221
Sceliphron (Sceliphron) laetum 221
Sceliphron (Sceliphron) laetum laetum 221
schach, Polistes (Megapolistes) 210
schach, Vespa 210
schmiedeknechti connexum, Trypoxylon (Trypoxylon) 265
schmiedeknechti, Trypoxylon (Trypoxylon) 265
schulthessi, Paragia 156
schulthessi, Paragia (Paragia) 156
Scolioidea 151
sculleni, Cerceris 296 , 297
sculpturata, Nitela 256
sculpturatus, Sericophorus 249
sculpturatus, Sericophorus (Sericophorus) 249
scutellatus, Psenulus 230

scutellatus, Psenulus carinifrons 230
secernendus, Argogorytes 281
secernendus, Arpactus 281
sedis, Incertae 303
sedula, Cerceris 302
semifossulatus, Sphex 225
semifossulatus, Sphex (Sphex) 225
semirufus, Paralastor 194
separatum, Pison (Pison) 262
separatus, Pison 262
septentrionalis, Pseudalastor tridentatus 200
serena, Liris (Leptolarra) 236
serena, Notogonia 236
sericeum, Zoyphium 253
sericeus godeffroyi, Sphex (Sphex) 226
sericeus, Sericophorus (Zoyphidium) 253
sericeus, Sphex (Sphex) 226
Sericogaster 151
Sericophorus 245
Sericophorus abnormis 255
Sericophorus aliceae 245
(Sericophorus) aliceae, Sericophorus 245
Sericophorus bicolor 245
(Sericophorus) bicolor, Sericophorus 245
Sericophorus brisbanensis 245
(Sericophorus) brisbanensis, Sericophorus 245
Sericophorus carinatus 245
(Sericophorus) carinatus, Sericophorus 245
Sericophorus castaneus 245
(Sericophorus) castaneus, Sericophorus 245
(Sericophorus) chalybaeus chalybaeus, Sericophorus 246
Sericophorus chalybaeus fulleri 246
(Sericophorus) chalybaeus fulleri, Sericophorus 246
(Sericophorus) chalybaeus, Sericophorus 246
Sericophorus chalybeus 246
Sericophorus claviger burnsiellus 246
(Sericophorus) claviger burnsiellus, Sericophorus 246
(Sericophorus) claviger claviger, Sericophorus 246
(Sericophorus) claviger, Sericophorus 246
Sericophorus cliffordi 246
(Sericophorus) cliffordi, Sericophorus 246
(Sericophorus) cockerelli, Sericophorus 246
Sericophorus cyanophilus 247
(Sericophorus) cyanophilus, Sericophorus 247
Sericophorus dipteroides 252
Sericophorus elegantior 247
(Sericophorus) elegantior, Sericophorus 247
Sericophorus froggatti 247
(Sericophorus) froggatti, Sericophorus 247
Sericophorus funebris 247
(Sericophorus) funebris, Sericophorus 247
Sericophorus gracilis 247
(Sericophorus) gracilis, Sericophorus 247
Sericophorus hackeri 246
Sericophorus inornatus 247
(Sericophorus) inornatus, Sericophorus 247
Sericophorus lilacinus 247
(Sericophorus) lilacinus, Sericophorus 247
Sericophorus littoralis 247
(Sericophorus) littoralis, Sericophorus 247
Sericophorus metallescens 247
(Sericophorus) metallescens, Sericophorus 247
Sericophorus minutus 248
(Sericophorus) minutus, Sericophorus 248
Sericophorus nigror 248

(Sericophorus) nigror, Sericophorus 248
Sericophorus niveifrons 248
(Sericophorus) niveifrons, Sericophorus 248
Sericophorus occidentalis 248
(Sericophorus) occidentalis, Sericophorus 248
Sericophorus patongensis 248
(Sericophorus) patongensis, Sericophorus 248
Sericophorus pescotti 248
(Sericophorus) pescotti, Sericophorus 248
(Sericophorus) raymenti, Sericophorus 248
Sericophorus relucens 248
Sericophorus relucens nigricornis 249
(Sericophorus) relucens nigricornis, Sericophorus 249
(Sericophorus) relucens relucens, Sericophorus 248
Sericophorus relucens ruficornis 249
(Sericophorus) relucens ruficornis, Sericophorus 249
(Sericophorus) relucens, Sericophorus 248
Sericophorus rufipes 248
Sericophorus rufobasalis 249
(Sericophorus) rufobasalis, Sericophorus 249
Sericophorus rufotibialis 249
(Sericophorus) rufotibialis, Sericophorus 249
Sericophorus rugosus 249
(Sericophorus) rugosus, Sericophorus 249
Sericophorus sculpturatus 249
(Sericophorus) sculpturatus, Sericophorus 249
Sericophorus (Sericophorus) 245
(Sericophorus), Sericophorus 245
Sericophorus (Sericophorus) aliceae 245
Sericophorus (Sericophorus) bicolor 245
Sericophorus (Sericophorus) brisbanensis 245
Sericophorus (Sericophorus) carinatus 245
Sericophorus (Sericophorus) castaneus 245
Sericophorus (Sericophorus) chalybaeus 246
Sericophorus (Sericophorus) chalybaeus chalybaeus 246
Sericophorus (Sericophorus) chalybaeus fulleri 246
Sericophorus (Sericophorus) claviger 246
Sericophorus (Sericophorus) claviger burnsiellus 246
Sericophorus (Sericophorus) claviger claviger 246
Sericophorus (Sericophorus) cliffordi 246
Sericophorus (Sericophorus) cockerelli 246
Sericophorus (Sericophorus) cyanophilus 247
Sericophorus (Sericophorus) elegantior 247
Sericophorus (Sericophorus) froggatti 247
Sericophorus (Sericophorus) funebris 247
Sericophorus (Sericophorus) gracilis 247
Sericophorus (Sericophorus) inornatus 247
Sericophorus (Sericophorus) lilacinus 247
Sericophorus (Sericophorus) littoralis 247
Sericophorus (Sericophorus) metallescens 247
Sericophorus (Sericophorus) minutus 248
Sericophorus (Sericophorus) nigror 248
Sericophorus (Sericophorus) niveifrons 248
Sericophorus (Sericophorus) occidentalis 248
Sericophorus (Sericophorus) patongensis 248
Sericophorus (Sericophorus) pescotti 248
Sericophorus (Sericophorus) raymenti 248
Sericophorus (Sericophorus) relucens 248
Sericophorus (Sericophorus) relucens nigricornis 249
Sericophorus (Sericophorus) relucens relucens 248
Sericophorus (Sericophorus) relucens ruficornis 249
Sericophorus (Sericophorus) rufobasalis 249
Sericophorus (Sericophorus) rufotibialis 249
Sericophorus (Sericophorus) rugosus 249
Sericophorus (Sericophorus) sculpturatus 249

Sericophorus (Sericophorus) spryi 249
Sericophorus (Sericophorus) subviridis 249
Sericophorus (Sericophorus) sydneyi 250
Sericophorus (Sericophorus) tallongensis 250
Sericophorus (Sericophorus) teliferopodus 250
Sericophorus (Sericophorus) teliferopodus okiellus 250
Sericophorus (Sericophorus) teliferopodus teliferopodus 250
Sericophorus (Sericophorus) victoriensis 250
Sericophorus (Sericophorus) violaceus 250
Sericophorus (Sericophorus) viridis 250
Sericophorus (Sericophorus) viridis roddi 251
Sericophorus (Sericophorus) viridis viridis 250
Sericophorus spryi 249
(Sericophorus) spryi, Sericophorus 249
Sericophorus subviridis 249
(Sericophorus) subviridis, Sericophorus 249
Sericophorus sydneyi 250
(Sericophorus) sydneyi, Sericophorus 250
Sericophorus tallongensis 250
(Sericophorus) tallongensis, Sericophorus 250
Sericophorus teliferopodus 250
Sericophorus teliferopodus okiellus 250
(Sericophorus) teliferopodus okiellus, Sericophorus 250
(Sericophorus) teliferopodus, Sericophorus 250
(Sericophorus) teliferopodus teliferopodus, Sericophorus 250
Sericophorus victoriensis 250
(Sericophorus) victoriensis, Sericophorus 250
Sericophorus violaceus 250
(Sericophorus) violaceus, Sericophorus 250
Sericophorus viridis roddi 251
(Sericophorus) viridis roddi, Sericophorus 251
(Sericophorus) viridis, Sericophorus 250
(Sericophorus) viridis viridis, Sericophorus 250
Sericophorus (Zoyphidium) 251
Sericophorus (Zoyphidium) affinis 251
Sericophorus (Zoyphidium) argyreus 251
Sericophorus (Zoyphidium) asperithorax 251
Sericophorus (Zoyphidium) cingulatus 251
Sericophorus (Zoyphidium) clypeatus 251
Sericophorus (Zoyphidium) collaris 251
Sericophorus (Zoyphidium) crassicornis 251
Sericophorus (Zoyphidium) dipteroides 252
Sericophorus (Zoyphidium) doddi 252
Sericophorus (Zoyphidium) emarginatus 252
Sericophorus (Zoyphidium) erythrosoma 252
Sericophorus (Zoyphidium) ferrugineus 252
Sericophorus (Zoyphidium) flavofasciatus 252
Sericophorus (Zoyphidium) frontalis 252
Sericophorus (Zoyphidium) fuscipennis 252
Sericophorus (Zoyphidium) hackeri 252
Sericophorus (Zoyphidium) humilis 253
Sericophorus (Zoyphidium) iridipennis 253
Sericophorus (Zoyphidium) kohlii 253
Sericophorus (Zoyphidium) laevigatus 253
Sericophorus (Zoyphidium) niger 253
Sericophorus (Zoyphidium) ornatus 253
Sericophorus (Zoyphidium) punctuosus 253
Sericophorus (Zoyphidium) pusillus 253
Sericophorus (Zoyphidium) rufoniger 253
Sericophorus (Zoyphidium) sericeus 253
Sericophorus (Zoyphidium) splendidus 254
Sericophorus (Zoyphidium) striatulus 254
Sericophorus (Zoyphidium) tuberculatus 254

severa, Bembex 293
severa, Bembix 293
sgarambus, Polistes 212
sgarambus, Polistes (Polistella) 212
siccata, Cerceris 302
signatus, Bembecinus 285
signatus, Stizus 285
similis, Alastor 193
simillimum, Pison (Pison) 262
simillimus, Paralastor 194
simillimus, Pison 262
simplex, Paralastor 194
simplicilamellatus, Eumenes 176
simplicilamellatus, Eumenes (Eumenes) 176
simulans, Pison (Parapison) 262
simulans, Pison (Pison) 262
simulator, Paralastor 194
smithianum, Macrocalymma 181
smithiensis, Williamsita (Androcrabro) 275
smithii, Alastor (Paralastor) 195
smithii, Paragia 156
smithii, Paragia (Paragia) 156
smithii, Paralastor 195
sobrina, Paragia 158
sobrina, Paragia (Paragiella) 158
socialistica, Icaria 216
socialistica, Ropalidia (Icariola) 216
Solenius 277
(Solenius), Lestica 277
(Solenius) manifestatus, Crabro 275
(Solenius) relicta, Lestica 277
solitarius, Paralastor 195
spargovillensis, Acarodynerus 164
spargovillensis, Odynerus (Stenodynerus) 164
spathulifera, Liris (Leptolarra) 236
spathulifera, Notogonia 236
spectrum, Acarodynerus 164
SPHECIDAE 151, 152, 160, 218
Sphecius 283
Sphecius lanio 283
Sphecius (Nothosphecius) 283
Sphecius (Nothosphecius) pectoralis 283
SPHECOIDEA 150, 151, 152
Sphex 222
(Sphex) ahasverus, Sphex 222
Sphex amator 223
Sphex argentatus 222
(Sphex) argentatus, Sphex 222
Sphex argentifrons 225
Sphex aurifex 226
Sphex aurifrons 226
Sphex bannitus 225
(Sphex) basilicus, Sphex 223
Sphex bilobatus 223
(Sphex) bilobatus, Sphex 223
Sphex caementarium 221
Sphex canescens 223
Sphex carbonaria 223
Sphex carbonicolor 223
(Sphex) carbonicolor, Sphex 223
Sphex clavigera 222
Sphex clavus 229
Sphex cognata 223
(Sphex) cognatus, Sphex 223
Sphex compressa 220

Sphex darwiniensis 223
(Sphex) darwiniensis, Sphex 223
Sphex decorata 224
(Sphex) decoratus, Sphex 224
(Sphex) dolichocerus, Sphex 225
Sphex ephippium 224
(Sphex) ephippium, Sphex 224
(Sphex) ermineus, Sphex 224
Sphex finschii 224
(Sphex) finschii, Sphex 224
Sphex formosa 223
Sphex formosellus 224
(Sphex) formosellus, Sphex 224
Sphex fumipennis 224
(Sphex) fumipennis, Sphex 224
Sphex gilberti 224
(Sphex) gilberti, Sphex 224
Sphex globosa 227
Sphex godeffroyi 226
Sphex gratiosa 225
Sphex gratiosissimus 225
(Sphex) imperialis, Sphex 226
Sphex (Isodontia) abditus nugenti 226
Sphex (Isodontia) albohirtus 226
Sphex luctuosa 224
(Sphex) luctuosus, Sphex 224
Sphex mimulus 225
(Sphex) mimulus, Sphex 225
Sphex modesta 225
(Sphex) modestus, Sphex 225
Sphex nigella 227
Sphex nitidiventris 225
Sphex obscurella 227
Sphex opulenta 223
Sphex praetexta 226
Sphex princeps 222
Sphex resplendens 225
(Sphex) resplendens, Sphex 225
(Sphex) rhodosoma, Sphex 225
(Sphex) rugifer, Sphex 225
Sphex semifossulatus 225
(Sphex) semifossulatus, Sphex 225
(Sphex) sericeus godeffroyi, Sphex 226
(Sphex) sericeus, Sphex 226
Sphex (Sphex) 222
(Sphex), Sphex 222
Sphex (Sphex) ahasverus 222
Sphex (Sphex) argentatus 222
Sphex (Sphex) basilicus 223
Sphex (Sphex) bilobatus 223
Sphex (Sphex) carbonicolor 223
Sphex (Sphex) cognatus 223
Sphex (Sphex) darwiniensis 223
Sphex (Sphex) decoratus 224
Sphex (Sphex) dolichocerus 225
Sphex (Sphex) ephippium 224
Sphex (Sphex) ermineus 224
Sphex (Sphex) finschii 224
Sphex (Sphex) formosellus 224
Sphex (Sphex) fumipennis 224
Sphex (Sphex) gilberti 224
Sphex (Sphex) imperialis 226
Sphex (Sphex) luctuosus 224
Sphex (Sphex) mimulus 225
Sphex (Sphex) modestus 225

Sphex (Sphex) resplendens 225
Sphex (Sphex) rhodosoma 225
Sphex (Sphex) rugifer 225
Sphex (Sphex) semifossulatus 225
Sphex (Sphex) sericeus 226
Sphex (Sphex) sericeus godeffroyi 226
Sphex (Sphex) staudingeri 226
Sphex (Sphex) vestitus 226
Sphex staudingeri 226
(Sphex) staudingeri, Sphex 226
Sphex umbrosus 222
Sphex vestita 226
(Sphex) vestitus, Sphex 226
Sphex vidua 227
Sphex violacea 220
Sphex wallacei 225
Sphodrotes 254
Sphodrotes acuticollis 254
Sphodrotes cygnorum 254
Sphodrotes dearmata 254
Sphodrotes marginalis 254
Sphodrotes nemoralis 254
Sphodrotes occidentalis 255
Sphodrotes ordinaria 255
Sphodrotes pilosellus 255
Sphodrotes prima 255
Sphodrotes punctuosa 255
Sphodrotes rubra 255
Sphodrotes rubricata 255
Sphodrotes rubricatus 255
Sphodrotes splendens 255
spilocephala, Icaria 215
spilocephala, Ropalidia (Icariola) gregaria 215
Spilomena 218 , 232
Spilomena australis 232
Spilomena bimaculata 232
Spilomena elegantula 232
Spilomena hobartia 232
Spilomena iridescens 232
Spilomena longiceps 232
Spilomena longifrons 233
Spilomena luteiventris 233
Spilomena rufitarsus 233
spilota, Cerceris 302
spinicornis, Arpactus 282
spinicornis, Austrogorytes 282
spiniger, Acanthostethus 278
spiniger, Nysson (Acanthostethus) 278
spinipleuris, Cerceris 302
Spinocrabro 272
spinolae, Pison 262
spinolae, Pison (Pison) 262
spinosiusculus, Stenodyneriellus 206
spinosus, Discoelius 203
spinosus, Pseudozethus 203
spinulifer, Rhopalum 272
splendens, Sphodrotes 255
splendida, Abispa (Abispa) 161
splendida, Abispa (Abispa) splendida 161
splendida australis, Abispa (Abispa) 161
splendida splendida, Abispa (Abispa) 161
splendidum, Zoyphium 254
splendidus, Odynerus 161
splendidus, Sericophorus (Zoyphidium) 254
spryi, Arpactus 282

spryi, Austrogorytes 282
spryi, Sericophorus 249
spryi, Sericophorus (Sericophorus) 249
staudingeri, Sphex 226
staudingeri, Sphex (Sphex) 226
steindachneri, Arpactophilus 231
steindachneri, Arpactophilus steindachneri 231
steindachneri deserticolus, Arpactophilus 231
steindachneri deserticolus, Harpactophilus 231
steindachneri steindachneri, Arpactophilus 231
Stenodyneriellus 205
Stenodyneriellus bicoloratus 205
Stenodyneriellus brisbanensis 205
Stenodyneriellus carnarvonensis 205
Stenodyneriellus darnleyensis 205
Stenodyneriellus novempunctatus 205
Stenodyneriellus pseudancistrocerus 206
Stenodyneriellus punctatissimus 206
Stenodyneriellus spinosiusculus 206
Stenodyneriellus tricoloratus 206
Stenodyneriellus turneriellus 206
Stenodyneriellus yanchepensis 206
Stenodyneriellus yanchepensis nigrithorax 206
Stenodyneriellus yanchepensis yanchepensis 206
(Stenodyneroides) subalaris, Odynerus 198
(Stenodynerus) exarmatus, Odynerus 163
(Stenodynerus) posttegulatus, Odynerus 163
(Stenodynerus) spargovillensis, Odynerus 164
Stenolabus fulvus 178
Stenolabus integer 178
Stenolabus vulneratus 179
Stenopolistes 210
(Stenopolistes) laevigatissimus, Polistes 211
(Stenopolistes), Polistes 210
(Stenopolistes) riekii, Polistes 211
stenopygus, Argogorytes 280
Stigmus queenslandensis 231
stimulator, Tachysphex 243
Stizus antipodum 284
Stizus egens 284
Stizus gorytoides 284
Stizus pectoralis 283
Stizus signatus 285
Stizus turneri 285
strenuum, Pison 262
strenuum, Pison (Pison) 262
striaticollis, Aphelotoma 219
striatula, Anacrucis 254
striatulus, Sericophorus (Zoyphidium) 254
subalaris, Flammodynerus 177
subalaris, Odynerus 177
subalaris, Odynerus (Stenodyneroides) 198
Subancistrocerus 206
Subancistrocerus monstricornis 207
(Subancistrocerus) monstricornis, Ancistrocerus 207
subhabilis, Paralastor 195
submersus, Paralastor 195
subobscurus, Paralastor 195
suboloris, Paralastor 195
subopacus, Tachysphex 243
subplebeius, Paralastor 195
subpunctulatus, Paralastor 195
subviridis, Sericophorus 249
subviridis, Sericophorus (Sericophorus) 249
succincta, Pseudepipona 202

succincta, Pseudepipona succincta 202
succincta purgata, Pseudepipona 202
succincta succincta, Pseudepipona 202
succinctus, Odynerus (Leionotus) 202
sulcatus, Arpactophilus 231
sulcatus, Harpactophilus 231
sulcatus, Tachytes 238
summus, Paralastor 195
superbum, Rhynchium 205
superbum, Rygchium 205
superbus, Pseudalastor 200
suspiciosa, Ammophila 228
suspiciosa, Podalonia tydei 228
sydneyi, Sericophorus 250
sydneyi, Sericophorus (Sericophorus) 250
Symphyta 160
synchromus, Paralastor 195
Syneuodynerus 207
(Syneuodynerus) aurantipilosella, Pseudepipona 207
Syneuodynerus aurantipilosellus 207
(Syneuodynerus) longebispinosa, Pseudepipona 207
Syneuodynerus longebispinosus 207
(Syneuodynerus) occidentata, Pseudepipona 207
Syneuodynerus occidentatus 207
synoecus, Polistes 212 , 213
synoecus, Polistes (Polistella) humilis 212

Tachyrhostus claviger 246
Tachyrrhostus 245
Tachyrrhostus cyaneus 246
tachyrrhostus, Tachytes 238
Tachyrrhostus viridis 250
Tachysphex 238
Tachysphex aborigenus 238
Tachysphex adelaidae 240
Tachysphex brevicornis 238
Tachysphex buccalis 239
Tachysphex circulans 239
Tachysphex contrarius 239
Tachysphex debilis 243
Tachysphex depressiventris 239
Tachysphex discrepans 239
Tachysphex eucalypticus 239
Tachysphex fanuiensis 239
Tachysphex fanuiensis corallinus 239
Tachysphex fanuiensis howeanus 239
Tachysphex foliaceus 240
Tachysphex fortior 240
Tachysphex galeatus 240
Tachysphex hypoleius 240
Tachysphex imbellis 240
Tachysphex mackayensis 240
Tachysphex maculipennis 240
Tachysphex maximus 241
Tachysphex multifasciatus 241
Tachysphex nefarius 241
Tachysphex novarae 241
Tachysphex pacificus 241
Tachysphex paucispina 241
Tachysphex persistans 241
Tachysphex pilosulus 241
Tachysphex platypus 242
Tachysphex pleuralis 242
Tachysphex politus 242

Tachysphex proteus 242
Tachysphex pubescens 242
Tachysphex pugnator 242
Tachysphex puncticeps 242
Tachysphex rhynchocephalus 243
Tachysphex rugidorsatus 242
Tachysphex stimulator 243
Tachysphex subopacus 243
Tachysphex tenuis 243
Tachysphex tenuisculptus 243
Tachysphex truncatifrons 243
Tachysphex vardyi 243
Tachysphex vividus 243
Tachysphex walkeri 244
Tachytes 237
Tachytes aestuans 237
Tachytes approximatus 237
Tachytes australis 236 , 240
Tachytes codonocarpi 237
Tachytes dispersus 237
Tachytes fatalis 237
Tachytes femoratus 233
Tachytes formosissimus 237
Tachytes hypoleius 240
Tachytes mitis 238
Tachytes novarae 241
Tachytes plutocraticus 238
Tachytes relucens 238
Tachytes rubellus 238
Tachytes sulcatus 238
Tachytes tachyrrhostus 238
taeniatum, Rhopalum (Corynopus) 268
taeniatum, Rhopalum (Rhopalum) 268
tallongensis, Sericophorus 250
tallongensis, Sericophorus (Sericophorus) 250
tamarinus, Epiodynerus 175
tamarinus, Epiodynerus tamarinus 175
tamarinus inviolatus, Epiodynerus 175
tamarinus, Odynerus (Leionotus) 175
tamarinus tamarinus, Epiodynerus 175
tarsatus, Austrogorytes 283
tarsatus, Gorytes 283
tasmaniae, Antamenes (Antamenes) 166
tasmaniae, Pachymenes 166
tasmanica, Aphelotoma 219
tasmanica, Aphelotoma tasmanica 219
tasmanica auriventris, Aphelotoma 219
tasmanica, Lyroda michaelseni 244
tasmanica tasmanica, Aphelotoma 219
tasmanica, Williamsita (Androcrabro) 276
tasmanicus, Acanthostethus 278
tasmanicus, Crabro 276
tasmanicus, Nysson (Acanthostethus) 278
tasmanicus, Pison 262
tasmaniensis, Alastor 196
tasmaniensis, Australozethus 169
tasmaniensis, Australozethus tasmaniensis 169
tasmaniensis, Epiodynerus 175
tasmaniensis montanus, Australozethus 169
tasmaniensis montanus, Pseudozethus 169
tasmaniensis, Odynerus (Leionotus) 175
tasmaniensis, Paralastor 196
tasmaniensis, Polistes 212
tasmaniensis tasmaniensis, Australozethus 169
teleporus, Delta incola 171

teleporus, Eumenes incola 171
teliferopodus okiellus, Sericophorus 250
teliferopodus okiellus, Sericophorus (Sericophorus) 250
teliferopodus, Sericophorus 250
teliferopodus, Sericophorus (Sericophorus) 250
teliferopodus, Sericophorus (Sericophorus) teliferopodus 250
teliferopodus teliferopodus, Sericophorus (Sericophorus) 250
tenebrosum, Pison (Parapison) 263
tenebrosum, Pison (Pison) 263
tenuicornis, Astaurus 280
tenuicornis, Clitemnestra 280
tenuis, Tachysphex 243
tenuisculptus, Tachysphex 243
tenuiventre, Rhopalum (Rhopalum) 266
tenuiventris, Crabro (Rhopalum) 266
tepicum, Rhopalum (Rhopalum) 266
tepida, Vespa 210
tepidus picteti, Polistes 210
tepidus, Polistes (Megapolistes) 210
tepidus, Polistes (Megapolistes) tepidus 210
tepidus tepidus, Polistes (Megapolistes) 210
territorialis, Pseudoturneria 272
testaceum, Rhopalum 266
testaceum, Rhopalum (Rhopalum) 266
thooma, Bembix 293
thoracicus, Clitemnestra 280
thoracicus, Miscothyris 280
tibiale, Pison (Pison) 263
tibialis, Pison 263
tibooburra, Bembix 293
tomentosum, Alyson 303
torresiana, Ropalidia (Icariola) mutabilis 216
townsvillensis austrinus, Polistes (Polistella) 213
townsvillensis, Polistes 211 , 212
townsvillensis, Polistes (Polistella) 212
townsvillensis, Polistes (Polistella) townsvillensis 212
townsvillensis townsvillensis, Polistes (Polistella) 212
transgrediens, Pseudalastor tridentatus 200
transiens, Crabro (Rhopalum) 266
transiens, Rhopalum (Rhopalum) 266
transmarinum, Delta 172
transmarinus, Eumenes 172
trepida, Bembex 293
trepida, Bembix 293
triangularis, Acanthostethus 279
triangularis, Nysson (Acanthostethus) 279
triangulum, Acarodynerus 164
triangulum, Odynerus 164
tricarinulatus, Paralastor 196
trichophthalma, Ropalidia (Icariola) 217
tricolor, Arpactophilus 232
tricolor, Crabro 271
tricolor, Harpactophilus 232
tricolor imbelle, Rhopalum 270
tricolor, Paragia 156
tricolor, Paragia (Paragia) 156
tricolor, Paralastor 196
tricolor, Podagritus (Echuca) 271
tricolor, Polistes 212
tricoloratus, Stenodyneriellus 206
tridentatus, Alastor (Paralastor) 200
tridentatus, Crabro 275
tridentatus paganus, Pseudalastor 200

tridentatus, Pseudalastor 200
tridentatus, Pseudalastor tridentatus 200
tridentatus septentrionalis, Pseudalastor 200
tridentatus transgrediens, Pseudalastor 200
tridentatus tridentatus, Pseudalastor 200
tridentifera, Bembex 292
trifida, Cerceris 302
truncatifrons, Tachysphex 243
Trypoxylon 264
(Trypoxylon) albitarsatum huonense, Trypoxylon 264
(Trypoxylon) albitarsatum, Trypoxylon 264
Trypoxylon bituberculatum 264
(Trypoxylon) bituberculatum, Trypoxylon 264
Trypoxylon connexum 265
Trypoxylon eximium 264
(Trypoxylon) eximium, Trypoxylon 264
Trypoxylon flavipes 264
(Trypoxylon) flavipes, Trypoxylon 264
Trypoxylon huonense 264
Trypoxylon mindanaonis 265
(Trypoxylon) mindanaonis, Trypoxylon 265
Trypoxylon papuanum 265
(Trypoxylon) papuanum, Trypoxylon 265
(Trypoxylon) schmiedeknechti connexum, Trypoxylon 265
(Trypoxylon) schmiedeknechti, Trypoxylon 265
(Trypoxylon), Trypoxylon 264
Trypoxylon (Trypoxylon) 264
Trypoxylon (Trypoxylon) albitarsatum 264
Trypoxylon (Trypoxylon) albitarsatum huonense 264
Trypoxylon (Trypoxylon) bituberculatum 264
Trypoxylon (Trypoxylon) eximium 264
Trypoxylon (Trypoxylon) flavipes 264
Trypoxylon (Trypoxylon) mindanaonis 265
Trypoxylon (Trypoxylon) papuanum 265
Trypoxylon (Trypoxylon) schmiedeknechti 265
Trypoxylon (Trypoxylon) schmiedeknechti connexum 265
tubarum, Rhopalum (Corynopus) 269
tubarum, Rhopalum (Rhopalum) 269
tuberculatum, Zoyphium 254
tuberculatus, Alastor 196
tuberculatus, Paralastor 196
tuberculatus, Sericophorus (Zoyphidium) 254
tuberculicorne, Rhopalum 266
tuberculicorne, Rhopalum (Rhopalum) 266
tuberculiventris, Bembex 294
tuberculiventris, Bembix 294
turneri, Bembecinus 285
turneri, Ropalidia (Icariola) 217
turneri, Stizus 285
turneriellus, Stenodyneriellus 206
Turneriola 272
tydei, Podalonia 228
tydei suspiciosa, Podalonia 228

uloola, Bembix 294
umbrosus, Sphex 222
undeneya, Bembix 294
unicolor, Ischnocoelia robusta 179
unispinosa, Cerceris 302

vardyi, Tachysphex 243

variabilis, Bembex 294
variabilis, Bembix 294
variabilis, Polistes (Polistella) 213
variabilis reginae, Polistes 212
variabilis, Vespa 213
variegata flavinoda, Ropalidia 215
variegata interrupta, Ropalidia 215
variitarse, Rhopalum 269
variitarse, Rhopalum (Corynopus) 269
varipes, Cerceris 302
vedetta, Williamsita 276
vedetta, Williamsita (Androcrabro) 276
veitchi, Crabro 273
veitchi, Neodasyproctus 273
venusta, Cerceris 299 , 302
venusta, Paragia 159
venusta, Paragia (Paragiella) 159
vernalis, Antamenes (Antamenes) 166
vernalis, Odynerus (Ancistrocerus) 166
verreauxii, Discoelius 203
verreauxii, Pseudozethus 203
verutum, Rhopalum (Rhopalum) 266
verutus, Dasyproctus 266
Vespa (Abispa) australiana 161
Vespa albifrons 182
Vespa arcuata 170
Vespa bicincta 198
Vespa campaniformis 171
Vespa dominula 209
Vespa ephippium 161
Vespa germanica 208
Vespa humilis 212
Vespa olivacea 210
Vespa rufipes 204
Vespa schach 210
Vespa tepida 210
Vespa variabilis 213
Vespa vulgaris 209
VESPIDAE 150 , 151 , 152 , 208
vespiformis, Bembex 294
vespiformis, Bembix 294
vespiformis, Paragia 155
vespiformis, Paragia (Cygnaea) 155
Vespinae 151
VESPOIDEA 150 , 151 , 152
Vespula 151 , 208
Vespula (Paravespula) 208
Vespula (Paravespula) germanica 208
Vespula (Paravespula) vulgaris 209
vestita, Sphex 226
vestitum, Pison (Pison) 263
vestitus, Pison 263
vestitus, Sphex (Sphex) 226
victor, Paralastor 196
victoriensis, Australodynerus merredinensis 168
victoriensis, Bembix 294
victoriensis, Sericophorus 250
victoriensis, Sericophorus (Sericophorus) 250
victrix, Cerceris 302
vidua, Isodontia 227
vidua, Sphex 227
viduus, Paralastor 196
violacea, Sphex 220
violaceus, Sericophorus 250
violaceus, Sericophorus (Sericophorus) 250

viridis roddi, Sericophorus 251
viridis roddi, Sericophorus (Sericophorus) 251
viridis, Sericophorus (Sericophorus) 250
viridis, Sericophorus (Sericophorus) viridis 250
viridis, Tachyrrhostus 250
viridis viridis, Sericophorus (Sericophorus) 250
virosum, Pison 263
virosum, Pison (Pison) 263
vividus, Tachysphex 243
vorticosus, Antamenes (Antamenes) 166
vorticosus, Pachymenes 166
vulgaris, Vespa 209
vulgaris, Vespula (Paravespula) 209
vulneratus, Alastor (Paralastor) 197
vulneratus, Paralastor 197
vulneratus, Stenolabus 179
vulpinus, Alastor (Paralastor) 197
vulpinus excisus, Paralastor 197
vulpinus, Paralastor 197
vulpinus, Paralastor vulpinus 197
vulpinus vulpinus, Paralastor 197

wadamiri, Bembix 294
walkeri, Paragia 159
walkeri, Paragia (Paragiella) 159
walkeri, Tachysphex 244
wallacei, Sphex 225
wangoola, Bembix 295
wanna, Bembix 295
warawara, Bembix 295
weema, Bembix 295
westwoodii, Pison 263
westwoodii, Pison (Pison) 263
wilcannia, Bembix 295
Williamsita 274
Williamsita (Androcrabro) 274
Williamsita (Androcrabro) bivittata 275
Williamsita (Androcrabro) bushiella 275
Williamsita (Androcrabro) manifestata 275
Williamsita (Androcrabro) neglecta 275
Williamsita (Androcrabro) ordinaria 275
Williamsita (Androcrabro) riekiella 275
Williamsita (Androcrabro) smithiensis 275
Williamsita (Androcrabro) tasmanica 276
Williamsita (Androcrabro) vedetta 276
Williamsita bushiella 275
Williamsita riekiella 275
Williamsita vedetta 276
wiluna, Bembix 295
windorum, Cerceris 303
wollowra, Bembix 295
wolpa, Bembix 295
wowine, Bembix 295

xanthochroma, Ischnocoelia 179
xanthochromus, Paralastor 197
xanthorrhoicus, Polistes humilis 212
xanthura, Cerceris 303
xanthura, Eumenes 172
xanthurum, Delta 172
xanthurum, Delta xanthurum 172
xanthurum xanthurum, Delta 172
xanthus, Paralastor 197

xenum, Rhopalum (Corynopus) 269
xenum, Rhopalum (Rhopalum) 269
xerophilus meesi, Paralastor 197
xerophilus, Paralastor 197
xerophilus, Paralastor xerophilus 197
xerophilus xerophilus, Paralastor 197

yalta, Bembix 296
yanchepensis, Australodynerus 206
yanchepensis nigrithorax, Australodynerus 206
yanchepensis nigrithorax, Stenodyneriellus 206
yanchepensis, Stenodyneriellus 206
yanchepensis, Stenodyneriellus yanchepensis 206
yanchepensis yanchepensis, Stenodyneriellus 206
yarrowi, Podagritus (Echuca) 272
yorki, Dasyproctus 274
yunkara, Bembix 296

ziegleri, Cerceris 296
Zoyphidium 251
(Zoyphidium) affinis, Sericophorus 251
(Zoyphidium) argyreus, Sericophorus 251
(Zoyphidium) asperithorax, Sericophorus 251
(Zoyphidium) cingulatus, Sericophorus 251
(Zoyphidium) clypeatus, Sericophorus 251
(Zoyphidium) collaris, Sericophorus 251
(Zoyphidium) crassicornis, Sericophorus 251
(Zoyphidium) dipteroides, Sericophorus 252
(Zoyphidium) doddi, Sericophorus 252
(Zoyphidium) emarginatus, Sericophorus 252
(Zoyphidium) erythrosoma, Sericophorus 252
(Zoyphidium) ferrugineus, Sericophorus 252
(Zoyphidium) flavofasciatus, Sericophorus 252
(Zoyphidium) frontalis, Sericophorus 252
(Zoyphidium) fuscipennis, Sericophorus 252
(Zoyphidium) hackeri, Sericophorus 252
(Zoyphidium) humilis, Sericophorus 253
(Zoyphidium) iridipennis, Sericophorus 253
(Zoyphidium) kohlii, Sericophorus 253
(Zoyphidium) laevigatus, Sericophorus 253
(Zoyphidium) niger, Sericophorus 253
(Zoyphidium) ornatus, Sericophorus 253
(Zoyphidium) punctuosus, Sericophorus 253
(Zoyphidium) pusillus, Sericophorus 253
(Zoyphidium) rufoniger, Sericophorus 253
(Zoyphidium) sericeus, Sericophorus 253
(Zoyphidium), Sericophorus 251
(Zoyphidium) splendidus, Sericophorus 254
(Zoyphidium) striatulus, Sericophorus 254
(Zoyphidium) tuberculatus, Sericophorus 254
Zoyphium 251
Zoyphium affine 251
Zoyphium argyreum 251
Zoyphium collare 251
Zoyphium crassicorne 251
Zoyphium doddi 252
Zoyphium emarginatum 252
Zoyphium erythrosoma 252
Zoyphium flavofasciatum 252
Zoyphium frontale 252
Zoyphium fuscipenne 252
Zoyphium hackeri 252
Zoyphium humile 253

Zoyphium iridipenne 253
Zoyphium kohlii 253
Zoyphium nigrum 253
Zoyphium ornatum 253
Zoyphium pusillum 253
Zoyphium rufipes 249

Zoyphium rufonigrum 253
Zoyphium sericeum 253
Zoyphium splendidum 254
Zoyphium tuberculatum 254
Zuba 214

Publication date of the previous volume

Volume 1   29 November 1983